NANOMATERIALS IN PLANTS, ALGAE, AND MICROORGANISMS

Nanomaterials in Plants, Algae, and Microorganisms

CONCEPTS AND CONTROVERSIES: VOLUME 1

Edited by

Durgesh Kumar Tripathi

Parvaiz Ahmad

Shivesh Sharma

Devendra Kumar Chauhan

Nawal Kishore Dubey

ELSEVIER

ACADEMIC PRESS

An imprint of Elsevier

Academic Press is an imprint of Elsevier
125 London Wall, London EC2Y 5AS, United Kingdom
525 B Street, Suite 1800, San Diego, CA 92101-4495, United States
50 Hampshire Street, 5th Floor, Cambridge, MA 02139, United States
The Boulevard, Langford Lane, Kidlington, Oxford OX5 1GB, United Kingdom

Library of Congress Cataloging-in-Publication Data
A catalog record for this book is available from the Library of Congress

British Library Cataloguing-in-Publication Data
A catalogue record for this book is available from the British Library

ISBN: 978-0-12-811487-2

For information on all Academic Press publications visit our website at https://www.elsevier.com/books-and-journals

Publisher: Candice Janco
Acquisition Editor: Anneka Hess
Editorial Project Manager: Emily Thompson
Production Project Manager: Surya Narayanan Jayachandran
Cover Designer: Mark Rogers

Typeset by TNQ Books and Journals

Contents

5. Plant Response to Engineered Nanoparticles

UDAY BURMAN AND PRAVEEN KUMAR

6. Nanoparticle-Induced Morphological Responses of Roots and Shoots of Plants

SEYED M. TALEBI

7. Recent Progress of Nanotoxicology in Plants

MUHAMMAD ZIA-UR-REHMAN, MUHAMMAD F. QAYYUM, FATIMA AKMAL, MUHAMMAD A. MAQSOOD, MUHAMMAD RIZWAN, MAQSODA WAQAR AND MUHAMMAD AZHAR

8. Exploring Plant-Mediated Copper, Iron, Titanium, and Cerium Oxide Nanoparticles and Their Impacts

ANUJA KOUL, ANIL KUMAR, VIVEK K. SINGH, DURGESH K. TRIPATHI AND SHARADA MALLUBHOTLA

9. Gold Nanomaterials to Plants: Impact of Bioavailability, Particle Size, and Surface Coating

NITIN KUMAR, PRANAV TRIPATHI AND SEEMA NARA

10. Responses of Plants to Iron Oxide Nanoparticles

MUHAMMAD ZIA-UR-REHMAN, ASIF NAEEM, HINNAN KHALID, MUHAMMAD RIZWAN, SHAFAQAT ALI AND MUHAMMAD AZHAR

11. Effects of Rare Earth Oxide Nanoparticles on Plants

HAMAAD R. AHMAD, MUHAMMAD ZIA-UR-REHMAN, MUHAMMAD I. SOHAIL, MUHAMMAD ANWAR UL HAQ, HINNAN KHALID, MUHAMMAD A. AYUB AND GOHAR ISHAQ

12. Influence of Titanium Dioxide Nanoparticles ($nTiO_2$) on Crop Plants: A Systematic Overview

ALESSANDRO MATTIELLO, DANIEL LIZZI AND LUCA MARCHIOL

13. Interaction of Copper Oxide Nanoparticles With Plants: Uptake, Accumulation, and Toxicity

NAMIRA ARIF, VASHALI YADAV, SWATI SINGH, DURGESH K. TRIPATHI, NAWAL KISHORE DUBEY, DEVENDRA KUMAR CHAUHAN AND LUCIA GIORGETTI

14. Impacts of Cerium Oxide Nanoparticles ($nCeO_2$) on Crop Plants: A Concentric Overview

DANIEL LIZZI, ALESSANDRO MATTIELLO AND LUCA MARCHIOL

15. Plant and Nanoparticle Interface at the Molecular Level: An Integrated Overview

GAUSIYA BASHRI, PARUL PARIHAR, RACHANA SINGH, ANURADHA PATEL AND SHEO M. PRASAD

16. Nanotechnology in Crop Protection

SUNITA OJHA, DEEPIKA SINGH, ARGHYA SETT, HASNAHANA CHETIA, DEBAJYOTI KABIRAJ AND UTPAL BORA

17. Impact of Nanoparticles on Oxidative Stress and Responsive Antioxidative Defense in Plants

VINAY KUMAR, MANSI SHARMA, TUSHAR KHARE AND SHABIR H. WANI

18. Nanoparticles and Organic Matter: Process and Impact

RAJENDRAN VELMURUGAN AND ARAN INCHAROENSAKDI

19. Ecological Risks of Nanoparticles: Effect on Soil Microorganisms

PRAVEEN KUMAR , UDAY BURMAN AND RAMESH K. KAUL

20. Application of Nanotechnology to Enhance the Nutrient Quality of Food Crops and Agricultural Production

PRAGYA MISHRA, ANITA SINGH, RAGHVENDRA R. MISHRA, SUYEAN ONG AND SHEO M. PRASAD

21. Potential Applications and Avenues of Nanotechnology in Sustainable Agriculture

KANCHAN VISHWAKARMA, NEHA UPADHYAY, NITIN KUMAR, DURGESH K. TRIPATHI, DEVENDRA KUMAR CHAUHAN, SHIVESH SHARMA AND SHIVENDRA SAHI

22. Nanoencapsulation of Essential Oils: A Possible Way for an Eco-Friendly Strategy to Control Postharvest Spoilage of Food Commodities From Pests

AKASH KEDIA AND NAWAL KISHORE DUBEY

List of Contributors

Muhammad A. Ayub University of Agriculture Faisalabad, Faisalabad, Pakistan

Fatima Akmal University of Agriculture, Faisalabad, Pakistan

Shafaqat Ali Government College University, Faisalabad, Pakistan

Muhammad Anwar ul Haq University of Agriculture Faisalabad, Faisalabad, Pakistan

Namira Arif Centre for Medical Diagnostic and Research, Motilal Nehru National Institute of Technology, Allahabad, India

Muhammad Azhar University of Agriculture, Faisalabad, Pakistan

Gausiya Bashri University of Allahabad, Allahabad, India

Utpal Bora Indian Institute of Technology Guwahati, Guwahati, India

Uday Burman ICAR-Central Arid Zone Research Institute, Jodhpur, India

Devendra Kumar Chauhan University of Allahabad, Allahabad, India

Hasnahana Chetia Indian Institute of Technology Guwahati, Guwahati, India

Nawal Kishore Dubey Banaras Hindu University, Varanasi, India

Somenath Ghatak University of Allahabad, Allahabad, India

Lucia Giorgetti National Research Council (IBBA-CNR), Pisa, Italy

Muhammad I. Sohail University of Agriculture Faisalabad, Faisalabad, Pakistan; University of Florida, Institute of Food and Agricultural Sciences, Indian River Research and Education Center, Fort Pierce, FL, United States

Aran Incharoensakdi Chulalongkorn University, Bangkok, Thailand

Gohar Ishaq University of Agriculture Faisalabad, Faisalabad, Pakistan

Prashant K. Rai University of Allahabad, Allahabad, India

Debajyoti Kabiraj Indian Institute of Technology Guwahati, Guwahati, India

Ramesh K. Kaul ICAR-Central Arid Zone Research Institute, Jodhpur, India

Akash Kedia Government General Degree College at Mangalkote, Burdwan, India

Hinnan Khalid University of Agriculture Faisalabad, Faisalabad, Pakistan

Tushar Khare Modern College of Arts, Science and Commerce, Savitribai Phule Pune University, Ganeshkhind, Pune, India

Anuja Koul Shri Mata Vaishno Devi University, Katra, India

Anil Kumar Shri Mata Vaishno Devi University, Katra, India

Deepak Kumar Banaras Hindu University, Varanasi, India

Nitin Kumar Centre for Medical Diagnostic and Research, Motilal Nehru National Institute of Technology, Allahabad, India

Praveen Kumar ICAR-Central Arid Zone Research Institute, Jodhpur, India

Vinay Kumar Modern College of Arts, Science and Commerce, Savitribai Phule Pune University, Ganeshkhind, Pune, India

Daniel Lizzi University of Udine, Udine, Italy; University of Trieste, Trieste, Italy

Sheo M. Prasad University of Allahabad, Allahabad, India

Sharada Mallubhotla Shri Mata Vaishno Devi University, Katra, India

Muhammad A. Maqsood University of Agriculture, Faisalabad, Pakistan

Luca Marchiol University of Udine, Udine, Italy

Alessandro Mattiello University of Udine, Udine, Italy

Pragya Mishra University of Allahabad, Allahabad, India

Asif Naeem Pakistan Atomic Energy Commission, Faisalabad, Pakistan

Seema Nara Motilal Nehru National Institute of Technology, Allahabad, India

Sunita Ojha Indian Institute of Technology Guwahati, Guwahati, India

SuYean Ong University Sains Malaysia, Penang, Malaysia

Sonika Pandey National Institute of Plant Genome Research, New Delhi, India

Parul Parihar University of Allahabad, Allahabad, India

Anuradha Patel University of Allahabad, Allahabad, India

Jainendra Pathak Banaras Hindu University, Varanasi, India

Jose R. Peralta-Videa The University of Texas at El Paso, El Paso, TX, United States

Muhammad F. Qayyum Bahaudin Zakariya University, Multan, Pakistan

Hamaad R. Ahmad University of Agriculture Faisalabad, Faisalabad, Pakistan

Raghvendra R. Mishra Banaras Hindu University, Varanasi, India

S. Rajeshkumar VIT University, Vellore, India

Rajneesh Banaras Hindu University, Varanasi, India

Muhammad Rizwan Government College University, Faisalabad, Pakistan

Shivendra Sahi Western Kentucky University, Bowling Green, KY, United States

Arghya Sett Indian Institute of Technology Guwahati, Guwahati, India

Gaurav Sharma Jawaharlal Nehru University, New Delhi, India

Mansi Sharma Modern College of Arts, Science and Commerce, Savitribai Phule Pune University, Ganeshkhind, Pune, India

Shivesh Sharma Centre for Medical Diagnostic and Research, Motilal Nehru National Institute of Technology, Allahabad, India

Shweta University of Allahabad, Allahabad, India

Anita Singh University of Allahabad, Allahabad, India

Deepika Singh Indian Institute of Technology Guwahati, Guwahati, India

Rachana Singh University of Allahabad, Allahabad, India

Shailendra P. Singh Banaras Hindu University, Varanasi, India

Swati Singh Centre for Medical Diagnostic and Research, Motilal Nehru National Institute of Technology, Allahabad, India

Vidya Singh Banaras Hindu University, Varanasi, India

Vivek K. Singh Shri Mata Vaishno Devi University, Katra, India

Rajeshwar P. Sinha Banaras Hindu University, Varanasi, India

Seyed M. Talebi Arak University, Arak, Iran

Durgesh K. Tripathi University of Allahabad, Allahabad, India; Centre for Medical Diagnostic and Research, Motilal Nehru National Institute of Technology, Allahabad, India; Banaras Hindu University, Varanasi, India

Pranav Tripathi Motilal Nehru National Institute of Technology, Allahabad, India

Neha Upadhyay Centre for Medical Diagnostic and Research, Motilal Nehru National Institute of Technology, Allahabad, India

Rajendran Velmurugan Chulalongkorn University, Bangkok, Thailand

S. Venkat Kumar VIT University, Vellore, India

Kanchan Vishwakarma Centre for Medical Diagnostic and Research, Motilal Nehru National Institute of Technology, Allahabad, India

Shabir H. Wani Mountain Research Centre for Field Crops, Khudwani, Sher-e-Kashmir University of Agricultural Sciences and Technology, Anantnag, India

Maqsoda Waqar University of Agriculture, Faisalabad, Pakistan

Geeta Watal University of Allahabad, Allahabad, India

Vashali Yadav Centre for Medical Diagnostic and Research, Motilal Nehru National Institute of Technology, Allahabad, India

Muhammad Zia-ur-Rehman University of Agriculture, Faisalabad, Pakistan

Preface

Nanotechnology is being widely used in different disciplines of science, which not only shows its applicability but also derives several benefits in medical science, agricultural sciences engineering, pharmaceuticals, and others. Being most valuable technique in the recent years its knowledge, updates, and related discoveries has not well documented in the form of books. Thus in the course of acquiring knowledge about the nanotechnological research in plants, algae, and microorganisms since the past up to the present we have found significant gap between the availability of books and emerging area of research. Thus, this book has been planned to collect the worldwide knowledge of scientists/researchers on nanotechnology, which may bridge the gap between researches being conducted from the past up to today, and the direction these researches might take in the future with respect to interactions of nanoparticles and plants, algae, and microorganisms.

The title of this book indicates that this book has collected the knowledge, discoveries, and the fruitful findings of nanoparticles and plants, algae, and microorganisms interactions from the past to recent. As several studies indicated the positive and some negative impacts of nanoparticle on living systems and therefore nanoparticles research somewhere is matter of controversies because of lack of concepts. Thus, we have collected the chapters discussing the wide knowledge of positive and negative impacts of nanoparticles on living system mainly plants, algae, and microorganisms. This book also included the chapters that briefly discussed the several aspects of nanoparticle research starting from basic knowledge to going on research.

Chapter 1 gives a concentric overview of availability and risk assessment of different metal and metal oxide nanoparticles on plants, microbes, and human beings. This chapter also includes the concepts of positive and negative impacts of nanoparticles on plants, microbes, and human beings and briefly discussed the future prospects. Chapter 2 discussed the brief account of plant-mediated metal and metal nanoparticle synthesis, characterization, and its wide impacts. The authors have also discussed the relevant plant parts for the appropriate synthesis of nanoparticles. Additionally Chapter 3 briefly discussed the potential use of spectroscopic techniques in characterization of green nanomaterials.

The next few chapters deal with the approaches and developments in nanotechnology; this chapter also discussed the challenges in the field of DNA nanotechnology and briefly described the possible way of its synthesis and its promising applications that could be developed. Authors also discussed the impact of engineered nanoparticles on different plant species. It has also discussed the exposure of nanoparticles intricacies of the different facets of their interaction at cellular, organ, and whole-plant level. It further discussed the impact of nanoparticles on morphology of root and shoot of plants and also on seed germinations, which depend on different factors such as chemical formula, size, reactivity, and the effective dosages.

In later chapters, authors have systematically arranged the knowledge of recent progress of nanotoxicology in plants and the way of plant-mediated cooper, iron, titanium, and cerium oxide nanoparticles synthesis and their possible impacts. Furthermore, bioavailability, particle size, surface coating, and impacts of gold nanomaterials to plants are briefly discussed. Additionally the response of iron oxide nanoparticles, rare earth oxide nanoparticles on plants, titanium dioxide nanoparticles, copper oxide nanoparticles, and cerium oxide nanoparticles in plants is briefly discussed. Further in the last few chapters, some very important issues such as phenomena of nanoparticles and molecular mechanisms in plants, nanoparticles, and organic matter, nanotechnology in crop protection, nanoparticles on oxidative stress and responsive antioxidative defense in plants, ecological risks of nanoparticles and their effect on soil microorganisms, nanoencapsulation of essential oils for the possible way of an eco-friendly strategy to control postharvest spoilage of food commodities from pests, application of nanotechnology to enhance the nutrient quality of food crop and agricultural production, and influence of nanomaterials on crops toward sustainable agriculture have been well discussed.

This book is a collection of global panorama that through different ways made us able to understand the possible interacting mechanism and impacts of nanoparticles with plants, algae, and microorganisms.

This book will be valuable for students, teachers, and researchers of diverse area related to botany, physiology, microbiology, environment, and nanotechnology, etc. To the best of our knowledge, we have made all the possible efforts to effectively compile this volume. However, we feel that there might be space for improvement; therefore, regarding this we seek involvement and the valuable suggestions from the readers, which would help us to make amendments for the upcoming volume.

Furthermore, we are highly grateful to the contributors of this volume, who have dedicated their worthy time and efforts in preparing and editing chapters as well as through their valuable suggestions. We owe our gratitude to Emily Joy Grace Thomson (Editorial Project Manager, Elsevier), Surya Narayanan Jayachandran (Project Manager, Reference Content Production), Swapna Praveen (Copyrights Coordinator, GR—Copyrights, Elsevier), Anneka Hess (Acquisitions Editor, Elsevier), and all the other staff members of Elsevier, who have been associated and helped in every possible ways to make this project worthy for publication of this volume.

Dr. Durgesh Kumar Tripathi
Dr. Parvaiz Ahmad
Dr. Shivesh Sharma
Prof. Devendra Kumar Chauhan
Prof. Nawal Kishore Dubey

CHAPTER

1

Availability and Risk Assessment of Nanoparticles in Living Systems: A Virtue or a Peril?

Shweta[1], *Durgesh K. Tripathi*[1,2,3], *Devendra Kumar Chauhan*[1], *Jose R. Peralta-Videa*[4]

[1]University of Allahabad, Allahabad, India; [2]Centre for Medical Diagnostic and Research, Motilal Nehru National Institute of Technology, Allahabad, India; [3]Banaras Hindu University, Varanasi, India; [4]The University of Texas at El Paso, El Paso, TX, United States

1.1 INTRODUCTION

The population of the world continues to increase at an alarming rate. Problems linked with overpopulation range from food and water scarcity to inadequacy of space for organisms. Overpopulation is also linked with several other demographic hazards. For instance, population explosion will not only result in exhaustion of natural repositories, but will also induce intense pressure on the world economy. At present, nanotechnology is a much discussed discipline of science because of its positive and negative aspects. Incidentally, because of industrialization and the ever-increasing population, nanopollution has been an emerging topic among scientists for investigation and debate. According to Wang and Wang (2014), nanotechnology is a multidisciplinary science that covers physical, chemical, biological, engineering, and electronic sciences. Hagens et al. (2007) defined nanotechnology as a measure of any substance at the macromolecular scale, molecular scale, and even the atomic scale. Moreover, according to Storrs (2005), nanotechnology is a branch of science that deals with the manipulation and control of any matter at the nanometer scale.

Nanotechnology and nanoparticles (NPs) play important roles in sustainable development and environmental challenges as well (Grieger et al., 2010; Peralta-Videa et al., 2011; Mukherjee et al., 2016). NPs possess both harmful as well as beneficial effects on the environment and its harboring components such as microbes, plants, and humans. For instance, it is demonstrated that TiO_2 NPs play beneficial roles in seed germination, enhanced root-shoot

Nanomaterials in Plants, Algae, and Microorganisms
http://dx.doi.org/10.1016/B978-0-12-811487-2.00001-3

length, and improved growth of seedlings in *Arabidopsis thaliana* (L.) (Szymańska et al., 2016; Lyu et al., 2017), cabbage (Andersen et al., 2016; Lyu et al., 2017), corn (Andersen et al., 2016; Lyu et al., 2017), lettuce (*Lactuca sativa* L.) (Andersen et al., 2016; Lyu et al., 2017), oat (*Avena sativa* L.) (Andersen et al., 2016; Lyu et al., 2017), and flax (*Linum usitatissium* L.) (Aghdam et al., 2016; Lyu et al., 2017). Conversely, Tripathi et al. (2017a) reported the toxicity of AgNPs on heterotrophic microbes as well as autotrophic plants. However, there are many other beneficial impacts also exerted by NPs including their roles in the management of wastewater and soil treatment (Tiwari et al., 2008; Chekli et al., 2013), cosmetics (Müller et al., 2002; Mu and Sprando, 2010; Mihranyan et al., 2012; Matranga and Corsi, 2012), food packaging and its protection from microbes (Duncan, 2011), electronics (Kachynski et al., 2008), agriculture and biomedicines (Hoet et al., 2004; Dong and Feng, 2007; Matranga and Corsi, 2012), in manufacturing pharmaceuticals (Matranga and Corsi, 2012), renewable energies (Pavasupree et al., 2006), and environmental remedies (Zhang, 2003; Tungittiplakorn et al., 2004). Moreover, Sharma et al. (2009) have described the green synthesis of AgNPs and demonstrated that they filter out the polluted air and water and also show antimicrobial activities.

On the other hand, they also show toxic effects on microbes (Matranga and Corsi, 2012; Tripathi et al., 2017a), plants (Stampoulis et al., 2009; Reddy et al., 2016; Tripathi et al., 2017b,c; Majumdar et al., 2015; Zuverza-Mena et al., 2016), as well as human beings (Handy and Shaw, 2007). It has been reported that the use of nanotechnological products leads to the increased accumulation of NPs in soil and aquatic ecosystems, which may be detrimental for living organisms (Lazareva and Keller, 2014; Keller and Lazareva, 2013). Similarly, Holden et al. (2012) demonstrated the toxic effect of NPs on microbes, invertebrates, aquatic organisms including algae, and whole ecosystems. However, several reports on plant and NP interfaces also demonstrated that when NPs interacted with plant cells the result was alterations in growth, biological function, gene expression, and ultimately hindrance to the growth and development of plants (Ochoa et al., 2017; López-Moreno et al., 2017; Tripathi et al., 2017b). Moreover, Majumdar et al. (2015) also described the impact of NPs on plants at proteomic and transcriptomic levels. Furthermore, Zezulka et al. (2013) demonstrated that NPs can lead to changes in the production of biomass, leaf area, and chlorophyll.

There are various NPs that are essential not only for the development of living organisms but also for the environment (Biswas and Wu, 2005). Therefore nanotechnology has been used to improve the quality of medicines (in making smart drugs), creams and cosmetic items, health supplements, textiles and clothing, air and water filters, household items, detergents, etc. (Donaldson et al., 2004; Xiao et al., 2005; Nel et al., 2006; Mihranyan et al., 2012).

Another aspect of nanotechnology is nanopollution, i.e., microbes, plants, and animals that directly or indirectly affect living organisms. Therefore, its intelligent use is necessary to check or control its negative impacts. Nanomaterials are big challenges for both nature and the ecosystem, because they lead to artificial evolution. They manipulate the matter at its elementary level and convert it to new matter that is not found in the realms of nature. Increasing the use of nanotechnological products leads to release of NPs in the environment, which requires assessment of environmental risk (Albrecht et al., 2006; Scheufele et al., 2007).

Therefore, the aim of this chapter is to critically review the application of NPs and toxicity imposed by them on microorganisms, plants, and humans, and to draw a conclusion, which could be useful for scientists working in nanotechnology.

1.2 SOURCES OF NPs IN THE ENVIRONMENT

The following are modes of sources from which NPs make their entry into the environment. A brief account of sources is depicted in Fig. 1.1.

1.2.1 Natural Sources

NPs are abundantly found in nature. They can be formed by various natural processes including eruptions of volcanoes, photochemical reactions, erosions, etc. In addition, they can also be obtained from plants and animals (Buzea et al., 2007). Air quality is affected by the liberation of vast quantities of NPs via cars, burning of charcoal, and volcanic eruptions, among others (Fig. 1.1). The presence of NPs is associated with human activities that can affect the environment worldwide. For instance, 10% of aerosols are generated by humans, while 90% of aerosols are produced naturally (Taylor, 2002). Aerosol particles can imbalance the whole energy of the planet by adsorbing huge quantities of solar insulation, and can scatter it back into the universe (Houghton, 2005).

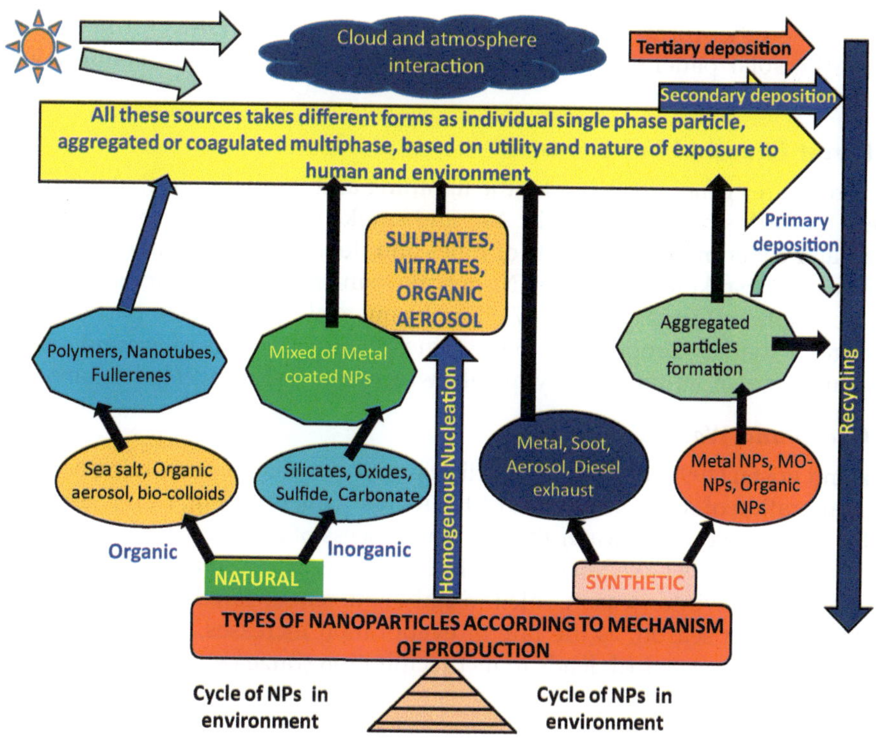

FIGURE 1.1 Cycle of nanoparticles in the environment. *Modified from Sajid, M., Ilyas, M., Basheer, C., Tariq, M., Daud, M., Baig, N., Shehzad, F., 2015. Impact of nanoparticles on human and environment: review of toxicity factors, exposures, control strategies, and future prospects. Environ. Sci. Pollut. R. 22 (6), 4122–4143.*

1.2.1.1 *Dust Storms*

Dust storms are considered a meteorological phenomenon. They are also known as sand storms, and are common in semiarid and arid zones (Buzea et al., 2007). Dust storms can move from one place to another by the process of suspension and saltation (Fig. 1.1).

1.2.1.2 *Terrestrial Dust Storms*

Terrestrial dust storms are unique and the largest source of NPs from migration of the anthropogenic as well as mineral dust pollutants. Shi et al. (2005) reported that 50% of aerosols present in the troposphere are mineral oriented and come from desert regions (Fig. 1.1). Their size varies around 100 nm or more. However, Taylor (2002) reported that the sizes of half of the dust particles are less than 2.5 μm (Fig. 1.1). Husar et al. (2001) also observed that in North America and Asia, especially in the Gobi Desert, the effects of terrestrial dust storms are very high and degrade the quality of air. Satellite imagery shows that the whole region of the Pacific is yellow in color because of terrestrial dust storms (Fig. 1.1).

1.2.1.3 *Extraterrestrial Dust Storms*

Extraterrestrial dust storms are broadly found in nature. Examples are dusts gathered from Mars and moons (Fig. 1.1). They adhere to the surfaces of equipment because they have magnetic NPs (Buzea et al., 2007). Lunar-grained particles are the smallest nanodust in comparison to other forms of terrestrial dusts. Extraterrestrial dust formed on Mars also accumulates on solar panel equipment, which can create problems in communication and movement (Fig. 1.1).

1.2.1.4 *Forest Fires and Volcanoes*

In the forest, fires generally occur by the lightning and it is believed that they have been a component of Earth's history (Fig. 1.1). Ash and smoke are two major products of fires that can spread and cover many thousands of miles and degrade the quality of air by enhancing particulate matter (Sapkota et al., 2005). Nanoscale to micron-sized particles are liberated during volcanic eruptions (Buzea et al., 2007; Strambeanu et al., 2015). These particles are actually ashes and gases that are propelled into the atmosphere to heights of around 18,000 m. This particulate matter also contains heavy metals, which may be harmful to humans (Fig. 1.1).

1.2.1.5 *Ocean and Water Evaporation*

Buseck and Pósfai (1999) reported huge distributions of aerosols involving oceans and seas through evaporation of water, with sizes of 100 to many microns (Fig. 1.1). These types of NPs can also be formed in oceanic water through the process of evaporation followed by precipitation. A good example is Lake Michigan, which contains an abundance of calcium carbonate (Fig. 1.1). During the winter season, calcium carbonate remains in the lake in high quantities, while in the summer season, as the rate of evaporation increases, the solubility of calcium carbonate also decreases, which leads to its precipitation (Buzea et al., 2007).

1.2.1.6 *Organisms*

Those bacteria and viruses that ranges from 30 to 700 nm and 10 to 400 nm respectively, are considered as small organisms having sizes of only a few microns (Bélanger et al., 2003; Buzea et al., 2007). However, there should be a clear distinction between particles such as NPs or microparticles and nanoorganisms or their components such as viruses, bacteria, cells, and other organelles (Fig. 1.1). The major distinction between organisms and nanoparticles is drawn on the basis of their energy levels. Energy can be depleted in the case of smaller organisms but in the case of NPs, energy does not deplete. Energy is not needed for the stabilization of NPs (Fig. 1.1). NPs can easily interact or change their energy via many chemical reactions with their surrounding environment (Fig. 1.1). Whether multicellular or unicellular, both types of organisms generate nanoorganic substances via two major processes: extracellular and intracellular (Ahmad et al., 2005). However, magnetic NPs, siliceous particles, and calcium carbonates are produced by magnetotactic bacteria, diatoms, and sulfur-layer bacteria, respectively (Ahmad et al., 2005). One property of magnetotactic bacteria is migration and orientation toward their favorable environment, with the help of nanometer size magnetic substances of (Fig. 1.1). Similarly, calcium phosphate shells are synthesized by a nanoorganism called a nanobacterium whose size varies from 20 to 300nm. They are porous in nature and hence slimy substances can be excreted through them (Kajander, 2006).

1.2.2 Anthropogenic Sources

Anthropogenic sources comprise emissions from combustion, smelting, refining of ore, welding, diesel vehicles, aeroplanes, and cooking, among others (Kajander, 2006). Engineered NPs are also the result of anthropogenic sources (Fig. 1.1).

1.2.2.1 *Diesel and Engine Exhaust NPs*

In urban areas, the major sources of NPs are gasoline and diesel engine exhaust emissions (Fig. 1.1). The size of NPs emitted from diesel and gasoline engines ranges from 20 to 130nm and 20 to 60nm, respectively, and these particles possess a spherical structure (Westerdahl et al., 2005). Anthropogenic sources of NPs include 90% diesel-generated NPs and 20% particle mass. Several studies showed that large numbers of NPs were found near freeways, which clearly indicate that vehicular pollution is a major source of nanopollution (Siegmann et al., 2008) (Fig. 1.1).

1.2.2.2 *Indoor Pollution and Buildings Demolition*

A report from the United States Environmental Protection Agency (http://www.epa.gov/iaq/index.html; www.epa.gov/airtrends/pmreport03/pmunderstand_2405) shows that in comparison to outdoor air, indoor air (houses) may be around 10 times more polluted (Fig. 1.1). Various human activities can lead to high quantities of indoor particulate matter (Fig. 1.1). Particles can also be liberated by activities such as smoking, cooking, cleaning of floors, combustion of candles and wood, etc. and from items such as textiles, spores, and smoke. It is also reported that in buildings, particles can infiltrate from outdoors to indoors, which may cause nanopollution (Fig. 1.1). Generally, humans spend most of their time performing indoor activities, which can ultimately affect their health. Older buildings that are likely to be reconstructed may be harmful too. The presence of harmful tiny particles of glasses, paper, and wood can interfere with respiration and lead to a negative impact on living organisms (Fig. 1.1).

1.2.2.3 *Cosmetics and Other Consumer Products*

Nowadays, NPs have also been used in cosmetics such as sunscreens, powders, lipsticks, etc. (Fig. 1.1). They can easily penetrate the skin and cover protective layers. In addition, for nurturing the skin, NPs are being used as delivery agents such as synthetic peptides, which help in the regeneration of cells (Fig. 1.1). Some NPs can aid the youthful appearance of skin cells and also possess antioxidant properties (Xiao et al., 2005). They have been proven effective in concealing wrinkles and small wounds, because of their smaller size and specific optical properties (Fig. 1.1). There is no way of distinguishing between harmful and beneficial NPs. However, NPs are biologically inert in animals as well as humans (Gurr et al., 2005). Currently, smaller particles of titanium oxide are used in everyday products such as food colorants, pigments, cosmetic creams, and sunscreens (Donaldson et al., 2004), and their negative impacts have also been reported in the literature (Kubota et al., 1994; Dunford et al., 1997; Serpone et al., 2001; Rehn et al., 2003; Buzea et al., 2007).

1.2.2.4 *Engineered Nanomaterials*

Manufacturing of NPs is wide and is a current field of study. NPs can be prepared through various processes such as plasma synthesis, evaporation at high temperature, flame pyrolysis, laser, thermal and electron evaporation (Peralta-Videa et al., 2016), and liquid phase methods, among others (Swihart, 2003). A unique property of engineered NPs is that they can be formed in any shape and size. Some NPs are firmly attached to the substrate and are safe for human health (Fig. 1.1).

1.3 GLOBAL EXTENSION AND ECONOMIC IMPACTS OF NATURAL AND ENGINEERED NPs

Natural and engineered nanomaterials (ENMs) provide a novel opportunity to boost the working operations of existing sectors. They are used predominantly in the production of industrial and commercial products (Colvin, 2003). However, because of their profound application in industrial and commercial fields, ENMs are also considered emerging toxicants that potentially induce adverse effects on the ecosystem's structure and function (Mohanty et al., 2014). Because of this, they have received increased attention from both scientific and regulatory communities (Miralles et al., 2012).

Thousands of ENM-related products have been discovered along with their consequent relative markets (Huang et al., 2015). Discharge of engineered NPs occurs deliberately or accidentally in the environment. Accidental release of engineered NPs occurs through atmospheric emission, leaching from sewage sludge, or most commonly from the erosion of engineered NPs containing nanomaterials (Buzea et al., 2007). On the other hand, deliberate release of NPs includes biosensors from medical and cosmetic applications. According to the "cradle-to-grave" approach, all of the positive or negative impacts of NPs are relevant to their sustainability (Huang et al., 2015). Predicting and assessing the environmental impact of ENMs is an important task that becomes complicated by shortage of information regarding their volumes of production and release.

Savolainen et al. (2010) showed safety-related issues of ENMs and nanotechnology, which have increased substantially (Fig. 1.2). In spite of this, ENMs also show many positive applications in people's everyday lives, such as in the formation of pure drinking water and clean energy (Savolainen et al., 2010). Several nano-based products are available in the market,

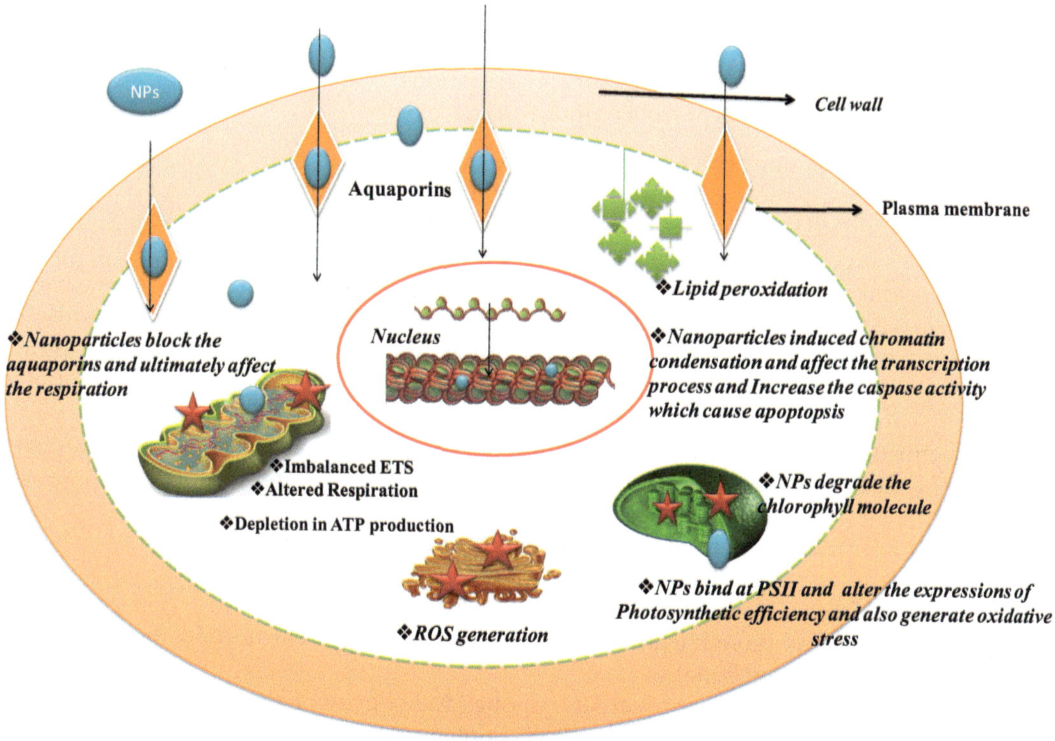

FIGURE 1.2 Various mechanisms by which nanoparticles (NPs) impose toxicity on cellular organelles. The figure describes the interaction of functionalized NPs with cellular organelles of plants. (1) Entry of NPs leads to mitochondrial dysfunction by unbalancing the electron transport system (ETS), altering respiration, and causing depletion in ATP production that eventually lead to the generation of depolarized mitochondria. (2) NPs also influence the critical size of the nucleus and convert the functioning of active DNA in its inactive form, which eventually leads to lysis of the nucleus. (3) By disrupting the usual pathway of electrons in the ETS, NPs lead to the generation of reactive oxygen species (ROS), which induce several adverse effects such as peroxidation of the lipid bilayer (LPO) and DNA damage. (4) Interaction of NPs with the chloroplast degrades the production of chlorophyll.

which may have few safety and health risks because at the workplace their exposure is very high in comparison to other surroundings (Savolainen et al., 2010). Knowledge of the environmental impact of NPs is limited and it is difficult to provide a reliable assessment. Many global issues related to environmental NPs are:

- Improved knowledge of environmental toxicity;
- Environmental NP exposure assessment, formation of a monitoring device;
- Knowledge of changes in environmental NP structure and formation of clumps at various concentrations in aerosols;
- Knowledge of environmental NP translocation in the human body;
- Identification of carcinogenic effects, pulmonary toxicity, genotoxicity, and effects on circulation;
- Generation of shattered approaches for safety of environmental NPs;

- Use of data for risk assessment of health with more emphasis on the occupational environment;
- Knowledge of environmentally relevant concentrations;
- Determination of the effects of soil components on toxicity.

Environmental NPs may enter any part of the body or organ and can cause pulmonary infections and fibrosis, and can even cause mesotheliomas in humans (Ferreira et al., 2013). Since NPs have unique properties, their values are very high in many technological areas. Many countries invest in them to generate new technologies for the betterment of the future. The US government is investing in nanotechnology with very steady growth. Furthermore, the President's 2017 Budget provided $1.4billion for the National Nanotechnology Initiative (NNI), with a collective entirety of almost $24billion since its founding in 2001 (including the 2017 request), establishing the key role that nanotechnology plays in the administration's innovation agenda (https://www.nano.gov/node/1573).

Manufacturing of functional nanofilters eliminates the problem of desalinization, reduces costs by 99%, and overcomes problems of water shortage in arid areas. Thus NPs have the ability to change the regional environment. With the advancement of nanotechnological applications, dependency on fossil fuels begins declining with the advent of solar cells and the invention of solar paints. These solar cells and solar paints are based on nanoengineered organic plastics. However, other engineered NPs are based on nanostructured aerogels. These NPs are very strong and have the ability to make hurricane- and earthquake-resistant structures, thus, provide societal benefits.

Delgado-Ramos (2014) discussed regulatory issues and policies of global trades and national implications. The economic protocol of nanotechnology suggests that expenditure channels are growing rapidly at the global level. In 2010, total global investment had been recorded at around US$70billion (Delgado-Ramos, 2014). According to Chad (2010), in 2008, private sectors had invested about two-third of total expenditure, i.e., around 60% of the 2010 investment.

In 2010, private sectors were major sources of investment in nanotechnology (Roco et al., 2011): 130% investment was recorded from 2004 to 2008 but it has now declined to 9.3% (Delgado-Ramos, 2014). After many surveys, it was found that in the context of research and development, some countries such as the United Kingdom, the United States, France, Japan, and Germany were at the forefront of nanotechnology, and other countries such as Russia and China were taking major steps to catch up (Liu et al., 2009). In the worldwide publications on nanoscience, China was in second place and before Japan; however, competition is intense between the United States, Europe, and Japan but the United States still holds the dominant place in the novel nanotechnological niche (Delgado-Ramos, 2014).

1.4 FORECASTING THE POTENTIAL RISK ASSOCIATED WITH NPs

Nanotechnology is at the forefront of science and technology, which has been advancing rapidly and has profound applications in human life, information technology, the environment, energy, and national security (Bai et al., 2010). Though the emergence of nanotechnology has been marked as the biggest engineering innovation since the dawn of the industrial revolution, at the same time it is also considered a double-edged sword, which not only has beneficial applications but also has adverse effects and hidden hazards on the environment, microorganisms, plants, and humans. In this regard, several governmental authorities and certification bodies such as Environment Health Services and various other nongovernmental organizations are recognizing the importance of risk assessment with regard to nanotechnologies all over the world (Sharma et al., 2012), and have made their own suggestions, views, and guidance (Dhawan et al., 2011).

In addition, nanotechnology also plays a key role in the food industries (Thul et al., 2013). In plants, NPs deal with pathological infections, and are used as growth adjuvants and nutrient supplements (Kim et al., 2009; Servin and White, 2016).

NPs can be linked to agrochemicals and another substance to deliver or control chemicals in plant cells and tissues, but in spite this, the adverse impacts of NPs cannot be denied, such as lipid peroxidation and DNA damage (Thul et al., 2013; Tripathi et al., 2017a; Medina-Velo et al., 2017; Barrios et al., 2017). A number of studies are available that clearly indicate NP toxicity on microbes, plants, and animals (Armendariz et al., 2004; Thul et al., 2013).

1.5 NP TOXICITIES IN MICROORGANISMS, PLANTS, AND HUMANS

1.5.1 Effects of NPs on Microorganisms

The detrimental effect of NPs on the microbial world could easily be manifested by inhibition of growth, wall formation, and cell morphological damage, which eventually exert negative impacts on the microbial community (Thul et al., 2013) (Table 1.1).

Akhavan and Ghaderi (2010) demonstrated that both graphene and graphene oxide nanowalls may lead to damage of the cell membrane of Gram-negative bacterium (*Escherichia coli*) and Gram-positive bacterium (*Staphylococcus aureus*), which may ultimately result in leakage of intracellular substances and thus cell death (Fig. 1.2; Table 1.1). Similarly, nano-TiO_2 and nano-ZnO also reduce the quantity and diversity of microorganisms in the soil and change compositions of the soil microbial community (Ge et al., 2011). After entering the microbial cell, NPs change the metabolism of microbes.

Thul et al. (2013) demonstrated the effect of NPs on bacteria in soil, and found that bacteria were responsible for enhancing productivity in the soil environment. Holden et al. (2012) showed the combined effects of NP hazards in different populations, communities, and ecosystems, and reported that bacteria can also assist the mobilization of NPs by adsorbing and accumulating other forms of NPs via the food chain, and can change many populations (Fig. 1.2; Table 1.1). Some NPs including AgNPs exert an antimicrobial effect at the ecosystem level.

Tilston et al. (2013) showed that soil microbes are a sufficient and skilled catalyst, which might either absorb or disperse the accumulated engineered NPs. Choi et al. (2008) reported that AgNPs after attachment to microbial cells can cause detrimental effects because of cell wall pitting (Fig. 1.2; Table 1.1). After 28 days of treatment with nano zerovalent iron (nZVI), aroclor-1242 can change bacterial aroclor congener profiles together with changes in the physicochemical properties of soil such as pH. Tilston et al. (2013) further showed that nZVI can change the bacterial colony composition and can decrease the chloroaromatic activity of mineralizing microorganisms. Yang et al. (2013) demonstrated the effect of 35 nm carbon-coated

TABLE 1.1 Impact of Nanoparticles (NPs) on Microorganisms

Types of NPs	Size of NPs (nm)	Dose	Source	Microorganisms	Effect of NPs on Microorganism(s)	References
Ag	14±6, 16, <100, 20, 8–10	9.3 µM, 100 µg/mL, 0.1 and 0.5 mg/L, 100 µg/mL, 1 mM	Silver nitrate	*Escherichia coli*, *Pseudomonas aeruginosa*, *Staphylococcus aureus*, *Salmonella typhimurium*, *Caenorhabditis elegans*	Antibacterial activity, inhibition of bacterial growth, bactericidal action, cell wall pitting, survival, growth, and reproduction reduced, reactive oxygen species (ROS) formation, glutathione *S*-transferase activity	Choi et al. (2008), Pal et al. (2007), Roh et al. (2009), Lim et al. (2012) and Sahu et al. (2013)
Au	20–50 nm	6–12 µg, 5–50 mg Au/kg dry mass	$HAuCl_4 \cdot 3H_2O$	*E. coli*, *S. aureus*, *Eisenia fetida*	Low toxicity to bacteria, growth, reproduction, and gene expression	Goodman et al. (2004) and Unrine et al. (2010)
CeO_2	15–45 nm	1 mg/L		Nitrogen-fixing bacteria, *C. elegans*	Decrease of N_2 fixation potentials, negative impact on survival, growth, reproduction, and gene expression	Priester et al. (2012) and Roh et al. (2010)
TiO_2, SiO_2, ZnO		Actual particle size range in suspension 175–810, 135–510, 420–640, 15–25 nm		*Bacillus subtilis*, *E. coli*	Mild toxicity caused by ROS production	Adams et al. (2006), Sapkota et al. (2011) and Li et al. (2011)
TiO_2, ZnO	15–20, 20–30 nm			Soil bacterial community	Reduced microbial biomass and diversity	Ge et al. (2011)
ZnO	50 mg/L, 40–100 nm	>10 mg/L 0.1–10,000 mg/L	$ZnSO_4$, $ZnHPO_4$	*Pseudomonas putida*, *E. fetida*, *Eisenia andrei*	Inhibition of bacterial growth, survival, and reproduction rate affected	Li et al. (2011) and Cañas et al. (2011)
TiO_2	35 nm	100, 250, 500, 750, 1000 mg/L	Nano-TiO_2	*Rhizobium leguminosarum*	Morphological changes to bacterial cells	Fan et al. (2014)

AgNPs or silver ions (Ag^+) on nitrifier *Nitrosomonas* species and nitrogen fixer *Azotobacter vinelandii*. The results showed that exposure of a sublethal dose of AgNPs or Ag^+ upregulated many nitrifying genes (*amoA1* and *amoC2*) in *Nitrosomonas europaea* (2.1 to 3.3-fold). NPs or nanomaterials can cause lipid peroxidation, which leads to damage in the cell membrane by accomplishing the generation of reactive oxygen species (ROS) (Manke et al., 2013; Huang et al., 2015) (Fig. 1.2; Table 1.1).

Cabiscol et al. (2000) illustrated consequences of ROS on bacteria that peroxidizes the lipid bilayer and leads to changes in the fluidity and permeability of the membrane, thus making cells more vulnerable to failure of nutrient uptake and osmotic stress. In the cell membrane, more peroxidized fatty acids result in DNA damage by releasing free radicals (Thul et al., 2013). Adams et al. (2006) compared the ecotoxicity of TiO_2, SiO_2, and ZnO NPs, which exert toxicity on bacteria in the presence of light.

The interaction of microbes and NPs for comparing the physicochemical properties and biological response of engineered and metal oxide NPs has been illustrated (Adams et al., 2006). It has been observed that toxicity of NPs is species specific and depends on size and shape of metallic NPs (Adams et al., 2006; Thul et al., 2013). Additionally, Zuverza-Mena et al. (2017) also reported that the behavior of NPs on plants also depends on the exposure conditions and characteristics of NPs. Moreover, the toxic behavior of NPs also depends on environmental effects. Du et al. (2017) reported that at the ambient concentration of CO_2 (370 μmol/mol), TiO_2 NPs possess toxicity in seedlings of *Oryza sativa*, while at a high concentration of CO_2 (570 μmol/mol), TiO_2 NPs induced visible signs of toxicity by declining the overall biomass of the plant.

Besides this, surface coating of NPs is also capable of enhancing microbial toxicity (Adams et al., 2006; Thul et al., 2013). However, Medina-Velo et al. (2017) reported that coated ZnO NPs, at concentrations of 250 and 500 mg/kg, showed a beneficial role, i.e., increased leaf length (~13%) and root length (~44%), compared to the control. Moreover, Ge et al. (2012) suggested that engineered NPs characteristically could change the bacterial colonies in a dose-dependent manner. Similarly, Priester et al. (2012) stated that engineered NPs such as CeO_2 can eliminate N_2 fixation and impair growth in soybean (Fig. 1.2; Table 1.1).

Aken (2015) reported the antimicrobial effect of AgNPs by transcriptional analysis of some bacteria (Fig. 1.2; Table 1.1). Transcriptional analysis was performed to show the pattern of gene expressions (Aken, 2015). Nagy et al. (2011) and McQuillan and Shaw (2014) studied the impact of AgNPs on model bacterium *E. coli* involving its whole-genome microarray and focused on patches of gene expressions in the regulation of metabolism of sulfur, oxidative balance, and homeostasis of Ag, Cu, and Fe. McQuillan et al. (2012) reviewed that exposure of AgNPs on *E. coli* leads to stress. Radzig et al. (2013) found that AgNPs can also result in oxidative damage of DNA, comparatively less resistance to porin synthesis, and disturb water homeostasis and redox balance (Fig. 1.2; Table 1.1). Du et al. (2011) showed adverse effects of TiO_2 and ZnO NPs on microbial communities in the soil. Yang et al. (2013) studied the impact of AgNPs and Ag^+ ions on nitrogen-cycling bacteria. The authors showed that a sublethal dose of 35 nm carbon-coated AgNPs or Ag^+ ions did not affect the expression of nitrogen-fixing genes *nifD*, *nifH*, *vnfD*, and *anfD* in *A. vinelandii* and denitrifying genes *narG*, *napB*, *nirH*, and *norB* in *Pseudomonas stutzeri*.

Pelletier et al. (2010) reported that exposure of CeO_2 NPs results in disturbance in cellular respiration, oxidative stress, and iron deficiency in *E. coli*. Similarly, Xie et al. (2011) also showed the effect of ZnO NPs on *Campylobacter jejuni* and reported upregulation of stress-responsive genes. Yang et al. (2012) reported the negative impact of quantum dots and other dissolved NPs in *Pseudomonas aeruginosa* by characterization of transporter genes and other stress-responsive genes. They reported that NPs induced the generation of ROS in spite of induced activities of catalases, peroxidases, etc. Sytar et al. (2013) showed that hydrogen peroxide, superoxide anion, and hydroxyl radical are examples of ROS that take oxygen from metal and cause cellular damage (Fig. 1.3; Table 1.1).

Hossain and Mukherjee (2012, 2013) stated that during the mitotic cycle in *E. coli*, CdS and CdO NPs inhibit the septum formation and downregulate important cell division proteins. Cui et al. (2012) showed that 4,6-diaminopyrimidine sulfhydryl-modified AuNPs affect ribosomal protein S10 (proteosynthesis), peroxidase, hydroperoxide reductase, and F-type ATP synthase. In *E. coli* and *S. aureus*, graphene and graphene oxide nanowalls lead to rupture of the cell membrane, which results in leakage of intracellular substances and thus cell death (Huang et al., 2007; Akhavan and Ghaderi, 2010).

1.5.2 Effects of NPs on Plants

The sizes of NPs are so small that they can easily penetrate through the cell wall of plant roots and may cause devastating effects (Rai et al., 2015). They can change expression of

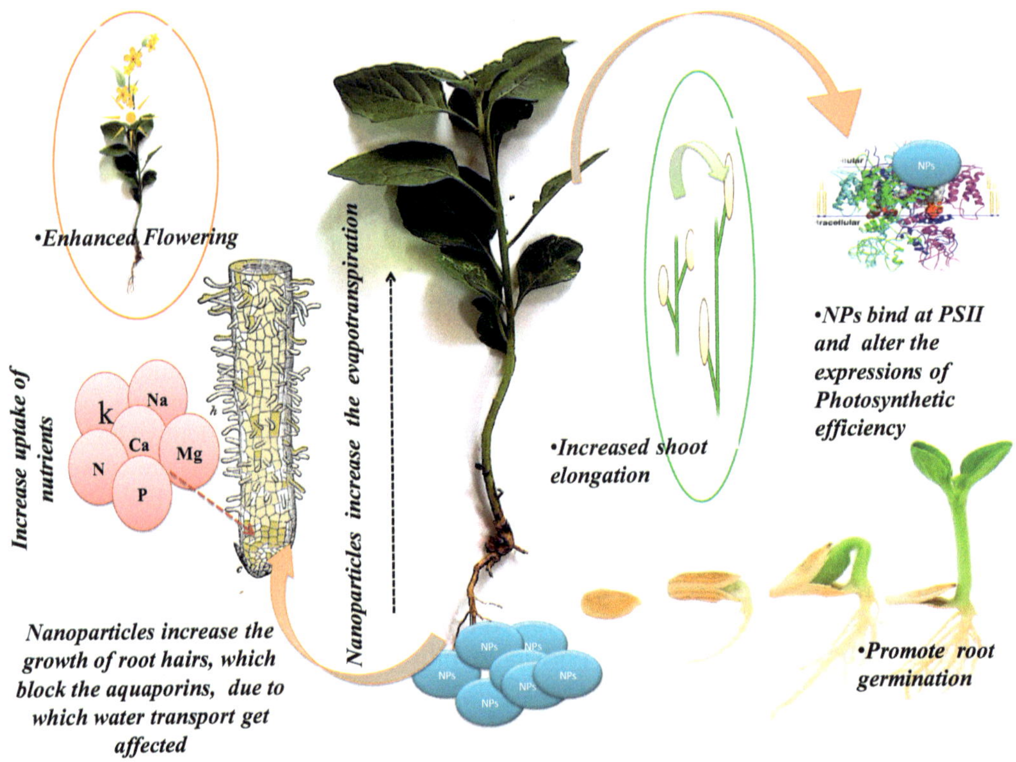

FIGURE 1.3 Nanoparticle toxicity at the whole plant level. *PSII*, photosystem II.

genes, biochemistry, physiology, and hence morphology of the plant (Siddiqui et al., 2015; Tripathi et al., 2016, 2017c,d,e; Singh et al., 2016; Shweta et al., 2016). DeRosa et al. (2010) demonstrated that engineered NPs such as ZnO NPs and carbon nanotubes have the ability to penetrate tissues of roots and seeds of *Solanum lycopersicum* through disrupting the plant surface (Tables 1.2 and 1.3). Furthermore, it has been demonstrated that NPs infiltrate into the cell wall and cell membrane of root epidermis accompanied by a multifaceted chain of

TABLE 1.2 Impact of Nanoparticles (NPs) on Plants

Types of NPs	Mechanism(s) of NP Toxicity	References
Fe_3O_4 and CoO	Disturb the respiration by blocking aquaporins	Wang et al. (2011) and Ghodake et al. (2011)
Ag	Acts as electron relay center and improves redox reactions In aquatic plants also, generation of reactive oxygen species is enhanced, which induces oxidative stress Enhances the production of secondary quinone electron acceptors in the electron transport chain, but inhibits the transport of electrons Modifies the nutritional value	Mallick et al. (2006), Oukarroum et al. (2013), Matorin et al. (2013) and Zuverza-Mena et al. (2016)
CeO_2	Changes or alters their redox state Ce^{3+} and Ce^{4+} to scavenge free radicals ($O_2^{\cdot -}$ and $\cdot OH$) and mimics superoxide dismutase activity Decreases light absorption and photochemical efficiency Increases the activity of the electron transport chain	Xie et al. (2008), Heckert et al. (2008), Horie et al. (2011), Boghossian et al. (2013) and Gomez-Garay et al. (2014)
TiO_2	In light and dark conditions, forms free radicals ($O_2^{\cdot -}$, $\cdot OH$ and $CO_2^{\cdot -}$); O_2^- / O_2^{2-} to O_2/H_2O_2)	Lei et al. (2008) and Fenoglio et al. (2009)
	Increases light absorption and quantum yield in photosystem II in *Spinacea olerocea*	Yang et al. (2007), Lei et al. (2007), Qi et al. (2013) and Mingyu et al. (2007)
ZnO	Traps electrons from —OH and produces $\cdot OH$	Lei et al. (2008) and Mclaren et al. (2009)
NiO	Produces $\cdot OH$ through Haber–Weiss reaction	Faisal et al. (2013)
CuO	Produces $\cdot OH$ through Fenton reaction	Fubini et al. (2007)
Fullerene, carbon nanotubes, graphene	Aggregation of root hairs and blocking of aquaporins	Liu et al. (2010), Qi et al. (2013), Tan et al. (2009) and Begum et al. (2011)
Au	Increases light absorption caused by plasmon resonance effect and decreases quantum yield because of chlorophyll to NP electron/energy transfer	Falco et al. (2011)
CdSe/ZnS	Enhance light absorption and improve the quality yield in photosystem I	Jung et al. (2010)
Mn	Improves phosphorylation activity in the electron transport chain	Lei et al. (2007)

TABLE 1.3 Impact of Nanoparticles (NPs) on Plants With Special Reference to Mechanisms of Action

NPs	Plant(s)	Size of NPs (nm)	Concentration	Effect of NPs on Plants With Their Mechanisms of Action	References
Ag (PVP- coated)	*Arabidopsis*	20	1, 2.5, 5, 10, 20 mg/L	Upregulated genes associated with the response to metal and oxidative stress, downregulated genes associated with the response to pathogen and hormonal stimuli	Kaveh et al. (2013)
Ag (polyethylene glycol coated)	*Arabidopsis*, *Populus tremula*	5–10	0.01–1 mg/L	Enhanced rate of evapotranspiration	Wang et al. (2013)
Al_2O_3	*Nicotiana tabacum*	<50	0.01–0.1 mg/L	Reduction in the activities of both dehydrogenase and oxidoreductase	Poborilova et al. (2013)
NiO	*Lycopersicum esculatum*	23.34	25, 50, 100, 250, 500, 1000, 1500, 2000 mg/L	Reduced catalase, glutathione, superoxide dismutase, and lipid peroxidation exhibit higher activities	Faisal et al. (2013)
Nano zerovalent iron	*Typha latifolia*	–	–	Reduced transpiration rate	Ma et al. (2013)
Quantum dot	*Arabidopsis*	6.3	5.8 nM	Ratio of reduced glutathione levels	Navarro et al. (2012)
Fullerene C_{60} and ZnO	*Arabidopsis*	–	–	Stress response of gene expression	Landa et al. (2012)
Single-walled carbon nanotube	*Oryza sativa*, *Arabidopsis*	1–2	25 mg/L	Induced chromatin condensation	Shen et al. (2010)
$C_{60}(OH)_{20}$	*Allium cepa*	1.12–1.74; 15.69–24.36	10, 110 mg/L	Plant cell damage	Chen et al. (2010)
TiO_2	*A. cepa*	10	0.25% mg/L	DNA damage	Ghosh et al. (2010)
	Zea mays	30	30, 1000 mg/L	Transpiration rate reduced	Asli and Neumann (2009)
Ag	*A. cepa*	<100	100 mg/L	Cell wall breaks, deformed chromosomes, indistinct disturbed metaphase, and mitosis	Kumari et al. (2009)
Multiwalled carbon nanotube	*O. sativa*	10–30	20, 40, 80 mg/L	Chromatin condensed inside the cytoplasm and caused cell death	Tan et al. (2009)
	Lycopersicon esculentum	10–35	50 mg/L	Upregulation of stress-related genes	Khodakovskaya et al. (2013)
Single-walled carbon nanotube (fluorescein isothiocyanate)	*N. tabacum*	Not reported	Not reported	Unnoticeable cell death	Liu et al. (2009)

events to penetrate the plant vascular bundle (xylem), and travel to the stele symplastically, to be translocated to the leaves. Similarly, Majumdar et al. (2014) clearly revealed the uptake mechanisms of CeO_2 NPs in kidney bean plants.

On the other hand, Ochoa et al. (2017) reported that the toxicity of NPs (CuO NPs) could be ameliorated by the phytohormone indole-3-acetic acid in *Pisum sativum*. Additionally, the toxicity of ZnO NPs and AgNPs is ameliorated by the supplementation of nitric oxide in *Triticum aesativum* and *P. sativum* (Tripathi et al., 2017c,d).

Nagy et al. (2011) reported upregulation of those genes that are involved in transport, reduction, and oxidation of isoelectronic NPs (Ag, Fe, and Cu), which may lead to oxidative stress by Fenton's reaction (Torney et al., 2007; Ševců et al., 2011). Generally, the analysis of transcription occurs on *Arabidopsis thaliana* plants and exposes the general changes in gene expressions in the presence of abiotic and biotic stimuli. Kaveh et al. (2013) showed that at higher doses (>5mg/L) of AgNPs, plant growth significantly decreases (Tables 1.2 and 1.3). Exposure of NPs (5mg/L) led to the upregulation of oxidative stress genes and downregulation of hormonal stimuli and pathogen genes (Kaveh et al., 2013). Yan et al. (2013) demonstrated that exposure of single-walled carbon nanotubes inhibits growth of root hairs in *Zea mays*. Landa et al. (2012) analyzed the whole genome using a microarray method in which they found that exposure of fullerene and ZnO NPs at 100mg/L on *A. thaliana* caused reduction in biomass or showed phytotoxic effects in association with upregulation of many genes involved in pathogens, and oxidative stress and in downregulation in genes related to cell organization, electron transport, energy pathways, and cell biosynthesis (Fig. 1.3; Tables 1.2 and 1.3).

Chu et al. (2012), Qian et al. (2013) and Nair and Chung (2014) demonstrated induction of genes in *A. thaliana* related to oxidative damage of DNA, oxidative stress, water homeostasis, and systemic acquired resistance (SAR) under NP stress. It has also been reported that NPs affect the major genes of photosynthetic pathways (Aken, 2015). Similarly, Ma et al. (2013) demonstrated the effect of indium oxide NPs in *A. thaliana* (L.) at a molecular level (Fig. 1.3; Tables 1.2 and 1.3). NPs also play a key role to trigger the oxidative stress genes and hence glutathione metabolism. Tassi et al. (2017) reported the application of CeO_2 NPs which could ameliorate the boron induced stress in *Helianthus annus* L.

Some NPs such as CdS quantum dots inhibited the growth and germination of plants by releasing Cd^{2+} ions (Fig. 1.3; Tables 1.2 and 1.3). Quantum dot NP exposure leads to over-expression of genes associated with pathogenesis, defense response, and SAR. In addition, Wang et al. (2014) reported the effect of graphene oxide NPs on translocation and their toxicity on *Arabidopsis* plants, and also showed that carbon NPs led to an increase in susceptibility of plants toward other abiotic stresses (Fig. 1.3; Tables 1.2 and 1.3).

Graphene NPs, in the presence of polyethylene glycol, negatively affect the development of seedlings in *A. thaliana* (Wang et al., 2014). Dimkpa et al. (2013) reported the disruption of growth of *Triticum aestivum* L. against AgNPs and demonstrated that in the presence of AgNPs, alterations in plant biomass and expression of genes related to oxidative stress occur (Figs. 1.1 and 1.4; Tables 1.2 and 1.3). Burklew et al. (2012) and Frazier et al. (2014) demonstrated that Al_2O_3 and TiO_2NPs declined plant biomass and germination rate because of upregulation of a number of microRNAs in *Nicotiana tabacum*. In the presence of TiO_2 NPs, the activities of glutathione reductase and ascorbate peroxidase in *Vicia faba* were found be decreased (Foltête et al., 2011; Rico et al., 2015).

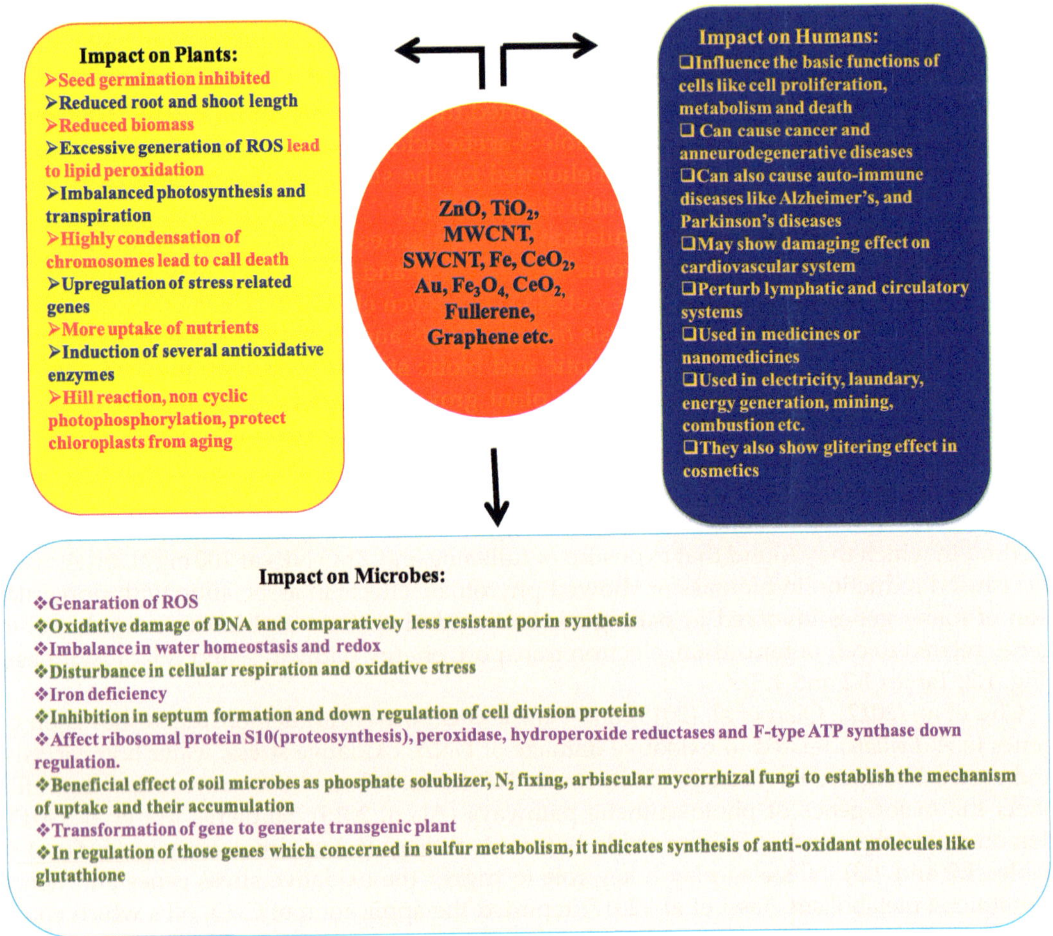

FIGURE 1.4 Overview of toxicity of nanoparticles on microorganisms, plants, and humans. *MWCNT*, multi-walled carbon nanotube; *SWCNT*, single-walled carbon nanotube.

Shen et al. (2010) observed that exposure of single-walled NPs to protoplast cells of *A. thaliana* and rice (*O. sativa*) resulted in ROS generation, DNA damage, and cell death (Fig. 1.4). Moreover, Yan et al. (2013) showed that exposure of single-walled carbon nanotubes hampers growth of root hairs in *Z. mays*. Introduction of NPs in plants induces a wide range of molecular responses, mostly related to transcription factors, which govern biotic and abiotic stresses (Aken, 2015). Lopez-Moreno et al. (2010) observed that exposure of ZnO and CeO_2 NPs resulted in homologous recombination, mutation, DNA damage, and/or deletions in soybean. In addition, Patlolla et al. (2012) demonstrated genotoxicity in bean plants (*V. faba*) after exposure of AgNPs, which manifested as increased micronuclei and chromosomal aberrations as well as decreased mitotic index (Fig. 1.3; Tables 1.2 and 1.3).

Stress in plants induces disturbance in photosynthetic activity because of oxidative stress. Stresses also lead to alteration in photosynthetic efficiency, photochemical fluorescence, and quantum yield. Experiments performed on *Spinacea olerocea* and *Lycopersicon esculentum* reveal alterations in light absorption and quantum yield of photosystem II (Yang et al., 2007; Lei et al., 2007; Qi et al., 2013; Rico et al., 2015). It was also observed that AuNPs and AgNPs affect the photosynthetic reaction center of photosystem I and quantum yield (Govorov and Carmeli, 2007; Shweta et al., 2016). Thus, previous studies indicate that NP exposure affects plants from morphology to gene expression (Figs. 1.1, 1.2 and 1.4).

1.5.3 Effects of NPs on Humans

In the present environment, everyone is being exposed to foreign NPs, which ultimately enter our bodies through inhalation, orally, or by other means. It is a fact that every organism encounters NP toxicity on a daily basis, which may have ill effects (Buzea et al., 2007) (Fig. 1.2).

There are various pathways through which NPs may enter in the human body:

- Through skin (by use of cosmetics containing NPs);
- Through inhalation (when NPs are present in the surrounding environment);
- Through ingestion (in the form of food products containing NPs);
- Through injection (in the form of nanomedicines).

The major toxic intruders are viruses made up of nucleoproteins. They are the most advanced type of toxicant and interact with biological systems as well as exploit the cellular machinery for their replication (Buzea et al., 2007). If the virus is benign, it may cause a common cold or flu-like symptoms.

NPs possess the capacity to cross the threshold level in humans, thus, cause damaging effects on cardiovascular systems (Buzea et al., 2007) (Fig. 1.2). Some NPs enter through dermal barriers (skin) or lungs (Table 1.4) and hence perturb the lymphatic and circulatory systems of both humans and animals (Oberdörster et al., 2005; Buzea et al., 2007) (Fig. 1.2).

After reaching the tissues or organs, NPs can potentially damage cellular processes (Fig. 1.2; Table 1.4). NP toxicity depends greatly on the arrangement of atoms (Buzea et al., 2007; Sutariya and Pathak, 2014). Asbestos NPs may cause lung cancer and other diseases (Buzea et al., 2007). Asbestos is found in various shapes and forms, hence, shows various ranges of toxicities (Buzea et al., 2007; Boulanger et al., 2014). NP synthesis, which is carried out by various physical processes ranging from combustion to erosion, may create health hazards both benign and lethal (Buzea et al., 2007; Kumar and Kumbhat, 2016). Industrial NPs are very tiny in appearance but can potentially increase pollution. Inhalation of NPs may cause several diseases such as lung cancer, asthma, bronchitis, emphysema, and neurodegenerative diseases such as Parkinson's and Alzheimer's diseases (Buzea et al., 2007; Kumar and Kumbhat, 2016) (Table 1.4) (Fig. 1.2).

1.6 ENVIRONMENTAL FATE OF NPs

In society, use of NPs such as Au, Ag, ZnO, and TiO_2 is increasing notably. Their application is so vast that they are also being used in the field of medicines (Fig. 1.4). Studies have shown that NPs present in water have lower concentrations, i.e., from mg/L to μg/L (Boxall et al., 2007; Mueller and Nowack, 2008; Neal et al., 2011), but because of a lack of sufficient data, their occurrence behavior is under debate (Fig. 1.4).

TABLE 1.4 Some Major Impacts of Nanoparticles (NPs) on Human

sNPs	Size (nm)	Doses	Cell Line	Major Effects	References
AgNPs	7–20	6.25–50 μg/mL	Skin carcinoma (A431) and fibrosarcoma (HT-1080)	Reduction in cell viability, increases oxidative stresses, DNA fragmentation and consequently increased caspase-3 activities	Arora et al. (2008)
	100	2.5–5.0 μg/mL	Human mesenchymal stem cells (hMSCs)	Proliferation of cells and chemotactive response declined; however, release of IL-8 inclined	Greulich et al. (2009)
	5–10	0.5 μg/mL	Hepatoma HepG2 cells	Increased cytotoxicity and oxidative stress	Kim et al. (2009)
	7–10	0.1–3.0 μg/mL	Hepatoma HepG2 cells	DNA damaged repair genes increased and induction of micronuclei reported. Hence shows nontoxic behavior	Kawata et al. (2009)
	25	0.46–2.32 ng/cm^2	Abdominal skin obtained from surgical waste	Silver nitrate has capacity to permeate the damaged skin cells	Larese et al. (2009)
	69	0–7.5 μg/mL	Acute monocytic leukemia cell line (THP-1)	Generation of reactive oxygen species and finally apoptosis of cell	Foldbjerg et al. (2009)
	20–80	0.34–1.7 μg/mL	Epidermal keratinocytes (HEKs)	Carbon-coated AgNPs: dose-dependent toxicity leads to decreases in viability of cell. Uncoated AgNPs: non toxic	Samberg et al. (2010)
	18	11 μg/mL	Baby hamster kidney (BHK21) and human colon adenocarcinoma (HT29) cells	p53-mediated apoptosis	Gopinath et al. (2010)
	100	10–200 μg/mL	Human embryonic stem cell (hESC)-derived neural stem/progenitor cells (NPCs)	Cell cycle arrested and induced cell death	Oh et al. (2016)
Citrate-coated colloidal AgNPs	30	100 μg/mL	Human skin HaCaT keratinocytes	Photo-, geno-, and cytotoxicity increased in dose-dependent manner	Lu et al. (2010)
Polyethylenimine-coated quantum dots (PEC-QD)	30–80	18 nM	Human foreskin fibroblast cells (HFF-1, ATCC SCRC-1041) and primary human dermal fibroblasts	Shown rate-determining cell death	Ware et al. (2014)

TABLE 1.4 Some Major Impacts of Nanoparticles (NPs) on Human—cont'd

sNPs	Size (nm)	Doses	Cell Line	Major Effects	References
$PS\text{-}NH_2$ or PS-COOH (amino-modified or carboxylated polystyrene NPs)	50 and 40	25 μg/mL	Lung cancer epithelial A549 cells	Disturbance in cyclin levels. Accumulated in lysosomal bodies, which lead to swelling and finally disintegration, then death of the cell	Kim and Ryu (2013)
Iron oxide NPs	30–65	1–100 mg/mL	Lung cancer cells (A549)	Toxic for lung cancerous cells	
	30–65	1–100 mg/mL	Fibroblast cells (IMR-90)	Nontoxic or selectively destroys only cancerous cells	
ZnO NPs	50–70	5 and 0.8 μg/mL	Epidermal cells	DNA damaging effect was induced by ZnO NP treatment	Sharma et al. (2009)
Mesoporous silica NPs	100	–	3T3-L1 cells and human mesenchymal stem cells (hMSCs)	Positively charged mesoporous silica NPs shown inhibitory effect on 3T3-L1 cells only	Chung et al. (2007)
TiO_2 NPs	20	30–150 μg/mL	Human lung epithelial cell line (H1299)	Shown to cause detrimental effect on H1299	Tedja et al. (2012)

Weinberg et al. (2011) showed that natural water containing engineered NPs in their aggregation and sedimentation phases may help in the removal of NPs from the suspension of water. Zhang et al. (2009) and Dobias and Bernier-Latmani (2013) showed that the aggregation of NPs occurs by neutralization of charge, electrical double-layer compression, bridging, and charge neutralization, which occur because of the combined effect of electrolyte species, pH, ionic strength, and other properties of water. All these factors together impart stability to NPs. In this way, it is possibly that NPs may enter in the food chain resulting in their biomagnifications or they can affect the organism at each trophic level (Rico et al., 2015; Unrine et al., 2012; Liu et al., 2013). Dobias and Bernier-Latmani (2013) stated that the cation of AgNPs may persist in the environment in the form of Ag^+ (Fig. 1.4). However, it is found that instead of high dilutions, the behavior and toxicity of engineered NPs in fresh water are different from sea water (Matranga and Corsi, 2012). For instance, suspended particulate matter and natural colloids show modified fate in estuarine waters (Dunphy Guzman et al., 2006; Hyung et al., 2007; Ju-Nam and Lead, 2008; Xie et al., 2008; Matranga and Corsi, 2012). Processes that affect the fate of engineered NPs are precipitation, agglomeration, and aggregation (Stolpe and Hassellöv, 2007; Matranga and Corsi, 2012). The bioavailability and transformation (physical, chemical, and biological reactions in living organisms) of engineered NPs generally depend on leaving their properties unchanged and the local chemistry of the environment (Ju-Nam and Lead, 2008; Matranga and Corsi, 2012).

1.7 CONCLUDING REMARKS AND FUTURE PERSPECTIVE

On the one hand, the increasing population in developing countries such as India and China is a major issue, which affect socioeconomic development. On the other hand, several sensitive problems are emerging globally because of lifestyle changes and environmental factors. Nanotechnology provides many applications in the field of medicines, energy, electronics, and life sciences. The objectives of humans are increasing day by day, which rely on the use of new technologies. In this connection, use of nanoscale products has increased abnormally and their negative impacts are still unknown.

In spite of several achievements in the field of nanotechnology, studies show that NPs adversely affect cellular, physiological, and molecular processes in microorganisms, plants, and humans, which is not a good indicator for the coming generation. Furthermore, recent findings will open new avenues of morphological, physiological, biochemical, and molecular investigations, which should appreciably advance our understanding of nanotoxicology. This will help to curtail the amount of NPs in the environment, and improve the environment for the survival of microorganisms, plants, and animals.

References

Adams, L.K., Lyon, D.Y., Alvarez, P.J., 2006. Comparative eco-toxicity of nanoscale TiO_2, SiO_2, and ZnO water suspensions. Water Res. 40, 3527–3532.

Aghdam, M.T.B., Mohammadi, H., Ghorbanpour, M., 2016. Effects of nanoparticulate anatase titanium dioxide on physiological and biochemical performance of *Linum usitatissimum* (Linaceae) under well-watered and drought stress conditions. Braz. J. Bot. 39 (1), 139–146.

Ahmad, A., Senapati, S., Khan, M.I., Kumar, R., Sastry, M., 2005. Extra-/intracellular biosynthesis of gold nanoparticles by an alkalotolerant fungus *Trichothecium* sp. J. Biomed. Nanotechnol. 1, 47–53.

Aken, V.B., 2015. Gene expression changes in plants and microorganisms exposed to nanomaterials. Curr. Opin. Biotechnol. 33, 206–219.

Akhavan, O., Ghaderi, E., 2010. Toxicity of graphene and graphene oxide nanowalls against bacteria. ACS Nano 4, 5731–5736.

Albrecht, M.A., Evans, C.W., Raston, C.L., 2006. Green chemistry and the health implications of nanoparticles. Green Chem. 8, 417–432.

Andersen, C.P., King, G., Plocher, M., Storm, M., Pokhrel, L.R., Johnson, M.G., Rygiewicz, P.T., 2016. Germination and early plant development of ten plant species exposed to TiO_2 and CeO_2 nanoparticles. Environ. Toxicol. Chem. 35 (9), 2223–2229.

Armendariz, V., Herrera, I., Jose-yacaman, M., Troiani, H., Santiago, P., Gardea-Torresdey, J.L., 2004. Size controlled gold nanoparticle formation by *Avena sativa* biomass: use of plants in nanobiotechnology. J. Nanopart. Res. 6 (4), 377–382.

Arora, S., Jain, J., Rajwade, J.M., Paknikar, K.M., 2008. Cellular responses induced by silver nanoparticles: in vitro studies. Toxicol. Lett. 179 (2), 93–100.

Asli, S., Neumann, P.M., 2009. Colloidal suspensions of clay or titanium dioxide nanoparticles can inhibit leaf growth and transpiration via physical effects on root water transport. Plant Cell Environ. 32 (5), 577–584.

Bai, J., Zhong, X., Jiang, S., Huang, Y., Duan, X., 2010. Graphene nanomesh. Nat. Nanotechnol. 5, 190–194.

Barrios, A.C., Medina-Velo, I.A., Zuverza-Mena, N., Dominguez, O.E., Peralta-Videa, J.R., Gardea-Torresdey, J.L., 2017. Nutritional quality assessment of tomato fruits after exposure to uncoated and citric acid coated cerium oxide nanoparticles, bulk cerium oxide, cerium acetate and citric acid. Plant Physiol. Biochem. 110, 100–107.

Begum, P., Ikhtiari, R., Fugetsu, B., 2011. Graphene phytotoxicity in the seedling stage of cabbage, tomato, red spinach, and lettuce. Carbon 49 (12), 3907–3919.

Bélanger, R.R., Benhamou, N., Menzies, J.G., 2003. Cytological evidence of an active role of silicon in wheat resistance to powdery mildew (*Blumeria graminis* f. sp. *tritici*). Phytopathology 93 (4), 402–412.

Biswas, P., Wu, C.Y., 2005. Nanoparticles and the environment. J. Air Waste Manag. Assoc. 55 (6), 708–746.
Boghossian, A.A., Sen, F., Gibbons, B.M., Sen, S., Faltermeier, S.M., Giraldo, J.P., Zhang, C.T., Zhang, J., Heller, D.A., Strano, M.S., 2013. Application of nanoparticle antioxidants to enable hyperstable chloroplasts for solar energy harvesting. Adv. Energy Mater. 3 (7), 881–893.
Boulanger, G., Andujar, P., Pairon, J.C., Billon-Galland, M.A., Dion, C., Dumortier, P., Brochard, P., Sobaszek, A., Bartsch, P., Paris, C., Jaurand, M.C., 2014. Quantification of short and long asbestos fibers to assess asbestos exposure: a review of fiber size toxicity. Environ. Health 13 (1), 1.
Boxall, A.B., Chaudhry, Q., Sinclair, C., Jones, A., Aitken, R., Jefferson, B., Watts, C., 2007. Current and Future Predicted Environmental Exposure to Engineered Nanoparticles. Central Science Laboratory, Department of the Environment and Rural Affairs, London, UK, p. 89.
Burklew, C.E., Ashlock, J., Winfrey, W.B., Zhang, B., 2012. Effects of aluminum oxide nanoparticles on the growth, development, and microRNA expression of tobacco (*Nicotiana tabacum*). PLoS One 7, e34783.
Buseck, P.R., Pósfai, M., 1999. Airborne minerals and related aerosol particles: effects on climate and the environment. Proc. Natl. Acad. Sci. U.S.A. 96, 3372–3379.
Buzea, C., Pacheco, I.I., Robbie, K., 2007. Nanomaterials and nanoparticles: sources and toxicity. Biointerphases 2, MR17–MR71.
Cabiscol, E., Tamarit, J., Ros, J., 2000. Oxidative stress in bacteria and protein damage by reactive oxygen species. Int. Microbiol. 3, 3–8.
Cañas, J.E., Qi, B., Li, S., Maul, J.D., Cox, S.B., Das, S., Green, M.J., 2011. Acute and reproductive toxicity of nano-sized metal oxides (ZnO and TiO_2) to earthworms (*Eisenia fetida*). J. Env. Mon. 13 (12), 3351–3357.
Chad, S., 2010. What Determines Productivity? No. w15712. National Bureau of Economic Research.
Chekli, L., Phuntsho, S., Roy, M., Shon, H.K., 2013. Characterisation of Fe-oxide nanoparticles coated with humic acid and Suwannee River natural organic matter. Sci. Total Environ. 461–462, 19–27.
Chen, R., Ratnikova, T.A., Stone, M.B., Lin, S., Lard, M., Huang, G., Hudson, J.S., Ke, P.C., 2010. Differential uptake of carbon nanoparticles by plant and mammalian cells. Small 6 (5), 612–617.
Choi, O., Deng, K.K., Kim, N.J., Ross, L., Surampalli, R.Y., Hu, Z., 2008. The inhibitory effects of silver nanoparticles, silver ions, and silver chloride colloids on microbial growth. Water Res. 42, 3066–3074.
Chu, H., Kim, H.J., Kim, J.S., Kim, M.S., Yoon, B.D., Park, H.J., Kim, C.Y., 2012. A nanosized Ag–silica hybrid complex prepared by γ-irradiation activates the defense response in *Arabidopsis*. Radiat. Phys. Chem. 81 (2), 180–184.
Chung, T.H., Wu, S.H., Yao, M., Lu, C.W., Lin, Y.S., Hung, Y., Mou, C.Y., Chen, Y.C., Huang, D.M., 2007. The effect of surface charge on the uptake and biological function of mesoporous silica nanoparticles in 3T3-L1 cells and human mesenchymal stem cells. Biomater. 28 (19), 2959–2966.
Colvin, V.L., 2003. The potential environmental impact of engineered nanomaterials. Nat. Biotechnol. 21, 1166–1170.
Cui, Y., Zhao, Y., Tian, Y., Zhang, W., Lü, X., Jiang, X., 2012. The molecular mechanism of action of bactericidal gold nanoparticles on *Escherichia coli*. Biomaterials 33, 2327–2333.
Delgado-Ramos, G.C., 2014. Nanotechnology in Mexico: global trends and national implications for policy and regulatory issues. Technol. Soc. 37, 4–15.
DeRosa, M.C., Monreal, C., Schnitzer, M., Walsh, R., Sultan, Y., 2010. Nanotechnology in fertilizers. Nat. Nanotechnol. 5, 91.
Dhawan, A., Shanker, R., Das, M., Gupta, K.C., 2011. Guidance for safe handling of nanomaterials. J. Biomed. Nanotechnol. 7, 218–224.
Dimkpa, C.O., Latta, D.E., McLean, J.E., Britt, D.W., Boyanov, M.I., Anderson, A.J., 2013. Fate of CuO and ZnO nano- and microparticles in the plant environment. Environ. Sci. Technol. 47, 4734–4742.
Dobias, J., Bernier-Latmani, R., 2013. Silver release from silver nanoparticles in natural waters. Environ. Sci. Technol. 47, 4140–4146.
Donaldson, K., Stone, V., Tran, C.L., Kreyling, W., Borm, P.J.A., 2004. Nanotoxicology. Occup Environ. Med. 61 (9), 727–728.
Dong, Y., Feng, S.S., 2007. Poly (D, L-lactide-co-glycolide)(PLGA) nanoparticles prepared by high pressure homogenization for paclitaxel chemotherapy. Int. J. Pharm. 342 (1), 208–214.
Du, W., Sun, Y., Ji, R., Zhu, J., Wu, J., Guo, H., 2011. TiO_2 and ZnO nanoparticles negatively affect wheat growth and soil enzyme activities in agricultural soil. J. Environ. Monit. 13, 822–828.
Du, W., Gardea-Torresdey, J.L., Xie, Y., Yin, Y., Zhu, J., Zhang, X., Ji, R., Gu, K., Peralta-Videa, J.R., Guo, H., 2017. Elevated CO_2 levels modify TiO_2 nanoparticle effects on rice and soil microbial communities. Sci. Total Environ. 578, 408–416.

Duncan, T.V., 2011. Applications of nanotechnology in food packaging and food safety: barrier materials, antimicrobials and sensors. J. Colloid Interface Sci. 363, 1–24.

Dunford, R., Salinaro, A., Cai, L., Serpone, N., Horikoshi, S., Hidaka, H., Knowland, J., 1997. Chemical oxidation and DNA damage catalysed by inorganic sunscreen ingredients. FEBS Lett. 418, 87–90.

Dunphy Guzman, K.A., Finnegan, M.P., Banfield, J.F., 2006. Influence of surface potential on aggregation and transport of titania nanoparticles. Environ. Sci. Technol. 40, 7688–7693.

Environmental Protection Agency. http://www.epa.gov/iaq/index.html; www.epa.gov/airtrends/pmreport03/pmunderstand_2405.pdf.

Faisal, M., Saquib, Q., Alatar, A.A., Al-Khedhairy, A.A., Hegazy, A.K., Musarrat, J., 2013. Phytotoxic hazards of NiO-nanoparticles in tomato: a study on mechanism of cell death. J. Hazard Mater. 250, 318–332.

Falco, W.F., Botero, E.R., Falcão, E.A., Santiago, E.F., Bagnato, V.S., Caires, A.R.L., 2011. In vivo observation of chlorophyll fluorescence quenching induced by gold nanoparticles. J. Photochem. Photobiol. A: Chem. 225 (1), 65–71.

Fan, R., Huang, Y.C., Grusak, M.A., Huang, C.P., Sherrier, D.J., 2014. Effects of nano-TiO_2 on the agronomically-relevant Rhizobium–legume symbiosis. Sci. Total Environ. 466, 503–512.

Fenoglio, I., Greco, G., Livraghi, S., Fubini, B., 2009. Non-UV-induced radical reactions at the surface of TiO_2 nanoparticles that may trigger toxic responses. Chem-A Eur. J 15 (18), 4614–4621.

Ferreira, A.J., Cemlyn-Jones, J., Cordeiro, C.R., 2013. Nanoparticles, nanotechnology and pulmonary nanotoxicology. Rev. Port. Pneumol. 19 (1), 28–37 (English Edition).

Foldbjerg, R., Olesen, P., Hougaard, M., Dang, D.A., Hoffmann, H.J., Autrup, H., 2009. PVP-coated silver nanoparticles and silver ions induce reactive oxygen species, apoptosis and necrosis in THP-1 monocytes. Toxicol. Lett. 190 (2), 156–162.

Foltête, A.S., Masfaraud, J.F., Bigorgne, E., Nahmani, J., Chaurand, P., Botta, C., Labille, J., Rose, J., Férard, J.F., Cotelle, S., 2011. Environmental impact of sunscreen nanomaterials: ecotoxicity and genotoxicity of altered TiO_2 nanocomposites on *Vicia faba*. Environ. Pollut. 159 (10), 2515–2522.

Frazier, T.P., Burklew, C.E., Zhang, B., 2014. Titanium dioxide nanoparticles affect the growth and microRNA expression of tobacco (*Nicotiana tabacum*). Funct. Integr. Genom. 14, 75–83.

Fubini, B., Fenoglio, I., Tomatis, M., 2007. Physicochemical characteristics of nanoparticles that determine potential toxicity. Nanotoxicology: characterization, dosing, and health effects. Informa Healthcare, New York, pp. 59–70.

Ge, Y., Schimel, J.P., Holden, P.A., 2011. Evidence for negative effects of TiO_2 and ZnO nanoparticles on soil bacterial communities. Environ. Sci. Technol. 45 (4), 1659–1664.

Ge, Y., Schimel, J.P., Holden, P.A., 2012. Identification of soil bacteria susceptible to TiO_2 and ZnO nanoparticles. Appl. Environ. Microbiol. 78 (18), 6749–6758.

Ghodake, G., Seo, Y.D., Lee, D.S., 2011. Hazardous phytotoxic nature of cobalt and zinc oxide nanoparticles assessed using. Allium cepa. J. Hazard Mater. 186 (1), 952–955.

Ghosh, M., Bandyopadhyay, M., Mukherjee, A., 2010. Genotoxicity of titanium dioxide (TiO_2) nanoparticles at two trophic levels: plant and human lymphocytes. Chemosphere 81 (10), 1253–1262.

Gomez-Garay, A., Pintos, B., Manzanera, J.A., Lobo, C., Villalobos, N., Martín, L., 2014. Uptake of CeO_2 nanoparticles and its effect on growth of *Medicago arborea* in vitro plantlets. Biol. Trace Elem. Res. 161 (1), 143–150.

Goodman, C.M., McCusker, C.D., Yilmaz, T., Rotello, V.M., 2004. Toxicity of gold nanoparticles functionalized with cationic and anionic side chains. Bioconjugate Chem. 15 (4), 897–900.

Gopinath, P., Gogoi, S.K., Sanpui, P., Paul, A., Chattopadhyay, A., Ghosh, S.S., 2010. Signaling gene cascade in silver nanoparticle induced apoptosis. Colloid. Surface B 77 (2), 240–245.

Govorov, A.O., Carmeli, I., 2007. Hybrid structures composed of photosynthetic system and metal nanoparticles: plasmon enhancement effect. Nano Lett. 7, 620–625.

Greulich, C., Kittler, S., Epple, M., Muhr, G., Köller, M., 2009. Studies on the biocompatibility and the interaction of silver nanoparticles with human mesenchymal stem cells (hMSCs). Langenbeck's Archives of Surgery 394 (3), 495–502.

Grieger, K.D., Baun, A., Owen, R., 2010. Redefining risk research priorities for nanomaterials. J. Nanopart. Res. 12 (2), 383–392.

Gurr, J.R., Wang, A.S., Chen, C.H., Jan, K.Y., 2005. Ultrafine titanium dioxide particles in the absence of photoactivation can induce oxidative damage to human bronchial epithelial cells. Toxicology 213, 66–73.

Hagens, W.I., Oomen, A.G., de Jong, W.H., Cassee, F.R., Sips, A.J., 2007. What do we (need to) know about the kinetic properties of nanoparticles in the body? Regul. Toxicol. Pharm. 49 (3), 217–229.

Handy, R.D., Shaw, B.J., 2007. Toxic effects of nanoparticles and nanomaterials: implications for public health, risk assessment and the public perception of nanotechnology. Health Risk Soc. 9 (2), 125–144.

Heckert, E.G., Karakoti, A.S., Seal, S., Self, W.T., 2008. The role of cerium redox state in the SOD mimetic activity of nanoceria. Biomaterials 29 (18), 2705–2709.

Hoet, P.H., Brüske-Hohlfeld, I., Salata, O.V., 2004. Nanoparticles-known and unknown health risks. J. Nanobiotechnol. 2, 12.

Holden, P.A., Nisbet, R.M., Lenihan, H.S., Miller, R.J., Cherr, G.N., Schimel, J.P., Gardea-Torresdey, J.L., 2012. Ecological nanotoxicology: integrating nanomaterial hazard considerations across the subcellular, population, community, and ecosystems levels. Acc. Chem. Res. 46 (3), 813–822.

Horie, M., Nishio, K., Kato, H., Fujita, K., Endoh, S., Nakamura, A., Miyauchi, A., Kinugasa, S., Yamamoto, K., Niki, E., Yoshida, Y., 2011. Cellular responses induced by cerium oxide nanoparticles: induction of intracellular calcium level and oxidative stress on culture cells. J. Biochem. 150 (4), 461–471.

Hossain, S.T., Mukherjee, S.K., 2012. CdO nanoparticle toxicity on growth, morphology, and cell division in *Escherichia coli*. Langmuir 28, 16614–16622.

Hossain, S.T., Mukherjee, S.K., 2013. Toxicity of cadmium sulfide (CdS) nanoparticles against *Escherichia coli* and HeLa cells. J. Hazard. Mater. 260, 1073–1082.

Houghton, J., 2005. Global warming. Rep. Prog. Phys. 68, 1343–1403. http://www.nature.com.

Huang, W.C., Tsai, P.J., Chen, Y.C., 2007. Functional Gold Nanoparticles as Photothermal Agents for Selective-Killing of Pathogenic Bacteria.

Huang, S., Wang, L., Liu, L., Hou, Y., Li, L., 2015. Nanotechnology in agriculture, livestock, and aquaculture in China. A review. Agron. Sustain. Dev. 35, 369–400.

Husar, R.B., Tratt, D.M., Schichtel, B.A., Falke, S.R., Li, F., Jaffe, D., Gasso, S., Gill, T., Laulainen, N.S., Lu, F., Reheis, M.C., 2001. Asian dust events of April 1998. J. Geophys. Res. Atmos. 106 (D16), 18317–18330.

Hyung, H., Fortner, J.D., Hughes, J.B., Kim, J.H., 2007. Natural organic matter stabilizes carbon nanotubes in the aqueous phase. Environ. Sci. Technol. 41 (1), 179–184.

Ju-Nam, Y., Lead, J.R., 2008. Manufactured nanoparticles: an overview of their chemistry, interactions and potential environmental implications. Sci. Total Environ. 400, 396–414.

Jung, H., Gulis, G., Gupta, S., Redding, K., Gosztola, D.J., Wiederrecht, G.P., Stroscio, M.A., Dutta, M., 2010. Optical and electrical measurement of energy transfer between nanocrystalline quantum dots and photosystem I. J. Phys. Chem. B 114 (45), 14544–14549.

Kachynski, A.V., Kuzmin, A.N., Nyk, M., Roy, I., Prasad, P.N., 2008. Zinc oxide nanocrystals for nonresonant nonlinear optical microscopy in biology and medicine. J. Phys. Chem. C 112, 10721–10724.

Kajander, E.O., 2006. Nanobacteria-propagating calcifying nanoparticles. Lett. Appl. Microbiol. 42, 549–552.

Kaveh, R., Li, Y.S., Ranjbar, S., Tehrani, R., Brueck, C.L., Aken, V.B., 2013. Changes in *Arabidopsis thaliana* gene expression in response to silver nanoparticles and silver ions. Environ. Sci. Technol. 47, 10637–10644.

Kawata, K., Osawa, M., Okabe, S., 2009. In vitro toxicity of silver nanoparticles at noncytotoxic doses to HepG2 human hepatoma cells. Environ. Sci. Technol. 43 (15), 6046–6051.

Keller, A.A., Lazareva, A., 2013. Predicted releases of engineered nanomaterials: from global to regional to local. Environ. Sci. Technol. Lett. 1 (1), 65–70.

Khodakovskaya, M.V., Kim, B.S., Kim, J.N., Alimohammadi, M., Dervishi, E., Mustafa, T., Cernigla, C.E., 2013. Carbon nanotubes as plant growth regulators: effects on tomato growth, reproductive system, and soil microbial community. Small 9 (1), 115–123.

Kim, S., Choi, J.E., Choi, J., Chung, K.H., Park, K., Yi, J., Ryu, D.Y., 2009. Oxidative stress-dependent toxicity of silver nanoparticles in human hepatoma cells. Toxicol. Vitro 23 (6), 1076–1084.

Kim, S., Ryu, D.Y., 2013. Silver nanoparticle-induced oxidative stress, genotoxicity and apoptosis in cultured cells and animal tissues. J. Appl. Toxicol. 33, 78–89.

Kim, S.W., Kim, K.S., Lamsal, K., Kim, Y.J., Kim, S.B., Jung, M., Sim, S.J., Kim, H.S., Chang, S.J., Kim, J.K., Lee, Y.S., 2009. An in vitro study of antifungal effect of silver nanoparticles on oak wilt pathogen *Raffaelea* sp. J. Microbiol. Biotechnol. 19 (8), 760–764.

Kubota, Y., Shuin, T., Kawasaki, C., Hosaka, M., Kitamura, H., Cai, R., Sakai, H., Hashimoto, K., Fujishima, A., 1994. Photokilling of T-24 human bladder cancer cells with titanium dioxide. Br. J. Cancer 70 (6), 1107.

Kumar, N., Kumbhat, S., 2016. Essentials in Nanoscience and Nanotechnology. John Wiley and Sons.

Kumari, M., Mukherjee, A., Chandrasekaran, N., 2009. Genotoxicity of silver nanoparticles in *Allium cepa. Sci.* Total Environ. 407 (19), 5243–5246.

Landa, P., Vankova, R., Andrlova, J., Hodek, J., Marsik, P., Storchova, H., White, J.C., Vanek, T., 2012. Nanoparticle-specific changes in *Arabidopsis thaliana* gene expression after exposure to ZnO, TiO_2, and fullerene soot. J. Hazard. Mater. 241–242, 55–62.

Larese, F.F., D'Agostin, F., Crosera, M., Adami, G., Renzi, N., Bovenzi, M., Maina, G., 2009. Human skin penetration of silver nanoparticles through intact and damaged skin. Toxicol. 255 (1), 33–37.

Lazareva, A., Keller, A.A., 2014. Estimating potential life cycle releases of engineered nanomaterials from wastewater treatment plants. Acs. Sustain. Chem. Eng. 2 (7), 1656–1665.

Lei, Z., Mingyu, S., Xiao, W., Chao, L., Chunxiang, Q., Liang, C., Hao, H., Xiaoqing, L., Fashui, H., 2007. Effects of nano-anatase on spectral characteristics and distribution of LHCII on the thylakoid membranes of spinach. Biol. Trace Elem. Res. 120 (1–3), 273–283.

Lei, Z., Mingyu, S., Xiao, W., Chao, L., Chunxiang, Q., Liang, C., Hao, H., Xiaoqing, L., Fashui, H., 2008. Antioxidant stress is promoted by nano-anatase in spinach chloroplasts under UV-B radiation. Biol. Trace Elem. Res. 121 (1), 69–79.

Li, M., Zhu, L., Lin, D., 2011. Toxicity of ZnO nanoparticles to *Escherichia coli*: mechanism and the influence of medium components. Environ. Sci. Technol. 45 (5), 1977–1983.

Lim, H.K., Asharani, P.V., Hande, M.P., 2012. Enhanced genotoxicity of silver nanoparticles in DNA repair deficient mammalian cells. Front Genet 3.

Liu, X., Zhang, P., Li, X., Chen, H., Dang, Y., Larson, C., Roco, M.C., Wang, X., 2009. Trends for nanotechnology development in China, Russia, and India. J. Nanopart. Res. 11 (8), 1845–1866.

Liu, Q., Zhao, Y., Wan, Y., Zheng, J., Zhang, X., Wang, C., Fang, X., Lin, J., 2010. Study of the inhibitory effect of water-soluble fullerenes on plant growth at the cellular level. ACS Nano 4 (10), 5743–5748.

Liu, Z., Wang, P., Zha, T., 2013. Land-price dynamics and macroeconomic fluctuations. Econometrica 81, 1147–1184.

Lopez-Moreno, M.L., De La Rosa, G., Hernandez-Viezcas, J.A., Peralta-Videa, J.R., Gardea-Torresdey, J.L., 2010. X-ray absorption spectroscopy (XAS) corroboration of the uptake and storage of CeO_2 nanoparticles and assessment of their differential toxicity in four edible plant species. J. Agric. Food Chem. 58, 3689–3693.

López-Moreno, M.L., de la Rosa, G., Cruz-Jiménez, G., Castellano, L., Peralta-Videa, J.R., Gardea-Torresdey, J.L., 2017. Effect of ZnO nanoparticles on corn seedlings at different temperatures; X-ray absorption spectroscopy and ICP/OES studies. Microchem. J. 134, 54–61

Lu, W., Senapati, D., Wang, S., Tovmachenko, O., Singh, A.K., Yu, H., Ray, P.C., 2010. Effect of surface coating on the toxicity of silver nanomaterials on human skin keratinocytes. Chem. Phys. Lett. 487 (1), 92–96.

Lyu, S., Wei, X., Chen, J., Wang, C., Wang, X., Pan, D., 2017. Titanium as a beneficial element for crop production. Front. Plant Sci. 8.

Ma, C., Chhikara, S., Xing, B., Musante, C., White, J.C., Dhankher, O.P., 2013. Physiological and molecular response of *Arabidopsis thaliana* (L.) to nanoparticle cerium and indium oxide exposure. ACS Sustain. Chem. 1, 768–778.

Majumdar, S., Peralta-Videa, J.R., Bandyopadhyay, S., Castillo-Michel, H., Hernandez-Viezcas, J.A., Sahi, S., Gardea-Torresdey, J.L., 2014. Exposure of cerium oxide nanoparticles to kidney bean shows disturbance in the plant defense mechanisms. J. Hazard. Mater. 278, 279–287.

Majumdar, S., Almeida, I.C., Arigi, E.A., Choi, H., VerBerkmoes, N.C., Trujillo-Reyes, J., Flores-Margez, J.P., White, J.C., Peralta-Videa, J.R., Gardea-Torresdey, J.L., 2015. Environmental effects of nanoceria on seed production of common bean (*Phaseolus vulgaris*): a proteomic analysis. Environ. Sci. Technol. 49 (22), 13283–13293.

Mallick, K., Witcomb, M., Scurrell, M., 2006. Silver nanoparticle catalysed redox reaction: an electron relay effect. Mater. Chem. Phys. 97 (2), 283–287.

Manke, A., Wang, L., Rojanasakul, Y., 2013. Mechanisms of nanoparticle-induced oxidative stress and toxicity. BioMed Res. Int.Article ID 942916, 15 pp. http://dx.doi.org/10.1155/2013/942916.

Matorin, D.N., Todorenko, D.A., Seïfullina, N., Zaiadan, B.K., Rubin, A.B., 2013. Effect of silver nanoparticles on the parameters of chlorophyll fluorescence and P700 reaction in the green alga. Chlamydomonas reinhardtii. Microbiol. 83 (1), 33–40.

Matranga, V., Corsi, I., 2012. Toxic effects of engineered nanoparticles in the marine environment: model organisms and molecular approaches. Mar. Environ. Res. 76, 32–40.

Mclaren, A., Valdes-Solis, T., Li, G., Tsang, S.C., 2009. Shape and size effects of ZnO nanocrystals on photocatalytic activity. J. Am. Chem. Soc. 131 (35), 12540–12541.

McQuillan, J.S., Shaw, A.M., 2014. Differential gene regulation in the Ag nanoparticle and Ag^+-induced silver stress response in *Escherichia coli*: a full transcriptomic profile. Nanotoxicology 8, 177–184.

McQuillan, J.S., Groenaga Infante, H., Stokes, E., Shaw, A.M., 2012. Silver nanoparticle enhanced silver ion stress response in *Escherichia coli K12*. Nanotoxicology 6, 857–866.

Medina-Velo, I.A., Barrios, A.C., Zuverza-Mena, N., Hernandez-Viezcas, J.A., Chang, C.H., Ji, Z., Zink, J.I., Peralta-Videa, J.R., Gardea-Torresdey, J.L., 2017. Comparison of the effects of commercial coated and uncoated ZnO nanomaterials and Zn compounds in kidney bean (*Phaseolus vulgaris*) plants. J. Hazard. Mater. 332, 214–222.

Mihranyan, A., Ferraz, N., Strømme, M., 2012. Current status and future prospects of nanotechnology in cosmetics. Prog. Mater. Sci. 57, 875–910.

Mingyu, S., Xiao, W., Chao, L., Chunxiang, Q., Xiaoqing, L., Liang, C., Hao, H., Fashui, H., 2007. Promotion of energy transfer and oxygen evolution in spinach photosystem II by nano-anatase TiO_2. Biol. Trace Elem. Res. 119 (2), 183–192.

Miralles, P., Church, T.L., Harris, A.T., 2012. Toxicity, uptake, and translocation of engineered nanomaterials in vascular plants. Environ. Sci. Technol. 46, 9224–9239.

Mohanty, A., Wu, Y., Cao, B., 2014. Impacts of engineered nanomaterials on microbial community structure and function in natural and engineered ecosystems. Appl. Microbiol. Biotechnol. 98 (20), 8457–8468.

Mu, L., Sprando, R.L., 2010. Application of nanotechnology in cosmetics. Pharm. Res. 27, 1746–1749.

Mueller, N.C., Nowack, B., 2008. Exposure modeling of engineered nanoparticles in the environment. Environ. Sci. Technol. 42, 4447–4453.

Mukherjee, A., Peralta-Videa, J.R., Gardea-Torresdey, J., White, J.C., 2016. Effects and uptake of nanoparticles in plants. Eng. Nanopart. Environ. 4.

Müller, R.H., Radtke, M., Wissing, S.A., 2002. Nanostructured lipid matrices for improved microencapsulation of drugs. Int. J. Pharm. 242, 121–128.

Nagy, A., Harrison, A., Sabbani, S., Munson Jr., R.S., Dutta, P.K., Waldman, W.J., 2011. Silver nanoparticles embedded in zeolite membranes: release of silver ions and mechanism of antibacterial action. Int. J. Nanomed. 6, 1833–1852.

Nair, P.M.G., Chung, I.M., 2014. Cell cycle and mismatch repair genes as potential biomarkers in *Arabidopsis thaliana* seedlings exposed to silver nanoparticles. Bull. Environ. Contam. Toxicol. 92, 719–725.

Navarro, D.A., Bisson, M.A., Aga, D.S., 2012. Investigating uptake of water-dispersible CdSe/ZnS quantum dot nanoparticles by *Arabidopsis thaliana* plants. J. Hazard. Mater. 211, 427–435.

Neal, C., Jarvie, H., Rowland, P., Lawler, A., Sleep, D., Scholefield, P., 2011. Titanium in UK rural, agricultural and urban/industrial rivers: geogenic and anthropogenic colloidal/sub-colloidal sources and the significance of within-river retention. Sci. Total Environ. 409, 1843–1853.

Nel, A., Xia, T., Mädler, L., Li, N., 2006. Toxic Potential of Materials at the Nanolevel. Science 311 (5761), 622–627.

Oberdörster, G., Oberdörster, E., Oberdörster, J., 2005. An emerging discipline evolving from studies of ultrafine particles supplemental web sections. Environ. Health Perspect. 113 (7), 823–839.

Ochoa, L., Medina-Velo, I.A., Barrios, A.C., Bonilla-Bird, N.J., Hernandez-Viezcas, J.A., Peralta-Videa, J.R., Gardea-Torresdey, J.L., 2017. Modulation of CuO nanoparticles toxicity to green pea (*Pisum sativum* Fabaceae) by the phytohormone indole-3-acetic acid. Sci. Total Environ. 598, 513–524.

Oh, J.H., Son, M.Y., Choi, M.S., Kim, S., Choi, A.Y., Lee, H.A., Kim, K.S., Kim, J., Song, C.W., Yoon, S., 2016. Integrative analysis of genes and miRNA alterations in human embryonic stem cells-derived neural cells after exposure to silver nanoparticles. Toxicol. Appl. Pharm. 299, 8–23.

Oukarroum, A., Barhoumi, L., Pirastru, L., Dewez, D., 2013. Silver nanoparticle toxicity effect on growth and cellular viability of the aquatic plant Lemna gibba. Environ. Toxicol. Chem. 32 (4), 902–907.

Pal, S., Tak, Y.K., Song, J.M., 2007. Does the antibacterial activity of silver nanoparticles depend on the shape of the nanoparticle? A study of the gram-negative bacterium. Escherichia coli. Appl. Environ. Microb 73 (6), 1712–1720.

Patlolla, A.K., Berry, A., May, L., Tchounwou, P.B., 2012. Genotoxicity of silver nanoparticles in *Vicia faba*: a pilot study on the environmental monitoring of nanoparticles. Int. J. Environ. Res. Public Health 9, 1649–1662.

Pavasupree, S., Ngamsinlapasathian, S., Nakajima, M., Suzuki, Y., Yoshikawa, S., 2006. Synthesis, characterization, photocatalytic activity and dye-sensitized solar cell performance of nanorods/nanoparticles TiO_2 with mesoporous structure. J. Photochem. Photobiol. A 184, 163–169.

Pelletier, D.A., Suresh, A.K., Holton, G.A., McKeown, C.K., Wang, W., Gu, B., Mortensen, N.P., Allison, D.P., Joy, D.C., Allison, M.R., Brown, S.D., 2010. Effects of engineered cerium oxide nanoparticles on bacterial growth and viability. Appl. Environ. Microbiol. 76 (24), 7981–7989.

Peralta-Videa, J.R., Zhao, L., Lopez-Moreno, M.L., de la Rosa, G., Hong, J., Gardea-Torresdey, J.L., 2011. Nanomaterials and the environment: a review for the biennium 2008–2010. J. Hazard. Mater. 186, 1–15.

Peralta-Videa, J.R., Huang, Y., Parsons, J.G., Zhao, L., Lopez-Moreno, L., Hernandez-Viezcas, J.A., Gardea-Torresdey, J.L., 2016. Plant-based green synthesis of metallic nanoparticles: scientific curiosity or a realistic alternative to chemical synthesis? Nan. Environ. Engin. 1 (1), 4.

Poborilova, Z., Opatrilova, R., Babula, P., 2013. Toxicity of aluminium oxide nanoparticles demonstrated using a BY-2 plant cell suspension culture model. Environ. Exp. Bot. 91, 1–11.
Priester, J.H., Ge, Y., Mielke, R.E., Horst, A.M., Moritz, S.C., Espinosa, K., Gelb, J., Walker, S.L., Nisbet, R.M., An, Y.J., Schimel, J.P., 2012. Soybean susceptibility to manufactured nanomaterials with evidence for food quality and soil fertility interruption. Proc. Natl. Acad. Sci. U.S.A. 109 (37), E2451–E2456.
Qi, M., Liu, Y., Li, T., 2013. Nano-TiO_2 improve the photosynthesis of tomato leaves under mild heat stress. Biol. Trace Elem. Res. 156, 323–328.
Qian, H., Peng, X., Han, X., Ren, J., Sun, L., Fu, Z., 2013. Comparison of the toxicity of silver nanoparticles and silver ions on the growth of terrestrial plant model *Arabidopsis thaliana*. J. Environ. Sci. 25, 1947–1956.
Radzig, M.A., Nadtochenko, V.A., Koksharova, O.A., Kiwi, J., Lipasova, V.A., Khmel, I.A., 2013. Antibacterial effects of silver nanoparticles on gram-negative bacteria: influence on the growth and biofilms formation, mechanisms of action. Colloid Surf. B 102, 300–306.
Rai, M., Ribeiro, C., Mattoso, L., Duran, N., 2015. Nanotechnologies in Food and Agriculture. Springer.
Reddy, P.V.L., Hernandez-Viezcas, J.A., Peralta-Videa, J.R., Gardea-Torresdey, J.L., 2016. Lessons learned: are engineered nanomaterials toxic to terrestrial plants? Sci. Total Environ. 568, 470–479.
Rehn, B., Seiler, F., Rehn, S., Bruch, J., Maier, M., 2003. Investigations on the inflammatory and genotoxic lung effects of two types of titanium dioxide: untreated and surface treated. Toxicol. Appl. Pharmacol. 189, 84–95.
Rico, C.M., Peralta-Videa, J.R., Gardea-Torresdey, J.L., 2015. Chemistry, biochemistry of nanoparticles, and their role in antioxidant defense system in plants. In: Siddiqui, M.H., Al-Whaibi, M.H., Mohammad, F. (Eds.), Nanotechnology and Plant Sciences: Nanoparticles and Their Impact on Plants. Springer, pp. 1–17 (Chapter 1).
Roco, M.C., Mirkin, C.A., Hersam, M.C., 2011. Nanotechnology research directions for societal needs in 2020, Vol. 1. Springer Netherlands, p. 690.
Roh, J.Y., Park, Y.K., Park, K., Choi, J., 2010. Ecotoxicological investigation of CeO_2 and TiO_2 nanoparticles on the soil nematode *Caenorhabditis elegans* using gene expression, growth, fertility, and survival as endpoints. Environ. Toxicol. Pharmacol. 29 (2), 167–172.
Roh, J.Y., Sim, S.J., Yi, J., Park, K., Chung, K.H., Ryu, D.Y., Choi, J., 2009. Ecotoxicity of silver nanoparticles on the soil nematode *Caenorhabditis elegans* using functional ecotoxicogenomics. Environ. Sci. Technol. 43 (10), 3933–3940.
Sahu, N., Soni, D., Chandrashekhar, B., Sarangi, B.K., Satpute, D., Pandey, R.A., 2013. Synthesis and characterization of silver nanoparticles using Cynodon dactylon leaves and assessment of their antibacterial activity. Bioproc. Biosyst. Eng. 36 (7), 999–1004.
Sajid, M., Ilyas, M., Basheer, C., Tariq, M., Daud, M., Baig, N., Shehzad, F., 2015. Impact of nanoparticles on human and environment: review of toxicity factors, exposures, control strategies, and future prospects. Environ. Sci. Pollut. R 22 (6), 4122–4143.
Samberg, M.E., Oldenburg, S.J., Monteiro-Riviere, N.A., 2010. Evaluation of silver nanoparticle toxicity in skin in vivo and keratinocytes in vitro. Environ. Health Persp. 118 (3), 407.
Sapkota, A., Anceno, A.J., Baruah, S., Shipin, O.V., Dutta, J., 2011. Zinc oxide nanorod mediated visible light photoinactivation of model microbes in water. Nanotechno. 22 (21), 215703.
Sapkota, A., Symons, J.M., Kleissl, J., Wang, L., Parlange, M.B., Ondov, J., Breysse, P.N., Diette, G.B., Eggleston, P.A., Buckley, T.J., 2005. Impact of the 2002 Canadian forest fires on particulate matter air quality in Baltimore City. Environ. Sci. Technol. 39 (1), 24–32.
Savolainen, K., Alenius, H., Norppa, H., Pylkkänen, L., Tuomi, T., Kasper, G., 2010. Risk assessment of engineered nanomaterials and nanotechnologies-a review. Toxicology 269, 92–104.
Scheufele, D.A., Corley, E.A., Dunwoody, S., Shih, T.J., Hillback, E., Guston, D.H., 2007. Scientists worry about some risks more than the public. Nat. Nanotechnol. 2, 732–734.
Serpone, N., Salinaro, A., Emeline, A., 2001. Deleterious effects of sunscreen titanium dioxide nanoparticles on DNA: efforts to limit DNA damage by particle surface modification. In: The International Symposium on Biomedical Optics, International Society for Optics and Photonics Published in Proceedings Nanoparticles and Nanostructured Surfaces: Novel Reporters with Biological Applications, vol. 4258, pp. 86–98.
Servin, A.D., White, J.C., 2016. Nanotechnology in agriculture: next steps for understanding engineered nanoparticle exposure and risk. NanoImpact 1, 9–12.
Ševců, A., El-Temsah, Y.S., Joner, E.J., Černík, M., 2011. Oxidative stress induced in microorganisms by zero-valent iron nanoparticles. Microbes Environ. 26 (4), 271–281.
Sharma, V.K., Yngard, R.A., Lin, Y., 2009. Silver nanoparticles: green synthesis and their antimicrobial activities. Adv. Colloid Interface Sci. 145, 83–96.

Sharma, V., Singh, P., Pandey, A.K., Dhawan, A., 2012. Induction of oxidative stress, DNA damage and apoptosis in mouse liver after sub-acute oral exposure to zinc oxide nanoparticles. Mutat. Res. 745 (1), 84–91.

Shen, C.X., Zhang, Q.F., Li, J., Bi, F.C., Yao, N., 2010. Induction of programmed cell death in *Arabidopsis* and rice by single-wall carbon nanotubes. Am. J. Bot. 97, 1602–1609.

Shi, Z., Shao, L., Jones, T.P., Lu, S., 2005. Microscopy and mineralogy of airborne particles collected during severe dust storm episodes in Beijing, China. J. Geophys. Res. 110.

Shweta, Tripathi, D.K., Singh, S., Singh, S., Dubey, N.K., Chauhan, D.K., 2016. Impact of nanoparticles on photosynthesis: challenges and opportunities. Mater. Focus 5 (5), 405–411.

Siddiqui, M.H., Al-Whaibi, M.H., Firoz, M., Al-Khaishany, M.Y., 2015. Role of nanoparticles in plants. Nanotechnol. Plant Sci. 19–35 Springer International Publishing.

Siegmann, P., Acevedo, F.J., Siegmann, K., Maldonado-Bascón, S., 2008. A probabilistic source attribution model for nanoparticles in air suspension applied on the main roads of Madrid and Mexico City. Atmos. Environ. 42 (17), 3937–3948.

Singh, S., Tripathi, D.K., Dubey, N.K., Chauhan, D.K., 2016. Effects of nano-materials on seed germination and seedling growth: striking the slight balance between the concepts and controversies. Mater. Focus 5 (3), 195–201.

Stampoulis, D., Sinha, S.K., White, J.C., 2009. Assay-dependent phytotoxicity of nanoparticles to plants. Environ. Sci. Technol. 43 (24), 9473–9479.

Stolpe, B., Hassellöv, M., 2007. Changes in size distribution of fresh water nanoscale colloidal matter and associated elements on mixing with seawater. Geochim. Cosmochim. Acta 71, 3292–3301.

Storrs, J., 2005. Nanofuture: What's Next for Nanotechnology. Prometheus Books.

Strambeanu, N., Demetrovici, L., Dragos, D., 2015. Natural sources of nanoparticles. In: Nanoparticles' Promises and Risks. Springer International Publishing, pp. 9–19.

Sutariya, V.B., Pathak, Y., 2014. Biointeractions of Nanomaterials. CRC Press.

Swihart, M.T., 2003. Vapor-phase synthesis of nanoparticles. Curr. Opin. Colloid Interface Sci. 8 (1), 127–133.

Sytar, O., Kumar, A., Latowski, D., Kuczynska, P., Strzałka, K., Prasad, M.N.V., 2013. Heavy metal-induced oxidative damage, defense reactions, and detoxification mechanisms in plants. Acta Physiol. Plant. 35, 985–999.

Szymańska, R., Kołodziej, K., Ślesak, I., Zimak-Piekarczyk, P., Orzechowska, A., Gabruk, M., Żądło, A., Habina, I., Knap, W., Burda, K., Kruk, J., 2016. Titanium dioxide nanoparticles (100–1000 mg/l) can affect vitamin E response in *Arabidopsis thaliana*. Environ. Pollut. 213, 957–965.

Tan, X.M., Lin, C., Fugetsu, B., 2009. Studies on toxicity of multi-walled carbon nanotubes on suspension rice cells. Carbon 47 (15), 3479–3487.

Tassi, E., Giorgetti, L., Morelli, E., Peralta-Videa, J.R., Gardea-Torresdey, J.L., Barbafieri, M., 2017. Physiological and biochemical responses of sunflower (*Helianthus annuus* L.) exposed to nano-CeO_2 and excess boron: modulation of boron phytotoxicity. Plant Physiol. Biochem. 110, 50–58.

Taylor, D.A., 2002. Dust in the wind. Environ. Health Perspect. 110, A80–A87.

Tedja, R., Lim, M., Amal, R., Marquis, C., 2012. Effects of serum adsorption on cellular uptake profile and consequent impact of titanium dioxide nanoparticles on human lung cell lines. ACS nano 6 (5), 4083–4093.

Thul, S.T., Sarangi, B.K., Pandey, R.A., 2013. Nanotechnology in Agroecosystem: implications on plant productivity and its soil environment. Expert Opin. Environ. Biol. 2, 1.

Tilston, E.L., Collins, C.D., Mitchell, G.R., Princivalle, J., Shaw, L.J., 2013. Nanoscale zerovalent iron alters soil bacterial community structure and inhibits chloroaromatic biodegradation potential in Aroclor 1242-contaminated soil. Environ. Pollut. 173, 38–46.

Tiwari, D.K., Behari, J., Sen, P., 2008. Application of Nanoparticles in Waste Water Treatment 1. Citeseer.

Torney, F., Trewyn, B.G., Lin, V.S., Wang, K., 2007. Mesoporous silica nanoparticles deliver DNA and chemicals into plants. Nat. Nanotechnol. 2, 295–300.

Tripathi, D.K., Singh, S., Singh, V.P., Prasad, S.M., Chauhan, D.K., Dubey, N.K., 2016. Silicon nanoparticles more efficiently alleviate arsenate toxicity than silicon in maize cultiver and hybrid differing in arsenate tolerance. Front. Environ. Sci. 4, 46.

Tripathi, D.K., Tripathi, A., Shweta, S.S., Singh, Y., Vishwakarma, K., Yadav, G., Sharma, S., Singh, V.K., Mishra, R.K., Upadhyay, R.G., Dubey, N.K., 2017a. Uptake, accumulation and toxicity of silver nanoparticle in autotrophic plants, and heterotrophic microbes: a concentric review. Front. Microbiol. 8.

Tripathi, D.K., Singh, S., Singh, S., Pandey, R., Singh, V.P., Sharma, N.C., Prasad, S.M., Dubey, N.K., Chauhan, D.K., 2017b. An overview on manufactured nanoparticles in plants: uptake, translocation, accumulation and phytotoxicity. Plant Physiol. Biochem. 110, 2–12.

Tripathi, D.K., Mishra, R.K., Singh, S., Singh, S., Vishwakarma, K., Sharma, S., Singh, V.P., Singh, P.K., Prasad, S.M., Dubey, N.K., Pandey, A.C., 2017c. Nitric oxide ameliorates zinc oxide nanoparticles phytotoxicity in wheat seedlings: implication of the ascorbate–glutathione cycle. Front. Plant Sci. 8.

Tripathi, D.K., Singh, S., Singh, S., Srivastava, P.K., Singh, V.P., Singh, S., Prasad, S.M., Singh, P.K., Dubey, N.K., Pandey, A.C., Chauhan, D.K., 2017d. Nitric oxide alleviates silver nanoparticles (AgNps)-induced phytotoxicity in *Pisum sativum* seedlings. Plant Physiol. Biochem. 110, 167–177.

Tripathi, D.K., Singh, S., Singh, V.P., Prasad, S.M., Dubey, N.K., Chauhan, D.K., 2017e. Silicon nanoparticles more effectively alleviated UV-B stress than silicon in wheat (*Triticum aestivum*) seedlings. Plant Physiol. Biochem. 110, 70–81.

Tungittiplakorn, W., Lion, L.W., Cohen, C., Kim, J.Y., 2004. Engineered polymeric nanoparticles for soil remediation. Environ. Sci. Technol. 38, 1605–1610.

Unrine, J.M., Shoults-Wilson, W.A., Zhurbich, O., Bertsch, P.M., Tsyusko, O.V., 2012. Trophic transfer of Au nanoparticles from soil along a simulated terrestrial food chain. Environ. Sci. Technol. 46 (17), 9753–9760.

Unrine, J.M., Tsyusko, O.V., Hunyadi, S.E., Judy, J.D., Bertsch, P.M., 2010. Effects of Particle Size on Chemical Speciation and Bioavailability of Copper to Earthworms Exposed to Copper Nanoparticles. J. Environ. Qual. 39 (6), 1942–1953.

Wang, H., Kou, X., Pei, Z., Xiao, J.Q., Shan, X., Xing, B., 2011. Physiological effects of magnetite (Fe_3O_4) nanoparticles on perennial ryegrass (*Lolium perenne* L.) and pumpkin (*Cucurbita mixta*) plants. Nanotoxicol 5 (1), 30–42.

Wang, J., Koo, Y., Alexander, A., Yang, Y., Westerhof, S., Zhang, Q., Schnoor, J.L., Colvin, V.L., Braam, J., Alvarez, P.J., 2013. Phytostimulation of poplars and Arabidopsis exposed to silver nanoparticles and Ag^+ at sublethal concentrations. Environ. Sci. Tech. 47 (10), 5442–5449.

Wang, Q., Zhao, S., Zhao, Y., Rui, Q., Wang, D., 2014. Toxicity and translocation of graphene oxide in *Arabidopsis* plants under stress conditions. RSC Adv. 4, 60891–60901.

Wang, E.C., Wang, A.Z., 2014. Nanoparticles and their applications in cell and molecular biology. Integr. Biol. 6 (1), 9–26.

Ware, M.J., Godin, B., Singh, N., Majithia, R., Shamsudeen, S., Serda, R.E., Meissner, K.E., Rees, P., Summers, H.D., 2014. Analysis of the influence of cell heterogeneity on nanoparticle dose response. ACS Nano 8 (7), 6693–6700.

Weinberg, H., Galyean, A., Leopold, M., 2011. Evaluating engineered nanoparticles in natural waters. TrAC 30, 72–83.

Westerdahl, D., Fruin, S., Sax, T., Fine, P.M., Sioutas, C., 2005. Mobile platform measurements of ultrafine particles and associated pollutant concentrations on freeways and residential streets in Los Angeles. Atmos. Environ. 39, 3597–3610.

Xiao, L., Takada, H., Maeda, K., Haramoto, M., Miwa, N., 2005. Antioxidant effects of water-soluble fullerene derivatives against ultraviolet ray or peroxylipid through their action of scavenging the reactive oxygen species in human skin keratinocytes. Biomed. Pharmacother. 59, 351–358.

Xie, B., Xu, Z., Guo, W., Li, Q., 2008. Impact of natural organic matter on the physicochemical properties of aqueous C60 nanoparticles. Environ. Sci. Technol. 42, 2853–2859.

Xie, Y., He, Y., Irwin, P.L., Jin, T., Shi, X., 2011. Antibacterial activity and mechanism of action of zinc oxide nanoparticles against *Campylobacter jejuni*. Appl. Environ. Microbiol. 77, 2325–2331.

Yan, S., Zhao, L., Li, H., Zhang, Q., Tan, J., Huang, M., He, S., Li, L., 2013. Single-walled carbon nanotubes selectively influence maize root tissue development accompanied by the change in the related gene expression. J. Hazard. Mater. 246–247, 110–118.

Yang, F., Liu, C., Gao, F., Su, M., Wu, X., Zheng, L., Hong, F., Yang, P., 2007. The improvement of spinach growth by nano-anatase TiO_2 treatment is related to nitrogen photoreduction. Biol. Trace Elem. Res. 119 (1), 77–88.

Yang, H., Fan, W., Vaneski, A., Susha, A.S., Teoh, W.Y., Rogach, A.L., 2012. Heterojunction engineering of CdTe and CdSe quantum dots on TiO_2 nanotube arrays: intricate effects of size-dependency and interfacial contact on photoconversion efficiencies. Adv. Funct. Mater. 22 (13), 2821–2829.

Yang, Y., Wang, J., Xiu, Z., Alvarez, P.J., 2013. Impacts of silver nanoparticles on cellular and transcriptional activity of nitrogen-cycling bacteria. Environ. Toxicol. Chem. 32, 1488–1494.

Zezulka, Š., Kummerová, M., Babula, P., Váňová, L., 2013. Lemna minor exposed to fluoranthene: growth, biochemical, physiological and histochemical changes. Aquat. Toxicol. 140–141, 37–47.

Zhang, W.X., 2003. Nanoscale iron particles for environmental remediation: an overview. J. Nanopart. Res. 5, 323–332.

Zhang, Y., Chen, Y., Westerhoff, P., Crittenden, J., 2009. Impact of natural organic matter and divalent cations on the stability of aqueous nanoparticles. Water Res. 43, 4249–4257.

Zuverza-Mena, N., Armendariz, R., Peralta-Videa, J.R., Gardea-Torresdey, J.L., 2016. Effects of silver nanoparticles on radish sprouts: root growth reduction and modifications in the nutritional value. Front. Plant Sci. 7.

Zuverza-Mena, N., Martínez-Fernández, D., Du, W., Hernandez-Viezcas, J.A., Bonilla-Bird, N., López-Moreno, M.L., Komárek, M., Peralta-Videa, J.R., Gardea-Torresdey, J.L., 2017. Exposure of engineered nanomaterials to plants: insights into the physiological and biochemical responses-a review. Plant Physiol. Biochem. 110, 236–264.

Further Reading

Alkilany, A.M., Murphy, C.J., 2010. Toxicity and cellular uptake of gold nanoparticles: what we have learned so far? J. Nanopart. Res. 12, 2313–2333.

Ankamwar, B., Chaudhary, M., Sastry, M., 2005. Gold nanotriangles biologically synthesized using tamarind leaf extract and potential application in vapor sensing. Syn. React. Inorg. Met. Org. Nano Met. Chem. 35, 19–26.

Apel, K., Hirt, H., 2004. Reactive oxygen species: metabolism, oxidative stress, and signal transduction. Annu. Rev. Plant Biol. 55, 373–399.

Austin, L.A., Kang, B., Yen, C.W., El-Sayed, M.A., 2011. Nuclear targeted silver nanospheres perturb the cancer cell cycle differently than those of nanogold. Bioconjug. Chem. 22, 2324–2331.

Austin, L.A., Mackey, M.A., Dreaden, E.C., El-Sayed, M.A., 2014. The optical, photothermal and facile surface chemical properties of gold and silver nanoparticles in biodiagnostics, therapy, and drug delivery. Arch. Toxicol. 88, 1391–1417.

Bergh, A., Craford, G., Duggal, A., Haitz, R., 2001. The promise and challenge of solid-state lighting. Phys. Today 54 (12), 42–47.

Biener, J., Farfan-Arribas, E., Biener, M., Friend, C.M., Madix, R.J., 2005. Synthesis of TiO_2 nanoparticles on the Au (111) surface. J. Chem. Phys. 123, 94705.

Cardona, C.M., Jannach, S.H., Huang, H., Itojima, Y., Leblanc, R.M., Gawley, R.E., Baker, G.A., Brauns, E.B., 2002. Spatially resolved derivatization of solid-phase-synthesis beads with fluorescent dendrimers: creation of localized microdomains. Helv. Chim. Acta 85 (10), 3532–3558.

Chang, A.L.S., Khosravi, V., Egbert, B., 2006. A case of argyria after colloidal silver ingestion. J. Cutan. Pathol. 33, 809–811.

Christou, P., McCabe, D.E., Swain, W.F., 1988. Stable transformation of soybean callus by DNA-coated gold particles. Plant Physiol. 87, 671–674.

de Jong, W.H., Borm, P.J., 2008. Drug delivery and nanoparticles: applications and hazards. Int. J. Nanomed. 3, 133–149.

de Jong, W.H., Roszek, B., Geertsma, R.E., 2005. Nanotechnology in Medical Applications: Possible Risks for Human Health RIVM Report 265001002., pp. 1–46.

Dos Santos, C.A., Seckler, M.M., Ingle, A.P., Gupta, I., Galdiero, S., Galdiero, M., Gade, A., Rai, M., 2014. Silver nanoparticles: therapeutical uses, toxicity, and safety issues. J. Pharm. Sci. 103 (7), 1931–1944.

Duncan, R., 2003. The drawing era of polymer therapeutics. Nat. Rev. Drug Discov. 2, 347–360.

Evelyn, A., Mannick, S., Sermon, P.A., 2003. Unusual carbon-based nanofibers and chains among diesel-emitted particles. Nano Lett. 3, 63–64.

Fabrega, J., Luoma, S.N., Tyler, C.R., Galloway, T.S., Lead, J.R., 2011. Silver nanoparticles: behaviour and effects in the aquatic environment. Environ. Int. 37 (2), 517–531.

Ferrari, M., 2005. Cancer nanotechnology: opportunities and challenges. Nat. Rev. Cancer 5, 161–171.

Gilbert, B., Huang, F., Zhang, H., Waychunas, G.A., Banfield, J.F., 2004. Nanoparticles: strained and stiff. Science 305 (5684), 651–654.

Goebel, H.H., Muller, J., DeMyer, W., 1973. Myopathy associated with Marfan's syndrome fine structural and histochemical observations. Neurology 23, 1257–1268.

Gottschalk, F., Sonderer, T., Scholz, R.W., Nowack, B., 2009. Modeled environmental concentrations of engineered nanomaterials (TiO_2, ZnO, Ag, CNT, fullerenes) for different regions. Environ. Sci. Technol. 43, 9216–9222.

Handford, C.E., Dean, M., Henchion, M., Spence, M., Elliott, C.T., Campbell, K., 2014. Implications of nanotechnology for the agri-food industry: opportunities, benefits and risks. Trends Food Sci. Technol. 40 (2), 226–241.

Hett, A., 2004. Nanotechnology. Small Matter, Many Unknowns. Risk Perception Swiss Re, Zurich, Switzerland, pp. 1–55.

Jianrong, C., Yuqing, M., Nongyue, H., Xiaohua, W., Sijiao, L., 2004. Nanotechnology and biosensors. Biotechnol. Adv. 22 (7), 505–518.

Karn, B., Kuiken, T., Otto, M., 2009. Nanotechnology and in situ remediation: a review of the benefits and potential risks. Environ. Health Perspect. 1823–1831.

Karuppanapandian, T., Moon, J.-C., Kim, C., Manoharan, K., Kim, W., 2011. Reactive oxygen species in plants: their generation, signal transduction, and scavenging mechanisms. Aust. J. Crop Sci. 5, 709–725.

Keller, A.A., McFerran, S., Lazareva, A., Suh, S., 2013. Global life cycle releases of engineered nanomaterials. J. Nanopart. Res. 15 (6), 1692.

Köhler, A.R., Som, C., Helland, A., Gottschalk, F., 2008. Studying the potential release of carbon nanotubes throughout the application life cycle. J. Clean. Prod. 16 (8), 927–937.

Lahiani, M.H., Dervishi, E., Chen, J., Nima, Z., Gaume, A., Biris, A.S., Khodakovskaya, M.V., 2013. Impact of carbon nanotube exposure to seeds of valuable crops. ACS Appl. Mater. Interfaces 5 (16), 7965–7973.

Loeschner, K., Hadrup, N., Qvortrup, K., Larsen, A., Gao, X., Vogel, U., Mortensen, A., Lam, H.R., Larsen, E.H., 2011. Distribution of silver in rats following 28 days of repeated oral exposure to silver nanoparticles or silver acetate. Part. Fibre Toxicol. 8 (1), 1.

Ma, X., Wang, C., 2010. Fullerene nanoparticles affect the fate and uptake of trichloroethylene in phytoremediation systems. Environ. Eng. Sci. 27, 989–992.

Ma, X., Geiser-Lee, J., Deng, Y., Kolmakov, A., 2010. Interactions between engineered nanoparticles (ENPs) and plants: phytotoxicity, uptake and accumulation. Sci. Total Environ. 408, 3053–3061.

Maier, L.A., 2002. Clinical approach to chronic beryllium disease and other nonpneumoconiotic interstitial lung diseases. J. Thorac. Imaging 17, 273–284.

Marmiroli, M., Pagano, L., Savo Sardaro, M.L., Villani, M., Marmiroli, N., 2014. Genome-wide approach in *Arabidopsis thaliana* to assess the toxicity of cadmium sulfide quantum dots. Environ. Sci. Technol. 48, 5902–5909.

Maurer, L.L., Meyer, J.N., 2016. A systematic review of evidence for silver nanoparticles-induced mitochondrial toxicity. Environ. Sci. 3 (2), 311–322.

McKendry, I.G., Hacker, J.P., Stull, R., Sakiyama, S., Mignacca, D., Reid, K., 2001. Long-range transport of Asian dust to the lower Fraser Valley, British Columbia, Canada. J. Geophys. Res. Atmos. 106, 18361–18370.

Michalet, X., Pinaud, F.F., Bentolila, L.A., Tsay, J.M., Doose, S., Li, J.J., Sundaresan, G., Wu, A.M., Gambhir, S.S., Weiss, S., 2005. Quantum dots for live cells, in vivo imaging, and diagnostics. Science 307 (5709), 538–544.

Moore, M.N., 2006. Do nanoparticles present ecotoxicological risks for the health of the aquatic environment? Environ. Int. 32, 967–976.

Nair, R., Varghese, S.H., Nair, B.G., Maekawa, T., Yoshida, Y., Kumar, D.S., 2010. Nanoparticulate material delivery to plants. Plant Sci. 179, 154–163.

Nekrasova, G.F., Ushakova, O.S., Ermakov, A.E., Uimin, M.A., Byzov, I.V., 2011. Effects of copper (II) ions and copper oxide nanoparticles on *Elodea densa* Planch. Russ. J. Ecol. 42 (6), 458–463.

Nemery, B., 1990. Metal toxicity and the respiratory tract. Eur. Respir. J. 3, 202–219.

Nowack, B., Bucheli, T.D., 2007. Occurrence, behavior and effects of nanoparticles in the environment. Environ. Pollut. 150, 5–22.

Ohbo, Y., Fukuzako, H., Takeuchi, K., Takeuchi, K., Takigawa, M., 1996. Argyria and convulsive seizures caused by ingestion of silver in a patient with schizophrenia. Psychiatry Clin. Neurosci. 50, 89–90.

Olcott, H.S., Fraenkel-Conrat, H., 1947. Specific group reagents for proteins. Chem. Rev. 41, 151–197.

Olcott, C.T., Richter, G.W., 1958. Experimental argyrosis. VI. Electron microscopic study of ingested silver in the kidney of the rat. Lab. Investig. 7, 103–109.

Ray, P.C., Yu, H., Fu, P.P., 2009. Toxicity and environmental risks of nanomaterials: challenges and future needs. J. Environ. Sci. Health Part C 27 (1), 1–35.

Roco, M.C., 2011. The Long View of Nanotechnology Development: The National Nanotechnology Initiative at 10 Years.

Rungby, J., 1986. Exogenous silver in dorsal root ganglia, peripheral nerve, enteric ganglia, and adrenal medulla. Acta Neuropathol. 69, 45–53.

Rungby, J., Danscher, G., 1983. Localization of exogenous silver in brain and spinal cord of silver exposed rats. Acta Neuropathol. 60, 92–98.

Ryan, K., Cousins, J.B. (Eds.), July 15, 2009. The SAGE International Handbook of Educational Evaluation. Sage.

Seok, J., Warren, H.S., Cuenca, A.G., Mindrinos, M.N., Baker, H.V., Xu, W., Richards, D.R., McDonald-Smith, G.P., Gao, H., Hennessy, L., Finnerty, C.C., 2013. Genomic responses in mouse models poorly mimic human inflammatory diseases. Proc. Natl. Acad. Sci. U.S.A. 110 (9), 3507–3512.

Service, R.F., 2004. Science policy. Nanotech forum aims to head off replay of past blunders. Science (N.Y.) 306 (5698), 955.

Seshan, K., 2012. Handbook of Thin Film Deposition. William Andrew.

Shah, C.P., Farkas, C.S., 1985. The health of Indians in Canadian cities: a challenge to the health care system. Can. Med. Assoc. J. 133 (9), 859.

Singh, S., Vishwakarma, K., Singh, S., Sharma, S., Dubey, N.K., Singh, V.K., Liu, S., Tripathi, D.K., Chauhan, D.K., March 24, 2017. Understanding the plant and nanoparticle interface at transcriptomic and proteomic level: a concentric overview. Plant Gene. http://dx.doi.org/10.1016/j.plgene.2017.03.006.

Song, G., Gao, Y., Wu, H., Hou, W., Zhang, C., Ma, H., 2012. Physiological effect of anatase TiO_2 nanoparticles on *Lemna minor*. Environ. Toxicol. Chem. 31, 2147–2152.

Stone, P.J., Pereira Jr., W., Biles, D., Snider, G.L., Kagan, H., Franzblau, C., 1977. Studies on the fate of pancreatic elastase in the hamster lung: 14C-guanidinated elastase 1–3. Am. Rev. Respir. Dis. 116, 49–56.

Tangaa, S.R., Selck, H., Winther-Nielsen, M., Khan, F.R., 2016. Trophic transfer of metal based nanoparticles in aquatic environments:a review and recommendations for future research focus. Environ. Sci. 3 (5), 966–981.

Trask, P.D., 1938. Relation of Salinity to the Calcium Carbonate Content of Marine Sediments.

Walker, F., 1971. Experimental argyria: a model for basement membrane studies. Br. J. Exp. Pathol. 52, 589–593.

Walker, F., 1972. The deposition of silver in glomerular basement membrane. Virchows Arch. B 11, 90–96.

Yu-Nam, Y., Lead, R., 2008. Manufactured nanoparticles: an overview of their chemistry, interactions and potential environmental implications. Sci. Total. Environ. 400, 396–414.

van der Zande, M., Vandebriel, R.J., Van Doren, E., Kramer, E., Herrera Rivera, Z., Serrano-Rojero, C.S., Gremmer, E.R., Mast, J., Peters, R.J., Hollman, P.C., Hendriksen, P.J., 2012. Distribution, elimination, and toxicity of silver nanoparticles and silver ions in rats after 28-day oral exposure. ACS Nano 6, 7427–7442.

Ze, Y., Liu, C., Wang, L., Hong, M., Hong, F., 2011. The regulation of TiO_2 nanoparticles on the expression of light-harvesting complex II and photosynthesis of chloroplasts of *Arabidopsis thaliana*. Biol. Trace Elem. Res. 143 (2), 1131–1141.

Zeliadt, N., 2010. Silver beware: antimicrobial nanoparticles in soil may harm plant life. Sci. Am. http://www.scientificamerican.com/article.cfm.

CHAPTER

2

Plant-Based Synthesis of Nanoparticles and Their Impact

S. Venkat Kumar, S. Rajeshkumar
VIT University, Vellore, India

2.1 INTRODUCTION

Nanotechnology is all about studying and applying very small items having sizes less than 100 nm. Since nanoparticles possess significant properties, which include size, biocompatibility, and selectivity, it makes nanotechnology a widely applicable field in innumerable sectors such as environmental engineering, energy and electronics, manufacturing and materials, chemical engineering, biotechnology, biopesticides, cosmetics, aquaculture technology, sensor applications, biological- or medicine-like fluorescent biological labels (Bruchez et al., 1998; Chan and Nie, 1998; Wang et al., 2002), drug and gene delivery (Mah et al., 2000; Panatarotto et al., 2003), biological warfare agent detection of pathogens (Edelstein et al., 2000), probing of DNA structures (Mahtab et al., 1995), detection of proteins (Nam et al., 2003), tissue engineering (Ma et al., 2003; De la Isla et al., 2003), tumor engineering, separation and purification of biological molecules and cells (Molday and MacKenzie, 1982), hyperthermia (Yoshida and Kobayashi, 1999), phagokinetic studies (Parak et al., 2002), magnetic resonance imaging contrast enhancement (Weissleder et al., 1990), and agriculture and plant sciences (Peralta-Videa et al., 2014, 2016; Tripathi et al., 2015, 2016, 2017a,b,c,d,e; Cox et al., 2016; Shweta et al., 2016; Singh et al., 2016, 2017; Tiwari et al., 2017).

Several techniques are utilized for the synthesis of silver nanoparticles, which includes chemical reduction of silver ions (Liz-Marzán and Lado-Tourino, 1996), thermal decomposition of organic solvents (Esumi et al., 1990), photoreduction in reverse micelles (Sun et al., 2001), etc.; but all these techniques are costly, and toxic and harmful chemicals are used posing hazards to the environment because of the properties in magnetic and optical polarizability, electrical conductivity (Chang and Yen, 1995), and antimicrobial activity and catalysis (Shiraishi and Toshima, 2000). Because of the side effects of chemical synthesis, biological methods incorporating microorganisms (Nanda and Saravanan, 2009; Pugazhenthiran et al., 2009; Gurunathan et al., 2009; Korbekandi et al., 2013a,b; Konishi and Uruga, 2007; Nair and

Nanomaterials in Plants, Algae, and Microorganisms
http://dx.doi.org/10.1016/B978-0-12-811487-2.00002-5

Pradeep, 2002; Klaus et al., 1999; Aziz et al., 2015, 2016; Prasad et al., 2016), enzymes (Willner et al., 2006; Talekar et al., 2016), and plants (Shankar et al., 2004; Chandran et al., 2006; Jae and Beom, 2009; Dubey et al., 2009; Bar et al., 2009) were used for the synthesis of silver nanoparticles providing eco-friendly, rapid, and low-cost attributes. Among these biological methods, plant-based nanoparticle synthesis is more appropriate because of its eco-friendly, low-cost, and safe nature for human therapeutic use (Kumar and Yadav, 2009).

Gold nanoparticles are the most stable metal nanoparticles having exclusive physical, magnetic, electronic, thermal, optical, biomedical, and biochemical applications. Gold nanoparticles are used for several applications because of their high absorption factor, huge surface area, bioadjustability, conjugation in the stable covalent link between molecules, noncytotoxicity, assembly of several types, application to catalysis and biology, and surface functionalization (Paciotti et al., 2004; Daniel and Astruc, 2004; Sokolov et al., 2003; Blanco et al., 2008; Murphy et al., 2008; Jinhwa et al., 2008; Das et al., 2012; Sardar et al., 2009; Talapin et al., 2009; Khan et al., 2014a,b; Guo et al., 2014; Lan et al., 2013; Mendoza et al., 2010; Bisker et al., 2012; Abouelkheir et al., 2016; Shedbalkar et al., 2014). Gold nanoparticles are synthesized from a diversity of plant species. Plant-mediated synthesis of gold nanoparticles needs less incubation time when compared with microbial systems and can be scaled up easily for commercial needs (Shankar et al., 2004; Ankamwar et al., 2005; Noruzi, 2015).

Zinc oxide (ZnO) nanoparticles are the main focus since they are a stable metal oxide, eco-friendly, and do not harm humans and animals. ZnO has a property that fights against infectious diseases because of its antimicrobial properties. Among various metal oxides such as TiO_2, MgO, CaO, and ZnO used for various purposes, ZnO is much more interesting because of its numerous optical, magnetic, high isoelectric point, biocompatible, fast electron transfer, piezoelectric, excellent thermally and chemically stable, and gas-sensing properties. Because of its various properties, ZnO exhibits adsorption ability, has high catalytic efficiency, and can be used as sensors, as a fungicide, in rubber processing, paints, wastewater treatment, catalysis, solar energy conversion, cosmetics, fibers, drug delivery, and as an antibacterial (Ramesh et al., 2014; Sangeetha et al., 2011; Vaseem and Umar, 2010; Ravichandrika et al., 2012; Iravani, 2011; Stoimenov et al., 2002).

2.2 PLANT-MEDIATED SYNTHESIS OF SILVER NANOPARTICLES

A wide range of resources are used for the synthesis of silver nanoparticles from bacterial strains such as *Bacillus licheniformis* (Kalishwaralal et al., 2008), *Enterococcus* sp. (Rajeshkumar et al., 2013a,b), *Bacillus* sp. (Pugazhenthiran et al., 2009; Malarkodi et al., 2013), *Serratia nematodiphila* (Malarkodi et al., 2013), *Klebsiella planticola* (Paulkumar et al., 2013a,b), *Bacillus subtilis* (Paulkumar et al., 2013a,b), *Escherichia coli* (Gurunathan et al., 2009), and *Vibrio alginolyticus* (Rajeshkumar et al., 2013a,b), fungal strains such as *Aspergillus niger* (Vanaja et al., 2015), *Pleurotus ostreatus* (Devika et al., 2012), *Aspergillus terreus* (Li et al., 2012), and *Penicillium fellutanum* (Kathiresan et al., 2009), and marine macroalgae such as *Chaetomorpha linum* (Ragupathi Raja Kannan et al., 2012), *Padina tetrastromatica* (Rajeshkumar et al., 2012a,b), *Sargassum tenerrimum* (Kumar et al., 2012), *Turbinaria conoides* (Rajeshkumar et al., 2012a,b), and *Sargassum longifolium* (Rajeshkumar et al., 2014). The various parts of plants that have been utilized for the synthesis of silver nanoparticles are shown in Fig. 2.1. The various steps involved in the

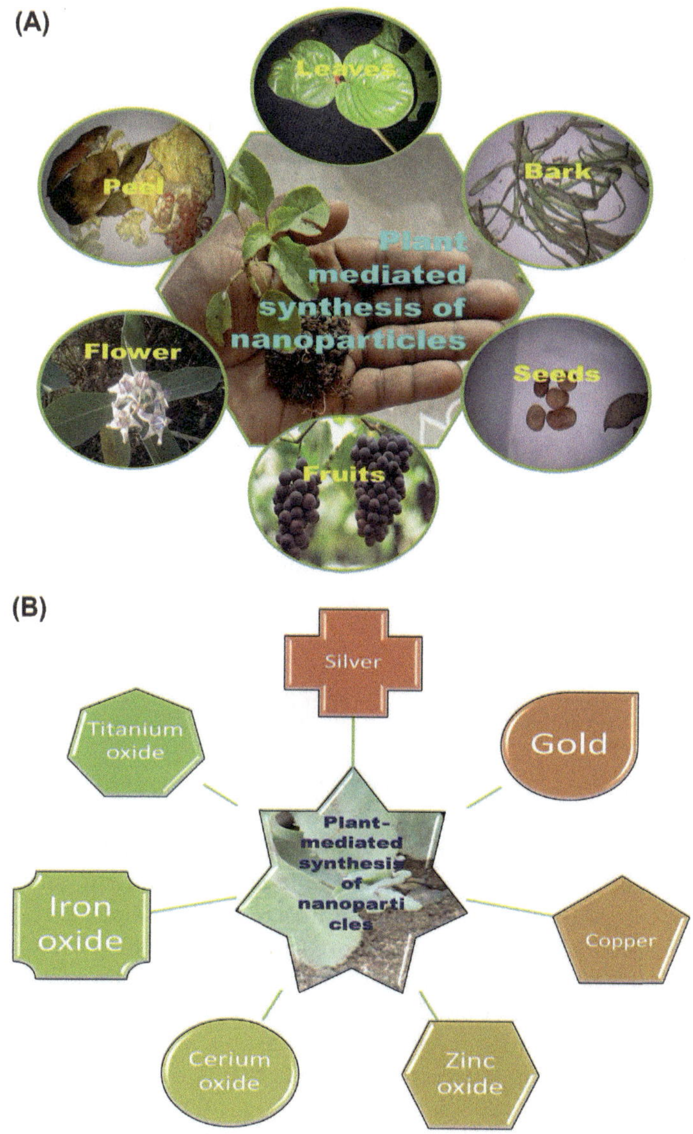

FIGURE 2.1 (A) Different parts of plants used in (B) plant-mediated synthesis of various types of nanoparticles.

plant-mediated synthesis of nanoparticles are shown in Fig. 2.2. The synthesized nanoparticles are characterized using various techniques such as UV-Vis spectroscopy, transmission electron microscopy, scanning electron microscopy, X-ray diffraction assay, energy dispersive analysis, and selected area electron diffraction pattern (Fig. 2.3).

The major portions of plants where silver nanoparticles are synthesized include leaves and roots, stems, fruits, and flowers. As far as the leaf is concerned, the following list gives the names of plants with their scientific names and sizes: rose *Catharanthus roseus* having size

FIGURE 2.2 Steps involved in the synthesis of nanoparticles by plants.

48–67 nm (Selva Bharath et al., 2012), Indian copperleaf *Acalypha indica* having size 20–30 nm (Krishnaraj et al., 2010), camphor tree *Cinnamomum camphora* having size 5–40 nm (Huang et al., 2008), Cuban oregano *Coleus aromaticus* with size 44 nm (Vanaja and Annadurai, 2013), coriander *Coriandrum sativum* leaf with size 26 nm (Sathyavathi et al., 2010), creeping tick trefoil *Desmodium triflorum* having size 10 nm (Ahmad et al., 2011), asthma weed *Euphorbia hirta* having size 40–50 nm (Elumalai et al., 2010), *Gliricidia sepium* with size 27 nm (Raut Rajesh et al., 2009), edible fig *Ficus carica* having size 10–20 nm (Singh and Bhakat, 2012), lemon *Citrus limon* having size >100 nm (Vankar and Shukla, 2012), herb louisa *Lippia citriodora* having size 15–30 nm (Cruza et al., 2010), bitterweed *Parthenium* having size 50 nm (Parashar et al., 2009), Paederia *Paederia foetida* having size 24 nm (Lavanaya et al., 2013), Elaeagnus *Elaeagnus indica* having size 30 nm (Natarajan et al., 2013), chanca piedra *Phyllanthus amarus* Schum. & Thonn having size 32–53 nm (Annamali et al., 2011), periwinkle *Catharanthus roseus* having size 35–55 nm (Kulkarni et al., 2012), some plant leaves such as *Alstonia scholars*, *Calotropis gigantia*, *Ficus religiosa* Linn., *Hevea brasiliensis*, *Musa paradisiaca*, *Achras sapota* with family scientific names Apocynaceae, Asclepiadaceae, Moraceae, Euphorbiaceae, Musaceae, Sapotaceae, respectively, all having sizes 401–434 nm (Mondal et al., 2011), strawberry *Arbutus unedo* having size 9–15 nm (Srinivas Naik et al., 2013), achiote *Bixa orellana* with size 35–65 nm (Thilagam et al., 2013), bastard oleaster *Elaeagnus latifolia* having size 30–50 nm (Phanjom et al., 2012), oleander of *Nerium oleander* having size 48–67 nm (Suganya et al., 2012), ocimum of *Ocimum bacillicum* having size 58–89 nm (Sivarangani and Meenakshisundaram, 2013),

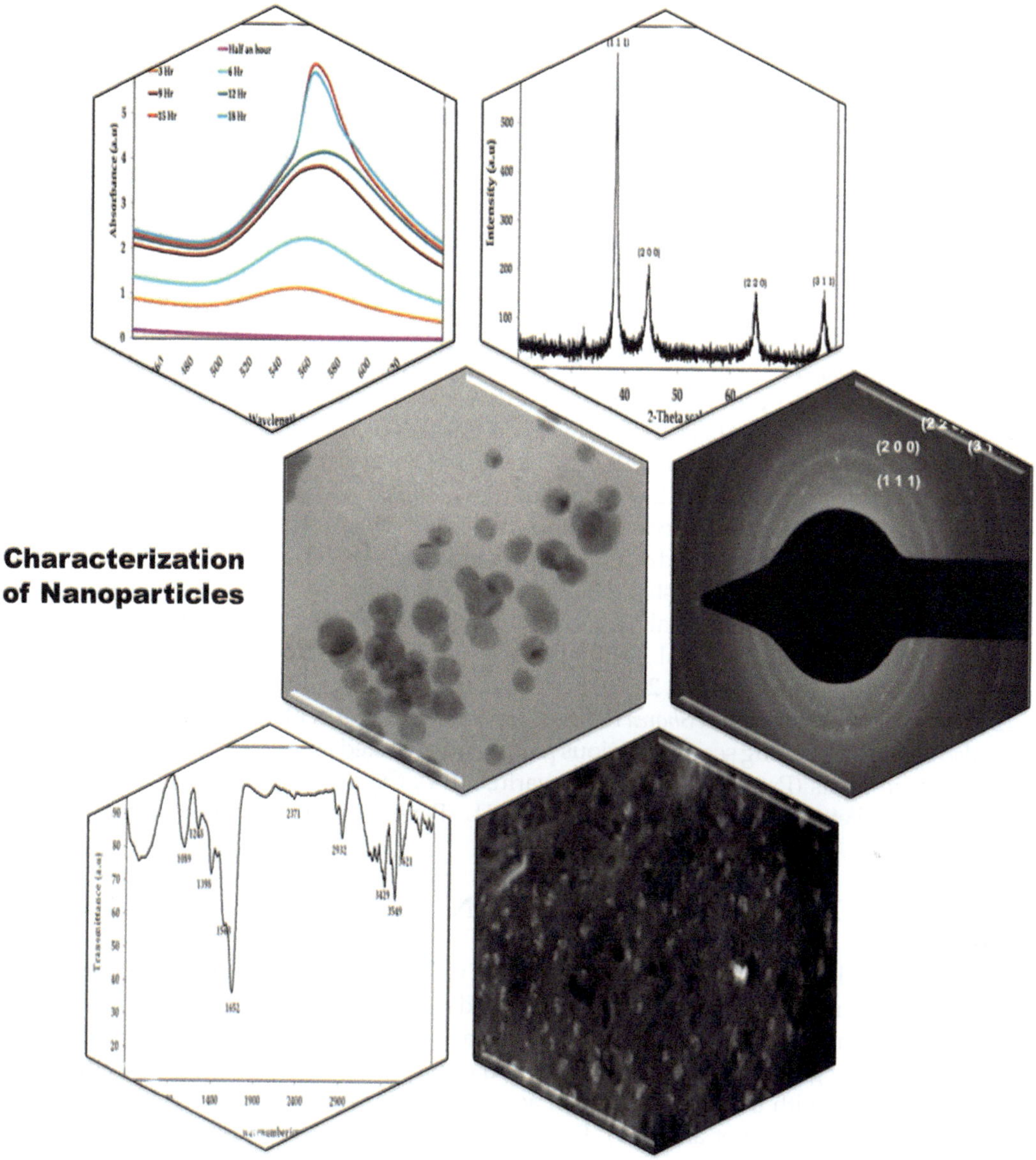

FIGURE 2.3 Characterization of nanoparticles using different analytical techniques.

wodier of *Odina wodier* having size 5–30 nm (Arun Kumar et al., 2013), carob of *Ceratonia siliqua* L. with size 5–40 nm (Awwad et al., 2013), walnut of *Juglans regia* L. having size 10–50 nm (Korbekandi et al., 2013c), sheesham of *Dalbergia sissoo* with size 5–55 nm (Singh et al., 2012a,b), asthma plant of *Euphorbia hirta* with size 40–50 nm (Elumalai et al., 2010), Tanner's Cassia of *Cassia auriculata* with size 420–435 nm (Parveen et al., 2012), Siam weed of *Chromolaena odorata* having size 40–70 nm (Geetha et al., 2012), Sanskrit of *Cephalandra indica* with size 40–90 nm (Celin Hemalatha and Renitta, 2011), thistle of *Sonchus asper* with size

2–100 nm (Verma et al., 2013), potato bush of *Phyllanthus reticulatus* with size 11–30 nm (Kudle et al., 2013), Margosa tree of *Azhadirachta indica* with size 21.07 nm (Lalitha et al., 2013).

Silver nanoparticles are also synthesized by using fruits and are as follows: yellow-berried nightshade with scientific name *Solanum xanthocarpum* having size 350–450 nm (Selva Bharath et al., 2012), grapefruit of *Vitis vinifera* (Korbekandi et al., 2013c), *V. vinifera* fruit extract (Gnanajobitha et al., 2013a,b), banana peel extract (Bankar et al., 2010), *Solanum trilobatum* fruit extract (Ramar et al., 2015), *Carambola* fruit extract (Mane Gavade et al., 2015), papaya fruit extract (Jaina et al., 2009), *Momordica cymbalaria* of size 15.5 nm (Swamy et al., 2015), *Piper longum* of size 46 nm (Reddy et al., 2014), and *Lycopersicon esculentum* (Maiti et al., 2014).

Silver nanoparticles are synthesized from flower extracts as well and are as follows: morning glory with scientific name *Ipomoea indica* having size 10–50 nm (Pavani et al., 2013), marvel of Peru of *Mirabilis jalapa* having size 60–70 nm (Vankar and Bajpai, 2010), frangipani of *Plumeria alba* Linn. having size 10–50 nm (Nagaraj et al., 2013), cork tree of *Millingtonia hortensis* with size 10–40 nm (Gnanajobitha et al., 2013a,b), Tanner's Cassia with scientific name *Cassia auriculata* having size 10–40 nm (Velavan et al., 2012), safflower of *Carthamus tinctorius* L. with size between 40 and 200 nm (Nagaraj et al., 2012a,b), balsam tree of *Gnidia glauca* having size 5–20 nm (Patil et al., 2013), and olibanum of *Boswellia serrate* with size 60–84 nm (Kudle et al., 2000).

Apart from leaves and fruit extracts, silver nanoparticles were synthesized from roots and stems of several plants and are listed as follows: *Portulaca oleracea* L. (Gholamreza et al., 2014), *Macrotyloma uniflorum* (Mutahir et al., 2016), *Glycyrrhiza glabra* (Dinesh et al., 2012), *Cochlospermum religiosum* (L.) Alston (Sasikala et al., 2015), *Boswellia ovalifoliolata* (Ankanna et al., 2010), *Thespesia populnea* (L.) Soland (Bhumi et al., 2013), *Svensonia hyderobadensis* (Linga Rao and Savithramma, 2012), *Boswellia ovalifoliolata* and *Shorea tumbuggaia* (Savithramma et al., 2011). Silver nanoparticles were also synthesized using seeds of various plants and are listed as follows: *Brassica nigra* (Pandit, 2015), soybean seeds (Prasad and Venkateswarlu, 2014), *Cyperus esculentus* and *Butyrospermum paradoxum* (Ajayi et al., 2015), and *Syzygium cumini* L. (Banerjee and Narendhirakannan, 2011).

2.3 GOLD NANOPARTICLE SYNTHESIS USING PLANTS

The concentrations of the gold (Au) atoms play a very important role in shaping the nanoparticles that have various applications in different fields. The reduction of the Au(III) ion to the Au atom involves binding of the atom to the cell surface, while other reduced Au also binds and aggregates to form gold nanoparticles. At the initial stages, Au ions have a greater opportunity to come in contact with the atom and form aggregates. Thus larger particles at small concentrations were formed (Cai et al., 2011). Rare metals have also paved their way in the field of medicine, diagnosis, and therapy or cancer treatment, and are also used as antiangiogenesis, antiarthritic, and antimalarial agents. They are successfully used as delivery molecules that assist in slowing down the cancer growth or even apoptosis of cancerous cells (Honary et al., 2012). The biosynthesis of nanoparticles can have specified morphology and redefined shapes such as hexagon, triangle, rods, hierarchical tubes, decahedrons, icosahedrons, and nodous ribbons as well as nanotriangles. The research focuses more on their shapes as each has potential applications in various fields such as optical coating, hyperthermia of tumors, and scanning the tunneling of microscopes, which can be used as conductive tips (Du et al., 2007). Because of its large application in a variety of fields, gold nanoparticles are synthesized using various parts of plants such as leaf, fruit, root, and stem and these are given in Table 2.1.

TABLE 2.1 Different Types of Plants and Their Parts Used for the Synthesis of Gold Nanoparticles

S. No.	Scientific Name	Plant Part Used	Size	References
1	*Ipomoea carnea*	Leaf, stem, and root	25–100 nm	Abbasi et al. (2015)
2	Geranium	Leaf	12 ± 3 nm	Franco-Romano et al. (2014)
3	*Morinda citrifolia*	Root	12.17–38.26 nm	Suman et al. (2014)
4	*Phoenix dactylifera*	Leaf	32 and 45 nm	Zayed and Eisa (2014)
5	*Salicornia brachiata*		25–35 nm	Ahmed et al. (2014)
6	*Cassia fistula*	Stem bark	55.2–98.4 nm	Daisy and Saipriya (2012)
7	*Eucommia ulmoides*	Bark	16.4 nm	Guo et al. (2015)
8	*Sesbania grandiflora*	Leaf	34.11 nm	Das and Velusamy (2014)
9	*Salix alba*	Leaves	50–80 nm	Islam et al. (2015)
10	*Pistia stratiotes* L.	Weed pistia	2–40 nm	Anuradha et al. (2015)
11	*Magnolia kobus* and *Diopyros kaki*	Leaf	5–300 nm	Song et al. (2009)
12	*Averrhoa bilimbi*	Fruit	75–150 nm	Rimal Isaac et al. (2013)
13	*Cacumen platycladi*	Leaf	2.2–42.8 nm	Zhan et al. (2011)
14	*Cinnamomum zeylanicum*	Leaf	25 nm	Smitha et al. (2009)
15	Coriander	Leaf	6.75 nm	Badrinarayanan and Sakthivel (2008)
16	*Gymnema sylvestre*	Leaves	72.8 nm	Arunachalam et al. (2014)
17	*Tagetes erecta*	Flower	30–50 nm	Krishnamurthy et al. (2012)
18	*Terminalia catappa*	Leaf	10–35 nm	Ankamwar (2010)
19	*Lansium domesticum*	Fruit	20–40 nm	Shankar et al. (2014)
20	*Nerium oleander*	Leaves	2–10 nm	Tahir et al. (2015a,b)
21	*Punica granatum*	Fruit peel	70 nm	Ganeshkumar et al. (2013)
22	*Euphorbia hirta*	Leaf	50 nm	Annamalai et al. (2013)
23	*Moringa oleifera*	Flower	3–5 nm	Anand et al. (2015)
24	*Terminalia arjuna*	Leaf	20–50 nm	Gopinath et al. (2013)
25	*Hibiscus sabdariffa*	Leaf and stem	10–60 nm	Mishra et al. (2016)
26	*Bacopa monnieri* (BLE)	Leaf	3–45 nm	Punuri et al. (2013)

(Continued)

TABLE 2.1 Different Types of Plants and Their Parts Used for the Synthesis of Gold Nanoparticles—cont'd

S. No.	Scientific Name	Plant Part Used	Size	References
27	*Terminalia catappa*	Leaf	10–35 nm	Ankamwar (2010)
28	*Citrus maxima*	Fruit	25.7710 nm	Yu et al. (2016)
29	*Olea europaea*	Leaf	50–100 nm	Khalil et al. (2012)
30	*Ficus religiosa*	Bark	20–30 nm	Wani et al. (2013)
31	*Mangifera indica*	Leaf	~20 nm	Philip (2010)
32	*Acalypha indica*	Leaves	20–30 nm	Krishnaraj et al. (2014)
33	*Ixora coccinea*	Flower	5–10 nm	Nagaraj et al. (2011)
34	*Cacumen platycladi*	Leaf	2–70 nm	Wua et al. (2012)
35	*Padina gymnospora*	Leaves	53–67 nm	Singh et al. (2012a,b)
36	*Citrus limon*, *Citrus reticulate*, and *Citrussinensis*	Fruits	15–80 nm	Sujitha and Kannan et al. (2013)
37	*Cassia auriculata*	Flowers	12–41 nm	Venkatachalam et al. (2013)
38	*Abutilon indicum*	Leaf	1–20 nm	Mata et al. (2016)
39	*Butea monosperma*	Leaf	20–80 nm	Patra et al. (2015)
40	*Coleus forskohlii*	Root	5–18 nm	Naraginti et al. (2016)
41	*Genipa americana*	Fruit	15–40 nm	Kumar et al. (2016)
42	*Suaeda monoica*	Leaf	14.5 nm	Arockiya Aarthi Rajathi et al. (2014)
43	*Nepenthes khasiana*	Leaves		Dhamecha et al. (2016)
44	*Spinacia oleracea*	Leaves	10–15 nm	Megarajan et al. (2016)
45	*Phoenix dactylifera* L.	Leaf	32 and 45 nm	Zayed and Eisa (2014)
46	*Mimosa pudica*	Leaf	12.5 nm	Uma Suganya et al. (2016)
47	*Acacia nilotica*	Bark	10–50 nm	Emmanuel et al. (2014)
48	*Salvadora persica*	Stem	10–30 nm	Tahir et al. (2015a,b)
49	*Cajanus cajan*	Seed	9–41 nm	Ashokkumar et al. (2014)
50	*Terminalia arjuna*	Fruit	Ta1 (60 nm) Ta3 (20 nm) Ta5 (14 nm)	Mohan Kumar et al. (2013a)

51	*Morinda citrifolia* L.	Root	12.17–38.26 nm	Suman et al. (2014)
52	*Nyctanthes arbor-tristis*	Flower	19.8 ± 5.0 nm	Das et al. (2011)
53	*Elettaria cardamomum*	Seed pod	432.3 nm	Pattanayak and Nayak (2013)
54	Olive	Leaf	50–100 nm	Khalil et al. (2012)
55	*Argemone mexicana*	Leaf	22–26 nm	Varun et al. (2015)
56	*Ocimum sanctum*	Leaves	10–300 nm	Lee et al. (2016)
57	*Erythrina variegata*	Leaves	20–50 nm	Ali et al. (2016b)
58	*Panicum maximum*	Root	14.28 nm	Agarwal and Srivastava (2014)
59	*Azadirachta indica* L.	Leaf	15–18 nm	Bindhani and Panigrahi (2014)
60	*Silybum marianum*	Leaves		Kosalai and Chandran (2016)
61	*Ficus benghalensis*	Leaf	2–100 nm	Francis et al. (2014)
62	*Diospyros ferrea*	Leaves	70–90 nm	Ramesh and Armash (2015)
63	Rosaceae	Flower	10–20 nm	Kahkha and Kahkha (2015)
64	*Bougainvillea glabra*	Leaf		Leema Rose et al. (2014)
65	*Costus igneus*	Leaves	54–62 nm	Velumani (2015)
66	*Aloe vera*	Leaf	50–350 nm	Chandran et al. (2006)
67	*Bauhinia purpurea*	Flower	20–50 nm	Radha et al. (2016)
68	*Hibiscus rosa-sinensis*	Leaves	16–30 nm	Yasmin et al. (2014)
69	*Plumeria alba* L.	Flower	20–30 nm	Nagaraj et al. (2012a,b)
70	*Hygrophila spinosa*	Leaf	50–80 nm	Koperuncholan (2015)
71	*Nepenthes khasiana*	Leaves	50–80 nm	Bhau et al. (2015)
72	*Amaranthus spinosus*	Leaf	10.74 nm	Das et al. (2012)
73	*Abelmoschus esculentus*	Seed	45–75 nm	Jayaseelan et al. (2013)
74	*Gnidia glauca*	Flower	5–20 nm	Ghosh et al. (2012)
75	*Vitis vinifera*	Leaves and seed	10–17 nm	Ismail et al. (2014)
76	*Ficus religiosa*	Bark	20–30 nm	Wani et al. (2013)
77	*Nitraria schoberi*	Fruits	30 and 40 nm	Rad et al. (2013a,b)
78	*Cicer arietinum* L.	Leaf	30–80 nm	Singh et al. (2013)

2.4 PLANT-ASSISTED SYNTHESIS OF ZINC OXIDE NANOPARTICLES

Several types of inorganic metal oxides have been synthesized in a number of studies, e.g., TiO_2, CuO, and ZnO. Of all these metal oxides, ZnO nanoparticles are of maximum interest because they are inexpensive to produce, safe, and can be prepared easily (Elumalai et al., 2015). Zinc oxide nanoparticles have gained interest in the past several years because of their wide range of applications in the fields of electronics, optics, and biomedical systems (Azizi et al., 2014). ZnO nanoparticles exhibit tremendous semiconducting properties because of their large band gap (3.37 eV) (Thema et al., 2015) and high exciton binding energy (60 meV) (Anbuvannan et al., 2015a,b), e.g., high catalytic activity and optic, UV filtering, antiinflammatory (Ramesh et al., 2015), and wound-healing properties (Nagajyothi et al., 2013). Because of their UV filtering properties, they have been extensively used in cosmetics such as sunscreen lotions. They have a wide range of biomedical applications such as drug delivery, anticancer, antidiabetic, antibacterial, antifungal, and agricultural properties (Ramesh et al., 2015). ZnO nanoparticles have a very strong antibacterial effect at very low concentrations of Gram-negative and Gram-positive bacteria as confirmed by studies, and have shown a stronger antibacterial effect than the ZnO nanoparticles synthesized chemically (Dobrucka and Dugaszewska, 2016). Apart from their protein adsorption properties, they have also been employed in rubber manufacturing, dental applications, paint, and for removing sulfur and arsenic from water (Ali et al., 2016a). ZnO nanoparticles have been reported in different morphologies such as nanoflakes, nanoflowers, nanobelts, nanorods, and nanowires (Anbuvannan et al., 2015a,b). Table 2.2 shows the green synthesis of ZnO nanoparticles isolated from various plant parts such as leaves, flowers, dried leaves, and fruit peels.

TABLE 2.2 Plant-Mediated Synthesis of ZnO Nanoparticles

S. No.	Plant (Family)	Part Taken for Extraction	Shape	References
1	*Pongamia pinnata* (legumes)	Fresh leaves	Spherical, hexagonal, nanorods	Sundrarajan et al. (2015)
2	*Trifolium pratense* (legumes)	Flower	Spherical	Dobrucka and Dugaszewska (2016)
3	*Solanum nigrum* (Solanaceae)	Leaf extract	Hexagonal wurtzite, quasispherical	Ramesh et al. (2015)
4	*Eichhornia crassipes* (Pontederiaceae)	Leaf extract	Spherical without aggregation	Vanathi et al. (2014)
5	*Ocimum basilicum* L. var. *purpurascens* (Lamiaceae)	Leaf extract	Hexagonal (wurtzite)	Salam et al. (2014)
6	*Anisochilus carnosus* (Lamiaceae)	Leaf extract	Hexagonal wurtzite, quasi spherical	Anbuvannan et al. (2015a,b)
7	*Azadirachta indica* (Meliaceae)	Leaf	Spherical	Bhuyan et al. (2015)
8	*Rosa canina* (Rosaceae)	Fruit extract	Spherical	Jafarirad et al. (2016)

TABLE 2.2 Plant-Mediated Synthesis of ZnO Nanoparticles—cont'd

S. No.	Plant (Family)	Part Taken for Extraction	Shape	References
9	*Aloe vera* (Liliaceae)	Freeze-dried leaf peel	Spherical, hexagonal	Qian et al. (2015)
10	*Azadirachta indica* (Meliaceae)	Fresh leaves	Spherical	Elumalai and Velmurugan (2015)
11	*Agathosma betulina* (Rutaceae)	Dry leaves	Quasispherical agglomerates	Thema et al. (2015)
12	*Aloe vera* (Liliaceae)	Leaf extract	Spherical, oval, hexagonal	Ali et al. (2016a)
13	*Coptidis rhizoma* (Ranunculaceae)	Dried rhizome	Spherical, rod shaped	Nagajyothi et al. (2014)
14	*Phyllanthus niruri* (Phyllanthaceae)	Leaf extract	Hexagonal wurtzite, quasi spherical	Anbuvannan et al. (2015a,b)
15	*Cocus nucifera* (Arecaceae)	Coconut water	Spherical and predominantly hexagonal without any agglomeration	Nithya Deva Krupa and Vimala (2016)
16	*Gossypium* (Malvaceae)	Cellulosic fiber	Wurtzite spherical, nanorod	Aladpoosh and Montazer (2015)
17	*Moringa oleifera* (Moringaceae)	Leaf	Spherical and granular nanosized shape with a group of aggregates	Elumalai et al. (2015)
18	*Nephelium lappaceum* (Sapindaceae)	Fruit peels	Spherical and hexagonal	Yuvakkumar et al. (2015)
19	*Nephelium lappaceum* L. (Sapindaceae)	Fruit peels	Needle shaped forming agglomerate	Yuvakkumar et al. (2014)
20	*Plectranthus amboinicus* (Lamiaceae)	Leaf extract	Rod-shaped nanoparticle with agglomerates	Fu and Fu (2015)
21	*Santalum album* (Santalaceae)	Leaves	Nanorods	Kavithaa et al. (2016)
22	*Vitex negundo* (Lamiaceae)	Leaf	Spherical	Ambika and Sundrarajan (2015)
23	*Vitex negundo* (Lamiaceae)	Flowers	Hexagonal	Ambika and Sundrarajan (2015)
24	*Azadirachta indica* (Meliaceae)	Fresh leaves	Hexagonal disk, nanobuds	Madan et al. (2016)
25	*Parthenium hysterophorus* L. (Asteraceae)	Leaf extract	Spherical, hexagonal	Rajiv et al. (2013)
26	*Calatropis gigantea* (Apocynaceae)	Fresh leaves	Spherical shaped forming agglomerates	Vidya et al. (2013)
27	*Sphathodea campanulata* (Bignoniaceae)	Leaf extract	Spherical	Ochieng et al. (2015)

2.5 OTHER NANOPARTICLES SYNTHESIZED USING PLANT SOURCES

Apart from silver, gold, and zinc oxide, other nanoparticles such as lead, copper, titanium oxide, indium oxide, selenium, platinum, palladium, etc. have been synthesized using plant sources (Table 2.3).

TABLE 2.3 Plant-Assisted Synthesis of Nanoparticles

Plant Source	Nanoparticles	Remarks	References
Aloe vera (*Aloe barbadensis* Miller)	Indium oxide	5–50 nm; spherical	Maensiri et al. (2008)
Diopyros kaki	Platinum	15–19 nm	Song et al. (2010)
Jatropha curcas L. latex (aqueous extract)	Lead	10–12.5 nm	Joglekar et al. (2011)
Euphorbiaceae latex	Copper	Very good antimicrobial activity	Valodkar et al. (2011)
Catharanthus roseus leaf extract	Titanium dioxide	Good effect against *Hippobosca maculata* and *Bovicola ovis*	Velayutham et al. (2011)
Capsicum annuum	Selenium	Selenium with protein nanocomposites	Li et al. (2007)
Sorghum bran extract	Iron and silver	Synthesized at room temperature	Njagi et al. (2011)
Annona squamosa L. peel and larvicidal agent	Palladium	Application as an acaricidal, insecticidal	Roopan et al. (2011)
Pulicaria glutinosa extract and supporting of poly(ethylene glycol)	Palladium	Applied for catalytic activity	Borah et al. (2015)
Pulicaria glutinosa extract	Palladium	Catalytic activity toward the Suzuki coupling reaction	Khan et al. (2014a,b)
Piper betle leaves broth	Palladium	Used for antifungal studies	Mallikarjuna et al. (2013)
Terminalia chebula aqueous extract	Palladium and iron nanoparticles		Mohan Kumar et al. (2013a,b)
Aqueous extracts of *Hordeum vulgare* and *Rumex acetosa* plants	Iron oxide		Makarov et al. (2014)
Ocimum sanctum electrolysis applications	Platinum	Water	Soundarrajan et al. (2012)
Dried *Anacardium occidentale* leaf extract	Platinum	Used for thermal and catalytic applications	Sheny et al. (2013)

2.6 CONCLUSION AND FUTURE PROSPECTS

The environment-responsive process for the synthesis of nanoparticles of size range between 1 and 100 nm is a most promising stage in the arena of nanotechnology. Because of environmental hazards that occur in the chemical synthesis of metal nanoparticles, research has concentrated on synthesizing nanoparticles using plant sources. Eco-friendly methods for the synthesis of nanoparticles are nontoxic and more applicable for biomedical applications. Microorganisms including bacteria, fungi, yeast, algae, actinomycetes, and plant sources from gymnosperms to angiosperms are extensively used for the green synthesis of metallic nanoparticles. Because of the phytochemical properties, variability of biomolecules contained in plant sources is the main reason for the synthesis of nanoparticles when compared with microbes. Among the various sources used for the synthesis of metallic nanoparticles, plant leaves are the best because of their stability and faster rate of synthesis. This chapter focused on the green synthesis of metallic nanoparticles by means of various plant sources. Various types of nanoparticles have been used in a number of applications in the biomedical field, hence using plant-based nanoparticles may increase the biocompatibility and reduce the toxicology of future applications.

Acknowledgments

Authors would like to express our sincere gratitude to Dean, SBST and management of VIT University for continuous encouragement.

References

Abbasi, T., Anuradha, J., Ganaie, S.U., Abbasi, S.A., 2015. Gainful utilization of the highly intransigent weed ipomoea in the synthesis of gold nanoparticles. J. King Saud Univ. Sci. 27, 15–22.

Abouelkheir, S.S., El-Sersy, N.A., Sabry, S.A.-F., 2016. Potential application of *Bacillus* sp. SDNS gold nanoparticles. Int. J. Curr. Microbiol. Appl. Sci. 5 (4), 546–552.

Agarwal, K., Srivastava, M.M., 2014. Synthesis and characterization of gold nanoparticles embedded with extract of the plant *Panicum maximum* with enhanced antioxidant. Behavior 3 (10).

Ahmad, N., Sharma, S., Singh, V.N., Shamsi, S.F., Fatma, A., Mehta, B.R., 2011. Biosynthesis of silver nanoparticles from *Desmodium triflorum*: a novel approach towards weed utilization. Biotechnol. Res. Int. 2011. http://dx.doi.org/10.4061/2011/454090.

Ahmed, K.B.A., Subramanian, S., Sivasubramanian, A., Veerappan, G., Veerappan, A., 2014. Preparation of gold nanoparticles using *Salicornia brachiata* plant extract and evaluation of catalytic and antibacterial activity. Spectrochim. Acta A Mol. Biomol. Spectrosc. 130, 54–58.

Ajayi, I.A., Raji, A.A., Ogunkunle, E.O., 2015. Green synthesis of silver nanoparticles from seed extracts of *Cyperus esculentus* and *Butyrospermum paradoxum*. Int. J. Pharm. Biol. Sci. 10 (1), 76–90.

Aladpoosh, R., Montazer, M., 2015. The role of cellulosic chains of cotton in biosynthesis of ZnO nanorods producing multifunctional properties: mechanism, characterizations and features. Carbohydr. Polym. 126, 122–129.

Ali, K., Dwivedi, S., Azam, A., Saquib, Q., Al-Said, M.S., Alkhedhairy, A.A., Musarrat, J., 2016a. *Aloe vera* extract functionalized zinc oxide nanoparticles as nanoantibiotics against multi-drug resistant clinical bacterial isolates. J. Colloid Interface Sci. 472, 145–156.

Ali, A.M., Jaber, H.H., Obald, J.B., 2016b. Green synthesis and characterization of gold nanoparticles from erythrina variegate plant extract. Int. J. Res. 3 (9), 657–664.

Ambika, S., Sundrarajan, M., 2015a. Green biosynthesis of ZnO nanoparticles using *Vitex negundo* L. extract: spectroscopic investigation of interaction between ZnO nanoparticles and human serum albumin. J. Photochem. Photobiol. B Biol. 149, 143–148.

Ambika, S., Sundrarajan, M., 2015b. Antibacterial behaviour of *Vitex negundo* extract assisted ZnO nanoparticles against pathogenic bacterial. J. Photochem. Photobiol. B Biol. 146, 52–57.

Anand, K., Gengan, R.M., Phulukdaree, A., Chuturgoon, A., 2015. Agroforestry waste *Moringa oleifera* petals mediated green synthesis of gold nanoparticles and their anti-cancer and catalytic activity. J. Ind. Eng. Chem. 21, 1105–1111.

Anbuvannan, M., Ramesh, M., Viruthagiri, G., Shanmugam, N., Kannadasan, N., 2015a. Synthesis, characterization and photocatalytic activity of ZnO nanoparticles prepared by biological method. Mol. Biomol. Spectrosc. 143, 304–308.

Anbuvannan, M., Ramesh, M., Viruthagiri, G., Shanmugam, N., Kannadasan, N., 2015b. *Anisochilus carnosus* leaf extract mediated synthesis of zinc oxide nanoparticles for antibacterial and photocatylitic activities. Mater. Sci. Semicond. Process. 39, 621–628.

Ankamwar, B., 2010. Biosynthesis of gold nanoparticles (Green-Gold) using leaf extract of *Terminalia catappa*. E-J. Chem. 7 (4), 1334–1339. http://dx.doi.org/10.1155/2010/745120.

Ankamwar, B., Chaudhary, M., Sastry, M., 2005. Gold nanotriangles biologically synthesized using tamarind leaf extract and potential application in vapor sensing. Synth. React. Inorg. Metal Org. Nano Metal Chem. 35 (1), 19–26.

Ankanna, S., Prasad, T.N.V.K.V., Elumalai, E.K., Savithramma, N., 2010. Production of biogenic silver nanoparticles using *Boswellia ovalifoliolata* stem bark. Digest J. Nanomater. Biostruct. 5, 369–372.

Annamalai, A., Christina, V.L.P., Sudha, D., Kalpana, M., Lakshmi, P.T.V., 2013. Green synthesis, characterization and antimicrobial activity of Au NPs using *Euphorbia hirta* L. leaf extract. Colloids Surf. B Biointerfaces 108, 60–65.

Annamali, A., Babu, S.T., Josa, N.A., sudha, D., Christina, V., 2011. Lyza biosynthesis and characterization of silver and gold nanoparticles using aqueous leaf extraction of *Phyllanthus amarus* Schum. & Thonn. World Appl. Sci. J. 13 (8), 1833–1840.

Anuradha, J., Abbasi, T., Abbasi, S.A., 2015. An eco-friendly method of synthesizing gold nanoparticles using an otherwise worthless weed pistia (*Pistia stratiotes* L.). J. Adv. Res. 6, 711–720.

Arockiya Aarthi Rajathi, F., Arumugam, R., Saravanan, S., Anantharaman, P., 2014. Phytofabrication of gold nanoparticles assisted by leaves of *Suaeda monoica* and its free radical scavenging property. J. Photochem. Photobiol. B Biol. 135, 75–80.

Arun Kumar, C., Astalakshmi, A., Nima, P., Ganesan, V., 2013. Plant mediated synthesis of silver nanoparticles using leaves of *Odina wodier* Roxb. Int. J. Adv. Res. 1 (5), 265–272.

Arunachalam, K.D., Arun, L.B., Annamalai, S.K., Aarrthy Arunachalam, M., 2014. Biofunctionalized gold nanoparticles synthesis from *Gymnema sylvestre* and its preliminary anticancer activity. Int. J. Pharm. Pharm. Sci.. ISSN: 0975-1491 6 (4).

Ashokkumar, T., Prabhu, D., Geetha, R., Govindaraju, K., Manikandan, R., Arulvasu, C., Singaravelu, G., 2014. Apoptosis in liver cancer (HepG2) cells induced by functionalized gold nanoparticles. Colloids Surf. B Biointerfaces 123, 549–556.

Awwad, A.M., Salem, N.M., Abdeen, A.O., 2013. Green synthesis of silver nanoparticles using carob leaf extract and its antibacterial activity. Int. J. Ind. Chem. 4, 29.

Aziz, N., Faraz, M., Pandey, R., Shakir, M., Fatma, T., Varma, A., et al., 2015. Facile algae-derived route to biogenic silver nanoparticles: synthesis, antibacterial, and photocatalytic properties. Langmuir 31, 11605–11612.

Aziz, N., Pandey, R., Barman, I., Prasad, R., 2016. Leveraging the attributes of *Mucor hiemalis*-derived silver nanoparticles for a synergistic broad-spectrum antimicrobial platform. Front. Microbiol. 7, 1984.

Azizi, S., Ahmad, M.B., Namvar, F., Mohamad, R., 2014. Green biosynthesis and charecterization of zinc oxide nanoparticles using brown marine microalga *Sargassum muticum* aqueous extract. Mater. Lett. 116, 5–277.

Badrinarayanan, K., Sakthivel, N., 2008. Coriander leaf mediated biosynthesis of gold nanoparticles. Mater. Lett. 62, 4588–4590.

Banerjee, J., Narendhirakannan, R.T., 2011. Biosynthesis of silver nanoparticles from *Syzygium cumini* (L). seeds extract and evaluation of their in vitro antioxidant activities. Digest J. Nanomater. Biostruct. 6 (3), 961–968.

Bankar, A., Joshi, B., Kumar, A.R., Zinjarde, S., 2010. Banana peel extract mediated novel route for the synthesis of silver nanoparticles. Colloids Surf. A Physicochem. Eng. Asp. 368, 58–63.

Bar, H., Bhui, D.K., Sahoo, G.P., Sarkar, P., Priyanka, S., 2009. Green synthesis of silver nanoparticles using seed extract of *Jatropha curcas*. Colloids Surf. A Physicochem. Eng. Asp. 348, 212–216.

Bhau, B.S., Ghosh, S., Puri, S., Borah, B., Sarmah, D.K., Khan, R., 2015. Green synthesis of gold nanoparticles from the leaf extract of *Nepenthes khasiana* and antimicrobial assay. Adv. Mater. Lett. 6 (1), 55–58.

Bhumi, G., Lingarao, M., Savithramma, N., 2013. Biological synthesis of silver nanoparticles from stembark of *Thespesia populnea* (L.) soland. Indian Streams Res. J. 3, 1–7.

Bhuyan, T., Mishra, K., Khanuja, M., Prasad, R., Varma, A., 2015. Biosynthesis of zinc oxide nanoparticles from *Azadirachta indica* for antibacterial and photocatylitic activities. Mater. Sci. Semicond. Process. 32, 55–61.

Bindhani, B.K., Panigrahi, A.K., 2014. Green synthesis of gold nanoparticles using neem (*Azadirachta indica* L. Leaf Extract Biomed. Appl. 5 (3), 457–464.

Bisker, G., Yeheskely-Hayon, D., Minai, L., Yelin, D., 2012. Controlled release of Rituximab from gold nanoparticles for phototherapy of malignant cells. J. Control Release 162, 303–309.

Blanco, M.C., Lourido, R., Vazquez-Vazquez, C., Pastor, E., Planes, G.A., Rivas, J., Lopez-Quintela, M.A., 2008. Synthesis of atomic gold clusters with strong electrocatalytic activities. Langmuir 24, 12690–12694.

Borah, R.K., Saikia, H.J., Mahanta, A., Das, V.K., Bora, U., Thakur, A.J., 2015. Biosynthesis of poly(ethylene glycol)-supported palladium nanoparticles using *Colocasia esculenta* leaf extract and their catalytic activity for Suzuki–Miyaura cross-coupling reactions. RSC Adv. 5, 72453–72457.

Bruchez, M., Moronne, M., Gin, P., Weiss, S., Alivisatos, A.P., 1998. Semiconductor nanocrystals as fluorescent biological labels. Science 281, 2013–2016.

Cai, F., Li, J., Sun, J., Ji, Y., 2011. Biosynthesis of gold nanoparticles by biosorption using *Magnetospirillum gryphiswaldense* MSR-1. Chem. Eng. J. 175, 70–75.

Celin Hemalatha, F., Renitta, P., 2011. Biomimetic synthesis and characterization of plant–mediated silver nanoparticles using *Cephalandra indica* extract and evaluation of their antibacterial activity. Elixir Chem. Phys. 40, 5186–5188.

Chan, W.C.W., Nie, S.M., 1998. Quantum dot bioconjugates for ultrasensitive nonisotopic detection. Science 281, 2016–2018.

Chang, L.T., Yen, C.C., 1995. Studies on the preparation and properties of conductive polymers. VIII. Use of heat treatment to prepare metallized films from silver chelate of PVA and PAN. J. Appl. Polym. Sci. 55 (2), 371–374.

Chandran, S.P., Chaudhary, M., Pasricha, R., Ahmad, A., Sastry, M., 2006. Synthesis of gold nanotriangles and silver nanoparticles using *Aloe vera* plant extract. Biotechnol. Prog. 22, 577–583.

Cox, A., Venkatachalam, P., Sahi, S., Sharma, N., 2016. Silver and titanium dioxide nanoparticle toxicity in plants: a review of current research. Plant Physiol. Biochem. 107, 147–163.

Cruza, D., Falea, P.L., Mouratoa, A., Vaza, P.D., Serralheiro, M.L., Linoa, A.R., 2010. Preparation and physicochemical characterization of Ag nanoparticles biosynthesized by *Lippia citriodora* (Lemon Verbena). Colloids Surf. B Biointerfaces 81, 67–73.

Daisy, P., Saipriya, K., 2012. Biochemical analysis of *Cassia fistula* aqueous extract and phytochemically synthesized gold nanoparticles as hypoglycemic treatment for diabetes mellitus. Int. J. Nanomed. 7, 1189–1202.

Daniel, M.C., Astruc, D., 2004. Gold nanoparticles: assembly, supramolecular chemistry, quantum-size-related properties, and applications toward biology, catalysis, and nanotechnology. Chem. Rev. 104 (1), 293–346.

Das, J., Velusamy, P., 2014. Catalytic reduction of methylene blue using biogenic gold nanoparticles from Sesbania grandiflora L. J. Taiwan Inst. Chem. Eng. 45, 2280–2285.

Das, R.K., Gogoi, N., Bora, U., 2011. Green synthesis of gold nanoparticles using *Nyctanthes arbortristis* flower extract. Bioprocess Biosyst. Eng. 34, 615–619.

Das, R.K., Gogoi, N., Babu, P.J., Sharma, P., Mahanta, C., Bora, U., 2012. The synthesis of gold nanoparticles using *Amaranthus spinosus* leaf extract and study of their optical properties. Adv. Mater. Phys. Chem. 2, 275–281.

Das, R.S., Singh, B., Mukhopadyay, S., Benerjee, R., 2012. Gold nano particles catalyzed oxidation of hydrazine by a metallo-superoxide complex: experimental evidences for surface activity of gold nano particles. Dalton Trans. 41, 4641–4648.

De la Isla, A., Brostow, W., Bujard, B., Estevez, M., Rodriguez, J.R., Vargas, S., Castano, V.M., 2003. Nanohybrid scratch resistant coating for teeth and bone viscoelasticity manifested in tribology. Mater. Res. Innov. 7, 110–114.

Devika, R., Elumalai, S., Manikandan, E., Eswaramoorthy, D., 2012. Biosynthesis of silver nanoparticles using the fungus *Pleurotus ostreatus* and their antibacterial activity. 1, 557. http://dx.doi.org/10.4172/scientificreports.557.

Dhamecha, D., Jalalpure, S., Jadhav, K., 2016. *Nepenthes khasiana* mediated synthesis of stabilized gold nanoparticles: characterization and biocompatibility studies. J. Photochem. Photobiol. B Biol. 154, 108–117.

Dinesh, S., Karthikeyan, S., Arumugam, P., 2012. Biosynthesis of silver nanoparticles from *Glycyrrhiza glabra* root extract. Arch. Appl. Sci. Res. 4 (1), 178–187.

Dobrucka, R., Dugaszewska, J., 2016. Biosynthesis and antibacterial activity of ZnO nanoparticles using Trifolium pratense flower extract. Saudi J. Biol. Sci. 23, 517–523.

Du, L., Jiang, H., Liu, X., Wang, E., 2007. Biosynthesis of gold nanoparticles assisted by *Escherichia coli* DH5α and its application on direct electrochemistry of hemoglobin. Electrochem. Commun. 9, 1165–1170.

Dubey, M., Bhadauria, S., Kushwah, B.S., 2009. Green synthesis of nanosilver particles from extract of *Eucalyptus hybrida* (Safeda) leaf. Digest J. Nanomater. Biostruct. 4 (3), 537–543.

Edelstein, R.L., Tamanaha, C.R., Sheehan, P.E., Miller, M.M., Baselt, D.R., Whitman, L.J., Colton, R.J., 2000. The BARC biosensor applied to the detection of biological warfare agents. Biosens. Bioelectron. 14, 805–813.

Elumalai, K., Velmurugan, S., 2015. Green synthesis, characterization and antimicrobial activities of zincoxide nanoparticles from the leaf extract of *Azadirachta indica* (L.). Appl. Surf. Sci. 345, 329–336.

Elumalai, E.K., Prasad, T.N.V.K.V., Hemachandran, J., Viviyan Therasa, S., Thirumalai, T., David, E., Elumalai, E.K., et al., 2010. Extracellular synthesis of silver nanoparticles using leaves of *Euphorbia hirta* and their antibacterial activities. J. Pharm. Sci. Res. 2 (9), 549–554.

Elumalai, K., Velmurugan, S., Ravi, S., Kathiravan, V., Ashok kumar, S., 2015. Green synthesized of zinc oxide nanoparticles using *Moringa oleifera* leaf extract and evaluation of its antimicrobial activity. Mol. Biomol. Spectrosc. 143, 158–164.

Emmanuel, R., Karuppiah, C., Chen, S.-M., Palanisamy, S., Padmavathy, S., Prakash, P., 2014. Green synthesis of gold nanoparticles for trace level detection of a hazardous pollutant (nitrobenzene) causing Methemoglobinaemia. J. Hazard. Mater. 279, 117–124.

Esumi, K., Tano, T., Torigoe, K., Meguro, K., 1990. Preparation and characterization of biometallic Pd-Cu colloids by thermal decomposition of their acetate compounds in organic solvents. J. Chem. Mater. 2, 564–567.

Francis, G., Thombre, R., Parekh, F., Leksminarayan, P., 2014. Bioinspired synthesis of gold nanoparticles using *Ficus benghalensis* (indian banyan) leaf extract. Chem. Sci. Trans. 3 (1), 470–474.

Franco-Romano, M., Gil, M.L.A., Palacios-Santander, J.M., Delgado-Jaén, J.J., Naranjo-Rodriguez, I., Hidalgo-Hidalgo de Cisneros, J.L., Cubillana-Aguilera, L.M., 2014. Sonosynthesis of gold nanoparticles from a geranium leaf extract. Ultrason. Sonochem. 21, 1570–1577.

Fu, L., Fu, Z., 2015. *Plectranthus amboinicus* leaf extract–assisted biosynthesis of ZnO nanoparticles and their photocatalytic activity. Ceram. Int. 41 (2), 2492–2496. http://dx.doi.org/10.1016/j.ceramint.2014.10.069.

Ganeshkumar, M., Sathishkumar, M., Ponrasua, T., Dinesh, M.G., Suguna, L., 2013. Spontaneous ultra-fast synthesis of gold nanoparticles using *Punica granatum* for cancer targeted drug delivery. Colloids Surf. B Biointerfaces 106, 208–216.

Geetha, N., Harini, K., Jerlin Showmya, J., Selva Priya, K., 2012. King and Robinson biofabrication of silver nanoparticles using leaf extract of *Chromolaena odorata* (L.). In: International Conference on Nuclear Energy, Environmental and Biological Sciences (ICNEEBS'2012), pp. 56–59.

Gholamreza, A., Varshosaz, J., Shahbazi, N., 2014. Synthesis of silver nanoparticle using *Portulaca oleracea* L. extracts. Nanomed. J. 1 (2), 94–99.

Ghosh, S., Patil, S., Ahire, M., Kitture, R., Gurav, D.D., Jabgunde, A.M., Kale, S., Pardesi, K., Shinde, V., Bellare, J., Dhavale, D.D., Chopade, B.A., 2012. *Gnidia glauca* flower extract mediated synthesis of gold nanoparticles and evaluation of its chemocatalytic potential. J. Nanobiotechnol. 10, 17.

Gnanajobitha, G., Paulkumar, K., Vanaja, M., Rajeshkumar, S., Malarkodi, C., Annadurai, G., Kannan, C., 2013a. Fruit-mediated synthesis of silver nanoparticles using *Vitis vinifera* and evaluation of their antimicrobial efficacy. J. Nanostruct. Chem. 3, 67.

Gnanajobitha, G., Vanaja, M., Paulkumar, K., Rajeshkumar, S., Malarkodi, C., Annadurai, G., Kannan, C., 2013b. Green synthesis of silver nanoparticles using *Millingtonia hortensis* and evaluation of their antimicrobial efficacy. Int. J. Nanomater. Biostruct. 3 (1), 21–25.

Gopinath, K., Venkatesh, K.S., Ilangovan, R., Sankaranarayanan, K., Arumugam, A., 2013. Green synthesis of gold nanoparticles from leaf extract of *Terminali aarjuna*, for the enhanced mitotic cell division and pollen germination activity. Ind. Crops Prod. 50, 737–742.

Guo, Q., Guo, Q., Yuan, J., Zeng, J., 2014. Biosynthesis of gold nanoparticles using a kind of flavonol: dihydromyricetin. Colloids Surf. A Physicochem. Eng. Asp. 441, 127–132.

Guo, M., Li, W., Yang, F., Liu, H., 2015. Controllable biosynthesis of gold nanoparticles from a *Eucommia ulmoides* bark aqueous extract. Spectrochim. Acta A Mol. Biomol. Spectrosc. 142, 73–79.

Gurunathan, S., Kalishwaralal, K., Vaidyanathan, R., Venkataraman, D., Pandian, S.R., 2009. Biosynthesis, purification and characterization of silver nanoparticles using *Escherichia coli*. Colloids Surf. B Biointerfaces 74, 328–335.

Honary, S., Gharaei-Fathabad, E., Paji, Z.K., Eslamifar, M., 2012. A novel biological synthesis of gold nanoparticle by Enterobacteriaceae family. Trop. J. Pharm. Res. 11 (6), 887–891.

Huang, J., Lin, L., Li, Q., Sun, D., Wang, Y., Lu, Y., He, N., Yang, K., Yang, X., Wang, H., Wang, W., Lin, W., 2008. Continuous-flow biosynthesis of silver nanoparticles by Lixivium of sundried *Cinnamomum camphora* leaf in tubular microreactors. Ind. Eng. Chem. Res. 47, 6081–6090.

Iravani, S., 2011. Green synthesis of metal nanoparticles using plants. Green Chem. 13 (10), 2638–2650.

Islam, N.U., Jalil, K., Shahid, M., Rauf, A., Muhammad, N., Khan, A., Shah, M.R., Khan, M.A., 2015. Green synthesis and biological activities of gold nanoparticles functionalized with Salix alba. Arabian J. Chem. http://dx.doi.org/10.1016/j.arabjc.2015.06.025.

Ismail, E.H., Khalil, M.M.H., Al Seif, F.A., El-Magdou, F., Jan 2014. Biosynthesis of gold nanoparticles using extract of grape (*Vitis vinifera*) leaves and seeds. 3 (1), 1–12.

Jae, Y.S., Beom, S.K., 2009. Rapid biological synthesis of silver nanoparticles using plant leaf extracts. Bioprocess Biosyst. Eng. 32, 79–84.

Jafarirad, S., Mehrabi, M., Divband, B., Kosari-Nasab, M., 2016. Biofabrication of zinc oxide nanoparticles using fruit extract of canina and their toxic potential against bacteria: a mechanistic approach. Mater. Sci. Eng. C Mater. Biol. Appl. 59, 296–302.

Jaina, D., Daimab, H.K., Kachhwahaa, S., Kotharia, S.L., 2009. Synthesis of plant-mediated silver nanoparticles using papaya fruit extract and evaluation of their anti microbial activities. Digest J. Nanomater. Biostruct. 4 (4), 723–727.

Jayaseelan, C., Ramkumar, R., Rahuman, A.A., Peruma, P., 2013. Green synthesis of gold nanoparticles using seed aqueous extract of *Abelmoschus esculentus* and its antifungal activity. Ind. Crops Prod. 45, 423–429.

Jinhwa, H., Kim, D.-S., Kim, Z.H., Lee, Y.W., Kim, D., Kim, M., Kwon, K., Park, H.J., Yun, W.S., Han, S.W., 2008. Controlled synthesis and characterization of the enhanced local field of octahedral Au nanocrystals. Chem. Commun. 46, 6120–6122.

Joglekar, S., Kodam, K., Dhaygude, M., Hudlikar, M., 2011. Novel route for rapid biosynthesis of lead nanoparticles using aqueous extract of *Jatropha curcas* L. latex. Mater. Lett. 65, 3170–3172.

Kahkha, M.R.R., Kahkha, H.R., 2015. Green synthesize of gold nanoparticle using rosacea flower extract. Int. Res. J. Appl. Basic Sci. 9 (10), 1722–1726.

Kalishwaralal, K., Deepak, V., Ramkumarpandian, S., Nellaiah, H., Sangiliyandi, G., 2008. Extracellular biosysnthesis of silver nanoparticles by the culture supernatant of *Bacillus licheniformis*. Mater. Lett. 626, 4411–4413.

Kathiresan, K., Manivannan, S., Nabeel, M.A., Dhivya, B., 2009. Studies on silver nanoparticles synthesized by a marine fungus, *Penicillium fellutanum* isolated from coastal mangrove sediment. Colloids Surf. B Biointerfaces 71, 133–137.

Kavithaa, K., Paulpandi, M., Ponraj, T., Murugan, K., Sumathi, S., 2016. Induction of intrinsic apoptotic pathway in human breast cancer (MCF-7) cells through facile biosynthesized ZnO nanorods. Karbala Int. J. Modern Sci. 2, 46–55.

Khalil, M.M.H., Ismail, E.H., El-Magdoub, F., 2012. Biosynthesis of Au nanoparticles using olive leaf extract: 1st nano updates. Arabian J. Chem. 5 (4), 431–437.

Khan, A.K., Rashid, R., Murtaza, G., Zahra, A., 2014a. Gold nanoparticles: synthesis and applications in drug delivery. Trop. J. Pharm. Res. 13 (7), 1169–1177.

Khan, M., Khan, M., Kuniyil, M., Adil, S.F., Al-Warthan, A., Alkhathlan, H.Z., Tremel, W., Tahir, M.N., Siddiqui, M.R.H., 2014b. Biogenic synthesis of palladium nanoparticles using *Pulicaria glutinosa* extract and their catalytic activity towards the Suzuki coupling reaction. Dalton Trans. 43, 9026–9031.

Klaus, T., Joerger, R., Olsson, E., Granqvist, C.G., 1999. Silver-based crystalline nanoparticles, microbially fabricated. Proc. Natl. Acad. Sci. U.S.A. 96, 13611–13614.

Konishi, Y., Uruga, T., 2007. Bioreductive deposition of platinum nanoparticles on the bacterium *Shewanella* algae. J. Biotechnol. 128, 648–653.

Koperuncholan, M., 2015. Bioreduction of chloroauric acid ($HAuCl_4$) for the synthesis of gold nanoparticles (GNPs): a special empathies of pharmacological activity. Int. J. Phytopharm. 5 (4), 72–80.

Korbekandi, H., Susan, G.A., Jalayer, S., Jalayer, M.S., Bandegani, M., 2013a. Nanosilver particle production using *Juhlands regia* L. (walnut) leaf extract. J. Nat. Pharm. Prod. 8 (1).

Korbekandi, H., Zeynab, A., Siavash, I., Sajjad, A., 2013b. Optimization of biological synthesis of silver nanoparticles using *Fusarium oxysporum*. Iran. J. Pharm. Res. 12 (3), 289–298.

Korbekandi, H., Asghari, G., Jalayer, S.S., Jalayer, M.S., Bandegani, M., 2013c. Nanosilver particle production using *Juglans regia* L. (Walnut) leaf extract. Jundishapur. J. Nat. Pharm. Prod. 8 (1), 20–26.

Kosalai, D., Chandran, M., 2016. Phytochemical analysis and anti-oxidant activity of gold nanoparticles synthesizing plant – *Silybum marianum*. Int. J. Curr. Microbiol. App. Sci. 5 (4), 469–475.

Krishnamurthy, N.B., Nagaraj, B., Malakar, B., Liny, P., Dinesh, R., 2012. Green synthesis of gold nanoparticles using *Tagetes erecta* L. (mari gold) flower extract & evaluation of their antimicrobial activities. Int. J. Pharma Bio Sci. 3 (1), 212–221.

Krishnaraj, C., Jagan, E.G., Rajasekar, S., Selvakumar, P., Kalaichelvan, P.T., Mohan, N., 2010. Synthesis of silver nanoparticles using *Acalypha indica* leaf extracts and its antibacterial activity against water borne pathogens. Colloids Surf. B Biointerfaces 76, 50–56.

Krishnaraj, C., Muthukumaran, P., Ramachandran, R., Balakumaran, M.D., Kalaichelvan, P.T., 2014. *Acalypha indica* Linn: biogenic synthesis of silver and gold nanoparticles and their cytotoxic effects against MDA-MB-231, human breast cancer cells. Biotechnol. Rep. 4, 42–49.

Kudle, K.R., Donda, M.R., Merugu, R., Kudle, M.R., Rudra, M.P.P., 2000. Microwave assisted green synthesis of silver nanoparticles using *Boswellia serrata* flower extract and evaluation of their antimicrobial activity. J. Med. Entomol. 5, 721–725.

Kudle, K.R., Donda, M.R., Alwala, J., Kudle, M.R., Sreedhar, B., Rudra, M.P.P., 2013. Synthesis of silver nanoparticles from *Phyllanthus reticulatus*: an investigation on the effect of broth (leaves and root) concentration in reduction mechanism and particle size. 2 (5), 2839–2849.

Kulkarni, A.P., Srivastava, A., Zunjarrao, R.S., 2012. Plant mediated synthesis of silver nanoparticles and their applications. Int. J. Pharm. Bio Sci. 3 (4), 121–127.

Kumar, V., Yadav, S.K., 2009. Plant-mediated synthesis of silver and gold nanoparticles and their applications. J. Chem. Technol. Biotechnol. 84, 151–157.

Kumar, P., Senthamil Selvi, S., Lakshmi Prabha, A., Prem, K.K., Ganeshkumar, R.S., Govindaraju, M., 2012. Synthesis of silver nanoparticles from *Sargassum tenerrimum* and screening phytochemicals for its antibacterial activity. Nano Biomed. Eng. 4, 12–16.

Kumar, B., Smita, K., Cumbal, L., Camacho, J., Hernandez- Gallegos, E., de Guadalupe Chavez-Lopez, M., Grijalva, M., Andrade, K., 2016. One pot phytosynthesis of gold nanoparticles using *Genipa americana* fruit extract and its biological applications. Mater. Sci. Eng. 62, 725–731.

Lalitha, A., Subbaiya, R., Ponmurugan, P., 2013. Green synthesis of silver nanoparticles from leaf extract *Azhadirachta indica* and to study its anti-bacterial and antioxidant property. Int. J. Curr. Microbiol. App. Sci. 2 (6), 228–235.

Lan, M.Y., Hsu, Y.B., Hsu, C.H., Ho, C.Y., Lin, J.C., Lee, S.W., 2013. Induction of apoptosis by high-dose gold nanoparticles in nasopharyngeal carcinoma cells. Auris Nasus Larynx 40, 563–568.

Lavanya, M., Veenavardhini, S.V., Gim, G.H., Kathiravan, M.N., Kim, S.W., 2013. Synthesis, characterization and evaluation of antimicrobial efficacy of silver nanoparticles using *Paederia foetida* L. leaf extract. Int. Res. J. Biol. Sci. 2 (3), 28–34.

Lee, S.Y., Sneha, K., Cho, C.-W., Yun, Y.-S., 2016. Biosynthesis of gold nanoparticles using ocimum sanctum extracts by solvents with different polarity. ACS Sustain. Chem. Eng. 4, 2651–2659.

Leema Rose, A., Vidhya, S., Ramagirija, S., Ramya, M., Ramya Mary, A., Reka, M.S., Roselin Ansilda, M., 2014. Kinetic study on green synthesis of gold nanoparticles using *Bougainvillea glabra* leaf extract. IJCPS 3 (5).

Li, S., Shen, Y., Xie, A., Yu, X., Zhang, X., Yang, L., 2007. Rapid, room-temperature synthesis of amorphous selenium/protein composites using *Capsicum annuum* L. extract. Nanotechnology 18 (40), 9–10.

Li, G., He, D., Qian, Y., Guan, B., Gao, S., Cui, Y., Yokoyama, K., Wang, L., 2012. Fungus-mediated green synthesis of silver nanoparticles using *Aspergillus terreus*. Int. J. Mol. Sci. 13, 466–476.

Linga Rao, M., Savithramma, N., 2012. Antimicrobial activity of silver nanoparticles synthesized by using stem extract of *Svensonia hyderobadensis* (Walp.) Mold—a rare medicinal plant. Res. Biotechnol. 3, 41–47.

Liz-Marzán, L.M., Lado-Tourino, I., 1996. Reduction and stabilization of silver nanoparticles in ethanol by nonionic surfactants. Langmuir 12, 3585–3589.

Ma, J., Wong, H., Kong, L.B., Peng, K.W., 2003. Biomimetic processing of nanocrystallite bioactive apatite coating on titanium. Nanotechnology 14, 619–623.

Madan, H.R., Sharma, S.C., Udayabhanu, Suresh, D., Vidya, Y.S., Nagabhushana, H., Rajanaik, H., Anantharaju, K.S., Prashantha, S.C., SadanandaMaiya, P., 2016. Facile green fabrication of nanostructure ZnO plates, bullets, flower, prismatic tips, closed pine cone: their antibacterial, antioxidant, photoluminescent and photocatylitic properties. Spectrochim. Acta A Mol. Biomol. Spectrosc. 152, 404–416.

Maensiri, S., Laokul, P., Klinkaewnarong, J., Phokha, S., Promarak, V., Seraphin, S., 2008. Indium Oxide (In_2O_3) Nanoparticles Using *Aloe vera* Plant Extract: Synthesis and Optical Properties. J. Optoelectron. Adv. Mater. 10, 161–165.
Mah, C., Zolotukhin, I., Fraites, T.J., Dobson, J., Batich, C., Byrne, B.J., 2000. Microsphere-mediated delivery of recombinant AAV vectors in vitro and in vivo. Mol. Ther. 1, S239. http://dx.doi.org/10.1006/mthe.2000.0174.
Mahtab, R., Rogers, J.P., Murphy, C.J., 1995. Protein-sized quantum dot luminescence can distinguish between "straight", "bent", and "kinked" oligonucleotides. J. Am. Chem. Soc. 117, 9099–9100.
Maiti, S., Krishnan, D., Barman, G., Ghosh, S.K., Laha, J.K., 2014. Antimicrobial activities of silver nanoparticles synthesized from *Lycopersicon esculentum* extract. J. Anal. Sci. Tech. 5, 40.
Makarov, V.V., Makarova, S.S., Love, A.J., Sinitsyna, O.V., Dudnik, A.O., Yaminsky, I.V., Taliansky, M.E., Kalinina, N.O., 2014. Biosynthesis of stable iron oxide nanoparticles in aqueous extracts of *Hordeum vulgare* and *Rumex acetosa* plants. Langmuir 30, 5982–5988.
Malarkodi, C., Rajeshkumar, S., Paulkumar, K., Vanaja, M., Gnana Jobitha, G., Annadurai, G., 2013. Bactericidal activity of bio mediated silver nanoparticles synthesized by *Serratia nematodiphila*. Drug Invention Today 5 (3), 1–7.
Mallikarjuna, K., Sushma, N.J., Subba Reddy, B.V., Narasimha, G., Deva Prasad Raju, B., 2013. Palladium nanoparticles: single-step plant mediated green chemical procedure using *Piper betle* leaves broth and their anti-fungal studies. Int. J. Chem. Anal. Sci. 4, 14–18.
Mane Gavade, S.J., Nikam, G.H., Dhabbe, R.S., Sabale, S.R., Tamhankar, B.V., Mulik, G.N., 2015. Green synthesis of silver nanoparticles by using carambola fruit extract and their antibacterial activity. Adv. Nat. Sci. Nanosci. Nanotechnol. 6, 1–6.
Mata, R., Nakkala, J.R., Sadras, S.R., 2016. Polyphenol stabilized colloidal gold nanoparticles from *Abutilon indicum* leaf extract induce apoptosis in HT-29 colon cancer cells. Colloids Surf. B Biointerfaces 143, 499–510.
Megarajan, S., Ayaz Ahmed, K.B., Rajendra Kumar Reddy, G., Suresh Kumar, P., Anbazhagan, V., 2016. Phytoproteins in green leaves as building blocks for photosynthesis of gold nanoparticles: an efficient electrocatalyst towards the oxidation ofascorbic acid and the reduction of hydrogen peroxide. J Photochem. Photobiol. B 155, 7–12.
Mendoza, K.C., McLane, V.D., Kim, S., Griffin, J.D., 2010. In vitro application of gold nanoprobes in live neurons for phenotypical classification, connectivity assessment, and electrophysiological recording. Brain Res. 1325, 19–27.
Mishra, P., Ray, S., Sinha, S., Das, B., Khan, M.I., Behera, S.K., Suraj, S.-I.Y., Tripathy, K., Mishra, A., 2016. Facile biosynthesis of gold nanoparticles by using extract of *Hibiscus sabdariffa* and evaluation of its cytotoxicity against U87 glioblastoma cells under hyperglycemic condition. Biochem. Eng. J. 105, 264–272.
Mohan Kumar, K., Mandal, B.K., Kiran Kumar, H.A., Maddinedi, S.B., 2013a. Green synthesis of size controllable gold nanoparticles. Spectrochim. Acta A Mol. Biomol. Spectrosc. 116, 539–545.
Mohan Kumar, K., Mandal, B.K., Siva Kumar, K., Sreedhara Reddy, P., Sreedhar, B., 2013b. Biobased green method to synthesise palladium and iron nanoparticles using *Terminalia chebula* aqueous extract. Spectrochim. Acta A Mol. Biomol. Spectrosc. 102, 128–133.
Molday, R.S., MacKenzie, D., 1982. Immunospecific ferromagnetic iron dextran reagents for the labeling and magnetic separation of cells. J. Immunol. Methods 52, 353–367.
Mondal, A.K., Parui, S.M., Samanta, S., Mallick, S., 2011. Synthesis of ecofriendly silver nanoparticle from plant latex used as an important taxonomic tool for phylogenetic interrelationship. Adv. Biores. 2 (1), 122–133.
Murphy, C.J., et al., 2008. Chemical sensing and imaging with metallic nanorods. Chem. Commun. 5, 544–557.
Mutahir, A., Chandra, H., Baunthiyal, M., 2016. Green synthesis of silver nanoparticles from different parts of the plant *Macrotyloma uniflorum*. Octa J. Biosci. 3 (2), 58–61.
Nagajyothi, P.C., Minh An, T.N., Sreekanth, T.V.M., Lee, J.-I., Lee, D.J., Lee, K.D., 2013. Green route biosynthesis: characterization and catalytic activity of ZnO nanoparticles. Mater. Lett. 108, 160–163.
Nagajyothi, P.C., Sreekanth, T.V.M., Tettey, C.O., Jun, Y.I., Mook, S.H., 2014. Characterization, antibacterial, antioxidant, and cytotoxic activities of ZnO nanoparticles using *Coptidis rhizoma*. Bioorg. Med. Chem. Lett. 24, 4298–4303.
Nagaraj, B., Krishnamurthy, N.B., Liny, P., Divya, T.K., Dinesh, R., 2011. Biosynthesis of gold nanoparticles of *Ixora coccinea* flower extract & their antimicrobial activities. Int. J. Pharma Bio Sci. 2 (4), 557–565.
Nagaraj, B., Malakar, B., Divya, T.K., Krishnamurthy, N.B., Liny, P., Dinesh, A., Iconaru, S.L., Ciobanu, C.S., 2012a. Synthesis of plant mediated gold nanoparticles using flower extracts of *Carthamus tinctorius* L. (safflower) and evaluation of their biological activities. Digest J. Nanomater. Biostruct. 7 (3), 1289–1295.
Nagaraj, B., Malakar, B., Divya, T.K., Krishnamurthy, N.B., Liny, P., Dinesh, R., 2012b. Environmental benign synthesis of gold nanoparticles from the flower extracts of *Plumeria alba* Linn. (Frangipani) and evaluation of their biological activities. Int. J. Drug Dev. Res. 4 (1), 144–150.

Nagaraj, B., Malakar, B., Divya, T.K., Krishnamurthy, N.B., Liny, P., Dinesh, R., 2013. Environmental benign synthesis of gold nanoparticles from the flower extracts of *Plumeria alba* Linn. (Frangipani) and evaluation of their biological activities. Int. J. Nanomater. Biostruct. 3 (1), 21–25.

Nair, B., Pradeep, T., 2002. Coalescence of nanoclusters and formation of submicron crystallites assisted by *Lactobacillus* strains. Cryst. Growth Design 2, 293–298.

Nam, J.M., Thaxton, C.C., Mirkin, C.A., 2003. Nanoparticles-based bio-bar codes for the ultrasensitive detection of proteins. Science 301, 1884–1886.

Nanda, A., Saravanan, M., 2009. Biosynthesis of silver nanoparticles from *Staphylococcus aureus* and its antimicrobial activity against MRSA and MRSE. Nanomedicine 5, 452–456.

Naraginti, S., Lakshmi Kumari, P., Das, R.K., Sivakumar, A., Patil, S.H., Andhalkar, V.V., 2016. Amelioration of excision wounds by topical application of green synthesized, formulated silver and gold nanoparticles in albino Wistar rats. Mater. Sci. Eng. C Mater. Biol. Appl. 62, 293–300.

Natarajan, R.K., Yagam, A.A.J.N., Gurunagarajan, S., Muthukumar, Ekambaram, N., Manimaran, A., 2013. *Elaeagnus indica* mediated green synthesis of silver nanoparticles and its potent toxicity against human pathogens. Global J. Pharmacol. 7 (3), 222–231. http://dx.doi.org/10.5829/idosi.wasj.2013.23.10.73.

Nithya Deva Krupa, A., Vimala, R., 2016. Evaluation of tetraethoxysilane(TEOS) sol-gel coatings, modified with green synthesized of zinc oxide nanoparticles for combating microfouling. Mater. Sci. Eng. C Mater. Biol. Appl. 61, 728–735.

Njagi, E.C., Huang, H., Stafford, L., Genuino, H., Galindo, H.M., Collins, J.B., 2011. Biosynthesis of iron and silver nanoparticles at room temperature using aqueous sorghum bran extracts. Langmuir 27, 264–271.

Noruzi, M., 2015. Biosynthesis of gold nanoparticles using plant extracts. Bioprocess Biosyst. Eng. 38 (1), 1–14.

Ochieng, P.E., Iwuoha, E., Michira, I., Masikini, M., Ondiek, J., Githira, P., Kamau, G.N., 2015. Green route synthesis and characterization of ZnO nanoparticles using *Spathodea campanulata*. Int. J. Biochemiphys. 23, 53–61.

Paciotti, G.F., Myer, L., Weinreich, D., Goia, D., Pavel, N., McLaughlin, R.E., Tamarkin, L., 2004. Colloidal gold: a novel nanoparticle vector for tumor directed drug delivery. Drug Deliv. 11, 169–183.

Panatarotto, D., Prtidos, C.D., Hoebeke, J., Brown, F., Kramer, E., Briand, J.P., Muller, S., Prato, M., Bianco, A., 2003. Immunization with peptide-functionalized carbon nanotubes enhances virus-specific neutralizing antibody responses. Chem. Biol. 10, 961–966.

Pandit, R., 2015. Green synthesis of silver nanoparticles from seed extract of *Brassica nigra* and its antibacterial activity. Nusant. Biosci. 7 (1), 15–19.

Parak, W.J., Boudreau, R., Gros, M.L., Gerion, D., Zanchet, D., Micheel, C.M., Willams, S.C., Alivisatos, A.P., Larabell, C.A., 2002. Cell motility and metastatic potential studies based on quantum dot imaging of phagokinetic tracks. Adv. Mater. 14, 882–885.

Parashar, V., Parashar, R., Avinash, B.S., Pandey, C., 2009. *Parthenium* leaf extract mediated synthesis of silver nanoparticles: a novel approach towards weed utilization. Digest J. Nanomater. Biostruct. 4 (1), 45–50.

Parveen, A., Roy, A.S., Rao, S., 2012. Biosynthesis and characterization of silver nanoparticles from *Cassia auriculata* leaf extract and in vitro evaluation of antimicrobial activity. Int. J. Appl. Biol. Pharm. Technol. 3 (2), 222–228.

Patil, S., Ahire, M., Kitture, R., Gurav, D.D., Jabgunde, A.M., Kale, S., Pardesi, K., Shinde, V., Bellare, J., Dhavale, D.D., Chopade, B.A., 2013. *Gnidia glauca* flower extract mediated synthesis of gold nanoparticles and evaluation of its chemocatalytic potential. Int. Res. J. Pharm. 4 (6).

Patra, S., Mukherjee, S., Barui, A.K., Ganguly, A., Sreedhar, B., Patra, C.R., 2015. Green synthesis, characterization of gold and silver nanoparticles and their potential application for cancer therapeutics. Mater. Sci. Eng. C Mater. Biol. Appl. 53, 298–309.

Pattanayak, M., Nayak, P.L., 2013. Green synthesis of gold nanoparticles using *Elettaria cardamomum* (ELAICHI) aqueous extract. World J. Nano Sci. Technol. 2 (1), 01–05.

Paulkumar, K., Rajeshkumar, S., GnanaJobitha, G., Vanaja, M., Malarkodi, C., Annadurai, G., 2013a. Eco-friendly synthesis of silver chloride nanoparticles using *Klebsiella planticola* (MTCC 2277). Int. J. Green Chem. Bioproc. 3 (1), 12–16.

Paulkumar, M., GnanaJobitha, G., Vanaja, M., RajeshKumar, S., Malarkodi, C., Annadurai, G., 2013b. Biosynthesis of silver chloride nanoparticles using *Bacillus subtilis* MTCC 3053 and assessment of its antifungal activity. ISRN Nanomater. 1–8. http://dx.doi.org/10.1155/2013/317963.

Pavani, K.V., Gayathramma, K., Banerjee, A., Suresh, S., 2013. Phyto-synthesis of silver nanoparticles using extracts of *Ipomoea indica* flowers. Am. J. Nanomater. 1 (1), 5–8.

Peralta-Videa, J.R., Hernandez-Viezcas, J.A., Zhao, L., Diaz, B.C., Ge, Y., Priester, J.H., Holden, P.A., Gardea-Torresdey, J.L., 2014. Cerium dioxide and zinc oxide nanoparticles alter the nutritional value of soil cultivated soybean plants. Plant Physiol. Biochem. 80, 128–135.

Peralta-Videa, J.R., Huang, Y., Parsons, J.G., Zhao, L., Lopez-Moreno, L., Hernandez-Viezcas, J.A., Gardea-Torresdey, J.L., 2016. Plant-based green synthesis of metallic nanoparticles: scientific curiosity or a realistic alternative to chemical synthesis? Nanotechnol. Environ. Eng. 1 (1), 4.

Phanjom, P., Sultana, A., Sarma, H., Ramchiary, J., Goswami, K., Baishya, P., 2012. Plant-mediated synthesis of silver nanoparticles using *Elaeagnus latifolia* leaf extract. Digest J. Nanomater. Biostruct 7 (3), 1117–1123.

Philip, D., 2010. Rapid green synthesis of spherical gold nanoparticles using *Mangifera indica* leaf. Spectrochim. Acta A Mol. Biomol. Spectrosc. 77, 807–810.

Prasad, C., Venkateswarlu, P., 2014. Soybean seeds extract based green synthesis of silver nanoparticles. Indian J. Adv. Chem. Sci. 2 (3), 208–211.

Prasad, R., Pandey, R., Barman, I., 2016. Engineering tailored nanoparticles with microbes: quo vadis? Wiley Interdiscip. Rev. Nanomed. Nanobiotechnol. 8, 316–330.

Pugazhenthiran, N., Anandan, S., Kathiravan, G., Prakash, N.K.U., Crawford, S., et al., 2009. Microbial synthesis of silver nanoparticles by *Bacillus* sp. J. Nanopart. Res. 11, 1811–1815.

Punuri, J.B., Sharma, P., Saranya, S., Bora, U., 2013. Synthesis of gold nanoparticles using ethonolic leaf extract of *Bacopa monnieri* and UV irradiation. Mater. Lett. 93, 431–434.

Qian, Y., Yao, J., Russel, M., Chen, K., Wang, X., 2015. Characterization of green synthesized nano formulation (ZnO–*A. vera*) and their antibacterial activity against pathogens. Environ. Toxicol. Pharmacol. 39, 736–746.

Rad, M.S., Rad, J.S., Heshmati, G.A., Miri, A., Sen, D.J., 2013a. Biological synthesis of gold and silver nanoparticles by *Nitraria schoberi* fruits. Am. J. Adv. Drug Deliv. ISSN:232-547X.

Rad, M.S., Rad, J.S., Heshmati, G.A., Miri, A., Malarkodi, C., Rajeshkumar, S., Paulkumar, K., Gnanajobitha, G., Vanaja, M., Annadurai, G., 2013b. Bacterial synthesis of silver nanoparticles by using optimized biomass growth of *Bacillus* sp. Nanosci. Nanotechnol. Int. J. 3 (2), 26–32.

Radha, R., Murugalakshmi, M., Kokial Rani, S., 2016. Eco friendly synthesis and characterization of gold nanoparticles from *Bauhinia purpurea* flower extract. Imperial J. Interdiscip. Res. 2 (6), 306–310.

Ragupathi Raja Kannan, R., Arumugam, R., Ramya, D., Manivannan, K., Anantharaman, P., 2012. Green synthesis of silver nanoparticles using marine macroalga *Chaetomorpha linum*. Appl. Nanosci. 3, 229–233.

Rajeshkumar, S., Kannan, C., Annadurai, G., 2012a. Green synthesis of silver nanoparticles using marine brown *Turbinaria conoides* and its antibacterial activity. Int. J. Pharma Bio Sci. 3 (4), 502–510.

Rajeshkumar, S., Kannan, C., Annadurai, G., 2012b. Synthesis and characterization of antimicrobial silver nanoparticles using marine brown seaweed *Padina tetrastromatica*. Drug Invention Today 4 (10), 511–513.

Rajeshkumar, S., GnanaJobitha, G., Malarkodi, C., Kannan, C., Annadurai, G., 2013a. Optimization of marine bacteria *Enterococcus* sp. biomass growth by using response surface methodology. J. Environ. Nanotechnol. 2 (1), 20–27.

Rajeshkumar, S., Malarkodi, C., Paulkumar, K., Vanaja, M., GnanaJobitha, G., Annadurai, G., 2013b. Intracellular and extracellular biosynthesis of silver nanoparticles by using marine bacteria *Vibrio alginolyticus*. Nanosci. Nanotechnol. Int. J. 3 (1), 21–25.

Rajeshkumar, S., Malarkodi, C., Paulkumar, K., Vanaja, M., GnanaJobitha, G., Annadurai, G., 2014. Algae mediated green fabrication of silver nanoparticles and examination of its antifungal activity against clinical pathogens. Int. J. Metals. 2014. http://dx.doi.org/10.1155/2014/692643.

Rajiv, P., Rajeshwari, S., Venckatesh, R., 2013. Bio-fabrication of zinc oxide nanoparticles using leaf extract of *Parthenium hysterophorus* L. and its size dependant antifungal activity against plant fungal pathogens. Spectrochim. Acta A Mol. Biomol. Spectrosc. 112, 384–387.

Ramar, M., Manikandan, B., Marimuthu, P.N., Raman, T., Maralinga, A., Subramanian, P., Karthick, S., Munusamy, A., 2015. Synthesis of silver nanoparticles using *Solanum trilobatum* fruits extract and its antibacterial, cytotoxic activity against human breast cancer cell line MCF 7. Spectrochim. Acta A Mol. Biomol. Spectrosc. 140 (5), 223–228.

Ramesh, V., Armash, A., 2015. Green synthesis of gold nanoparticles against pathogens and cancer cells. IJPR 5 (10), 250–256.

Ramesh, P., Rajendran, A., MeenakshiSundaram, M., 2014. Green syntheis of zinc oxide nanoparticles using flower extract *Cassia auriculata*. J. Nanosci. Nanotechnol. 2 (1), 41–45.

Ramesh, M., Anbuvannan, M., Viruthagiri, G., 2015. Green synthesis of ZnO nanoparticles using *Solanum nigrum* leaf extract and their antibacterial activity. Spectrochim. Acta A Mol. Biomol. Spectrosc. 136, 864–870.

Raut Rajesh, W., Lakkakula Jaya, R., Kolekar Niranjan, S., Mendhulkar Vijay, Kashid Sahebrao, 2009. Phytosynthesis of silver nanoparticle using *Gliricidia sepium* (Jacq.). Curr. Nanosci. 5, 117–122.

Ravichandrika, K., Kiranmayi, P., Ravikumar, R.V.S.S.N., 2012. Synthesis, characterization and antibacterial activity of ZnO nanoparticles. Int. J. Pharm. Pharm. Sci. 4 (4), 336–338.

Reddy, N.J., Vali, D.N., Rani, M., Rani, S.S., 2014. Evaluation of antioxidant, antibacterial and cytotoxic effects of green synthesized silver nanoparticles by *Piper longum* fruit. Mat. Sci. Eng. C 34, 115–122.

Rimal Isaac, R.S., Sakthivel, G., Murthy, C., 2013. Green synthesis of gold and silver nanoparticles using *Averrhoa bilimbi* fruit extract. J. Nanotechnol. 2013. http://dx.doi.org/10.1155/2013/906592.

Roopan, S.M., Bharathi, A., Kumar, R., Khanna, V.G., Prabhakarn, A., 2011. Acaricidal, insecticidal, and larvicidal efficacy of aqueous extract of *Annona squamosa* L. peel as biomaterial for the reduction of palladium salts into nanoparticles. Colloids Surf. B Biointerfaces 92, 209–212.

Salam, H.A., Sivaraj, R., Venckatesh, R., 2014. Green synthesis and characterization of zinc oxide nanoparticles from *Ocimum basilicum* L. var. purpurascens Benth.-*Lamiaceae* leaf extract. Mater. Lett. 131, 16–18.

Sangeetha, G., Rajeshwari, S., Venckatesh, R., 2011. Green synthesis of zinc oxide nanoparticles by *Aloe barbadensis* miller leaf extract: structure and optical properties. Mater. Res. Bull. 46, 2560–2566.

Sardar, R., Funston, A.M., Mulvaney, P., Murray, R.W., 2009. Gold nanoparticles: past, present, and future. Langmuir 25, 13840–13851.

Sasikala, A., Linga Rao, M., Savithramma, N., Prasad, T.N.V.K.V., 2015. Synthesis of silver nanoparticles from stem bark of *Cochlospermum religiosum* (L.) Alston: an important medicinal plant and evaluation of their antimicrobial efficacy. Appl. Nanosci. 5 (7), 827–835.

Sathyavathi, R., Balamurali Krishna, M., Venugopal Rao, S., Saritha, R., Narayana Rao, D., 2010. Biosynthesis of silver nanoparticles using *Coriandrum sativum* leaf extract and their application in nonlinear optics. Adv. Sci. Lett. 3 (2), 138–143.

Savithramma, N., Lingarao, M., Suvarnalatha Devi, P., 2011. Evaluation of antibacterial efficacy of biologically synthesized silver nanoparticles using stem barks of *Boswellia ovalifoliolata* and *Shorea tumbuggaia*. J. Biol. Sci. 11, 39–45.

Selva Bharath, M., Packia Lekshmi, N.C.J., Dinesh Kumar, P., Raja Brindha, J., Jeeva, S., 2012. Synthesis of plant mediated silver nanoparticles using *Solanum xanthocarpum* fruit extract and evaluation of their anti microbial activities. J. Pharm. Res. 5 (9), 4888–4892.

Shankar, S.S., Ahmed, A., Akkamwar, B., Sastry, M., Rai, A., Singh, A., 2004. Biological synthesis of triangular gold nanoprism. Nat. Mater. 3, 482–488.

Shankar, S., Jaiswal, L., Aparna, R.S.L., Prasad, R.G.S.V., 2014. Synthesis, characterization, in vitro bio compatibility, and antimicrobial activity of gold, silver and gold silver alloy nanoparticles prepared from *Lansium domesticum* fruit peel extract. Mater. Lett. 137, 75–78.

Shedbalkar, U., Singh, R., Wadhwani, S., Gaidhani, S., Chopade, B.A., 2014. Microbial synthesis of gold nanoparticles: current status and future prospects. Adv. Colloid Interf. Sci. 209, 40–48.

Sheny, D., Philip, D., Mathew, J., 2013. Synthesis of platinum nanoparticles using dried *Anacardium occidentale* leaf and its catalytic and thermal applications. Spectrochim. Acta A Mol. Biomol. Spectrosc. 114, 267–271.

Shiraishi, Y., Toshima, N., 2000. Oxidation of ethylene catalyzed by colloidal dispersions of poly (sodium acrylate)-protected silver nanoclusters. Colloids Surf. A Physicochem. Eng Asp. 169, 59–66.

Shweta, Tripathi, D.K., Singh, S., Singh, S., Dubey, N.K., Chauhan, D.K., 2016. Impact of nanoparticles on photosynthesis: challenges and opportunities. Mater. Focus 5 (5), 405–411.

Singh, P.P., Bhakat, C., 2012. Green synthesis of gold nanopoarticles and silver nanoparticles from leaves and bark of *Ficus carica* for nanotechnological application. Int. J. Sci. Res. Publ. 2 (5), 1–5.

Singh, C., Baboota, R.K., Naik, P., Singh, H., 2012a. Biocompatible synthesis of silver and gold nanoparticles using leaf extract of *Dalbergia sissoo*. Adv. Mater. Lett. 3 (4), 279–285.

Singh, M., Kalaivani, R., Manikandan, S., Sangeetha, N., Kumaraguru, A.K., 2012b. Facile green synthesis of variable metallic gold nanoparticle using *Padina gymnospora*, a brown marine macroalga. Appl. Nanosci. http://dx.doi.org/10.1007/s13204-012-0115-7.

Singh, A., Sharma, M.M., Batra, A., 2013. Synthesis of gold nanoparticles using chick pea leaf extract using green chemistry. J. Optoelectron. Biomed. Mater. 5 (2), 27–32.

Singh, S., Tripathi, D.K., Dubey, N.K., Chauhan, D.K., 2016. Effects of nano-materials on seed germination and seedling growth: striking the slight balance between the concepts and controversies. Mater. Focus 5 (3), 195–201.

Singh, S., Vishwakarma, K., Singh, S., Sharma, S., Dubey, N.K., Singh, V.K., Liu, S., Tripathi, D.K., Chauhan, D.K., 2017. Understanding the plant and nanoparticle interface at transcriptomic and proteomic level: a concentric overview. Plant Gene. http://dx.doi.org/10.1016/j.plgene.2017.03.006.

Sivarangani, K., Meenakshisundaram, M., 2013. Biological synthesis of silver nanoparticals using *Ocimum bacillicum* leaf extract and their antimicrobial activity. IRJP 4 (1), 225–229.

Smitha, S.L., Philip, D., Gopchandran, K.G., 2009. Green synthesis of gold nanoparticles using *Cinnamomum zeylanicum* leaf broth. Spectrochim. Acta A Mol. Biomol. Spectrosc. 74, 735–739.

Sokolov, K., Follen, M., Aaron, J., Pavlova, I., Malpica, A., Lotan, R., Richards-Kortum, R., 2003. Real-time vital optical imaging of precancer using anti-epidermal growth factor receptor antibodies conjugated to gold nanoparticles. Cancer Res. 63, 1999–2004.

Song, J.Y., Jang, H.-K., Kim, B.S., 2009. Biological synthesis of gold nanoparticles using *Magnolia Kobus* and *Diopyros kaki* leaf extracts. Process Biochem. 44, 1133–1138.

Song, J.Y., Kwon, E.Y., Kim, B.S., 2010. Biological synthesis of platinum nanoparticles using *Diopyros kaki* leaf extract. Bioprocess Biosyst. Eng. 33, 159–164.

Soundarrajan, C., Sankari, A., Dhandapani, P., Maruthamuthu, S., Ravichandran, S., Sozhan, G., Palaniswamy, N., 2012. Rapid biological synthesis of platinum nanoparticles using *Ocimum sanctum* for water electrolysis applications. Bioprocess Biosyst. Eng. 35, 827–833.

Srinivas Naik, L., Paul Marx, K., Sree Vennela, P., Devi, V.R., 2013. Green synthesis of silver nanoparticles using strawberry leaf extract (*Arbutus unedo*) and evaluation of its antimicrobial activity-a novel study. Int. J. Nanomater. Biostruct. 3 (3), 47–50.

Stoimenov, P.K., Klinger, R.L., Marchin, G.L., Klabunde, K.J., 2002. Metal oxide nanoparticles as bactericidal agents. Langmuir 18, 6679–6686.

Suganya, R.S., Priya, K., Sumi Roxy, B., 2012. Phytochemical screening and antibacterial activity from *Nerium oleander* and evaluate their plant mediate nanoparticle synthesis. IRJP 3 (5), 285–288.

Sujitha, M.V., Kannan, S., 2013. Green synthesis of gold nanoparticles using *Citrus* fruits (*Citrus limon*, *Citrus reticulata* and *Citrus sinensis*) aqueous extract and its characterization. Spectrochim. Acta A Mol. Biomol. Spectrosc. 102, 15–23.

Suman, T.Y., Radhika Rajasree, S.R., Ramkumar, R., Rajthilak, C., Perumal, P., 2014. The Green synthesis of gold nanoparticles using an aqueous root extract of *Morinda citrifolia* L. Spectrochim. Acta A Mol. Biomol. Spectrosc. 118, 11–16.

Sun, Y.P., Atorngitjawat, P., Meziani, M.J., 2001. Preparation of silver nanoparticles via rapid expansion of water in carbon dioxide microemulsion into reductant solution. Langmuir 17, 5707–5710.

Sundrarajan, M., Ambika, S., Bharathi, K., 2015. Plant-extract mediated synthesis of ZnO nanoparticles using *Pongamia pinnata* and their activity against pathogenic bacteria. Adv. Powder Technol. 26, 1294–1299.

Swamy, M.K., Akhtar, M.S., Mohanty, S.K., Sinniah, U.R., 2015. Synthesis and characterization of silver nanoparticles using fruit extract of *Momordica cymbalaria* and assessment of their in vitro antimicrobial, antioxidant and cytotoxicity activities. Spectrochim. Acta A Mol. Biomol. Spectrosc. 151, 939–944.

Tahir, K., Li, B., Khan, S., Nazir, S., Khan, Z.U.H., Khan, A.U., Islam, R.U., 2015a. Enhanced chemocatalytic reduction of aromatic nitro compounds by biosynthesized gold nanoparticles. J. Alloys Compd. 651, 322–327.

Tahir, K., Nazir, S., Li, B., Khan, A.U., Khan, Z.U.H., Gong, P.Y., Khan, S.U., Ahmad, A., 2015b. *Nerium oleander* leaves extract mediated synthesis of gold nano- particles and its antioxidant activity. Mater. Lett. 156, 198–201.

Talapin, D.V., Lee, J.S., Kovalenko, M.V., Shevchenko, E.V., 2009. Prospects of colloidal nanocrystals for electronic and optoelectronic applications. Chem. Rev. 110, 389–458.

Talekar, S., Asavari, J., Radhika, C., Arya, N., Rahul, B., 2016. Immobilized enzyme mediated synthesis of silver nanoparticles using cross-linked enzyme aggregates (CLEAs) of NADH-dependent nitrate reductase. Nano Struct. Nano Objects 6, 23–33.

Thema, F.T., Manikandan, E., Dhlamini, M.S., Maaza, M., 2015. Green synthesis of ZnO nanoparticles via *Agathosma betulina* natural extract. Mater. Lett. 161, 124–127.

Thilagam, M.A., Tamilselvi, B., Chandrasekeran, C., 2013. Rose phytosynthesis of silver nanoparticles using medicinal and dye yielding plant of *Bixa orellana* L. Leaf extract. JPSI 2 (4), 9–13.

Tiwari, M., Sharma, N.C., Fleischmann, P., Burbage, J., Venkatachalam, P., Sahi, S.V., 2017. Nanotitania exposure causes alterations in physiological, nutritional and stress responses in tomato (*Solanum lycopersicum*). Front. Plant Sci. 8.

Tripathi, D.K., Singh, V.P., Prasad, S.M., Chauhan, D.K., Dubey, N.K., 2015. Silicon nanoparticles (SiNp) alleviate chromium (VI) phytotoxicity in *Pisum sativum* (L.) seedlings. Plant Physiol. Biochem. 96, 189–198.

Tripathi, D.K., Singh, S., Singh, V.P., Prasad, S.M., Chauhan, D.K., Dubey, N.K., 2016. Silicon nanoparticles more efficiently alleviate arsenate toxicity than silicon in maize cultiver and hybrid differing in arsenate tolerance. Front. Environ. Sci. 4, 46.

Tripathi, D.K., Singh, S., Singh, S., Pandey, R., Singh, V.P., Sharma, N.C., Prasad, S.M., Dubey, N.K., Chauhan, D.K., 2017a. An overview on manufactured nanoparticles in plants: uptake, translocation, accumulation and phytotoxicity. Plant Physiol. Biochem. 110, 2–12.

Tripathi, D.K., Mishra, R.K., Singh, S., Singh, S., Vishwakarma, K., Sharma, S., Singh, V.P., Singh, P.K., Prasad, S.M., Dubey, N.K., Pandey, A.C., 2017b. Nitric oxide ameliorates zinc oxide nanoparticles phytotoxicity in wheat seedlings: implication of the ascorbate–glutathione cycle. Front. Plant Sci. 8.

Tripathi, D.K., Singh, S., Singh, S., Srivastava, P.K., Singh, V.P., Singh, S., Prasad, S.M., Singh, P.K., Dubey, N.K., Pandey, A.C., Chauhan, D.K., 2017c. Nitric oxide alleviates silver nanoparticles (AgNps)-induced phytotoxicity in *Pisum sativum* seedlings. Plant Physiol. Biochem. 110, 167–177.

Tripathi, D.K., Tripathi, A., Shweta, S.S., Singh, Y., Vishwakarma, K., Yadav, G., Sharma, S., Singh, V.K., Mishra, R.K., Upadhyay, R.G., Dubey, N.K., 2017d. Uptake, accumulation and toxicity of silver nanoparticle in autotrophic plants, and heterotrophic microbes: a concentric review. Front. Microbiol. 8.

Tripathi, D.K., Singh, S., Singh, V.P., Prasad, S.M., Dubey, N.K., Chauhan, D.K., 2017e. Silicon nanoparticles more effectively alleviated UV-B stress than silicon in wheat (*Triticum aestivum*) seedlings. Plant Physiol. Biochem. 110, 70–81.

Uma Suganya, K., Govindaraju, S., Ganesh Kumar, K., Prabhu, V., Arulvasu, D., Stalin Dhas, C., Karthick, T., Niranjan Changmai, V., 2016. Anti-proliferative effect of biogenic gold nanoparticles against breastcancer cell lines (MDA-MB-231 & MCF-7). Appl. Surf. Sci. 37, 415–424.

Valodkar, M., Nagar, P.S., Jadeja, R.N., Thounaojam, M.C., Devkar, R.V., Thakore, S., 2011. Euphorbiaceae latex induced green synthesis of non-cytotoxic metallic nanoparticle solutions: a rational approach to antimicrobial applications. Colloids Surf. A Physicochem. Eng Asp. 384, 337–344.

Vanaja, M., Annadurai, G., 2013. *Coleus aromaticus* leaf extract mediated synthesis of silver nanoparticles and its bactericidal activity. Appl. Nanosci. 3 (3), 217–223.

Vanaja, M., Paulkumar, K., Rajeshkumar, S., Gnana Jobitha, G., Malarkodi, C., Chitra, M.K., Annadurai, G., 2015. Fungal assisted intracellular and enzyme based synthesis of silver nanoparticles and its characterization and bactericidal efficiency. Int. Res. J. Pharm. Biosci. 2 (3), 8–19.

Vanathi, P., Rajiva, P., Narendhran, S., Rajeshwari, S., Rahman, P.K.S.M., Venckatesh, R., 2014. Biosynthesis and characterization of phyto mediated zinc oxide nanoparticles: a green chemistry approach. Mater. Lett. 134, 13–15.

Vankar, P.S., Bajpai, D., 2010. Preparation of gold nanoparticles from *Mirabilis jalapa* flowers. Indian J. Biochem. Biophys. 47, 157–160.

Vankar, P.S., Shukla, D., 2012. Biosynthesis of silver nanoparticles using lemon leaves extract and its application for antimicrobial finish on fabric. Appl. Nanosci. 2 (2), 163–168.

Varun, S., Sellappa, S., Khan, M.R., Vijayakumar, S., 2015. Green synthesis of gold nanoparticles using *Argemone mexicana* L. Leaf extract and its characterization. Int. J. Pharm. Sci. Rev. Res. 32 (2), 42–44.

Vaseem, M., Umar, A., 2010. ZnO nanoparticles: growth, properties, and applications. Am. Sci. Publ. 5, 2–3.

Velavan, S., Arivoli, P., Mahadevan, K., 2012. Biological reduction of silver nanoparticles using *cassia auriculata* flower extract and evaluation of their in vitro antioxidant activities. Nanosci. Nanotechnol. Int. J. 2 (4), 30–35.

Velayutham, K., Rahuman, A.A., Rajakumar, G., Santhoshkumar, T., Marimuthu, S., Jayaseelan, C., 2011. Evaluation of *Catharanthus roseus* leaf extract-mediated biosynthesis of titanium dioxide nanoparticles against *Hippobosca maculata* and *Bovicola ovis*. Parasitol Res. 111 (6), 2329–2337.

Velumani, S., 2015. Green synthesis of gold nanoparticles from *Costus igneus*. IJARIIE 1 (5), 2395–4396 ISSN (O).

Venkatachalam, M., Govindaraju, K., Mohamed Sadiq, A., Tamilselvan, S., Ganesh Kumar, V., Singaravelu, G., 2013. Functionalization of gold nanoparticles as antidiabetic nanomaterial. Spectrochim. Acta Mol. Biomol. Spectrosc. 116, 331–338.

Verma, A., Joshi, P., Arya, A., 2013. Synthesis of plant-mediated silver nanoparticles using plant extract of *Sonchus asper*. Int. J. Nanotechnol. Appl. 3 (4), 11–18.

Vidya, C., Hirematha, S., Chandraprabha, M.N., Lourdu Antonyraj, M.A., Gopal, I.V., Jain, A., Bansal, K., 2013. Green synthesis of zinc oxide nanoparticles by *Calatropis gigantea*. Int. J. Curr. Eng. Technol.. ISSN: 2277-4106.

Wang, S., Mamedova, N., Kotov, N.A., Chen, W., Studer, J., 2002. Antigen/antibody immunocomplex from CdTe nanoparticle bioconjugates. Nano Lett. 2, 817–822.

Wani, K., Choudhari, A., Chikate, R., Kaul-Ghanekar, R., 2013. Synthesis and Characterization of Gold Nanoparticles Using Ficus Religiosa ExtractApplied Science Innovations Pvt. Ltd., India Carbon – Sci. Tech. 5/1, pp. 203–210.

Weissleder, R., Elizondo, G., Wittenburg, J., Rabito, C.A., Bengele, H.H., Josephson, L., 1990. Ultrasmall superparamagnetic iron oxide: characterization of a new class of contrast agents for MR imaging. Radiology 175, 489–493.

Willner, I., Baron, R., Willner, B., 2006. Growing metal nanoparticles by enzymes. J. Adv. Mater. 18, 1109–1120.

Wua, W., Huang, J., Wu, L., Sun, D., Lin, L., Zhou, Y., Wang, H., Li, Y., Wu, T.-Y., Chen, S.-M., Ajmal Ali, M., Fahad Al Hemaid, M.A., 2012. Green synthesis and electrochemical characterizations of gold nanoparticles using leaf extract of *Magnolia kobus*. Int. J. Electrochem. Sci. 7, 12742–12751.

Yasmin, A., Ramesh, K., Rajeshkumar, S., 2014. Optimization and stabilization of gold nanoparticles by using herbal plant extract with microwave heating. Yasmin et al. Nano Converg. 2014 (1), 12.

Yoshida, J., Kobayashi, T., 1999. Intracellular hyperthermia for cancer using magnetite cationic liposomes. J. Magn. Magn. Mater. 194, 176–184.

Yu, J., Xu, D., Guan, H.N., Wang, C., Huang, L.K., Chi, D.F., 2016. Facileone-step green synthesis of gold nanoparticles using *Citrus maxima* aqueous extract sandits catalytic activity. Mater. Lett. 166, 110–112.

Yuvakkumar, R., Suresh, J., Joseph Nathanael, A., Sundrarajan, M., Hong, S.I., 2014. Novel green synthetic strategy to prepare ZnO nanocrystals using rambutan (*Nephelium lappaceum* L.) peel extract and its antibacterial applications. Mater. Sci. Eng. C Mater. Biol. Appl. 41, 17–27.

Yuvakkumar, R., Suresh, J., Saravanakumar, B., Nathanael, A.J., Hong, S.I., Rajendran, V., 2015. Rambutan peels promoted biomimetic synthesis of bioinspired zinc oxide nanochains for biomedical applications. Mol. Biomol. Spectrosc. 137, 250–258.

Zayed, M.F., Eisa, W.H., 2014. *Phoenix dactylifera* L. leaf extract phytosynthesized gold nanoparticles; controlled synthesis and catalytic activity. Spectrochim. Acta A Mol. Biomol. Spectrosc. 121, 238–244.

Zhan, G., Huang, J., Lin, L., Lin, W., Emmanuel, K., Li, Q., 2011. Synthesis of gold nanoparticles by *Cacumen platycladi* leaf extract and its simulated solution: toward the plant-mediated biosynthetic mechanism. J. Nanopart. Res. 13, 4957–4968.

Further Reading

Lee, H.J., Lee, G., Jang, N.R., Yun, J.H., Song, J.Y., Kim, B.S., 2011. Biological synthesis of copper nanoparticles using plant extract. Nanotechnology 1, 371–374.

Manoj, L., Vishwakarma, V., 2015. Green synthesis and spectroscopic characterisations of gold nanoparticles using in vitro grown hypericin rich shoot cultures of *Hypericum hookerianum*. Int. J. ChemTech Res. 8 (11), 194–199.

CHAPTER

3

Potential of Spectroscopic Techniques in the Characterization of "Green Nanomaterials"

Gaurav Sharma[1], *Sonika Pandey*[2], *Somenath Ghatak*[3], *Geeta Watal*[3], *Prashant K. Rai*[3]

[1]Jawaharlal Nehru University, New Delhi, India; [2]National Institute of Plant Genome Research, New Delhi, India; [3]University of Allahabad, Allahabad, India

3.1 INTRODUCTION

Nanoparticles, defined as a miniaturized particle of range 10^{-9} with the dimensional size of 1–100 nm, are the building blocks of nanotechnology (McNeil, 2005). Nanoparticle-derived materials are related to numerous scientific and technological applications in the fields of pharmaceuticals, chemistry, electronic sensors, energy generation, environment monitoring, and other areas of biosciences (McDonald et al., 2005; Esparza et al., 2008; Tripathi et al., 2016, 2017a,b,c,d,e; Singh et al., 2017; Tiwari et al., 2017) and have shown several positive and negative impacts (Peralta-Videa et al., 2014; Tripathi et al., 2015; Cox et al., 2016; Shweta et al., 2016; Singh et al., 2016; Peralta-Videa et al., 2016). For the synthesis of nanoparticles, various conventional chemical and physical techniques such as laser ablation, sol–gel, pyrolysis, and lithography electrodeposition are available (Davar and Salavati-Niasari, 2011). However, they are not only expensive, but also require high energy input and use toxic chemical/organic solvents as reducing agents. This results in production of hazardous by-products. As a result, biologically synthesized nanoparticles known as green nanoparticles/nanomaterials are gaining importance. A large number of living organisms have already been known as a source of green nanoparticles, such as cyanobacteria, bacteria, fungi, plants, and their bioderivatives (MubarakAli et al., 2011; Aziz et al., 2015, 2016; Prasad et al., 2016). In addition, biomolecules such as proteins, carbohydrates, polyphenols, flavonoids, and their bioconjugates can also be utilized efficiently for the capping of metallic nanoparticles.

Nanomaterials in Plants, Algae, and Microorganisms
http://dx.doi.org/10.1016/B978-0-12-811487-2.00003-7

The wide range of applications is mainly because of the unique physicochemical and crystallographic properties of nanomaterial that are highly different from their bulk counterparts; their uniqueness is associated with dependency on the size of the nanoparticle. Luminescent semiconductor crystal-derived nanosensors known as quantum dot' nanoparticles exhibit massive change in electronic and optical properties, such as a size-tunable light emission, symmetric emission spectra, and broad absorption spectra, which is a function of size, arising because of quantum confinement effects (Alivisatos, 1996; Kairdolf et al., 2013). Because of changes in the density of electronic states, the excitation process is limited to a few transitions, and can be understood on the basis of the relation between position and momentum in free and confined particles. Furthermore, the efficacy in chemical reactivity of metal nanocrystals used as a catalyst is also governed by size. Nanoscale gold/copper catalysts synthesized in different shapes, such as irregular nanoparticles, nanobelts, and nanoplatelets, showed a remarkable increase in surface area that provides better molecular selectivity to catalyze a specific reaction (Zhou et al., 2006).

The characterization of nanomaterial is generally based on two methods: microscopic imaging and spectroscopic analysis. The imaging methods involve different types of microscopy such as electron microscopy, ion microscopy, and atomic force microscopy. Although microscopic methods have paved the way for scientists to see these objects at the nanoscale, most of the chemical, structural, and optical properties cannot be characterized simply by imaging. To overcome these problems, spectroscopic methods for characterization of nanoparticles have gained importance. This has opened new doors for exploring nanomaterial that mainly provides information supporting elemental and structural analysis.

By definition, spectroscopy pertains to the interaction between matter and electromagnetic radiations (gamma, X-ray, ultraviolet, visible, infrared, microwave radiation and radio waves) that are categorized into eight groups on the basis of the frequency. This interaction results in the emission of characteristic spectra that are influenced by wave properties such as their energy, frequency, phase angle, velocity, amplitude, polarization, and direction of propagation that are altered during the interaction with matter/nanomaterials, and form the basis of various topological analyses. In this chapter, a brief account of various spectroscopic techniques has been described that are commonly used to study the physicochemical characteristics of nanomaterial.

3.2 OVERVIEW OF METHODS FOR SYNTHESIS OF NANOPARTICLES

Nanoparticles can be synthesized by using various routes: physical, chemical, and biological methods as shown in Fig. 3.1. There are numerous approaches for physical methods of nanomaterial synthesis such as the arc discharge method, evaporation, condensation, spray pyrolysis, vapor phase synthesis, laser beam ablation, and inert gas condensation.

In the chemical method, nanoparticles are mainly synthesized by reduction in the presence of inorganic or organic agents. For example, reduction of silver ions (Ag^{+}) in aqueous or nonaqueous solutions can be carried out in the presence of various reducing agents such as sodium citrate, elemental hydrogen, ascorbate, sodium borohydride, Tollens' reagent,

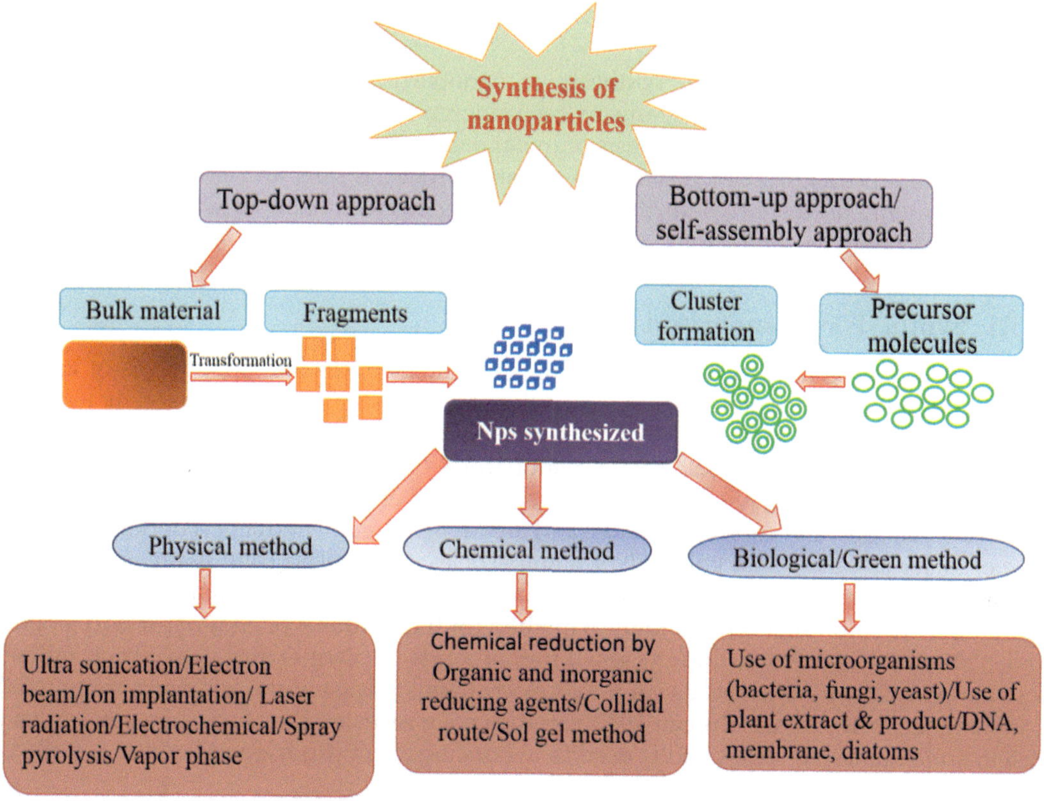

FIGURE 3.1 Illustration of different methods involved in the synthesis of nanoparticles.

N,N-dimethylformamide, and poly(ethylene glycol)-block copolymers. Although both physical and chemical methods can produce nanoparticles in a very short duration they lead to environmental contamination by producing various hazardous wastes as by-products. This problem can be overcome by adopting a biological method using biotechnological tools that are considered safe and ecologically supported.

Nowadays, the concept of green nanotechnology has attracted scientists all over the world. Gold nanoparticles synthesized from green and zimbro tea were simple and cost-effective (Geraldes et al., 2016). Moreover, gold nanoparticles generated through green synthesis agglomerate, which is caused by the presence of numerous phytochemicals such as thearubigins, theaflavins, and catechins that act as both reducing and stabilizing agents. Halophytes grow well in extreme saline environments and hence produce very useful secondary metabolites. Silver nanoparticles biosynthesized by different plant parts (leaves, bark, and root) of *Avicenna marina* mangrove plant were found to show antimicrobial activity (Sankar and Abideen, 2015). The green synthesis of nanoparticles can also minimize the environmental challenges, scarcity of energy resources, and dependency on chemically derived nanoparticles.

3.3 SOURCE FOR GREEN SYNTHESIS OF NANOMATERIALS

The biological approach for synthesis of nanoparticles has increased tremendously. Molecules derived from living organisms can undergo highly controlled assembly that provides a suitable source for the synthesis of metal nanoparticles, which was found to be cost-effective and eco-friendly. Green synthesis of nanoparticles is safer and easy to handle. Some of the commonly used sources for the manufacturing of green nanoparticles are plants, seaweed, algae, fungi, bacteria, etc.

3.3.1 Synthesis of Nanomaterial From Plants

Plants are active reservoirs of a broad range of metabolites that are sources of antioxidants, polymers, and biomolecules. Most of the plants or their products are easily available throughout the year and can act as a source for synthesizing nanoparticles. The main mechanism involved during the production of nanoparticles is plant-assisted reduction caused by phytochemicals. Some of the common phytochemicals are polyphenols, flavones, ketones, aldehydes, amides, and carboxylic acids, and are involved in the reduction of the metal ions. One of major advantages of gold/silver nanoparticles synthesized from plant extract is that the process can be carried out at ambient temperature, atmospheric pressure, and performed in water, which represents an economical method for the synthesis of nanoparticles (Geraldes et al., 2016). During the past few years, different kinds of plant extracts have been utilized for nanoparticle synthesis because of their low cost, renewable nature, and nontoxicity (Nath and Banerjee, 2013; Rajesh Kumar, 2012). Studies have revealed that *Rumex hymenosepalus* is rich in antioxidant molecules and is used as a reducing agent (Rodríguez-León et al., 2013).

3.3.2 Synthesis of Nanomaterial From Algae

Algae are the other abundant source for the green synthesis of nanoparticles. Because of cosmopolitan distribution these can be utilized for production of large-scale and eco-friendly metallic nanoparticles. In the last few years, the demand for gold and silver nanoparticles synthesized using algae has increased tremendously because of easy handling of raw material, the presence of bioactive phytochemicals, and high stability. Some of the commonly used algae from which gold nanoparticles have been synthesized are *Sargassum wightii* (Singaravelu et al., 2007), *Turbinaria conoides* (Rajeshkumar et al., 2013), *Laminaria japonica* (Ghodake and Lee, 2011), and *Stoechospermum marginatum* (Rajathi et al., 2012).

3.3.3 Synthesis of Nanomaterial From Fungi

Fungi are also an important source for the production of larger amounts of nanoparticles. This is because fungal mycelium has a tendency to secrete large amounts of proteins that can be used for higher production of nanomaterial, which makes it a better candidate as compared to bacteria. Some of common examples of fungus are *Pediococcus pentosaceus* (Shahverdi et al., 2007), *Fusarium oxysporum* (Ummartyotin et al., 2012), and *Aspergillus fumigatus* (Bhainsa and D'Souza, 2006) reported for synthesis of silver nanoparticles using the reduction method.

3.3.4 Synthesis of Nanomaterial From Bacteria

Microorganisms such as bacteria constitute a large domain of prokaryotic organisms, and have an ability to survive in a wide range of environmental conditions ranging from optimal to highly extreme. However, the simplest microbes have a remarkable capability to mobilize or immobilize and reduce metal ions at the nanometer scale. In the last few decades there has been tremendous advancement in the biosynthesis of nanoparticles using bacteria with desired morphologies/sizes, fast and clean. Different bacterial species such as *Bacillus cereus, Escherichia coli, Bacillus subtilis,* and *Pseudomonas aeruginosa* have been reported for removing Ag^{+}, Cd^{2+}, Cu^{2+}, and La^{3+} from solution and also have a binding capacity to large quantities of metallic cations (Mullen et al., 1989). Cell biomass and cell extracts of bacteria can be potential sources of synthesis of nanoparticles such as gold, silver, titanium, platinum, palladium, cadmium sulfide, titanium dioxide, magnetite, and so forth (Iravani, 2014). Magnetotactic bacteria and S-layer bacteria are some common examples of bacteria-synthesizing inorganic materials.

3.4 FACTORS GOVERNING SYNTHESIS OF GREEN NANOPARTICLES AND THEIR ANALYSIS

In general, the size of nanomaterials is closer to single atoms and molecules, and as a result their mass becomes extremely small leading to negligible gravitational force. As a consequence, nanoparticles exhibit either additional or different properties, which arise because of a variety of reasons including quantum confinement effects, which are totally different from the bulk material made up of the same elemental constituents (Wang and Wang, 2014). Furthermore, because the size of the particle is decreased, the surface area per unit mass increases because these nanoparticles possess a relatively large surface area. In addition, surface area of a particular material depends on the size and shape. Therefore when evaluating the toxicity and behavior of nanoparticles, it is not only the concentration and composition that are important determinants but also the physicochemical properties of nanomaterials must be considered (Tiede et al., 2008). Some of the major determinants that govern the biosynthesis of green nanoparticles are mentioned next.

3.4.1 Technique for Particle Synthesis

Nanoparticles can be synthesized from methods based on physical synthesis, chemical synthesis, and biological synthesis. These methods mostly comprise two approaches, namely, bottom up and top down, as shown in Fig. 3.1. When smaller components are arranged together to form more complex aggregates this is referred to as the bottom-up approach. This method requires either a chemical or physical force that operates at the nanoscale to assemble smaller units into larger structures. Organic synthesis of homogeneous carbon nanotubes and quantum dots made of indium gallium arsenide are some of the common examples (Jasti and Bertozzi, 2010). However, the top-down approach leads to the formation of smaller nanoscale particles that are derived from larger ones. Lithographic patterning using a short wavelength optical source is one of the commonest examples

(Pease and Chou, 2008). Although each procedure has its own specific benefits and drawbacks, biological methods are widely accepted for the synthesis of nanoparticles because they use biodegradable and naturally occurring materials in conjunction with green technology and are therefore environmentally sustainable and more acceptable than traditional methods (Vadlapudi and Kaladhar, 2014).

3.4.2 pH Effect on Aggregation of Nanoparticles

Variations in the pH of solution can lead to drastic changes in size, shape, or aggregates of synthesized nanoparticles. Previous studies have shown that the size of silver nanoparticles synthesized at different pH values was inversely proportional (Alqadi et al., 2014). It was found that at high pH, the surface plasmon resonance peak becomes narrower and shifted toward the shorter wavelength region, while boarding of the surface plasmon resonance peak leads to a wider range of nanoparticles size in the solution. Thus the smaller size of nanoparticles can be synthesized by increasing the solution pH and vice versa. Many studies have also revealed that at high pH, spherical nanoparticles can be obtained, while cylindrical and triangular patterns can be obtained at low pH (Zhou et al., 2013).

3.4.3 Temperature

Temperature is another important determinant that controls the synthesis of nanoparticles with different morphological patterns leading to unique functional properties. Various chemical methods for synthesizing nanoparticles such as the photoinduced method, electrochemical method, ultrasonic-assistant method, solvothermal method, and templating method are highly dependent on the reaction temperature for the formation and growth of gold/silver nanoplates. Jiang et al. (2011) reported that silver nanoparticles of different patterns, such as plates and spheres, can be generated through a synergetic reduction approach using two or three reducing agents simultaneously. Different methods require different temperature conditions. Generally, a higher temperature >350°C is needed for physical methods and a lower temperature is required for chemical methods (Patra and Baek, 2014), while nanoparticles based on green synthesis require an ambient temperature around 100°C (Rai et al., 2006).

3.4.4 Pressure

Pressure is another important parameter affecting the size/shape of magnetite nanoparticles synthesized by the coprecipitation technique. In a study based on solution algorithm, it was observed that the formation of nanoparticles with different compositions can be carried out using mechanisms such as binary homogeneous nucleation, coagulation, and binary heterogeneous cocondensation, by changing only the saturation pressure of one material in thermal plasma fabrication (Shigeta and Watanabe, 2016). As the difference in saturation pressure increases, there is an increase in the time lag of cocondensation of two material vapors during the collective growth of the metal silicide nanopowder resulting in generation of a wide range of nanopowders. In addition, the reduction of metal ions using biological agents has been found to be accelerated at ambient pressure (Abhilash, 2012).

3.4.5 Stabilizing Agent

Stabilization of synthesized nanoparticles is necessary for various industrial usages, medical purposes, drug delivery, and manufacturing technologies. A highly stable nanoparticle can be synthesized via stabilization of the nanoparticle surface through a coating of a dispersant layer over the surface of the nanoparticle. 3-(Trimethoxysilyl)propyl methacrylate (TPM) is one such example that has been used as the stabilizing agent for the preparation of TPM-modified ZnO nanoparticles (Hung and Whang, 2005). It was observed that the presence of TPM on the surface of nanoparticles effectively stimulates the stability of colloidal ZnO nanoparticles and the compatibility between the organic matrix in the solid nanohybrid and inorganic nanoparticles. The thickness of the dispersant layer is a key factor that determines the stability of suspensions containing highly concentrated nanoparticles. In a study it was reported that a thick layer of adherent results in an excluded volume around the nanoparticles, while a thin layer causes particles to aggregate (Studart et al., 2007). Polymeric substances such as starch can also work efficiently as protecting agent because they contain size-confined, nanosized pools of inter- and intramolecular origin, which can be used for the synthesis of nanoparticles. Polyhydroxylated macromolecules can form supramolecular aggregations, which form a base for the nucleation/growth of nanoparticles because of inter- and intramolecular hydrogen bonding resulting in molecular level (Studart et al., 2007).

3.4.6 Particle Size Distribution/Surface Area

There is tremendous demand for nanoparticles of increasingly smaller particle size and different material composition. The particle size and particle size distributions of nanoparticles are two important factors governing the properties of suspensions, such as rheology, surface area, film gloss, and packing density. The application of many nanomaterials such as TiO_2, carbon nanotubes, and dendrimers synthesized for industrial, environmental, and toxicity investigation is often determined by the properties of nanomaterials, such as particle size, surface area, crystal structures, and morphological features (Tiwari et al., 2008). Moreover particle size also determines the physical and chemical properties of nanoparticles, e.g., the melting point of nanoparticles has been reported to decrease dramatically when the size of the nanoparticles reached the atomic scale.

3.4.7 Particle Proximity Effect

The physicochemical properties of nanomaterials are influenced not only by size/shape but matrix for dispersion or particle proximity is also a big constraint. This is because when these either individual or aggregate miniaturized particles come in contact with or are near to the surface of other nanoparticles, a change in properties is observed. This proximity effect can be utilized for generating more advanced nanoparticles. In a study it was reported that the proximity effect controlled the superconducting behavior of novel biphasic Pb–Sn nanoparticles that were embedded in an Al matrix that played a key role in attaining the effective transition temperature of the system (Bose et al., 2008). Also different-shaped nanoparticles can be evolved because of the proximity effect. Yang et al. (2016) studied the electrochemical dynamics in chains of silver particles on a silver ion solid electrolyte, which is a size-dependent process. It was observed that when the grid spacing is small as compared to the size of the formed silver particles, anomalous chains of unequally sized particles leading to an irregular chain are formed because of the proximity effect during the tip-induced electrochemical process. Moreover, the proximity effect of nanoparticles can be implemented in the capping of particles, substrate–substance interaction, and magnetic properties of the nanoparticles. Particle size and size distribution are crucial for the characterization of nanoparticle systems because they are involved in the determination of in vivo distribution, toxicity, biological fate, and the targeting ability of nanoparticles in various applications (Geraldes et al., 2016).

3.4.8 Other Factors

Nanoparticles synthesized from living systems are highly dynamic in nature. Plant-extracted metabolites that can act as reducing and stabilizing agents for the synthesis of nanoparticles are highly dependent on plant species, plant parts (root, stem, leaf, seed, etc.) and methods/conditions used for extraction. In addition, the different types of nanoparticles require a different purification method that also controls the quality and quantity of synthesized nanoparticles. Moreover, the behavior of synthesized nanoparticles is highly affected by experimental conditions (Ghorbani et al., 2011). Previous studies have shown that stability, reactivity, and physicochemical properties are influenced by reaction conditions. In a study it was observed that the crystalline nature of zinc sulfide nanoparticles changed immediately when its environment was changed from a wet to a dry condition.

Although there are numerous factors that must be considered while synthesizing nanoparticles, three main steps remain that must be evaluated for the formation of green nanoparticles. These include solvent medium used for nanoparticle synthesis, selection of reducing agent, and capping of nanoparticles with a nontoxic stabilizing agent.

3.5 OVERVIEW OF SPECTROSCOPIC TECHNIQUES APPLICABLE TO NANOPARTICLE ANALYSIS

Nowadays a wide range of characterization techniques is available for the detection of nanoparticles that explore the physical and chemical characteristics of the synthesized nanomaterials used. Some of the commonly used methods are described in this chapter as shown in Fig. 3.2.

3.5.1 Nuclear Magnetic Resonance

Nuclear magnetic resonance (NMR) is a widely used spectroscopic technique in which nuclei in a magnetic field absorb and then reemit electromagnetic radiation, a phenomenon utilized for the characterization of conformational dynamics of nanomaterial at the atomic level. This method is applicable for 3D structure determination of a solid compound and a suspension/colloidal solution. Unlike microscopic imaging and diffraction techniques that can provide a long-range order of structural evidence of crystallinity, NMR can be used to determine the structures of various biomolecules, polymers, and amorphous substances that are generally devoid of long-range ordered arrangement (Wang et al., 2001). This method

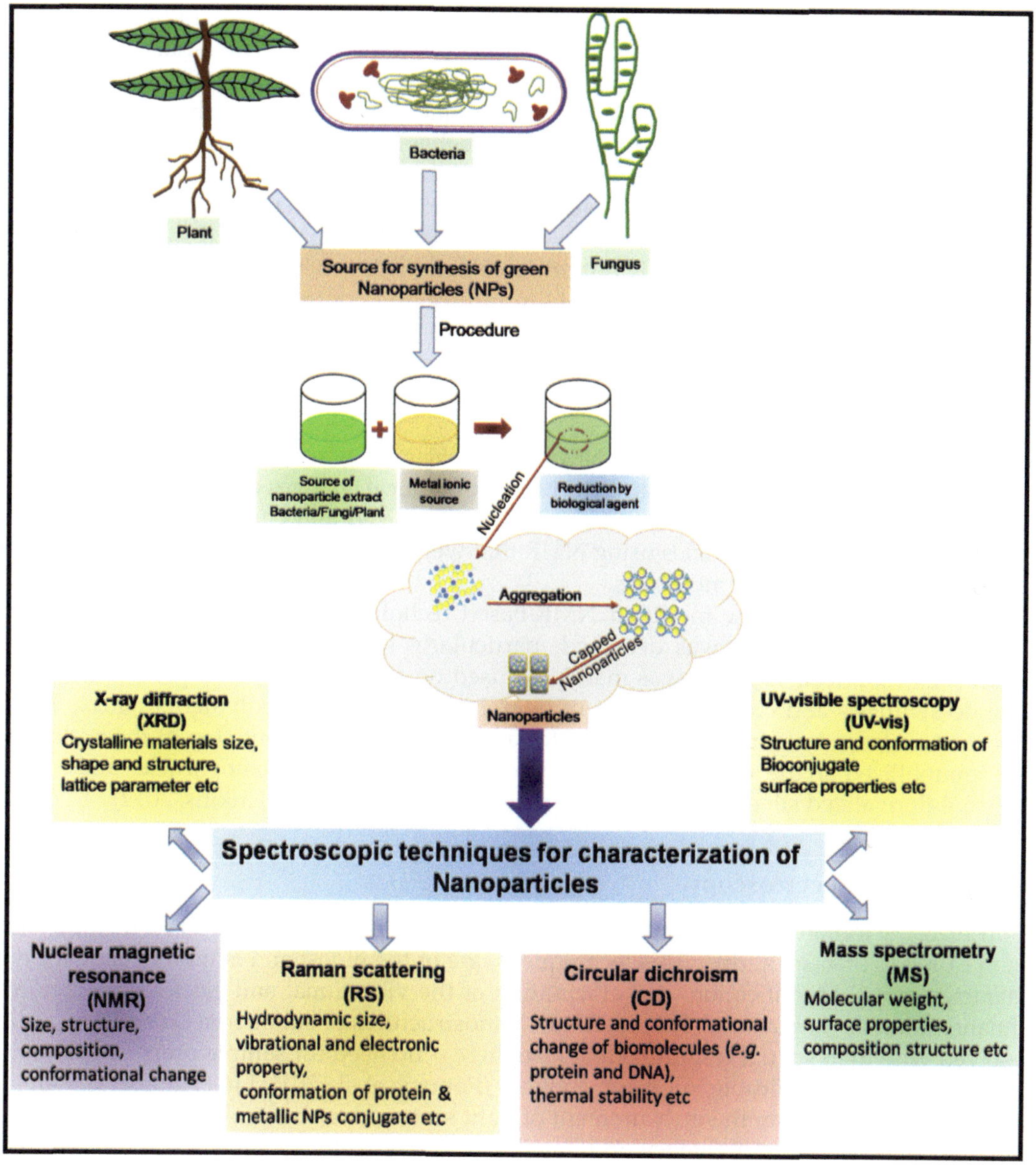

FIGURE 3.2 Schematic representation of different biological sources for the synthesis of nanoparticles and their characterization using spectroscopic techniques.

is gaining importance because of a wide range of applications that involve not only structural characterization but also the evaluation of dynamic interactions of the species in different conditions. Carter et al. (2005) used NMR screening for air- and water-stable silicon nanocrystals prepared by the bromine oxidation of porous silicon nanoparticles. NMR can be easily applied for characterization of the phase behavior and of the surface adsorption phenomena of colloids in a suspension. One such example is the pulse gradient spin-echo NMR technique. Based on diffusion NMR spectroscopy, large colloidal materials, particularly vesicles formed from macromolecular amphiphiles and nanoparticles, can be easily resolved (Valentini et al., 2004; Carter et al., 2005). This technique can also facilitate the in situ localization and visualization of spatiotemporal metabolic behavior in the tissues of the living system in their native form (Gholipour et al., 2015).

In the past few decades, NMR-based characterization of nanoparticles has gained importance because of its noninvasiveness, high throughput, decreased sample preparation, high reproducibility, and easily quantifiable data. This is because low radio-frequency excitation energy is used, which is unable to induce any conformational changes associated with molecular structures (Lupoi et al., 2014). However, because of low detection sensitivity toward biological samples, spectra often show considerable peak overlap over a long time and the samples are required for analysis through NMR (Sapsford et al., 2011; Lupoi et al., 2014).

There is an immense requirement of NMR in exploring the potential of nanomaterial-based biodiagnostic fingerprinting that can detect the biochemical changes associated with disease conditions or drug delivery. Crooks et al. (2012) synthesized and characterized dendrimer-encapsulated metal nanoparticles using NMR that were used as catalysts for the hydrogenation of unsaturated organic molecules.

In the mid-1990s, a new modified NMR-based technique called high-resolution magic angle spinning (HR-MAS) was designed, particularly for nonsolid materials (Alam and Jenkins, 2012). HR-MAS NMR was mainly focused on studying the solid-phase organic or peptide synthesis and characterization of ligands attached to the polymeric matrix. These modified surfaces were further employed for analyzing each synthetic step of the cyclopeptide immobilized on the surface of poly(vinylidene fluoride)-based nanoparticles and studying thiol-derivatized silver clusters (Deshayes et al., 2010; Alam and Jenkins, 2012).

3.5.2 Raman Spectroscopy

In 1928, the Indian scientist Raman discovered Raman spectroscopy based on the phenomenon called Raman scattering. This technique is one of the oldest and is widely used for the characterization, identification, and elucidation of the vibrational and electronic structures of nanomaterials (Pathak and Thassu, 2016). Nanostructures of submicron size can be easily resolved with high spatial resolution and also there is no need for sample preparation, making this technique applicable for in situ analysis (Popovic et al., 2011; Lin et al., 2014). Raman spectroscopy is based on the measurement of light scattered from a molecule when a monochromatic light of single frequency originated from a laser source interacts with the sample. When a sample is irradiated with a light source then the light is either absorbed, transmitted, or scattered. When the laser interacts with the sample, the electron cloud is perturbed and excites the molecule to a next higher state, resulting in either lower Stokes scattering or higher/anti-Stokes scattering, a frequency that is detected by the photodetector. In general, Raman spectra are produced when the energy of scattered photons is different from incident photons, showing the occurrence of energy transfer from one molecule to another, which is known as Raman shift. This is usually expressed in terms of wave number and the spectrum is denoted in terms of the intensity of the Raman-scattered light as a function of Raman shift.

The scattering process seen during light interaction with matter can be divided into two groups, i.e., elastic and inelastic, and Raman spectroscopy uses the inelastic scattering of light with the analyte. In general, nanoparticles are weak in Raman scattering because of their nanoscale size resulting in a very low frequency of about 1 in 10^7 photons. This drawback of a Raman signal of low sensitivity was overcome by developing a more advanced form known as surface-enhanced Raman scattering (SERS) and tip-enhanced Raman spectroscopy. These were ultrasensitive characterization techniques that could work even to a single molecule level (Li et al., 2010). Nanomaterials made of Ag, Au, and Cu had been characterized by SERS on the basis of particle size and distribution, electron–photon coupling, phonon dispersion, etc. Applications of Raman spectroscopy to biomass have previously been reviewed and can be used to determine quantitatively the temperature shifts in frequency of the Raman peaks associated with carbon nanotubes (Casimir et al., 2015).

3.5.3 X-Ray Diffraction

The X-ray diffraction (XRD) technique is a well-known powerful nondestructive technique, first described by German physicist Von Laue in 1912 for characterizing crystalline materials. By examining the XRD peaks, a lot of information can be extracted from the XRD data related to crystalline nature, nature of phase present, and grain size (Liu et al., 1995). Moreover, this technique is widely used for resolving the tertiary structures of various forms of crystalline samples (powder, bulk, and thin film) at the atomic scale (Cantor and Schimmel, 1980; Sapsford et al., 2011) but is limited to amorphous materials. When X-rays are passed through a sample, a diffraction peak is produced resulting from constructive interference of a monochromatic beam scattered at specific angles from each set of lattice planes. This interaction can be explained by Bragg's law as follows (Cantor and Schimmel, 1980):

Bragg's equation:

$$n\lambda = 2d \sin (\theta)$$

where λ is the wavelength of the incident rays, θ is the angle between incident rays and the surface of the crystal, d is the spacing between layers of the atoms, and n is an integer (1, 2, 3, 4, etc.) showing constructive interference.

Using this equation, the relationship between the diffraction angles (Bragg angle), X-ray wavelength, and interplanar spacing of the crystal planes is known. Consequently, one can depict the crystalline phase of the material. For example, scattering of X-rays through a small angle gives the average interparticle distance; on the other hand, diffraction through a wide angle can be used for defining the atomic structure of nanoclusters (Alivisatos, 1996). In addition, the diffraction line widths are closely related to the size, size distribution, defects, and strain in nanocrystals. As the size of the nanocrystals decreases, the line width is broadened because of loss of long-range order relative to the bulk (Gesquiere, 2010).

There are different variants of X-ray-based spectroscopic techniques for determining the chemical composition of nanocrystals. X-ray absorption spectroscopy such as extended X-ray absorption fine structure and X-ray absorption near edge structure, X-ray fluorescence spectroscopy, energy dispersive X-ray spectroscopy, and X-ray photoelectron spectroscopy are some of the techniques that are based on detecting and analyzing emitted or absorbed radiations from a sample after excitation with X-rays (Barr, 1994; Koningsberger and Prins, 1988).

Studies have revealed that X-ray-based characterization techniques can play an important role in nanomaterial-based catalyst research. A highly stable gold and silver nanoparticle using arabinoxylan from *Plantago ovate* seed husk was characterized by the XRD method (Amin et al., 2013). Gajendran et al. (2014) synthesized and studied the silver nanoparticles to prevent the apoptotic effect on human breast cancer obtained from *Datura anoxia*. Although XRD is frequently used for nanomaterial characterization, such as estimation of active surface area and material structural property, the method is not capable of determining the microstructures and getting results only from a single conformation/binding state of the sample (Cao, 2004). Also the low intensity of diffracted X-rays obtained from low atomic number materials limits further usage (Lin et al., 2014).

3.5.4 Circular Dichroism

Chiral molecules have different groups of atoms attached, resulting in nonsuperimposable mirror images. This phenomenon of molecular asymmetry is generally possessed by many biomolecules such as protein, nucleotides, and amino acids, and hence they show chirality (Hammes, 2005). Based on the interaction of the chiral molecule with plane polarized light, a technique called circular dichroism (CD) was developed. CD is a very effective technique for characterizing a wide range of molecules from macromolecular scale to nanoscale particles (Fasman, 1996). Also various conformational changes associated with biomolecules (protein and nucleic acids) and interactions of donor–acceptor (protein–protein, protein–DNA, protein–ligand, and DNA–ligand interactions) can be ideally studied by using CD (Ranjbar and Gill, 2009). With the advancement in technology and need, different types of CD techniques have been developed for structural evaluation of materials, such as electronic CD, magnetic CD, fluorescence-detected CD, magnetic vibrational CD, near-infrared CD, vibrational CD, stopped-flow CD, and synchrotron radiation CD (Ranjbar and Gill, 2009).

Interaction of nanoparticles with biomolecules such as protein is an essential process that can influence cellular uptake and transport, drug delivery, and bioimaging, which is the basis of nanoparticle bioreactivity. Nanoparticles are achiral but in conjugation with biomolecules show chiroptical properties. Bioconjugates of the hemoproteins, myoglobin, and hemoglobin have been synthesized by their adsorption on spherical gold and silver nanoparticles and gold nanorods, and have been characterized by using CD. Jiang et al. (2011) investigated the adsorption-induced conformational changes for cyt *c* adsorbed on gold nanoparticles by analysis of CD spectra and observed that the different interaction forces between cyt *c* and gold nanoparticles resulted in different conformational changes. In addition, oriental CD was also used for evaluating the alignment of peptide membrane along with confirmation changes associated in the lipid bilayers (Bürck et al., 2016).

CD spectroscopy is a nondestructive extensively used technique for identification of the detailed structure of biomolecules such as protein, carbohydrates, and nucleic acid adsorbed on nanomaterials. Purdie et al. (1989) applied CD for analysis of enantiomeric mixtures of nicotine and cocaine, vitamins such as D and E, and cholesterol and other steroids. It provides an easy detection procedure and rapid diagnostic test for drug assay or other substances in complex mixtures/certain structural features relative to the helicity and stereochemistry of native structures. In addition, this method can be used to measure the equilibrium dissociation constant, K_d, of binding interactions between a host macromolecule and a ligand. In spite of numerous applications, there are several disadvantages of this technique. CD cannot be used to obtain specific structural information such as the contribution of a single amino acid residue in a protein-type biomolecule resulting in CD spectrum (Ranjbar and Gill, 2009). The method also requires relatively concentrated and purified samples that cannot be used for mixtures of samples (Lanucara et al., 2014). For CD measurement, molecules have to be chiral otherwise weak spectra of CD are obtained if only nonoptical chromophores are present. This CD spectropolarimeter is an expensive instrument and suffers from numerous experimental errors during measurement.

3.5.5 Mass Spectroscopy

Mass spectroscopy is a powerful analytical technique for elucidating compounds on the basis of elemental/chemical composition and identification of unknown compounds in a sample and mass of a molecule. The basic principle involves the identification of charged particles or molecular ions with different masses based on their mass-to-charge ratio separation (McNaught and Wilkinson, 1997; Sparkman, 2000). The mass spectroscopy instrument contains three major components: an ion source for generating gaseous ions, an analyzer for determining ions in the mass-to-charge ratio, and a detector system for recording the different ionic species present in the sample (Tiede et al., 2008; Gmoshinski et al., 2013).

In mass spectroscopy prior to detection, biomolecules are first ionized and then converted into their respective volatile components without thermal decomposition. Based on the types, ionization techniques that are coupled with mass spectrophotometer analyzer required to ionize different molecules (liquid, solid, and metal) are generally grouped into three categories. The most commonly used methods comprise matrix-assisted laser desorption/ionization and electrospray ionization, which are commonly used to ionize and volatilize the thermally labile liquid/solid biological derivatives. However, for analysis of metallic samples such as nanomaterials, inductively coupled plasma ionization is used (Tiede et al., 2008; Gmoshinski et al., 2013).

Since the use of convectional mass spectrometry there have been various modifications for only resolving unknown compounds, molecular mass evaluation, and identification of purity of known substances, etc. Laser-induced fluorescence and ion trap mass spectrometry are newly developed techniques widely used for analyzing nanoparticles that possess intrinsic fluorescence or are extrinsically labeled with dye molecules (Cai et al., 2003; Peng et al., 2003). For identification of the size of single particles, single particle mass spectrometry is employed (Tiede et al., 2008). Mass spectroscopy is commonly utilized for characterizing nanoparticles because it is a highly sensitive technique with a great degree of accuracy, which results in very low sample requirement in the range of 10^{-9} to 10^{-12} mol (Lin et al., 2014). Carbon-based nanomaterials that include carbon nanotubes, carbon nanodots, and graphene or their derivatives have been characterized using mass spectrometry. Chen et al. (2015) utilized the mass spectrometry technique for imaging the distribution of carbon nanomaterials in mice for drug delivery. Also therapeutic efficacy of cerium oxide-based nanoparticles in various tissues of mice was assayed by inductively coupled plasma-mass spectrometry analysis. Although mass spectrometry has been used for both sizing and quantitative assessment of the various nanocrystals, it has limited applications because of expensive instrumentation and unavailability of databases for molecular species characterization, especially nanomaterial bioconjugates and sample fragmentation caused during ionization (Sapsford et al., 2011; Lavigne et al., 2013).

3.5.6 Visible (UV-Vis) Spectroscopy

UV-Vis is one of the oldest and routinely used techniques for quantitative analysis of nano-sized molecules. Different types of analytes such as transition metal ions and bioconjugates of organic and inorganic nanoparticles can be easily characterized by UV-Vis spectroscopy (Perkampus, 1992). The operating principle of UV-Vis spectroscopy comprises irradiating the sample to be analyzed with electromagnetic radiation, resulting in electronic transitions that are obtained in the form absorption spectra (Clark et al., 1993). A UV-Vis spectrophotometer is one of the simplest techniques and is made of components comprising the source, optical bench, sample and reference beam lines, monochromatic light, and a detector. UV-Vis spectroscopy is based on the Beer–Lambert law, which says that the absorbance of a solution is directly proportional to the concentration of the absorbing species in the solution and the path length, so the greater number of molecules absorbing light of a particular wavelength will result in the greater extent of absorption of light (Zhang, 2009a,b). The formula of the Beer–Lambert law is:

$$A = log_{10}\left(I_0/I\right) = \varepsilon * c * L$$

where A is absorbance, I_0 is intensity of light incident upon the sample cell, I is intensity of light leaving the sample cell, c is molar concentration of solute, L is length of sample cell (cm), and ε is molar absorptivity.

UV-Vis offers the possibility of characterizing metallic nanomaterials, which utilizes the color properties, seen because of absorption of light of a particular wavelength. Quantum dots and organic-conjugated nanomaterials such as fullerenes and carbon nanotubes of different sizes have been studied by this method. In addition, molecules can be resolved on the basis of size-dependent properties ranging from nano- to atomic scales; they can also be observed in a UV-Vis spectrum. Andrievsky et al. (2002) performed the comparative study of two types of fullerene–water colloidal systems: molecular-colloidal C_{60} solution in water (C_{60}FWS) and typical monodisperse C_{60} hydrosol with the aid of UV–Vis spectroscopy. The relationship between absorbance spectra and particle size distributions for ZnO quantum-sized nanocrystals was studied by applying UV-Vis spectroscopy (Pesika et al., 2003). Although this technique is quick and easy to perform, errors in equipment design and other technical problems will lead to false detection of spectra. Moreover, accuracy in measurement can be affected by electronic circuit design and quality of detector circuit that can further reduce the sensitivity of the measurement.

3.5.7 Dynamic Light Scattering

Dynamic light scattering (DLS) is a spectroscopy-based simple and quick technique for determining the particle size distribution of synthesized nanoparticles. DLS utilizes Rayleigh light scattering phenomenon (Bar-Ziv et al., 1997). The principle is that when a light beam falls on the particle a Rayleigh scattering phenomenon is observed (Zhang, 2009a,b). Therefore motion of a particle can be determined by observing the intensity of light scattered by the sample at some fixed angle. This is a very useful technique for determining nanoparticle shape and hydrodynamic size of small particles, polymers, and aggregates in a solution or suspension by scattering light of monochromatic wavelength (Sapsford et al., 2011). It also measures the temporal fluctuation that usually contains a mixture of the constructive and destructive interferences of the scattered light, through which the particle size can be calculated using the Stokes–Einstein equation (Lin et al., 2014; Brar and Verma, 2011; Sapsford et al., 2011).

Although it is a very simple and easily applicable technique, sample analysis of varying sizes or nonspherical nanoparticles cannot be accurately measured (Uskokovic, 2012). Also it gives inaccurate measurements when aggregations are present in the composition of nanoparticles or the size of particles is of a relatively small range (1 nm to 3 μm) (Bootz et al., 2004).

3.6 SUMMARY

Nanomaterials are generally composed of two basic components: a metallic part and organic molecules. In addition, nanoparticles possess unique optical properties that arise because of the nanoscale size of the particle and very large surface-to-volume ratio that differs widely from the macroscopic materials. Thus the final size and structure of nanoparticles are important factors that govern the working efficiency. Green nanomaterials are often used to treat a number of diseases in various traditional systems of medicine. Hence it is deemed important to characterize and quality control the complex mixtures such as formulations or therapeutic preparations. In this context, spectroscopy is an important method for the characterization of nanomaterials in formulations or nanodrugs and provides an important insight into the size, shape, structure, and chemistry. Spectroscopic techniques play an important role in the characterization of biologically synthesized nanoparticles on the basis of physicochemical properties and provide noteworthy information on use the nanoparticles with a range of applications. Thus spectroscopy has motivated the upsurge in nanotechnology research by characterization and evaluation of nanoparticles.

References

Abhilash, B.P., 2012. Synthesis of zinc-based nanomaterials: a biological perspective. IET Nanobiotechnol. 6 (4), 144–148.

Alam, T.M., Jenkins, J.E., 2012. HR-MAS NMR Spectroscopy in Material Science. INTECH Open Access Publisher.

Alivisatos, A.P., 1996. Semiconductor clusters, nanocrystals, and quantum dots. Science 271 (5251), 933.

Alqadi, M., Abo Noqtah, O., Alzoubi, F., Alzouby, J., Aljarrah, K., 2014. pH effect on the aggregation of silver nanoparticles synthesized by chemical reduction. Mater. Sci.-Pol. 32 (1), 107–111.

Amin, M., Iram, F., Iqbal, M.S., Saeed, M.Z., Raza, M., Alam, S., 2013. Arabinoxylan-mediated synthesis of gold and silver nanoparticles having exceptional high stability. Carbohydr. Polym. 92 (2), 1896–1900.

Andrievsky, G.V., Klochkov, V.K., Bordyuh, A.B., Dovbeshko, G.I., 2002. Comparative analysis of two aqueous-colloidal solutions of C-60 fullerene with help of FTIR reflectance and UV–Vis spectroscopy. Chem. Phys. Lett. 364, 8–17.

Aziz, N., Faraz, M., Pandey, R., Shakir, M., Fatma, T., Varma, A., et al., 2015. Facile algae-derived route to biogenic silver nanoparticles: synthesis, antibacterial, and photocatalytic properties. Langmuir 31, 11605–11612.

Aziz, N., Pandey, R., Barman, I., Prasad, R., 2016. Leveraging the attributes of *Mucor hiemalis*-derived silver nanoparticles for a synergistic broad-spectrum antimicrobial platform. Front. Microbiol. 7, 1984. http://dx.doi.org/10.3389/fmicb.2016.01984.

Barr, T.L., 1994. Modern ESCA: The Principles and Practice of X-Ray Photoelectron Spectroscopy. CRC Press, New York, p. 384.

Bar-Ziv, R., Meller, A., Tlusty, T., Moses, E., Stavans, J., Safran, S.A., 1997. Localized dynamic light scattering: probing single particle dynamics at the nanoscale. Phys. Rev. Lett. 78 (1), 154.

Bhainsa, K.C., D'Souza, S.F., 2006. Extracellular biosynthesis of silver nanoparticles using the fungus *Aspergillus fumigatus*. Colloids Surf. B Biointerfaces 47, 160–164.

Bootz, A., Vogel, V., Schubert, D., Kreuter, J., 2004. Comparison of scanning electron microscopy, dynamic light scattering and analytical ultracentrifugation for the sizing of poly (butyl cyanoacrylate) nanoparticles. Eur. J. Pharm. Biopharm. 57 (2), 369–375.

Bose, S., Bhattacharya, V., Chattopadhyay, K., Ayyub, P., 2008. Proximity effect controlled superconducting behavior of novel biphasic Pb–Sn nanoparticles embedded in an Al matrix. Acta Mater. 56 (16), 4522–4528.

Brar, S.K., Verma, M., 2011. Measurement of nanoparticles by light-scattering techniques. Trends Anal. Chem. 30 (1), 4–17.

Bürck, J., Wadhwani, P., Fanghänel, S., Ulrich, A.S., 2016. Oriented circular dichroism: a method to characterize membrane-active peptides in oriented lipid bilayers. Acc. Chem. Res. 49 (2), 184–192.

Cai, Y., Peng, W.P., Chang, H.C., 2003. Ion trap mass spectrometry of fluorescently labeled nanoparticles. Anal. Chem. 75, 1805–1811.

Cantor, C.R., Schimmel, P.R., 1930. Biophysical Chemistry. Freeman, New York.

Cao, G., 2004. Nanostructures and Nanomaterials: Synthesis, Properties and Applications. Imperial College Press.

Carter, R.S., Harley, S.J., Power, P.P., Augustine, M.P., 2005. Use of NMR spectroscopy in the synthesis and characterization of air- and water-stable silicon nanoparticles from porous silicon. Chem. Mater. 17, 2932–2939.

Casimir, D., Garcia-Sanchez, R., Misra, P., 2015. Raman spectroscopy, modeling and simulation studies of carbon nanotubes. In: Applied Spectroscopy and the Science of Nanomaterials. Springer, Singapore, pp. 1–20.

Chen, S., Xiong, C., Liu, H., Wan, Q., Hou, J., He, Q., Badu-Tawiah, A., Nie, Z., 2015. Mass spectrometry imaging reveals the sub-organ distribution of carbon nanomaterials. Nat. Nanotechnol. 10 (2), 176–182.

Clark, B.J., Frost, T., Russell, M.A., 1993. UV Spectroscopy: Techniques, Instrumentation, Data Handling. Chapman & Hall, London, UK.

Cox, A., Venkatachalam, P., Sahi, S., Sharma, N., 2016. Silver and titanium dioxide nanoparticle toxicity in plants: a review of current research. Plant Physiol. Biochem. 107, 147–163.

Crooks, R.M., Zhao, M., Sun, L., Chechik, V., Yeung, L.K., 2001. Dendrimer-encapsulated metal nanoparticles: synthesis, characterization, and applications to catalysis. Acc. Chem. Res. 34 (3), 181–190.

Davar, F., Salavati-Niasari, M., 2011. Synthesis and characterization of spinel-type zinc aluminate nanoparticles by a modified sol–gel method using new precursor. J. Alloys Compd. 509 (5), 2487–2492.

Deshayes, S., Maurizot, V., Clochard, M.C., Berthelot, T., Baudin, C., Déléris, G., 2010. Synthesis of specific nanoparticles for targeting tumor angiogenesis using electron-beam irradiation. Radiat. Phys. Chem. 79 (3), 208–213.

Esparza, R., Rosas, G., Valenzuela, E., Gamboa, S., Pal, U., Pérez, R., 2008. Structural analysis and shape-dependent catalytic activity of Au, Pt and Au/Pt nanoparticles. Matéria (Rio de Janeiro) 13 (4), 579–586.

Fasman, G.D., 1996. Circular Dichroism and the Conformational Analysis of Biomolecules. Plenum Publishing Corp, New York, USA.

Gajendran, B., Chinnasamy, A., Durai, P., Raman, J., Ramar, M., 2014. Biosynthesis and characterization of silver nanoparticles from *Datura inoxia* and its apoptotic effect on human breast cancer cell line MCF7. Mater. Lett. 122, 98–102.

Geraldes, A.N., da Silva, A.A., Leal, J., Estrada-Villegas, G.M., Lincopan, N., Katti, K.V., Lug., atildeo, A.B., 2016. Green nanotechnology from plant extracts: synthesis and characterization of gold nanoparticles. Adv. Nanoparticles 5 (03), 176.

Gesquiere, A.J., 2010. Optical properties and spectroscopy of nanomaterials. J. Am. Chem. Soc. 132 (10), 3637–3638.

Ghodake, G., Lee, D.S., 2011. Biological synthesis of gold nanoparticles using the aqueous extract of the brown algae *Laminaria japonica*. J. Nanoelectron. Optoelectron. 6, 1–4.

Gholipour, S., Davoodnia, A., Nakhaei-Moghaddam, M., 2015. Synthesis, characterization, and antibacterial evaluation of new alkyl 2-amino-4-aryl-4H-chromene-3-carboxylates. Chem. Heterocycl. Comp. 51 (2), 808.

Ghorbani, H.R., Safekordi, A.A., Attar, H., Sorkhabadi, S.M., 2011. Biological and non-biological methods for silver nanoparticles synthesis. Chem. Biochem. Eng. Q. 25 (3), 317–326.

Gmoshinski, I.V., Khotimchenko, S.A.E., Popov, V.O., Dzantiev, B.B., Zherdev, A.V., Demin, V.F., Buzulukov, Y.P., 2013. Nanomaterials and nanotechnologies: methods of analysis and control. Russ. Chem. Rev. 48 (1).

Hammes, G.G., 2005. Spectroscopy for the Biological Sciences. John Wiley & Sons, Inc Circular Dichorism, New York, USA.
Hung, C.H., Whang, W.T., 2005. Effect of surface stabilization of nanoparticles on luminescent characteristics in ZnO/poly (hydroxyethyl methacrylate) nanohybrid films. J. Mater. Chem. 15 (2), 267–274.
Iravani, S., 2014. Bacteria in nanoparticle synthesis: current status and future prospects. Int. Sch. Res. Notices 2014.
Jasti, R., Bertozzi, C.R., 2010. Progress and challenges for the bottom-up synthesis of carbon nanotubes with discrete chirality. Chem. Phys. Lett. 494 (1), 1–7.
Jiang, X.C., Chen, W.M., Chen, C.Y., Xiong, S.X., Yu, A.B., 2011. Role of temperature in the growth of silver nanoparticles through a synergetic reduction approach. Nanoscale Res. Lett. 6 (1), 32.
Kairdolf, B.A., Smith, A.M., Stokes, T.H., Wang, M.D., Young, A.N., Nie, S., 2013. Semiconductor quantum dots for bioimaging and biodiagnostic applications. Annu. Rev. Anal. Chem. (Palo Alto Calif.) 6 (1), 143.
Koningsberger, D.C., Prins, R. (Eds.), 1988. A Series of Monographs on Analytical and its Applications. Wi, New York.
Lavigne, J.P., Espinal, P., Dunyach-Remy, C., Messad, N., Pantel, A., Sotto, A., 2013. Mass spectrometry: a revolution in clinical microbiology. Clin. Chem. Lab. Med. 51, 257–270.
Li, J.F., Huang, Y.F., Ding, Y., Yang, Z.L., Li, S.B., Zhou, X.S., Feng, R.F., Zang, W., Zhou, Z.Y., Wu, D.Y., Ren, B., Wang, Z.L., Tian, Z.Q., 2010. Shell-isolated nanoparticle-enhanced Raman spectroscopy. Nature 464 (7287), 392–395.
Lin, P.C., Lin, S., Wang, P.C., Sridhar, R., 2014. Techniques for physicochemical characterization of nanomaterials. Biotechnol. Adv. 32 (4), 711–726.
Liu, D., Wang, Q., Chang, H.L.M., Chen, H., 1995. Variant structure in metal-organic-chemical-vapor-deposition-derived SnO_2 thin films on sapphire (0001). J. Mater. Res. 10 (06), 1516–1522.
Lupoi, J.S., Singh, S., Simmons, B.A., Henry, R.J., 2014. Assessment of lignocellulosic biomass using analytical spectroscopy: an evolution to high-throughput techniques. Bioenergy Res. 7 (1), 1–23.
McDonald, S.A., Konstantatos, G., Zhang, S., Cyr, P.W., Klem, E.J., Levina, L., Sargent, E.H., 2005. Solution-processed PbS quantum dot infrared photodetectors and photovoltaics. Nat. Mater. 4 (2), 138–142.
McNaught, A.D., Wilkinson, A., 1997. Compendium of Chemical Terminology. IUPAC recommendations.
McNeil, S.E., 2005. Nanotechnology for the biologist. J. Leukoc. Biol. 78 (3), 585–594.
MubarakAli, D., Thajuddin, N., Jeganathan, K., Gunasekaran, M., 2011. Plant extract mediated synthesis of silver and gold nanoparticles and its antibacterial activity against clinically isolated pathogens. Colloids Surf. B 85 (2), 360–365.
Mullen, M.D., Wolf, D.C., Ferris, F.G., Beveridge, T.J., Flemming, C.A., Bailey, G.W., 1989. Bacterial sorption of heavy metals. Appl. Environ. Microbiol. 55 (12), 3143–3149.
Nath, D., Banerjee, P., 2013. Green nanotechnology–a new hope for medical biology. Environ. Toxicol. Pharmacol. 36 (3), 997–1014.
Pathak, Y., Thassu, D. (Eds.), 2016. Drug Delivery Nanoparticles Formulation and Characterization, vol. 191. CRC Press.
Patra, J.K., Baek, K.H., 2014. Green nanobiotechnology: factors affecting synthesis and characterization techniques. J. Nanomater. 2014, 219.
Pease, R.F., Chou, S.Y., 2008. Lithography and other patterning techniques for future electronics. Proc. IEEE 96 (2), 248–270.
Peng, W.P., Cai, Y., Lee, Y.T., Chang, H.C., 2003. Laser induced fluorescence/ion trap as a detector for mass spectrometric analysis of nanoparticles. Int. J. Mass Spectrom. 229, 67–76.
Peralta-Videa, J.R., Hernandez-Viezcas, J.A., Zhao, L., Diaz, B.C., Ge, Y., Priester, J.H., Holden, P.A., Gardea-Torresdey, J.L., 2014. Cerium dioxide and zinc oxide nanoparticles alter the nutritional value of soil cultivated soybean plants. Plant Physiol. Biochem. 80, 128–135.
Peralta-Videa, J.R., Huang, Y., Parsons, J.G., Zhao, L., Lopez-Moreno, L., Hernandez-Viezcas, J.A., Gardea-Torresdey, J.L., 2016. Plant-based green synthesis of metallic nanoparticles: scientific curiosity or a realistic alternative to chemical synthesis? Nanotechnol. Environ. Eng. 1 (1), 4.
Perkampus, H.H., 1992. UV-VIS Spectroscopy and its Applications. Springer-Verlag, Berlin, Germany.
Pesika, N.S., Stebe, K.J., Searson, P.C., 2003. Relationship between absorbance spectra and particle size distributions for quantum-sized nanocrystals. J. Phys. Chem. B 107 (38), 10412–10415.
Popovic, Z.V., Dohcevic-Mitrovic, Z., Scepanovic, M., Grujic-Brojcin, M., Askrabic, S., 2011. Raman scattering on nanomaterials and nanostructures. Ann. Phys. 523 (1–2), 62–74.
Prasad, R., Pandey, R., Barman, I., 2016. Engineering tailored nanoparticles with microbes: quo vadis? Wiley Interdiscip. Rev. Nanomed. Nanobiotechnol. 8, 316–330. http://dx.doi.org/10.1002/wnan.1363.

Purdie, N., Swallows, K.A., Murphy, L.H., Purdie, R.B., 1989. Analytical applications of circular dichroism. J. Pharm. Biomed. Anal. 7 (12), 1519–1526.

Rai, A., Singh, A., Ahmad, A., Sastry, M., 2006. Role of halide ions and temperature on the morphology of biologically synthesized gold nanotriangles. Langmuir 22 (2), 736–741.

Rajathi, F.A.A., Parthiban, C., Ganesh Kumar, V., Anantharaman, P., 2012. Biosynthesis of antibacterial gold nanoparticles using brown alga, *Stoechospermum marginatum* (kützing). Spectrochim. Acta A 99, 166–173.

Rajesh Kumar, S., Kannan, C., Annadurai, G., 2012. Green synthesis of silver nanoparticles using marine brown algae *Turbinaria conoides* and its antibacterial activity. Int. J. Pharm. Bio. Sci. 3 (4), 502–510.

Rajeshkumar, S., Malarkodi, C., Gnanajobitha, G., Paulkumar, K., Vanaja, M., Kannan, C., Annadurai, G., 2013. Seaweed-mediated synthesis of gold nanoparticles using *Turbinaria conoides* and its characterization. J. Nanostruct. Chem. 3 (1), 1–7.

Ranjbar, B., Gill, P., 2009. Circular dichroism techniques: biomolecular and nanostructural analyses — a review. Chem. Biol. Drug Des. 74, 101–120.

Rodríguez-León, E., Iñiguez-Palomares, R., Navarro, R.E., Herrera-Urbina, R., Tánori, J., Iñiguez-Palomares, C., Maldonado, A., 2013. Synthesis of silver nanoparticles using reducing agents obtained from natural sources (*Rumex hymenosepalus* extracts). Nanoscale Res. Lett. 8 (1), 318.

Sankar, M.V., Abideen, S., 2015. Pesticidal effect of green synthesized silver and lead nanoparticles using Avicennia marina against grain storage pest Sitophilus oryzae. Int. J. Nanomater. Biostruct. 5 (3), 32–39.

Sapsford, K.E., Tyner, K.M., Dair, B.J., Deschamps, J.R., Medintz, I.L., 2011. Analyzing nanomaterial bioconjugates: a review of current and emerging purification and characterization techniques. Anal. Chem. 83 (12), 4453–4488.

Shahverdi, A.R., Minaeian, S., Shahverdi, H.R., Jamalifar, H., Nohi, A.A., 2007. Rapid synthesis of silver nanoparticles using culture supernatants of *Enterobacteria*: a novel biological approach. Process Biochem. 42, 919–923.

Shigeta, M., Watanabe, T., 2016. Effect of saturation pressure difference on metal–silicide nanopowder formation in thermal plasma fabrication. Nanomaterials 6 (3), 43.

Shweta, Tripathi, D.K., Singh, S., Singh, S., Dubey, N.K., Chauhan, D.K., 2016. Impact of nanoparticles on photosynthesis: challenges and opportunities. Mater. Focus 5 (5), 405–411.

Singaravelu, G., Arockiyamari, J., Ganesh Kumar, V., Govindaraju, K., 2007. A novel extracellular biosynthesis of monodisperse gold nanoparticles using marine algae, Sargassum wightii Greville. Colloids Surf. B Biointerfaces 57, 97–101.

Singh, S., Tripathi, D.K., Dubey, N.K., Chauhan, D.K., 2016. Effects of nano-materials on seed germination and seedling growth: striking the slight balance between the concepts and controversies. Mater. Focus 5 (3), 195–201.

Singh, S., Vishwakarma, K., Singh, S., Sharma, S., Dubey, N.K., Singh, V.K., Liu, S., Tripathi, D.K., Chauhan, D.K., 2017. Understanding the plant and nanoparticle interface at transcriptomic and proteomic level: a concentric overview. Plant Gene. https://doi.org/10.1016/j.plgene.2017.03.006.

Sparkman, O.D., 2000. Mass spectrometry desk reference. J. Am. Soc. Mass Spectrom. 11 (12), 1144.

Studart, A.R., Amstad, E., Gauckler, L.J., 2007. Colloidal stabilization of nanoparticles in concentrated suspensions. Langmuir 23 (3), 1081–1090.

Tiede, K., Boxall, A.B., Tear, S.P., Lewis, J., David, H., Hassellöv, M., 2008. Detection and characterization of engineered nanoparticles in food and the environment. Food Addit. Contam. 25 (7), 795–821.

Tiwari, V., Jiang, J., Sethi, V., Biswas, P., 2008. One-step synthesis of noble metal–titanium dioxide nanocomposites in a flame aerosol reactor. Appl. Catal. A 345 (2), 241–246.

Tiwari, M., Sharma, N.C., Fleischmann, P., Burbage, J., Venkatachalam, P., Sahi, S.V., 2017. Nanotitania exposure causes alterations in physiological, nutritional and stress responses in tomato (*Solanum lycopersicum*). Front. Plant Sci. 8.

Tripathi, D.K., Singh, V.P., Prasad, S.M., Chauhan, D.K., Dubey, N.K., 2015. Silicon nanoparticles (SiNp) alleviate chromium (VI) phytotoxicity in *Pisum sativum* (L.) seedlings. Plant Physiol. Biochem. 96, 189–198.

Tripathi, D.K., Singh, S., Singh, V.P., Prasad, S.M., Chauhan, D.K., Dubey, N.K., 2016. Silicon nanoparticles more efficiently alleviate arsenate toxicity than silicon in maize cultiver and hybrid differing in arsenate tolerance. Front. Environ. Sci. 4, 46.

Tripathi, D.K., Singh, S., Singh, S., Pandey, R., Singh, V.P., Sharma, N.C., Prasad, S.M., Dubey, N.K., Chauhan, D.K., 2017a. An overview on manufactured nanoparticles in plants: uptake, translocation, accumulation and phytotoxicity. Plant Physiol. Biochem. 110, 2–12.

Tripathi, D.K., Mishra, R.K., Singh, S., Singh, S., Vishwakarma, K., Sharma, S., Singh, V.P., Singh, P.K., Prasad, S.M., Dubey, N.K., Pandey, A.C., 2017b. Nitric oxide ameliorates zinc oxide nanoparticles phytotoxicity in wheat seedlings: implication of the ascorbate–glutathione cycle. Front. Plant Sci. 8.

Tripathi, D.K., Singh, S., Singh, S., Srivastava, P.K., Singh, V.P., Singh, S., Prasad, S.M., Singh, P.K., Dubey, N.K., Pandey, A.C., Chauhan, D.K., 2017c. Nitric oxide alleviates silver nanoparticles (AgNps)-induced phytotoxicity in Pisum sativum seedlings. Plant Physiol. Biochem. 110, 167–177.

Tripathi, D.K., Tripathi, A., Shweta, S.S., Singh, Y., Vishwakarma, K., Yadav, G., Sharma, S., Singh, V.K., Mishra, R.K., Upadhyay, R.G., Dubey, N.K., 2017d. Uptake, accumulation and toxicity of silver nanoparticle in autotrophic plants, and heterotrophic microbes: a concentric review. Front. Microbiol. 8.

Tripathi, D.K., Singh, S., Singh, V.P., Prasad, S.M., Dubey, N.K., Chauhan, D.K., 2017e. Silicon nanoparticles more effectively alleviated UV-B stress than silicon in wheat (*Triticum aestivum*) seedlings. Plant Physiol. Biochem. 110, 70–81.

Ummartyotin, S., Bunnak, N., Juntaro, J., Sain, M., Manuspiya, H., 2012. Synthesis of colloidal silver nanoparticles for printed electronics. C. R. Chim. 15, 539–544.

Uskoković, V., 2012. Dynamic light scattering based microelectrophoresis: main prospects and limitations. J. Dispers. Sci. Technol. 33 (12), 1762–1786.

Vadlapudi, V., Kaladhar, D.S.V.G.K., 2014. Review: green synthesis of silver and gold nanoparticles. Middle East J. Sci. Res. 19 (6), 834–842.

Valentini, M., Vaccaro, A., Rehor, A., Napoli, A., Hubbell, J.A., Tirelli, N., 2004. Diffusion NMR spectroscopy for the characterization of the size and interactions of colloidal matter: the case of vesicles and nanoparticles. J. Am. Chem. Soc. 126, 2142–2147.

Wang, E.C., Wang, A.Z., 2014. Nanoparticles and their applications in cell and molecular biology. Integr. Biol. 6 (1), 9–26.

Wang, L.Q., Exarhos, G.J., Liu, J., 2001. Characterization of nanophase materials. In: Nuclear Magnetic Resonance — Characterization of Self-Assembled Nanostructural Materials. Wiley, V.C.H., Verlag., Gmb, H, pp. 243–260.

Yang, S.M., Paranthaman, M.P., Noh, T.W., Kalinin, S.V., Strelcov, E., 2016. Nanoparticle shape evolution and proximity effects during tip-induced electrochemical processes. ACS Nano 10 (1), 663–671.

Zhang, J.Z., 2009a. Optical Properties and Spectroscopy of Nanomaterials. World Scientific, Singapore.

Zhang, J.Z., 2009b. Spectroscopic techniques for studying optical properties of nanomaterials. In: Optical Properties and Spectroscopy of Nanomaterials, pp. 11–46.

Zhou, K., Wang, R., Xu, B., Li, Y., 2006. Synthesis, characterization and catalytic properties of CuO nanocrystals with various shapes. Nanotechnology 17 (15), 3939.

Zhou, D., Ji, Z., Jiang, X., Dunphy, D.R., Brinker, J., Keller, A.A., 2013. Influence of material properties on TiO_2 nanoparticle agglomeration. PLoS One 8 (11), e81239.

CHAPTER

4

DNA in Nanotechnology: Approaches and Developments

Rajneesh, Jainendra Pathak, Vidya Singh, Deepak Kumar, Shailendra P. Singh, Rajeshwar P. Sinha

Banaras Hindu University, Varanasi, India

4.1 INTRODUCTION

DNA and RNA (in some viruses) are the genetic material of all living things residing on Earth. They consist of nitrogenous bases adenine, guanine, cytosine, and thymine. The hydrogen-bonding ability of DNA makes it an important material for assembling into predictable nanoscale structures (Seeman, 2010). The molecular recognition property of DNA was utilized for creating artificial DNA nanostructures for technological purposes in DNA nanotechnology (Kumar et al., 2016a,b). DNA has an inherent potential of self-assembly and it interacts with a wide array of molecules, hence it is an important biomolecule (Zahid et al., 2013).

Nanobiotechnology is the sum of two words: "nano," which means study or synthesis of structures having a size range from 1 to 100 nm, while "biotechnology" refers to the use of living things or biological molecules for the development of technological tools. By utilizing biological systems, nanobiotechnologists have developed several innovative nanodevices (Sekhon, 2005). Over the last three decades, one-, two-, and even three-dimensioned branched junction structures having distinct and intricate geometries were constructed by manipulating single- and double-stranded DNAs. The "bottom-up" approach has been used by a majority of researchers for the synthesis of dynamic DNA nanostructures, which results in the development of several macroscopic structures with nanometer-size features (Kallenbach et al., 1983; Aldaye et al., 2008; Shih and Lin, 2010; Pinheiro et al., 2011).

Accurate spatially controlled DNA nanostructures have allowed workers to explore their novel applications in fields such as structural biology, directed material assembly, nanorobotics, biocatalysis, DNA computing, drug delivery, and disease diagnosis (Zhang et al., 2014a,b). Advancement in DNA nanotechnology has resulted in the production of various reprogrammable functionalized devices and sensors. For the first time in the early

Nanomaterials in Plants, Algae, and Microorganisms
http://dx.doi.org/10.1016/B978-0-12-811487-2.00004-9

1990s a method was reported by Seeman and coworkers in which DNA underwent hybridization in various ways to synthesize self-assembling nanostructures. DNA tiles having sticky ends were synthesized and allowed to hybridize for formation of a cube-like structure (Seeman, 1982, 2003).

Structures of various sizes and shapes can be created using nanoconstruction. The Rothemund model of DNA origami enabled scientists to develop various 2D shapes by folding long strands of DNA (Rothemund, 2006; Wei et al., 2012). This methodology is not only restricted to produce 2D nanostructures; at the same time 3D structures were also produced by this approach (Sparreboom et al., 2005; Reichert and Wenger, 2008; Acharya et al., 2009; Bharali et al., 2009). Through strand displacement cycles, DNA strands may undergo multiple hybridizations, utilizing a hinge of the DNA itself (Yurke et al., 2000). Yurke et al. (2000) showed that changes in the structure of DNA at the molecular level are carried out through hybridization and strand displacement instead of using biomolecules such as proteins or other biosupportive molecules. This was achieved by connecting two double helical strands of DNA through another short DNA sequence, which acted as a "hinge." Repeated cycling of the two strands into a closed and opened state occurred through this "hinge" by the addition of two single-stranded DNA molecules consecutively (Yurke et al., 2000). Based on this method, various types of nanostructures were assembled in a sequence-specific manner. Previously, for doing this there was reliance on the effect of variations in the surrounding environment of the DNA such as temperature, salt, and pH (Mao et al., 1999; Kay et al., 2007). For generation and modification of DNA nanostructures, DNA-modifying enzymes were also used. These DNA-modifying enzymes are specific in their action, depending upon biological processes, and work as miniature nanofactories (Keller and Marx, 2012). Ease of manipulation of DNA and sequence specificity of nanoarchitectural structures enable them to organize into different biological molecules such as proteins, peptides, and viral capsids (Hemminga et al., 2010), as well as complex structures such as carbon nanotubules and other nanoparticles. DNA nanostructures possessing a self-assembling property have enhanced the activity of enzyme cascades, and shifted surface plasmon resonance wavelengths rely on their custom-controlled arrangement (Park et al., 2005; Lund et al., 2005; Liu et al., 2005; Sönnichsen et al., 2005; Li et al., 2006; Erben et al., 2006; Chhabra et al., 2007; Williams et al., 2007; Saccà et al., 2010; Stephanopoulos et al., 2010).

Despite all major breakthroughs in DNA nanotechnology, it is still in its initial phase and only about 30% of the synthesized DNA structures show similarity to the original design (Glotzer and Solomon, 2007). In the DNA computational area, establishment of technology for fabricating modern DNA nanostructures still remains a great challenge. Scientists compare this process with the development of computers, electronics, and automobiles. Apart from errors in designing genetic sequence, another drawback is the prolonged thermal cycling for up to 24 h, which is needed for producing useful nanodevices. Over a decade has passed since production of the first functional prototype in automobiles. Canonical complementary base pair interactions, which give DNA the unique property of self-assembly, form the basis of DNA nanotechnology. This also result in the formation of DNA nanostructures with programmable designs of desired functional properties and geometries (Kumar et al., 2016a,b). High biocompatibility and low cytotoxicity of DNA inside cells make it a suitable candidate for applications in pharmaceutics (Li et al., 2013; Chao et al., 2014). Properties of DNA such as complementary base pairing, excellent molecular recognition characteristics, along with stability, mechanical rigidity, custom synthesis with variable length of strands, and nanodimensions of the repeating unit, allow the formation of various shapes of nanostructures (Jones et al., 2015; Kumar et al., 2016a,b). Nanotechnology has a wide range of applications in agriculture, plant science, and various disciplines of biology such as diagnostics, health, electronics, etc. (Peralta-Videa et al., 2014, 2016; Aziz et al., 2016; Tripathi et al., 2015, 2016, 2017a,b,c,d,e; Prasad et al., 2016; Shweta et al., 2016; Singh et al., 2016, 2017). Because various types of nanoparticles are being synthesized frequently these days their deleterious effect needs to be studied prior to their release into the environment (Aziz et al., 2015, 2016; Prasad et al., 2016). DNA nanostructures have selective advantages over other nanoparticles because they are not detrimental to the environment. The major problem in structural DNA nanotechnology is the cost of DNA synthesis, which is used for synthesizing DNA nanostructures. Hence there is a need to develop new approaches for minimizing the production cost of DNA for its judicious and efficient use in DNA nanotechnology.

This chapter deals with applications of DNA nanostructures in various fields of biology and other interdisciplinary applications and challenges in the field of DNA nanotechnology.

4.2 SYNTHESIS OF DNA NANOSTRUCTURES

The term "scaffolded DNA origami" was first introduced by Paul Rothemund in 2006 at the California Institute of Technology. This finding formed the basis of structural DNA nanotechnology by intensifying the size and complexity of DNA nanostructures (self-assembled) in a simple "one-pot" reaction (Andersen et al., 2009). The process of DNA origami synthesis constitutes the folding of a single-stranded scaffold of long circular viral DNA (bacteriophage M13mp18, 7.249 kb) with hundreds of short staple strands of DNA into a desired shape. A wide array of 2D and 3D nanostructures were successfully developed by utilizing the DNA origami technique. This technique is also being used for the fabrication of a nanoscale pattern of proteins, nanoparticles, and various other functional molecular components into well-defined positions (Rothemund, 2006).

In the designed DNA nanostructures, different types of chemically modified staples were inserted at various predefined locations, which resulted in different additional functionalities in DNA nanostructures. Computer software called "caDNAno," based on the design of a DNA origami, was used for finding the placement of individual staple strands (Douglas et al., 2009a). Because of the Watson–Crick base pairing ability of DNA, each staple strand binds to a specific position in the template DNA. The nanostructures were synthesized by using a one-pot reaction in this scaffold strand and mixing the staple strands, followed by rapid heating coupled with slow cooling, which allows different staple strands to pull the long scaffold strand into a desired shape (Kumar et al., 2016a,b).

Many researchers have proposed a flexible approach for manipulation of DNA shape and conformation for designing 3D and 2D curved and twisted nanostructures such as ellipsoid shells, spherical shells, and nanoflasks (Douglas et al., 2009b; Dietz et al., 2009; Han et al., 2011; Tian et al., 2017). Furthermore, multilayer 3D nanostructures and curved objects were formed by using a series of four-arm junctions for creating gridiron-like DNA nanostructures (Han et al., 2013; Jones et al., 2015). A combination of merits of tile and origami assembly resulted in a new strategy, in which single-stranded blocks having concatenated sticky ends were used for developing 2D and 3D DNA canvases (Ke et al., 2012a,b; Wei et al., 2012). In a different line, a highly automated method was developed utilizing a routing algorithm based on relaxation simulation and graph theory that traces scaffold strands through the target structures. Such structures contain one helix per edge and display considerable stability under the ionic condition of biological assays (Benson et al., 2015). The design concept of DNA origami, its assembly principles, and shape were discussed by Linko and Dietz (2013) and Castro et al. (2011). For the purposes of design, the double helical structure of DNA is represented as domains of a double helix and these two domains are connected via multiple interhelix connections, which comprise immobilized Holliday junctions (Seeman, 2003).

4.3 CHARACTERIZATION

DNA nanostructures have been characterized by various techniques such as polyacrylamide gel electrophoresis (PAGE) or agarose gel electrophoresis (AGE), transmission electron microscopy (TEM), atomic force microscopy (AFM), and dynamic light scattering (DLS) (Fig. 4.1). In PAGE/AGE techniques, migration of DNA occurs through a polyacrylamide or agarose matrix under the influence of an applied electric field, which results in separation of nanostructures based on their size, and are further analyzed under UV after staining with UV-sensitive dyes such as ethidium bromide. The nanostructures were recovered from the

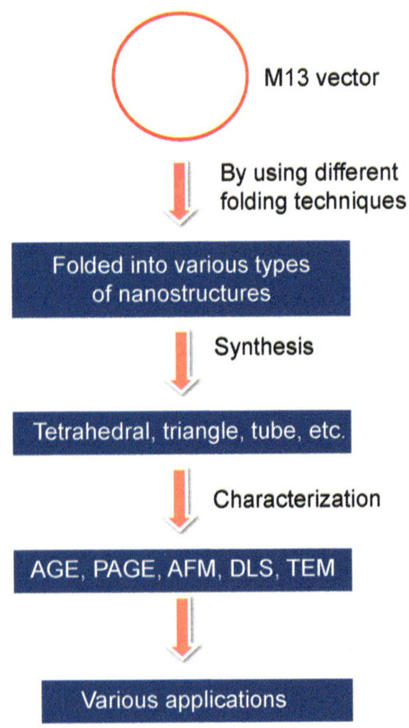

FIGURE 4.1 Steps during the synthesis and characterization of DNA nanostructures. *AFM*, Atomic force microscopy; *AGE*, agarose gel electrophoresis; *DLS*, dynamic light scattering; *PAGE*, polyacrylamide gel electrophoresis; *TEM*, transmission electron microscopy.

gel by utilizing suitable extraction techniques. By running a polyacrylamide or agarose gel followed by comparison of the bands' intensity to a known standard, estimation of intensity of assembled DNA nanostructures can be done (Chang et al., 2011; Lee et al., 2012). There was a decrease in folding quality of DNA nanostructures loaded with drug, while a sharp band observed before the loading of the drug and also DNA nanostructures loaded with drug show lower mobility indicating intercalation of the drug into the DNA (Zhao et al., 2012; Jiang et al., 2012).

The assembly of DNA, the shape, the size, and the monodispersity of nanostructures were confirmed by TEM and AFM. These techniques were used for studying the morphology of nanostructures over and before drug loading (Chang et al., 2011; Zhao et al., 2012; Jiang et al., 2012). TEM images confirmed that the doxorubicin (doxo)-loaded twisted DNA nanostructures were more compact, elongated, and straight than the twisted DNA folded in the absence of doxo (Zhao et al., 2012), while by using AFM images, Jiang et al. (2012) showed that after intercalation of the drug, the morphology of the nanostructures was retained before and after doxo loading. Various states as well as conformations of DNA nanorobots were analyzed by utilizing TEM (Douglas et al., 2012). The hydrodynamic diameter of the nanostructures was determined by using the DLS technique, which detects the fluctuations in scattered light intensity caused by diffusing particles. The type of ions in medium, their concentration, size of surface structure, and particle core affect the measurement. Commonly, DNA nanostructures show a narrow size distribution as evidenced by DLS analysis, representing a monodisperse and uniform hydrodynamic size (Chang et al., 2011; Lee et al., 2012; Kim et al., 2013; Zhang et al., 2014a,b). Tables 4.1A and 4.1B summarize DNA nanostructures and their properties as well as applications. All these techniques were used in a particular order for demonstrating the dimensions (shape and size) of nanostructures.

4.4 CORRECTION OF SEQUENCE MISMATCH

Rapid advancements in DNA-based nanobiotechnology caused an increase in the demand for synthetic DNA. We are now able to synthesize short molecules such as single-stranded DNA up to entire viral genomes by using nucleotides. The high cost and large time duration needed for the synthesis of DNA make it somewhat limited for use as a research tool in nanotechnology. Another major problem is the significant higher error rate in synthetic DNA sequences (Coelho et al., 2013). The synthesis of long DNA molecules needs synthetic machinery that has a low error rate and, if errors occur, it should have a suitable system for correction. Correction of errors is expensive and requires a long time duration too. There are various ways to develop error-free sequences of DNA. This method involves physical separation, in which metals are used to chelate partially denatured purine bases that result in the removal of errors, but this method is also limited to physical separation (Gilleron et al., 2013). In the hairpin polymerase chain reaction (PCR) approach there is complete separation of genuine mutations from misincorporations occurring from polymerase. In hairpin PCR, all DNA sequences are converted into hairpin followed by ligation oligonucleotide caps to DNA ends. Conditions of such reactions are optimized in such a way that hairpin DNA is amplified efficiently and DNA strands are copied in a single pass

TABLE 4.1A Nanostructures and Their Properties

STRUCTURAL CHARACTERIZATION

DNA Nanostructure	Characterization	Size (nm)	Cargo (Loading)	Stability	Targeting	Loading Efficiency	Release	References
Aptamer-tethered DNA nanotrain	AFM, AGE (3%), TEM	AFM: length 100 nm	Doxo, epirubicin, daunorubicin	>45 h Solution: PBS + 5 mM Mg^{2+} Temperature: N/A	Protein tyrosin kinase 7	N/A	Endocytosis	Kim et al. (2013)
Hexagonal barrel with aptamer-based lock (antigen keys)	AFM, AGE (2%), DLS, TEM	DLS: 90 nm TEM: 35 nm × 35 nm × 45 nm	Fab antibody fragments	N/A	CD33, CDw32 8	N/A	N/A	Douglas et al. (2012)
Icosahedral: five- (120 bp) and six- (144 bp) (aptamer) point-star structure	AGE (2%), DLS, TEM	DLS: five stars (28.2 ± 3 nm) six stars (28.6 ± 5 nm) TEM: 25 nm	Doxo	>30 min Solution: cell culture medium Temperature: 37°C	Mucin 1. Tumor surface marker	1200 molecules/structure Time: 1 h Incubation temperature: room temperature	N/A	Chang et al. (2011)
Open-caged DNA (pyramidal), 408 bp	PAGE (7%)	N/A	Doxo	35 h (half-life) Solution: cell culture medium + 10% FBS Temperature: 37°C	N/A	172 molecules doxo/structure (at 15% of loading efficiency)	50% of doxo release from py-doxo in PBS in 5 h and 3 h in FBS. Free doxo in 20 min	Kumar et al. (2016a,b)
Tetrahedral	PAGE (3.5%)	N/A	Streptavidin and CpGs	5 h Solution: FBS 50% Temperature: room temperature	N/A	N/A	N/A	Liu et al. (2012)
Tetrahedral (30 bp/edge, total 6 edges)	AFM, DLS, PAGE (5%)	AFM: height 7.5 nm DLS: 28.6 ± 2.38 nm	siRNA against GFP gene DNA structure conjugated with folic acid	N/A	Folic acid receptor (FAR)	1–6 siRNA per tetrahedral (1 siRNA/edge)	N/A	Lee et al. (2012)
Tetrahedral	AFM, DLS, PAGE (6%)	AFM: height 2–3 nm DLS: 9.08 ± 3 nm	Doxo	>4 to < 23 h Solution: 10% human serum Temperature: 37°C	N/A	26 molecules/structure Time: 1 h, Temperature: N/A	100% in 10 h. <10% in 3 h. Solution: PBS pH 7.4 Temperature: 37°C 60% in 10 h Solution: PBS pH 5 Temperature: 37°C	Kim et al. (2013)

Triangle and tube	AFM, AGE (1%)	AFM: triangle: edge ≈ 150 nmtube: length ≈ 183 nm	Doxo	N/A	N/A	>200,000 molecules (calculated by our group) Time: 24 h Temperature: room temperature	≈15% pH 7.4, 35% pH 5.5 in 48 h ≈ 25% in MCF-7 cell lysate >40% with 50 U DNase I Solution: PBS Temperature: 37°C	Jiang et al. (2012)
Tube	AGE (2%), TEM	TEM: length 80 nm, diameter ≈20 nm	62 CpG sequence specific for mouse Toll-like receptor 9	6 h Solution: cell culture medium Temperature: 37°C	Toll-like receptor 9	62 binding sites per tube	N/A	Schüller et al. (2011)
Tube (tile)	AGE (2%), TEM	TEM: length 27 nm, diameter 8 nm	siRNA (GFP) 8 h Solution:	8 h Solution: cell culture medium Temperature: 37°C (degradation depends on Mg^{2+} concentration, oligonucleotide sequences, salt concentration, structural extension)	FAR	N/A	N/A	Kocabey et al. (2014)
Tube with different global twist [straight (S) 10.5 bp/turn and twist (T) tube 12 bp/turn]	AGE (2%), TEM	TEM: length 138 nm, diameter 13 nm	Doxo	48 h (T-tube) Solution: 10% FBS Temperature: 37°C	N/A	N/A	80% (T-tube) and 90% (S-tube) in 10 h Solution: PBS pH 7.4 Temperature: 37°C	Zhao et al. (2012)
Tube, triangle and square	AFM, AGE (1%), DLS	AFM:-tube: height 7 nm, diameter 380 nm, triangle: edge 120 nm, square: length × width 90 nm × 60 nm DLS: triangle: 59 nm square: 80.9 nm tube: 98.6 nm	Doxo	24 h Solution: serum Temperature: 37°C	N/A	>200,000 molecules (calculated by our group) Time: 24 h Temperature: room temperature	≈20% of doxo is released in 48 h at pH 7.4, ≈35% is released at pH 5.5 Solution: PBS Temperature: 37°C	Zhang et al. (2014a,b)

AFM, Atomic force microscopy; *AGE*, agarose gel electrophoresis; *CpG*, cytosine–phosphate–guanosine; *DLS*, dynamic light scattering; *doxo*,doxorubicin; *FBS*, fetal bovine serum; *N/A*, not applicable; *PAGE*, polyacrylamide gel electrophoresis; *PBS*, phosphate-buffered saline; *siRNA*, short interfering RNA; *TEM*, transmission electron microscopy.

Adapted from Kumar, V., Bayda, S., Hadla, M., Caligiuri, I., Russo Spena, C., Palazzolo, S., Kempter, S., Corona, G., Toffoli, G., Rizzolio, F., 2016a. Enhanced chemotherapeutic behavior of open-caged DNA@Doxorubicin nanostructures for cancer cells. J. Cell Physiol. 231 (1), 106–110; Kumar, V., Palazzolo, S., Bayda, S., Corona, G., Toffoli, G., Rizzolio, F., 2016b. DNA nanotechnology for cancer therapy. Theranostics 6 (5), 710.

TABLE 4.1B Nanostructures and Their Biological Applications

BIOLOGICAL ACTIVITY						
		In Vitro Tests		In Vivo Tests		
DNA Nanostructure	Internalization	Cells	Activity	Cell	Activity	References
Aptamer-tethered DNA nanotrain	CEM cells (human T-cell acute lymphocytic leukemia $PTK7^+$) and Ramos (human B lymphocyte Burkitt's lymphoma $PTK7^-$)	Drug-aptamer-DNA-drug is more cytotoxic on $PTK7^+$ cells than free drug. Same efficacy on $PTK7^-$ cells	Drug-aptamer-DNA-drug is more cytotoxic on $PTK7^+$ cells than free drug. Same efficacy on $PTK7^-$ cells	CEM ($PTK7^+$) xenograft mouse model	Increased antitumor efficacy and reduced side effects of doxo delivered via aptamer-DNA nanotrain. Inj: I.V.	Kim et al. (2013)
Hexagonal barrel with aptamerbased lock (antigen keys)	N/A	NKL	Increase apoptosis	N/A	N/A	Douglas et al. (2012)
Icosahedral: five-(120bp) and six-(144bp) (aptamer) point-star structure	Dynamin-dependent and lathrin-mediated endocytosis. Degradation in lysosomes	Breast cancer MCF-7 ($MUC1^+$), CHO-K1 ($MUC1^-$)	Doxo-aptamer six-point star structure is more effective on cell viability of $MUC1^+$ cell than free doxo. Same efficacy on MUC^- cells	N/A	N/A	Chang et al. (2011)
Open-caged DNA (pyramidal), 408bp	py-Doxo is able to penetrate inside MDA-MB-231, release Doxo in the nucleus	MDA-MB-231, HepG2	Decrease cell viability compared to free doxo	N/A	N/A	Kumar et al. (2016a,b)
Tetrahedral	Endocytosis. Antigen localized in lysosomes after 2h	RAW264.7 (macrophage, Abelson murine leukemia virus transformed)	Increased internalization of the complex tetrahedron-STV-CpG by APC cells	BALB/c immunocompetent mice	Mice immunized with the complex tetrahedron-STV-CpG developed a stronger and longer immunitary response	Liu et al. (2012)
Tetrahedral (30bp/edge, total 6 edges)	N/A	HeLa cells (LUC^+). KB cells (HeLa cell contaminant overexpressing folate receptor)	HeLa: LUC expression <50%. KB: GFP expression <40%	KB xenograft tumors	siRNA no effect on LUC. siRNA-DNA tetrahedral: IC_{50} LUC expression: 1.8mg/kg. Inj: intratumor or tail-vein Stability: siRNA 6min; siRNA- DNA tetrahedral	Lee et al. (2012)

Tetrahedral	Macropinocytosis and caveolae-mediated endocytosis pathways	Breast cancer MCF-7 (drug sensitive) and MCF-7-ADR (doxo resistant)	Doxo-tetrahedral is more effective on cell viability of MCF-7-ADR than free doxo. Same efficacy on MCF-7 cells	N/A	N/A	Kim et al. (2013)
Triangle and tube	Endocytosis and localization of origami in lysosomes after 6 h of treatment	Breast cancer MCF-7 and MCF-7 resistant	MCF-7: 2.5 μM and MCF-7 resistant >100 μM. DNA origami loaded with doxo enhanced the cells' death compared to free doxo on MCF-7 doxo-resistant cells. No differences between free doxo and origami doxo in regular MCF-7	N/A	N/A	Jiang et al. (2012)
Tube	N/A	Splenocytes from female C57BL/6 mice	N/A Immunoresponse through the TLR9. Nontoxic	N/A	N/A	Schüller et al. (2011)
Tube (tile)	Endosomal trapping (no release)	N/A	No effects on GFP expression	N/A	N/A	Kocabey et al. (2014)
Tube with different global twist [straight (S) 10.5 bp/turn and twist (T) tube 12 bp/turn]	N/A	Breast cancer MDA-MB-231; MDA-MB-468; MCF-7	In all the cell lines tested the T-DNA IC_{50} is ≈2 times lower than free doxo (Kumar et al., 2016a,b)	N/A	N/A	Zhao et al. (2012)
Tube, triangle, and square	N/A	Breast cancer MDA-MB-231	No significant difference compared to the free doxo	MDA-MB-231 cells orthotopic breast cancer model	Increased EPR effect of DNA origami. Significant tumor reduction in mice treated with doxo-origami compared to free doxo Inj: I.V.	Zhang et al. (2014a,b)

doxo, doxorubicin; *EPR*, Enhanced permeability and retention; *I.V.*, intravenous; *Inj*, Injection; *Luc*, Luciferase; *N/A*, not applicable; *siRNA*, short interfering RNA.

Adapted from Kumar, V., Bayda, S., Hadla, M., Caligiuri, I., Russo Spena, C., Palazzolo, S., Kempter, S., Corona, G., Toffoli, G., Rizzolio, F., 2016a. Enhanced chemotherapeutic behavior of open-caged DNA@Doxorubicin nanostructures for cancer cells. J. Cell Physiol. 231 (1), 106–110; Kumar, V., Palazzolo, S., Bayda, S., Corona, G., Toffoli, G., Rizzolio, F., 2016b. DNA nanotechnology for cancer therapy. Theranostics 6 (5), 710.

(Wooddell et al., 2013). "Selective destruction" is one method for the removal for sequential errors. The mismatch repair proteins MutH, MutL, and MutS were utilized under double-strand cleavage conditions, followed by size selection of uncleaved product for their isolation (Smith and Modrich, 1997). All these techniques have been used for reducing errors and decreasing mutations in PCR products (Ke et al., 2012a).

Overlap extension PCR and combined dual asymmetrical PCR were used for synthesizing error-free DNA sequences (Young and Dong, 2004). The PCR-dependent purification procedures also involve cloning and gel electrophoresis. However, all these existing methods do not fully remove errors in long synthetic DNA sequences (Agrawal et al., 2009). In new approaches, proteins responsible for self-assembly and natural error correction are being used for error-free DNA production. Among these proteins, the nuclease I enzyme, which is an endonuclease, is very useful (Karukstis et al., 1998). The CEL I protein is very effective because it reduces the sixfold error in synthetic DNA (Hughes et al., 2012). This enzyme is added to the earlier amplified PCR product and then subjected to a second cycle of amplification. Thereafter, analysis by gel electrophoresis and further quantification and cloning were performed. CEL I is an enzyme occurring naturally that cleaves mismatched DNA sequences (Haran et al., 1993; Fritze et al., 2006; Perrault and Shih, 2014). Common deletion and insertion errors are removed by CEL I, and it serves as an effective tool for the removal of errors that may take place during DNA synthesis (Sahay et al., 2013). PCR-based technologies are used for the synthesis of synthetic DNA, but currently microarrays are also utilized (Hong et al., 1999). In microarray technology, typical DNA synthesis depends upon spatial confinement of reactions on chips. Controlling the locations on the chip can be done to minimize error rates, where the reagents eventually end up (Zahid et al., 2013). Photochemistry is utilized for error minimization. In this, light can be used as a tool for blocking reaction at the error site. In another approach, redox reactions are directed only at desired sites. These strategies can result in reduction of error rates to 1 in 600 bases (Marcato, 2014).

4.5 DNA NANOSTRUCTURES IN BIOLOGICAL APPLICATIONS

Advancements in DNA nanotechnology have opened new avenues for its applications in the field of biomedicine. Nanomaterials such as semiconductors, polymers, and metals present various advantages for in vivo applications because of their physical and chemical properties (Dietz et al., 2009). Nanobiotechnology has paved a way for the development of an advanced field called nanomedicine because it gives a new outlook on therapy and analytics in medicine and pharmacology (Davis et al., 2008).

In the future, DNA nanotechnology will have to face several challenges for its advancement. Complex material of nanoscale level is the result of millions of years of natural evolution. Advancement in the field of structural DNA nanotechnology is anticipated to allow important progress in the field of nanotechnology (Zahid et al., 2013). DNA nanotechnology has emerged as an interdisciplinary research field, with participants from chemistry, physics, computer science, materials science, and biology collaborating to answer the future challenges in nanotechnology (Zahid et al., 2013). Fig. 4.2 explains the interdisciplinary approaches of DNA nanotechnology and its diverse applications.

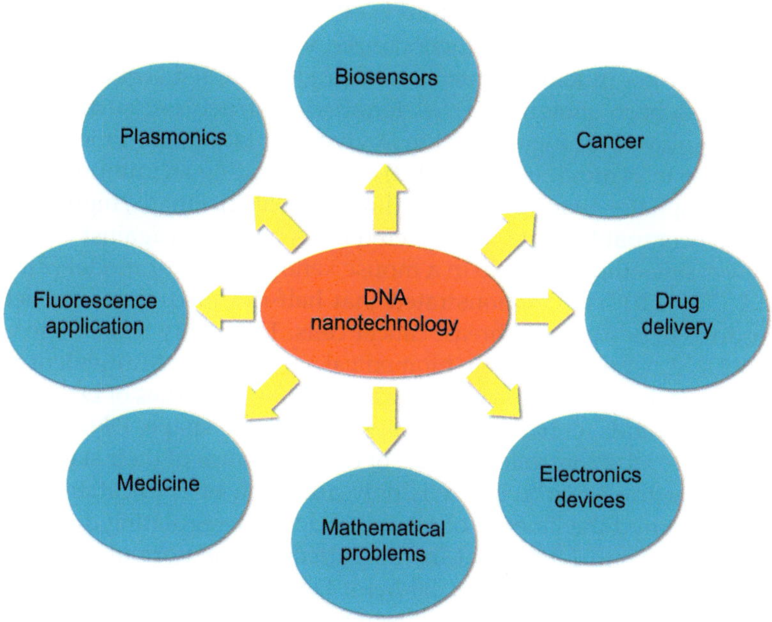

FIGURE 4.2 Applications of DNA nanotechnology in various fields.

4.6 DRUG DELIVERY APPLICATIONS

Until now, several DNA-based nanostructures, namely, tetrahedral, icosahedral (Zhang et al., 2008), nanotube (Yan et al., 2003; Rothemund et al., 2004; Liu et al., 2004; Mathieu et al., 2005), square, and triangle (Rothemund, 2006), have been synthesized for in vitro and in vivo drug delivery applications. The DNA nanostructures are taken up by the cells efficiently without application of transfection agents in comparison to double-stranded DNA (Walsh et al., 2011) and, when these are packaged densely, they could also be efficiently used for drug delivery (Li et al., 2011). Fluorescence-based assays were used for detecting cellular uptake of DNA origami. They have fluorescent labels, which are a drawback of fluorescence-based assays. In another instance, cellular uptake of DNA origami structures was quantified utilizing quantitative real-time PCR of M13 vector amplicons (Okholm et al., 2014). To increase the efficiency of cellular uptake of a DNA-based nanostructure, the DNA nanostructure was coated with virus capsid proteins and it was believed that self-assembly and binding of virus capsids on the surface of origami occur *via* electrostatic interaction, and packaging of DNA nanostructures occurs inside the capsids (Mikkilä et al., 2014). Intercalators based on acridine derivatives were used for modifying the surface characteristics of DNA nanostructures, which resulted in an increase in the cellular uptake of DNA nanostructures in comparison to unchanged origami (Brglez et al., 2015).

Nanostructures such as DNA tiles, tetrahedral (Kim et al., 2013) and icosahedral (Chang et al., 2011), were found to be efficient for delivering drugs such as doxo to breast cancer cells.

One drug-resistant cell tetrahedral structure is effective. The icosahedral structure delivers the doxo in a targeted way once it is functionalized with aptamer sequences specific against the tumor surface marker mucin 1. Protection of single-stranded sequences against nuclease degradation has also been shown by DNA tetrahedral structures; particularly, these structures have been employed for increasing the in vivo circulation half-time of short interfering RNA (siRNA) from 6 to 24 min (Lee et al., 2012) and for eliciting an immunoresponse by delivery of cytosine–phosphate–guanosine (CpG) (Liu et al., 2012). Finally, high antitumor efficacy was shown by a biocompatible aptamer-tethered DNA nanotrain against a folic acid receptor and showed fewer side effects of doxo in a mouse xenograft tumor model (Zhu et al., 2013).

Kumar et al. (2016a,b) have demonstrated that half-icosahedral nanostructures can efficiently deliver doxo to hepatic and breast cancer cells. This study reveals the value of shape and structure of DNA nanostructures in biomedical applications. Multivalent and multifunctional drug carriers were synthesized by using the unique properties of DNA origami, such as dense packaging of strands, enhanced size, and controllable shape. DNA origami structures have great potential for the controlled release of drugs because of their stability in cell lysates and can be slowly degraded in living cells only after 72 h of application (Mei et al., 2011). Zhao et al. (2012) reported the kinetics of release of doxo by regulating the global twist of the DNA origami structure and found that in comparison to the normal structure, the twisted structure form released doxo slowly. A 30-helix DNA origami nanotube conjugated with CpG oligonucleotides was constructed and its immunostimulatory effectiveness analyzed on mouse spleen cells (Kumar et al., 2016a,b). The DNA nanotubes conjugated with CpG have more internalization efficiency than CpG alone and also CpG-conjugated DNA nanotubes show a superior immune response (Schüller et al., 2011). Construction of DNA barrels has been done in a structure that was capable of selectively interfacing with cells for delivering signaling molecules to the surfaces of the cell (Douglas et al., 2012). Tables 4.1a and 4.1b show the details of various types of synthesized DNA nanostructures and their drug delivery applications in in vivo and in vitro experiments (Jiang et al., 2012; Zhang et al., 2014b). From in vivo experiments it was observed that they efficiently delivered doxo to both resistant and normal cancer cells (Zhang et al., 2014b). The triangular DNA nanostructures have enhanced retention time in comparison to the square and tube structures. Most studies show that the intercalation attribute of doxo within DNA duplex is important for drug loading in DNA nanostructures (Zhao et al., 2012; Jiang et al., 2012; Zhang et al., 2014b). Doxo shows higher loading efficiency with DNA nanostructures in comparison to doxo alone. More than 70% efficiency was observed in the case of the DNA tube (Jiang et al., 2012). Through covalent modification, there is virtually no limit to changing DNA with various functional groups. Large arrays of DNA oligomers having a large variety of functional end groups are commercially available nowadays. DNA having an amino group interacts with aminocarboxylic groups and DNA with thiol modification binds to maleimide groups (Srinivasan et al., 2006).

In a new approach, quantum dot (QD) has been used for labeling plasmid DNA (Srinivasan et al., 2006). In this approach, plasmid DNA was conjugated with a core made up of cadmium selenide/zinc sulfide encapsulated with phospholipid/polyethylene oxide, using a linker molecule peptide nucleic acid–*N*-succinimidyl-3-(2-pyridylthio) propionate. These do not affect the function and also promote tagging. This QD-tagged DNA has high efficiency of transfection, which can be monitored. It also retains its functionality upon internalization in the nucleus. Apart from covalent modification, DNA negative charge is also utilized, which can bind positive charge on the surface of another molecule (such as gold nanoparticles along with quaternary ammonium) (Wang and Murray, 2004). Pellegrino et al. (2007) showed that DNA incubated in high stoichiometry excess with gold nanoparticles results in its nonspecific adsorption. For binding DNA with nanoparticles a noncovalent receptor–ligand system such as avidin–biotin is used (Levy et al., 2005; Niemeyer et al., 1998).

4.7 DNA NANOTECHNOLOGY IN CANCER

Millions of people develop certain forms of cancer every year. In cancer, normal cells lose their control of growth, which not only affects the particular organ, but also disturbs the physiology of the entire body. So, one of the major endeavors of nanomedicine is to find a way to diagnose and treat cancers (Zou, 2005; Reichert and Wenger, 2008). Usually, an optimum concentration of drug is used in cancer therapies at the tumor site to kill the cancer cells, which causes certain effects on normal cells as well. Nanotechnology presents an alternative way that prevents the killing of normal cells in chemotherapy.

The current theme of research in nanotechnology is the development of techniques such as nanocarriers and ligand-targeted delivery of therapeutic drugs, having a size range of 10–100nm. In 2009, the Food and Drug Administration (United States) approved albumin-based nanoparticles or liposomes for trials in clinics (Sparreboom et al., 2005; Bharali et al., 2009). Because of their lipid composition, liposomes easily cross the cell membrane to deliver therapeutic products to cells. Albumin-bound nanostructures can be used in breast cancer therapy for enhancing the permeability of the endoplasmic reticulum, whereas the liposome-based nanoparticles are used for delivering microscopic amounts of chemotherapeutic drugs to cancer sites (Sparreboom et al., 2005). Most nanostructures are not sufficient for the treatment of cancer. This resulted in the development of a fullproof "nano-system," which is comprised of four basic steps:

1. Designing of nanosystems for carrying cargo to the tumor site, which may be therapeutics or diagnostics.
2. The nanosystems contain more than one ligand attached to them, which increase their affinity and specificity in the affected cells.
3. More than one drug is loaded on these nanosystems at one time, which is important in multivalent drug therapy.
4. Because biomolecules such as DNA or RNA are used for development of these nanosystems, they may evade most drug-resistance mechanisms.

Nanosystems that have these properties reduce damage to surrounding noncancerous cells by efficient delivery of high concentrations of drugs to cancer cells (Acharya et al., 2009). A detailed study is required to define the fate of various DNA nanostructures within cells.

The mechanisms of excretion, recycling, and subcellular redistribution of the DNA nanostructures could be studied using the RNA interference approach (Kumar et al., 2016a,b). The relative instability of DNA nanostructures makes them unsuitable for in vivo application. Certain case studies reported that the pharmacological potential of the cargo is enhanced by DNA nanostructures. From the point of view of chemotherapeutic drugs, DNA tiles and DNA origami were found to enhance doxo effectiveness, which was confirmed by the tumor growth assay in xenograft mouse models. Initial studies on biodistribution show that liver, kidney, and spleen are the main organs where DNA nanostructures are accumulated (Lee et al., 2012; Zhang et al., 2014a,b). Until now, little research has been done on pharmacokinetics and pharmacodynamics profiling and understanding these aspects is very important for the applicability of DNA nanostructures in the field of therapy (Kumar et al., 2016a,b). The PEGylated liposomal doxo has two times lower potency than doxo alone. Drug is released slowly in the cell culture because of the highly stable Doxil, and this turns out to be an advantage in vivo by enhancing the enhanced permeability and retention (EPR) effect in the tumor cells (Hong et al., 1999). However, further investigations are needed on designing specific drug doses to evaluate in vivo effects of the origami over an extended period of time.

DNA nanostructures are considered as foreign material inside the cell, and were identified by DNA sensing machinery on the immune system, which led to their degradation. DNA normally resides in the nucleus. When it accumulates in other parts of the cell except the nucleus, it is recognized as "anomalous" material. A Toll-like receptor (TLR) pathway is important for triggering immune responses against the B-form of DNA. An anomalous DNA structure activates TLR9 and shows type I interferon response in dendritic cells of endosomes. Cells are placed with deoxyribonucleases (DNases) in the cytoplasm, which degrade foreign DNA (Paludan and Bowie, 2013). RNA polymerase III, interferon-inducible gene 16, and the Mre11–Rad50–Nbs1 complex are a group of sensors present in both nucleus and cytoplasm. DNA hydrolyzing autoantibodies, DNase II, neutral DNase I, and phosphodiesterase I are the DNA-degrading enzymes responsible for 90% of DNase activity of blood plasma that DNA encounters first (Tamkovich et al., 2006). DNase I, which acts on double-stranded DNA, single-stranded DNA, and chromatin producing fragments of various lengths, is a protein secreted into the alimentary tract and bloodstream (Samejima and Earnshaw, 2005). The compact DNA structure has less sensitivity to enzymatic degradation in comparison to linear DNA. A tetrahedron DNA structure was developed by Keum and Bermudez (2009) with edges of about 7 nm, having an enzyme restriction site, CTNAG (DdeI). The tetrahedron structure is less sensitive to DdeI cleavage in comparison to the linear DNA (having the same restriction site), because of increased mechanical stability of ligated DNA. The nonspecific degradation ability of DNase I was also tested and it was observed that the linear fragment was more prone to digestion as compared to the tetrahedron. This difference in the properties could be explained by the fact that the local and global helix geometries vary between linear and tetrahedral structures. To overcome the host immune surveillance, a new system was developed, which was based on the envelope of viral particles (Perrault and Shih, 2014).

The encapsulated DNA drastically decreased the inflammatory response that was demonstrated in in vitro experiments. The circulation half-life of the DNA nanostructures was found to be increased by a three-stage drug delivery systems (membrane–DNA–drug) in the blood to finely control the drug release. The outer membrane was modified by adding ligands, specific receptors, or antibodies for developing target-specific cells (Perrault and Shih, 2014). For enhancing performance as a delivery system, it is key to understand the interaction between DNA origami and DNA sensing machinery. In an efficient delivery system the recycling pathway is a limiting factor as seen in siRNA delivered with lipid nanoparticles (Sahay et al., 2013), and DNA immunosurveillance also plays a fundamental role. Targeted drug transport could be achieved by using aptamer-tethered DNA nanotrains as carriers in cancer therapy (Zhu et al., 2013).

4.8 ROLE IN SOLVING MATHEMATICAL PROBLEMS

DNA-based nanostructures were used for solving various biological and mathematical problems. A wide array of mathematical problems can be solved by utilizing DNA molecules, considering Adelman's experiment as a basis. Qian et al. (2011) used "seesaw" logic gates for calculating square roots. These gates were based on the concept that pairing of a single stretch of DNA occurs with different molecules, which results in competition for binding sites. The moment when a molecule is attached to DNA, it can be simultaneously allowed for other molecules to speed up themselves to the resident sequences; it can be displaced again. During this process these "gates" were loaded with different input molecules and as a result developed logical output molecules.

The number represented by various DNA strands, of which output can yield the square root, serve as answers (Wickham et al., 2012). Researchers have also developed an artificial neural network for mimicking smart biological computation (Qian et al., 2011; Mailloux et al., 2015). Four neurons were used for developing this network. A neuron is prone to any incoming stimuli in its natural environment, and after reaching a certain threshold it either "reacts" or "fires." Boolean logic circuits and other circuits were developed on logic gates, which may store memories. By using DNA logic circuits one can recall memory from incomplete information and also determine when conflicting answers are obtained (Qian et al., 2011). The independent sets of problem were solved by using sticker-based DNA (Taghipour et al., 2012). This model had a random access memory contrary to the earlier sticker-based DNA systems, thus no further extension of its strands and enzymes is required (Roweis et al., 1998). Utilizing memory complexes consisting of DNA, a solution space was developed. Then, various independent sets of problems were solved by using sticker-based parallel algorithms in polynomial time.

Biomolecules other than DNA have also been used for computational purposes. RNA was used to solve an assortment of chess problems through DNA computing (Faulhammer et al., 2000). By applying the same concept, Bandyopadhyay et al. (2010) used 2,3-dichloro-5,6-dicyano-*p*-benzoquinone, which has the ability to transform itself between four different states to mimic natural phenomena such as detection of cancer growth and diffusion of heat. For design problems, a DNA molecule encoding all possible solutions is synthesized and attached to supportive surfaces. The repeated cycles of hybridization and action of exonuclease enzymes are required for digesting, identifying, and eliminating nonsolution strands of DNA. Thereafter, to amplify remaining molecules, PCR is used and most of them are hybridized to a range of molecules (Liu et al., 2000).

However, these methods may be far too advanced for today's computers to handle. So, the future perspective of this current research would be DNA computing and consistent perfection. The development of circuits and electrical components by using nanotechnology comes under DNA physical applications. Nanoscale electronic devices have been developed at the molecular level and are commonly referred to as molecular electronics (Petty et al., 1995). Complementary metal-oxide-semiconductor (CMOS) technology has an important role in nanoelectronics. CMOS is necessary in analog circuits (for example, data convertors and image sensors) and logic-based devices (for example, microcontrollers, digital logic circuits, and microprocessors) (Baker, 2008). For miniaturization and increasing processing speed, CMOS changes according to demand. CMOS circuitry has a constraint that can significantly affect the shape and size of computers and various other electronics.

DNA provides an answer to these problems. The self-guided assembly method is utilized for developing carbon nanotube devices and wires (Patwardhan et al., 2004). Electronic devices (nanowire and transistors) are synthesized by using these materials (Bachtold et al., 2001; DeHon, 2003), thus their characteristics are very similar to a CMOS circuit. Such devices have the advantage that DNA can be easily assembled in very high numbers and densities as compared to a normal circuit in a typical electrical system. Despite this, the efficiency of DNA in terms of power consumption cost makes it a material of interest for this purpose (Patwardhan et al., 2004).

4.9 BIOSENSORS

Biosensors are devices in which biological molecules are used for detecting different targets and by utilizing output elements these recognition events are converted into electrical, optical, or mass-sensitive signals to obtain quantity (Borisov and Wolfbeis, 2008; Suginta et al., 2013; Azimzadeh et al., 2017). The biosensor in which DNA is utilized as a probe for detection is an important subarea of biosensors.

Nowadays, biosensors based on DNA provide a rapid, easy, and economical alternative to detection. However, there are problems associated with these DNA-based biosensors, such as instability and reproducibility in comparison to other available biosensors. Sensors are in demand because of their wide range of applications in different areas (such as in manufacturing, medicine, innovation, and environmental sectors) (Chao et al., 2016; Azimzadeh et al., 2017). The use of different nanomaterials derived from various sources increased the demand for DNA-based biosensors toward the development of fast, easy, and economical detection of analytes. The available DNA-based biosensors retain their abilities but there is still room for improvement in different directions: (1) increase in sensitivity, (2) decrease the detection limit by using various magnification methods, and (3) new output signals to fulfill the detection style.

DNA-based nanodevices are important in research because humans like the ability to control the world at the nanoscale (Krishnan and Simmel, 2011; Chao et al., 2014; Arts et al., 2016; Pandian et al., 2017). DNA or RNA (DNA tweezers, PX-JX2-based devices, and DNA walkers) sequences with overhang are used as triggers because transformation of the devices can initiate branch migration (Yurke et al., 2000; Yan et al., 2002; Sherman and Seeman, 2004; Shin and Pierce, 2004; Omabegho et al., 2009; Gu et al., 2009, 2010; Engelen et al., 2017). Deoxyribozymes (DNAzymes) were the first walking device to be triggered by enzymes in which DNA ligase was used to combine two DNA strands together; to break their connections, DNA restriction enzymes were used (Tian et al., 2005).

Parallel theories were later demonstrated by using a DNA nicking enzyme. The DNAzymes or ribozymes have the capability of catalyzing biochemical reactions, and may also be able to drive DNA motors to walk on double-stranded DNA tracks. This theory was used to develop a molecular spider, which walks on a 100-nm origami scaffold platform (Lund et al., 2010; Wickham et al., 2012). Certain complexes [hexamine cobalt(III)] $(Co(NH_3)_6{}^{3+}$ are capable of making the transition from the B-form of double helical DNA (right handed) to the Z-form (left handed). By using these properties of complexes the first rotary nanomechanical (B–Z) device was constructed (Mao et al., 1999).

4.10 TECHNICAL CHALLENGES

Custom cookie-cutter shapes could be created through DNA origami, with a mass of ~5 MDa each (Pinheiro et al., 2011). For DNA origami, the main issue is the dependencies on the M13 vector, having a genome size of 7 kb, as the sole source of the scaffold. Bigger structures could be achieved through two ways, i.e., either by folding a longer unique scaffold or by folding multiple scaffolds with distinct sequences (Pinheiro et al., 2011). Furthermore, it was considered that the M13 vector does not have all sequences needed for high-yield folding for various DNA nanostructures. There is the possibility that in the future more complex DNA nanostructures will be synthesized. The two major prominent hurdles are high error rate in the self-assembly of DNA nanostructures and high cost of synthetic DNA. To solve this problem, a selective destruction method was employed. In this method a mismatch repair protein such as Mut H and Mut S is used in a double-strand cleavage environment coupled with isolation of uncleaved products (Smith and Modrich, 1997). About US$700 is spent on constructing a new M13-based origami, whereas oligonucleotide synthesis on the 25-nmol scale costs US$0.10 per base.

Emerging commercial availability provides key technological opportunities, which results in the availability of arrays at affordable rates. Reproduction of small amounts of each of the tens of thousands of unique oligonucleotide sequences is done for less than US$0.001 per base. There is the possibility of further development of reliable low-cost methods for enzymatic amplification of subsets of strands from these arrays (Cheng et al., 2009; Tian et al., 2009). Advances in technologies give the hope that development of custom-designed DNA nanostructures that are 1 GDa in mass (~100 times more complex than M13-based origami) could be achieved for ~US$1000. Lowering the cost of enzymatic amplification reactions by rapid development in technologies has enabled the production of desired amounts of complex DNA nanostructures, which is very important for their applications. Economic feasibility is the major concern in the field of DNA nanotechnology, but array-printed DNA-assisted gene synthesis may be an answer to the challenges in this field (Cheng et al., 2009; Tian et al., 2009). Thus we can generate many unique scaffold sequences, each designed for optimized folding into a particular origami shape, or at least a large number of distinct generic scaffold sequences that can help independent foldings in a single pot. All these discussions raise the question: what are the rules for effective sequence design? The answer requires much work in this direction, involving interplay between experiment and theory (Pinheiro et al., 2011).

4.11 CONCLUSION AND FUTURE PERSPECTIVES

Nowadays, DNA is an important engineering material available in nanotechnology. DNA has enormous potential for self-assembly, which results in the formation of programmable nanostructures. It can also be used as a platform for the synthesis for various mechanical, chemical, and physical devices. A mismatch repair mechanism coupled with synthesis gives a better way of producing error-free DNA. Mismatch, excision, and homologous recombination repair are representations of an interstrand repair mechanism (Wu et al., 2005).

Nanotechnology provides an important platform for the development of detectors in the field of chemical and biological detection, which are highly sensitive and efficient. Sophisticated architectures need to be developed urgently for diverse applications in terms of DNA-based nanostructures (Zahid et al., 2013). Presently, utilizing the DNA origami technique, attempts are being made to transform DNA into different shapes, but the real need is to synthesize refined and intelligent structures that have viable chemical, physical, and biological applications. Presently, DNA computation is in its budding phase and future development of novel techniques is required in which DNA would be utilized to have a more important role in biological computation and data storage (Zahid et al., 2013). DNA nanotechnology provides a better way for the development of novel vehicles for drug delivery, having required biocompatibility along with the enhanced cellular uptake kinetics of DNA nanostructures loaded with drugs (Kumar et al., 2016a,b).

Judicious selection of the functionalities, size, and shape of DNA nanostructures, because of the size and shape-dependent EPR effect, enables optimization and tuning of DNA nanotechnology for their best performance in in vitro and in vivo experiments. To avoid any misinterpretation, before in vivo applications of any prospective DNA nanostructures, very watchful and strict in vitro testing is needed (Kocabey et al., 2014). Advances in this area have brought new hopes. RCA-based scaffolds and several single-stranded DNA approaches provide a means to synthesize various-sized DNA nanostructures, while in Rothemund's origami process only limited-sized nanostructures were synthesized (Iinuma et al., 2014; Marchi et al., 2014). Utilization of DNA triangular origami coupled with the lithography technique resulted in the generation of large-area spatially ordered arrays of gold nanoparticles, which may help them to bind uniformly to the surfaces, e.g., like an electrode (Hung et al., 2010). Precise placement of guest molecules such as fluorophores, proteins, or nanoparticles, up to nanometer range, could be achieved through DNA nanostructures. These assemblies' reactions may be utilized in the study of biological processes with precise control over spacing, number, and molecular identity (Schlichthaerle et al., 2016).

References

Acharya, S., Dilnawaz, F., Sahoo, S.K., 2009. Targeted epidermal growth factor receptor nanoparticle bioconjugates for breast cancer therapy. Biomaterials 30 (29), 5737–5750.

Agrawal, P., Barthwal, S.K., Barthwal, R., 2009. Studies on self-aggregation of anthracycline drugs by restrained molecular dynamics approach using nuclear magnetic resonance spectroscopy supported by absorption, fluorescence, diffusion ordered spectroscopy and mass spectrometry. Eur. J. Med. Chem. 44 (4), 1437–1451.

Aldaye, F.A., Palmer, A.L., Sleiman, H.F., 2008. Assembling materials with DNA as the guide. Science 321 (5897), 1795–1799.

Andersen, E.S., Dong, M., Nielsen, M.M., Jahn, K., Subramani, R., Mamdouh, W., Golas, M.M., Sander, B., Stark, H., Oliveira, C.L., Pedersen, J.S., 2009. Self-assembly of a nanoscale DNA box with a controllable lid. Nature 459 (7243), 73–76.

Arts, R., Den Hartog, I., Zijlema, S.E., Thijssen, V., van der Beelen, S.H., Merkx, M., 2016. Detection of antibodies in blood plasma using bioluminescent sensor proteins and a smartphone. Anal. Chem. 88 (8), 4525–4532.

Azimzadeh, M., Rahaie, M., Nasirizadeh, N., Daneshpour, M., Naderi-Manesh, H., 2017. Electrochemical *miRNA* biosensors: the benefits of nanotechnology. Nanomed. Res. J. 2 (1), 36–48.

Aziz, N., Faraz, M., Pandey, R., Shakir, M., Fatma, T., Varma, A., et al., 2015. Facile algae-derived route to biogenic silver nanoparticles: synthesis, antibacterial, and photocatalytic properties. Langmuir 31, 11605–11612.

Aziz, N., Pandey, R., Barman, I., Prasad, R., 2016. Leveraging the attributes of *Mucor hiemalis*-derived silver nanoparticles for a synergistic broad-spectrum antimicrobial platform. Front. Microbiol. 7, 1984. http://dx.doi.org/10.3389/fmicb.2016.01984.

Bachtold, A., Hadley, P., Nakanishi, T., Dekker, C., 2001. Logic circuits with carbon nanotube transistors. Science 294 (5545), 1317–1320.

Baker, R.J., 2008. CMOS: Circuit Design, Layout, and Simulation, second ed. Wiley, New York.
Bandyopadhyay, A., Pati, R., Sahu, S., Peper, F., Fujita, D., 2010. Massively parallel computing on an organic molecular layer. Nat. Phys. 6 (5), 369–375.
Benson, E., Mohammed, A., Gardell, J., Masich, S., Czeizler, E., Orponen, P., Högberg, B., 2015. DNA rendering of polyhedral meshes at the nanoscale. Nature 523 (7561), 441–444.
Bharali, D.J., Khalil, M., Gurbuz, M., Simone, T.M., Mousa, S.A., 2009. Nanoparticles and cancer therapy: a concise review with emphasis on dendrimers. Int. J. Nanomed. 4, 1.
Borisov, S.M., Wolfbeis, O.S., 2008. Optical biosensors. Chem. Rev. 108 (2), 423–461.
Brglez, J., Nikolov, P., Angelin, A., Niemeyer, C.M., 2015. Designed intercalators for modification of DNA origami surface properties. Chem. Eur. J. 21 (26), 9440–9446.
Castro, C.E., Kilchherr, F., Kim, D.N., Shiao, E.L., Wauer, T., Wortmann, P., Bathe, M., Dietz, H., 2011. A primer to scaffolded DNA origami. Nat. Methods 8 (3), 221–229.
Chang, M., Yang, C.S., Huang, D.M., 2011. Aptamer-conjugated DNA icosahedral nanoparticles as a carrier of doxorubicin for cancer therapy. ACS Nano 5 (8), 6156–6163.
Chao, J., Liu, H., Su, S., Wang, L., Huang, W., Fan, C., 2014. Structural DNA nanotechnology for intelligent drug delivery. Small 10 (22), 4626–4635.
Chao, J., Zhu, D., Zhang, Y., Wang, L., Fan, C., 2016. DNA nanotechnology-enabled biosensors. Biosens. Bioelectron. 76, 68–79.
Cheng, W., Campolongo, M.J., Cha, J.J., Tan, S.J., Umbach, C.C., Muller, D.A., Luo, D., 2009. Free-standing nanoparticle superlattice sheets controlled by DNA. Nat. Mater. 8 (6), 519–525.
Chhabra, R., Sharma, J., Ke, Y., Liu, Y., Rinker, S., Lindsay, S., Yan, H., 2007. Spatially addressable multi protein nanoarrays templated by aptamer-tagged DNA nanoarchitectures. J. Am. Chem. Soc. 129 (34), 10304–10305.
Coelho, T., Adams, D., Silva, A., Lozeron, P., Hawkins, P.N., Mant, T., Perez, J., Chiesa, J., Warrington, S., Tranter, E., Munisamy, M., 2013. Safety and efficacy of RNAi therapy for transthyretin amyloidosis. N. Engl. J. Med. 369 (9), 819–829.
Davis, M.E., Chen, Z.G., Shin, D.M., 2008. Nanoparticle therapeutics: an emerging treatment modality for cancer. Nat. Rev. Drug Discov. 7 (9), 771–782.
DeHon, A., 2003. Array-based architecture for FET-based, nanoscale electronics. IEEE Trans. Nanotechnol. 2 (1), 23–32.
Dietz, H., Douglas, S.M., Shih, W.M., 2009. Folding DNA into twisted and curved nanoscale shapes. Science 325 (5941), 725–730.
Douglas, S.M., Dietz, H., Liedl, T., Högberg, B., Graf, F., Shih, W.M., 2009b. Self-assembly of DNA into nanoscale three-dimensional shapes. Nature 459 (7245), 414–418.
Douglas, S.M., Marblestone, A.H., Teerapittayanon, S., Vazquez, A., Church, G.M., Shih, W.M., 2009a. Rapid prototyping of 3D DNA-origami shapes with caDNAno. Nucleic Acids Res. 436.
Douglas, S.M., Bachelet, I., Church, G.M., 2012. A logic-gated nano-robot for targeted transport of molecular payloads. Science 335 (6070), 831–834.
Engelen, W., Meijer, L.H., Somers, B., De Greef, T.F., Merkx, M., 2017. Antibody-controlled actuation of DNA-based molecular circuits. Nat. Commun. 8, 14473.
Erben, C.M., Goodman, R.P., Turberfield, A.J., 2006. Single-molecule protein encapsulation in a rigid DNA cage. Angew. Chem. 118 (44), 7574–7577.
Faulhammer, D., Cukras, A.R., Lipton, R.J., Landweber, L.F., 2000. Molecular computation: RNA solutions to chess problems. Proc. Natl. Acad. Sci. U.S.A. 97 (4), 1385–1389.
Fritze, A., Hens, F., Kimpfler, A., Schubert, R., Peschka-Süss, R., 2006. Remote loading of doxorubicin into liposomes driven by a transmembrane phosphate gradient. Biochim. Biophys. Acta 1758 (10), 1633–1640.
Gilleron, J., Querbes, W., Zeigerer, A., Borodovsky, A., Marsico, G., Schubert, U., Manygoats, K., Seifert, S., Andree, C., Stöter, M., Epstein-Barash, H., 2013. Image-based analysis of lipid nanoparticle-mediated siRNA delivery, intracellular trafficking and endosomal escape. Nat. Biotechnol. 31 (7), 638–646.
Glotzer, S.C., Solomon, M.J., 2007. Anisotropy of building blocks and their assembly into complex structures. Nat. Mater. 6 (8), 557–562.
Gu, H.Z., Chao, J., Xiao, S.J., Seeman, N.C., 2009. Dynamic patterning programmed by DNA tiles captured on a DNA origami substrate. Nat. Nanotechnol. 4 (4), 245–248.
Gu, H., Chao, J., Xiao, S.J., Seeman, N.C., 2010. A proximity-based programmable DNA nanoscale assembly line. Nature 465 (7295), 202–205.

Han, D., Pal, S., Nangreave, J., Deng, Z., Liu, Y., Yan, H., 2011. DNA origami with complex curvatures in three-dimensional space. Science 332 (6027), 342–346.

Han, D., Pal, S., Yang, Y., Jiang, S., Nangreave, J., Liu, Y., Yan, H., 2013. DNA gridiron nanostructures based on four-arm junctions. Science 339 (6126), 1412–1415.

Haran, G., Cohen, R., Bar, L.K., Barenholz, Y., 1993. Transmembrane ammonium sulfate gradients in liposomes produce efficient and stable entrapment of amphipathic weak bases. Biochim. Biophys. Acta 1151 (2), 201–215.

Hemminga, M.A., Vos, W.L., Nazarov, P.V., Koehorst, R.B., Wolfs, C.J., Spruijt, R.B., Stopar, D., 2010. Viruses: incredible nanomachines. New advances with filamentous phages. Eur. Biophys. J. 39 (4), 541–550.

Hong, R.L., Huang, C.J., Tseng, Y.L., Pang, V.F., Chen, S.T., Liu, J.J., Chang, F.H., 1999. Direct comparison of liposomal doxorubicin with or without polyethylene glycol coating in C-26 tumor-bearing mice: is surface coating with polyethylene glycol beneficial? Clin. Cancer Res. 5 (11), 3645–3652.

Hughes, R.A., Miklos, A.E., Ellington, A.D., 2012. Enrichment of error-free synthetic DNA sequences by CEL I nuclease. Curr. Protoc. Mol. Biol. 99, 1–10 3.24.

Hung, A.M., Micheel, C.M., Bozano, L.D., Osterbur, L.W., Wallraff, G.M., Cha, J.N., 2010. Large-areas partially ordered arrays of gold nanoparticles directed by lithographically confined DNA origami. Nat. Nanotechnol. 5 (2), 121–126.

Iinuma, R., Ke, Y., Jungmann, R., Schlichthaerle, T., Woehrstein, J.B., Yin, P., 2014. Polyhedral self-assembled from DNA tripods and characterized with 3D DNA- paint. Science 344 (6179), 65–69.

Jiang, Q., Song, C., Nangreave, J., Liu, X., Lin, L., Qiu, D., Wang, Z.G., Zou, G., Liang, X., Yan, H., Ding, B., 2012. DNA origami as a carrier for circumvention of drug resistance. J. Am. Chem. Soc. 134 (32), 13396–13403.

Jones, M.R., Seeman, N.C., Mirkin, C.A., 2015. Programmable materials and the nature of the DNA bond. Science 347 (6224), 1260901.

Kallenbach, N.R., Ma, R.I., Seeman, N.C., 1983. An immobile nucleic acid junction constructed from oligonucleotides. Nature 305 (5937), 829–831.

Karukstis, K.K., Thompson, E.H., Whiles, J.A., Rosenfeld, R.J., 1998. Deciphering the fluorescence signature of daunomycin and doxorubicin. Biophys. Chem. 73 (3), 249–263.

Kay, E.R., Leigh, D.A., Zerbetto, F., 2007. Synthetic molecular motors and mechanical machines. Angew. Chem. Int. Ed. 46 (1–2), 72–191.

Ke, Y., Bellot, G., Voigt, N.V., Fradkov, E., Shih, W.M., 2012a. Two design strategies for enhancement of multilayer-DNA-origami folding: underwinding for specific intercalator rescue and staple-break positioning. Chem. Sci. 3 (8), 2587–2597.

Ke, Y., Ong, L.L., Shih, W.M., Yin, P., 2012b. Three-dimensional structures self-assembled from DNA bricks. Science 338 (6111), 1177–1183.

Keller, S., Marx, A., 2012. The use of enzymes for construction of DNA-based objects and assemblies. Chem. Soc. Rev. 40 (12), 5690–5697.

Keum, J.W., Bermudez, H., 2009. Enhanced resistance of DNA nanostructures to enzymatic digestion. Chem. Commun. (Camb.) 45, 7036–7038.

Kim, K.R., Kim, D.R., Lee, T., Yhee, J.Y., Kim, B.S., Kwon, I.C., Ahn, D.R., 2013. Drug delivery by a self-assembled DNA tetrahedron for overcoming drug resistance in breast cancer cells. Chem. Commun. (Camb.) 49 (20), 2010–2012.

Kocabey, S., Meinl, H., MacPherson, I.S., Cassinelli, V., Manetto, A., Rothenfusser, S., Liedl, T., Lichtenegger, F.S., 2014. Cellular uptake of tile-assembled DNA nanotubes. Nanomaterials 5 (1), 47–60.

Krishnan, Y., Simmel, F.C., 2011. Nucleic acid based molecular devices. Angew. Chem. Int. Ed. 50 (14), 3124–3156.

Kumar, V., Bayda, S., Hadla, M., Caligiuri, I., Russo Spena, C., Palazzolo, S., Kempter, S., Corona, G., Toffoli, G., Rizzolio, F., 2016a. Enhanced chemotherapeutic behavior of open-caged DNA@Doxorubicin nanostructures for cancer cells. J. Cell Physiol. 231 (1), 106–110.

Kumar, V., Palazzolo, S., Bayda, S., Corona, G., Toffoli, G., Rizzolio, F., 2016b. DNA nanotechnology for cancer therapy. Theranostics 6 (5), 710.

Lee, H., Lytton-Jean, A.K., Chen, Y., Love, K.T., Park, A.I., Karagiannis, E.D., Sehgal, A., Querbes, W., Zurenko, C.S., Jayaraman, M., Peng, C.G., 2012. Molecularly self-assembled nucleic acid nanoparticles for targeted in vivo siRNA delivery. Nat. Nanotechnol. 7 (6), 389–393.

Levy, M., Cater, S.F., Ellington, A.D., 2005. Quantum-dot aptamer beacons for the detection of proteins. Chem. Biochem. 6 (12), 2163–2166.

Li, H., LaBean, T.H., Kenan, D.J., 2006. Single-chain antibodies against DNA aptamers for use as adapter molecules on DNA tile arrays in nanoscale materials organization. Org. Biomol. Chem. 4 (18), 3420–3426.

Li, J., Pei, H., Zhu, B., Liang, L., Wei, M., He, Y., Chen, N., Li, D., Huang, Q., Fan, C., 2011. Self-assembled multivalent DNA nanostructures for noninvasive intracellular delivery of immunostimulatory CpG oligonucleotides. ACS Nano 5 (11), 8783–8789.

Li, J., Fan, C., Pei, H., Shi, J., Huang, Q., 2013. Smart drug delivery nano-carriers with self-assembled DNA nanostructures. Adv. Mater. 25 (32), 4386–4396.

Linko, V., Dietz, H., 2013. The enabled state of DNA nanotechnology. Curr. Opin. Biotechnol. 24 (4), 555–561.

Liu, Q., Wang, L., Frutos, A.G., Condon, A.E., Corn, R.M., Smith, L.M., 2000. DNA computing on surfaces. Nature 403 (6766), 175–179.

Liu, D., Park, S.H., Reif, J.H., LaBean, T.H., 2004. DNA nanotubes self-assembled from triple-crossover tiles as templates for conductive nanowires. Proc. Natl. Acad. Sci. U.S.A. 101 (3), 717–722.

Liu, Y., Lin, C., Li, H., Yan, H., 2005. Aptamer-directed self-assembly of protein arrays on a DNA nanostructure. Angew. Chem. 117 (28), 4407–4412.

Liu, X., Xu, Y., Yu, T., Clifford, C., Liu, Y., Yan, H., Chang, Y., 2012. A DNA nanostructure platform for directed assembly of synthetic vaccines. Nano Lett. 12 (8), 4254–4259.

Lund, K., Liu, Y., Lindsay, S., Yan, H., 2005. Self-assembling a molecular pegboard. J. Am. Chem. Soc. 127 (50), 17606–17607.

Lund, K., Manzo, A.J., Dabby, N., Michelotti, N., Johnson-Buck, A., Nangreave, J., Taylor, S., Pei, R., Stojanovic, M.N., Walter, N.G., Winfree, E., Yan, H., 2010. Molecular robots guided by prescriptive landscapes. Nature 465 (7295), 206–210.

Mailloux, S., Gerasimova, Y.V., Guz, N., Kolpashchikov, D.M., Katz, E., 2015. Bridging the two worlds: a universal interface between enzymatic and DNA computing systems. Angew. Chem. Int. Ed. 54 (22), 6562–6566.

Mao, C., Sun, W., Shen, Z., Seeman, N.C., 1999. A nanomechanical device based on the B-Z transition of DNA. Nature 397 (6715), 144–146.

Marcato, P.D., 2014. Pharmacokinetics and pharmacodynamics of nanomaterials. In: Durán, N., Guterres, S.S., Alves, O.L. (Eds.), Nanotoxicology Materials, Methodologies and Assessments. Springer, NY, USA, pp. 97–110.

Marchi, A.N., Saaem, I., Vogen, B.N., Brown, S., LaBean, T.H., 2014. Toward larger DNA origami. Nano Lett. 14 (10), 5740–5747.

Mathieu, F., Liao, S., Kopatsch, J., Wang, T., Mao, C., Seeman, N.C., 2005. Six-helix bundles designed from DNA. Nano Lett. 5 (4), 661–665.

Mei, Q., Wei, X., Su, F., Liu, Y., Youngbull, C., Johnson, R., Lindsay, S., Yan, H., Meldrum, D., 2011. Stability of DNA origami nanoarrays in cell lysate. Nano Lett. 11 (4), 1477–1482.

Mikkila¨, J., Eskelinen, A.P., Niemela¨, E.H., Linko, V., Frilander, M.J., To¨rma¨, P., Kostiainen, M.A., 2014. Virus-encapsulated DNA origami nanostructures for cellular delivery. Nano Lett. 14 (4), 2196–2200.

Niemeyer, C.M., Bürger, W., Peplies, J., 1998. Covalent DNA-streptavidin conjugates as building blocks for novel biometallic nanostructures. Angew. Chem. Int. Ed. 37 (16), 2265–2268.

Okholm, A.H., Nielsen, J.S., Vinther, M., Sørensen, R.S., Schaffert, D., Kjems, J., 2014. Quantification of cellular uptake of DNA nanostructures by qPCR. Methods 67 (2), 193–197.

Omabegho, T., Sha, R., Seeman, N.C., 2009. A bipedal DNA Brownian motor with coordinated legs. Science 324 (5923), 67–71.

Paludan, S.R., Bowie, A.G., 2013. Immune sensing of DNA. Immunity 38 (5), 870–880.

Pandian, R.K.S., Yuan, C.J., Lin, C.C., Wang, W.H., Chang, C.C., 2017. DNA-based nanowires and nanodevices. Adv. Phys. X 2 (1), 22–34.

Park, S.H., Yin, P., Liu, Y., Reif, J.H., LaBean, T.H., Yan, H., 2005. Programmable DNA self-assemblies for nanoscale organization of ligands and proteins. Nano Lett. 5 (4), 729–733.

Patwardhan, J.P., Dwyer, C., Lebeck, A.R., Sorin, D.J., 2004. Circuit and system architecture for DNA-guided self-assembly of nanoelectronics. In: Kyoto, R.J. (Ed.), Foundations of Nanoscience: Self-assembled Architectures and Devices. Sci. Technica. Inc., pp. 344–358.

Pellegrino, T., Sperling, R.A., Alivisatos, A.P., Parak, W.J., 2007. Gel electrophoresis of gold-DNA nanoconjugates. J. Biomed. Biotechnol. 26796.

Peralta-Videa, J.R., Hernandez-Viezcas, J.A., Zhao, L., Diaz, B.C., Ge, Y., Priester, J.H., Holden, P.A., Gardea-Torresdey, J.L., 2014. Cerium dioxide and zinc oxide nanoparticles alter the nutritional value of soil cultivated soybean plants. Plant Physiol. Biochem. 80, 128–135.

Peralta-Videa, J.R., Huang, Y., Parsons, J.G., Zhao, L., Lopez-Moreno, L., Hernandez-Viezcas, J.A., Gardea-Torresdey, J.L., 2016. Plant-based green synthesis of metallic nanoparticles: scientific curiosity or a realistic alternative to chemical synthesis? Nanotechnol. Environ. Eng. 1 (1), 4.

Perrault, S.D., Shih, W.M., 2014. Virus-inspired membrane encapsulation of DNA nanostructures to achieve *in vivo* stability. ACS Nano 8 (5), 5132–5140.

Petty, M.C., Bryce, M.R., Bloor, D., 1995. An Introduction to Molecular Electronics, first ed. Oxford University Press, London.

Pinheiro, A.V., Han, D., Shih, W.M., Yan, H., 2011. Challenges and opportunities for structural DNA nanotechnology. Nat. Nanotechnol. 6 (12), 763–772.

Prasad, R., Pandey, R., Barman, I., 2016. Engineering tailored nanoparticles with microbes: quo vadis? Wiley Interdiscip. Rev. Nanomed. Nanobiotechnol. 8, 316–330. http://dx.doi.org/10.1002/wnan.1363.

Qian, L., Winfree, E., Bruck, J., 2011. Neural network computation with DNA strand displacement cascades. Nature 475 (7356), 368–372.

Reichert, J.M., Wenger, J.B., 2008. Development trends for new cancer therapeutics and vaccines. Drug Discov. Today 13 (1–2), 30–37.

Rothemund, P.W., Ekani-Nkodo, A., Papadakis, N., Kumar, A., Fygenson, D.K., Winfree, E., 2004. Design and characterization of programmable DNA nanotubes. J. Am. Chem. Soc. 126 (50), 16344–16352.

Rothemund, P.W.K., 2006. Folding DNA to create nanoscale shapes and patterns. Nature 440 (7082), 297–302.

Roweis, S., Winfree, E., Burgoyne, R., Chelyapov, N.V., Goodman, M.F., Rothemund, P.W., Adleman, L.M., 1998. A sticker-based model for DNA computation. J. Comput. Biol. 5 (4), 615–629.

Saccà, B., Meyer, R., Erkelenz, M., Kiko, K., Arndt, A., Schroeder, H., Rabe, K.S., Niemeyer, C.M., 2010. Orthogonal protein decoration of DNA origami. Angew. Chem. Int. Ed. 49 (49), 9378–9383.

Sahay, G., Querbes, W., Alabi, C., Eltoukhy, A., Sarkar, S., Zurenko, C., Karagiannis, E., Love, K., Chen, D., Zoncu, R., Buganim, Y., Schroeder, A., Langer, R., Anderson, D.G., 2013. Efficiency of siRNA delivery by lipid nanoparticles is limited by endocytic recycling. Nat. Biotechnol. 31 (7), 653–658.

Samejima, K., Earnshaw, W.C., 2005. Trashing the genome: the role of nucleases during apoptosis. Nat. Rev. Mol. Cell Biol. 6 (9), 677–688.

Schlichthaerle, T., Strauss, M.T., Schueder, F., Woehrstein, J.B., Jungmann, R., 2016. DNA nanotechnology and fluorescence applications. Curr. Opin. Biotechnol. 30 (39), 41–47.

Schüller, V.J., Heidegger, S., Sandholzer, N., Nickels, P.C., Suhartha, N.A., Endres, S., Bourquin, C., Liedl, T., 2011. Cellular immunostimulation by CpG-sequence-coated DNA origami structures. ACS Nano 5 (12), 9696–9702.

Seeman, N.C., 1982. Nucleic acid junctions and lattices. J. Theor. Biol. 99 (2), 237–247.

Seeman, N.C., 2003. DNA in a material world. Nature 421 (6921), 427–431.

Seeman, N.C., 2010. Nanomaterials based on DNA. Annu. Rev. Biochem. 79, 65–87.

Sekhon, B.S., 2005. Nanobiotechnology: an overview of drug discovery, delivery and development. Pharmacol. Ther. 69, 13.

Sherman, W.B., Seeman, N.C., 2004. A precisely controlled DNA biped walking device. Nano Lett. 4 (7), 1203–1207.

Shih, W.M., Lin, C., 2010. Knitting complex weaves with DNA origami. Curr. Opin. Struct. Biol. 20 (3), 276–282.

Shin, J.S., Pierce, N.A., 2004. A synthetic DNA walker for molecular transport. J. Am. Chem. Soc. 126 (35), 10834–10835.

Shweta, Tripathi, D.K., Singh, S., Singh, S., Dubey, N.K., Chauhan, D.K., 2016. Impact of nanoparticles on photosynthesis: challenges and opportunities. Mater. Focus 5 (5), 405–411.

Singh, S., Tripathi, D.K., Dubey, N.K., Chauhan, D.K., 2016. Effects of nano-materials on seed germination and seedling growth: striking the slight balance between the concepts and controversies. Mater. Focus 5 (3), 195–201.

Singh, S., Vishwakarma, K., Singh, S., Sharma, S., Dubey, N.K., Singh, V.K., Liu, S., Tripathi, D.K., Chauhan, D.K., 2017. Understanding the plant and nanoparticle interface at transcriptomic and proteomic level: a concentric overview. Plant Gene. http://doi.org/10.1016/j.plgene.2017.03.006.

Smith, J., Modrich, P., 1997. Removal of polymerase-produced mutant sequences from PCR products. Proc. Natl. Acad. Sci. U.S.A. 94 (13), 6847–6850.

Sönnichsen, C., Reinhard, B.M., Liphardt, J., Alivisatos, A.P., 2005. A molecular ruler based on plasmon coupling of of single gold and silver nanoparticles. Nat. Biotechnol. 23 (6), 741–745.

Sparreboom, A., Scripture, C.D., Trieu, V., Williams, P.J., De, T., Yang, A., Beals, B., Figg, W.D., Hawkins, M., Desai, N., 2005. Comparative preclinical and clinical pharmacokinetics of a cremophor-free, nanoparticle albumin-bound paclitaxel (ABI-007) and paclitaxel formulated in Cremophor (Taxol). Clin. Cancer Res. 11 (11), 4136–4143.

Srinivasan, C., Lee, J., Papadimitrakopoulos, F., Silbart, L.K., Zhao, M., Burgess, D.J., 2006. Labeling and intracellular tracking of functionally active plasmid DNA with semiconductor quantum dots. Mol. Ther. 14 (2), 192–201.

Stephanopoulos, N., Liu, M., Tong, G.J., Li, Z., Liu, Y., Yan, H., Francis, M.B., 2010. Immobilization and one-dimensional arrangement of virus capsids with nanoscale precision using DNA origami. Nano Lett. 10 (7), 2714–2720.

Suginta, W., Khunkaewla, P., Schulte, A., 2013. Electrochemical biosensor applications of polysaccharides chitin and chitosan. Chem. Rev. 113 (7), 5458–5479.

Taghipour, H., Taghipour, A., Rezaei, M., Esmaili, H., 2012. Solving the independent set problem by sticker based DNA computers. Am. J. Mol. Biol. 2 (2), 153–158.

Tamkovich, S.N., Cherepanova, A.V., Kolesnikova, E.V., Rykova, E.Y., Pyshnyi, D.V., Vlassov, W., Laktionov, P.P., 2006. Circulating DNA and DNase activity in human blood. Ann. N. Y. Acad. Sci. 1075, 191–196.

Tian, Y., He, Y., Chen, Y., Yin, P., Mao, C., 2005. A DNAzyme that walks processively and autonomously a long a one-dimensional track. Angew. Chem. Int. Ed. 44 (28), 4355–4358.

Tian, J., Ma, K., Saaem, I., 2009. Advancing high-throughput gene synthesis technology. Mol. Biosyst. 5 (7), 714–722.

Tian, C., Kim, H., Sun, W., Kim, Y., Yin, P., Liu, H., 2017. DNA nanostructures-mediated molecular imprinting lithography. ACS Nano 11, 227–238.

Tripathi, D.K., Singh, V.P., Prasad, S.M., Chauhan, D.K., Dubey, N.K., 2015. Silicon nanoparticles (SiNp) alleviate chromium (VI) phytotoxicity in *Pisum sativum* (L.) seedlings. Plant Physiol. Biochem. 96, 189–198.

Tripathi, D.K., Singh, S., Singh, V.P., Prasad, S.M., Chauhan, D.K., Dubey, N.K., 2016. Silicon nanoparticles more efficiently alleviate arsenate toxicity than silicon in maize cultiver and hybrid differing in arsenate tolerance. Front. Environ. Sci. 4, 46.

Tripathi, D.K., Singh, S., Singh, S., Pandey, R., Singh, V.P., Sharma, N.C., Prasad, S.M., Dubey, N.K., Chauhan, D.K., 2017a. An overview on manufactured nanoparticles in plants: uptake, translocation, accumulation and phytotoxicity. Plant Physiol. Biochem. 110, 2–12.

Tripathi, D.K., Mishra, R.K., Singh, S., Singh, S., Vishwakarma, K., Sharma, S., Singh, V.P., Singh, P.K., Prasad, S.M., Dubey, N.K., Pandey, A.C., 2017b. Nitric oxide ameliorates zinc oxide nanoparticles phytotoxicity in wheat seedlings: implication of the ascorbate–glutathione cycle. Front. Plant Sci. 8.

Tripathi, D.K., Singh, S., Singh, S., Srivastava, P.K., Singh, V.P., Singh, S., Prasad, S.M., Singh, P.K., Dubey, N.K., Pandey, A.C., Chauhan, D.K., 2017c. Nitric oxide alleviates silver nanoparticles (AgNps)-induced phytotoxicity in *Pisum sativum* seedlings. Plant Physiol. Biochem. 110, 167–177.

Tripathi, D.K., Tripathi, A., Shweta, S.S., Singh, Y., Vishwakarma, K., Yadav, G., Sharma, S., Singh, V.K., Mishra, R.K., Upadhyay, R.G., Dubey, N.K., 2017d. Uptake, accumulation and toxicity of silver nanoparticle in autotrophic plants, and heterotrophic microbes: a concentric review. Front. Microbiol. 8.

Tripathi, D.K., Singh, S., Singh, V.P., Prasad, S.M., Dubey, N.K., Chauhan, D.K., 2017e. Silicon nanoparticles more effectively alleviated UV-B stress than silicon in wheat (*Triticum aestivum*) seedlings. Plant Physiol. Biochem. 110, 70–81.

Walsh, A.S., Yin, H., Erben, C.M., Wood, M.J., Turberfield, A.J., 2011. DNA cage delivery to mammalian cells. ACS Nano 5 (7), 5427–5432.

Wang, G., Murray, R.W., 2004. Controlled assembly of monolayer-protected gold clusters by dissolved DNA. Nano Lett. 4 (1), 95–101.

Wei, B., Dai, M., Yin, P., 2012. Complex shapes self-assembled from single-stranded DNA tiles. Nature 485 (7400), 623–626.

Wickham, S.F.J., Bath, J., Katsuda, Y., Endo, M., Hidaka, K., Sugiyama, H., Turberfield, A.J., 2012. A DNA-based molecular motor that can navigate a network of tracks. Nat. Nanotechnol. 7 (3), 169–173.

Williams, B.A.R., Lund, K., Liu, Y., Yan, H., Chaput, J.C., 2007. Self-assembled peptide nanoarrays: an approach to studying protein-protein interactions. Angew. Chem. 119 (17), 3111–3114.

Wooddell, C.I., Rozema, D.B., Hossbach, M., John, M., Hamilton, H.L., Chu, Q., Hegge, J.O., Klein, J.J., Wakefield, D.H., Oropeza, C.E., Deckert, J., 2013. Hepatocyte-targeted RNAi therapeutics for the treatment of chronic hepatitis B virus infection. Mol. Ther. 21 (5), 973–985.

Wu, Q., Christensen, L.A., Legerski, R.J., Vasquez, K.M., 2005. Mismatch repair participates in error-free processing of DNA interstrand crosslinks in human cells. EMBO Rep. 6 (6), 551–557.

Yan, H., Zhang, X., Shen, Z., Seeman, N.C., 2002. A robust DNA mechanical device controlled by hybridization topology. Nature 415 (6867), 62–65.

Yan, H., Park, S.H., Finkelstein, G., Reif, J.H., LaBean, T.H., 2003. DNA-templated self-assembly of protein arrays and highly conductive nanowires. Science 301 (5641), 1882–1884.

Young, L., Dong, Q., 2004. Two-step total gene synthesis method. Nucleic Acids Res. 32 (7), e59.

Yurke, B., Turberfield, A.J., Mills, A.P., Simmel, F.C., Neumann, J.L., 2000. A DNA-fuelled molecular machine made of DNA. Nature 406 (6796), 605–608.

Zahid, M., Kim, B., Hussain, R., Amin, R., Park, S.H., 2013. DNA nanotechnology: a future perspective. Nanoscale Res. Lett. 8 (1), 1.

Zhang, C., Su, M., He, Y., Zhao, X., Fang, P.A., Ribbe, A.E., Jiang, W., Mao, C., 2008. Conformational flexibility facilitates self-assembly of complex DNA nanostructures. Proc. Natl. Acad. Sci. U.S.A. 105 (31), 10665–10669.

Zhang, F., Nangreave, J., Liu, Y., Yan, H., 2014a. Structural DNA nanotechnology: state of the art and future perspective. J. Am. Chem. Soc. 136 (32), 11198–11211.

Zhang, Q., Jiang, Q., Li, N., Dai, L., Liu, Q., Song, L., Wang, J., Li, Y., Tian, J., Ding, B., Du, Y., 2014b. DNA origami as an *in vivo* drug delivery vehicle for cancer therapy. ACS Nano 8 (7), 6633–6643.

Zhao, Y.X., Shaw, A., Zeng, X., Benson, E., Nyström, A.M., Högberg, B., 2012. DNA origami delivery system for cancer therapy with tunable release properties. ACS Nano 6 (10), 8684–8691.

Zhu, G., Zheng, J., Song, E., Donovan, M., Zhang, K., Liu, C., Tan, W., 2013. Self-assembled, aptamer-tethered DNA nanotrains for targeted transport of molecular drugs in cancer theranostics. Proc. Natl. Acad. Sci. U.S.A. 110 (20), 7998–8003.

Zou, W., 2005. Immunosuppressive networks in the tumor environment and their therapeutic relevance. Nat. Rev. Cancer 5 (4), 263–274.

CHAPTER

5

Plant Response to Engineered Nanoparticles

Uday Burman, Praveen Kumar
ICAR-Central Arid Zone Research Institute, Jodhpur, India

5.1 INTRODUCTION

The term "nano" means "dwarf" and particles with one dimension less than 100 nm qualify to be nanoparticles. Nanoparticles have always existed in our environment from both natural and anthropogenic sources in all domains, including soil (as clay), water (as colloids), and air (as ultrafine particles), clearly demonstrating their ubiquitous nature. The production of various types of metal nanoparticles primarily used for commercial purposes has been reported to be more than 10,000 tonnes (Hendren et al., 2011; Piccinno et al., 2012). They are being widely used in consumer products and hence engineered in abundance. According to an estimate, 63%–91% of engineered nanoparticles produced globally end up in landfills, 8%–28% in soils, 0.4%–7% in water bodies, and 0.1%–1.5% in the atmosphere. The magnitude of potential engineered nanoparticle mass released to the environment in India, however, remains obscure (Sharma, 2015). Therefore concerns over possible adverse effects on plants and animals are raised especially for those nanoparticles that are likely to find their way into the environment. These nanoparticles are more likely to influence growth and metabolism of plants because they are sessile. However, any toxic effect on plants is transferred to other plants in the food chain and web and ultimately influences the whole ecosystem.

The response of plants to nanomaterials varies with the type of plant species, their growth stages, and the nature of nanomaterials. Engineered nanoparticles exhibit various characters including size, shape, and charge that influence their functionality. Additionally, settings and dissolute agglomeration in aqueous medium are also important though difficult to measure accurately. Toxicity could be expressed as depression in seed germination and root elongation, and reduction in biomass accumulation. Though some attempts have been made to elucidate the detoxification mechanism (Tripathi et al., 2017a) all aspects are not fully understood and many issues related to anatomical and histoenzymological reactions need to be studied in more detail to elucidate the sequence of events involved.

Nanomaterials in Plants, Algae, and Microorganisms
http://dx.doi.org/10.1016/B978-0-12-811487-2.00005-0

5.2 SIZE IS NOT THE ONLY CRITERION

Nanoparticles can be made by a top-down approach from a variety of macro-sized particles, and their actions depend on chemical composition and the size and/or shape of the particles as well as their chemical characteristics. Focus has been on correlating the properties of nanomaterials, such as size, shape, chemical functionality, surface charge, and composition with biomolecular signaling, biological kinetics, transportation, and toxicity. Correlation studies that were helpful is assigning a particular biological response to a definitive characteristic by systematically eliminating other parameters one by one are still meager.

Reactive oxygen species (ROS) activity, reflecting adverse biological effects, was highest for 30nm TiO_2 particles and as the size decreased to 10nm it decreased progressively but remained constant for particles smaller than 10nm (Jiang et al., 2008). However, root growth of radish and rapeseed incubated in nano-Zn (35nm) suspension was significantly inhibited; however, such an inhibition was not detected when soaked in nano-ZnO (15–25nm) suspension probably because of the selective permeability of the seed coat. Results of the research as mentioned earlier do not throw light on the mode of action. Thus there is a need for scientific debate supported by experimental evidence that clearly brings out mass and total surface area or any other parameter as the key players responsible for an observed biological response to nanoparticle exposure.

Shape is also important because this influences delivery efficiency and consequently response pattern. Even for particles >100nm size, rods showed higher uptake than spheres, cylinders, or cubes (Gratton et al., 2008), but for particles <100nm in size, spheres reflected increased uptake than rods because they experienced less interaction with cell surface receptors because of the short axis compared to the long axis of the rods (Chithrani and Chan, 2007). Cytotoxicity is altered when nanoparticles interact differently because of the difference in characteristics of the proteins coated on them, especially when they are interacting with the same receptor. Pure alumina nanoparticles (13nm) without any modifications reduced root elongation in corn (*Zea mays*), cucumber (*Cucumis sativus*), soybean (*Glycine max*), and carrot (*Daucus carota*) plants. When nanoparticles were loaded with phenanthrene their toxicity significantly decreased and consequently no adverse effects on the roots of plants was observed (Yang and Watts, 2005).

5.3 METHOD OF APPLICATION AND ENTRY OF NANOPARTICLES INTO PLANTS

Application of nanoparticles in soil invariably results in their adsorption and this consequently influences their bioavailability. Nanoparticles may also be applied to plants using injection or spraying methods. One of the main drawbacks of the injection method is that it cannot be used for large-scale field application. Attempts have also been made to concentrate the particles at the desired area in plants by using magnets. This opens an area of smart delivery to target locations. An observation of bioferrofluids along with metallic nanoparticles in the xylem vessels (Melendi et al., 2008) is one such attempt. Additionally, spraying seems to be a viable option because nanoparticles have been reported to penetrate plant tissue and reach regions distant from the point of application even though movements over short distances are favored. It is important to note here that most of the studies in the literature provide limited information because they have been performed at the germination stage when the vascular system is not fully developed.

TiO_2 nanoparticles of 30 mm size could not pass through the pore size (6.6 mm) of maize roots and hence did not register any entry (Asli and Neumann, 2009). In addition, the selective permeability determined by the pore diameter of the cell wall ranging from 5 to 20 nm as well as interaction of biomolecules encountered in the pathway are also important. However, once the nanoparticles have entered the cell wall it also exercises a certain amount of specificity as well as regulating the entry on the basis of the size of their pores (Fleischer et al., 1999; Navarro et al., 2008). After entering they may move apoplastically and symplastically. Nanoparticles come in contact with the plant cell surface where they influence the physical and chemical nature of the plant organs. Additionally, they may also influence subcellular organization. Aggregation of ZnO nanoparticles at the root surface has been observed in sequence through scanning electron microscopy (Lin and Xing, 2008). Initially, they aggregate in adjacent parenchyma cells reflecting cell-to-cell movements. In another study, through transmission electron microscopy it was confirmed that ZnO nanoparticles could pass through the plant cell wall and were present in the roots of *Schoenoplectus tabernaemontani*, but the potential for its translocation to the shoot was limited (Zhang et al., 2015). The nanoparticles are internalized by wrapping a membrane around them to form a closed vesicular structure primarily because of a localized decrease in Gibb's free energy (Gao et al., 2005). This may be caused by interaction between ligands on nanoparticles and receptors on the cell surface wherein shortage of either the receptor or the ligands does not result in optimum endocytosis (Yuan et al., 2010). The net accumulation of nanoparticles in the cell shall always be the sum total effect of endocytosis and exocytosis, the two processes that have been reported to be influenced by both size and shape (Chithrani and Chan, 2007).

Such a movement from parenchymatous to vascular tissue has been reported for TiO_2 nanoparticles in wheat (Larue et al., 2011). Entry at the junction point of lateral roots to finally reach xylem through the cortex and pericycle (Dietz and Herth, 2011) is also possible. Such cell-to-cell transport may be through plasmodesmata (Rico et al., 2011). An organic moiety attached to nanoparticles often facilitates their penetration and subsequent internalization into cells. Smaller particles are reported to enter plant cells easily, whereas larger nanoparticles are usually unable to enter cells and cannot consequently influence cell metabolism (Moore, 2006; Verano-Braga et al., 2014). However, the possibility of entry of these large nanoparticles exists from hydathodes or flower stigmas.

Pumpkin plants exposed to nanoparticles of mixed size allowed differential movement depending on their size. At cells close to the application point, nanoparticles of mixed size were observed, while at a distance far from the application point particles of more or less homogeneous size below 46 mm were recorded (Corredor et al., 2009). This clearly suggests the role of an appropriate size for long-distance transport within plants. The net accumulation of a specific shape of nanoparticles inside the cell also depends on the net uptake (facilitated through endocytosis) and exocytosis. Furthermore, the time taken and cellular energy required to wrap the particles for either endo- or exocytosis depends upon the length-to-width aspect ratio of the nanoparticle.

Furthermore, nanoparticles that are positively charged are taken up at a faster rate than neutral or negatively charged ones because membranes are invariably slightly negatively charged (Thorek and Tsourkas, 2008). This is probably caused by strong electrostatic attraction between the biomembrane and positively charged nanoparticle. Once in the cytosol, nanoparticles may disrupt mitochondrial function resulting in the production of ROS and consequent activation of the cascade of oxidative stress-mediated reactions (AshaRani et al., 2008). Nanoparticles may also either remain in the cytosol or may be exocytosed. In the former case, upon mitosis they are also distributed within daughter cells (Rees et al., 2011).

5.4 BIOTRANSFORMATION OF NANOPARTICLES IN PLANTS

No biotransformation was observed in the case of nanoparticles of Fe_3O_4 in pumpkin (Zhu et al., 2008) and CeO_2 in soybean (Lopez-Moreno et al., 2010). However, some studies revealed that nanoparticles are transformed in/on the root surface, as in the case of Zn in mesquite (Rico et al., 2011), Ni in mesquite (Parsons et al., 2010), Ag in zucchini (Stampoulis et al., 2009), Al in ryegrass, and Cu in mungbean and wheat, but no test was done to determine whether Ag, Al, or Cu in the plant existed as nanoparticles or ions; thus no definite conclusion could be made based on the reported studies. Elaborate studies are required to understand the mechanism of ionization of nanoparticles and the factors influencing them. In the case of ZnO nanoparticles, however, transformation probably occurs at the root surface where they were present inside the soybean plants at Zn^{2+} oxidation, even though ZnO nanoparticles were not detected inside the plant, and also the release of Zn^{2+} ions was very low from ZnO nanoparticles in aqueous hydroponic solution.

The uptake and translocation of nanoparticles across root cells that also involve both active and passive transport processes depend on the characteristics of moiety and plant species. Adherence of nanoparticles sprayed onto the leaf surface and their stability depend on leaf morphology and the presence of trichomes as well as the nature and quantity of waxy coatings on the leaf. The presence of thin water film further improves the hydraulic connection between the leaf surface and its interiors. This coupled with microorganisms at the leaf surface provides a favorable phyllosphere for facilitating absorption and subsequent transformation of nanoparticles. Entry could thus be through either a cuticular or stomatal path, as also recognized for Ag nanoparticles in lettuce leaves (Larue et al., 2014).

5.5 EFFECTS OF NANOPARTICLES

5.5.1 On Legume–Rhizobium Symbiosis

Exposure to TiO_2 nanoparticles adversely influences interaction between pea and rhizobium leading to poor nodule development and N_2 fixation that could be caused by alteration in polysaccharide composition of the cell wall leading to change in host response. This reflects the possible negative potential of TiO_2 nanoparticles with respect to rhizobium–legume interaction (Fan et al., 2014). In our laboratory, too, nitrogen fixation was differentially influenced depending on the concentration of ZnO nanoparticles: 10 ppm was deleterious, while 1.5 ppm improved in clusterbean, cowpea, and green gram. In mothbean, however, either concentration was deleterious for nitrogen fixation. Additionally, an interactive effect of concentration and duration of exposure time was also observed in green gram (Kumar et al., 2015).

Nanoparticle-induced changes in nitrogenase (N_2ase) may be caused by alterations in conformity of tertiary and quaternary configuration of the protein, mainly as a result of the pattern and intensity of interaction because of the presence of various electrostatic, hydrophobic, and hydrophilic bonds (Mahmoudi et al., 2011). This also accounts for the observed variation in response of nanoparticles in different plant species because the proteins of different species also differ in their physicochemical characteristics, which play a key role in the expression of the response as discussed earlier. Thus, this interaction could be either beneficial or harmful. Nanoparticle-induced changes in the microenvironment of N_2ase may therefore be caused by the interaction of protein and/or protein complex that depends on the surface and physicochemical characteristics of both particle and protein (Lynch and Dawson, 2008; Saptarshi et al., 2013).

5.5.2 On Growth

There have been publications on the effects of engineered nanomaterials on higher plants wherein both positive and negative effects were reported (Monica and Cremonini, 2009). Alumina nanoparticles that were found to inhibit root elongation in several plants including corn, cucumber, soybean, cabbage, and carrot at relatively low dosage was one of the first reports of negative effects of nanoparticles (Yang and Watts, 2005). In addition, Lee et al. (2013) also reported that the biomass of buckwheat seedlings was significantly reduced in response to ZnO nanoparticles compared to ZnO macroparticles at concentrations between 10 and 2000 mg/L. Retardation in growth and deformation of epidermal cells because of exposure to CuO nanoparticles has also been reported to be associated with changes in levels of indole acetic acid in wheat (Adams et al., 2017).

Some publications on ZnO nanoparticles also showed both positive and negative effects (Lin and Xing, 2008; Priester et al., 2012; Prasad et al., 2012). ZnO nanoparticles negatively affected the developmental stages and reproduction of soybean plants in a soil microcosm (Yoon et al., 2014). Our results have also demonstrated that the effect of nano-ZnO could either be beneficial or toxic and this depends on plant species, plant part, and concentration used (Burman et al., 2013a, 2013b; Kumar et al., 2015).

Seed germination and root growth of zucchini seeds were adversely affected when grown in hydroponic solution containing ZnO nanoparticles (Stampoulis et al., 2009), whereas the seed germination of ryegrass and corn was inhibited by nano-Zn (35 nm) and nano-ZnO (15–25 nm), respectively (Lin and Xing, 2007). It is important to note here that because seed germination guarantees plant survival, any variation in the effect of nanoparticles on this process shall have a bearing on the species composition in a habitat and hence their relative abundance and the ecosystem services dependent on them. In radish and rape, application of nano-Zn suspension inhibited root growth but when seeds were soaked such an adverse effect was not observed because of selective permeability of the seed coat.

No significant effect on seed emergence of wheat, radish, and veltch was observed upon exposure to either nano-ZnO or bulk ZnO probably because of its lower dissolution/solubility (Lopez-Moreno et al., 2010). However, upon application of 1000 ppm $ZnCl_2$, increased Zn ion permeability through the seed coat of veltch compared to that of wheat and radish also accounted for inhibition of germination. Additionally, enhanced dissolution because of interaction with root exudates may also result in toxicity as well as disruption of structural and functional characteristics of roots in the case of *Brassica napus* (Kouhi et al., 2015). Perusal of the literature reveals a dose-dependent response wherein the concentration that registers inhibitory response varies with the plant species and the chemical identity of the moiety (Lin and Xing, 2007–2008; Wang et al., 2012; Yang et al., 2015): 100 and 50mg/L nano-ZnO were nontoxic for root growth in maize and rice, respectively. Application at higher doses reflected toxicity symptoms and a 50% inhibition (IC_{50}) was recorded at 1500 and 40mg/L in maize and rice, respectively. However, IC_{50} for nano-CuO was much lower at 60mg/L, while inhibitory symptoms were observed only at 1000mg/L with Al_2O_3 in maize. The size of seed also plays an important role in the expression of nanoparticle response. Small-seeded species such as rapeseed and lettuce are likely to reflect an increased effect because of their large surface/volume ratio compared to wheat. However, seed size is not the only deciding factor for phytotoxicity. The chemical nature of the seed coat that imparts selective permeability and consequently differential entry to various nanoparticles also influences a response. Pore size on the surface of the seed coat could also be responsible for inhibition of root growth after seed soaking.

The role of solubility in reflecting the nanoparticle-mediated response has also been indicated in another study wherein 10ppm was more effective than 1.5ppm; also $ZnSO_4$ was more effective than bulk ZnO followed by nano-ZnO (Burman et al., 2013b). This was probably because of the difference in solubility and consequent bioavailability of zinc because $ZnSO_4$ readily dissociated to Zn^{++}, while the solubility of bulk ZnO in water was very low (Mortvedt, 1992). The small nanoparticle size of zinc oxide is likely to have a higher dissolution rate.

Nano-TiO_2 treatment, up to 4%, has been reported to accelerate the germination of the aged seeds of spinach in comparison to bulk TiO_2 (Zheng et al., 2005). TiO_2 nanoparticles in low concentration (10ppm) promoted seed germination of wheat compared to bulk-sized TiO_2, but at 100ppm they were inhibitory (Hassan et al., 2012). In fennel, 40ppm TiO_2 nanoparticles resulted in reduction in mean germination time by 31.8% and also positively influenced other indices such as germination vigor, vigor index, and mean daily germination (Hassan et al., 2013). These responses could be caused by the influence on permeability of the seed coat to water and subsequent uptake, as also observed in tomato when exposed to carbon nanotubes (CNTs) (Khodakovskaya et al., 2009).

Phytotoxic responses of Al and Cu oxide nanoparticles were also found to be dependent on concentration in the case of rice and maize (Yang et al., 2015). In comparison to the control, ZnO nanoparticle treatments at 200 and 300mg/L reduced *Arabidopsis* growth by about 20% and 80%, respectively (Lee et al., 2010). The higher dose of ZnO nanoparticles in this case resulted in drastic reduction of chlorophyll *a*, *b* and carotenoids as well as net photosynthesis and transpiration rate. Additionally, in another study on *Arabidopsis*, deleterious response of ZnO nanoparticles on growth was attributed to the induction of oxidative stresses (Wang et al., 2016).

In other studies, laboratory-synthesized nanoparticles (0.1%, w/v) showed substantial growth promontory effects on chickpea seed germination, seedling length, and fresh and dry weight. Nanoparticle-treated seedlings also showed a remarkable increase of chlorophyll in chickpea (Sathiyanarayanan and Banu, 2016). It has been reported that nanoparticle treatment improves the photosynthetic rate by improving the synthesis of photosynthetic pigments (Xie et al., 2012; Priyadarshini et al., 2012; Siddiqui and Al-Whaibi, 2014). The reason for increased photosynthetic pigment might also be because of enhanced electron transport rates, as suggested by Giraldo et al. (2014).

5.5.3 On Metabolites

However, to understand the molecular mechanisms of plant response to nanoparticles, links need to be established between up- and down regulation of genes with cell metabolism. Relative dominance of the region of the genome results in regulation of the concentration of specific protein, which in turn may be responsible for plants' adaption to the nanoparticles' exposure.

Perusal of the literature related to the response of plants to nanoparticle exposure reveals chemical interactions that involve disturbance of ion cell membrane transport activity (Auffan et al., 2008; Nel et al., 2006), oxidative damage (Foyer et al., 2002), and lipid peroxidation (Kamat et al., 2000) in most cases. Toxicity may also be caused by the characteristic physicochemical properties and subsequent interaction at the cellular level rather than by the ion released, as in the case of *Eruva sativa* roots exposed to Ag nanoparticles (Vannini et al., 2013). Often the toxic effects on photosynthesis, lipid peroxidation, and level of antioxidants could be partially alleviated by treating nanoparticle-exposed plants to nitric oxide, as in the case of pea treated with Ag nanoparticles (Tripathi et al., 2017b) and wheat treated with ZnO nanoparticles (Tripathi et al., 2017c). This provides a better insight into the possible mode of action of some nanoparticles on plants.

The positive effect of nano-TiO_2 on spinach growth is a perfect example reflecting regulation at multiple points of different pathways (Zheng et al., 2005; Hong et al., 2005; Yang et al., 2006; Gao et al., 2006). The positive effect of TiO_2 nanoparticles on spinach growth could be because of ROS generation that may be responsible for antimicrobial activity leading to improved growth and development of spinach. This photosterilization by nano-TiO_2 could be responsible for improving the growth and development of plants. Nano-TiO_2 could also increase light absorbance, accelerate the transport and transformation of light energy, protect chloroplasts from aging, and prolong the photosynthetic time of the chloroplasts. Nano-TiO_2 might also enter the chloroplast and accelerate its oxidation–reduction reactions leading to efficient electron transport.

There are reports that in some cases the metabolism could be influenced by magnetic rather than chemical characteristics, as in the case of Fe nanoparticles that are likely to induce both phenotypic and genotypic effects, subsequently leading to change in productivity. Paramagnetic properties of Fe nanoparticles lodged at any place in the matrix of the biomembrane could totally influence transmembrane ion flow and could also possibly influence the redox reaction of photosystems that ultimately affects photosynthetic yield. Additionally, subjecting these nanoparticles to an electromagnetic field could also result in a stimulatory effect on nucleic acid biosynthesis. In tissues, temperature may rise as a consequence of electromagnetic energy adsorbed by metallic nanoparticles.

The technology of magnetic nanoparticles could, through controlled release and target delivery of agrochemicals, be used for improving plant disease resistance, increasing efficient nutrient utilization, and enhancing plant growth. Magnetic nanoparticles could be used as capping agents for porous carriers of various chemicals. In this case the localization and uncapping process carried out using external magnets could subsequently release the payload at the target site. Thus, these magnetic particles could be used to concentrate moiety in a locality by magnets, in addition to encapsulation that provides better biocompatibility and larger absorption surface that in turn facilitates "smart" delivery of nanoparticles. Concentration of bioferrofluids in *Cucurbita pepo* is one such example (Corredor et al., 2009).

Roots of *Brassica nigra* showed an 80% increase in 2,2-diphenyl-1-picrylhydrazyl radical scavenging activity, total antioxidant, and reducing power potential in the presence of 10 mg/L CuO nanoparticles in the media. Furthermore, though nonenzymatic antioxidant molecules, phenols, and flavonoids were also increased, it was concentration dependent (Zafar et al., 2016). Further biomass accumulation in the ZnO nanoparticle-treated chickpea seedlings compared to those treated with bulk ZnO could be attributed to low ROS levels that resulted in less lipid peroxidation, as also evident from the low content of malondialdehyde (Burman et al., 2013b).

Thus the mechanism of action in the case of exposure of plants to ZnO nanoparticles could be through (1) creation of new holes and subsequent increase in permeability, (2) the result of savings of plant energy otherwise utilized by plants in scavenging free radicals (ROS), or (3) part of the enhanced C accumulated in the shoot having enhanced photosynthesis possibly because of protection of the cellular membrane from the UV radiation, as evident from less malondialdehyde content in the treated plants in our experiments on pearl millet and clusterbean (Burman et al., 2013a). However, reasons for higher C fixation in the presence of nano-ZnO are not clear. Two indirect effects of ZnO on plants may influence photosynthesis, i.e.: (1) high biocompatibility and fast electron transfer kinetics of nano-ZnO (Kumar and Shen-Ming, 2008), which may increase the rate of some biochemical reactions, and (2) protection of the plant membrane from damage by UV radiation-induced lipid peroxidation. In our experiment, too, malondialdehyde content, an indicator of lipid peroxidation, was 4.3% and 10.7% less in nano-ZnO-treated pearl millet and clusterbean leaves, respectively, than the corresponding controls (Burman et al., 2013a). In contrast, deleterious effects could be caused by aggregation of nanoparticles resulting in the blocking of pores and channels. Lin and Xing (2007) also reported severe damage to epidermal and cortical cells and even impairment of the endodermal and vascular cells of the ryegrass by nano-Zn, leading to inhibition of root growth.

5.5.4 On Different Enzymes

Improvement of spinach growth and increase of the chlorophyll and protein contents by nano-TiO_2 treatment are not only related to the promotion of photosynthesis as detailed in the preceding paragraphs but may be caused by the enhancement of nitrogen metabolism, otherwise synthesis of chlorophyll, amino acid, protein (enzyme), and nucleic acid would have been suffocated by the lack of nitrogen. Thus nano-TiO_2 improved plant growth by enhanced nitrogen metabolism. Nano-anatase TiO_2 treatment could also accelerate the enhanced transformation of ammonium nitrogen to organic nitrogen (e.g., protein and chlorophyll) in nano-anatase TiO_2-treated spinach plants that resulted in improved growth. This was associated with high nitrate reductase, glutamate dehydrogenase, glutamine synthase, and glutamate pyruvate transaminase activities.

In many studies, besides the increase in enzyme activity including α,β-amylase, ascorbate peroxidase, peroxidase, and catalase, higher malondialdehyde content indicates that a better protective mechanism might exist under the application of nanoparticles. A similar protective mechanism is reported also to result in better fertilizer and water utilization as well as an increase in nitrate reductase activity in soybean because of the combined application of nano-sized SiO_2 and TiO_2 (Lu et al., 2002).

Adsorption of enzymatic proteins to metallic nanoparticles coupled with the potential of specific conjugate formation because of surface specificity may be responsible for the observed increase in activity. However, further research is required to correlate these changes with the concomitant enhancement of expression of messenger RNA of these enzymes. Furthermore, when these nanoparticles and biomolecules reflect the potential to self-conjugate to form a hybrid moiety they can be exploited to construct nanodevices and nanosensors for use in agriculture.

5.6 EFFECT ON ABIOTIC AND BIOTIC STRESS

Under drought, soybean germination percentage and rate over control increased after application of 1 g/L of nano-ZnO. The treatments were effective in mobilization of a seed reservoir toward growth (Sedghi et al., 2013). TiO_2 nanoparticles at 0.02% also increased almost all agronomic traits including gluten and starch content in wheat subjected to water stress (Jaberzadeh et al., 2013). In contrast, under flooding Ag nanoparticles reflected a size-specific response wherein 15 nm facilitated soybean growth compared to the larger (50–50 nm) and smaller (2 nm) nanoparticles. This was caused by change in content and pattern of protein under Ag nanoparticle exposure, which was related to stress, signaling, and cell metabolism. Watson et al. (2015) reported a higher soluble Zn content in the acidic soil, as well as higher concentration in wheat shoots, compared to the alkaline soil upon amendment with ZnO nanoparticles. Nanoparticles of the same metal may behave differently in different environments, e.g., amendment of alkaline soil with 500 mg/kg resulted in lower Zn content in wheat plant compared to those grown in acidic soil. Furthermore, the former condition led to increased lateral root production, while those in the acidic soil recorded decreased root growth.

Silicon, the second most abundant mineral (Epstein, 1999), reflects a positive effect on the growth of plants when they are subjected to abiotic and biotic stresses. On the other hand, 2 mM nano-Si under 50 mM NaCl showed a reduction in the germination properties of tomato, while 1 mM nanoSi under 25 mM showed enhancement of germination characteristics such as germination rate, root length, and dry weight (Haghighi et al., 2012). Siddiqui and Whaibi (2014) also reported improvement in germination and seedling vigor in tomato by application of 8 g/L SiO_2 of 12 nm size. Interestingly, nanoscale silica at 15 kg/ha reflected a positive response in maize compared to bulk silica in terms of protein, chlorophyll, and phenol content (Suriyaprabha et al., 2012). Si nanoparticles have also been reported to protect wheat seedlings from UV-B radiation stress by maintaining high levels of antioxidants (Tripathi et al., 2017d).

One study reveals that adverse effects of salt stress on cultivated crops could be reduced by CeO_2 nanoparticle application probably by both regulation of protein synthesis and photosynthetic mechanism (Rossi et al., 2016).

Surface-modified hydrophobic nanosilica has been successfully used to control a range of agricultural insect pests (Barik et al., 2008; Rahman et al., 2009). Properly functionalized lipophilic nanosilica is absorbed into the cuticular lipids of insects because of physiosorption. This results in uncontrolled evaporation of water and consequent death by desiccation (Athanassiou et al., 2007; Mewis and Ulrichs, 2001).

The potential dual use of nanoscale amendments to both suppress disease and promote nutrient status is also being advocated (Giannousi et al., 2013). The foliar application of 100 mg/L inhibited development of powdery mildew in field-cultivated cucumber and pumpkin (Lamsal et al., 2011) and reflected a dose-dependent increase in fungal control up to 100 mg/L foliar spray; furthermore, the nanoparticle treatment was about 25% more effective than the corresponding bulk Ag amendment.

5.7 EFFECTS OF CARBON-BASED NANOMATERIALS

Carbon-based nanomaterials (CNTs) also reflected both positive and negative effects on plants. CNTs influenced plant growth but graphene nanocarbon structures did not. This clearly demonstrates the fact that size is not the only criterion of response. These nanomoieties differ in their ability to penetrate plant tissue and hence differ in their response. CNTs induce activation of genes (LeAqp2) responsible for regulation of activity of protein and movement of water through channels that further mellow the stress-related effects in tomato (Khodakovskaya et al., 2011). Multiwall CNTs (MWCNTs) in tomato also increased germination and plant growth by stimulating aquaporins and subsequently cell division and extension (Khodakovskaya et al., 2013). The interaction of MWCNTs with protein and polysaccharides generates a cascade of signals resulting in accumulation of compounds leading to cell wall thickening and subsequent growth. However, in other studies, MWCNTs have been reported to cause cell death by high oxidative stress and subsequent accumulation of ROS. Detachment of plasma membrane from cells was also reported to be responsible for cell death in this case.

5.8 NANOBIOTECHNOLOGY

The field of nanobiotechnology, a subfield of nanotechnology, has not emerged as a well-defined, single research discipline such as chemistry, physics, or biology. The development and growth of nanobiotechnology has been shaped by the convergence of many scientific and technological revolutions of the 20th century. Properly functionalized nanomaterials serve as a platform to transport large numbers of genes as well as chemicals that trigger gene expression in plants. Focus on the study of the plant–nanoparticle interface at the transcript level in most cases has been topical. Selective modulation of gene expression and nanofiber-induced delivery of genetic material because of surface modification by attachment of plasmid DNA manipulates cell biochemistry (McKnight et al., 2003, 2004). This technology could be upscaled for its commercial application in the delivery of important agrochemicals (pesticides, regulators, etc.) in addition to improvement in crop efficiency and environment monitoring.

The successful delivery and integration of plasmid DNA has been confirmed from gene expression. This process is similar to the microinjection method of gene delivery and is hence possible with plant cells in which the treated cells could be regenerated into whole plants that would express the introduced trait. It is possible to make DNA tethered on carbon nanofibers without allowing them to integrate into host genome and still allow some transcriptions of the tethered genes. Hence this technique has an additional advantage in that it disallows the transmission of modified traits to next generations. This differentiates it from the existing genetic engineering methods and a one-time modification of the cells is possible through this technique. A gene and its chemical inducer with its ends capped with Au nanoparticles could also be loaded with mesoporous silica nanoparticles to protect the molecules from leaching out. Uncapping of Au nanoparticles through an external energy source enabled release of chemicals and triggered gene expression in plants under controlled-release conditions.

5.9 PRACTICAL POSSIBILITIES AND THE WAY FORWARD

Nanotechnology that has been revolutionizing many important scientific fields, ranging from biology to medicine, provides a new interdisciplinary avenue into agriculture and food sciences by converging science and engineering. Nanoparticles, because of their high surface area, sorption capacity, and the possibility of controlled release at target sites in response to environmental and biological demand, are likely to be a part of future smart delivery systems. Furthermore, their capacity to modify gene expression and consequently biological pathways also gives them immense potential (Nair et al., 2010). "Environment-responsive" properties such as pH and oxygen concentration, coupled with external stimuli such as near-infrared light, probably generates heat that in turn excites the trigger mechanism for localized chemical/nutrient release (Agarwal et al., 2011; Wu et al., 2008). In today's era of widespread imbalance in nutrient availability, such an approach wherein nanoparticles of nutrients themselves or their encapsulation in polymeric nanoparticle shells or simply loading of nutrients (by adsorption) on nanoparticles could bring the world a step nearer to nutritional security. Protection of pesticides from UV-induced degradation by encapsulation inside a porous hollow silica shell, formation of nanocomposites to achieve regulated and timely release of fertilizer, and reduction in vascular blockage coupled with inhibition of microbial growth for the longer shelf-life of vase flowers such as gerbera are a few more examples of the potential use of nanoscience in the field of agriculture.

The last decade has witnessed an upsurge in publications related to engineered nanoparticles. Often, these publications are from laboratories that are frequently restricted to a "spray and observe" concept. Hence it is difficult to deduce any conceptual outline of the mode of action of the nanoparticle more so in the wake of lack of complete characterization of nanoparticles. Incomplete information about the characteristics of nanoparticles often leads to incorrect or very general conclusions because of dilution of information. We are currently unable to draw specific or general conclusions regarding the impact of size, shape, and surface chemistry-dependent interactions. This also does not throw any light on the relationship of different parameters with functionality. However, a general idea of the whole scenario right from entry in the plant to the possible responses may be drawn, as depicted in Fig. 5.1. Thus all relevant aspects that help to elucidate the mechanism involved in the response of plants to nanoparticle exposure needs to be studied step by step through well-executed fundamental studies that also draw their strength from the recent advancement in the technology and instrumentation involved. Such studies may help in designing and developing strategies involving the use of nanoparticles for agriculture. However, as the domain is closely linked to the food chain and web there is an urgent need to put in place proper regulations/legislation related to the use of nanomaterials, their confinement, and disposal, thereby creating awareness among all stakeholders for the biosafety and ethical issues involved in nanotechnology and its application.

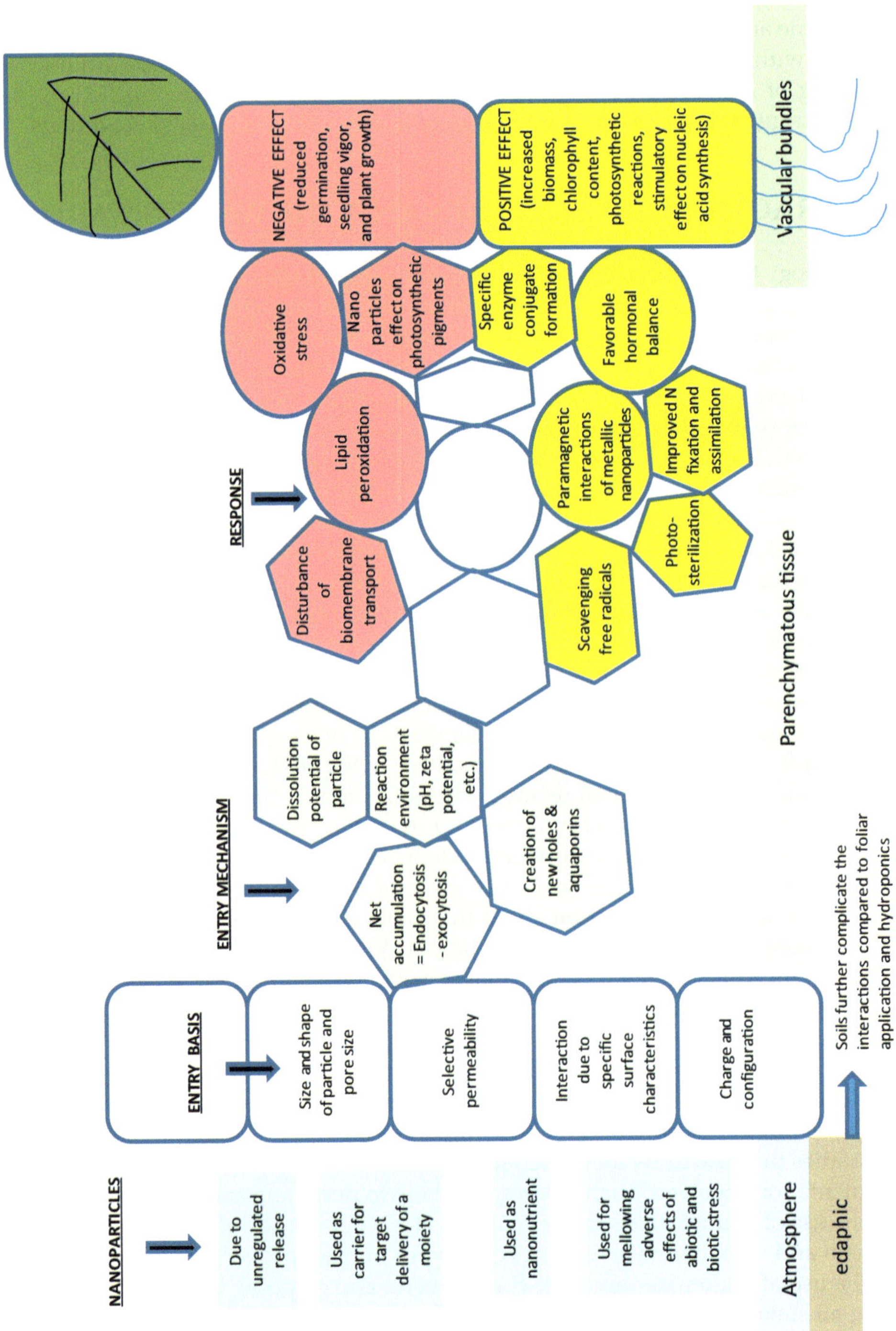

FIGURE 5.1 Schematic representation of different facets of interaction of nanoparticles and plants.

References

Adams, J., Wright, M., Wagner, H., Valiente, J., Britt, D., Anderson, A., 2017. Cu from dissolution of CuO nanoparticles signals changes in root morphology. Plant Physiol. Biochem. 110, 108–117.

Agarwal, A., Mackey, M.A., El-Sayed, M.A., Bellamkonda, R.V., 2011. Remote triggered release of doxorubicin in tumors by synergistic application of thermo-sensitive liposomes and gold nanorods. ACS Nano 5, 4919–4926.

AshaRani, P.V., Low Kah Mun, G., Hande, M.P., Valiyaveettil, S., 2008. Cytotoxicity and genotoxicity of silver nanoparticles in human cells. ACS Nano 3, 279–290.

Asli, S., Neumann, M., 2009. Colloidal suspensions of clay or titanium dioxide nanoparticles can inhibit leaf growth and transpiration via physical effects on root water transport. Plant Cell Environ. 32, 577–584.

Athanassiou, C.G., Kavallieratos, N.G., Meletsis, C.M., 2007. Insecticidal effect of three diatomaceous earth formulations, applied alone or in combination, against three stored-product beetle species on wheat and maize. J. Stored Prod. Res. 43, 330–334.

Auffan, M., Achouak, W., Rose, J., Roncato, M.A., Chanéac, C., Waite, D.T., Masion, A., Woicik, J.C., Wiesner, M.R., Bottero, J.Y., 2008. Relation between the redox state of iron-based nanoparticles and their cytotoxicity toward *Escherichia coli*. Environ. Sci. Technol. 42, 6730–6735.

Barik, T.K., Sahu, B., Swain, V., 2008. Nanosilica-from medicine to pest control. Parasitol. Res. 103, 253–258.

Burman, U., Tarafdar, J.C., Kaul, R.K., Saini, M., Kumar, K., Praveen-Kumar, 2013a. Changes in carbon partitioning in pearl millet (*Pennisetum glaucum*) and clusterbean (*Cyamopsis tetragonoloba*) in response to ZnO nanoparticle application. Indian J. Agri. Sci. 83, 352–354.

Burman, U., Saini, M., Kumar, P., 2013b. Effect of zinc oxide nanoparticles on growth and antioxidant system of chickpea seedlings. Toxicol. Environ. Chem. 95, 605–612.

Chithrani, B.D., Chan, W.C., 2007. Elucidating the mechanism of cellular uptake and removal of protein coated gold nanoparticles of different sizes and shapes. Nano Lett. 7, 1542–1550.

Corredor, E., Testillano, P.S., Coronado, M.J., Melendi, P.G., Pacheco, R.F., Marquina, C., Ibarra, M.R., Fuente, J.M.D.L., Rubiales, D., Luque, A.P., Risueño, M.C., 2009. Nanoparticle penetration and transport in living pumpkin plants: in situ subcellular identification. BMC Plant Biol. 9 (45). http://dx.doi.org/10.1186/1471-2229-9-45.

Dietz, K.J., Herth, S., 2011. Plant nanotoxicology. Trends Plant Sci. 16, 582–589.

Epstein, E., 1999. Silicon. Annu. Rev. Plant Physiol. Plant Mol. Biol. 50, 641–664.

Fan, R., Huang, Y.C., Michael, A., Grusak, C.P., Huang, D., Sherrier, J., 2014. Effects of nano-TiO_2 on the agronomically-relevant *Rhizobium*–legume symbiosis. Sci. Total Environ. 466, 503–512. http://dx.doi.org/10.1016/j.scitotenv.2013.07.032.

Fleischer, A., O'Neill, M.A., Ehwald, R., 1999. The pore size of non-graminaceous plant cell walls is rapidly decreased by borate ester cross-linking of the pectic polysaccharide rhamnogalacturonan II. Plant Physiol. 121, 829–838.

Foyer, S., Crowley, C., Smaihi, M., Bonfils, C., Erlanger, B.F., Seta, P., Larroque, C., 2002. Cellular localisation of a water-soluble fullerene derivative. Biochem. Biophys. Res. Commun. 294, 116–119.

Gao, F., Hong, F., Liu, C., Zhang, L., Su, M., Wu, X., Yang, F., Wu, C., Yang, P., 2006. Mechanism of nano-anatase TiO_2 on promoting photosynthetic carbon reaction of spinach. Biol. Trace Elem. Res. 111, 239–253.

Gao, H., Shi, W., Freund, L.B., 2005. Mechanics of receptor-mediated endocytosis. Proc. Natl. Acad. Sci. U.S.A. 102, 9469–9474.

Giannousi, K., Avramidis, I., Dendrinou-Samara, C., 2013. Synthesis, characterization and evaluation of copper based nanoparticles as agrochemicals against *Phytophthora infestans*. RSC Adv. 3, 21743–21752.

Giraldo, J.P., Landry, M.P., Faltermeier, S.M., McNicholas, T.P., Iverson, N.M., Boghossian, A.A., Reuel, N.F., Hilmer, A.J., Sen, F., Brew, J.A., Strano, M.S., 2014. Plant nanobionics approach to augment photosynthesis and biochemical sensing. Nat. Mater. 13, 400–408. http://dx.doi.org/10.1038/nmat3890.

Gratton, S.E., Ropp, P.A., Pohlhaus, P.D., Luft, J.C., Madden, V.J., 2008. The effect of particle design on cellular internalization pathways. Proc. Natl. Acad. Sci. U.S.A. 105, 11613–11618.

Haghighi, M., Afifipour, Z., Mozafarian, M., 2012. The effect of N-Si on tomato seed germination under salinity levels. J. Biol. Environ. Sci. 6, 87–90.

Hassan, F., Kamali, M., Jafari, L., Moghaddam, P.R., 2013. Phytotoxicity and stimulatory impacts of nanosized and bulk titanium dioxide on fennel (*Foeniculum vulgare* Mill). Chemosphere 91, 506–511.

Hassan, F., Moghaddam, P.R., Shahtahmassebi, N., Fotovat, A., 2012. Impact of bulk and nanosized titanium dioxide (TiO_2) on wheat seed germination and seedling growth. Biol. Trace Elem. Res. 146, 101–106.

Hendren, C.O., Mesnard, X., Dro¨ge, J., Weisner, M.R., 2011. Estimating production data for five engineered nanomaterials as a basis for exposure assessment. Environ. Sci. Technol. 45, 2562–2569.

Hong, F.S., Zhou, J., Liu, C., 2005. Effect of nano-TiO_2 on photochemical reaction of chloroplasts of spinach. Biol. Trace Elem. Res. 105, 269–280.

Jaberzadeh, A., Moaveni, P., TohidiMoghadam, H.R., Zahedi, H., 2013. Influence of bulk and nanoparticles titanium foliar application on some agronomic traits, seed gluten and starch contents of wheat subjected to water deficit stress. Not. Bot. Horti Agrobot. 41, 201–207.

Jiang, J., Oberdoerster, G., Elder, A., Gelein, R., Mercer, P., Biswas, P., 2008. Does nanoparticle activity depend upon size and crystal phase? Nanotoxicology 2, 33–42.

Kamat, J.P., Devasagayam, T.P., Priyadarsini, K.I., Mohan, H., 2000. Reactive oxygen species mediated membrane damage induced by fullerene derivatives and its possible biological implications. Toxicology 155, 55–61.

Khodakovskaya, M., Dervishi, E., Mahmood, M., Xu, Y., Li, Z., Watanabe, F., Biris, A.S., 2009. Carbon nanotubes are able to penetrate plant seed coat and dramatically affect seed germination and plant growth. ACS Nano 3, 3221–3227.

Khodakovskaya, M.V., Khodakovskaya, M.V., de Silva, K., Nedosekin, D.A., Dervishi, E., Biris, A.S., Shashkov, E.V., Galanzha, E.I., Zharov, V.P., 2011. Complex genetic, photothermal and photoacoustic analysis of nanoparticle-plant interactions. Proc. Natl. Acad. Sci. U.S.A. 108, 1028–1033.

Khodakovskaya, M.V., Kim, B.S., Kim, J.N., Alimohammadi, M., Dervishi, E., Mustafa, T., Cernigla, C.E., 2013. Carbon nanotubes as plant growth regulators: effects on tomato growth, reproductive system, and soil microbial community. Small 9, 115–123.

Kouhi, S.M.M., Lahouti, M., Ganjeali, A., Entezari, M.H., 2015. Comparative effects of ZnO nanoparticles, ZnO bulk particles, and Zn^{2+} on *Brassica napus* after long-term exposure: changes in growth, biochemical compounds, antioxidant enzyme activities, and Zn bioaccumulation. Water Air Soil Pollut. 226, 364–374. http://dx.doi.org/10.1007/s11270-015-2628-7.

Kumar, P., Burman, U., Santra, P., 2015. Effect of nano-zinc oxide on nitrogenase activity in legumes: an interplay of concentration and exposure time. Int. Nano Lett. 5,191–198. http://dx.doi.org/10.1007/s40089-015-0155-6.

Kumar, S.A., Shen, M.C., 2008. Nano structured zinc oxide particles in chemically modified electrodes for biosensor applications. Anal. Lett. 41, 141–158.

Lamsal, K., Kim, S.W., Jung, J.H., Kim, Y.S., Kim, K.S., Lee, Y.S., 2011. Inhibition effects of silver nanoparticles against powdery mildews on cucumber and pumpkin. Mycobiology 39, 26–32.

Larue, C., Castillo-Michel, H., Sobanska, S., Cécillon, L., Bureau, S., Barthès, V., Ouerdane, L., Carrière, M.S.G., 2014. Foliar exposure of the crop *Lactuca sativato* silver nanoparticles: evidence for internalization and changes in Ag speciation. J. Hazard Mater. 264, 98–106.

Larue, C., Khodja, H., Herlin-Boime, N., Brisset, F., Flank, A.M., Fayard, B., Chaillou, S., Carrie're, M., 2011. Investigation of titanium dioxide nanoparticles toxicity and uptake by plants. J. Phys. Conf. Ser. 304, 012057.

Lee, C.W., Mahendra, S., Zodrow, K., Li, D., Tsai, Y.C., Braam, J., Alvarez, P.J.J., 2010. Developmental phytotoxicity of metal oxide nanoparticles to *Arabidopsis thaliana*. Environ. Toxicol. Chem. 29, 669–675.

Lee, S., Kim, S., Kim, S., Lee, I., 2013. Assessment of phytotoxicity of ZnO NPs on a medicinal plant, *Fagopyrum esculentum*. Environ. Sci. Pollut. Res. 20, 848–854.

Lin, D.H., Xing, B.S., 2007. Phytotoxicity of nanoparticles: inhibition of seed germination and root elongation. Environ. Pollut. 150, 243–250.

Lin, D., Xing, B., 2008. Root uptake and phytotoxicity of ZnO nanoparticles. Environ. Sci. Technol. 42, 5580–5585.

Lopez-Moreno, M., de la Rosa, G., Hernández-Viezcas, J., Castillo-Michel, H., Botez, C.E., Peralta-Videa, J.R., Gardea-Torresdey, J.L., 2010. Evidence of the differential biotransformation and genotoxicity of ZnO and CeO_2 nanoparticles on soybean (*Glycine max*) plants. Environ. Sci. Technol. 44, 7315–7320.

Lu, C., Zhang, C., Wen, J., Wu, G., Tao, M., 2002. Research of the effect of nanometer materials on germination and growth enhancement of *Glycine max* and its mechanism. Soybean Sci. 21 (3), 168–171.

Lynch, I., Dawson, K.A., 2008. Protein–nanoparticle interactions. Nano Today 3, 40–47.

Mahmoudi, M., Lynch, I., Ejtehadi, M.R., Monopoli, M.P., Bombelli, F.B., Laurent, S., 2011. Protein- nanoparticle interactions:opportunities and challenges. Chem. Rev. 111, 5610–5637. http://dx.doi.org/10.1021/cr100440g.

McKnight, T.E., Melechko, A.V., Griffin, G.D., Guillorn, M.A., Merkulov, V.I., Serna, F., Hensley, D.K., Doktycz, M.J., Lowndes, D.H., Simpson, M.L., 2003. Intracellular integration of synthetic nanostructures with viable cells for controlled biochemical manipulation. Nanotechnology 14, 551–556.

McKnight, T.E., Melechko, A.V., Hensley, D.K., Mann, D.G.J., Griffin, G.D., Simpson, M.L., 2004. Tracking gene expression after DNA delivery using spatially indexed nanofiber arrays. Nano Lett. 4, 1213–1219.

Melendi, P.G.L., Ndez-Pacheco, R.F., Coronado, M.J., Corredor, E., Testillano, P.S., Risuen˘, O.M.C., Marquina, C., Ibarra, M.R., Rubiales, D., Pe´ Rez-De-Luque, A., 2008. Nanoparticles as smart treatment-delivery systems in plants: assessment of different techniques of microscopy for their visualization in plant tissues. Ann. Bot. 101, 187–195. http://dx.doi.org/10.1093/aob/mcm283.

Mewis, I., Ulrichs, C., 2001. Action of amorphous diatomaceous earth against different stages of the stored product pests *Tribolium confusum, Tenebrio molitor, Sitophilus granaries* and *Plodiainter punctella*. J. Stored Prod. Res. 37 (2), 153–164.

Monica, R.C., Cremonini, R., 2009. Nanoparticles and higher plants. Caryologia 62, 161–165.

Moore, M., 2006. Do nanoparticles present ecotoxicological risks for the health of the aquatic environment? Environ. Int. 32, 967–976.

Mortvedt, J.J., 1992. Crop response to level of water soluble zinc in granular zinc fertilizers. Fertil. Res. 33, 249–255.

Nair, R., Varghese, S.H., Nair, B.G., Maekawa, T., Yoshida, Y., Kumar, D.S., 2010. Nanoparticulate material delivery to plants. Plant Sci. 179, 154–163.

Navarro, E., Baun, A., Behra, R., Hartmann, N.B., Filser, J., Miao, A., Quigg, A., Santschi, P.H., Sigg, L., 2008. Environmental behaviour and ecotoxicity of engineered nanoparticles to algae, plants, and fungi. Ecotoxicology 17, 372–386.

Nel, A., Xia, T., Mädler, L., Li, N., 2006. Toxic potential of materials at the nano level. Science 311, 622–627.

Parsons, J.G., Lòpez, M.L., Gonzales, C.M., Peralta-Videa, J.R., Garrea-Torresdey, J.L., 2010. Toxicity and biotransformation of uncoated and coated nickel hydroxide nanoparticles on mesquite plants. Environ. Toxicol. Chem. 29, 1146–1154.

Piccinno, F., Gottschalk, F., Seeger, S., Nowack, B., 2012. Industrial production quantities and uses of ten engineered nanomaterials in Europe and the world. J. Nanopart. Res. 14, 1–11. http://dx.doi.org/10.1007/s11051-012-1109-9.

Prasad, T.N.V.K.V., Sudhakar, P., Sreenivasulu, Y., Latha, P., Munaswamy, V., Reddy, K.R., Sreeprasad, T.S., Sajanlal, P.R., Pradeep, T., 2012. Effect of nanoscale zinc oxide particles on the germination, growth and yield of peanut. J. Plant Nutr. 35, 905–927.

Priester, J.H., Ge, Y., Mielke, R.E., Horst, A.M., Moritz, S.C., Espinosa, K., Gelb, J., Walker, S.L., Nisbet, R.M., An, Y.-J., Schime, J.P., Palmer, R.G., Hernandez-Viezcas, J.A., Zhao, L., Gardea-Torresdey, J.L., Holden, P.A., 2012. Soybean susceptibility to manufactured nanomaterials with evidence for food quality and soil fertility interruption. Proc. Natl. Acad. Sci. U.S.A. http://dx.doi.org/10.1073/pnas.1205431109.

Priyadarshini, S., Deepesh, B., Zaidi, M.G.H., PardhaSaradhi, P., Khanna, P.K., Arora, S., 2012. Silver nanoparticle-mediated enhancement in growth and antioxidant status of *Brassica juncea*. Appl. Biochem. Biotechnol. 167, 2225–2233.

Rahman, A., Seth, D., Mukhopadhyaya, S.K., Brahmachary, R.L., Ulrichs, C., Goswami, A., 2009. Surface functionalized amorphous nanosilica and microsilica with nanopores as promising tools in biomedicine. Naturwissenschaften 96, 31–38.

Rees, P., Brown, M.R., Summers, H.D., Holton, M.D., Errington, R.J., 2011. A transfer function approach to measuring cell inheritance. BMC Syst. Biol. 5, 31–37.

Rico, C.M., Majumdar, S., Duarte-Gardea, M., Peralta-Videa, J.R., Gardea-Torresdey, J.L., 2011. Interaction of nanoparticles with edible plants and their possible implications in the food chain. J. Agric. Food Chem. 59, 3485–3498.

Rossi, L., Zhang, W., Lombardini, L., Ma, X., 2016. The impact of cerium oxide nanoparticles on the salt stress responses of *Brassica napus* L. Environ. Pollut. 219, 28–36.

Saptarshi, S.R., Dusch, A., Lopata, A.L., 2013. Interaction of nanoparticles with proteins: relation to bio-reactivity of the nanoparticle. J. Nanobiotechnol. 11, 26–27. http://www.J.Nanobiotechnology.com/content/11/1/26.

Sathiyanarayanan, A.N., Banu, K.N., 2016. Silver-chitosan nanoparticles induced biochemical variations of chickpea (*Cicer arietinum* L.). Biocatal. Agric. Biotechnol. 8, 39–44.

Sedghi, M., Hadi, M., Toluie, S.G., 2013. Effect of nano zinc oxide on the germination parameters of soybean seeds under drought stress. Ann. West Univ. Timiş. Ser. Biol. 24, 73–78.

Sharma, P., 2015. Nanomaterials from food packaging and commercial products into ecological and soil environment. Curr. Sci. 109, 1223–1224.

Siddiqui, M.H., Al-Whaibi, M.H., 2014. Role of nano-SiO_2 in germination of tomato (*Lycopersicum esculentum* seeds Mill.). Saudi J. Biol. Sci. 21, 13–17.

Stampoulis, D., Sinha, S.K., White, J.C., 2009. Assay dependent phytotoxicity of nanoparticles to plants. Environ. Sci. Technol. 43, 9473–9479.

Suriyaprabha, R., Karunakaran, G., Yuvakkumar, R., Prabu, P., Rajendran, V., Kannan, N., 2012. Growth and physiological responses of maize (*Zea mays* L.) to porous silica nanoparticles in soil. J. Nanopart. Res. 14, 1294–1307.

Thorek, D.L.J., Tsourkas, A., 2008. Size, charge and concentration dependent uptake of iron oxide particles by non-phagocytic cells. Biomaterials 29, 3583–3590.

Tripathi, D.K., Singh, S., Singh, S., Pandey, R., Singh, V.P., Sharma, N.C., Prasad, S.M., Dubey, N.K., Chauhan, D.K., 2017a. An overview on manufactured nanoparticles in plants: uptake, translocation, accumulation and phytotoxicity. Plant Physiol. Biochem. 110, 2–12. http://dx.doi.org/10.1016/j.plaphy.2016.07.030.

Tripathi, D.K., Singh, S., Singh, S., Srivastava, P.K., Singh, V.P., Singh, S., Singh, S., Prasad, S.M., Singh, P.K., Dubey, N.K., Pandey, A.C., Chauhan, D.K., 2017b. Nitric oxide alleviates silver nanoparticles (AgNps)-induced phytotoxicity in *Pisum sativum* seedlings. Plant Physiol. Biochem. 110, 167–177. http://dx.doi.org/10.1016/j.plaphy.2016.06.015.

Tripathi, D.K., Mishra, R.K., Singh, S., Singh, S., Vishwakarma, K., Sharma, S., Singh, V.P., Singh, P.K., Prasad, S.M., Dubey, N.K., Pandey, A.C., Sahi, S., Chauhan, D.K., 2017c. Nitric oxide ameliorates zinc oxide nanoparticles phytotoxicity in wheat seedlings: implication of the ascorbate–glutathione cycle. Front. Plant Sci. 8, 1–10. http://dx.doi.org/10.3389/fpls.2017.00001.

Tripathi, D.K., Singh, S., Singh, V.P., Prasad, S.M., Dubey, N.K., Chauhan, D.K., 2017d. Silicon nanoparticles more effectively alleviated UV-B stress than silicon in wheat (*Triticum aestivum*) seedlings. Plant Physiol. Biochem. 110, 70–81.

Vannini, C., Domingo, G., Onelli, E., Prinsi, B., Marsoni, M., Espen, L., Bracale, M., 2013. Morphological and proteomic responses of *Eruca sativa* exposed to silver nanoparticles or silver nitrate. PLoS One 8, e68752. http://dx.doi.org/10.1371/journal.pone.0068752.

Verano-Braga, T., Miethling-Graff, R., Wojdyla, K., Rogowska-Wrzesinska, A., Brewer, J.R., Erdmann, H., Kjeldsen, F., 2014. Insights into the cellular response triggered by silver nanoparticles using quantitative proteomics. ACS Nano 8, 2161–2175.

Wang, Z., Xie, X., Zhao, J., Liu, X., Feng, W., White, J.C., Xing, B., 2012. Xylem- and phloem-based transport of CuO nanoparticles in maize (*Zea mays* L.). Environ. Sci. Technol. 46, 4434–4441.

Wang, X.Y., Chen, S., Li, Q., Wang, W., Hou, C., Gao, X., Wang, L., Wang, S., 2016. Zinc oxide nanoparticles affect biomass accumulation and photosynthesis in *Arabidopsis*. Front. Plant Sci. 6, 1243. http://dx.doi.org/10.3389/fpls.2015.01243.

Watson, J.L., Fang, T., Dimpka, C.O., Britt, D.W., McLean, J.E., Jacobson, A., Anderson, A.J., 2015. The phytotoxicity of ZnO nanoparticles on wheat varies with soil properties. Biometals 28, 101–112.

Wu, G., Mikhailovsky, A., Khant, H.A., Fu, C., Chiu, W., Zasadzinski, J.A., 2008. Remotely triggered liposome release by near-infrared light absorption via hollow gold nanoshells. J. Am. Chem. Soc. 130, 1875–1877.

Xie, Y., Li, B., Zhang, Q., Zhang, C., 2012. Effects of nano-silicon dioxide on photosynthetic fluorescence characteristics of *Indocalamus barbatus* McClure. J. Nanjing For. Univ. (Nat. Sci. Ed.) 2, 59–63.

Yang, Z., Chen, J., Dou, R., Gao, X., Mao, C., Wang, L., 2015. Assessment of the phytotoxicity of metal oxide nanoparticles on two crop plants, maize (*Zea mays* L.) and rice (*Oryza sativa* L.). Int. J. Environ. Res. Public Health 12, 15100–15109. http://dx.doi.org/10.3390/ijerph121214963.

Yang, F., Hong, F., You, W., Liu, C., Gao, F., Wu, C., Yang, P., 2006. Influences of nano-anatase TiO_2 on the nitrogen metabolism of growing spinach. Biol. Trace Elem. Res. 110, 179–190.

Yang, L., Watts, D.J., 2005. Particle surface characteristics may play important role in phytotoxicity of alumina nanoparticles. Toxicol. Lett. 158, 122–132.

Yoon, S.-J., Kwak, J.I., Lee, W.-M., Holden, P.A., An, Y.-J., 2014. Zinc oxide nanoparticles delay soybean development: a standard soil microcosm study. Ecotoxicol. Environ. Saf. 100, 131–137.

Yuan, H.Y., Li, J., Bao, G., Zhang, S.L., 2010. Variable nanoparticle-cell adhesion strength regulates cellular uptake. Phys. Rev. Lett. 105, 1381011–1381014.

Zafar, H., Ali, A., Zia, M., 2016. CuO Nanoparticles inhibited root growth from *Brassica nigra* seedlings but induced root from stem and leaf explants. Appl. Biochem. Biotechnol. 18, 365–378. http://dx.doi.org/10.1007/s12010-016-2217-2.

Zhang, D., Hua, T., Xiao, F., Chen, C., Gersberg, R.M., Liu, Y., Stuckey, D., Ng, W.J., 2015. Phytotoxicity and bioaccumulation of ZnO nanoparticles in *Schoenoplectus tabernaemontani* Soon Keat Tan. Chemosphere 120, 211–219.

Zheng, L., Hong, F., Lu, S., Liu, C., 2005. Effect of nano-TiO_2 on strength of naturally aged seeds and growth of spinach. Biol. Trace Elem. Res. 105, 83–91.

Zhu, H., Han, J., Xiao, J.Q., Jin, Y., 2008. Uptake, translocation and accumulation of manufactured iron oxide nanoparticles by pumpkin plants. J. Environ. Monit. 10, 713–717.

CHAPTER

6

Nanoparticle-Induced Morphological Responses of Roots and Shoots of Plants

Seyed M. Talebi
Arak University, Arak, Iran

6.1 INTRODUCTION

From the last few decades, nanotechnology has been recognized as a novel and popular field. Nanotechnology allows advanced study in different areas, and its findings could reveal new usages in biotechnology as well as agriculture fields. In the wide ranges of disciplines such as energy, medicine, and biology, nanotechnology proposes an expanding investigation similar to reproductive science and technology, alternation of agricultural and food wastages to energy and other beneficial products via enzymatic nanobioprocessing, chemical sensors, purification of water, disease debarment, and treatment in plants using different nanocides (Carmen et al., 2003; Nair et al., 2010).

Though fertilizer compounds are essential for growth and development of plants, most of the compounds used are presently unavailable to crops because of various parameters such as leaching, decay by photolysis, hydrolysis, and parsing. Therefore there is a need to decrease nutrient casualties in fertilization, and to amplify the plant product by the operation of novel uses with the assistance of nanotechnology and nanomaterials. These types of fertilizers might bear traits, which are impressive to plants, deliver the nutrients on request, and control the use of chemicals fertilizers, which regulate growth and development of plants and raise activity of the target (DeRosa et al., 2010; Nair et al., 2010).

Giraldo et al. (2014) stated that higher plants have a prominent ability to develop mechanism to perform better under suitable and unsuitable conditions. Investigators have decided to expand novel methods, which could benefit crops by raising their native operations. It is important to know that nanoparticles have unrivaled physicochemical characteristics and the potential to enhance the metabolism of plants.

Nanomaterials in Plants, Algae, and Microorganisms
http://dx.doi.org/10.1016/B978-0-12-811487-2.00006-2

Investigators (e.g., Galbraith, 2007; Torney et al., 2007) have stated that nanoparticles are capable of penetrating into the cells of herbs; in addition, they can carry DNA and other chemical compounds into the cells of plants. Therefore the nanotechnology research field presents novel possibilities in plant biotechnology to target special gene manipulation and expression in the particular cells of plants (Singh et al., 2017). Researchers have increased the ability of plants to capture more energy from the sun by inserting carbon nanotubes into chloroplasts. These structures could act as dummy antennae, which could permit chloroplasts to receive light wavelengths that are not in their normal limited area, for example, near-infrared, green, and also ultraviolet (Cossins, 2014; Giraldo et al., 2014). Moreover, engineered carbon nanotubes increase the germination of seeds and the growth and development of plants (Lahiani et al., 2013; Siddiqui and Al-Whaibi, 2014). However, most of the studies on nanoparticles are concerned with toxicity, while some studies that have been executed on nanoparticles are of benefit to herbs. Study in the area of nanotechnology is needed to discover new utilizations to target the special delivery of chemical molecules, proteins, and nucleotides for genetic variations of crop plants (Torney et al., 2007; Scrinis and Lyons, 2007).

The field of science has great potential for those studying plant science and other areas and to expand novel tools for interpolation of nanoparticles into plants, which could increase available functions and create new ones (Cossins, 2014). In this chapter, studies related to the effects of some important nanoparticles on growth and development of roots and shoots have been summarized.

6.1.1 Functions of Roots and Shoots

Almost all green plants are able to manufacture energy, which they need to grow. Through photosynthesis reactions, a leaf produces assimilates from carbon dioxide, which is taken up with the ambient air through a plant's stomata. This source of carbohydrate is very important for plant growth and development. Carbohydrates are used at the shoot level but are also transferred to reach other plant parts. However, it is important to know that roots are heterotrophic organs for carbon sources (Freixes et al., 2002).

A root does not yield biomass by itself, and its growth is completely dependent on assimilates coming from the leaf. It is observed by various experiments in different plant taxa, for example, Aguirrezabal et al. (1994) in sunflower, and Muller et al. (1998) in maize. Bingham et al. (1996) reported that defoliating a plant causes a rapid decline in the primary and secondary rate of root elongation. Roots provide anchorage in the soil, but they are also important to extract water and nutrients from the soil. For example, soil water shortage is known to have a strong effect on the growth of shoots (Reymond et al., 2003). Therefore shoots and roots are connected and their respective growth is integrated and coordinated at the entire plant level.

Different mechanisms such as hormonal (Werner et al., 2003; Teale et al., 2006) and metabolic signals (Freixes et al., 2002) allow this coordination. Production of various phytohormones such as cytokinins, auxins, or abscisic acid is strictly controlled, and changes in their comparative affluence at various stages of development or in response to the environment drive the growth of the different organs.

6.1.2 Environmental Conditions Alter the Model of Biomass Allocation Between Shoots and Roots

Plants are also able to change their growth and to regulate allocation models between the organs in response to ecological conditions. When a resource is confined, an amplified biomass allocation can be seen toward the organ responsible for the limiting resource acquisition. The name of this theory is "functional equilibrium theory," formalized first by Brouwer in 1962, and presumes that the part preoccupied in the receipt of a resource has primacy on that resource. This opinion is a foundation of many other theories in all domains of plant biology, from plant ecology and evolution to modeling and ecophysiology (Lacointe, 2000).

Bloom et al. (1985) reported that an implicit hypothesis of this is that there are tradeoffs in allotment among leaves, stems, and roots of plants. A plant changes its appropriation to shoots if the carbon reach of the shoot is ruined by a small amount of upper-ground resources, such as sun rays and carbon dioxide. Consequently, the growth of leaves is confined by water, nutrients, and lower assimilates that are registered above land. The surplus assimilates are next carried to the roots, boosting growth of roots related to that of shoots (Grechi et al., 2007). This integration of ecological conditions at the entire plant level can also be illustrated by split root experiments.

6.2 EFFECTS OF DIVERSE NANOPARTICLES ON GROWTH AND DEVELOPMENT OF PLANTS

Nanoparticle compound interaction with plants creates different changes in structural and also physiological features, pertaining to the attributes of nanoparticles (Tripathi et al., 2015, 2016, 2017a,b,c,d,e; Singh et al., 2016; Shweta et al., 2016; Cox et al., 2017; Du et al., 2017). The effects of nanoparticles depend on various parameters such as nanoparticle chemical combination, scale, surface covering, reactivity, and most significantly the dosage at which these compounds are efficient (Khodakovskaya et al., 2012).

Ma et al. (2010a) stated that both affirmative and negative effects of nanoparticles on herb growth and development and the engineered nanoparticle effect on herbs pertain to the combination, doping, scale, and physical and chemical qualities of engineered nanoparticles as well as plant type. Engineered nanoparticles' influence belongs to their doping and differs from species to species. However, in this section the effects of some nanoparticles on various aspects of plant phenomena such as seed germination, shoot and roots morphology, growth, and development are discussed.

6.2.1 Rare Earth Oxide Nanoparticles

Hu et al. (2004) reported that much progress in the study of the consequence of rare earth cations (RE^{3+}) on metabolite compounds of plants' and crops' efficiency and modalities have been built. However, the biochemical and physiological impacts of rare earth oxide nanoparticles on plants have not been investigated yet. Different studies (e.g., Hong et al., 2002; Xie et al., 2002) confirmed that rare earth ions have affirmative consequences on the growth of plants if used correctly. The responses to rare earth ions pertain to the kind of plant, growth phase, status of soil, and amount of these elements. These elements have an affirmative outcome for plant growth at light condensations, but a negative result has been observed at greater concentrations. However, this concept is yet unfounded. This mechanism is illustrated by "hormesis effects," that is, a dose–response event determined by a low amount of incitement and a high quantity of deterrence (Calabrese and Baldwin, 2002).

Ma et al. (2010b) examined the toxicity of some rare earth oxide nanoparticles (CeO_2, La_2O_3, nano-Gd_2O_3, and Yb_2O_3) on the root growth of some important crops of higher plants, namely, *Brassica napus*, *Raphanus sativus*, *Triticum aestivum*, *Lactuca sativa*, *Brassica oleracea*, *Lycopersicon esculentum*, as well as *Cucumis sativus*. The impacts of examined nanoparticles on root elongation differed significantly among various nanoparticles and also plant types. Results showed that application of 2000 mg/L nano-CeO_2 had no impact on the growth of roots of all studied plants, barring lettuce. However, nano-La_2O_3, nano-Gd_2O_3, and nano-Yb_2O_3, in a concentration of 2000 mg/L, prevented root growth of all studied taxa. Inhibitory efficacy of nano-La_2O_3, nano-Gd_2O_3, and nano-Yb_2O_3 also varied in several growth activities of taxa. For wheat, the deterrence chiefly happens within the seed incubation procedure, but lettuce and rape were prevented on both seed soaking and incubation proceedings. The 50% inhibitory dopings differed greatly between the studied plants, for example, the concentration for rape was nearly 40 mg/L of nano-La_2O_3, 20 mg/L of nano-Gd_2O_3, and 70 mg/L of nano-Yb_2O_3, respectively. In the concentration domains, the RE^{3+} ion liberated from the nanoparticles had poor traces on the growth of roots.

6.2.2 Nanoparticles of Silicon Dioxide

The growth and development of plants start from seed germination followed by elongation of roots and stems egress as the primitive indications of plant growth and development. Hence to comprehend the cycle of plant phenology with relevance to nanoparticles is very important. The results of former investigations proved that the efficacy of nanomaterials on the germination of seeds is related to the concentrations of these compounds.

For example, Siddiqui and Al-Whaibi (2014) found that the lower amounts of nano-SiO_2 increased germination of seeds in tomato (Fig. 6.1). The lower concentration of silicon dioxide nanoparticles increased seed germination of *Lycopersicum esculentum.* The concentration of 8 g/Lof silicon dioxide nanoparticles enhanced seed germination by 22.16%, mean time of germination by 3.98%, seedling vigor index by 507.82%, and seed germination index by 22.15% compared to control plants. Concentration of 8 g/L of silicon dioxide nanoparticles increased the fresh weight of seedlings by 116.58% and seedling dry weight by 117.46% compared to control seedlings, and also silicon dioxide nanoparticles play a very important role in root and shoot growth (Siddiqui and Al-Whaibi, 2014).

In addition, nano-SiO_2 enhanced germination of seed by preparing further nutrients' accessibility to *Zea mays* L. seeds, pH, and electrical conductivity to the culture medium (Suriyaprabha et al., 2012). In an experimental study, Baoshan et al. (2004) used exogenous nanoparticles of silicon dioxide on seedlings of *Larix olgensis.* Their results showed that nano-SiO_2 amplified various factors of the growth and condition of seedlings, such as mean

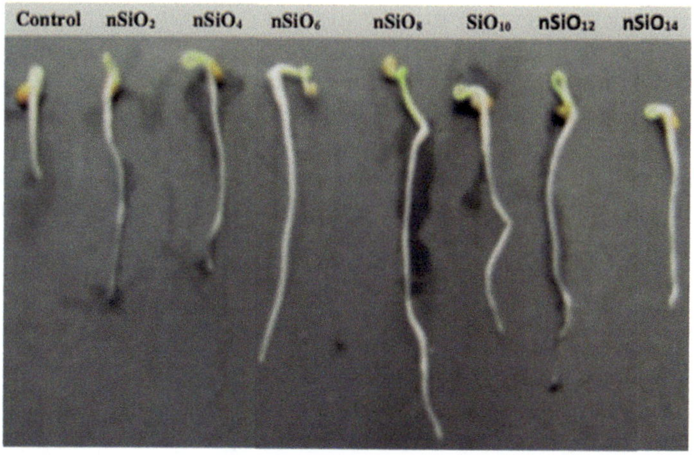

FIGURE 6.1 Effect of silicon dioxide nanoparticles on growth of seedlings in tomato. (1) SiO_0 (control), (2) SiO_2, (3) SiO_4, (4) SiO_6, (5) SiO_8, (6) SiO_{10}, (7) SiO_{12}, and (8) SiO_{14} [the concentration (in g/L for nanoparticles) is shown as a subscript]. *Reprinted from Siddiqui, M.H., Al-Whaibi, M.H., 2014. Role of nano-SiO_2 in germination of tomato (Lycopersicum esculentum seeds Mill.). Saudi J. Biol. Sci., 21, 13–17, Copyright (2014), with permission from Elsevier.*

height, diameter of root collar, main length of roots, seedling lateral root number, and also induction of chlorophyll synthesis.

Under abiotic stress, nano-SiO_2 increases the germination of seeds. This condition was seen in different species such as tomato (Haghighi et al., 2012) and squash (Siddiqui et al., 2014). They observed that, under NaCl stress application, nano-SiO_2 increased the germination of seeds and motivated the system of antioxidants.

Shah and Belozerova (2009) examined the effect of various nanoparticles (e.g., silica, palladium, gold, and copper) and found that all of these nanoparticles had a significant effect on the seeds of lettuce. Lu et al. (2002) reported that external usage of silicon dioxide and titanium dioxide nanoparticles developed soybean seed germination by developing nitrate reductase associated with an increasing capability of seeds to uptake and use water and nutrients (Zheng et al., 2005).

In saline conditions, nano-SiO_2 maximized the fresh as well as dry weight of leaves, content of chlorophyll, and proline agglomeration. Different investigations (Haghighi et al., 2012; Li et al., 2012; Kalteh et al., 2014; Siddiqui et al., 2014) confirmed that an increment in the proline accumulation, unattached amino acids, nutrient quantity, and also activity of antioxidant enzymes are caused by nano-SiO_2, therefore increasing the plant's endurance to environmental stresses.

Salinity stress significantly decreased growth and development of *Ocimum basilicum*. The leaf dry and fresh weight of this plant declined with an increment in concentration of NaCl, while they were significantly enhanced with an application of silicon nanoparticles. The content of chlorophyll was minimized in salt-stressed plant, while it developed with treatment with silicon nanoparticles (Siddiqui et al., 2007). Application of 1 mM silicon dioxide nanoparticles could prominently alleviate the side effect of salt stress on percentage of germination, length of root and shoot, weight of seedling, mean germination time, seedling vigor index, and cotyledon reserve mobilization of *Lens culinaris* (Janmohammadi et al., 2015).

Wang et al. (2014) studied the effect of quantum dots and silica-coated quantum dot treatment on rice plants. They found that silica-coated quantum dots significantly promoted root growth of this plant, compared to control samples. Nano-SiO_2 developed the growth of the plant by enhancing exchange of gases, net rate of photosynthesis, level of transpiration, conductance of stomata, activities of photosystem II potential, rate of electron transportation, and photochemical quell (Xie et al., 2011; Siddiqui et al., 2014).

6.2.3 Zinc Oxide Nanoparticles

Experiments confirmed that nanoparticles of zinc oxide maximize the growth and development of plants. Lower concentration of zinc oxide nanoparticles had useful efficacy on the seed germination of different plant species such as peanut (Prasad et al., 2012), soybean (*Glycine max*) (Sedghi et al., 2013), wheat (*T. aestivum*), and onion (*Allium cepa*) (Raskar and Laware, 2014). However, higher doses of zinc oxide nanoparticles decreased germination of seeds.

The nanoparticles' effect on germination of seeds relates to concentrations of nanoparticles and differs from species to species. de la Rosa et al. (2013) applied various concentrations of zinc oxide nanoparticles to plants such as alfalfa, cucumber, and tomato; they observed that only seed pullulating cucumber was increased.

Zinc oxide nanoparticles had considerable effects on many physiological traits of *Cyamopsis tetragonoloba*, such as plant biomass, growth of foliage and roots, area of roots, synthesis of chlorophyll, and also protein, population of rhizospheric microorganisms, acid phosphatase, and activities of various enzymes such as phytase as well as alkaline phosphatase in cluster bean rhizosphere (Raliya and Tarafdar, 2013). Examinations with light and scanning electron microscopes and inductive-coupled plasma/atomic emission spectroscopy showed that roots of *Vigna radiata* and *Cicer arietinum* take up zinc oxide nanoparticles and develop the length of roots and shoots as well as root and shoot biomass (Mahajan et al., 2011).

Moreover, zinc oxide nanoparticles supplemented with MS medium increased somatic embryogenesis, shooting, regeneration of plantlets, and induced synthesis of proline, activity of superoxide dismutase, and catalase as well as peroxidase, therefore ameliorating endurance to various stresses (Helaly et al., 2014).

Application of zinc oxide nanoparticles significantly enhanced the formation of lateral root and also variations of element amounts in wheat seedlings. Furthermore, zinc oxide nanoparticles changed the morphology of roots by increasing the density of roots through the formation of lateral roots. These conditions maximize uptake of nutrient for the plants (Watson et al., 2015). In addition, many investigations have also proved that exposure to other engineered nanoparticles results in a change in plant root growth patterns (Miralles et al., 2012; Dimkpa et al., 2013).

Prakash et al. (2017) treated seedlings of *Arabidopsis thaliana* with five concentrations of zinc oxide nanoparticles (0, 20, 50, 100, and 200 mg/L) and analyzed morphological changes. The obtained data confirmed that the fresh weight as well as length of primary root was decreased after exposure to most of the concentrations (except for 20 mg/L) of zinc oxide nanoparticles. However, treatment with 20 mg/L of zinc oxide nanoparticles triggered a 9% increase in the formation of lateral root. No root-to-shoot translocation of Zn was seen in seed lines and accumulation was mainly located in the tips of root, primary–lateral root junctions, and root–shoot junctions (Fig. 6.2).

The effects of zinc oxide nanoparticles were examined on *Gossypium hirsutum* L. plants. The results confirmed that the shoot and root lengths and total biomass significantly maximized with an increment of nanoparticle concentrations compared to control. The maximum growth rate observed was 130.0% and 115.2% for shoots and roots, respectively, in the presence of 200 mg/L nanoparticles. It is very important to know that the growth rate of shoots is significantly higher than root tissues following zinc oxide nanoparticle exposure with respect to control plants (Venkatachalam et al., 2017).

6.2.4 Manganese Oxide Nanoparticles

Manganese oxide nanoparticles significantly increased elongation of root in seedlings of *Lactuca sativa* by more than 50%. It confirmed that manganese oxide nanoparticles could be applied as an Mn micronutrient fertilizer or enhancer of plant growth. Although, these nanoparticles improving agronomic productions, they had no or little phytotoxicity at levels more than 50 ppm (Liu et al., 2016).

6.2.5 Iron Oxide Nanoparticles

Sergey et al. (2015) examined the various effects of iron oxide nanoparticles on *A. thaliana* physiology at two different concentrations: 3 and 25 mg/L. In the treated samples with a concentration of 3 mg/L, no effect was seen on seedlings or root length, while 25 mg/L-treated plants had reduced seedlings and root length. In the treated samples, no observable phenotypic variations were reported in the size of the plant or its total structure.

These conditions confirmed that environmental pollution with this nanoparticle could go unnoticed. Charged iron oxide nanoparticles are transferred all over the plant tissues and can be identified in different tissues of roots, leaves, floral organs, and stems. This nanoparticle possesses prominent effects on the length of root in *A. thaliana* seedlings.

Iron oxide nanoparticles significantly maximized elongation of root in *Lactuca sativa* seedlings by 12%–26%. This showed that iron oxide nanoparticles could be applied as iron fertilizer in concentrations of 5–20 ppm (Srivastava et al., 2014).

Iron pyrite nanoparticle-treated seeds of *Spinacia oleracea* produced significantly broader leaves and larger leaf numbers with increased biomass when compared to the plant samples developed from control seeds. The results showed that the possible mechanism resulting in the increment of biomass is an amplified stored starch breakdown in the treated seeds with iron pyrite resulting in prominently better growth (Srivastava et al., 2014).

Askary et al. (2016a) examined the effects of iron oxide nanoparticles on stem and leaf trichomes in *Mentha piperita* (Labiatae) under saline stress. The obtained results proved

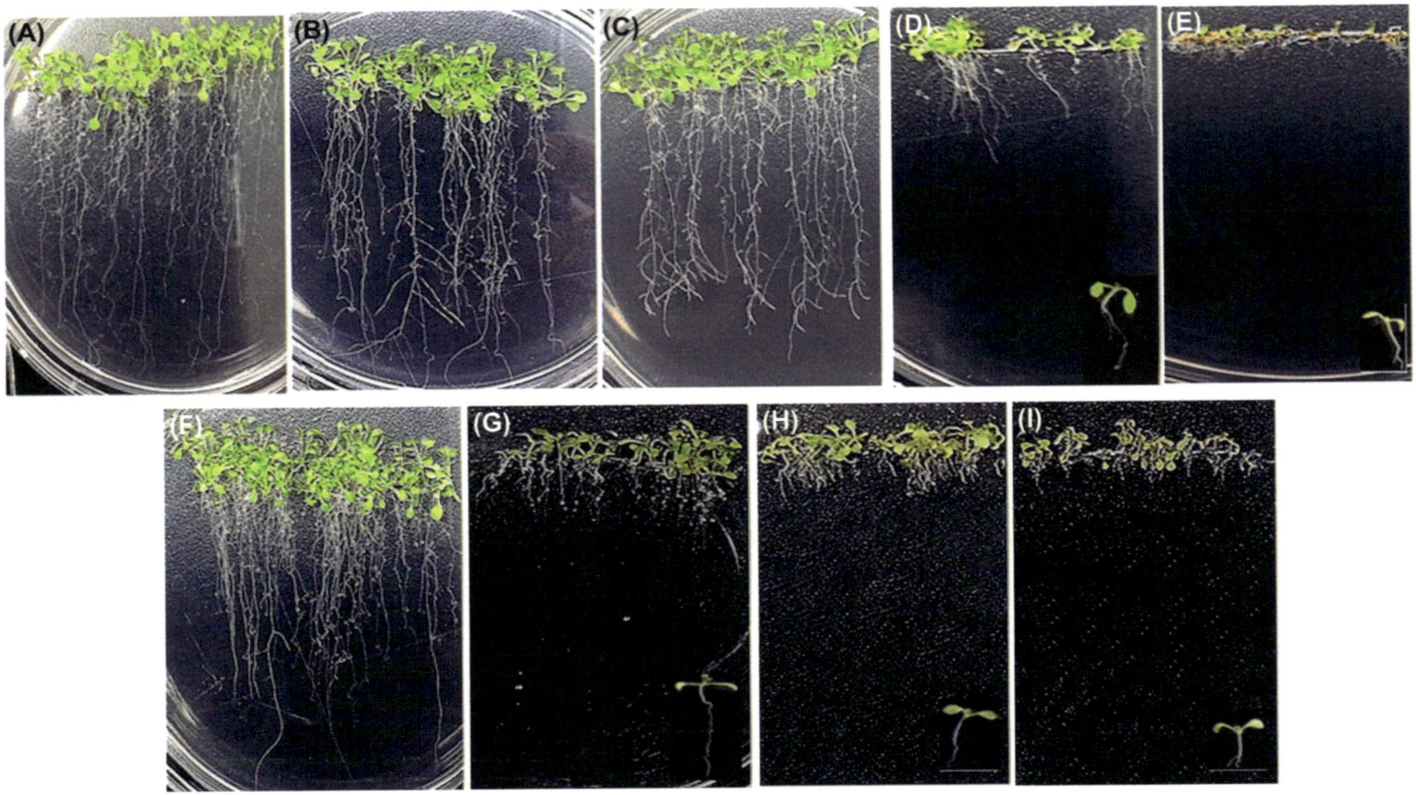

FIGURE 6.2 Phenotypes of *Arabidopsis thaliana* seedlings exposed to various concentrations of zinc and zinc oxide nanoparticles for 14days. (A) Control and seedlings grown in the presence of 20, 50,100, and 200mg/L of (B–E) Zn and (F–I) zinc oxide nanoparticles. *Reprinted from Prakash, M., Nair, G., Chung, M., 2017. Regulation of morphological, molecular and nutrient status in Arabidopsis thaliana seedlings in response to ZnO nanoparticles and Zn ion exposure. Sci. Total Environ. 575, 187–198, Copyright (2017), with permission from Elsevier.*

that these nanoparticles had a negative effect on the density of trichomes. It seems that iron nanoparticles had significant effects on plant morphological and anatomical traits.

Various mechanisms were proposed for decreasing the density of trichomes. First, iron nanoparticles could have negative effects on trichome formation. Second, Fe nanoparticles could remove salt stress effects. Davar Zareii et al. (2014) demonstrated that the side effects of drought stress can be declined by iron nanoparticle application. In addition, different studies (e.g., Eichert et al., 2008; Basiuk et al., 2011) confirmed that foliar absorption of metal nanoparticles is linked to the difference in leaf morphological and anatomical characteristics, particularly of trichomes, stomata, and hypodermis.

The effects of various concentrations of iron nanoparticles were examined on salt-treated plants of *Lallemantia royleana* (Askary et al., 2016b). Significant negative and positive correlations were observed between the densities of trichomes with concentrations of salt and iron oxide nanoparticles. Significant negative correlations ($P \leq .05$) were recorded between concentrations of iron nanoparticles with simple four-celled and sessile hairs, while these nanoparticles had significant positive correlations ($P \leq .05$) with peltate and simple one-celled types. Pariona et al. (2017) studied the effect of citrate-coated magnetite nanoparticles on the germination as well as early growth of *Quercus macdougallii* (oak). They used two types of magnetite nanoparticles, namely, NP1 and NP2. The morphology of NP1 was quasispherical, with sizes of 6–10 nm, but the NP2 had sizes between 65 and 160 nm.

The obtained results confirmed that application of nanoparticles amplified the germination up to 33% in respect of the control samples. Moreover, the magnetite nanoparticle treatments maximized growth, dry biomass, and chlorophyll concentration. Therefore treatments with magnetite iron nanoparticles could be potentially used to improve conservation and reforestation of threatened forestry species.

In a study the effect of six concentrations (0, 5, 10, 20, 30, and 40 μM) of iron nanoparticles was examined on growth parameters of *Catharanthus roseus* plants over 70 days. The findings showed that the nanoparticles prominently developed leaf dimensions in respect of control plants, but had no effect on the stomata aggregate of leaves. The largest amount of leaf growth parameters was acquired in a 30 μL concentration of these nanoparticles; however, the lowest quantities of the mentioned parameters were seen in 0 μL iron oxide nanoparticles (Askary et al., 2014).

6.2.6 Carbon Nanotubes

Between different nanoparticles, carbon nanotubes have obtained a significant position, because they have inimitable electrical, mechanical, thermal, as well as chemical attributes. Most investigations on carbon nanotubes have been exemplified on animals and humans (Ke et al., 2011; Tiwari et al., 2014). Therefore the available data on carbon nanotubes and their connection with the cells of plant metabolism are scarce.

Srinivasan and Saraswathi (2010) stated that because of the unmatched properties of these nanotubes, carbon nanotubes possess the ability to interpenetrate the wall of cells as well as their membrane and also prepare adequate delivery of chemicals to them. Single-walled carbon nanotubes operate as nanotransporters for DNA delivery and also paint molecules into plant cells. However, different investigators have stated that many-walled carbon nanotubes have a unique strength of impressing the germination of seeds and growth of plants, and act as a system of delivery of DNA and chemical compounds to plant cells.

Multiwalled carbon nanotubes encourage water, essential calcium, as well as iron nutrients to absorb yield, which could maximize the germination of seeds with growth and development of plants (Villagarcia et al., 2012; Tiwari et al., 2014). Multiwalled carbon nanotubes added to agar culture medium stimulated seed germination of some important crop plants such as soybean, barley, and corn due to their ability to penetrate the coats of seed as the nanotubes agglomerates were seen inside the seed coats using various instruments such as Raman spectroscopy and transmission electron microscopy (Lahiani et al., 2013). Moreover, these investigators stated that multiwalled carbon nanotubes regulated expression of genes that coded different kinds of water channel proteins in the seed coats of soybean, barley, and corn.

The largest germinations in tomato, hybrid *Bt* cotton (Nalwade and Neharkar, 2013), *Brassica juncea* (Mondal et al., 2011), *Phaseolus mungo* (Gajanan et al., 2010), and rice were recorded with multiwalled carbon nanotubes (Nair et al., 2010). While, various studies confirmed the positive role of carbon nanotubes in seed germination and plant growth and development.

Khodakovskaya et al. (2012) believed that multiwalled carbon nanotubes act as regulators for seed germination and growth. In addition, these scientists showed that these compounds have the ability to increase tobacco cell growth of culture by up regulating the marker genes for cell divisions (*CycB*), formation of cell wall (*NtLRX1*), and transport of water (aquaporin, *NNtPIP1*).

Wang et al. (2012) stated that oxidized multiwalled carbon nanotubes significantly maximized elongation of cells in the root system and amplified the activity of dehydrogenase. However, other investigators believed that multiwalled carbon nanotubes do not show a desirable impact on germination of seed in different species. These conditions hold true when plants were given a high concentration of multiwalled carbon nanotubes (Lin and Xing, 2007; Husen and Siddiqi, 2014).

These nanotubes amplified the growth of shoots and activity of peroxidase and dehydrogenase, which may be caused first by absorption and agglomeration of multiwalled carbon nanotubes by roots, followed by the translocation from root to leaf (Smirnova et al., 2012), which could induce the expression of genes (Khodakovskaya et al., 2012; Wang et al., 2012; Lahiani et al., 2013).

The existence of water-soluble carbon nanotubes within *T. aestivum* plants was proved by using a scanning electron and fluorescence microscope, which induced the seedlings' growth in light and dark conditions (Tripathi and Sarkar, 2014). Moreover, multiwalled carbon nanotubes increase the capacity of water retention and biomass, flowering and fruit manufacture, and enhance the medical qualities of plant species (Khodakovskaya et al., 2013; Husen and Siddiqi, 2014).

The inhibitory effect of multiwalled carbon nanotubes on plant growth has been shown in many experiments (Ikhtiar et al., 2013; Begum and Fugetsu, 2012; Begum et al., 2014; Tiwari et al., 2014). Thus the effects of carbon nanotubes on plants differ from species to species, in their growth stages, and in the nature of nanoparticles.

6.2.7 Gold Nanoparticles

There are few investigations on the interplay of gold nanoparticles with different plant taxa. Shah and Belozerova (2009) believed that gold nanoparticles encourage phytotoxicity in plants by hampering the function of aquaporins. They are a class of phytoproteins, which aid in the transport of a broad range of molecules such as water.

The effects of gold nanoparticles in the improvement of seed germination, leaf number, leaf area, height of plant, and chlorophyll content were reported in various plants such as lettuce and cucumber (Barrena et al., 2009), *B. juncea* (Arora et al., 2012), *Boswellia ovalifoliolata* (Savithramma et al., 2012), and *Gloriosa superba* (Gopinath et al., 2014), which advances the production of yields (Arora et al., 2012; Gopinath et al., 2014).

Furthermore, gold nanoparticles possess a key function in the germination of seed as well as the antioxidant system in *A. thaliana* and can change microRNA expression amount, which regulates various structural, metabolic, and physiological phenomena in plants (Kumar et al., 2013).

Physiological examination confirmed a low increment in the mean length of shoots with a total decline in the middle length of roots. With different amounts of gold nanoparticles, a small increment in the mean length of shoots ranging from 6.81 to 7.36 cm was seen in *B. juncea* seedlings. A prominent growth in the mean length of shoots of *B. juncea* in MS medium free of gold nanoparticles was seen beyond the growth of seedlings in the MS medium completed with gold nanoparticles. With gold nanoparticle doping, a small enhancement in the mean length of shoots occurred. This is in agreement with a concurrent increment in the middle length of roots of *B. juncea* ranging from 9.42 to 11.62 cm up to the complement of 100 ppm of gold nanoparticles.

Additional enhancement in gold nanoparticle concentration has shown a decrease in the mean length of roots of about 9.43 cm. Such decrement in the mean length of seedlings roots might be caused by the diversion of metabolic energy and metabolites from the processes involved in the normal growth and development to the processes involved in alleviating the toxic effects imposed by gold nanoparticles (Woyro and Stelzer, 1988).

6.2.8 Silver Nanoparticles

Although many investigations exist on the effects of silver nanoparticles on prokaryotic and animal cells, a number of examinations were performed on plant species (Monica and Cremonini, 2009; Krishnaraj et al., 2012).

Examinations proved that nanoparticles have both favorable and unfavorable results on growth and development of plants. Examinations on the effect of biologically synthesized silver nanoparticles on hydroponically grown *Bacopa monnieri* showed that these nanoparticles had a prominent outcome on the germination of seed and induced protein and saccharide synthesis, while minimizing total phenol amounts and activities of catalase and peroxidase (Krishnaraj et al., 2012).

Biologically synthesized silver nanoparticles increased germination of seeds as well as growth of seedlings of *B. ovaliofoliolata* (Savithramma et al., 2012). Silver nanoparticles amplified shoot and root length, leaf area, chlorophyll, contents of carbohydrates and proteins, as well as antioxidant enzymes in different species such as *B. juncea*, common bean, and corn (Salama, 2012; Sharma et al., 2012).

Studies confirmed that the positive and negative effects of silver nanoparticles on elongation of roots depend on the type of plants (Gruyer et al., 2013). They demonstrated that the length of root was amplified in barley, while it declined in lettuce. Effects of silver nanoparticles on the germination of some wetland weed taxa (*Lolium multiflorum, Panicum virgatum, Carex lurida, Carex scoparia, Carex vulpinoidea, Carex crinita, Eupatorium fistulosum, Phytolaca americana, Scirpus cyperinus, Lobelia cardinalis,* and *Juncus effusus*) were examined.

The results showed that silver nanoparticles declined the rate of germination in one species, namely, *E. fistulosum* (Yin et al., 2012).

Rezvani et al. (2012) stated that silver nanoparticles induce elongation of roots by obstruction signaling of ethylene in *Crocus sativus*. However, the impact of silver nanoparticles on plant morphology and physiology depends on two main factors: the size and shape of nanoparticles. Examinations on the effect of three different shapes of silver nanoparticles on physiological as well as molecular response of *Arabidopsis* plants confirmed that decahedral silver nanoparticles promoted the greatest grade of root elongation; however, the globular-shaped silver nanoparticles had no effect on root growth, and a triggered shape created the highest amount of anthocyanin accumulation in seedlings of this species.

Silver nanoparticles in two shapes, decahedral and globular, gave the lowest and greatest amounts for Cu/Zn superoxide dismutase, respectively. However, all of examined size and shape of silver nanoparticles regulated accumulations of different proteins such as cell-division-cycle kinase 2, protochlorophyllide oxidoreductase, and fructose-1,6-bisphosphate aldolase. In addition, they induced genes expression involved in cellular events; for instance, these nanoparticles induced the gene expression of indoleacetic acid protein 8, 9-*cis*-epoxy-carotenoid dioxygenase, and dehydration-responsive RD22. Silver nanoparticles activated the aminocyclopropane-1-carboxylic acid (ACC)-derived inhibition of root elongation in seedlings of *Arabidopsis*, and minimized the expression of ACC synthase 7 and ACC oxidase (Syu et al., 2014).

Hojjat (2016) studied the effect of silver nanoparticles on various morphological and physiological features in *Lens culinaris* Medik. It was observed that the percentage of germination was not affected significantly between the different treatments. Its seedlings under silver nanoparticles had higher root and shoot length, dry mass, and germination speed with respect to control plants. However, in nanoparticle-treated plants, the mean time of germination was lower than in control plants. The obtained results confirmed that the effect of silver nanoparticles was significant on germination percentage ($P \leq .05$).

Treatment of seedlings of *Trigonella foenum-graecum* L. with biosynthesized silver nanoparticles showed that this compound has significant impact on the various growth parameters of seedlings, such as leaf number, root length, shoot length, and wet weight. Moreover, high-performance liquid chromatography analysis showed that in silver nanoparticle-treated seedlings a decline was seen in the production of the major phytochemical diosgenin. However, untreated plants yielded more diosgenin; therefore the recorded phytochemical enhancement effect induced by the nanoparticles was very significant.

Silver nanoparticles play important roles in enhancement of both plant growth and diosgenin synthesis. Therefore nanoparticles can increase the biosynthesis of natural medicinal products in plants (Jasim et al., 2016).

Salama (2012) surveyed the results of silver nanoparticles on the growth factors, such as lengths of shoots and roots, area of leaf surface, carbohydrate content, chlorophyll, and proteins, of some edible crops: common bean (*Phaseolus vulgaris* L.) and corn (*Z. mays* L.). He used a number of concentrations of silver nanoparticles (20, 40, 60, 80, and 100 ppm). The data confirmed that lower amounts of silver nanoparticles had a stimulating outcome on plant growth but the amplified levels created an inhibitory result. In addition, incremented levels of silver nanoparticles from 20 to 60 ppm caused an increase in the lengths of stems and roots, area of leaf surface, chlorophyll, carbohydrates, as well as protein amounts of these crops. However, the minimum value of these factors occurred with untreated samples, and increasing the amount of silver nanoparticles decreased these components.

6.2.9 Nanoaluminum

Burklew et al. (2012) stated that aluminum has many uses in various industries, e.g., it can be used in different materials with high energy, as an explosive material, and as a propellant, because it has prominent electrical properties, resistance to heat shock, and high mechanical resistance in high temperatures.

Nanoparticles of aluminum have different results on shoot development and growth. Riahi-Madvar et al. (2012) observed that nanoparticles of aluminum inhibited root elongation of *T. aestivum*. They stated that aluminum nanoparticles induced oxidative stress in roots cells of this plant. It is very important to know that the toxic effect is probably not nano-specified but is caused by the dissolution of aluminum oxide nanoparticles. However, the effects of nanoparticles on plants depend on the plant type and concentration of nanoparticles (Stewart et al., 2015).

Zhao et al. (2013) examined the results of aluminum nanoparticles on root and shoot growth of wheat. They observed that by amplifying nanoparticle concentration the length of shoot and root was minimized. However, these nanoparticles had a slight positive effect at 100 mg/L concentration. While the use of nanotitanium dioxide at a concentration of 100 mg/L had a positive effect on growth factor it could alleviate the negative effects of nano-aluminum in combined effects.

6.2.10 Titanium Dioxide Nanoparticles

The commercial name of titanium dioxide is Titania and its chemical formula is TiO_2. When this compound is used as a pigment it is called white titanium or white pigment. Titanium dioxide has various applications, e.g., it can remove organic pollution, and can be used in polychlorobiphenyl, surfactants production, insecticides as well as organic colors. Furthermore, titanium dioxide is useful for manufacturing self-cleaning hydrophobic products and can be used in sunscreen. It can also be used as a photocatalyst in purifying wastes and cleaning air (Karunakaran et al., 2014).

Different investigators have centralized the effects of titanium dioxide nanoparticles on various organisms such as plants and animals. However, studies on titanium dioxide nanoparticle use with herbs are inconclusive. Titanium dioxide nanoparticles increased the germination of seed and improved the growth of the radicle and plumule of canola seedlings (Mahmoodzadeh et al., 2013).

Palmqvist et al. (2015) studied different consequences of titanium dioxide nanoparticles on plants and observed that this compound can improve yields by up to 30% by enhancing photosynthesis through influencing the absorption of light. Titanium dioxide nanoparticles are encouraging as a useful source of nutrients for plants to maximize production of biomass because of increased metabolic activities, and local nutrient usage by promoting microbial activities.

Various examinations have confirmed the effects of nanoparticles on plant growth factors. For instance, spinach seeds treated with nanotitanium dioxide in low concentrations improved the shoot fresh/dry weight. In addition, chlorophyll content was enhanced by 17% and photosynthesis rate by 29% as compared with untreated plants (Zheng et al., 2005).

However, Song et al. (2013) believed that titanium and nanotitanium were more effective on roots than the shoots and increased the root fresh/dry weight. Studies on soybean confirmed that titanium movement is very slow and is more effective in treated organs. For this reason, use of nanotitanium dioxide solution in the environment stimulated roots more than other shoots (Calza et al., 2014).

The findings of Karunakaran et al. (2014) proved that nanotitanium dioxide in optimal concentrations stimulated the root and shoot growth of *T. aestivum*, while in high concentrations it has an inhibiting effect on the growth of these organs. Their results also confirmed that the root fresh and dry weights were significantly affected by nanotitanium dioxide.

Jaberzadeh et al. (2013) demonstrated that titanium dioxide nanoparticles maximized growth and yielded components in water deficit-stressed plants of *T. aestivum*. The nanoparticles regulate activity of different enzymes involved in the metabolism of nitrogen, such as glutamate dehydrogenase, glutamic–pyruvic transaminase, nitrate reductase, and glutamine synthase, which favor the plants to absorb nitrate and also helps the inorganic nitrogen conversion to organic ones in the shape of protein and chlorophyll, which can maximize the fresh and dry weight of plant (Yang et al., 2006; Mishra et al., 2014).

Crabtree (1998) reported that titanium dioxide nanoparticles act like a photocatalyst and induce reduction–oxidation reactions. Also Yang et al. (2006) demonstrated that the nanoparticles noticeably promote aged seed's vigor, chlorophyll formation, stimulates activity of Rubisco and enhances photosynthesis, thereby increasing growth and development of plant.

Nanoparticles maximize the absorbance of light, speed up the forwarding and alternation of the energy of sunlight, preserve chloroplast senility, and increase the photosynthetic period of chloroplasts (Yang et al., 2006). All of these are caused by titanium dioxide nanoparticles, which protect the chloroplasts from extreme radiation by maximizing the antioxidant enzyme's activities, such as catalase peroxidase, dismutase, and superoxide (Hong et al., 2005).

Castiglionea et al. (2016) employed two kinds of titanium dioxide nanoparticles to assess if various responses in function of different particle size and features were elicited. Furthermore, as corresponding bulk material, considered for decades an inert and safe material, but it has been recently classified as possibly carcinogenic to humans (Group 2B carcinogen, IARC 2010), they also tested the effects of the same concentration of this form. To evaluate the hypothesized effects of these materials on seeds of *Vicia faba*, they considered various physiological, histochemical, cytological, and biochemical end-points.

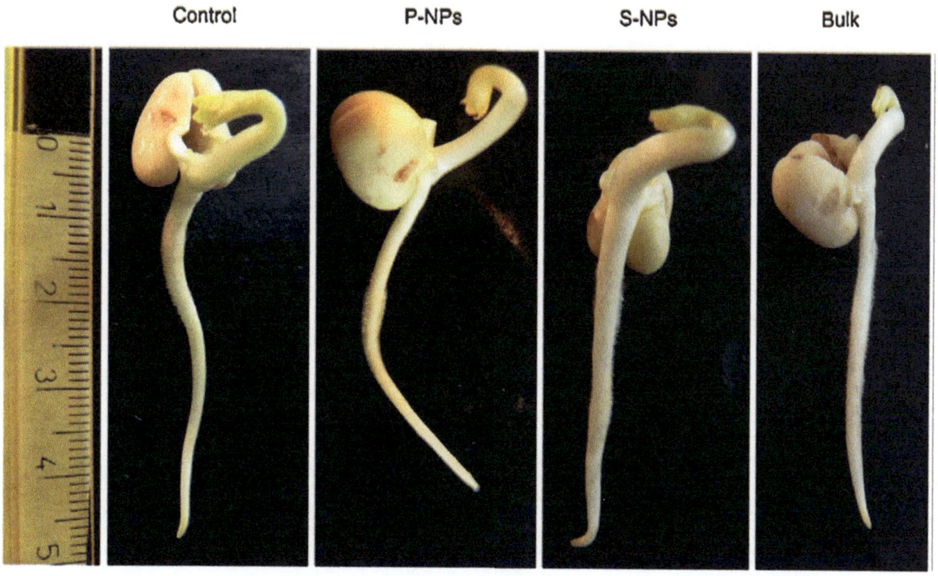

FIGURE 6.3 Seedlings of *Vicia faba*. Representative samples after 72h germination in water (control) in the presence of 50mg/L of TiO_2 P-NPs, S-NPs, and bulk counterpart (bulk). *Reprinted from Castiglionea, R.M., Giorgettib, L., Bellanib, L., Mucciforac, S., Bottegaa, S., Spanòa, C., 2016. Root responses to different types of TiO_2 nanoparticles and bulk counterpart in plant model system Vicia faba L. Environ. Exp. Bot. 130, 11–21, Copyright (2016), with permission from Elsevier.*

Given that nanoparticles can influence growth and development of plants directly entering plant cells by means of various penetration mechanisms or even without being internalized, a further aim of their work was to understand how these materials can affect the root ultrastructure of this species with regard to the function of the shape and the size of the nanoparticles. The percentage of germination did not show significant differences between control and treated plants. After 72h germination, roots from B-treated seeds were significantly shorter than P-NP-treated (nanoparticles sample P: spherical shape, produced by pulsed laser ablation in liquids) roots but not significantly different from control and S-NP (nanoparticles sample S: tetragonal crystals, from Sigma-Aldrich, USA)-treated materials. The highest vigor index was detected in P-NP-treated roots, while the minimum value was characteristic of B-(bulk counterpart) treated samples (Fig. 6.3).

6.2.11 Copper Nanoparticles

Hafeez et al. (2015) examined effects of copper nanoparticles for enhancing growth and yield of wheat cultivar Millat-2011. Their findings showed that germination of seed was not affected with 0.2–0.8ppm of copper nanoparticles, while it decreased significantly at 1.0ppm. In solution culture, concentration of copper nanoparticles of more than 2ppm was detrimental to this plant. However, application of low concentrations of copper nanoparticles (0.2, 0.4, 0.6, 0.8, and 1.0ppm) conspicuously amplified leaf area, chlorophyll content, leaf freshness, as well as dry weight, and also arid weight of the root in comparison to control samples.

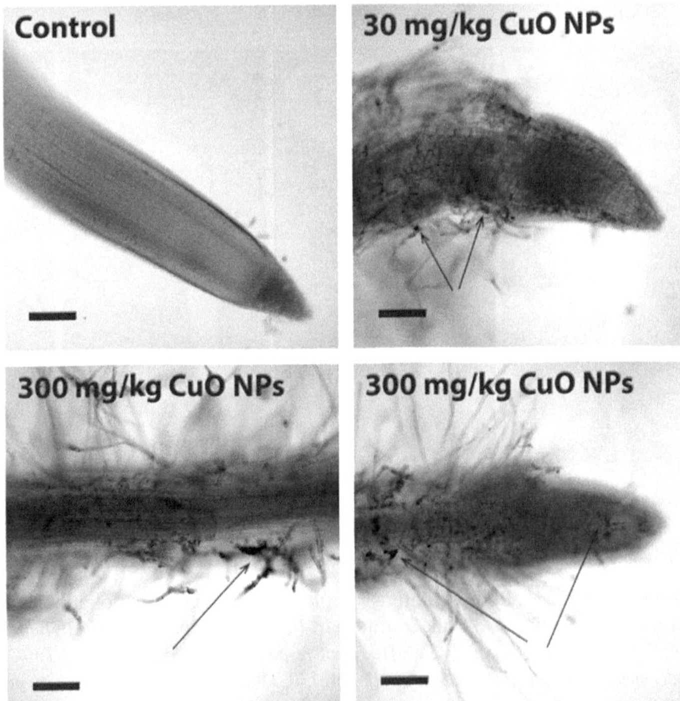

FIGURE 6.4 Presence of aggregates of copper oxide nanoparticles on roots of plantlets grown for 7 days in sand amended with 30 mg Cu and 300 mg/kg from copper oxide nanoparticles. Aggregates are not apparent from roots grown without the nanoparticles. Images taken with the LSM 710 are representative of at least 10 different root samples. Bar is 200 mm. The *arrows* show aggregates of nanoparticles on the root hairs and cell surface. *Reproduced from Adams, J., Wright, M, Wagner, H, Valiente, J., Britt, D., Anderson, A. 2017. Cu from dissolution of CuO nanoparticles signals changes in root morphology. Plant Physiol. Biochem. 110, 108–117. Copyright © 2017. Elsevier Masson SAS. All rights reserved.*

Furthermore, application of different concentrations of copper nanoparticles (10, 20, 30, 40, and 50 ppm) to pot soil significantly maximized wheat growth as well as yield as compared with control plants. It is important to know that 30 ppm copper nanoparticles manufactured significantly higher content of chlorophyll, leaf area, spikes/pot number, grains/spike number, weight of 100 grains, and also grain yield (Hafeez et al., 2015). Treatment of *Oryza sativa*, var. Jyoti with copper(II) oxide nanoparticles decreased seed germination rate, length of shoot and root, and also biomass. Results confirmed that the toxic effect of Cu accumulation in roots and shoots led to reduced photosynthesis (Da Costa and Sharma, 2016).

The presence of copper oxide nanoparticles (>10 mg Cu/kg) prevented wheat growth in sand. In this condition, different morphological variations occur such as root hair proliferation, shortening of the division zones, and elongation. Moreover, the epidermal cells in the compressed root tip became abnormal in shape (Figs. 6.4 and 6.5). Variations in root morphology caused by the copper oxide nanoparticles were likely due to copper release from dissolution at the surface of the root, because resembling alternations happened with copper ions (6 mg/kg) (Adams et al., 2017). The studies of Peng et al. (2015) demonstrated that copper oxide nanoparticles induced cell death in the germinated seedlings of rice in the meristematic as well as elongation zone and is proposed to be related to stunted growth.

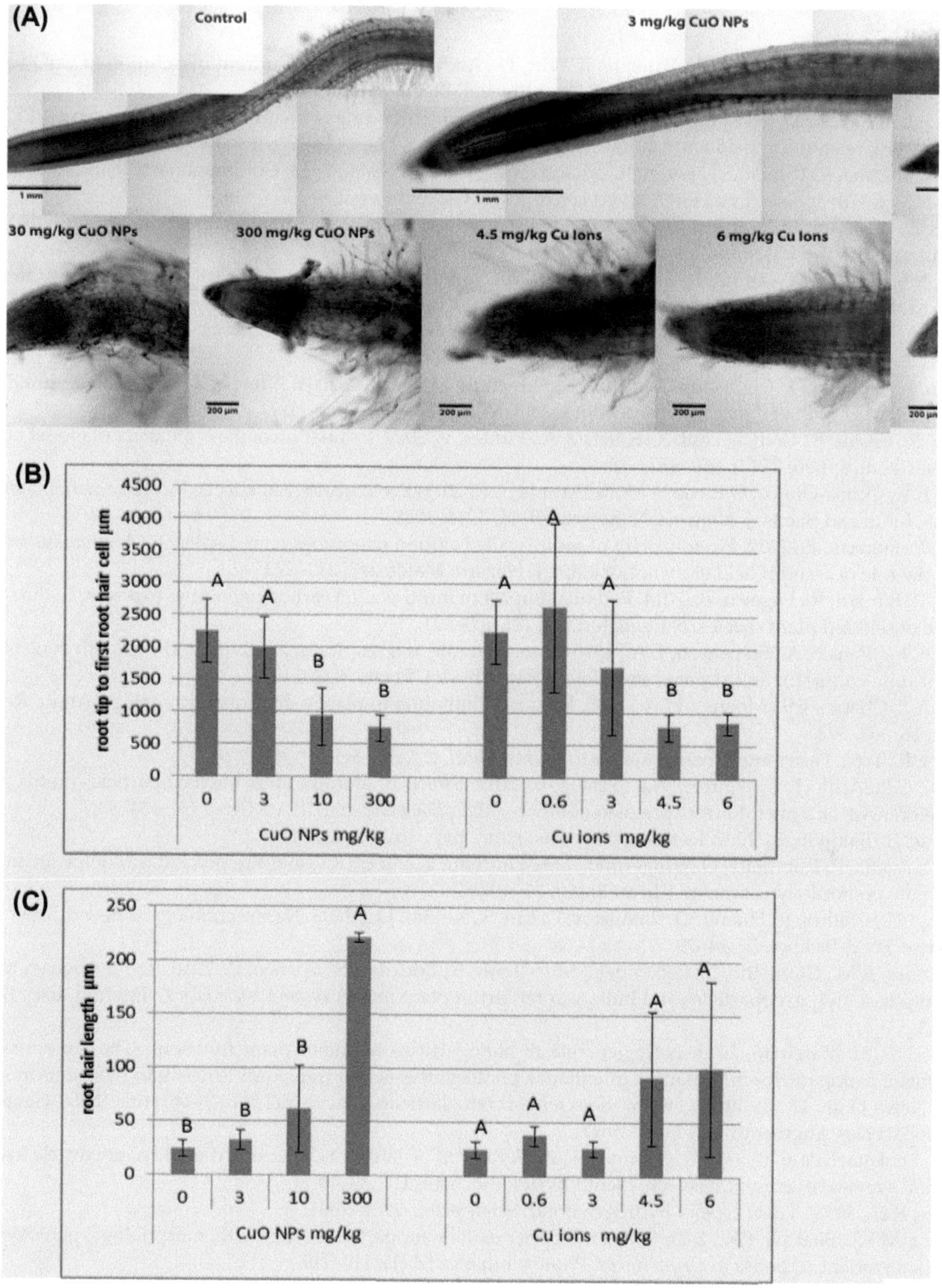

FIGURE 6.5 Images of root tips from plants grown with copper oxide nanoparticles, Cu ions, or without challenge. Images are typically of three different roots from three independent studies all grown for 7 days ($n=9$). (A) Root tips grown as a control with no Cu additions, with 3, 30, and 300 mg/kg Cu from copper oxide nanoparticles and with 4.5 and 6 mg/kg Cu ions. (B) The relationship between distance to the initial formation of root hairs and Cu treatments. Data are based on at least three roots from three different studies. (C) The relationship between lengths of initial root hairs detected from the root tip and Cu treatments. Values are for 30 root hairs/treatment. Means and standard deviations are shown. Values that differ from the control are designated with a different letter. *Reproduced from Adams, J., Wright, M, Wagner, H, Valiente, J., Britt, D., Anderson, A. 2017. Cu from dissolution of CuO nanoparticles signals changes in root morphology. Plant Physiol. Biochem. 110, 108–117.*

References

Adams, J., Wright, M., Wagner, H., Valiente, J., Britt, D., Anderson, A., 2017. Cu from dissolution of CuO nanoparticles signals changes in root morphology. Plant Physiol. Biochem. 110, 108–117.

Aguirrezabal, L.A.N., Deleens, E., Tardieu, F., 1994. Root elongation rate is accounted for by intercepted PPFD and source-sink relations in field and laboratory-grown sunflower. Plant Cell Environ. 17, 443–450.

Arora, S., Sharma, P., Kumar, S., Nayan, R., Khanna, P.K., Zaidi, M.G.H., 2012. Gold-nanoparticle induced enhancement in growth and seed yield of *Brassica juncea*. Plant Growth Regul. 66, 303–310.

Askary, M., Amirjani, M.R., Saberi, T., 2014. Evaluation of the effects of iron nanofertilizer on leaf growth, antioxidants and carbohydrate contents of *Catharanthus roseus*. JPPF 3 (7), 43–56.

Askary, M., Talebi, S.M., Amini, F., Bangan, A.D.B., 2016a. Effects of stress on foliar trichomes plasticity in *Mentha piperita*. Nusant. Biosci. 8, 32–38.

Askary, M., Talebi, S.M., Amini, F., Sarabadani, B., 2016b. Trichomes plasticity in *Lallemantia royleana* under different stress. Nusant. Biosci. 8, 128–133.

Baoshan, L., Shao-qi, D., Chun-hui, L., Li-Jun, F., Shu-chun, Q., Min, Y., 2004. Effect of TMS (nanostructured silicon dioxide) on growth of *Changbai larch* seedlings. J. For. Res. 15, 138–140.

Barrena, R., Casals, E., Coln̄, J., Font, X., SJnchez, A., Puntes, V., 2009. Evaluation of the ecotoxicity of model nanoparticles. Chemosphere 75 (7), 850–857.

Basiuk, E.V., Ochoa-Olmos, O.E., de la Mora-Estrada, L.F., 2011. Ecotoxicological effects of carbon nanomaterials on algae, fungi and plants. J. Nanosci. Nanotechnol. 11, 3016–3038.

Begum, P., Fugetsu, B., 2012. Phytotoxicity of multi-walled carbon nanotubes on red spinach (*Amaranthus tricolor* L) and the role of ascorbic acid as an antioxidant. J. Hazard. Mater. 243, 212–222.

Begum, P., Ikhtiari, R., Fugetsu, B., 2014. Potential impact of multi-walled carbon nanotubes exposure to the seedling stage of selected plant species. Nanomaterials 4 (2), 203–221.

Bingham, I.J., Panico, A., Stevenson, E.A., 1996. Extension rate and respiratory activity in the growth zone of wheat roots: time-course for adjustments after defoliation. Physiol. Plant. 98, 201–209.

Bloom, A.J., Chapin, F.S., Mooney, H.A., 1985. Resource limitation in plants- an economic analogy. Annu. Rev. Ecol. Syst. 16, 363–392.

Brouwer, R., 1962. Distribution of dry matter in plants. Neth. J. Agric. Sci. 10, 361–376.

Burklew, C.E., Ashlock, J., Winfrey, W.B., Zhang, B., 2012. Effects of aluminium oxide nanoparticles on the growth, development, and microRNA expression of tobacco (*Nicotiana tabacum*). PLoS One 7 (5), e34783.

Calabrese, E., Baldwin, L., 2002. Defining hormesis. Hum. Exp. Toxicol. 21, 91–97.

Calza, P., Avetta, P., Rubulotta, G., Sangermano, M., Laurenti, E., 2014. TiO_2-soybean peroxidase composite materials as a new photocatalytic system. Chem. Eng. J. 239 (0), 87–92.

Carmen, I.U., Chithra, P., Huang, Q., Takhistov, P., Liu, S., Kokini, J.L., 2003. Nanotechnology: a new frontier in food science. Food Technol. 57, 24–29.

Castiglionea, R.M., Giorgettib, L., Bellanib, L., Mucciforac, S., Bottegaa, S., Spanòa, C., 2016. Root responses to different types of TiO_2 nanoparticles and bulk counterpart in plant model system *Vicia faba* L. Environ. Exp. Bot. 130, 11–21.

Cossins, D., (3), March 16, 2014. Next generation: nanoparticles augment plant functions. The incorporation of synthetic nanoparticles into plants can enhance photosynthesis and transform leaves into biochemical sensors. Sci. News Opin. 28 (3). http://www.the-scientist.com/?articles.view/articleNo/39440/title/Next-Generation–Nanoparticles-Augment-Plant-Functions/.

Cox, A., Venkatachalam, P., Sahi, S., Sharma, N., 2017. Reprint of: silver and titanium dioxide nanoparticle toxicity in plants: a review of current research. Plant Physiol. Biochem. 110, 33–49.

Crabtree, R.H., 1998. A new type of hydrogen bond. Science 282, 2000–2001.

Da Costa, M.V.J., Sharma, P.K., 2016. Effect of copper oxide nanoparticles on growth, morphology, photosynthesis, and antioxidant response in *Oryza sativa*. Photosynthetica 54 (1), 110–119.

de la Rosa, G., Lopez-Moreno, M.L., de Haro, D., Botez, C.E., Peralta-Videa, J.R., Gardea-Torresdey, J.L., 2013. Effects of ZnO nanoparticles in alfalfa, tomato, and cucumber at the germination stage: root development and X-ray absorption spectroscopy studies. Pure Appl. Chem. 85 (12), 2161–2174.

DeRosa, M.C., Monreal, C., Schnitzer, M., Walsh, R., Sultan, Y., 2010. Nanotechnology in fertilizers. Nat. Nanotechnol. 5, 91. http://dx.doi.org/10.1038/nnano.2010.2.

Dimkpa, C.O., McLean, J.E., Britt, D.W., Anderson, A.J., 2013. Antifungal activity of ZnO nanoparticles and their interactive effect with a biocontrol bacterium on growth antagonism of the plant pathogen *Fusarium graminearum*. Biometals 26, 913–924.

Du, W., Tan, W., Peralta-Videa, J.R., Gardea-Torresdey, J.L., Ji, R., Yin, Y., Guo, H., 2017. Interaction of metal oxide nanoparticles with higher terrestrial plants: physiological and biochemical aspects. Plant Physiol. Biochem. 110, 210–225.

Eichert, T., Kurtz, A., Steiner, U., Goldbach, H.E., 2008. Size exclusion limits and lateral heterogeneity of the stomatal foliar uptake pathway for aqueous solutes and water-suspended nanoparticles. Physiol. Plant. 134, 151–160.

Freixes, S., Thibaud, M., Tardieu, F., Muller, B., 2002. Root elongation and branching is related to local hexose concentration in *Arabidopsis thaliana* seedlings. Plant Cell Environ. 25, 1357–1366.

Gajanan, G., Deuk, S.Y., Donghee, P., Sung, L.D., 2010. Phytotoxicity of carbon nanotubes assessed by *Brassica juncea* and *Phaseolus mungo*. J. Nanoelectron. Optoelectron. 5, 157–160.

Galbraith, D.W., 2007. Nanobiotechnology: silica breaks through in plants. Nat. Nanotechnol. 2, 272–273.

Giraldo, J.P., Landry, M.P., Faltermeier, S.M., McNicholas, T.P., Iverson, N.M., Boghossian, A.A., Reuel, N.F., Hilmer, A.J., Sen, F., Brew, J.A., Strano, M.S., 2014. Plant nanobionics approach to augment photosynthesis and biochemical sensing. Nat. Mater. http://dx.doi.org/10.1038/nmat3890.

Gopinath, K., Gowri, S., Karthika, V., Arumugam, A., 2014. Green synthesis of gold nanoparticles from fruit extract of *Terminalia arjuna*, for the enhanced seed germination activity *of Gloriosa superba*. J. Nanostruct. Chem. 4, 1–11.

Grechi, I., Hilbert, G., Milin, S., Robert, T., Gaudillere, J.P., 2007. Effect of light and nitrogen supply on internal C:N balance and control of root to shoot biomass allocation in grapevine. Environ. Exp. Bot. 59, 139–149.

Gruyer, N., Dorais, M., Bastien, C., Dassylva, N., Triffault-Bouchet, G., 2013. Interaction between sliver nanoparticles and plant growth. In: International Symposium on New Technologies for Environment Control, Energy-Saving and Crop Production in Greenhouse and Plant Factory–Greensys, Jeju, Korea, 6–11 October 2013.

Hafeez, A., Razzaq, A., Mahmood, T., Jhanzab, H.M., 2015. Potential of copper nanoparticles to increase growth and yield of wheat. J. Nanosci. Adv. Technol. 1 (1), 1–6.

Haghighi, M., Afifipour, Z., Mozafarian, M., 2012. The effect of N-Si on tomato seed germination under salinity levels. J. Biol. Environ. Sci. 6, 87–90.

Helaly, M.N., El-Metwally, M.A., El-Hoseiny, H., Omar, S.A., El-Sheery, N.I., 2014. Effect of nanoparticles on biological contamination of in vitro cultures and organogenic regeneration of banana. Aust. J. Crop Sci. 8, 612–624.

Hojjat, S.S., 2016. The effect of silver nanoparticle on lentil seed germination under drought stress. Int. J. Farm. Alli. Sci. 5 (3), 208–212.

Hong, F., Wang, L., Meng, X., Wei, Z., Zhao, G., 2002. The effect of cerium (III) on the chlorophyll formation in spinach. Biol. Trace Elem. Res. 89, 259–271.

Hong, F., Zhou, J., Liu, C., Yang, F., Wu, C., Zheng, L., Yang, P., 2005. Effect of nano-TiO_2 on photochemical reaction of chloroplasts of spinach. Biol. Trace Elem. Res. 105 (1–3), 269–279.

Hu, Z., Richter, H., Sparovek, G., Schnug, E., 2004. Physiological and biochemical effects of rare earth elements on plants and their agricultural significance: a review. J. Plant Nutr. 27, 183–220.

Husen, A., Siddiqi, K.S., 2014. Carbon and fullerene nanomaterials in plant system. J. Nanotechnol. 12, 1–10.

Ikhtiar, R., Begum, P., Watari, F., Fugetsu, B., 2013. Toxic effect of multiwalled carbon nanotubes on lettuce (*Lactuca Sativa*). Nano Biomed. 5, 18–24.

Jaberzadeh, A., Moaveni, P., Moghadam, H.R.T., Zahedi, H., 2013. Influence of bulk and nanoparticles titanium foliar application on some agronomic traits, seed gluten and starch contents of wheat subjected to water deficit stress. Not. Bot. Horti Agrobot. Cluj Napoca 41, 201–207.

Janmohammadi, M., Sabaghnia, N., Ahadnezhad, A., 2015. Impact of silicon dioxide nanoparticles on seedling early growth of lentil (*Lens culinaris* medik.) genotypes with various origins. Agric. For. 61 (3), 19–33.

Jasim, B., Thomas, R., Mathew, J., Radhakrishnan, E.K., 2016. Plant growth and diosgenin enhancement effect of silver nanoparticles in Fenugreek (*Trigonella foenum-graecum* L.). Saudi Pharm. J. http://dx.doi.org/10.1016/j.jsps.2016.09.012.

Kalteh, M., Alipour, Z.T., Ashraf, S., Aliabadi, M.M., Nosratabadi, A.F., 2014. Effect of silica nanoparticles on basil (*Ocimum basilicum*) under salinity stress. J. Chem. Health Risks 4, 49–55.

Karunakaran, G., Suriyaprabha, R., Manivasakan, P., Yuvakkumar, R., Rajendran, V., Mahmoodzadeh, H., Aghili, R., 2014. Effect on germination and early growth characteristics in wheat plants (*Triticum aestivum* L.) seeds exposed to TiO_2 nanoparticles. J. Chem. Health Risks 4 (1), 29–36.

Ke, P.C., Lin, S., Reppert, J., Rao, A.M., Luo, H., 2011. Uptake of carbon-based nanoparticles by mammalian cells and plants. In: Sattler, K.D. (Ed.), Handbook of Nanophysics: Nanomedicine and Nanorobotics. CRC Press, New York, pp. 1–30.

Khodakovskaya, M.V., de Silva, K., Biris, A.S., Dervishi, E., Villagarcia, H., 2012. Carbon nanotubes induce growth enhancement of tobacco cells. ACS Nano 6 (3), 2128–2135.

Khodakovskaya, M.V., Kim, B.S., Kim, J.N., Alimohammadi, M., Dervishi, E., Mustafa, T., Cernigla, C.E., 2013. Carbon nanotubes as plant growth regulators: effects on tomato growth, reproductive system, and soil microbial community. Small 9 (1), 115–123.

Krishnaraj, C., Jagan, E.G., Ramachandran, R., Abirami, S.M., Mohan, N., Kalaichelvan, P.T., 2012. Effect of biologically synthesized silver nanoparticles on *Bacopa monnieri* (Linn.) Wettst. plant growth metabolism. Process Biochem. 47 (4), 51–658.

Kumar, V., Guleria, P., Kumar, V., Yadav, S.K., 2013. Gold nanoparticle exposure induces growth and yield enhancement in *Arabidopsis thaliana*. Sci. Total Environ. 461, 462–468.

Lacointe, A., 2000. Carbon allocation among tree organs: a review of basic processes and representation in functional-structural tree models. Ann. For. Sci. 57, 521–533.

Lahiani, M.H., Dervishi, E., Chen, J., Nima, Z., Gaume, A., Biris, A.S., Khodakovskaya, M.V., 2013. Impact of carbon nanotube exposure to seeds of valuable crops. ACS Appl. Mater. Interfaces 5, 7965–7973.

Li, B., Tao, G., Xie, Y., Cai, X., 2012. Physiological effects under the condition of spraying nano-SiO_2 onto the *Indocalamus barbatus* McClure leaves. J. Nanjing For. Univ. 36, 161–164 (Natural Science Edition).

Lin, D., Xing, B., 2007. Phytotoxicity of nanoparticles: inhibition of seed germination and root growth. Environ. Pollut. 150 (2), 243–250.

Liu, R., Zhang, H., Lal, R., 2016. Effects of stabilized nanoparticles of copper, zinc, manganese, and iron oxides in low concentrations on lettuce (*Lactuca sativa*) seed germination: nanotoxicants or nano-nutrients. Water Air Soil Pollut. 227, 42–54.

Lu, C.M., Zhang, C.Y., Wen, J.Q., Wu, G.R., Tao, M.X., 2002. Research on the effect of nanometer materials on germination and growth enhancement of *Glycine max* and its mechanism. Soybean Sci. 21, 68–172.

Ma, X., Geiser-Lee, J., Deng, Y., Kolmakov, A., 2010a. Interactions between engineered nanoparticles (ENPs) and plants: phytotoxicity, uptake and accumulation. Sci. Total Environ. 408 (16), 3053–3061.

Ma, Y., Kuang, L., He, X., Bai, W., Ding, Y., Zhang, Z., Zhao, Y., Chai, Z., 2010b. Effects of rare earth oxide nanoparticles on root elongation of plants. Chemosphere 78, 273–279.

Mahajan, P., Dhoke, S.K., Khanna, A.S., 2011. Effect of nano-ZnO particle suspension on growth of mung (*Vigna radiata*) and gram (*Cicer arietinum*) seedlings using plant agar method. J. Nanotechnol. 2011, 1–7. http://dx.doi.org/10.1155/2011/696535.

Mahmoodzadeh, H., Nabavi, M., Kashefi, H., 2013. Effect of nanoscale titanium dioxide particles on the germination and growth of canola (*Brassica napus*). J. Ornam. Hortic. Plants 3, 25–32.

Miralles, P., Johnson, E., Church, T.L., Harris, A.T., 2012. Multiwalled carbon nanotubes in alfalfa and wheat: toxicology and uptake. J. R. Soc. Interface 9 (77), 3514–3527.

Mishra, V., Mishra, R.K., Dikshit, A., Pandey, A.C., 2014. Interactions of nanoparticles with plants: an emerging prospective in the agriculture industry. In: Ahmad, P., Rasool, S. (Eds.)Ahmad, P., Rasool, S. (Eds.), Emerging Technologies and Management of Crop Stress Tolerance: Biological Techniques, vol. 1, pp. 159–180.

Mondal, A., Basu, R., Das, S., Nandy, P., 2011. Beneficial role of carbon nanotubes on mustard plant growth: an agricultural prospect. J. Nanopart. Res. 13, 4519–4528.

Monica, R.C., Cremonini, R., 2009. Nanoparticles and higher plants. Caryologia 62 (2), 161–165.

Muller, B., Stosser, M., Tardieu, F., 1998. Spatial distributions of tissue expansion and cell division rates are related to irradiance and to sugar content in the growing zone of maize roots. Plant Cell Environ. 21, 149–158.

Nair, R., Varghese, S.H., Nair, B.G., Maekawa, T., Yoshida, Y., Kumar, D.S., 2010. Nanoparticulate material delivery to plants. Plant Sci. 179, 154–163.

Nalwade, A.R., Neharkar, S.B., 2013. Carbon nanotubes enhance the growth and yield of hybrid Bt cotton Var. ACH-177-2. Int. J. Adv. Sci. Tech. Res. 3, 840–846.

Palmqvist, N.G., Bejai, S., Meijer, J., Seisenbaeva, G.A., Kessler, V.G., 2015. Nano titania aided clustering and adhesion of beneficial bacteria to plant roots to enhance crop growth and stress management. Sci. Rep. 5, 10146.

Pariona, N., Martínez, A.I., Hernandez-Flores, H., Clark-Tapia, R., 2017. Effect of magnetite nanoparticles on the germination and early growth of *Quercus macdougallii*. Sci. Total Environ. 575, 869–875.

Peng, C., Zhang, H., Fang, H., Xu, C., Huang, H., Wang, Y., Sun, L., Yuan, X., Chen, Y., Shi, J., 2015. Natural organic matter-induced alleviation of the phytotoxicity to rice (*Oryza sativa* L.) caused by copper oxide nanoparticles. Environ. Toxicol. Chem. 34, 1996–2003.

Prakash, M., Nair, G., Chung, M., 2017. Regulation of morphological, molecular and nutrient status in *Arabidopsis thaliana* seedlings in response to ZnO nanoparticles and Zn ion exposure. Sci. Total Environ. 575, 187–198.

Prasad, T.N.V.K.V., Sudhakar, P., Sreenivasulu, Y., Latha, P., Munaswamy, V., Reddy, K.R., Sreeprasad, T.S.P., Sajanlal, R., Pradeep, T., 2012. Effect of nanoscale zinc oxide particles on the germination, growth and yield of peanut. J. Plant Nutr. 35 (6), 905–927.

Raliya, R., Tarafdar, J.C., 2013. ZnO nanoparticle biosynthesis and its effect on phosphorous-mobilizing enzyme secretion and gum contents in cluster bean (*Cyamopsis tetragonoloba* L.). Agric. Res. 2, 48–57.

Raskar, S.V., Laware, S.L., 2014. Effect of zinc oxide nanoparticles on cytology and seed germination in onion. Int. J. Curr. Microbiol. App. Sci. 3, 467–473.

Reymond, M., Muller, B., Leonardi, A., Charcosset, A., Tardieu, F., 2003. Combining quantitative trait loci analysis and an ecophysiological model to analyze the genetic variability of the responses of maize leaf growth to temperature and water deficit. Plant Physiol. 131, 664–675.

Rezvani, N., Sorooshzadeh, A., Farhadi, N., 2012. Effect of nano-silver on growth of saffron in flooding stress. World Acad. Sci. Eng. Technol. 1, 517–522.

Riahi-Madvar, A., Rezaee, F., Jalili, V., 2012. Effects of alumina nanoparticles on morphological properties and antioxidant system of *Triticum aestivum*. IJPP 3 (1), 595–603.

Salama, H.M.H., 2012. Effects of silver nanoparticles in some crop plants, common bean (*Phaseolus vulgaris* L.) and corn (*Zea mays* L.). Int. Res. J. Biotechnol. 3 (10), 190–197.

Savithramma, N., Ankanna, S., Bhumi, G., 2012. Effect of nanoparticles on seed germination and seedling growth of *Boswellia ovalifoliolata* an endemic and endangered medicinal tree taxon. Nano Vis. 2, 61–68.

Scrinis, G., Lyons, K., 2007. The emerging nano-corporate paradigm: nanotechnology and the transformation of nature, food and agri-food systems. Int. J. Sociol. Food Agric. 15, 22–44.

Sedghi, M., Hadi, M., Toluie, S.G., 2013. Effect of nano zinc oxide on the germination of soybean seeds under drought stress. Ann. West Univ. Timişoara Ser. Biol. 2, 73–78.

Sergey, B., LeFebvre, M., Sherwood, J., Xu, Y., Bao, Y., Ramonell, K.M., 2015. Developmental and reproductive effects of iron oxide nanoparticles in *Arabidopsis thaliana*. Int. J. Mol. Sci. 16, 24174–24193.

Shah, V., Belozerova, I., 2009. Influence of metal nanoparticles on the soil microbial community and germination of lettuce seeds. Water Air Soil Pollut. 197, 143–148.

Sharma, P., Bhatt, D., Zaidi, M.G., Saradhi, P.P., Khanna, P.K., Arora, S., 2012. Silver nanoparticle mediated enhancement in growth and antioxidant status of *Brassica juncea*. Appl. Biochem. Biotechnol. 167, 2225–2233.

Shweta, Tripathi, D.K., Singh, S., Singh, S., Dubey, N.K., Chauhan, D.K., 2016. Impact of nanoparticles on photosynthesis: challenges and opportunities. Mater. Focus 5 (5), 405–411.

Siddiqui, M.H., Al-Whaibi, M.H., 2014. Role of nano-SiO_2 in germination of tomato (*Lycopersicum esculentum* seeds Mill.). Saudi J. Biol. Sci. 21 (1), 13–17.

Siddiqui, M.H., Govorov, A.O., Carmeli, I., 2007. Hybrid structures composed of photosynthetic system and metal nanoparticles: plasmon enhancement effect2Lycopersicum esculentum. Nano Lett. 7 (3), 620–625.

Siddiqui, M.H., Al-Whaibi, M.H., Faisal, M., Al Sahli, A.A., 2014. Nano-silicon dioxide mitigates the adverse effects of salt stress on *Cucurbita pepo* L. Environ. Toxicol. Chem. 33 (11), 2429–2437. http://dx.doi.org/10.1002/etc.2697.

Singh, S., Tripathi, D.K., Dubey, N.K., Chauhan, D.K., 2016. Effects of nano-materials on seed germination and seedling growth: striking the slight balance between the concepts and controversiesCucurbita pepo. Mater. Focus 5 (3), 195–201.

Singh, S., Vishwakarma, K., Singh, S., Sharma, S., Dubey, N.K., Singh, V.K., Liu, S., Tripathi, D.K., Chauhan, D.K., 2017. Understanding the plant and nanoparticle interface at transcriptomic and proteomic level: a concentric overview. Plant Gene. http://dx.doi.org/10.1016/j.plgene.2017.03.006.

Smirnova, E., Gusev, A., Zaytseva, O., Sheina, O., Tkachev, A., Kuznetsova, E., Lazareva, E., Song, G., Gao, Y., Wu, H., Hou, W., Zhang, C., Ma, H., 2012. Physiological effect of anatase TiO_2 nanoparticles on *Lemna minor*. Environ. Toxicol. Chem. 31 (9), 2147–2152.

Song, U., Shin, M., Lee, G., Roh, J., Kim, Y., Lee, E.J., Lazareva, E., Song, G., Gao, Y., Wu, H., Hou, W., Zhang, C., Ma, H., 2013. Functional analysis of TiO_2 nanoparticle toxicity in three plant speciesLemna minor. Biol. Trace Elem. Res. 155 (9), 93–103.

Srinivasan, C., Saraswathi, R., Lee, G., Roh, J., Kim, Y., Lee, E.J., 2010. Nano-agriculture-carbon nanotubes enhance tomato seed germination and plant growth2. Curr. Sci. 99, 273–275.

Srivastava, G., Das, C.K., Das, A., Singh, S.K., Roy, M., Kim, H., Sethy, N., Kumar, A., Sharma, S.K., Singh, R.K., Philip, D., Das, M., 2014. Seed treatment with iron pyrite (FeS_2) nanoparticles increases the production of spinach. RSC Adv. 4 (102), 58495–58504.

Stewart, J., Hansen, T., McLean, J.E., McManus, P., Das, S., Britt, D.W., Anderson, A.J., Dimkpa, C.O., Sharma, S.K., Singh, R.K., Philip, D., Das, M., 2015. Salts affect the interaction of ZnO or CuO nanoparticles with wheat2. Environ. Toxicol. Chem. 34 (9), 2116–2125.

Suriyaprabha, R., Karunakaran, G., Yuvakkumar, R., Rajendran, V., Kannan, N., Britt, D.W., Anderson, A.J., Dimkpa, C.O., 2012. Silica nanoparticles for increased silica availability in maize (*Zea mays* L) seeds under hydroponic conditions. Curr. Nanosci. 8 (9), 902–908.

Syu, Y.Y., Hung, J.H., Chen, J.C., Chuang, H.W., Kannan, N., 2014. Impacts of size and shape of silver nanoparticles on *Arabidopsis* plant growth and gene expression. Plant Physiol. Biochem. 8, 57–64.

Teale, W.D., Paponov, I.A., Palme, K., Chuang, H.W., 2006. Auxin in action: signaling, transport and the control of plant growth and developmentArabidopsis. Nat. Rev. Mol. Cell Biol. 7, 847–859.

Tiwari, D.K., Dasgupta-Schubert, N., Villaseo̊r-Cendejas, L.M., Villegas, J., Carreto-Montoya, L., Borjas-Garcia, S.E., 2014. Interfacing carbon nanotubes (CNT) with plants: enhancement of growth, water and ionic nutrient uptake in maize (*Zea Mays*) and implications for nanoagriculture. Appl. Nanosci. 4, 577–591.

Torney, F., Trewyn, B.G., Lin, V.S.Y., Wang, K., Carreto-Montoya, L., Borjas-Garcia, S.E., 2007. Mesoporous silica nanoparticles deliver DNA and chemicals into plantsZea Mays. Nat. Nanotechnol. 2, 295–300.

Tripathi, S., Sarkar, S., Lin, V.S.Y., Wang, K., 2014. Influence of water soluble carbon dots on the growth of wheat plant. Appl. Nanosci. 2 http://dx.doi.org/10.1007/s13204-014-0355-9 295–300.

Tripathi, D.K., Singh, V.P., Prasad, S.M., Chauhan, D.K., Dubey, N.K., 2015. Silicon nanoparticles (SiNp) alleviate chromium (VI) phytotoxicity in *Pisum sativum* (L.) seedlings. Plant Physiol. Biochem. 96, 189–198http://dx.doi.org/10.1007/s13204-014-0355-9.

Tripathi, D.K., Singh, S., Singh, V.P., Prasad, S.M., Chauhan, D.K., Dubey, N.K., 2016. Silicon nanoparticles more efficiently alleviate arsenate toxicity than silicon in maize cultiver and hybrid differing in arsenate tolerancePisum sativum. Front. Environ. Sci. 4, 46–198.

Tripathi, D.K., Singh, S., Singh, S., Pandey, R., Singh, V.P., Sharma, N.C., Prasad, S.M., Dubey, N.K., Chauhan, D.K., 2017a. An overview on manufactured nanoparticles in plants: uptake, translocation, accumulation and phytotoxicity. Plant Physiol. Biochem. 110, 2–12.

Tripathi, D.K., Mishra, R.K., Singh, S., Singh, S., Vishwakarma, K., Sharma, S., Singh, V.P., Singh, P.K., Prasad, S.M., Dubey, N.K., Pandey, A.C., 2017b. Nitric oxide ameliorates zinc oxide nanoparticles phytotoxicity in wheat seedlings: implication of the ascorbate–glutathione cycle. Front. Plant Sci. 8, 2–12.

Tripathi, D.K., Singh, S., Singh, S., Srivastava, P.K., Singh, V.P., Singh, S., Prasad, S.M., Singh, P.K., Dubey, N.K., Pandey, A.C., Chauhan, D.K., 2017c. Nitric oxide alleviates silver nanoparticles (AgNps)-induced phytotoxicity in *Pisum sativum* seedlings. Plant Physiol. Biochem. 110, 167–177.

Tripathi, D.K., Tripathi, A., Shweta, S.S., Singh, Y., Vishwakarma, K., Yadav, G., Sharma, S., Singh, V.K., Mishra, R.K., Upadhyay, R.G., Dubey, N.K., 2017d. Uptake, accumulation and toxicity of silver nanoparticle in autotrophic plants, and heterotrophic microbes: a concentric reviewPisum sativum. Front. Microbiol. 8, 167–177.

Tripathi, D.K., Singh, S., Singh, V.P., Prasad, S.M., Dubey, N.K., Chauhan, D.K., Sharma, S., Singh, V.K., Mishra, R.K., Upadhyay, R.G., Dubey, N.K., 2017e. Silicon nanoparticles more effectively alleviated UV-B stress than silicon in wheat (*Triticum aestivum*) seedlings. Plant Physiol. Biochem. 110, 70–81.

Venkatachalam, P., Priyanka, N., Manikandan, K., Ganeshbabu, I., Indiraarulselvi, P., Geetha, N., Muralikrishna, K., Bhattacharya, R.C., Tiwari, M., Sharma, N., Sahi, S.V., 2017e. Enhanced plant growth promoting role of phycomolecules coated zinc oxide nanoparticles with P supplementation in cotton (*Gossypium hirsutum* L.). Plant Physiol. Biochem. 110, 118–127.

Villagarcia, H., Dervishi, E., Silva, K., Biris, A.S., Khodakovskaya, M.V., Geetha, N., Muralikrishna, K., Bhattacharya, R.C., Tiwari, M., Sharma, N., Sahi, S.V., 2012. Surface chemistry of carbon nanotubes impacts the growth and expression of water channel protein in tomato plantsGossypium hirsutum. Small 8, 2328–2334.

Wang, X., Han, H., Liu, X., Gu, X., Chen, K., Lu, D., 2012. Multi-walled carbon nanotubes can enhance root elongation of wheat (*Triticum aestivum*) plants. J. Nanopart. Res. 14 (6), 1–10.

Wang, A., Zheng, Y., Peng, F., Gu, X., Chen, K., Lu, D., 2014. Thickness-controllable silica coating of CdTe QDs by reverse microemulsion method for the application in the growth of riceTriticum aestivum. J. Spectrosc. 14 (6) http://dx.doi.org/10.1155/2014/169245 1–10.

Watson, J.L., Fang, T., Dimkpa, C.O., Britt, D.W., McLean, J.E., Jacobson, A., Anderson, A.J., 2015. The phytotoxicity of ZnO nanoparticles on wheat varies with soil properties. Biometals 28, 101–112. http://dx.doi.org/10.1007/s10534-014-9806-8.

Werner, T., Motyka, V., Laucou, V., Smets, R., van Onckelen, H., Schmülling, T., Anderson, A.J., 2003. Cytokinin-deficient transgenic Arabidopsis plants show multiple developmental alteration indicating opposite functions of cytokinins in the regulation of shoot and root meristem activity. Plant Cell 15, 2532–2550. http://dx.doi.org/10.1007/s10534-014-9806-8.

Woyro, K.H., Stelzer, R., Laucou, V., Smets, R., van Onckelen, H., Schmülling, T., 1988. Ion concentrations in the cytoplasm and vacuoles of rhizodermal cells from NaCl treated *Sorghum*, *Spartina* and *Puccinellia* plants. J. Plant Physiol. 133, 441–446.

Xie, Z., Zhu, J., Chu, H., Zhang, Y., Zeng, Q., Ma, H., Cao, Z., 2002. Effect of lanthanum on rice production, nutrient uptake, and distributionSorghumSpartinaPuccinellia. J. Plant Nutr. 25, 2315–2331.

Xie, Y., Li, B., Zhang, Q., Zhang, C., Lu, K., Tao, G., Cao, Z., 2011. Effects of nano-TiO_2 on photosynthetic characteristics of *Indocalamus barbatus*. J. Northeast For. Univ. 39, 22–25.

Yang, F., Hong, F., You, W., Liu, C., Gao, F., Wu, C., Yang, P., 2006. Influence of nano-anatase TiO_2 on the nitrogen metabolism of growing spinachIndocalamus barbatus. Biol. Trace Elem. Res. 110 (2), 179–190.

Yin, L., Colman, B.P., McGill, B.M., Wright, J.P., Bernhardt, E.S., Wu, C., Yang, P., 2012. Effects of silver nanoparticle exposure on germination and early growth of eleven wetland plants2. PLoS One 7 (2), 1–7.

Zareii, D.F., Roozbahani, A., Hosnamidi, A., Wright, J.P., Bernhardt, E.S., 2014. Evaluation the effect of water stress and foliar application of Fe nanoparticles on yield, yield components and oil percentage of safflower (*Carthamus tinctorious* L.). Int. J. Adv. Biol. Biomed. Res. 2 (4), 1150–1159.

Zhao, L., Sun, Y., Hernandez-Viezcas, J.A., Servin, A.D., Hong, J., Niu, G., Peralta-Videa, J.R., Duarte-Gardea, M., Gardea-Torresdey, J.L., 2013. Influence of CeO_2 and ZnO nanoparticles on cucumber physiological markers and bioaccumulation of Ce and Zn: a life cycle studyCarthamus tinctorious. J. Agric. Food Chem. 61 (49), 11945–11951.

Zheng, L., Hong, F., Lu, S., Liu, C., Hong, J., Niu, G., Peralta-Videa, J.R., Duarte-Gardea, M., Gardea-Torresdey, J.L., 2005. Effect of nano-TiO_2 on strength of naturally aged seeds and growth of spinach. Biol. Trace Elem. Res. 104 (1), 83–91.

CHAPTER

7

Recent Progress of Nanotoxicology in Plants

Muhammad Zia-ur-Rehman[1], *Muhammad F. Qayyum*[2], *Fatima Akmal*[1], *Muhammad A. Maqsood*[1], *Muhammad Rizwan*[3], *Maqsoda Waqar*[1], *Muhammad Azhar*[1]

[1]University of Agriculture, Faisalabad, Pakistan; [2]Bahaudin Zakariya University, Multan, Pakistan; [3]Government College University, Faisalabad, Pakistan

7.1 INTRODUCTION

Advancement in nanotechnology and nanoparticle (NP) investigations has successfully aggravated the craze to use these particles in agriculture and assess their impact on the food chain and environmental sustainability (Mura et al., 2013). Because of their smaller size, catalytic properties, and highly active surface, these particles can easily pass through cell membranes and cause toxicity by affecting the intracellular biomolecules. NP-associated phytotoxicity might be initiated by NP–soil component interactions that can change nutrient absorption by plants; in addition, NPs are also absorbed and stored in plants where they interact and can negatively affect cellular structures (Masarovičová and Kráľová, 2013). The toxic effects of NPs include restrictions in seed emergence, decreased number of seedlings, inhibition of stem and root growth, and postponement in flowering, hence decreased yield (Li et al., 2015; El-Temsah and Joner, 2012; Lin et al., 2009; Lee et al., 2008). Moreover, NPs are also responsible for cell wall breakage and reproductive damage including chromatin condensation, seize mitosis by disturbing the metaphase, and most importantly inhibiting genetic expression (Kaveh et al., 2013; Lin et al., 2009; Kumari et al., 2009). Higher applications of NPs have forced researchers to investigate their toxic effects and understand possible toxicity mechanisms, especially in plants for safe and sustainable food production. This chapter will explain NP types, characteristics, possible phytotoxic effects and mechanisms, progress in phytotoxicity, and highlight plant efforts and mechanisms to detoxify NPs.

Nanomaterials in Plants, Algae, and Microorganisms
http://dx.doi.org/10.1016/B978-0-12-811487-2.00007-4

7.2 ROLE OF NANOPARTICLES IN AGRICULTURE

To feed the increasing global population the scientific community is making continuous efforts to increase the yield of food crops. For this purpose, different technologies are recommended. Nanotechnology is one of the emerging popular technologies in agriculture and other fields because of the unique physicochemical properties of NPs. The positive and toxic effects of NPs on plant growth and development that have been documented are totally dependent on NP type, chemical composition, size, surface activity, and plant species, but the focus is mainly on concentration (Ma et al., 2010; Khodakovskaya et al., 2012; Tripathi et al., 2015, 2016a,b,c, 2017a,b,c,d). NPs are used to improve seed germination, control fertilizer losses (nanofertilizers or nanoencapsulated nutrients), boost plant metabolism, act as a carrier for DNA and chemicals inside cells, improve genetic manipulation and expression in targeted cells, enhance photosynthesis by the introduction of carbon nanotubes (CNTs) in chloroplast that capture more light energy of higher wavelengths (UV, green, and near infrared), and improve growth and yield of crop plants (Galbraith, 2007; Torney et al., 2007; de la Rosa et al., 2013; Nair et al., 2011; Lahiani et al., 2013; Giraldo et al., 2014; Cossins, 2014; Siddiqui and Al-Whaibi, 2014; Shweta et al., 2016). These NPs are used as carriers of herbicides, pesticides, fertilizers, and genes to target specific intracellular organelles, hence they are called "magic bullets" (Siddiqui et al., 2015). These are also used for the purification of water and conversion of agricultural waste into energy and other valuable by-products through enzymatic nanobioprocessing (Nair et al., 2011; Carmen et al., 2003). Under abiotic stresses, NPs-treatment enhances seed germination and biomass of crop plants, e.g., NPs improved germination and antioxidant enzymes in tomato and squash while they were grown under NaCl stress (Haghighi et al., 2012; Siddiqui et al., 2014). Under NaCl, stress germination and antioxidant enzymes were improved. For purified and sustainable food production it is necessary to consider both positive and toxic impacts of NPs. Beneficial and toxic effects are totally dependent on the dose/concentration of NPs applied.

7.3 TYPES AND CHARACTERISTICS OF TOXIC NANOPARTICLES

7.3.1 Carbon-Based Nanoparticles

Carbon-based NPs include CNTs (single-walled CNTs and multiwalled CNTs) and fullerenes. Research on carbon NPs reports the damaging effect of CNTs and fullerenes to plants (Begum and Fugetsu, 2012; Chen et al., 2010; Liu et al., 2010; Shen et al., 2010). Fullerene (C70) delayed the blossoming of rice for at least 1 month and reduced the seed-setting rate up to 4.6% compared to control (Lin et al., 2009).

Liu et al. (2010) found that water-soluble fullerene causes loss of gravitropism and shortening of seedling roots. Also single-walled CNTs induced reactive oxygen species (ROS) production, which leads to programmed cell death in *Arabidopsis* leaves and protoplasts. Similarly, multiwalled CNTs induced ROS production and necrotic lesions in red spinach leaf cells and tissues (Begum and Fugetsu, 2012). Despite the remarkable difference in plant and mammalian cell structure, both show similar response toward fullerenes (Chen et al., 2010). The presence of C60 fullerene in food crops such as tomato, soybean, and zucchini increased the bioaccumulation of the persistent pollutant dichlorodiphenyldichloroethylene (De La Torre-Roche et al., 2012). CNTs can penetrate the plant cell membrane into protoplasts (Serag et al., 2011). Additional investigations measured the uptake of C60 fullerene in plants and found that plants can gradually take up 14C-labeled C60 and persist in soil for a long time (Avanasi et al., 2014).

7.3.1.1 *Carbon Nanotubes*

CNTs have an important position among NPs because they can supply chemicals to cells by penetrating the cell membrane and cell wall of plants. Multiwalled CNTs showed positive influence on seed germination and plant growth by improving the nutrient and water uptake efficiency of plants (Villagarcia et al., 2012; Tiwari et al., 2014), by regulating the gene expression of various water channel proteins (Lahiani et al., 2013), by regulating marker genes for water transport, cell division, and cell wall formation (Khodakovskaya et al., 2012), and by enhancing dehydrogenase activity in root systems (Wang et al., 2012a). The multiwalled CNTs also exhibit an influential role in improving the plants' root, stem growth (Smirnova et al., 2012), biomass, flowering, fruiting, water retention capacity, and medicinal properties (Khodakovskaya et al., 2013; Husen and Siddiqi, 2014).

However, the inhibitory effect of multiwalled CNTs is also receiving attention (Tiwari et al., 2014; Ikhtiar et al., 2013; Begum and Fugetsu, 2012; Begum et al., 2014). Some researchers have found the negative role of multiwalled CNTs on plant growth because of their involvement in the enhancement of ROS production (Tan et al., 2009), reduction of dry weight, superoxide dismutase (SOD) activities, chlorophyll content (Lin et al., 2009), and inhibition of root growth (Doshi et al., 2008). Miralles et al. (2012) reported the comparative effect of the penetrability of multiwalled CNTs on the germination rate of wheat and alfalfa. The germination rate of wheat was improved compared to alfalfa plants possibly because of softer wheat seed penetrability of multiwalled CNTs. Multiwalled CNTs reduced the biomass of zucchini by 38% (Stampoulis et al., 2009), reduced the root length of lettuce (Lin and Xing, 2007), induced chromatin condensation, cell death, detachment of plasma membrane from the cell wall, and cell shrinkage of rice (Tan et al., 2009), and inhibited growth and cell death of red spinach (Begum and Fugetsu, 2012).

Single-walled CNTs act as effective nanotransporters for DNA and dye molecule delivery into the intact cells of plants (Srinivasan and Saraswathi, 2010). Single-walled CNTs revealed increased cucumber and onion root length, but also reported a reduction in lettuce root growth, no effect on cabbage and carrot root elongation (Cañas et al., 2008), and also delayed the flowering and yield of rice (Lin et al., 2009). Single-walled CNTs showed adverse effects on plant cell viability, induced chromatin condensation, and also induced ROS production that consequently lead to programmed cell death (Shen et al., 2010).

7.3.1.2 *Fullerene*

Kole et al. (2013) treated bitter melon-seed with fullerene and their results show significant increase in the biomass, phytomedicine content, fruit number and fresh mass, and overall yield of plant with respect to control but the effect diminished at high concentrations. While in contrast, De La Torre-Roche et al. (2013) reported reduced biomass of corn and soybean treated with C60 fullerene.

Fullerene is involved in the inhibition of cell growth because its adsorption leads to cell wall and cell membrane disruption and ROS production, thereby proving to be phytotoxic to tobacco plants (Liu et al., 2013). Fullerene $C_{70}(C(COOH)_2)_4$ reduced the root length and root gravitropism in *Arabidopsis* (Liu et al., 2010) and $C_{60}(OH)_{20}$ caused cell death of *Allium cepa* (Chen et al., 2010). Fullerene C60 is also involved in the upregulation of stress-related genes in *Arabidopsis* (Landa et al., 2012).

7.3.2 Metal and Metal Oxide Nanoparticles

The metals and metal-oxide NPs hold important position in NPs group due to their broad application. Nowadays, their inhibitory role in plant growth is discussed comprehensively (Ma et al., 2015a; Hossain, 2015; Gardea-Torresdey et al., 2014). CuO NPs showed an inhibitory effect on radish and annual and perennial ryegrass by damaging their DNA (Atha et al., 2012). The phytotoxic effect of CuO and ZnO NPs exhibited alteration in root morphology and reduced root growth with shortened length; in addition, their genotoxic effects showed interference in cellular homeostasis, damaged plant DNA, and thereby altered gene expression levels of buckwheat plant (Lee et al., 2013b). Similarly, reduction in wheat root length was observed when grown in CuO NP-treated sand (Dimkpa et al., 2013a). Previous studies demonstrated the phytotoxicity of metal oxide NPs in a dose-dependent manner (Watson et al., 2015; Zhang et al., 2015b; López-Moreno et al., 2016). Exposure to CuO NPs in sand, soil, and hydroponic conditions caused root shortening in an array of plants (Wang et al., 2012a,b,c,d; Dimkpa et al., 2013b; Shi et al., 2014a,b; Peng et al., 2015; Ebbs et al., 2016; Nair and Chung, 2015).

TiO_2 NPs have wide application in cosmetics, wastewater, and antibacterial products, while some controversial results show TiO_2 to have opposite results on plants. Gao et al. (2013) demonstrated that exposure of *Ulmus elongata* seedlings to TiO_2 NPs leads to reduced photosynthetic rate via nonstomatal regulation, increased chlorosis of leaf, defoliation, and caused disrupted growth. In contrast, TiO_2 application provided positive effects on plants, especially on root elongation, and showed no toxic effects on seed germination, enzymatic activities, chlorophyll content, total antioxidant capacity, SOD activities, and physiological differences even when exposed at high concentrations (Song et al., 2013a,b). Similarly, TiO_2 application explicated the positive influence on germination rate and germination time of fennel seeds (Feizi et al., 2013).

ZnO NPs also proved to be phytotoxic to plants. Lee et al. (2013a) gave a clear illustration of these NPs. They tested ZnO NP application to *Fagopyrum esculentum* and their results showed significant reduction in biomass, shortened roots, and stunted growth with the generation of ROS. The inhibitory effects of ZnO-NPs are strictly dependent to plant species and soil conditions as no biomass reduction as in ZnO NPs treated plants was observed by Zhao et al. (2013a,b). The detrimental effect of Ag NPs on plants is widely discussed. Ag NPs inhibit plant growth by ROS production (Qian et al., 2013), by functional impairment (Pokhrel and Dubey, 2013), by altering sulfhydryl groups of the photosynthetic enzymes (Xu et al., 2010), or by altering specific cell functions (Vannini et al., 2013).

Similarly, Oukarroum et al. (2013) reported ROS-associated cell variability and growth inhibition in plants. Ag NP application resulted in biomass reduction, reduced shoot elongation, lower chlorophyll content, higher SOD activities, and low fruit productivity in tomato plants (Song et al., 2013a,b). Au NP exposure enhances the defense system, antioxidant potential, germination rate, and vegetative growth of plants (Kumar et al., 2013). NiO NPs proved phytotoxic because they induced ROS generation, apoptosis, and necrosis in tomato root seedlings. These NPs are a potential hazard for plants because of damage to the DNA and finally cell death (Faisal et al., 2013).

7.3.3 Rare Earth Oxide Nanoparticles

Rare earth oxide (REO) NPs are used in various industries such as polishing powder, paint coating, automobile exhaust catalysts etc. due to their optic, catalytic, and magnetic properties (Kosynkin et al., 2000; Fu et al., 2003; Gordon et al., 2004; Karakoti et al., 2008). The REO-NPs are released into environment following different ways but their potential effects in the environment are still unknown. Among all REO NPs, CeO_2 NPs are widely used; they are among the 10 most commonly produced engineered nanomaterials worldwide (Marchiol et al., 2016). The phytotoxic effects of REO NPs are rarely discussed.

The phytotoxic effects of four REO NPs (nano-CeO_2, nano-La_2O_3, nano-Gd_2O_3, and nano-Yb_2O_3) on six different plants showed that nano-La_2O_3, nano-Gd_2O_3, and nano-Yb_2O_3 inhibited root elongation and seedling lengths of all test species, while CeO_2 NPs reduced root elongation of lettuce (Ma et al., 2010). Similarly, the phytotoxic effects of nano-CeO_2 and nano-La_2O_3 showed the inhibitory effect of nLa_2O_3 on root shoot elongation and biomass, while nCeO_2 showed no effects (Ma et al., 2015b). The application of CeO_2 NPs reduced the number of tillers, leaf area, and number of spikelets per plant of barley (Marchiol et al., 2016). López-Moreno et al. (2010) reported reduced germination, reduced shoot and root growth as well as biomass in response to CeO_2. Arabidopsis growth was decreased up to 85% in CeO_2 presence at 1000–2000 mg/L (Huang et al., 2012).

The toxicity of REO NPs changes in a dose-dependent manner. Corral-Diaz et al. (2014) demonstrated the stimulatory effect of REO NPs on seed germination at low concentration and vice versa as concentration increased. Similar findings showed that mild concentration treatment of CeO_2 had a stimulatory effect on plant growth, while high concentration showed adverse effects on the photosystem and antioxidant system of plants (Yang et al., 2016). Gui et al. (2015) reported the low concentration stimulation and high concentration inhibition of plant growth and biomass because of CeO_2 NP exposure. They found significant increase in plant biomass at 250 ppm and on 500–2000 ppm of (CeO2) NPs treatments, while the plant growth was retarded up to 85%, the chlorophyll production reduced and MDA concentration increased in dose-dependent fashion.

7.3.4 Quantum Dots (Artificial Atoms)

Quantum dots (QDs), also known as "artificial atoms," are the first nanotechnology to be used in biological sciences and have found extensive applications in many clinical and commercial products (Klimov, 2007; Valizadeh et al., 2012). QDs have unique electronic and fluorescent characteristics, such as narrow emission spectra, high photochemical stability, and continuous absorption spectra (Bruchez et al., 1998). These extraordinary optical and electrical characteristics associated with small size and high surface area make them more ideal for biomedical and biotechnological use (Kuzyniak et al., 2014; Niemeyer, 2001; Whitesides, 2005).

Before we take advantage of artificial atoms in biological applications, we should assess the ecotoxicological data of QDs. The toxicity of Cd-based QDs are mainly due to the release of Cd^{2+} upon surface oxidation of particles (Derfus et al., 2004). QDs are responsible for DNA damage and suppression of cell proliferation (Hoshino et al., 2004). Mercaptoundecanoic acid QDs can induce cell damage and even cell death (Shiohara et al., 2004). Alimohammadi et al. (2011) added QDs to nanotubules and they significantly changed the viability of tomato plants by dramatically accelerating root inhibition and leaf senescence.

Navarro et al. (2012) reported oxidative stress in Arabidopsis roots after exposure to QDs. Adsorption of water-soluble CdSe/ZnS QDs to the cell surface of the algae *Chlamydomonas* reduced their photosynthetic capabilities (Lin et al., 2009). The fate and transport of QDs in soil, plants, and insects showed that QDs can be transported across the environment (Al-Salim et al., 2011).

7.4 FACTORS AFFECTING PHYTOTOXICITY OF NANOPARTICLES

NPs have shown diverse impacts on plants and soil organisms based on their size, chemical composition, functionality, concentration, and especially type (Ge et al., 2011; Rico et al., 2011; Shoults-Wilson et al., 2011; de la Rosa et al., 2013; Yang et al., 2015; Andersen et al., 2016). However, there is a lack of research knowledge regarding NP phytotoxicity in relation to different NP characteristics and the environment.

7.4.1 Physicochemical Characteristics of Nanoparticles

The specific mechanism of NP toxicity is difficult to determine because of uncertainty in the contribution of elements toward toxicity. There are as many as 22 aspects that could cause change in the biological effects of NPs (Oberdorster, 2010). Some of the crucial factors may be the concentration of NPs, size and specific area of NPs, the physicochemical properties and stability of NPs, the species of plant and their developmental stage, soil conditions, the solution of the nanomaterials, etc. (Siddiqui et al., 2015).

7.4.1.1 Size and Surface Properties

Upon interaction with plants, NPs may affect plants based on their size, surface coating, chemical structure, etc. (Khodakovskaya et al., 2012). NPs encourage various anatomical and morphological changes that can be determined by their concentration and behavior (Siddiqui et al., 2015). Hoshino et al. (2004) illustrated an example whereby QD toxicity was determined by the composition of the capping material. Therefore QD chemical composition is a key characteristic in plant toxicity. Similar findings regarding ZnS coating on QDs led to the fact that ZnS coating enhanced the biocompatibility of QDs by shielding Cd^{2+} ion release from inner layers, thereby effectively reducing the toxicity of QDs (Su et al., 2009). The surface chemistry of CdSe and CdSe/ZnS with varied surface modifications has an important role in Cd^{2+} ion release along with a role in cytotoxicity (Kirchner et al., 2005). Green and Howman (2005) also observed that CdSe QDs cause the formation of free radicals, and this process was enhanced when UV light or air oxidation triggered QDs.

The mechanism of toxicity is modified with surface modification because this changes the cellular interactions of NPs. The toxicity of CuO NPs is found to be caused by solubilization of particles into toxic metal ions, while the mechanism of CuO NP toxicity changes by polymer coating in *Lemna gibba*. This surface modification results in reduced plant growth and increased ROS formation even at low concentrations (Perreault et al., 2014). In QDs, surface modification alters their physicochemical properties. Additionally, QD cytotoxicity depends on their surface molecules. The properties of QDs are not linked to those of the QD core materials but to molecules covering the surface of QDs (Shiohara et al., 2004). In addition, Au NPs cause leaf necrosis based on particle size; Au NPs of 18 nm size do not induce necrosis, but 3.5 nm-sized Au NPs do (Sabo-Attwood et al., 2012).

7.4.1.2 Dissolution

For determination of the toxicity of NPs, especially metal NPs that are both essential and toxic for plant functions, dissolution of the metal-containing NPs and their bioavailability play important roles (Gajjar et al., 2009). The transport of intact CuO NPs into roots and their redistribution via xylem and subsequently phloem in maize plants is documented (Wang et al., 2013a,b). However, in wheat seedlings, NP dissolution occurs when grown with CuO NPs based on the detection of copper sulfide deposits in leaf tissue (Dimkpa et al., 2013b). Similar results are reported for rice seedlings for CuO NPs uptake and their transformation into other Cu-complex compounds in the rhizosphere (Peng et al., 2015; Misra et al., 2012).

Certain organic compounds in the soil such as humic acids released through decomposition and/or produced by soil organisms and released through roots may also influence the release of metals, their bioavailability, and biological response (Peng et al., 2015; Grillo et al., 2015; Martineau et al., 2014; Calder et al., 2012). CuO NP dissolution in the rhizosphere changes the morphology of roots determined by the release of Cu from contact. This dissolution may be caused by protons, secretion, and metal ion chelates released from root cells into the rhizosphere. Dissolution of CuO NPs causes root inhibition by altering indole-3-acetic acid (IAA) distribution (Adams et al., 2016). The retardation of root elongation with CuO NPs is similar to the findings for work with Cu ions in *Arabidopsis*, where root elongation inhibition by Cu ions was correlated with modification in IAA levels because of altered transport (Yuan et al., 2013a,b).

7.4.1.3 Concentration of Nanoparticles

The application of NPs is concentration dependent. The concentration is one of the most important factor determining the phtotoxicity of NPs. In the majority of cases, increases in concentration lead to a sharp increase in toxicity threshold. A higher dosage of CuO NPs showed toxic effects (root shortening) by ion exposure on plants (Shi et al., 2014a,b; Wang et al., 2015; Nair and Chung, 2015; Pena et al., 2015). *Arabidopsis thaliana* showed increased growth at 0.5 g/L concentration by nanoscale zero-valent iron (nZVI) (Kim et al., 2014, 2015) and *Cucumis sativus* showed reduced root development within a 100–5000 mg/L concentration of Fe_3O_4 engineered NPs (ENPs). The effective concentration of Fe_3O_4 ENPs in lettuce and radish is more than 5000 mg/L and for cucumber it is 1682 mg/L (Wu et al., 2012). NPs of $nTiO_2$ inhibit plant growth, germination, root length, and biomass in a dose-dependent manner (Frazier et al., 2014). Rafique et al. (2014) reported that different concentrations of $nTiO_2$ ($\leq$ 60 mg/kg and >60 mg/kg) increased and reduced in root/shoot length and biomass, respectively.

El-Temsah and Joner (2012) investigated the potential toxic effects of nZVI on *Linum usitatissimum*, *Lolium perenne*, and *Hordeum vulgare* and reported complete inhibition of seed germination at 2 and 5 g/L of nZVI, while there were no detrimental effects at concentrations of 250 mg/L. Similarly, Fe NP exposure changed the number of lateral roots, internal CO_2 concentration, root and shoot lengths, and photosynthetic pigments of four rice cultivars as concentration increased from 4 to 9 mM (Pereira et al., 2013). Plant biomass and growth were reduced at a higher concentration of nZVI, i.e., 200 mg/L (Ma et al., 2013a,b,c). Mukherjee et al. (2014a,b,c) investigated the effect of ZnO ENPs in *Pisum sativum* (L.). They observed variation in physiological and biochemical responses in relation to ROS induction, plant growth, and chlorophyll content.

CeO_2 NPs showed differential responses in four edible plants, alfalfa, corn, cucumber, and tomato (López-Moreno et al., 2010). CeO_2 NPs inhibited the germination rate of tomato, corn, and cucumber with 2000 mg/L CeO_2 concentration but did not cause a noteworthy reduction for alfalfa. Moreover, the stem and root growth were differentially inhibited by CeO_2 NPs in these four plants. Though cucumber germination was not significantly affected by CeO_2 NPs, its root and stem growth were significantly inhibited (López-Moreno et al., 2010). Likewise, $nCeO_2$ at 250 mg/kg reduced the biomass of cilantro plants (Morales et al., 2013).

7.4.2 Effect of Soil-Type and Environmental Factors

The level of NP toxicity on the aquatic environment also depends on matrix properties such as concentration and composition of dissolved organic carbon, pH, ionic strength, and type of salts (Ju-Nam and Lead, 2008). Nevertheless, limited studies have focused on NP phytotoxicity in relation to environmental and soil variable conditions (El-Temsah and Joner, 2012; Du et al., 2011; Kim et al., 2011; Lee et al., 2012).

Previously, studies conducted in relation to heavy metals showed that NP phytotoxicity may be modified because of factors such as cation exchange capacity, pH, and NOM content (Kabata-Pendias and Pendias, 1993). In a study, Lee et al. (2012) reported the phytotoxicity of silver (Ag) NPs to sorghum (*Sorghum bicolor*) and mung bean (*Phaseolus radiatus*) when grown in agar medium. However, the bioavailability of Ag NPs was lower in soil medium and did not cause toxicity even at 100–2000 mg/kg.

In soils, multiple components affect the behavior of NPs, especially soil colloids, soil organic matter, and nutrients. These components may regulate the behavior of NPs by affecting NP aggregation in the environment. The susceptibility of homoaggregation of NPs leads to flagging mobility of these structures, and ultimately a toxic effect is pronounced and their bioavailability is reduced (Phenrat et al., 2009; Hotze et al., 2010). There are some factors that affect the process of aggregation, e.g., organic matter, ionic strength, type of cations present, and pH of soil (Hotze et al., 2010).

Because all these factors varied with type of soil, they have an indirect effect on the phytotoxicity of NPs. Fang et al. (2009) described that the susceptibility of $nTiO_2$ aggregation in soil increased because soil solution had a low dissolved organic carbon content and high ionic strength. Similarly, $nTiO_2$ homoaggregations increased in the presence of bivalent ions compared to monovalent ions (French et al., 2009). In most cases, increase in temperature and aging caused a decrease in the phytotoxicity of NPs, while light conditions augmented the toxic effect of NPs (Jośko and Oleszczuk, 2013). The phytotoxic effect of Si NPs is pH dependent (Geisler-Lee et al., 2012).

Light also induced various physiological responses of plants and induced NP toxicity. Zoua et al. (2016) checked the effect of post- and preillumination conditions on *Wolffia globosa* to Ag NP exposure. Results showed that light induced Ag NP toxicity by altering photosynthetic pigments and protein content.

7.5 PHYTOTOXIC EFFECTS OF NANOPARTICLES

The phytotoxicity of NPs in plants is best assessed and predicted by determining their toxic effects on plants, such as potential growth retardation, biomass and surface area reduction, seed germination inhibition, poor quality, etc. (Dietz and Herth, 2011; Aslani et al., 2014).

7.5.1 Reduced Seed Germination

NPs induced significant inhibition in seed germination (Shaw and Hossain, 2013; Frazier et al., 2014; Raliya et al., 2015; Singh et al., 2016). Researchers exhibited that multiwalled CNTs did not show a positive relation with seed germination in many plants even at high concentration exposure (Husen and Siddiqi, 2014; Lin and Xing, 2007). This is similar to the germination of ryegrass inhibited by Zn NPs at 2000 mg/L concentration and corn at 2000 mg/L Zn NPs (Lin and Xing, 2007). Ryegrass, barley, and flax showed inhibited germination at 10 mg/L for Ag NP suspension and 250 mg/L by aqueous suspensions of zero-valent iron NPs (Temsah and Joner, 2010).

Kumar et al. (2015) compared the effect of nano-ZnO, nano-Fe_2O_3, and fullerene NPs on seed germination in different plant species. Among NPs, ZnO NPs caused considerable reduction in seed germination in all plant species, thus cucumber was found most susceptible and rice least susceptible to NP phytotoxicity. However, nano-Fe_2O_3 and fullerene NPs showed no effect on seed germination. Seeds of lettuce, radish, and cucumber were incubated with different metal oxide NPs (CuO, NiO, TiO_2, Fe_2O_3, Co_3O_4) to assess their phytotoxic effect. Results showed that NiO and CuO NPs have deleterious impacts on the seed activities of all three. The inhibitory impact is caused by the surface adsorption of metal oxide NPs on seeds, which leads to free metal ion release near the seeds (Wu et al., 2012).

Another study showed that seed germination in rice was stimulated in various concentrations of bulk TiO_2 particles, while seed germination decreased when exposed to TiO_2 NP concentrations of 10, 1, and 0.01 mg/mL (Jalill and Yousef, 2015). On the other hand, Boonyanitipong et al. (2011) showed contradictory results claiming that TiO_2 NPs had no adverse influence on rice seed germination. These varied results may be caused by the difference in the variety of rice seeds. Feizi et al. (2013) reported that increasing concentrations of TiO_2 NPs decreased germination of sage seeds. Likewise, Pokhrel and Dubey (2013) reported inhibition of cabbage in a dose-dependent manner. CeO_2 NPs also depicted reduced germination of maize, tomato, and cucumber in aqueous suspension (López-Moreno et al., 2010). CuO NPs inhibit seed germination and root elongation in cucumber (Moon et al., 2014). Likewise, Da Costa and Sharma (2016) reported reduced seed germination in rice to nCuO exposure.

7.5.2 Effect on Morphology and Plant Growth

The toxic effects of NPs on plant growth and morphology are discussed in several reviews (Huang et al., 2012; López-Moreno et al., 2010; Musante and White, 2012; Begum and Fugetsu, 2012; Zhang et al., 2015a; Ebbs et al., 2016; Tripathi et al., 2017c). Studies revealed that high concentrations of oxidized multiwalled CNTs had a damaging effect on plant growth and biomass production (Mondal et al., 2011). Similarly, octahedral molybdenum NP uptake caused phytotoxicity that restricted plant growth (Aubert et al., 2012). A range of phytotoxicity effects was observed in several crop species such as *Lycopersicum esculentum*, *C. sativus*, *Lactuca sativa*, *Oryza sativa*, *Brassica oleracea*, and *Raphanus sativus* when exposed to single-walled and functionalized CNTs (Cañas et al., 2008). Biomass reduction by about 60% was observed in *Cucurbita pepo* multiwalled CNTs at 1000 mg/L (Stampoulis et al., 2009). Likewise, $C_{70}(C(COOH)_2)_{4\text{-}8}$ NPs stimulated negative geotropism in *A. thaliana* roots (Liu et al., 2010). Moreover, flowering of *O. sativa* was delayed by about a month in the presence of multiwalled CNTs and C70 NPs (Lin et al., 2009). Similarly, multiwalled CNTs and C60 fullerenes reduce the biomass of *Glycine max* and *Zea mays* by 36.5%–45% (De La Torre-Roche et al., 2013). Priester et al. (2012) found that the growth of *G. max* was reduced after exposure of CeO_2 NPs and Yb_2O_3, Gd_2O_3, and La_2O_3 NPs at 2000 mg/L.

The phototoxic exposure of 0.5 mg/L Ag NPs on *O. sativa* decreased the root and shoot biomass significantly and on further increase in concentration also affected chlorophyll, carotenoids, and leaf area (Nair and Chung, 2014a,b). Hawthorne et al. (2012) showed that in *C. pepo*, 49%–91% biomass reduction was reported because of 205–750 mg/L of Ag NPs. Pokhrel and Dubey (2013), who reported that the presence of Ag and ZnO NPs reduced the size of vacuoles resulting in reduced cell turgidity and size in *Z. mays* L. and *B. oleracea* var. *capitata* L., found similar results.

The phytotoxic effects of NPs also showed that ZnO NPs collapsed the structure of vacuoles, cortical cells, epidermis, and roots at a concentration of 1000 mg/L (Zhao et al., 2013a,b). NPs of 30 nm TiO_2 favored leaf growth in *Z. mays* (Asli and Neumann, 2009), while 35 nm TiO_2 NPs reduced the number of lateral roots in peas (Fan et al., 2014). Single-walled CNTs delayed or postponed yield and flowering at 400 mg/L in *O. sativa* (Lin et al., 2009). Root growth and length reduction are observed in various studies regarding NP phytotoxicity (López-Moreno et al., 2010; Liu et al., 2010; Atha et al., 2012; Yoon et al., 2014; Mukherjee et al., 2014a,b,c; Watson et al., 2015; Raliya et al., 2015; Tripathi et al., 2017a,b,c,d).

7.5.3 Effect on Quality and Grain Yield

ENPs have been reported to influence protein levels (Gao et al., 2008; Patra et al., 2013), carbohydrate contents (Salama, 2012; Mirzajani et al., 2013; Mondal et al., 2011), radical scavenging ability, phenolic content (Krystofova et al., 2013; Krishnaraj et al., 2012), and elemental concentrations in plants (Speranza et al., 2010; Gao et al., 2013). All these studies were executed for a short exposure time and at early growth stages; the effect on the quality of fruits or grains is still unfamiliar. Au NPs showed alteration in oil content and reducing sugars in mustard seeds (Arora et al., 2012). Likewise, Ag NPs increased the total soluble solids in cucumber (Shams et al., 2013a,b), but did not change the polyphenol content in borage (Seif et al., 2011). Still, studies on the effect of NPs on the quality of seeds are also needed.

Rico et al. (2013a,b,c) reported the effect of CeO_2 NPs at 0 and 500 mg/kg soil on the quality of rice grains. Their results showed that rice grains in plants treated with $nCeO_2$ were deficient in glutelin, lauric and valeric acids, prolamin, starch, Fe, and S. In addition, $nCeO_2$ decreased all antioxidant values in grains except flavonoids. The results clearly showed that the quality of rice is compromised in $nCeO_2$ treatment. Thus $nCeO_2$ showed significant modifications in rice nutritional value, which may have a long-term negative effect on food quality. Similar findings demonstrated that $nCeO_2$ and $nTiO_2$ treatments showed negative effects on barley kernels by changing the composition and nutritional quality of barley kernels (Pošćić et al., 2016).

7.6 PHYTOTOXIC MECHANISM OF NANOPARTICLES

7.6.1 Uptake, Accumulation, and Translocation of Nanoparticles in Plants

NP internalization into the reproductive and edible parts of plants is of specific concern because many NPs are harmful to biotic life (Anjum et al., 2016). Plants uptake NPs either through the soil or by foliar contact (Anjum et al., 2015; Shi et al., 2014a,b) and their uptake and transportation in plants varies with growth stages, cultivars, and plant species (Schwabe et al., 2015). NPs are adhered to plant surfaces and pass through plant tissues via natural micrometer- or nanometer-scale openings (Koo et al., 2015; Dan et al., 2015). The association between NPs and plant uptake is explained by various pathways, which are discussed next.

7.6.1.1 Pathways of Internalization of Nanoparticles

The NP uptake mechanism is considered to be an "active-transport mechanism," which consists of various other processes occurring in cells such as plasma membrane regulation, recycling, and signaling (Wang et al., 2012a,b,c,d). NPs adhere to the root surface and then enter the epidermis and cortex cells by using an apoplastic pathway to pass through plant cells (Anjum et al., 2013, 2015; Le at al., 2014).

When exposed to plants, NPs pierce the cell membrane and cell wall of root epidermal cells and are escorted by various complex events to cross xylem (vascular bundle), transfer to the stele via a symplastic pathway, and then translocate to the leaves. NPs travel via cell membrane pores to cross the integral cell membrane, which is specific to the size of the nanomaterial (Tripathi et al., 2016a,b,c). Before NPs reach the stele, they must be assimilated through a passive pathway via endodermal apoplast (Judy et al., 2016). Xylem is the significant vehicle that distributes and translocates NPs in plants (Aslani et al., 2014). The cell wall, through which acquisition of water molecules and other solutes occurs, is usually a porous network of polysaccharide fiber matrices (Held et al., 2015).

7.6.1.1.1 EXAMPLES

Carbon NPs penetrate through roots in *Triticum aestivum* (wheat), *Solanum lycopersicum* L. (tomato), *Helianthus annuus* L. (sunflower), and *P. sativum* L. (pea) and then translocate toward vascular bundles and reach the aerial parts by transpiration in xylem (Cifuentes et al., 2010; Aslani et al., 2014). The same mechanism is used for the uptake of Ag NPs in rice (Thuesombat et al., 2014).

Plant cells also adopt another strategy to uptake ENPs, which is called the endocytic pathway (Tonigold and Mailänder, 2016). The endocytic pathway is both dependent and independent of clathrin (Miralles et al., 2012). Clathrin-dependent passageways include the development of sheathed or folded structures on the cell membrane, which later become clathrin-coated vesicles (Ng et al., 2015). The clathrin-free pathway comes about by three processes: (1) phagocytosis, (2) caveolae, and (3) fluid phase endocytosis (Monti et al., 2015). *Nicotiana tabacum* has both alleyways (clathrin dependent and independent) to internalize Au NPs (Onelli et al., 2008). Si NPs also adopt the endocytosis pathway (Walker et al., 2016).

Furthermore, internalization of engineered multiwalled CNTs occurs via a distinct endosomal pathway (Majzoub et al., 2016) because the cell wall of plants acts as a semipermeable obstacle that only enables specific materials to pass via pores (Pauly and Keegstra, 2016).

7.6.1.1.2 EXAMPLES

In cells of *Medicago sativa*, QD bioaccumulation takes place precisely in the nucleus and cytoplasm (Rocha et al., 2016). Engineered nanomaterials might enter by parenchymatous intercellular space that aids in diffusion of liquid material to cotyledons (Ma et al., 2015a,b).

In the seed coat, aquaporins (another type of gate) regulate the entry of NPs (Yulianti et al., 2016).

7.6.1.2 *Foliar Application*

In top to downward translocation, NPs enter the plant leaves by two pathways, namely, penetration through stomatal openings or entrapment of NPs in the cuticle; the NPs are then redistributed in the stem and reach the root cells through phloem (Larue et al., 2014; Deepa et al., 2014; Hong et al., 2014). ZnO and TiO_2 NPs enter through the leaves when the size of particles is 25 ± 3.5 nm in tomato plants (Raliya et al., 2015) and CuO NPs enter through the leaves when their size is <50 nm in maize (Adhikari et al., 2016).

Foliar-applied CaO NPs in *Arachis hypogea* L. (groundnut) makes the Ca pass through the leaves, shoots, and roots by phloem (Deepa et al., 2014).

7.6.1.3 *Factors Affecting Uptake*

Various physicochemical differences among plants, such as variations in cell wall pore size and hydraulic conductivity, etc., may affect NP transport and accumulation (Judy et al., 2016). It seems that NP uptake, accumulation, and translocation are dependent on composition, size, shape, type and NP agglomeration, exposure concentration of NPs, and plant type (Zhang et al., 2015; Li et al., 2014a,b; Rico et al., 2015a,b; Gorczyca et al., 2015).

7.6.2 Alteration of Mineral Absorption and Assimilation and Biotransformation of Nanoparticles in Plant Cells

7.6.2.1 *Alteration in Mineral Uptake and Assimilation*

Absorption of sufficient quantities of mineral nutrients, primarily at early plant growth stages, is needed for the development and growth of crops. Plant nutritional status is negatively affected by NPs, i.e., plant uptake and accumulation of mineral nutrients are reduced on exposure to NP stress (Antisari et al., 2015; Dimkpa et al., 2015; Li et al., 2014a,b; Hong et al., 2015). Toxic metal ions significantly decrease plant mineral uptake (Zaheer et al., 2015; Rizwan et al., 2016a,b), and this decrease in plant mineral uptake may be caused by discharge of soluble metal ions from NPs (Mahmoodzadeh et al., 2013; Nair et al., 2014). However, there is a need to investigate how NPs decrease nutrient uptake through roots involving absorption, adsorption, and translocation to xylem, stem, and grains from the root surface.

7.6.2.1.1 EXAMPLES

Foliar application of Ag NPs indicated a sign of nutrient deficiency in tomato plants because Ag NPs reduce essential elements (Mg, Na, Mn, Cu, Fe, K, P, Ca, Zn, and B) uptake and translocation in plants (Shams et al., 2013a,b). CeO_2 NPs interrupted nutrients in Bt cotton; these NPs reduce Zn, P, Fe, and Mg concentrations in xylem sap because of the rupturing of chloroplasts and destruction of the structure of vascular bundles (Li et al., 2014a,b). SiO_2 and CeO_2 NPs changed nutrient elements (Na, Zn, Mg, Cu, and Fe) in roots and shoots of transgenic cotton (Corral-Diaz et al., 2014; Li et al., 2014a,b; Le et al., 2014), whereas CeO_2 NPs reduced Ni in different tissues of radish (Corral-Diaz et al., 2014). The concentration of Ca, Zn, and Fe decreased in bean plants on exposure of CuO NPs (Dimkpa et al., 2015). In a similar manner, CuO NPs also reduced Fe, Mg, Ca, P, and Mn in lettuce plants (Trujillo–Reyes et al., 2014; Hong et al., 2015). SnO_2 NPs decreased K, Ca, Mg, and S content in leaves and stems of tomato. SnO_2 nanomaterials reduced S, Mg, Ca, and K contents in stems and leaves of tomato (Antisari et al., 2015).

7.6.2.2 Biotransformation of Nanoparticles

Transformation is the biochemical modification through living entities, and either detoxification or toxicity is enhanced during this process (Peng et al., 2015). Biogenic reducing materials and root exudates such as organic acids as well as PO_4O_3 are the key factors convoluted in the process of biotransformation (Zhang et al., 2012). ENPs are modified through biological media and ambient environmental systems and their toxicity and fate may be changed (MacLeod and Johnson, 2015). Macromolecular modification, phosphorylation, sulfidation, and redox reactions are typical nanomaterial transformations (Son et al., 2015; Miclăuş et al., 2016).

For example, sulfidation of Ag NPs significantly reduced their toxic effects on *Caenorhabditis elegans* because of reduced silver sulfide solubility (Starnes et al., 2015). In plants, there is a biotransformation of $Ni(OH)_2$ to Ni^{2+} in leaves and shoots (Parsons et al., 2010). In rice, Cu(II) NPs reacted with phosphate, citrate, and cysteine ligands and reduced to Cu(I) (Peng et al., 2015). $NaYF_4$ NPs transformed into clusters of Y-phosphate in roots of maize plants (Yin et al., 2011). La_2O_3 NPs combined with carboxylic or phosphate groups, whereas CeO_2 NPs were transformed to a Ce(III)–carboxyl complex in cucumber plants (Ma et al., 2015a,b). It has been reported that CuO NPs were reduced from Cu_2S to Cu_2O in maize (Wang et al., 2012a,b,c,d). Two REO NPs (Yb_2O_3 and La_2O_3) biotransformed from REO NPs to rare earth phosphates in roots of cucumber. Increased dissolution of these NPs by root exudates (organic acid) plays a considerable role in their biotransformation and phytotoxicity (Zhang et al., 2012).

Modification of NPs through biotransformation also counteracts the beneficial effects of NPs by altering their toxicity and fate. Thereby the extent and mechanism of such biotransformations needs to be addressed in the future not only for appropriate NP applications but also for risk assessment.

7.6.3 Genotoxicity of Nanoparticles

Genotoxicity is caused when NPs interact with plant cells through the changing of gene expression in plants (Vannini et al., 2013; Lee et al., 2013a,b,c; Singh et al., 2017). NPs cause genotoxicity in many important agricultural crops (Golbamaki et al., 2015; Nagaonkar et al., 2015; Nair and Chung, 2015).

7.6.3.1 Aluminum Oxide Nanoparticles

These particles gradually lower the mitotic index in root cells of onion (Rajeshwari et al., 2015). Ag NPs instigate cytological changes in roots of germinated seed of barley and wheat. Pretreatment of seed with Ag NPs triggers deletion of chromosomes, binucleate cells, chromosomal aneuploidy, chromosomal aberrations, sticky chromosomes, deformed chromosomal fragments, micronuclei, and nuclei at anaphase and metaphase (Abou-Zeid and Baraka, 2014), and also reduces the mitotic index in tips of roots in faba bean (Patlolla et al., 2012). Seed priming of Ag NPs changes expressions of various proteins, which are involved in the cell defense system and primary metabolism in roots and shoots of wheat (Vannini et al., 2014). Different ZnO and Ag NPs (<100 nm size) impair phases of cell division, disturb metaphase, cause multiple breaks of chromosomes, and disintegrate the root cells of onion (Ma et al., 2011; Vannini et al., 2014; Kumari et al., 2009, 2011; Wang et al., 2015). In another study, Ag NP-facilitated differential transcription of genes was observed and was related to tolerance of oxidative stress in shoots and roots of rice seedlings (Nair and Chung, 2015). NPs change transcription of aquaporin and antioxidant genes in *A. thaliana* (Nair and Chung, 2014a,b; Qian et al., 2013). In *A. thaliana*, these nanomaterials galvanize the expression of genes linked to cellular events involving metabolism, cell proliferation, and hormonal pathways (Syu et al., 2014).

7.6.3.2 Copper Nanoparticles

CuO NPs also change gene expressions related to root growth in rice and destroy DNA in ryegrass and radish (Atha et al., 2012; Wang et al., 2015); they also decline the mitotic index and enhance the abnormality index in roots of onion (Nagaonkar et al., 2015). In *Brassica juncea* (Indian mustard) CuO NPs significantly reduced APX and CAT gene expression in roots, but did not alter these genes' expression in shoots (Nair and Chung, 2015). In a similar manner, Cu NPs upregulate gene expression of CuZn SOD in roots but do not change APX gene expression in roots and shoots of chickpea (Nair and Chung, 2015). When soybean is exposed to CuO NPs they upregulated gene expressions of CAD, C4H, and PAL in root cells and this NP stress stimulates genes that are related to the biosynthesis of proline and glutathione, assimilation of sulfur, and oxidative stress response (Nair et al., 2014).

7.6.3.3 Other Nanoparticles

The CeO_2 and TiO_2 NPs changed the regulation of 142 and 204 genes in *Arabidopsis thaliana*, respectively (Tumburu et al., 2015). In (maize), TiO_2 NPs significantly reduced mitotic activity and enhanced chromosomal aberrations in meristem of root tips (Castiglione et al., 2011). ZnO NPs prompted various types of root apical meristematic chromosomal sickness, mitotic aberrations, breakages, and bridges, which are time and concentration dependent in garlic and onion (Shaymurat et al., 2012; Raskar and Laware, 2014; Taranath et al., 2015). NPs of zero-valent iron considerably enhance H^+-ATPase isoform levels, which are involved in stomatal opening.

These studies showed in-depth molecular mechanisms underlying plant responses to metal and metal oxide NPs. However, genotoxicity studies of metal and metal oxide NPs in crop plants are limited. In addition, most of the studies were conducted in vitro and in vivo and might be helpful to understand the mechanisms behind the genotoxic effects of NPs in crop plants.

7.6.4 Increase in Reactive Oxygen Species

ROS are derivatives of oxygen that arise in oxygenated conditions and comprise hydrogen peroxide (H_2O_2), hydroxyl ions (OH^-), and superoxide anions (O^{2-}) (Yang et al., 2009). ROS production and oxidative stress initiation is the crucial biomarker of NP toxicity (Rizwan et al., 2016a,b) and is measured by direct ROS quantification or antienzyme activities such as catalase (CAT) and SOD (Thwala et al., 2013; Shaw and Hossain, 2013; Perreault et al., 2014; Oukarroum et al., 2012; Lin et al., 2012; Faisal et al., 2013; Begum and Fugetsu, 2012; Chang et al., 2012). Large quantities of ROS might be generated even at very low exposure of NPs to plants (Toduka et al., 2012). NPs are capable of reacting with biomolecules because of their high specific area, which provides NPs with high electronic density and reactive activity (Pisanic et al., 2009). Throughout this process, chemical reactions that enhance the formation of radicals (O^{2-}, OH^{-1}, etc.) interact with various protein groups, leading to accumulation of ROS and oxidative stress (Shaw et al., 2014a,b; Wang et al., 2011; Gorczyca et al., 2015; Mohammadi et al., 2013; Rao and Shekhawat, 2014).

7.6.4.1 Factors

ROS stress results in programmed cell death and this oxidative stress depends upon the dose of NPs rather than bulk metal material (Rajeshwari et al., 2015; Wang et al., 2011; Cui et al., 2014a,b). The time period of stress applied, e.g., stress of CuO NPs, does not change the chlorophyll contents of barley leaves after 10 days of growth, but after 20 days there was a sudden decrease in chlorophyll contents, regardless of NP concentrations (Shaw et al., 2014a,b). Plants can tolerate the toxicity of NPs up to a certain period, whereas extended exposure of NPs to plants imparts toxic effects because of the harmful effects of NPs on the plant defense mechanism, such as antioxidant enzymes (Zhao et al., 2015).

7.6.4.2 Interaction of Reactive Oxygen Species with Nanoparticles

NPs stimulate ROS production and release NP ions on exposure to acidic conditions of lysosomes (Nohl and Gille, 2013) or interaction with oxidative organelles such as mitochondria and redox active proteins in cells (Chang et al., 2012). ROS, after interacting with biomolecules, instigate an imbalance between reactive oxygen productions and the ability of biological systems to heal, resulting in damage and decontamination of reactive intermediates (Yang et al., 2009). NP-stimulated production of ROS leads to a variety of biological responses, which are dependent upon cellular pathway types, comparative abundance of ROS, and antioxidant response components, which belong to oxidative stress (Xia et al., 2015).

OH^{-1} is included in the category of highly toxic ROS species; it is capable of oxidizing nearly all the components of cells (Yamakoshi et al., 2003). Metal oxide NPs generate extracellular OH^{-1} that might stimulate oxidative damage of the plasma membrane, which exhibits toxic impacts in organisms because NP-induced oxidative stress is an authenticating cell-damaging mechanism (Pulskamp, 2007; Yang et al., 2009).

7.6.4.3 *Effect of Reactive Oxygen Species*

Extreme oxidative stress modifies nucleic acid, lipids, and proteins, which may lead to cell death or excite antioxidant defense mechanisms. In the interim, with enhanced production of ROS, NPs induce DNA damage by point mutations or breaking its strands and enhances expression of death receptor genes (Singh and Lillard, 2009; Yang et al., 2009). Furthermore, NP ions inactivate the functional proteins by dislodging metal ions in precise metalloproteins or chelating with biomolecules (Material et al., 2012). Another response of oxidative stress damage is the release of intracellular Ca^{2+} ions, leading to perturbation of mitochondria and cell death (Xia et al., 2015), Various diseases including autoimmune diseases, cardiovascular disease, lung disease, and aging are connected to oxidative stress (Yamakoshi et al., 2003).

In plants, oxidative stress interrupts biochemical reactions and decreases gas exchange and photosynthetic rate probably because of increased ROS production (Adrees et al., 2015; Das et al., 2015), and the impact on photosynthesis is extensively reported (Shaw et al., 2014a,b; Mirzajani et al., 2013; Abou-Zeid and Baraka, 2014; Rao and Shekhawat, 2014; Das et al., 2015). Seed priming of barley, soybean, and wheat with Ag NPs quenched photosynthetic pigments and killed chlorophyll fluorescence (Gorczyca et al., 2015; Zhao et al., 2013a,b; Abou-Zeid and Baraka, 2014). In a similar manner, CeO_2 NPs affected cucumber and radish plants (Zhao et al., 2013a,b; Corral-Diaz et al., 2014).

7.6.4.3.1 EXAMPLES

Ag NPs enhance electrolyte leakage in wheat and reduce viable algal cells and chlorophyll contents in green algae (Oukarroum et al., 2012). CeO_2 NPs increase lipid peroxidation and H_2O_2 production in rice and maize (Zhao et al., 2012a,b; Rico et al., 2013a,b). Similarly, CuO NPs enhance MDA contents in barley, chickpea, and rice leading to oxidative damage (Shaw and Hossain, 2013). ZnO and CuO NPs simulate oxidative stress by glutathione oxidation and lipid peroxidation in wheat (Dimkpa et al., 2012a,b). NiO NPs significantly induce cellular toxicity by increasing production of ROS in tomato roots (Faisal et al., 2013). Al_2O_3 NPs revealed phytotoxicity in tobacco only by reactive oxygen generation in a time- and concentration-dependent manner (Poborilova et al., 2013). CNT-induced phytotoxicity also endorsed ROS production leading to oxidative stress (Begum and Fugetsu, 2012).

From these studies it is obvious that NPs cause oxidative stress in plants in the form of increased production of ROS. Defense systems in plants can scavenge the production of ROS when stress is low to mild. However, at higher stress, ROS generation by plants is unavoidable. Similarly, the increase or decrease in the production of antioxidant enzymes in plants upon exposure to NPs is evident depending on certain factors such as size and type of NPs, time, and conditions of exposure.

7.6.5 Reduction in Antioxidative Enzymes

Plants have developed an antioxidant defense mechanism as a first line of defense against the generation of ROS, which contains both nonenzymatic and enzymatic antioxidants (Le et al., 2014; Zhao et al., 2012a,b, 2015; Rico et al., 2013c; Regier et al., 2015). Under slight NP stress, the actions of the main antioxidant enzymes are enhanced in various crops (Faisal et al., 2013). Conversely, under high NP stress, antioxidant enzymes activities are reduced possibly because of the oxidative eruption in plant tissues (Rao and Shekhawat, 2014). For instance, activities of CAT, ascorbate peroxidase (APX), and SOD were enhanced in seedlings of wheat on exposure to aluminum oxide NPs at a dosage of 200 ppm for 5 days, and at increased concentrations of NPs, activities of these enzymes reduced or did not change as compared to control plants (Riahi-Madvar et al., 2013).

Additionally, Ag NPs repressed SOD and CAT activities in roots and eaves of wheat (Gorczyca et al., 2015). CeO_2 NPs at the rate of 500 mg/L for 48 days reduced the activities of glutathione reductase (GR), CAT, and dehydroascorbate reductase (DHAR) in barley leaves (Rico et al., 2015a,b). In *Coriandrum sativum* L. (cilantro) plants, CeO_2 NPs enhanced APX activities in roots and CAT in shoots (Morales et al., 2013). In the same way, these NPs (at 500 ppm for 5 days) changed the activities of SOD in roots of lettuce (Cui et al., 2014a,b). CuO NPs in barley increased the activities of GR, SOD, and APX and reduced monodehydroascorbate reductase and DHAR activities in shoots (Shaw et al., 2014a,b). Similarly, these NPs (at 500 ppm) significantly enhanced the activity of SOD in shoots and roots of Indian mustard, whereas the same concentration of NPs causes a reasonable reduction in the activities of APX in roots and shoots (Nair and Chung, 2015). ZnO and CuO NPs (100 mg/L) considerably enhance the activities of CAT, peroxidase, and SOD in cucumber roots (Kim et al., 2012). NiO NPs (2 ppm) enhance the activities of SOD, GR, and CAT 1.7, 3.7, and 6.8 times higher, correspondingly, in roots of tomato (Faisal et al., 2013). TiO_2 NPs reduce the activity of GR at the dosage of 50 ppm and of APX at the rate of 25 and 5 ppm in roots of faba bean (Foltête et al., 2011).

7.7 DETOXIFICATION OF NANOPARTICLES IN PLANTS

7.7.1 Antioxidant Defense Mechanism

Exposure of plants to different NPs-stress results in the cellular production of ROS and the possible mechanism is interaction of NPs with cellular moiety (Tripathi et al., 2015, 2016a,b,c; Rico et al., 2015a,b).

Produced radicals of ROS behave as signaling molecules, thus stimulating the antioxidant defense mechanism to inhibit toxicity caused by free ROS radicals. This antioxidant defense mechanism involves both enzymatic antioxidants such as GR, DHAR, guaiacol peroxidase (GPX), APX, CAT, and SOD, and nonenzymatic agents such as phenolics, thiols, glutathione, etc. (Tripathi et al., 2016a,b,c). Of these, ascorbate and thiols (GSH or GSSG) are the most significant antioxidants of low molecular weight. Both GPX and CAT have key importance in the quenching of peroxy and ROS radicals, whereas SOD catalyzes the dismutation of the superoxide radical into H_2O_2 (Rico et al., 2015a,b). Types, chemical properties, exposure media, and concentration of NPs are key factors, which inspires the antioxidative defense response in plants (Rizwan et al., 2016a,b).

Enzymes such as GR, DHAR, and APX are involved in the homeostasis of complex redox actions in the ascorbate-glutathione cycle, which limits ROS overproduction or the oxidative status of plants. In nanomaterial-stimulated generation of ROS, APX reduced H_2O_2, which is generated by SOD into H_2O molecules. DHAR regenerates ascorbate, which for a second time is exploited by the activity of APX for H_2O_2 diminution (Rico et al., 2015a,b). Antioxidant enzymes have a potential against oxidative stress induced by NPs in plants (Wang et al., 2015).

Activation of antioxidant enzymes is stimulated by various NPs such as nCo_3O_4, nFe_3O_4, and $nCeO_2$ that prompt CAT activation; nAu, nCuO, $nMnO_2$, nCo_3O_4, nFe_3O_4, and $nCeO_2$ stimulate GPX, and fullerene, nPt, and $nCeO_2$ persuade activation of SOD. It is hard to distinguish these types of mimetic actions in trials using NP exposure to a single plant. NPs have a direct influence on antioxidant enzymatic activities (Tripathi et al., 2016a,b,c; Rico et al., 2015a,b) but currently there is still no confirmation that the NPs' chemical properties can be correlated with antioxidant enzymatic activities. In fact, there are erratic and uneven impacts of NPs on enzymatic actions of defense mechanisms.

CuO NPs improved the activities of GPX, APX, CAT, and SOD in rice (Da Costa and Sharma, 2016), soyabean (Nair and Chung, 2014a), and *A. thaliana* (Nair and Chung, 2014b). Decreased activities of APX were indicated in Indian mustard (Nair and Chung 2015) and CAT activity in alfalfa (Hong et al., 2015). Thus it is difficult to validate negative or positive impacts of NPs on different enzymatic activities. CeO_2 NPs induce phytotoxicity in *Phaseolus vulgaris* by distressing machinery of enzymatic antioxidant defense (Majumdar et al., 2014).

Thereby, correlation between toxicity of NPs and antioxidant enzymatic systems displayed in plants is still unclear and this point needs additional exploration. Nitric oxide has been indicated to have an enriching role in NP-induced phytotoxicity in rice (Chen et al., 2015). Equally, sodium nitroprusside and hematin play key roles in the upregulation of seed germination rate under silver and silver nitroprusside NP stress in *Brassica nigra* (Amooaghaie et al., 2015). However, there is a need for further studies to deliver high-quality results in this area.

7.7.2 Root Exudate Detoxification of Nanoparticles

NPs carry out significant changes in metabolic processes in leaf tissues and also influence root exudates. This change in metabolomics of root exudates indicates an active defense pathway against nanostress, i.e., amino acid upregulation to exclude/sequester NPs, citric acid downregulation to decrease ions of NPs, regulation of ascorbic acid against ROS, and phenolic compound upregulation to stimulate antioxidant systems (Zhao et al., 2016). Almost 30%–40% of fixed carbon by photosynthesis is transferred to the rhizosphere (Marschner, 2011) in the form of root exudates, involving CO_2, phenolic compounds, proteins, sugars, amino acids, and organic acids. These compounds play a significant role in stress tolerance of plants and exclusion of pollutants (Calvo et al., 2016). Plants upregulate organic acids, comprising amino acids, to complex or chelate toxic metal elements (e.g., Cu, Fe, Zn, Cd, Al) to inhibit translocation of these toxic metals to aboveground parts of plants (Dong et al., 2007).

7.7.2.1 *Amino Acids*

Eleven types of amino acids such as valine, threonine, serine, proline, phenylalanine, lysine, leucine, isoline, glycine, beta-alanine, and alanine are significantly upregulated as a response to nCu stress in a dose-dependent manner, and this increase is probably a defense response of plants (Zhao et al., 2016).

7.7.2.2 *Organic Acids*

Organic acids strongly chelate ions, have a key role in aiding plant mineral uptake, and exclude or sequester toxic metal ions. Succinic acid, oxalic acid, and citric acid detoxify the different dissolved metal ions (Meier et al., 2012). Organic acids hinder the passage of metals across the root's plasma membrane because of their strong ability to form extracellular stable complexes with Cd, Al, and Cu (Meier et al., 2012).

7.7.2.3 *Phenolic Compounds*

Phenolic compounds are secondary metabolites, which have an important function against stress (Shalaby et al., 2015). Benzoic and salicylic acid concentrations are significantly upregulated in root exudates on exposure of Cu/nano-Cu particles (Zhao et al., 2016). Both these have antifungal and antioxidant activities and have a crucial action in plant defense to combat various abiotic and biotic stressors (Borochov-Neori et al., 2015). Salicylic acid acts as a signaling substance in plant defense activation after attack by pathogens (Quentin et al., 2016) and considerably decreases H_2O_2 and MDA in leaves and roots of *O. sativa* L. (rice) (Mostofa and Fujita, 2013). Dehydroascorbic acid participates in the plant defense mechanism to combat oxidative stress because of NP toxicity (Johansen et al., 2015).

Root exudates are a hidden part of plant defense mechanisms against nontoxicity and have been ignored previously. However, they have a significant detoxification mechanism in some plants against NP-induced stress. Further studies including different proteomic techniques are required to elaborate the importance of root exudates in hindering phytotoxicity.

References

Abou-Zeid, L., Baraka, H.N., 2014. Combating oxidative stress as a hallmark of cancer and aging: computational modeling and synthesis of phenylene diamine analogs as potential antioxidant. Saudi Pharm. J. 22, 264–272.

Adams, J., Wright, M., Wagner, H., Valiente, J., Britt, D., Anderson, A., 2016. Cu from dissolution of CuO nanoparticles signals changes in root morphology. Plant Physiol. Biochem. 1–10.

Adhikari, T., Sarkar, D., Mashayekhi, H., Xing, B., 2016. Growth and enzymatic activity of maize (*Zea mays* L.) plant: solution culture test for copper dioxide nano particles. J. Plant Nutr. 39, 99–115.

Adrees, M., Ali, S., Rizwan, M., Zia-ur-Rehman, M., Ibrahim, M., Abbas, F., Farid, M., Qayyum, M.F., Irshad, M.K., 2015. Mechanisms of silicon-mediated alleviation of heavy metal toxicity in plants: a review. Ecotoxicol. Environ. Saf. 119, 186–197.

Alimohammadi, M., Xu, Y., Wang, D.Y., Biris, A.S., Khodakovskaya, M.V., 2011. Physiological responses induced in tomato plants by a two-component nanostructural system composed of carbon nanotubes conjugated with quantum dots and its in vivo multimodal detection. Nanotechnology 22, 295101.

Al-Salim, N., Barraclough, E., Burgess, E., Clothier, B., Deurer, M., Green, S., Malone, L., Weir, G., 2011. Quantum dot transport in soil, plants, and insects. Sci. Total Environ. 409, 3237–3248.

Amooaghaie, R., Tabatabaei, F., Ahadi, A.M., 2015. Role of hematin and sodium nitroprusside in regulating *Brassica nigra* seed germination under nanosilver and silver nitrate stresses. Ecotoxicol. Environ. Saf. 113, 259–270.

Andersen, C.P., King, G., Plocher, M., Storm, M., Pokhrel, L.R., Johnson, M.G., Rygiewicz, P.T., 2016. Germination and early plant development of ten plant species exposed to TiO_2 and CeO_2 nanoparticles. Environ. Toxicol. Chem. 35, 2223–2229.

Anjum, N.A., Gill, S.S., Duarte, A.C., Pereira, E., Ahmad, I., 2013. Silver nanoparticles in soil–plant systems. J. Nanopart. Res. 15, 1–26.

Anjum, N.A., Adam, V., Kizek, R., Duarte, A.C., Pereira, E., Iqbal, M., Lukatkin, A.S., Ahmad, I., 2015. Nanoscale copper in the soil–plant system – toxicity and underlying potential mechanisms. Environ. Res. 138, 306–325.

Anjum, N.A., Rodrigo, M.A.M., Moulick, A., Heger, Z., Kopel, P., Zítka, O., Adam, V., Lukatkin, A.S., Duarte, A.C., Pereira, E., Kizek, R., 2016. Transport phenomena of nanoparticles in plants and animals/humans. Environ. Res. 151, 233–243.

Antisari, L.V., Carbone, S., Gatti, A., Vianello, G., Nannipieri, P., 2015. Uptake and translocation of metals and nutrients in tomato grown in soil polluted with metal oxide (CeO_2, Fe_3O_4, SnO_2, TiO_2) or metallic (Ag, Co, Ni) engineered nanoparticles. Environ. Sci. Pollut. Res. 22, 1841–1853.

Arora, S., Sharma, P., Kumar, S., Nayan, R., Khanna, P.K., Zaidi, M.G.H., 2012. Gold nanoparticle induced enhancement in growth and seed yield of *Brassica juncea*. Plant Growth Regul. 66, 303–310.

Aslani, F., Bagheri, S., Julkapli, N.M., Juraimi, A.S., Hashemi, F.S.G., Baghdadi, A., 2014. Effects of engineered nanomaterials on plants growth: an overview. Sci. World J. 2014.

Asli, S., Neumann, P.M., 2009. Colloidal suspensions of clay or titanium dioxide nanoparticles can inhibit leaf growth and transpiration via physical effects on root water transport. Plant Cell Environ. 32, 577–584.

Atha, D.H., Wang, H., Petersen, E.J., Cleveland, D., Holbrook, R.D., Jaruga, P., Dizdaroglu, M., Xing, B., Nelson, B.C., 2012. Copper oxide nanoparticle mediated DNA damage in terrestrial plant models. Environ. Sci. Technol. 46, 1819–1827.

Aubert, T., Esnault, M.A., Cordier, S., Grasset, F., Cabello-Hurtado, F., 2012. Phytotoxicity of nanosized molybdenum octahedral clusters. In: XVII Congress of the FESPB.

Avanasi, R., Jackson, W.A., Sherwin, B., Mudge, J.F., Anderson, T.A., 2014. C60 fullerene soil sorption, biodegradation, and plant uptake. Environ. Sci. Technol. 48, 2792–2797.

Begum, P., Fugetsu, B., 2012. Phytotoxicity of multi-walled carbon nanotubes on red spinach (*Amaranthus tricolor* L.) and the role of ascorbic acid as an antioxidant. J. Hazard. Mater. 243, 212–222.

Begum, P., Ikhtiari, R., Fugetsu, B., 2014. Potential impact of multi-walled carbon nanotubes exposure to the seedling stage of selected plant species. Nanomaterials 4, 203–221.

Boonyanitipong, P., Kositsup, B., Kumar, P., Baruah, S., Dutta, J., 2011. Toxicity of ZnO and TiO_2 nanoparticles on germinating rice seed *Oryza sativa* L. Int. J. Biosci. Biochem. Bioinform. 1, 282–285.

Borochov-Neori, H., Lazarovitch, N., Judeinstein, S., Patil, B., Holland, D., 2013. Climate and salinity effects on color and health promoting properties in the pomegranate (*Punica granatum* L.) fruit arils. In: Tropical and Subtropical Fruits: Flavors, Color, and Health Benefits. ACS Symposium Series.

Bruchez Jr., M., Moronne, M., Gin, P., Weiss, S., Alivisatos, A.P., 1998. Semiconductor nanocrystals as fluorescent biological labels. Science 28, 2013–2016.

Calder, A.J., Dimkpa, C.O., McLean, J.E., Britt, D.W., Johnson, W., Anderson, A.J., 2012. Soil components mitigate the antimicrobial effects of silver nanoparticles towards a beneficial soil bacterium, *Pseudomonas chlororaphis* O6. Sci. Total Environ. 429, 215–222.

Calvo, O.C., Franzaring, J., Schmid, I., Müller, M., Brohon, N., Fangmeier, A., 2017. Atmospheric CO_2 enrichment and drought stress modify root exudation of barley. Global Change Biol. 23, 1292–1304.

Cañas, J.E., Long, M., Nations, S., Vadan, R., Dai, L., Luo, M., Ambikapathi, R., Lee, E.H., Olszyk, D., 2008. Effects of functionalized and non functionalized single walled carbon nanotubes on root elongation of select crop species. Environ. Toxicol. Chem. 27, 1922–1931.

Carmen, I.U., Chithra, P., Huang, Q., Takhistov, P., Liu, S., Kokini, J.L., 2003. Nanotechnology: a new frontier in food science. Food Technol. 57, 24–29.

Castiglione, M.R., Giorgetti, L., Geri, C., Cremonini, R., 2011. The effects of nano-TiO_2 on seed germination, development and mitosis of root tip cells of *Vicia narbonensis* L. and *Zea mays* L. J. Nanopart. Res. 13, 2443–2449.

Chang, Y., Zhang, M., Xia, L., Zhang, J., Xing, G., 2012. The toxic effects and mechanisms of CuO and ZnO nanoparticles. Materials 5, 2850–2871.

Chen, R., Ratnikova, T.A., Stone, M.B., Lin, S., Lard, M., Huang, G., Hudson, J.S., Ke, P.C., 2010. Differential uptake of carbon nanoparticles by plant and mammalian cells. Small 6, 612–617.

Chen, J., Liu, X., Wang, C., Yin, S.S., Li, X.L., Hu, W.J., Simon, M., Shen, Z.J., Xiao, Q., Chu, C.C., Peng, X.X., 2015. Nitric oxide ameliorates zinc oxide nanoparticles- induced phytotoxicity in rice seedlings. J. Hazard. Mater. 297, 173–182.

Cifuentes, Z., Custardoy, L., de la Fuente, J.M., Marquina, C., Ibarra, M.R., Rubiales, D., Pérez-de-Luque, A., 2010. Absorption and translocation to the aerial part of magnetic carbon-coated nanoparticles through the root of different crop plants. J. Nanobiotechnol. 8, 1.

Corral-Diaz, B., Peralta-Videa, J.R., Alvarez-Parrilla, E., Rodrigo-García, J., Morales, M.I., Osuna-Avila, P., Niu, G., Hernandez-Viezcas, J.A., Gardea-Torresdey, J.L., 2014. Cerium oxide nanoparticles alter the antioxidant capacity but do not impact tuber ionome in *Raphanus sativus* (L.). Plant Physiol. Biochem. 84, 277–285.

Cossins, D., March 16, 2014. Next generation: nanoparticles augment plant functions. The incorporation of synthetic nanoparticles into plants can enhance photosynthesis and transform leaves into biochemical sensors. Sci. News Opin. http://www.the-scientist.com/?articles.view/articleNo/39440/title/Next-Generation–Nanoparticles-Augment-Plant-Functions/.

Cui, D., Zhang, P., Ma, Y., He, X., Li, Y., Zhang, J., 2014a. Effect of cerium oxide nanoparticles on *Asparagus lettuce* cultured in an agar medium. Environ. Sci. Nano 1, 459–465.

Cui, W., Chen, H., Zhu, K., Jin, Q., Xie, Y., Cui, J., Xia, Y., Zhang, J., Shen, W., 2014b. Cadmium-induced hydrogen sulfide synthesis is involved in cadmium tolerance in *Medicago sativa* by reestablishment of reduced (*Homo*) glutathione and reactive oxygen species homeostases.

Da Costa, M.V.J., Sharma, P.K., 2016. Effect of copper oxide nanoparticles on growth, morphology, photosynthesis, and antioxidant response in *Oryza sativa*. Photosynthetica 54, 110–119.

Dan, Y., Zhang, W., Xue, R., Ma, X., Stephan, C., Shi, H., 2015. Characterization of gold nanoparticle uptake by tomato plants using enzymatic extraction followed by single-particle inductively coupled plasma–mass spectrometry analysis. Enviro. Sci. Technol. 49, 3007–3014.

Das, I., Krzyzosiak, A., Schneider, K., Wrabetz, L., D'Antonio, M., Barry, N., Sigurdardottir, A., Bertolotti, A., 2015. Preventing proteostasis diseases by selective inhibition of a phosphatase regulatory subunit. Science 348 (6231), 239–242.

de la Rosa, G., López-Moreno, M.L., de Haro, D., Botez, C.E., Peralta-Videa, J.R., Gardea-Torresdey, J.L., 2013. Effects of ZnO nanoparticles in alfalfa, tomato, and cucumber at the germination stage: root development and X-ray absorption spectroscopy studies. Pure Appl. Chem. 85, 2161–2174.

De La Torre-Roche, R., Hawthorne, J., Deng, Y., Xing, B., Cai, W., Newman, L.A., Wang, C., Ma, X., White, J.C., 2012. Fullerene-enhanced accumulation of p, p'-DDE in agricultural crop species. Environ. Sci. Technol. 46, 9315–9323.

De La Torre-Roche, R., Hawthorne, J., Deng, Y., Xing, B., Cai, W., Newman, L.A., Wang, Q., Ma, X., Hamdi, H., White, J.C., 2013. Multiwalled carbon nanotubes and C60 fullerenes differentially impact the accumulation of weathered pesticides in four agricultural plants. Environ. Sci. Technol. 471, 12539–12547.

Deepa, K., Singha, S., Panda, T., 2014. Doxorubicin nanoconjugates. J. Nanosci. Nanotechnol. 14, 892–904.

Derfus, A.M., Chan, W.C.W., Bhatia, S.N., 2004. Probing the cytotoxicity of semiconductor quantum dots. Nano Lett. 4, 11–18.

Dietz, K.J., Herth, S., 2011. Plant nanotoxicology. Trends Plant Sci. 16, 582–589.

Dimkpa, C.O., McLean, J.E., Latta, D.E., Manangon, E., Britt, D.W., Johnson, W.P., 2012a. CuO and ZnO nanoparticles: phytotoxicity, metal speciation, and induction of oxidative stress in sand-grown wheat. J. Nanopart. Res. 14, 1–15.

Dimkpa, C.O., McLean, J.E., Britt, D.W., Anderson, A.J., 2012b. Nano-CuO and interaction with nano-ZnO or soil bacterium provide evidence for the interference of nanoparticles in metal nutrition of plants. Ecotoxicology 24, 119–129.

Dimkpa, C.O., Latta, D.E., McLean, J.E., Britt, D.W., Boyanov, M.I., Anderson, A.J., 2013a. Fate of CuO and ZnO nano- and microparticles in the plant environment. Environ. Sci. Technol. 47, 4734–4742.

Dimkpa, C.O., McLean, J.E., Latta, D.E., Manangon, E., Britt, D.W., Johnson, W.P., Boyanov, M.I., Anderson, A.J., 2013b. CuO and ZnO nanoparticles: phytotoxicity, metal speciation and induction of oxidative stress in sand-grown wheat. J. Nanopart. Res. 14, 1125–1129.

Dimkpa, C.O., Hansen, T., Stewart, J., McLean, J.E., Britt, D.W., Anderson, A.J., 2015. ZnO nanoparticles and root colonization by a beneficial pseudomonad influence essential metal responses in bean (*Phaseolus vulgaris*). Nanotoxicology 9, 271–278.

Dong, J., Mao, W.H., Zhang, G.P., Wu, F.B., Cai, Y., 2007. Root excretion and plant tolerance to cadmium toxicity – a review. Plant Soil Environ. 53, 193–200.

Doshi, R., Braida, W., Christodoulatos, C., Wazne, M., O'Connor, G., 2008. Nano-aluminum: transport through sand columns and environmental effects on plants and soil communities. Environ. Res. 106, 296–303.

Du, W., Sun, Y., Ji, R., Zhu, J., Wu, J., Guo, H., 2011. TiO_2 and ZnO nanoparticles negatively affect wheat growth and soil enzyme activities in agricultural soil. J. Environ. Monit. 13, 822–828.

Ebbs, S.D., Bradfield, S.J., Kumar, P., White, J.C., Musante, C., Ma, X., 2016. Accumulation of zinc, copper, or cerium in carrot (*Daucus carota*) exposed to metal oxide nanoparticles and metal ions. Environ. Sci. Nano 3, 114–126.

El-Temsah, Y.S., Joner, E.J., 2012. Impact of Fe and Ag nanoparticles on seed germination and differences in bioavailability during exposure in aqueous suspension and soil. Environ. Toxicol. 27, 42–49.

Faisal, M., Saquib, Q., Alatar, A.A., Al-Khedhairy, A.A., Hegazy, A.K., Musarrat, J., 2013. Phytotoxic hazards of NiO-nanoparticles in tomato: a study on mechanism of cell death. J. Hazard. Mater. 250–251, 318–332.

Fan, R., Huang, Y.C., Grusak, M.A., Huang, C.P., Sherrier, D.J., 2014. Effects of nano-TiO_2 on the agronomically-relevant *Rhizobium legume* symbiosis. Sci. Total Environ. 466, 503–512.

Fang, J., Shan, X., Wen, B., Lin, J., Owens, G., 2009. Stability of titania nanoparticles in soil suspensions and transport in saturated homogeneous soil columns. Environ. Pollut. 157, 1101–1109.

Feizi, H., Kamali, M., Jafari, L., Moghaddam, P.R., 2013. Phytotoxicity and stimulatory impacts of nanosized and bulk titanium dioxide on fennel (*Foeniculum vulgare* Mill). Chemosphere 91, 506–511.

Foltête, A.S., Masfaraud, J.F., Bigorgne, E., Nahmani, J., Chaurand, P., Botta, C., Labille, J., Rose, J., Férard, J.F., Cotelle, S., 2011. Environmental impact of sunscreen nanomaterials: ecotoxicity and gentoxicity of altered TiO_2 nanocomposites on *Vicia faba*. Environ. Pollut. 159, 2515–2522. http://dx.doi.org/10.1016/j.envpol.2011.06.020.

Frazier, T.P., Burklew, C.E., Zhang, B.H., 2014. Titanium dioxide nanoparticles affect the growth and microRNA expression of tobacco (*Nicotiana tabacum*). Funct. Integr. Genom. 14, 75–83.

French, R.A., Jacobson, A.R., Kim, B., Isley, S.L., Penn, L., Baveye, P.C., 2009. Influence of ionic strength, pH, cation valence on aggregation kinetics of titanium dioxide nanoparticles. Environ. Sci. Technol. 43, 1354–1359.

Fu, Q., Saltsburg, H., Flytzani-Stephanopoulos, M., 2003. Active nonmetallic Au and Pt species on ceria-based water–gas shift catalysts. Science 301, 935–938.

Gajjar, P., Pettee, B., Britt, D.W., Huang, W., Johnson, W.P., Anderson, A.J., 2009. Antimicrobial activities of commercial nanoparticles against an environmental soil microbe, *Pseudomonas putida* KT2440. J. Biol. Eng. 2, 9.

Galbraith, D.W., 2007. Nanobiotechnology: silica breaks through in plants. Nat. Nanotechnol. 2, 272–273.

Gao, F., Liu, C., Qu, C., Zheng, L., Yang, F., Su, M., Hong, F., 2008. Was improvement of spinach growth by nano-TiO_2 treatment related to the changes of rubisco activase? Biometals 21, 211–217.

Gao, J., Xu, G., Qian, H., Liu, P., Zhao, P., Hu, Y., 2013. Effects of nano-TiO_2 on photosynthetic characteristics of *Ulmus elongata* seedlings. Environ. Pollut. 176, 63–70.

Gardea-Torresdey, J.L., Rico, C.M., White, J.C., 2014. Trophic transfer, transformation, and impact of engineered nanomaterials in terrestrial environments. Environ. Sci. Technol. 48, 2526–2540.

Ge, Y., Schimel, J.P., Holden, P.A., 2011. Evidence for negative effects of TiO_2 and ZnO nanoparticles on soil bacterial communities. Environ. Sci. Technol. 45, 1659–1664.

Geisler-Lee, J., Wang, Q., Yao, Y., Zhang, W., Geisler, M., Li, K., Huang, Y., Chen, Y., Kolmakov, A., Ma, X., 2012. Phytotoxicity,accumulation and transport of silver nanoparticles by *Arabidopsis thaliana*. Nanotoxicology 7, 323–337.

Giraldo, J.P., Landry, M.P., Faltermeier, S.M., McNicholas, T.P., Iverson, N.M., Boghossian, A.A., Reuel, N.F., Hilmer, A.J., Sen, F., Brew, J.A., Strano, M.S., 2014. Plant nanobionics approach to augment photosynthesis and biochemical sensing. Nat. Mater. 13, 400–408.

Golbamaki, N., Rasulev, B., Cassano, A., Robinson, R.l.M., Benfenati, E., Leszczynski, J., Cronin, M.T., 2015. Genotoxicity of metal oxide nanomaterials: review of recent data and discussion of possible mechanisms. Nanoscale 7, 2154–2198.

Gorczyca, A., Pociecha, E., Kasprowicz, M., Niemiec, M., 2015. Effect of nanosilver in wheat seedlings and *Fusarium culmorum* culture systems. Eur. J. Plant Pathol. 142, 251–261.

Gordon, W., Carter, J., Tissue, B., 2004. Long-lifetime luminescence of lanthanide doped gadolinium oxide nanoparticles for immunoassays. J. Lumin. 108, 339–342.

Green, M., Howman, E., 2005. Semiconductor quantum dots and free radical induced DNA nicking. Chem. Commun. 1, 121–123.

Grillo, R., Rosa, A.H., Fraceto, L.F., 2015. Engineered nanoparticles and organic matter: a review of state of the art. Chemosphere 119, 608–619.

Gui, X., Zhang, Z., Liu, S., Ma, Y., Zhang, P., He, X., Li, Y., Zhang, J., Li, H., Rui, Y., Liu, L., Cao, W., 2015. Fate and phytotoxicity of CeO_2 nanoparticles on lettuce cultured in the potting soil environment. PLoS One 10, e0134261.

Haghighi, M., Afifipour, Z., Mozafarian, M., 2012. The effect of N-Si on tomato seed germination under salinity levels. J. Biol. Environ. Sci 6, 87–90.

Hawthorne, J., Musante, C., Sinha, S.K., White, J.C., 2012. Accumulation and phytotoxicity of engineered nanoparticles to *Cucurbita pepo*. Int. J. Phytoremediat. 14, 429–442.

Held, M.A., Jiang, N., Basu, D., Showalter, A.M., Faik, A., 2015. Plant Cell Wall Polysaccharides: Structure and Biosynthesis. Polysaccharides: Bioactivity and Biotechnology, pp. 3–54.

Hong, G., Zou, Y., Antaris, A.L., Diao, S., Wu, D., Cheng, K., Zhang, X., 2014. Ultrafast fluorescence imaging in vivo with conjugated polymer fluorophores in the second near-infrared window. Nat. Commun. 5.

Hong, J., Rico, C.M., Zhao, L., Adeleye, A.S., Keller, A.A., Peralta-Videa, J.R., Gardea-Torresdey, J.L., 2015. Toxic effects of copper-based nanoparticles or compounds to lettuce (*Lactuca sativa*) and alfalfa (*Medicago sativa*). Environ. Sci. Process. Impacts 17, 177–185.

Hoshino, A., Fujioka, K., Oku, T., Suga, M., Sasaki, Y.F., Ohta, T., Yasuhara, M., Suzuki, K., Yamamoto, K., 2004. Physicochemical properties and cellular toxicity of nanocrystal quantum dots depend on their surface modification. Nano Lett. 4, 2163–2169.

Hossain, Z., Mustafa, G., Komatsu, S., 2015. Plant responses to nanoparticle stress. Int. J. Mol. Sci. 16, 26644–26653.

Hotze, E.M., Phenrat, T., Lowry, G.V., 2010. Nanoparticle aggregation: challenges to understanding transport and reactivity in the environment. J. Environ. Qual. 39, 1909–1924.

Huang, P., Ju, H.W., Min, J.H., Zhang, X., Chung, J.S., Cheong, H.S., Kim, C.S., 2012. Molecular and physiological characterization of the *Arabidopsis thaliana* oxidation-related zinc finger 2, a plasma membrane protein involved in ABA and salt stress response through the ABI2-mediated signaling pathway. Plant Cell Physiol. 53, 193–203.

Husen, A., Siddiqi, K.S., 2014. Carbon and fullerene nanomaterials in plant system. J. Nanotechnol. 12, 1–10.

Ikhtiar, R., Begum, P., Watari, F., Fugetsu, B., 2013. Toxic effect of multiwalled carbon nanotubes on lettuce (*Lactuca sativa*). Nano Biomed. 5, 18–24.

Jalill, R.D.H.A., Yousef, A.M., 2015. Comparison the phytotoxicity of TiO_2 nanoparticles with bulk particles on Amber 33 variety of rice (*Oryza sativa*) in vitro. Sch. Acad. J. Biosci. 3, 254–262.

Johansen, A., Pedersen, A.L., Jensen, K.A., Karlson, U., Hansen, B.M., Scott-Fordsmand, J.J., Winding, A., 2008. Effects of C60 fullerene nanoparticles on soil bacteria and protozoans. Environ. Toxicol. Chem. 27, 1895–1903.

Jośko, I., Oleszczuk, P., 2013. Manufactured nanomaterials–the connection between environmental fate and toxicity. Crit. Rev. Environ. Sci. Technol. 43, 2581–2616. http://dx.doi.org/10.1080/10643389.2012.694329.

Judy, J.D., Kirby, J.K., McLaughlin, M.J., Cavagnaro, T., Bertsch, P.M., 2016. Gold nanomaterial uptake from soil is not increased by arbuscular mycorrhizal colonization of *Solanum lycopersicum* (tomato). Nanomaterials 6, 68.

Ju-Nam, Y., Lead, J.R., 2008. Manufactures nanoparticles: an overview of their chemistry, interactions and potential environmental implications. Sci. Total Environ. 400, 396–414.

Kabata-Pendias, A., Pendias, H., 1993. Biogeochemia pierwiastków s´ladowych. PWN, S.A. Warszawa.

Karakoti, A., Monteiro-Riviere, N., Aggarwal, R., Davis, J., Narayan, R., Self, W., McGinnis, J., Seal, S., 2008. Nanoceria as antioxidant: synthesis and biomedical applications. JOM 60, 33–37.

Kaveh, R., Li, Y.S., Ranjbar, S., Tehrani, R., Brueck, C.L., Van Aken, B., 2013. Changes in *Arabidopsis thaliana* gene expression in response to silver nanoparticles and silver ions. Environ. Sci. Technol. 47, 10637–10644.

Khodakovskaya, M.V., de Silva, K., Biris, A.S., Dervishi, E., Villagarcia, H., 2012. Carbon nanotubes induce growth enhancement of tobacco cells. ACS Nano 6, 2128–2135.

Khodakovskaya, M.V., Kim, B.S., Kim, J.N., Alimohammadi, M., Dervishi, E., Mustafa, T., Cernigla, C.E., 2013. Carbon nanotubes as plant growth regulators: effects on tomato growth, reproductive system, and soil microbial community. Small 9, 115–123.

Kim, S., Kim, J., Lee, I., 2011. Effects of Zn and ZnO nanoparticles and Zn^{2+} on soil enzyme activity and bioaccumulation of Zn in *Cucumis sativus*. Chem. Ecol. 27, 49–57.

Kim, S., Lee, S., Lee, I., 2012. Alteration of phytotoxicity and oxidant stress potential by metal oxide nanoparticles in Cucumis sativus. Water Air Soil Pollut. 223, 2799–2806.

Kim, J.H., Lee, Y., Kim, E.J., Gu, S., Sohn, E.J., Seo, Y.S., An, H.J., Chang, Y.S., 2014. Exposure of iron nanoparticles to *Arabidopsis thaliana* enhances root elongation by triggering cell wall loosening. Environ. Sci. Technol. 48, 3477–3485.

Kim, J.H., Oh, Y., Yoon, H., Hwang, I., Chang, Y.S., 2015. Iron nanoparticle-induced activation of plasma membrane H(þ)-ATPase promotes stomatal opening in *Arabidopsis thaliana*. Environ. Sci. Technol. 4, 1113–1119.

Kirchner, C., Liedl, T., Kudera, S., Pellegrino, T., Javier, A.M., Gaub, H.E., Stölzle, S., Fertig, N., Parak, W.J., 2005. Cytotoxicity of colloidal CdSe and CdSe/ZnS nanoparticles. Nano Lett. 5, 331–338.

Klimov, V.I., 2007. Spectral and dynamical properties of multiexcitons in semiconductor nanocrystals. Annu. Rev. Phys. Chem. 58, 635–673.

Kole, C., Kole, P., Randunu, K.M., Choudhary, P., Podila, F., Ke, P.C., Rao, A.M., Marcus, R.K., 2013. Nanobiotechnology can boost crop production and quality: first evidence from increased plant biomass, fruit yield and phytomedicine content in bitter melon (*Momordica charantia*). BMC Biotechnol. 13, 37.

Koo, Y., Wang, J., Zhang, Q., Zhu, H., Chehab, E.W., Colvin, V.L., Alvarez, P.J., Braam, J., 2015. Fluorescence reports intact quantum dot uptake into roots and translocation to leaves of *Arabidopsis thaliana* and subsequent ingestion by insect herbivores. Environ. Sci. Technol. 49, 626–632.

Kosynkin, V., Arzgatkina, A., Ivanov, E., Chtoutsa, M., Grabko, A., Kardapolov, A., Sysina, N., 2000. The study of process production of polishing powder based on cerium dioxide. J. Alloys Compd. 303, 421–425.

Krishnaraj, C., Jagan, E.G., Ramachandran, R., Abirami, S.M., Mohan, N., Kalaichelvan, P.T., 2012. Effect of biologically synthesized silver nanoparticles on *Bacopa monnieri* (Linn.)Wettst. plant metabolism. Process Biochem. 47, 651–658.

Krystofova, O., Sochor, J., Zitka, O., Babula, P., Kudrle, V., Adam, V., Kizek, R., 2013. Effect of magnetic nanoparticles on tobacco BY-2 cell suspension culture. Int. J. Environ. Res. Public Health 10, 47–71.

Kumar, V., Guleria, P., Kumar, V., Yadav, S.K., 2013. Gold nanoparticle exposure induces growth and yield enhancement in *Arabidopsis thaliana*. Sci. Total Environ. 461, 462–468.

Kumar, S., Patra, A.K., Datta, S.C., Rosin, K.G., Purakayastha, T.J., 2015. Phytotoxicity of nanoparticles to seed germination of plants. IJAR 3, 854–865.

Kumari, M., Mukherjee, A., Chandrasekaran, N., 2009. Genotoxicity of silver nanoparticles in Allium cepa. Sci. Total Environ. 407, 5243–5246.

Kumari, M., Khan, S.S., Pakrashi, S., Mukherjee, A., Chandrasekaran, N., 2011. Cytogenetic and genotoxic effects of zinc oxide nanoparticles on root cells of *Allium cepa*. J. Hazard. Mater. 190, 613–621.

Kuzyniak, W., Adegoke, O., Sekhosana, K., D'souza, S., Tshangana, S.C., Hoffmann, B., Ermilov, E.A., Nyokong, T., Höpfner, M., 2014. Synthesis and characterization of quantum dots designed for biomedical use. Int. J. Pharm. 466, 382–389.

Lahiani, M.H., Dervishi, E., Chen, J., Nima, Z., Gaume, A., Biris, A.S., Khodakovskaya, M.V., 2013. Impact of carbon nanotube exposure to seeds of valuable crops. ACS Appl. Mater. Interfaces 5, 7965–7973.

Landa, P., Vankova, R., Andrlova, J., Hodek, J., Marsik, P., Storchova, H., White, J.C., Vanek, T., 2012. Nanoparticle-specific changes in *Arabidopsis thaliana* gene expression after exposure to ZnO, TiO_2, and fullerene soot. J. Hazard. Mater. 241–242, 55–62.

Larue, C., Castillo-Michel, H., Sobanska, S., Cécillon, L., Bureau, S., Barthès, V., Ouerdane, L., Carrière, M., Sarret, G., 2014. Foliar exposure of the crop *Lactuca sativa* to silver nanoparticles: evidence for internalization and changes in Ag speciation. J. Hazard. Mater. 264, 98–106.

Le, V.N., Rui, Y., Gui, X., Li, X., Liu, S., Han, Y., 2014. Uptake, transport, distribution and bio-effects of SiO_2 nanoparticles in Bt-transgenic cotton. J. Nanobiotechnol 12, 50.

Lee, S., Cha, E.J., Park, K., Lee, S.Y., Hong, J.K., Sun, I.C., Kim, S.Y., Choi, K., Kwon, I.C., Kim, K., Ahn, C.H., 2008. A near-infrared-fluorescence-quenched gold-nanoparticle imaging probe for in vivo drug screening and protease activity determination. Angew. Chem. Int. Ed. Engl. 47, 2804–2807.

Lee, W., Kwak, J., An, Y., 2012. Effect of silver nanoparticles in crop plants *Phaseolus radiatus* and *Sorghum bicolor*: media effect on phytotoxicity. Chemosphere 86, 491–499.

Lee, S., Kim, S., Kim, S., Lee, I., 2013a. Assessment of phytotoxicity of ZnO NPs on a medicinal plant, *Fagopyrum esculentum*. Environ. Sci. Pollut. Res. 20, 848–854.

Lee, S., Chung, H., Kim, S., Lee, I., 2013b. The genotoxic effect of ZnO and CuO nanoparticles on early growth of buckwheat, *Fagopyrum esculentum*. Water Air Soil Pollut. 224, 1668–1679.

Lee, S., Chung, H., Kim, S., Lee, I., 2013c. The genotoxic effect of ZnO and CuO nanoparticles on early growth of buckwheat, *Fagopyrum esculentum*. Water Air Soil Pollut. 224, 1–11.

Li, X., Gui, X., Rui, Y., Ji, W., Yu, Z., Peng, S., 2014a. Bt-transgenic cotton is more sensitive to CeO_2 nanoparticles than its parental non-transgenic cotton. J. Hazard. Mater. 274, 173–180.

Li, J., He, Y., Sun, W., Luo, Y., Cai, H., Pan, Y., Shen, M., Xia, J., Shi, X., 2014b. Hyaluronic acid-modified hydrothermally synthesized iron oxide nanoparticles for targeted tumor MR imaging. Biomaterials 35, 3666–3677.

Li, X., Schirmer, K., Bernard, L., Sigg, L., Pillai, S., Behra, R., 2015. Silver nanoparticle toxicity and association with the alga *Euglena gracilis*. Environ. Sci. Nano 2, 594–602.

Lin, D., Xing, B., 2007. Phytotoxicity of nanoparticles: inhibition of seed germination and root growth. Environ. Pollut. 150, 243–250.

Lin, S., Reppert, J., Hu, Q., Hudson, J.S., Reid, M.L., Ratnikova, T.A., Rao, A.M., Luo, H., Ke, P.C., 2009. Uptake, translocation, and transmission of carbon nanomaterials in rice plants. Small 5, 1128–1132.

Lin, A., Wang, Y., Tang, J., Xue, P., Li, C., Liu, L., Hu, B., Yang, F., Loake, G.J., Chu, C., 2012. Nitric oxide and protein *S*-nitrosylation are integral to hydrogen peroxide-induced leaf cell death in rice. Plant Physiol. 158, 451–464.

Liu, Q., Zhao, Y., Wan, Y., Zheng, J., Zhang, X., Wang, C., Fang, X., Lin, J., 2010. Study of the inhibitory effect of water-soluble fullerenes on plant growth at the cellular level. ACS Nano 4, 5743–5748.

Liu, Q., Zhang, X., Zhao, Y., Lin, J., Shu, C., Wang, C., Fang, X., 2013. Fullerene-induced increase of glycosyl residue on living plant cell wall. Environ. Sci. Technol. 47, 7490–7498.

López-Moreno, M.L., de la Rosa, G., Hernandez-Viezcas, J.A., Peralta-Videa, J.R., Gardea-Torresdey, J.L., 2010. X-ray absorption spectroscopy (XAS) corroboration of the uptake and storage of CeO_2 nanoparticles and assessment of their differential toxicity in four edible plant species. J. Agric. Food Chem. 58, 3689–3693.

López-Moreno, M.L., Avilés, L.L., Pérez, N.G., Irizarry, B.A., Perales, O., Cedeno-Mattei, Y., Román, F., 2016. Effect of cobalt ferrite ($CoFe_2O_4$) nanoparticles on the growth and development of *Lycopersicon lycopersicum* (tomato plants). Sci. Total Environ. 550, 45–52.

Ma, X., Geiser-Lee, J., Deng, Y., Kolmakov, A., 2010. Interactions between engineered nanoparticles (ENPs) and plants: phytotoxicity, uptake and accumulation. Sci. Total Environ. 408, 3053–3061.

Ma, Y., He, X., Zhang, P., Zhang, Z., Guo, Z., Tai, R., Xu, Z., Zhang, L., Ding, Y., Zhao, Y., 2011. Phytotoxicity and biotransformation of La_2O_3 nanoparticles in a terrestrial plant cucumber (*Cucumis sativus*). Nanotoxicology 5, 743–753.

Ma, C., Chhikara, S., Xing, B., Musante, C., White, J.C., Dhankher, O.P., 2013a. Physiological and molecular response of *Arabidopsis thaliana* (L.) to nanoparticle cerium and indium oxide exposure. ACS Sustain. Chem. Eng. 1, 768–778.

Ma, X., Gurung, A., Deng, Y., 2013b. Phytotoxicity and uptake of nanoscale zero-valent iron (nZVI) by two plant species. Sci. Total Environ. 443, 844–849.

Ma, R., Levard, C., Judy, J.D., Unrine, J.M., Durenkamp, M., Martin, B., Jefferson, B., Lowry, G.V., 2013c. Fate of zinc oxide and silver nanoparticles in a pilot wastewater treatment plant and in processed biosolids. Environ. Sci. Technol. 48, 104–112.

Ma, C.X., White, J.C., Dhankher, O.P., Xing, B., 2015a. Metal-based nanotoxicity and detoxification pathways in higher plants. Environ. Sci. Technol. 49, 7109–7122.

Ma, Y.H., Zhang, P., Zhang, Z.Y., He, X., Li, Y.Y., Zhang, J., Zheng, L.R., Chu, S.Q., Yang, K., Zhao, Y.L., Chai, Z.F., 2015b. Origin of the different phytotoxicity and biotransformation of cerium and lanthanum oxide nanoparticles in cucumber. Nanotoxicology 9, 262–270.

MacLeod, M.J., Johnson, J.A., 2015. PEGylated *N*-heterocyclic carbene anchors designed to stabilize gold nanoparticles in biologically relevant media. J. Am. Chem. Soc. 137, 7974–7977.

Mahmoodzadeh, H., Aghili, R., Nabavi, M., 2013. Physiological effects of TiO_2 nanoparticles on wheat (*Triticum aestivum*). Tech. J. Eng. Appl. Sci. 3, 1365–1370.

Majumdar, S., Peralta-Videa, J., Bandyopadhyay, S., Castillo-Michel, H., Hernandez- Viezcas, J., Sahi, S., Gardea-Torresdey, J., 2014. Exposure of cerium oxide nanoparticles to kidney bean shows disturbance in plant defense mechanism. J. Hazard. Mater. 278, 279–287.

Majzoub, R.N., Wonder, E., Ewert, K.K., Kotamraju, V.R., Teesalu, T., Safinya, C.R., 2016. Rab11 and lysotracker markers reveal correlation between endosomal pathways and transfection efficiency of surface-functionalized cationic liposome–DNA nanoparticles. J. Phys. Chem. B120, 6439–6453.

Marchiol, L., Mattiello, A., Pošćić, F., Fellet, G., Zavalloni, C., Carlino, E., Musetti, R., 2016. Changes in physiological and agronomical parameters of barley (*Hordeum vulgare*) exposed to cerium and titanium dioxide nanoparticles. Int. J. Environ. Res. Public Health 13, 332.

Marschner, H., 2011. Marschner's Mineral Nutrition of Higher Plants. Academic Press.

Martineau, N., McLean, J.E., Dimkpa, C.O., Britt, D.W., Anderson, A.J., 2014. Components from wheat roots modify the bioactivity of ZnO and CuO nanoparticles in a soil bacterium. Environ. Pollut. 187, 65–72.

Masarovičová, E., Kráľová, K., 2013. Metal nanoparticles and plants/Nanocząstki Metaliczne I Rośliny. Eco. Chem. Engineer. S 20, 9–22.

Meier, S., Alvear, M., Borie, F., Aguilera, P., Ginocchio, R., Cornejo, P., 2012. Influence of copper on root exudate patterns in some metallophytes and agricultural plants. Ecotoxicol. Environ. Saf. 75, 8–15.

Miclăuş, T., Beer, C., Chevallier, J., Scavenius, C., Bochenkov, V.E., Enghild, J.J., Sutherland, D.S., 2016. Dynamic protein coronas revealed as a modulator of silver nanoparticle sulphidation in vitro. Nat. Commun. 7, 11770. http://dx.doi.org/10.1038/ncomms11770.

Miralles, P., Johnson, E., Church, T.L., Harris, A.T., 2012. Multiwalled carbon nanotubes in alfalfa and wheat: toxicology and uptake. J. R. Soc. Interface 9, 3514–3527.

Mirzajani, F., Askari, H., Hamzelou, S., Farzaneh, M., Ghassempour, A., 2013. Effect of silver nanoparticles on *Oryza sativa* L. and its rhizosphere bacteria. Ecotoxicol. Environ. Saf. 88, 48–54.

Misra, S.B., Dybowska, A., Berhanu, D., Luoma, S.N., Valsami-Jones, E., 2012. The complexity of nanoparticle dissolution and its importance in nanotoxicological studies. Sci. Total Environ. 438, 225–232.

Mohammadi, M.T., Amini, R., Jahanbakhsh, Z., Shekarforoush, S., 2013. Effects of atorvastatin on the hypertension-induced oxidative stress in the rat brain. Iran. Biomed. J. 17, 152–157.

Mondal, A., Basu, R., Das, S., Nandy, P., 2011. Beneficial role of carbon nanotubes on mustard plant growth: an agricultural prospect. J. Nanopart. Res. 13, 4519–4528.

Monti, D.M., Guarnieri, D., Napolitano, G., Piccoli, R., Netti, P., Fusco, S., Arciello, A., 2015. Biocompatibility, uptake and endocytosis pathways of polystyrene nanoparticles in primary human renal epithelial cells. J. Biotechnol. 193, 3–10.

Moon, Y.S., Park, E.S., Kim, T.O., Lee, H.S., Lee, S.E., 2014. SELDI-TOF MS-based discovery of a biomarker in *Cucumis sativus* seeds exposed to CuO nanoparticles. Environ. Toxicol. Pharm. 38, 922–931.

Morales, M.I., Rico, C.M., Hernandez-Viezcas, J.A., Nunez, J.E., Barrios, A.C., Tafoya, A., Flores-Marges, J.P., Peralta-Videa, J.R., Gardea-Torresdey, J.L., 2013. Toxicity assessment of cerium oxide nanoparticles in cilantro (*Coriandrum sativum* L.) plants grown in organic soil. J. Agric. Food Chem. 61, 6224–6230.

Mostofa, M.G., Fujita, M., 2013. Salicylic acid alleviates copper toxicity in rice (*Oryza sativa* L.) seedlings by up-regulating antioxidative and glyoxalase systems. Ecotoxicology 22, 959–973.

Mukherjee, A., Peralta-Videa, J.R., Bandyopadhyay, S., Rico, C.M., Zhao, L., Gardea-Torresdey, J.L., 2014a. Physiological effects of nanoparticulate ZnO in green peas (*Pisum sativum* L.) cultivated in soil. Metallomics 6, 132–138.

Mukherjee, A., Pokhrel, S., Bandyopadhyay, S., Mädler, L., Peralta-Videa, J.R., Gardea- Torresdey, J.L., 2014b. A soil mediated phyto-toxicological study of iron doped zinc oxide nanoparticles (Fe at ZnO) in green peas (*Pisum sativum* L.). Chem. Eng. J. 258, 394–401.

Mukherjee, N., Mukherjee, S., Saini, P., Roy, P., Babu, S.P.S., 2014c. Antifilarial effects of polyphenol rich ethanolic extract from the leaves of *Azadirachta indica* through molecular and biochemical approaches describing reactive oxygen species (ROS) mediated apoptosis of *Setaria cervi*. Exp. Parasitol. 136, 41–58.

Mura, S., Seddaiu, G., Bacchini, F., Roggero, P.P., Greppi, G.F., 2013. Advances of nanotechnology in agro-environmental studies. Italian J. Agron. 8, 18.

Musante, C., White, J.C., 2012. Toxicity of silver and copper to *Cucurbitapepo*: differential effects of nano and bulk-size particles. Environ. Toxicol. 27, 510–517.

Nagaonkar, D., Shende, S., Rai, M., 2015. Biosynthesis of copper nanoparticles and its effect on actively dividing cells of mitosis in *Allium cepa*. Biotechnol. Prog. 31, 557–565.

Nair, P.M.G., Chung, I.M., 2014a. A mechanistic study on the toxic effect of copper oxide nanoparticles in soybean (*Glycine max* L.) root development and lignification of root cells. Biol. Trace Elem. Res. 162, 342–352.

Nair, P.M.G., Chung, I.M., 2014b. Physiological and molecular level effects of silver nanoparticles exposure in rice (*Oryza sativa* L.) seedlings. Chemosphere 112, 105–113.

Nair, P.M., Chung, I.M., 2015. Study on the correlation between copper oxide nanoparticles induced growth suppression and enhanced lignification in Indian mustard (*Brassica juncea* L.). Ecotoxicol. Environ. Saf. 113, 302–313.

Nair, R., Poulose, A., Nagaoka, Y., Yoshida, Y., Maekawa, T., Kumar, D., 2011. Uptake of FITC labeled silica nanoparticles and quantum dots by rice seedlings: effects on seed germination and their potential as biolabels for plants. J. Fluoresc. 21, 2057–2068.

Nair, P.M.G., Kim, S.H., Chung, I.M., 2014. Copper oxide nanoparticle toxicity in mung bean (*Vigna radiata* L.) seedlings: physiological and molecular level responses of in vitro grown plants. Acta Physiol. Plant 36, 2947–2958.

Navarro, D.A., Bisson, M.A., Aga, D.S., 2012. Investigating uptake of water-dispersible CdSe/ZnS quantum dot nanoparticles by *Arabidopsis thaliana* plants. J. Hazard. Mater. 211–212, 427–435.

Ng, C.T., Tang, F.M.A., Li, J.J.E., Ong, C., Yung, L.L.Y., Bay, B.H., 2015. Clathrin-mediated endocytosis of gold nanoparticles in vitro. Anat. Rec. 298, 418–427.

Niemeyer, C.M., 2001. Nanoparticles, proteins, and nucleic acids: biotechnology meets materials science. Angew. Chem. Int. Ed. 40, 4128–4158.

Nohl, H., Gille, L., 2013. Lysosomal ROS formation. Redox Rep 10, 199–205.

Oberdorster, G., 2010. Safety assessment for nanotechnology and nanomedicine: concepts of nanotoxicology. J. Intern. Med. 267, 89–105.

Onelli, E., Prescianotto-Baschong, C., Caccianiga, M., Moscatelli, A., 2008. Clathrin-dependent and independent endocytic pathways in tobacco protoplasts revealed by labelling with charged nanogold. J. Exp. Bot. 59, 3051–3068.

Oukarroum, A., Bras, S., Perreault, F., Popovic, R., 2012. Inhibitory effects of silver nanoparticles in two green algae, *Chlorella vulgaris* and *Dunaliella tertiolecta*. Ecotoxicol. Environ. Saf. 78, 80–85.

Oukarroum, A., Barhoumi, L., Pirastru, L., Dewez, D., 2013. Silver nanoparticle toxicity effect on growth and cellular viability of the aquatic plant *Lemna gibba*. Environ. Toxicol. Chem. 32, 902–907.

Parsons, J.G., Lopez, M.L., Gonzalez, C.M., Peralta-Videa, J.R., Gardea-Torresdey, J.L., 2010. Toxicity and biotransformation of uncoated and coated nickel hydroxide nanoparticles on mesquite. Plants. Environ. Toxicol. Chem. 29, 1146–1154.

Patlolla, A.K., Berry, A., May, L.B., Tchounwou, P.B., 2012. Genotoxicity of silver nanoparticles in *Vicia faba*: a pilot study on the environmental monitoring of nanoparticles. Int. J. Environ. Res. Public Health 9, 1649–1662.

Patra, P., Choudhury, S.R., Mandal, S., Basu, A., Goswani, A., Gogoi, R., Srivastava, C., Kumar, R., Gopal, M., 2013. Effect of sulfur and ZnO nanoparticles on stress physiology and plant (*Vigna radiata*) nutrition. Adv. Nanomater. Nanotechnol. 31, 299–307.

Pauly, M., Keegstra, K., 2016. Biosynthesis of the plant cell wall matrix polysaccharide xyloglucan. Annu. Rev. Plant Biol. 67, 235–259.

Pena, L.B., Mendez, A.E., Matayoshi, C.L., Zawoznik, M.S., Gallego, S.M., 2015. Early response of wheat seminal roots growing under copper excess. Plant Physiol. Biochem. 87, 115–123.

Peng, C., Zhang, H., Fang, H., Xu, C., Huang, H., Wang, Y., Sun, L., Yuan, X., Chen, Y., Shi, J., 2015. Natural organic matter-induced alleviation of the phytotoxicity to rice (*Oryza sativa* L.) caused by copper oxide nanoparticles. Environ. Toxicol. Chem. 34, 1996–2003.

Pereira, E.G., Oliva, M.A., Rosado-Souza, L., Mendes, G.C., Colares, D.S., Stopato, C.H., Almeida, A.M., 2013. Iron excess affects rice photosynthesis through stomatal and non-stomatal limitations. Plant Sci. 201–202, 81–92.

Perreault, F., Popovic, R., Dewez, D., 2014. Different toxicity mechanisms between bare and polymer-coated copper oxide nanoparticles in *Lemna gibba*. Environ. Pollut. 185, 219–227.

Phenrat, T., Long, T., Lowry, G., Veronesi, B., 2009. Partial oxidation ("Aging") and surface modification decrease the toxicity of nanosized zerovalent iron. Environ. Sci. Technol. 43, 195–200.

Pisanic, T.R., Jin, S., Shubayev, V.I., 2009. Iron oxide magnetic nanoparticle nanotoxicity: incidence and mechanisms. In: Nanotoxicity. John Wiley & Sons, Ltd, Chichester, UK, pp. 397–425. http://dx.doi.org/10.1002/9780470747803.ch20.

Poborilova, Z., Opatrilova, R., Babula, P., 2013. Toxicity of aluminium oxide nanoparticles demonstrated using a BY-2 plant cell suspension culture model. Environ. Exp. Bot. 91, 1–11.

Pokhrel, L.R., Dubey, B., 2013. Evaluation of developmental responses of two crop plants exposed to silver and zinc oxide nanoparticles. Sci. Total Environ. 452–453, 321–332.

Pošćić, F., Mattiello, A., Fellet, G., Miceli, F., Marchiol, L., 2016. Effects of cerium and titanium oxide nanoparticles in soil on the nutrient composition of barley (*Hordeum vulgare* L.) kernels. Int. J. Environ. Res. Public Health 13, 577.

Priester, J.H., Ge, Y., Mielke, R.E., Horst, A.M., Moritz, S.C., Espinosa, K., Gelb, J., Walker, S.L., Nisbet, R.M., An, Y.J., Schimel, J.P., 2012. Soybean susceptibility to manufactured nanoparticles with evidence for food quality and soil fertility interruption. Proc. Natl. Acad. Sci. Lett. 109, 2451–2456.

Pulskamp, K., Diabaté, S., Krug, H.F., 2007. Carbon nanotubes show no sign of acute toxicity but induce intracellular reactive oxygen species in dependence on contaminants. Toxicol. Lett. 168, 58–74.

Qian, H., Peng, X., Han, X., Ren, J., Sun, L., Fu, Z., 2013. Comparison of the toxicity of silver nanoparticles and silver ions on the growth of terrestrial plant model *Arabidopsis thaliana*. J. Environ. Sci. 25, 1947–1956.

Quentin, M., Baurès, I., Hoefle, C., Caillaud, M.C., Allasia, V., Panabières, F., Abad, P., Hückelhoven, R., Keller, H., Favery, B., 2016. The Arabidopsis microtubule-associated protein MAP65-3 supports infection by filamentous biotrophic pathogens by down-regulating salicylic acid-dependent defenses. J. Exp. Bot. 67, 1731–1743.

Rafique, R., Arshad, M., Khokhar, M.F., Qazi, I.A., Hamza, A., Virk, N., 2014. Growth response of wheat to titania nanoparticles application. NUST J. Eng. Sci. 7, 42–46.

Rajeshwari, A., Kavitha, S., Alex, S.A., Kumar, D., Mukherjee, A., Chandrasekaran, N., Mukherjee, A., 2015. Cytotoxicity of aluminum oxide nanoparticles on *Allium cepa* root tip—effects of oxidative stress generation and biouptake. Environ. Sci. Pollut. Res. 14, 11057–11066.

Raliya, R., Nair, R., Chavalmane, S., Wang, W.N., Biswas, P., 2015. Mechanistic evaluation of translocation and physiological impact of titanium dioxide and zinc oxide nanoparticles on the tomato (*Solanum lycopersicum* L.) plant. Metallomics 7, 1584–1594.

Rao, S., Shekhawat, G.S., 2014. Toxicity of ZnO engineered nanoparticles and evaluation of their effect on growth, metabolism and tissue specific accumulation in *Brassica juncea*. J. Environ. Chem. Eng. 2, 105–114.

Raskar, S.V., Laware, S.L., 2014. Effect of zinc oxide nanoparticles on cytology and seed germination in onion. Int. J. Curr. Microbiol. Appl. Sci. 3, 467–473.

Regier, N., Cosio, C., Von Moos, N., Slaveykova, V.I., 2015. Effects of copper-oxide nanoparticles dissolved copper and ultraviolet radiation on copper bioaccumulation, photosynthesis and oxidative stress in the aquatic macrophyte *Elodea nuttallii*. Chemosphere 128, 56–61.

Riahi-Madvar, A., Rezaee, F., Jalali, V., 2013. Effects of alumina nanoparticles on morphological properties and antioxidant system of *Triticum aestivum*. Iran. J. Plant Physiol. 3, 595–603.

Rico, C.M., Majumdar, S., Duarte-Gardea, M., Peralta-Videa, J.R., Gardea-Torresdey, J.L., 2011. Interaction of nanoparticles with edible plants and their possible implications in the food chain. J. Agric. Food Chem. 59, 3485–3498.

Rico, C.M., Hong, J., Morales, M.I., Zhao, L., Barrios, A.C., Zhang, J.Y., 2013a. Effect of cerium oxide nanoparticles on rice: a study involving the antioxidant defense system and in vivo fluorescence imaging. Environ. Sci. Technol. 47, 5635–5642.

Rico, C.M., Morales, M.I., Barrios, A.C., McCreary, R., Hong, J., Lee, W.-Y., Nunez, J., Peralta-Videa, J.R., Gardea-Torresdey, J.L., 2013b. Effect of cerium oxide nanoparticles on the quality of rice (*Oryza sativa* L.) grains. J. Agric. Food Chem. 61, 11278–11285.

Rico, C.M., Morales, M.I., McCreary, R., Castillo-Michel, H., Barrios, A.C., Hong, J., 2013c. Cerium oxide nanoparticles modify the antioxidative stress enzyme activities and macromolecule composition in rice seedlings. Environ. Sci. Technol. 47, 14110–14118.

Rico, C.M., Peralta-Videa, J.R., Gardea-Torresdey, J.L., 2015a. Chemistry, biochemistry of nanoparticles, and their role in antioxidant defense system in plants. In: Nanotechnology and Plant Sciences, pp. 1–17.

Rico, C.M., Barrios, A.C., Tan, W., Rubenecia, R., Lee, S.C., Varela-Ramirez, A., 2015b. Physiological and biochemical response of soil-grown barley (*Hordeum vulgare* L.) to cerium oxide nanoparticles. Environ. Sci. Pollut. Res. 22.

Rizwan, M., Ali, S., Adrees, M., Rizvi, H., Rehman, M.Z., Hannan, F., Qayyum, M.F., Hafeez, F., Ok, Y.S., 2016a. Cadmium stress in rice: toxic effects, tolerance mechanisms and management: a critical review. Environ. Sci. Pollut. Res. 23.

Rizwan, M., Ali, S., Abbas, T., Rehman, M.Z., Hannan, F., Keller, C., Al-Wabel, M.I., Ok, Y.S., 2016b. Cadmium minimization in wheat: a critical review. Ecotoxicol. Environ. Saf. 130, 43–53.

Rocha, T.L., Gomes, T., Mestre, N.C., Cardoso, C., Bebianno, M.J., 2015. Tissue specific responses to cadmium-based quantum dots in the marine mussel *Mytilus galloprovincialis*. Aqua. Toxicol. 1, 6910–6918.

Sabo-Attwood, T., Unrine, J.M., Stone, J.W., Murphy, C.J., Ghoshroy, S., Blom, D., Bertsch, P.M., Newman, L.A., 2012. Uptake, distribution and toxicity of gold nanoparticles in tobacco (*Nicotiana xanthi*) seedlings. Nanotoxicology 6, 353–360.

Salama, H.M.H., 2012. Effects of silver nanoparticles in some crop plants, common bean (*Phaseolus vulgaris* L.) and corn (*Zea mays* L.). Int. Res. J. Biotechnol. 3, 190–197.

Schwabe, F., Tanner, S., Schulin, R., Rotzetter, A., Stark, W., von Quadt, A., Nowack, B., 2015. Dissolved cerium contributes to uptake of Ce in the presence of differently sized CeO_2-nanoparticles by three crop plants. Metallomics 7, 466–477.

Seif, S.M., Sorroshzadeh, A., Rezazadeh, H., Naghdibadi, H.A., 2011. Effect of nano silver and silver nitrate on seed yield of borage. J. Med. Plant Res. 5, 171–175.

Serag, M.F., Kaji, N., Gaillard, C., Okamoto, Y., Terasaka, K., Jabasini, M., Tokeshi, M., Mizukami, H., Bianco, A., Baba, Y., 2011. Trafficking and subcellular localization of multiwalled carbon nanotubes in plant cells. ACS Nano 5, 493–499.

Shalaby, S., Horwitz, B.A., 2015. Plant phenolic compounds and oxidative stress: integrated signals in fungal–plant interactions. Curr. Genet. 61, 347–357.

Shams, G., Ranjbar, M., Amiri, A.A., Khodarahmpour, Z., 2013a. The effect of 35 nm silver nanoparticles on antagonistic and synergistic mineral elements in leaves and fruit of tomato (*Lycopersicon esculentum* Mill.). IJACS. 5.

Shams, G., Ranjbar, M., Amiri, A., 2013b. Effect of silver nanoparticles on concentration of silver element and growth indexes in cucumber (*Cucumis sativus* L. negeen). J. Nanopart. Res. 15, 1630–1642.

Shaw, A.K., Hossain, Z., 2013. Impact of nano-CuO stress on rice (*Oryza sativa* L.) seedlings. Chemosphere 93, 906–915.

Shaw, A.K., Ghosh, S., Kalaji, H.M., Bosa, K., Brestic, M., Zivcak, M., Hossain, Z., 2014a. Nano-CuO stress induced modulation of antioxidative defense and photosynthetic performance of Syrian barley (*Hordeum vulgare* L.). Environ. Exp. Bot. 102, 37–47.

Shaw, P.X., Werstuck, G., Chen, Y., 2014b. Oxidative stress and aging diseases. Oxid. Med. Cell. Longev. 2014.

Shaymurat, T., Gu, J., Xu, C., Yang, Z., Zhao, Q., Liu, Y., Liu, Y., 2012. Phytotoxic and genotoxic effects of ZnO nanoparticles on garlic (*Allium sativum* L.): a morphological study. Nanotoxicology 6, 241–248.

Shen, C.X., Zhang, Q.F., Li, J., Bi, F.C., Yao, N., 2010. Induction of programmed cell death in *Arabidopsis* and rice by single-wall carbon nanotubes. Am. J. Bot. 97, 1602–1609.

Shi, J.Y., Peng, C., Yang, Y.Q., Yang, J.J., Zhang, H., Yuan, X.F., Chen, Y.X., Hu, T., 2014a. Phytotoxicity and accumulation of copper oxide nanoparticles to the Cu tolerant plant *Elsholtzia splendens*. Nanotoxicology 8, 179–188.

Shi, Y., Pan, Y., Zhang, H., Zhang, Z., Li, M.J., Yi, C., Yang, M., 2014b. A dual-mode nanosensor based on carbon quantum dots and gold nanoparticles for discriminative detection of glutathione in human plasma. Biosens. Bioelectron. 56, 39–45.

Shiohara, A., Hoshino, A., Hanaki, K., Suzuki, K., Yamamoto, K., 2004. On the cyto-toxicity caused by quantum dots. Microbiol. Immunol. 48, 669–675.

Shoults-Wilson, W.A., Reinsch, B.C., Tsyusko, O.V., Bertsch, P.M., Lowry, G.V., Unrine, J.M., 2011. Role of particle size and soil type in toxicity of silver nanoparticles to earthworms. Soil Sci. Soc. Am. J. 75, 365–377.

Shweta, Tripathi, D.K., Singh, S., Singh, S., Dubey, N.K., Chauhan, D.K., 2016. Impact of nanoparticles on photosynthesis: challenges and opportunities. Mater. Focus 5 (5), 405–411.

Siddiqui, M.H., Al-Whaibi, M.H., 2014. Role of nano-SiO_2 in germination of tomato (*Lycopersicum esculentum* seeds Mill.). Saudi J. Biol. Sci. 21, 13–17.

Siddiqui, M.H., Al-Whaibi, M.H., Mohammad, F. (Eds.), 2015. Nanotechnology and Plant Sciences: Nanoparticles and Their Impact on Plants. Springer.

Siddiqui, M.H., Al-Whaibi, M.H., Faisal, M., Al-Sahli, A.A., 2014. Nano-silicon dioxide mitigates the adverse effects of salt stress on *Cucurbita pepo* L. Environ. Toxicol. Chem. 33, 2429–2437. http://dx.doi.org/10.1002/etc.2697.

Singh, R., Lillard, J.W., 2009. Nanoparticle-based targeted drug delivery. Exp. Mol. Pathol. 86, 215–223.

Singh, S., Tripathi, D.K., Dubey, N.K., Chauhan, D.K., 2016. Effects of nano-materials on seed germination and seedling growth: striking the slight balance between the concepts and controversies. Mater. Focus 5 (3), 195–201.

Singh, S., Vishwakarma, K., Singh, S., Sharma, S., Dubey, N.K., Singh, V.K., Liu, S., Tripathi, D.K., Chauhan, D.K., 2017. Understanding the plant and nanoparticle interface at transcriptomic and proteomic level: a concentric overview. Plant Gene. http://dx.doi.org/10.1016/j.plgene.2017.03.006.

Smirnova, E., Gusev, A., Zaytseva, O., Sheina, O., Tkachev, A., Kuznetsova, E., Lazareva, E., Onishchenko, G., Feofanov, A., Kirpichnikov, M., 2012. Uptake and accumulation of multiwalled carbon nanotubes change the morphometric and biochemical characteristics of *Onobrychis arenaria* seedlings. Front. Chem. Sci. Eng. 6, 132–138.

Son, B., Kim, H.D., Kim, M., Kim, J.A., Lee, J., Shin, H., Hwang, N.S., Park, T.H., 2015. Physical stimuli – induced chondrogenic differentiation of mesenchymal stem cells using magnetic nanoparticles. Adv. Healthcare Mater. 4, 1339–1347.

Song, U., Jun, H., Waldman, B., Roh, J., Kim, Y., Yi, J., Lee, E.J., 2013a. Functional analyses of nanoparticle toxicity: a comparative study of the effects of TiO_2 and Ag on tomatoes (*Lycopersicon esculentum*). Ecotoxicol. Environ. Saf. 93, 60–67.

Song, U., Shin, M., Lee, G., Roh, J., Kim, Y., Lee, E.J., 2013b. Functional analysis of TiO_2 nanoparticle toxicity in three plant species. Biol. Trace Elem. Res. 155, 93–103.

Speranza, A., Leopold, K., Maier, M., Taddei, A.R., Scoccianti, V., 2010. Pd-nanoparticles cause increased toxicity to kiwifruit pollen compared to soluble Pd(II). Environ. Pollut. 158, 873–882.

Srinivasan, C., Saraswathi, R., 2010. Nano-agriculture-carbon nanotubes enhance tomato seed germination and plant growth. Curr. Sci. 99, 273–275.

Stampoulis, D., Sinha, S.K., White, J.C., 2009. Assay-dependent phytotoxicity of nanoparticles to plants. Environ. Sci. Techol. 43, 9473–9479.

Starnes, D.L., Unrine, J.M., Starnes, C.P., Collin, B.E., Oostveen, E.K., Ma, R., Lowry, G.V., Bertsch, P.M., Tsyusko, O.V., 2015. Impact of sulfidation on the bioavailability and toxicity of silver nanoparticles to *Caenorhabditis elegans*. Environ. Pollut. 196, 239–246.

Su, Y.Y., He, Y., Lu, H.T., Sai, L.M., Li, Q.N., Li, W.X., Wang, L.H., Shen, P.P., Huang, Q., Fan, C.H., 2009. The cytotoxicity of cadmium based, aqueous phase–synthesized, quantum dots and its modulation by surface coating. Biomaterials 30, 19–25.

Syu, Y.Y., Hung, J.H., Chen, J.C., Chuang, H.W., 2014. Impacts of size and shape of silver nanoparticles on *Arabidopsis* plant growth and gene expression. Plant Physiol. Biochem. 83, 57–64.

Tan, X.-M., Lin, C., Fugetsu, B., 2009. Studies on toxicity of multi-walled carbon nanotubes on suspension rice cells. Carbon 47, 3479–3487.

Taranath, T.C., Patil, B.N., Santosh, T.U., Sharath, B.S., 2015. Cytotoxicity of zinc nanoparticles fabricated by *Justicia adhatoda* L. on root tips of *Allium cepa* L.—a model approach. Environ. Sci. Pollut. Res 22, 8611–8617.

Temsah, Y.S., Joner, E.J., 2010. Impact of Fe and Ag nanoparticles on seed germination and differences in bioavailability during exposure in aqueous suspension and soil. Environ. Toxicol. 27, 42–49.

Thuesombat, P., Hannongbua, S., Akasit, S., Chadchawan, S., 2014. Effect of silver nanoparticles on rice (*Oryza sativa* L. cv. KDML 105) seed germination and seedling growth. Ecotoxicol. Environ. Saf. 104, 302–309.

Thwala, M., Musee, N., Sikhwivhilu, L., Wepener, V., 2013. The oxidative toxicity of Ag and ZnO nanoparticles towards the aquatic plant *Spirodela punctuta* and the role of testing media parameters. Environ. Sci. Process. Impacts 15, 1830–1843.

Tiwari, D.K., Dasgupta-Schubert, N., Villaseñor-Cendejas, L.M., Villegas, J., Carreto-Montoya, L., Borjas-García, S.E., 2014. Interfacing carbon nanotubes (CNT) with plants: enhancement of growth, water and ionic nutrient uptake in maize (*Zea mays*) and implications for nanoagriculture. Appl. Nanosci. 4, 577–591.

Toduka, Y., Toyooka, T., Ibuki, Y., 2012. Flow cytometric evaluation of nanoparticles using side-scattered light and reactive oxygen species-mediated fluorescence-correlation with genotoxicity. Environ. Sci. Technol. 46, 7629–7636.

Tonigold, M., Mailänder, V., 2016. Endocytosis and intracellular processing of nanoparticles in dendritic cells: routes to effective immunonanomedicines.

Torney, F., Trewyn, B.G., Lin, V.S.Y., Wang, K., 2007. Mesoporous silica nanoparticles deliver DNA and chemicals into plants. Nat. Nanotechnol 2, 295–300.

Tripathi, D.K., Singh, V.P., Prasad, S.M., Chauhan, D.K., Dubey, N.K., 2015. Silicon nanoparticles (SiNp) alleviate chromium (VI) phytotoxicity in *Pisum sativum* (L.) seedlings. Plant Physiol. Biochem. 96, 189–198.

Tripathi, D.K., Singh, S., Singh, V.P., Prasad, S.M., Chauhan, D.K., Dubey, N.K., 2016a. Silicon nanoparticles more efficiently alleviate arsenate toxicity than silicon in maize cultivar and hybrid differing in arsenate tolerance. Front. Environ. Sci. 4, 46.

Tripathi, D.K., Singh, S., Singh, S., Srivastava, P.K., Singh, V.P., Singh, S., Singh, P.K., Prasad, S.M., Pandey, A.C., Chauhan, D.K., Dubey, N.K., 2016b. Nitric oxide alleviates silver nanoparticles (AgNps)-induced phytotoxicity in *Pisum sativum* seedlings. Plant Physiol. Biochem. 110.

Tripathi, D.K., Singh, S., Singh, V.P., Prasad, S.M., Chauhan, D.K., Dubey, N.K., 2016c. Silicon nanoparticles more efficiently alleviate arsenate toxicity than silicon in maize cultiver and hybrid differing in arsenate tolerance. Front. Environ. Sci. 4, 46.

Tripathi, D.K., Singh, S., Singh, S., Pandey, R., Singh, V.P., Sharma, N.C., et al., 2017a. An overview on manufactured nanoparticles in plants: uptake, translocation, accumulation and phytotoxicity. Plant Physiol. Biochem. 110, 2–12.

Tripathi, D.K., Tripathi, A., Guar, S., Singh, S., Singh, Y., Vishwakarma, K., et al., 2017b. Uptake, accumulation and toxicity of silver nanoparticle in autotrophic plants, and heterotrophic microbes: a concentric review. Front. Microbiol. 8, 7.

Tripathi, D.K., Mishra, R.K., Singh, S., Singh, S., Vishwakarma, K., Sharma, S., Singh, V.P., Singh, P.K., Prasad, S.M., Dubey, N.K., Pandey, A.C., 2017d. Nitric oxide ameliorates zinc oxide nanoparticles phytotoxicity in wheat seedlings: implication of the ascorbate–glutathione cycle. Front. Plant Sci. 8.

Tripathi, D.K., Singh, S., Singh, V.P., Prasad, S.M., Dubey, N.K., Chauhan, D.K., 2017e. Silicon nanoparticles more effectively alleviated UV-B stress than silicon in wheat (*Triticum aestivum*) seedlings. Plant Physiol. Biochem. 110, 70–81.

Trujillo-Reyes, J., Majumdar, S., Botez, C.E., Peralta-Videa, J.R., Gardea-Torresdey, J.L., 2014. Exposure studies of core–shell Fe/Fe3O4 and Cu/CuO NPs to lettuce (*Lactuca sativa*) plants: are they a potential physiological and nutritional hazard? J. Hazard. Mater. 267, 255–263.

Tumburu, L., Andersen, C.P., Rygiewicz, P.T., Reichman, J.R., 2015. Phenotypic and genomic responses to titanium dioxide and cerium oxide nanoparticles in Arabidopsis germinants. Environ. Toxicol. Chem. 34, 70–83.

Valizadeh, A., Mikaeili, H., Samiei, M., Farkhani, S.M., Zarghami, N., Kouhi, M., Akbarzadeh, A., Davaran, S., 2012. Quantum dots: synthesis, bioapplications, and toxicity. Nanoscale Res. Lett. 7, 480.

Vannini, C., Domingo, G., Onelli, E., Prinsi, B., Marsoni, M., Espen, L., Bracale, M., 2013. Morphological and proteomic responses of *Eruca sativa* exposed to silver nanoparticles or silver nitrate. PLoS One 8, e68752.

Vannini, C., Domingo, G., Onelli, E., De Mattia, F., Bruni, I., Marsoni, M., Bracale, M., 2014. Phytotoxic and genotoxic effects of silver nanoparticles exposure on germinating wheat seedlings. J. Plant Physiol. 171, 1142–1148.

Villagarcia, H., Dervishi, E., Silva, K., Biris, A.S., Khodakovskaya, M.V., 2012. Surface chemistry of carbon nanotubes impacts the growth and expression of water channel protein in tomato plants. Small 8, 2328–2334.

Walker, W.A., Tarannum, M., Vivero-Escoto, J.L., 2016. Cellular endocytosis and trafficking of cholera toxin B-modified mesoporous silica nanoparticles. J. Mater. Chem. B 4, 1254–1262.

Wang, Y.C., Lee, C.M., Lee, L.C., Tung, L.C., Hsieh-Li, H.M., Lee-Chen, G.J., Su, M.T., 2011. Mitochondrial dysfunction and oxidative stress contribute to the pathogenesis of spinocerebellar ataxia type 12 (SCA12). J. Biol. Chem. 286, 21742–21754.

Wang, X., Han, H., Liu, X., Gu, X., Chen, K., Lu, D., 2012a. Multi-walled carbon nanotubes can enhance root elongation of wheat (*Triticum aestivum*) plants. J. Nanopart. Res. 14, 1–10.

Wang, M., Chen, L., Chen, S., Ma, Y., 2012b. Alleviation of cadmium-induced root growth inhibition in crop seedlings by nanoparticles. Ecotoxicol. Environ. Saf. 79, 48–54.

Wang, A.Z., Langer, R., Farokhzad, O.C., 2012c. Nanoparticle delivery of cancer drugs. Annu. Rev. Med. 63, 185–198.

Wang, Z., Xie, X., Zhao, J., Liu, X., Feng, W., White, J.C., Xing, B., 2012d. Xylem-and phloem-based transport of CuO nanoparticles in maize (*Zea mays* L.). Environ. Sci. Technol. 46, 4434–4441.

Wang, J., Koo, Y., Alexander, A., Yang, Y., Westerhof, S., Zhang, Q.B., Schnoor, J.L., Colvin, V.L., Braam, J., Alvarez, P.J.J., 2013a. Phytostimulation of poplars and Arabidopsis exposed to silver nanoparticles and Ag+ at sublethal concentrations. Environ. Sci. Technol. 47, 5442–5449.

Wang, Q., Ebbs, S.D., Chen, Y., Ma, X., 2013b. Trans-generational impact of cerium oxide nanoparticles on tomato plants. Metallomics 5, 753–759.

Wang, S.L., Liu, H.Z., Zhang, Y.X., Xin, H., 2015. The effect of CuO NPs on reactive oxygen species and cell cycle gene expression in roots of rice. Environ. Toxicol. Chem. 34, 554–561.

Watson, J.L., Fang, T., Dimkpa, C.O., Britt, D.W., McLean, J.E., Jacobson, A., Anderson., A.J., 2015. The phytotoxicity of ZnO nanoparticles on wheat varies with soil properties. Biometals 28, 101–112.

Whitesides, G.M., 2005. Nanoscience, nanotechnology and chemistry. Small 1, 172–179.

Wu, S.G., Huang, L., Head, J., Chen, D.R., Kong, I.C., Yinjie, J., Tang, Y.G., 2012. Phytotoxicity of metal oxide nanoparticles is related to both dissolved metals ions and adsorption of particles on seed surfaces. J. Pet. Environ. Biotechnol. 3, 1–6.

Xia, X.J., Zhou, Y.H., Shi, K., Zhou, J., Foyer, C.H., Yu, J.Q., 2015. Interplay between reactive oxygen species and hormones in the control of plant development and stress tolerance. J. Exp. Bot. 66, 2839–2856.

Xu, Q.S., Hu, J.Z., Xie, K.B., Yang, H.Y., Du, K.H., Shi, G.X., 2010. Accumulation and acute toxicity of silver in *Potamogeton crispus* L. J. Hazard. Mater. 173, 186–193.

Yamakoshi, Y., Umezawa, N., Ryu, A., Arakane, K., Miyata, N., Goda, Y., Masumizu, T., Nagano, T., 2003. Active oxygen species generated from photoexcited fullerene (C60) as potential medicines: $O_2-\bullet$ versus $1O_2$. J. Am. Chem. Soc. 125, 12803–12809.

Yang, H., Liu, C., Yang, D., Zhang, H., Xi, Z., 2009. Comparative study of cytotoxicity, oxidative stress and genotoxicity induced by four typical nanomaterials: the role of particle size, shape and composition. J. Appl. Toxicol. 29, 69–78.

Yang, Z., Chen, J., Dou, R.Z., Gao, X., Mao, C.B., Wang, L., 2015. Assessment of the phytotoxicity of metal oxide nanoparticles on two crop plants, maize (*Zea mays* L.) and rice (*Oryza sativa* L. Int. J. Environ. Res. Public Health 12, 15100–15109.

Yang, X., Pana, H., Wanga, P., Zhao, F., 2016. Particle-specific toxicity and bioavailability of cerium oxide (CeO_2) nanoparticles to *Arabidopsis thaliana*. J. Hazard. Mater. http://dx.doi.org/10.1016/j.jhazmat.2016.03.054.

Yin, L., Cheng, Y., Espinasse, B., Colman, B.P., Auffan, M., Wiesner, M., Rose, J., Liu, J., Bernhardt, E.S., 2011. More than the ions: the effects of silver nanoparticles on *Lolium multiflorum*. Environ. Sci. Technol. 45, 2360–2367.

Yoon, S.J., Kwak, J.I., Lee, W.M., Holden, P.A., An, Y.J., 2014. Zinc oxide nanoparticles delay soybean development: a standard soil microcosm study. Ecotoxicol. Environ. Saf. 100, 131–137.

Yuan, H.M., Liu, W.C., Jin, Y., Lu, Y.T., 2013a. Role of ROS and auxin in plant response to metal-mediated stress. Plant Signal Behav. 8, e24671.

Yuan, H.M., Xu, H.H., Liu, W.C., Lu, Y.T., 2013b. Copper regulates primary root elongation through PIN1-mediated auxin redistribution. Plant Cell Physiol. 54, 766–778.

Yulianti, L., Bramono, K., Mardliyati, E., Freisleben, H.J., 2016. Effects of *Centella asiatica* ethanolic extract encapsulated in chitosan nanoparticles on proliferation activity of skin fibroblasts and keratinocytes, type I and III collagen synthesis and aquaporin 3 expression in vitro. J. Pharm. and Biomed. Sci. 6, 5.

Zaheer, I.E., Ali, S., Rizwan, M., Farid, M., Shakoor, M.B., Gill, R.A., Najeeb, U., Iqbal, N., Ahmad, R., 2015. Citric acid assisted phytoremediation of copper by *Brassica napus* L. Ecotoxicol. Environ. Saf. 120, 310–317.

Zhang, P., Ma, Y., Zhang, Z., He, X., Guo, Z., Tai, R., Ding, Y., Zhao, Y., Chai, Z., 2012. Comparative toxicity of nanoparticulate/bulk Yb_2O_3 and $YbCl_3$ to cucumber (*Cucumis sativus*). Environ. Sci. Technol. 46, 1834–1841.

Zhang, D., Hua, T., Xiao, F., Chen, C., Gersberg, R.M., Liu, Y., Stuckeyd, D., Nga, W.J., Tana, S.K., 2015a. Phytotoxicity and bioaccumulation of ZnO nanoparticles in *Schoenoplectus tabernaemontani*. Chemosphere 120, 211–219.
Zhang, R., Zhang, H., Tu, C., Hu, X., Li, L., Luo, Y., Christie, P., 2015b. Phytotoxicity of ZnO nanoparticles and the released Zn (II) ion to corn (*Zea mays* L.) and cucumber (*Cucumis sativus* L.) during germination. Environ. Sci. Pollut. Res. 22, 11109–11117.
Zhang, J., Zhao, Z., Xia, Z., Dai, L., 2015c. A metal-free bifunctional electrocatalyst for oxygen reduction and oxygen evolution reactions. Nat. Nanotechnol. 10, 444–452.
Zhao, L., Peng, B., Hernandez-Viezcas, J.A., Rico, C., Sun, Y., Peralta-Videa, J.R., et al., 2012a. Stress response and tolerance of *Zea mays* to CeO_2 nanoparticles: cross talk among H_2O_2, heat shock protein, and lipid peroxidation. ACS Nano 6, 9615–9622.
Zhao, Y., Swierk, J.R., Megiatto, J.D., Sherman, B., Youngblood, W.J., Qin, D., Lentz, D.M., Moore, A.L., Moore, T.A., Gust, D., Mallouk, T.E., 2012b. Improving the efficiency of water splitting in dye-sensitized solar cells by using a biomimetic electron transfer mediator. Proc. Natl. Acad. Sci. U.S.A. 109, 15612–15616.
Zhao, L., Hernandez-Viezcas, J.A., Peralta-Videa, J.R., Bandyopadhyay, S., Peng, B., Munoz, B., Keller, A.A., Gardea-Torresdey, J.L., 2013a. ZnO nanoparticle fate in soil and zinc bioaccumulation in corn plants (*Zea mays*) influenced by alginate. Environ. Sci. Process. Impacts 15, 260–266.
Zhao, L., Sun, Y., Hernandez-Viezcas, J.A., Servin, A.D., Hong, J., Niu, G., 2013b. Influence of CeO_2 and ZnO nanoparticles on cucumber physiological markers and bioaccumulation of Ce and Zn: a life cycle study. J. Agric. Food Chem. 61, 11945–11951.
Zhao, L., Sun, Y., Hernandez-Viezcas, J.A., Hong, J., Majumdar, S., 2015. Monitoring the environmental effects of CeO_2 and ZnO nanoparticles through the life cycle of corn (*Zea mays*) plants and in situ-XRF mapping of nutrients in kernels. Environ. Sci. Technol. 49, 2921–2928.
Zhao, H., Jin, Q., Wang, Y., Chu, L., Li, X., Xu, Y., 2016. Effects of nitric oxide on alleviating cadmium stress in *Typhaan gustifolia*. Plant Growth Regul. 78, 243–251.
Zoua, X., Lia, P., Huangb, Q., Zhanga, H., 2016. The different response mechanisms of *Wolffia globosa*: light-induced silver nanoparticle toxicity. Aquat. Toxicol. 176, 97–105.

Further Reading

Duressa, D., Soliman, K., Taylor, R., Senwo, Z., 2011. Proteomic analysis of soybean roots under aluminum stress. Int. J. Plant Genom. 2011.
Khodakovskaya, M.V., de Silva, K., Nedosekin, D.A., Dervishi, E., Biris, A.S., Shashkov, E.V., Galanzha, E.I., Zharov, V.P., 2011. Complex genetic, photothermal, and photoacoustic analysis of nanoparticle-plant interactions. Proc. Natl. Acad. Sci. U.S.A. 108, 1028–1033.
Kopittke, P.M., Menzies, N.W., 2006. Effect of Cu toxicity on growth of cowpea (*Vigna unguiculata*). Plant Soil 279, 287–296.
Lopes, I., Ribeiro, R., Antunes, F.E., Rocha-Santos, T.A.P., Rasteiro, M.G., Soares, A.M.V.M., Gonçalves, F., Pereira, R., 2012. Toxicity and genotoxicity of organic and inorganic nanoparticles to the bacteria *Vibrio fischeri* and *Salmonella typhimurium*. Ecotoxicology 21, 637–648.
Lowry, G.V., Gregory, K.B., Apte, S.C., Lead, J.R., 2012. Transformations of nanomaterials in the environment. Environ. Sci. Technol. 46, 6893–6899.
Michaud, A.M., Chappellaz, C., Hinsinger, P., 2008. Copper phytotoxicity affects root elongation and iron nutrition in durham wheat (*Triticum turgidum* durum L.). Plant Soil 310, 151–165.
Nigam, R., Srivastava, S., Prakash, S., Srivastava, M., 2001. Cadmium mobilisation and plant availability–the impact of organic acids commonly exuded from roots. Plant Soil 230, 107–113.
Taggart, L.E., McMahon, S.J., Currell, F.J., Prise, K.M., Butterworth, K.T., 2014. The role of mitochondrial function in gold nanoparticle mediated radiosensitisation. Cancer Nanotechnol. 5, 1–12.
Xu, J., Yang, L., Wang, Y., Wang, Z., 2004. Advances in the study uptake and accumulation of heavy metal in rice (*Oryza sativa*) and its mechanisms. Chin. Bull. Bot. 22, 614–622.
Yang, F., Hong, F., You, W., Liu, C., Gao, F., Wu, C., Yang, P., 2006. Influence of nano-anatase TiO_2 on the nitrogen metabolism of growing spinach. Biol. Trace Elem. Res. 110, 179–190.
Yang, M., Zhang, M., Tahara, Y., Chechetka, S., Miyako, E., Iijima, S., Yudasaka, M., 2014. Lysosomal membrane permeabilization: carbon nanohorn-induced reactive oxygen species generation and toxicity by this neglected mechanism. Toxicol. Appl. Pharm. 280, 117–126.

CHAPTER

8

Exploring Plant-Mediated Copper, Iron, Titanium, and Cerium Oxide Nanoparticles and Their Impacts

Anuja Koul[1], *Anil Kumar*[1], *Vivek K. Singh*[1], *Durgesh K. Tripathi*[2,3,4], *Sharada Mallubhotla*[1]

[1]Shri Mata Vaishno Devi University, Katra, India; [2]University of Allahabad, Allahabad, India; [3]Centre for Medical Diagnostic and Research, Motilal Nehru National Institute of Technology, Allahabad, India; [4]Banaras Hindu University, Varanasi, India

8.1 INTRODUCTION

Nanotechnology is a rapidly emerging area of applied science and engineering, which has laid down a strong foundation for the technological revolution in various sectors of humankind. It is a fascinating area of modern-day research, which comprises a choice of materials for the synthesis of particles and a strategy and operation of structural size (below 100 nm) (Kannan and Subbalaxmi, 2011; Mallikarjuna et al., 2012). Nanomaterials and nanoparticles are swiftly attaining renovation in new fronts because of their novel enhanced attributes. The field of nanotechnology has empowered and benefited sectors including healthcare, agriculture, biomedical, food technology, environment, drug therapeutics, cosmetics, mechanics, chemical industries, optics, electron transistors, dyes, pigments, etc. (Curtis and Wilkinson, 2001).

For thousands of years, humankind has used nanotechnology as an applicable technique; however, the idea and concept of nanotechnology was conceived by Professor R. Feynman in 1959 in his lecture entitled "There's Plenty of Room at the Bottom," delivered at an American Physical Society meeting where he demonstrated a process to control individual atoms and molecules (Feynman, 1960). Further development took place in 1974 at the International Conference on Industrial Production in Tokyo, where the word nanotechnology was used for the first time by N. Taniguchi to explain the super-thin processing of materials of nanometer precision leading to the conception of nanosized mechanisms. The field of nanotechnology

Nanomaterials in Plants, Algae, and Microorganisms
http://dx.doi.org/10.1016/B978-0-12-811487-2.00008-6

witnessed major breakthroughs with the invention of scanning tunneling microscopy in 1981 that could actually "see" individual atoms and that ultimately led to the beginning of modern-day nanotechnology. The National Science Foundation initiated the first nanotechnology program in the United States in 1991. In 2001, the United States approved the National Nanotechnology Initiative for prioritizing developments incurred in the field of nanotechnology. Subsequently, scientific research, technical programs, workshops, and other developments have been captivating research labs all over the world in this accepted and emerging area of technology as a core area for "the next industrial revolution" (Tolochko, 2009).

Distinguishing properties such as solubility, absorption, emission, conductivity, optical behavior, and reactivity of synthesized nanoparticles depends upon the size and surface of the particle (Theivasanthi and Alagar, 2010). These properties can be altered by varying the initial size of the producing matter (Kelly et al., 2003), while agglomeration can be prevented by altering the size and surface charge (Jiang et al., 2009). Depending on the type of nanoparticles and their sizes, they absorb light of particular wavelengths. They transmit different colors because of surface plasmon resonance, i.e., the interaction of electromagnetic radiation and electrons in the conduction band around the nanoparticles (Mulvaney, 1996; Park et al., 2006). For example, this distinguishing color phenomenon is exhibited by gold nanoparticles that are red in color, whereas original gold is yellow in color (Xu and Cortie, 2006).

The mechanical strength of nanoparticles can be increased by tailoring the size, structure, and surface area of the metal, and these particles show characteristic and specific electromagnetic behavior, which enables their usage in various electrical devices, transistors, etc. (Niihara et al., 1991).

Further expansion of this technology is a promising approach in the area of medicinal plant biotechnology (Dahl et al., 2007). Plants are an incredible source of novel products of medicinal importance for drug development. Among the commercially available therapeutic medicines in the market, many of them are simply modifications of plant-derived natural products. The advancement in biotechnological approaches provides several ways for designing, synthesizing, producing, and delivering these naturally occurring bioactive metabolites. It is pertinent to mention that plant-mediated synthesis of nanoproducts has gained considerable importance because of their diversifying attributes and features (Leela and Vivekanandan, 2008).

The application of nanotechnology when used for drug deliverance is based on the concept and capability of maneuvering synthesized molecules for the production of drugs/molecules that possess the requisite programmed application. Drugs are considered successful when they fulfill the requirements laid down by the theories of pharmacokinetics and dynamics of drug application. Drug delivery into the human body is either by the oral, skin epidermal, or transmucosal route, or through inhalation. Drugs ought to be absorbed, dispersed, easily metabolized, nontoxic, and must have efficacy toward absolute selectivity. Nanotechnology drug delivery methods refer to systems comprising microchips and microneedle transdermal remedial assemblies. These techniques also help in treating tumor cells in a body by targeting specific cancerous cells (Jong and Borm, 2008). They also find applications in dental implants, which are improved by coating nanotubes with antiinflammatory drugs that can be applied to areas of the implant region (Kaushik and Joshi, 2015). Success has also been achieved to control diabetes with nanoparticles that release insulin when the level of glucose in the body rises (Wang et al., 2009). Mesoporous nanoparticles containing specific enzymes have also been developed for antiaging skin creams (Pandey and Pandey, 2013).

Nanosponges with a coating of red blood cells have been synthesized and are reported to have the ability to absorb and remove toxins from the bloodstream (Hu et al., 2013). Nanotubes have been developed and widely used as sensors for monitoring inflammation and other disease responses (Ma et al., 2016). Nanotechnology has shown wide applications in the treatment of wounds, staph infestations, and bacteriostatic and bactericidal conditions (Mittal et al., 2013).

Nanoparticle-based applications have also been developed for the treatment of wastewater, groundwater bodies, filtration and purification of industrial solvents, water desalination, and other environmental clean-up processes. Other promising outcomes of nanotechnology include construction and applications of nanowires, nanorods, nanofibers, military goods, layers of graphene, etc. (Xu and Cortie, 2006). Traditional synthesis of nanosized particles involves the use of hazardous materials and high-cost and high-energy inputs. Therefore there is a need to search and develop improved methods for the synthesis of these particles. Development of biosynthetic protocols employing plant extracts emerged as a viable approach for the synthesis of nanoparticles (Baker et al., 2005), which is in accordance with the principles of green technology (Inbathamizh et al., 2013). Because of its rapid, cost-effective, and environment friendly properties, nanoparticles fabricated from plants have attracted the interest of researchers. The other reason for using plants for the synthesis of metal nanoparticles is because of the presence of a variety of reducing agents such as flavonoids, organic acids, dehydrogenases, and electron shuttles (Roy and Das, 2015).

Two distinct approaches have been adopted for the synthesis and fabrication of nanoparticles. These are the "top-down approach" and the "bottom-up approach" (Iqbal et al., 2012).

In the top-down approach, transfer of pattern is done repeatedly by adopting extreme lithographic, photolithographic, beam lithography, UV treatment, grinding, mulling, mining, and sputtering techniques and laser radiations. As reported, nanoparticles can also be synthesized by subjecting the substrate to atmospheric pressure and the evaporation–condensation principle (Ahmed et al., 2016). Different types of nanoparticles have been prepared using this technique (Iqbal et al., 2012). Despite great degrees of success there are several drawbacks associated with this approach such as requirements of large space, high input of energy, and power consumption, in addition to its being time consuming.

On the other hand, the bottom-up approach employs chemical/biological treatments to generate new nuclei, which may grow into the specific nanosize. Chemical reduction using organic/inorganic reducing agents such as ascorbate, sodium citrate, sodium borohydride, Tollens' reagent, polyethylene glycol, etc. are commonly used under this approach (Tran et al., 2013; Iravani et al., 2014). It is pertinent to mention that capping agents such as phyllanthin, sodium citrate, etc. may also be obligatory for the stability of nanoparticles in many cases. This approach of synthesis is generally considered better because of time economy, circumvention of stringent conditions, low space, and temperature requirements.

Despite these advantages, the chemicals used during this approach are generally toxic and can also lead to hazardous by-products. Therefore plant-mediated synthesis of nanoparticles has proved to be the changing face of nanotechnology in accordance with the rigorous principles of green synthetic technology.

The incorporation of plant-mediated nanoparticles has diverse applications in several branches of science because of their characteristic physical and chemical attributes. Synthesis of metallo-nanoparticles can easily be carried out through biological reduction of metals using

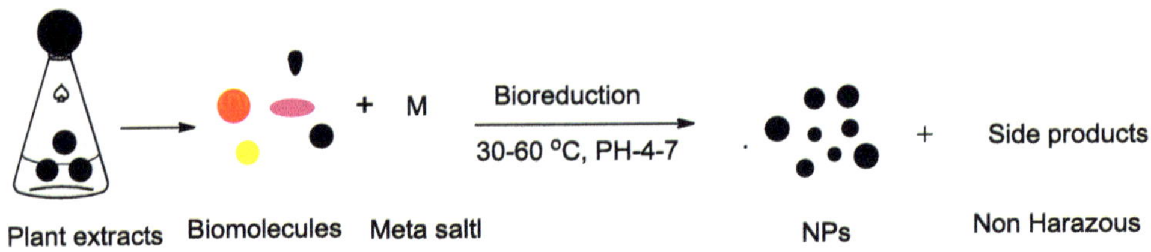

FIGURE 8.1 Synthesis of nanoparticles using plant extracts. *NPs*, nanoparticles.

plant parts such as stem, leaf, root, fruit, callus, or flower (Fig. 8.1). Tailoring of nanoparticles can be done by varying the concentration of plant extract used for the reduction of metal (Chandran et al., 2006; Dubey et al., 2010).

The stem of the plant contains high amounts of bioactive molecules such as phenolics, flavonoids, amines, and secondary metabolites having medicinally acclaimed properties; the nanoparticles thus synthesized are rich in medicinal attributes (Sasikala et al., 2015).

Fruits of the *Tribulus terrestris* species were used for ecofriendly production of silver nanoparticles with sustainable antimicrobial activities (Gopinath et al., 2012). Similarly, palladium nanoparticles were synthesized using grapefruits containing polyphenolic compounds that act efficiently against several bacterial and fungal diseases (Amarnath et al., 2012).

Seeds are also known to contain natural components such as lignin, vitamin, saponin, flavonoid, etc. and accordingly the synthesized nanoparticles were observed to contain parent components and deliverable attributes, for example, fenugreek seed-mediated gold nanoparticles possess antioxidant properties (Mittal et al., 2013).

The flower petals of several medicinally important plant species such as *Catharanthus roseus*, *Hibiscus rosa-sinensis*, *Mirabilis jalapa*, *Helianthus annuus*, etc. have also been used for the synthesis of metallic nanoparticles. These petals are rich sources of phenols, sugars, and proteins and therefore gold and silver nanoparticles thus synthesized possess pharmacologically important properties (Vankar and Bajpai, 2010; Das et al., 2011; Noruzi et al., 2011). The functional bioactive groups present in flower petals not only aid in bioreduction but also behave as effective antipathogenic agents.

Leaves are widely exploited for the synthesis of nanoparticles because of the presence of high therapeutic properties in the leaf part as compared to other plant parts of medicinally valuable plant species. Leaf extracts of medicinal plants such as *Murraya koenigii*, *Solanum nigrum*, *Alternanthera sessilis*, *Stevia rebaudiana*, etc. have been widely studied and used for the synthesis of silver and gold nanoparticles for several therapeutic attributes (Ahmed et al., 2016).

The extracts of plant parts are commonly prepared by macerating the selected plant tissue followed by preparation of its solution in a suitable solvent. The metal solution of a proper ratio is then added and finally the extract containing the embedded particles is harvested. These synthesized nanoparticles have attributes of both metal and plant tissue (Umesh and Vishwas, 2013). Several plant species have been employed for the synthesis of nanoparticles with benefiting attributes.

Nanoparticles have attracted the attention of many researchers because of their astonishing characteristics such as shape, size, and reactivity. Nanoparticles in the form of fertilizers, pesticides, and herbicides have been used in agriculture sectors to enhance plant growth and yield. Nanoparticles affect the plants on physiological, metabolic, and biochemical levels

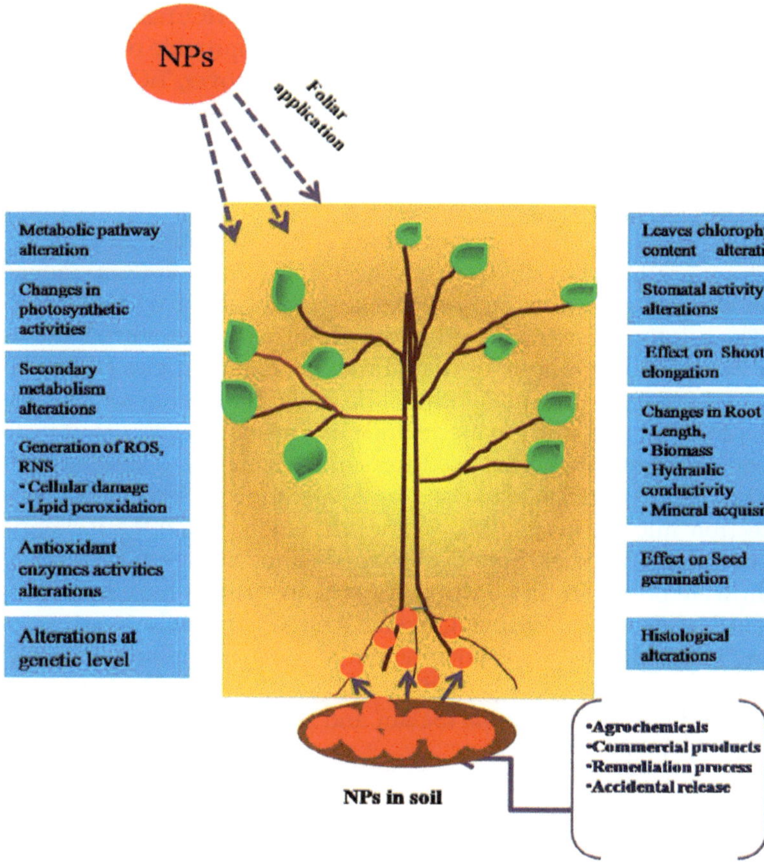

FIGURE 8.2 Schematic representation of different impacts of nanoparticles on terrestrial plants. *NPs*, nanoparticles; *RNS*, reactive nitrogen species; *ROS*, reactive oxygen species.

(Singh et al., 2016; Shweta et al., 2016; Tripathi et al., 2016, 2017a,b,c,d,e; Cox et al., 2017; Du et al., 2017; Fig. 8.2). The impacts of nanoparticles on crops vary differentially and are dependent on species of plants, their growth period, growth conditions, dose, and extent of exposure (Tripathi et al., 2015, 2016, 2017a,b,c,d,e; Arif et al., 2016; Singh et al., 2016; Shweta et al., 2016; Cox et al., 2017; Du et al., 2017;). For sustainable development of agricultural practices an understanding of the impacts of nanoparticles is of critical importance.

8.2 PLANT-MEDIATED TITANIUM DIOXIDE NANOPARTICLES AND THEIR IMPACT ON PLANTS AND OTHER LIVING SYSTEMS

Greener synthesis of titanium oxide nanoparticles is becoming one of the promising methodologies. Titanium dioxide (TiO_2) nanoparticles of varying sizes and structures can be produced by green methodologies using plants and microbes. Many researchers reported the production of TiO_2 nanoparticles from different plant extracts. TiO_2 nanoparticle synthesis via reaction between ethanolic leaf extracts and titanium tetraisoproxide of *Nyctanthes arbor-tristis* has been reported by Sundrarajan and Gowri (2011). Moreover, TiO_2 nanoparticles of size 100–200nm have been synthesized from titanium hydroxide using latex of *Jatropha curcas* L. (Hudlikar et al., 2012). Titanium hydroxide along with aqueous extract of *Eclipta prostrata* as biotemplating agent has been used for the synthesis of TiO_2 nanoparticles (Rajakumar et al., 2012). Spherical nanoparticles are obtained with sizes varying from 36 to 68nm (Kirthi et al., 2011). Rice straw in powder form has been utilized as biotemplating agent for synthesizing TiO_2 nanoparticles using tetraisopropoxide and acetic acid reaction (Ramimoghadam et al., 2014). In another study, aqueous extract of *Psidium guajava* along with titanium hydroxide has been used. The solution of extract and titanium hydroxide was stirred and kept at room temperature for 24h for the synthesis of titanium oxide nanoparticles (Kumar et al., 2014). An economic, simple method for the synthesis of rutile TiO_2 nanoparticles was proposed by Roopan et al. (2012). In this study, water extract of *Annona squamosa* fruit peel along with titanium hydroxide as a precursor was used for synthesizing TiO_2 nanoparticles. The characterization of these nanoparticles was done by UV, scanning electron microscopy (SEM), transmission electron microscopy (TEM), X-ray diffraction (XRD), and energy dispersive spectroscopy (EDS) methodologies. Spherical-shaped nanoparticles in aggregated form with sizes of 23 ± 2nm were observed. In addition, TiO_2 nanoparticles have been biosynthesized by using *C. roseus* leaf extract as reducing agent, which formed clustered nanoparticles of 25–110nm having irregular shapes (Velayutham et al., 2012). Similarly, the green synthesis of TiO_2 nanoparticles has also been reported by using *Ocimum basilicum, Calotropis gigantea, Azadirachta indica,* and *Clonorchis sinensis* (Salam and Sivaraj, 2014; Rao et al., 2015; Purkait et al., 2015; Marimuthu et al., 2013).

TiO_2 nanoparticles are the most widely produced nanoparticles and have been exploited in agriculture-relevant practices. These nanoparticles have been utilized in products protecting plants from different hazards. TiO_2 effects are some of the most studied and TiO_2 nanoparticles are well used even though their effects on plants and other living organisms are little known. Studies of TiO_2 nanoparticles on plants have provided mixed outcomes with positive as well as negative effects on plant physiology and metabolism (Asli and Neumann, 2009; Seeger et al., 2009; Ghosh et al., 2010). In comparison to other nanoparticles these nanoparticles are reported to have fewer toxic effects (Boonyanitipong et al., 2011; Kim et al., 2007). The study carried out by Chao and Choi (2005) explicated the efficient role of TiO_2 nanoparticles in protecting rice, olive, and maize crops from different diseases such as *Curvularia* leaf spot and bacteria leaf blight diseases. In addition, the severity of rice blast and tomato spray mold was found to be reduced after the application of TiO_2. The growth of plants was also found to increase because of the activity of these nanoparticles. Another work instigating the disease-controlling activities of TiO_2 nanoparticles was reported by Bowen et al. (1992). The combined effect of TiO_2, silica, and aluminum was observed to have protective activity against powdery mildew and downy diseases, which may be caused by induction of plant defense mechanisms. TiO_2 nanoparticles have photocatalytic and UV protective activities, and are used in plant protection products to enhance their effectiveness and reduce the application amounts or to decompose persistent compounds faster (Gogos et al., 2012).

The effects of TiO_2 nanoparticles (5nm) on seed germination have been reported by many researchers. Acceleration in spinach seed germination by treatment of TiO_2 nanoparticles was observed because of enhancement in water absorption (Zheng et al., 2005). However, no substantial changes in germination of wheat grains after inoculating with TiO_2 nanoparticles were found in a study carried out by Feizi et al. (2012). Reduced rate of germination of *Vicia narbonensis* seeds was observed to be inflicted by TiO_2 nanoparticles (Castiglione et al., 2011).

The alternation in biochemical conditions in nearby areas of roots of wheat because of accumulation of nanoparticles of TiO_2 was observed by Du et al. (2011). These alternations further affected the availability of nutrients and ultimately plant growth. Impacts of TiO_2 nanoparticles on photosynthetic activities were also deciphered in many studies. Different studies deciphering the impact of these nanoparticles on the photosynthetic process because of their photocatalytic and thermal conductivity are also reported (Rezaei et al., 2015; Rico et al., 2015b). The alternations in photosynthetic mechanisms led to the generation of oxidative stress, which further induced stress-neutralizing defense mechanisms (Rico et al., 2015b). By modulating chlorophyll *a*-mediated light absorption, electron transfer, and rate of and release of oxygen, these nanoparticles enhanced the Hill reaction and activity of chloroplast (Hong et al., 2005; Wang et al., 2008; Monica and Cremonini, 2009). In 2012 Yin et al. demonstrated the photocytotoxicity of TiO_2 nanoparticles in an array of cell systems and established that the resulting toxicity was led by reactive oxygen species (ROS).

8.3 PLANT-MEDIATED IRON OXIDE NANOPARTICLES AND THEIR IMPACT ON PLANTS AND OTHER LIVING SYSTEMS

Iron oxide (Fe_2O_3) nanoparticles are highly explored and utilized nanomaterials. Fe_2O_3 nanoparticles are widely used in diagnostics, controlled drug delivery systems, environmental engineering, and remediation processes because of their high surface and volume ratio, magnetic properties, and compatibility to biological systems (He and Zhao, 2005; Wang et al., 2014). Plant-based synthesis of Fe_2O_3 nanoparticles is one of the simple, economical, and efficient methods. *Eucalyptus globulus* leaf extract as reducing and capping agent along with ferric chloride has been used for the synthesis of crystalline Fe_2O_3 nanoparticles of 100 nm size. Characterization of nanoparticles was done by XRD, SEM, UV spectroscopy, TEM, and Fourier transform infrared spectroscopy (Balamurugan et al., 2014).

The biosynthesis of Fe_2O_3 nanoparticles from garlic vine leaf extract was reported by Prasad (2016). In this biogenic method $FeSO_4{\cdot}7H_2O$ was used as a precursor and octahedral nanoparticles of size 18.22 nm were obtained. Stabilization of nanoparticles was attributed by biomolecules present in plant extract as observed by thermogravimetric analysis. Pure hematite nanoparticles were reportedly synthesized by leaf extract of green tea utilizing the hydrothermal method (Ahmmad et al., 2013). Porous nanoparticles of average diameter 60 nm having spherical shape were observed. A comparative study of photocatalytic nanoparticles synthesized by green methodology and commercially available ones was done in this study. Plant-based nanoparticles were found to have higher surface area as well as higher photocatalytic activities than commercially available Fe_2O_3 nanoparticles. Leaf extract of *Ocimum sanctum* was used for the synthesis of Fe_2O nanoparticles of irregular shape having 20–47 nm size (Balamurugan et al., 2014). The reduction reaction was supposed to be caused by the presence of phenolic compounds and proteins in the extract.

With the rapid development of nanotechnology, developing nano-iron fertilizer is an important strategy to alleviate Fe deficiency and elevate the Fe fertilization effect in agricultural applications. Nanoparticles of Fe_2O_3 affect plants at the physiological, biochemical, and molecular level in a dose-dependent manner. Positive and negative effects of Fe_2O_3 nanoparticles on plants have been deciphered in many studies. The impact of Fe_2O_3 nanoparticles on germination and growth of roots was studied by Li et al. (2013) and Ren et al. (2011) in watermelon and mung bean. These nanoparticles enhanced germination of seeds, growth of seedlings, root elongation, and chlorophyll contents, ultimately improving physiological functions of plants and their resistivity under adverse conditions. In *Zea mays* L., Fe_2O_3 nanoparticles were found to have inducible effects on germination, enzyme activities, as well increased chlorophyll and malondialdehyde (MDA) contents (Li et al., 2016). In addition, the effect of Fe_2O_3 nanoparticles on physiological functions of *Solanum lycopersicum* was investigated by Shankramma et al. (2016). Substantial improvement in germination of seedlings and growth of plants was observed in Fe_2O_3 nanoparticle (<10 nm)-treated plants because of slow diffusion of these nanoparticles through nanoholes present in the seed coat. This study also concluded that a 50–200 mg/L dose of Fe_2O_3 nanoparticles could be beneficial from an agricultural perspective. Similar types of results were also observed in pumpkin after being treated with Fe_2O_3 nanoparticles (Zhu et al., 2008). Regardless of these positive effects of hematite nanoparticles, several researchers investigated the negative effects of these nanoparticles in different plant species. *Arabidopsis thaliana*, when exposed to Fe_2O_3 nanoparticles, displayed a reduction in chlorophyll content as well as in growth (Marusenko et al., 2013). Nanoparticles also alter the hydraulics of roots, which causes perturbations in nutrient and water uptake. The reduction in root hydraulic conductivity of *Helianthus annuus* treated with Fe_2O_3 nanoparticles was observed by Martínez-Fernández et al. (2015). It was also concluded that the amount of calcium, potassium, magnesium, and sulfur is reduced in shoots along with a decrease in chlorophyll content. Liu et al. (2016) demonstrated the effect of γ-Fe_2O_3 on the germination of lettuce seeds. Nanoparticles (5–20 mg/L) were found to be less toxic than their ionic forms and stimulated the elongation of roots. A study was performed by Gui et al. (2015) for analyzing the effects of γ-Fe_2O_3 on transgenic and wild-type rice plants. These nanoparticles influenced the production of abscisic acid, indole-3-acetic acid, and root phytohormones in a concentration-dependent manner. Higher activities of antioxidant enzymes in transgenic plants were observed with low concentrations of nanoparticles and lower activities were observed at higher concentrations.

8.4 PLANT-MEDIATED CERIUM OXIDE NANOPARTICLES AND THEIR IMPACTS ON PLANTS AND OTHER LIVING SYSTEMS

Biosynthesis of cerium oxide (CeO_2) by plant extracts is one of the beneficial means of green methodology, with spherical-shaped CeO_2 nanoparticles having low toxicity being produced. Synthesis of CeO_2 nanoparticles from different plant extracts has been reported in many studies. For instance, leaf extracts of *Aloe vera*, *Acalypha indica*, and *Gloriosa superba* acting as stabilizing as well as capping agents have been used (Priya et al., 2014; Kannan and Sundrarajan, 2014; Arumugam et al., 2015).

A. vera leaf extracts along with cerium nitrate as precursor was used by Dutta et al. (2016) resulting in the synthesis of nanoparticles of 2–3 nm size. These nanoparticles have the potential to neutralize the hazardous impacts of hydrogen peroxide and therefore could be used as an antioxidant agent for stress-generated diseases. In addition, spherical-shaped CeO_2 nanoparticles of 63.3 nm size were synthesized by using *Aloe barbadensis* gel reacting with $Ce(NO_3)_3{\cdot}6H_2O$ (Priya et al., 2014). Arumugam et al. (2015) used leaf extract of *Gloriosa superba* for biosynthesis of CeO_2 nanoparticles having antibacterial properties. A fast and effective method for synthesizing CeO_2 nanoparticles was proposed by Surendra and Roopan (2016). In this method, peel of *Moringa oleifera* was used as stabilizing and reducing agent for the synthesis of these nanoparticles by a microwave-assisted method. In addition, the aqueous extract of *Centella asiatica* along with cerium ammonium nitrate was used for the biosynthesis of CeO_2 nanoparticles. The synthesized nanoparticles were characterized by UV–visible spectroscopy, dynamic light scattering, SEM, high-resolution transmission electron microscopy, and EDS and found to have free radical scavenging potential.

Different studies elucidating the implications of CeO_2 nanoparticles on plants at the physiological level have been carried out by researchers. Wang et al. (2012) investigated the positive effects of CeO_2 nanoparticles on tomato plants. The exposure of these nanoparticles enhanced plant growth as well as its productivity (Wang et al., 2012). In an important study the role of CeO_2 nanoparticles in salt stress tolerance was explicated. The physiological and biochemical alternations caused by these nanoparticles in *Brassica napus* L. (canola) cv. under salt stress were elucidated. This study concluded that CeO_2 nanoparticles improved salt stress tolerance by inducing certain mechanisms that attributed an increase in total biomass. In addition, the chlorophyll content of plants was increased, which resulted in improved photosynthetic activities in salt stress conditions (Rossi et al., 2016).

Rico et al. (2015a,b) investigated the impacts of CeO_2 nanoparticles on the physiological parameters of barley. Effective improvements in height, biomass, and chlorophyll content of plants were observed in CeO_2 nanoparticle-treated barley plants. The accumulation of cerium with other nutrients (potassium, phosphorus, calcium, magnesium, iron, zinc, and copper) in barley grains was also observed (Rico et al., 2015a,b).

Various studies have been carried out for determining the impacts of nanoparticles on different plant species when they are cultivated in different conditions. The effect of varied dosages of CeO_2 nanoparticles on *Raphanus sativus* L. germination was investigated by Corral-Diaz et al. (2014). After 4 days of treatment it was observed that 500 mg CeO_2 nanoparticles per kg decreased the rate of seed germination, while after 15 days the effect was found to be insignificant. Therefore it has been concluded that these nanoparticles affect germination at an early stage. The interactions of CeO_2 nanoparticles with nutrients present in soil may also affect their accumulation in and availability to plants (Peralta-Videa et al., 2014). The stable nature of CeO_2 nanoparticles in varied biological conditions made them ideal material for studying the phytotoxicity of nanoparticles (Yang et al., 2017). Phytotoxicity may be attributed to the generation of ROS in plants exposed to nanoparticles. Ma et al. (2015) demonstrated the toxic effects of different types of CeO_2 nanoparticles on *Lactuca* plants. The increased level of MDA instigated cellular membrane damage in roots. MDA production is directly related to lipid peroxidation caused by ROS (Halliwell and Gutteridge, 2015). It has also been concluded that responses of plants to CeO_2 exposure are dependent on size as well as growth stages (Ma et al., 2015).

8.5 EXPLORING PLANT-MEDIATED COPPER NANOPARTICLES AND THEIR IMPACTS ON PLANTS AND OTHER LIVING SYSTEMS

Although silver and gold are widely used for the synthesis of plant-mediated nanoparticles, copper is also known to be employed for bioreduction for the synthesis of copper nanoparticles. Copper nanoparticles have gained importance because of their cost-effective nature compared to silver and gold. Bioreduction using copper metal is easier as compared to others because of their ease of preparation with polymers (Mallik et al., 2001). The stability of copper nanoparticles is caused by its higher conductive nature. Copper has a wide range of applications from catalytic activity to electronics (Sun et al., 2000; Maier et al., 2001; Wunder et al., 2010).

The most common approach for the synthesis of copper nanoparticles is the microemulsion method, which involves the chemical synthesis approach. Physical methods used for the synthesis of copper nanoparticles are radiolysis, microwave radiation, aerosol methods, etc.; however, these processes include the use of costly instruments and high inputs of energy, and in many cases by-product formation also takes place.

Copper nanoparticles have been reported to be synthesized using extracts of bacterial and fungal species such as *Morganella* sp., *Pseudomonas stutzeri*, *Penicillium aurantiogriseum*, and *Penicillium citrinum* (Varshney et al., 2010; Honary et al., 2012), but it has been demonstrated that the process of reduction by these microbial strains is very slow and limited. In view of these limitations a need to develop an alternative, safer, and quicker approach for the synthesis of copper nanoparticles has been perceived (Thakkar et al., 2010).

The green approach is thus employed for the synthesis of copper nanoparticles for a safer environment. Bark extracts of *Terminalia arjuna* have been used for the synthesis of copper nanoparticles (Yallappa et al., 2013). Copper nanoparticles were synthesized in an aluminum vessel using copper sulfate as a precursor in the presence of butter, curd, lime, and tamarind juice acting as capping reagents (Sastry et al., 2013). *Artabotrys odoratissimus* extract has been used as reducing reagent for the preparation of copper nanoparticles from copper sulfate at a temperature of 95°C resulting in particle sizes of 110–130 nm (Kathad and Gajera, 2014). Copper nanoparticles synthesized with the extract of leaves of the *Mangolia* tree act as stabilizing and reducing agents (Lee et al., 2011). Water bodies mainly infested with a number of microbes and prolonged use of disinfectants make these microbes resistant to these agents. Synthesized copper nanoparticles have been used as protective polyurethane foam over these water bodies, which act as a protective antibacterial sheath (Ruparelia et al., 2008). These particles also act as catalysts for the reduction of nitrobenzene (Kaur et al., 2014) dye reduction and aggregation, fluorescence quenching and enhancement (Mandal and De, 2015), and destabilization of tumor cells (Din and Rehan, 2016). Copper nanoparticles have wider applications than those of silver nanoparticles but they cannot be used in the drug industry because of their high toxicity as compared to silver. Considerable research into decreasing its toxicity needs to be done and alternative modes for its use should be explored.

8.6 CONCLUSION AND FUTURE PROSPECTS

Although the past decade has seen remarkable development in many of the different aspects of nanotechnology, plant-mediated nanoparticles is one of the emerging areas that have diverse applications in several branches of science because of their characteristic physical and chemical attributes. Interest in plant-mediated synthesis of nanoparticles is growing rapidly as it is considered as an alternative green synthetic approach strictly because of the green principles. These hybrid nanoparticles have revolutionized newer aspects in nanotechnology with regard to nanomedicine, nanobiosensors, nanoelectronics, nanooptics, catalytic chemistry, synthetic dye industry, cancer therapeutics, drug target delivery systems, nanodrug therapeutics, agriculture, nanoengineering, etc.

In this chapter, we have summarized the published literature regarding plant-mediated synthesis of metallo-nanoparticles utilizing various plant parts such as stem, leaf, root, fruit, callus, or flower through the biological reduction of different metals. We have emphasized different synthetic protocols and potential applications of important metallo-nanoparticles such as silver, gold, and copper. It has been well established in the literature that plant-mediated silver nanoparticles have gained sufficient evidence and support to be proven as a therapeutic agent for several diseases, thus benefiting society.

With the advancement in nanotechnology and vast diversity of plants available there is huge scope for further exploration of characteristically designed nanoparticles. To the best of our knowledge, this review of the literature is the first of its kind to assess and report plant-mediated synthesis of nanoparticles along with their broad diversified applications, which could be useful for students, researchers, academicians, etc. working in the field of nanotechnology.

Acknowledgment

Ms. Anuja Koul is thankful to Shri Mata Vaishno Devi University, Katra, J&K, and the University Grants Commission, Govt. of India, for fellowship.

References

Ahmed, S., Ahmad, M., Swami, B.L., Ikram, S., 2016. A review on plants extract mediated synthesis of silver nanoparticles for antimicrobial applications: a green expertise. J. Adv. Res. 7, 17–28.

Ahmmad, B., Leonard, K., Islam, M.S., Kurawaki, J., Muruganandham, M., Ohkubo, T., Kuroda, Y., 2013. Green synthesis of mesoporous hematite (α-Fe_2O_3) nanoparticles and their photocatalytic activity. Adv. Powder Technol. 24, 160–167.

Amarnath, K., Kumar, J., Reddy, T., Mahesh, V., Ayyappan, S.R., Nellore, J., 2012. Synthesis and characterization of chitosan and grape polyphenols stabilized palladium nanoparticles and their antibacterial activity. Colloids Surf. B 94, 254–261.

Arif, N., Yadav, V., Singh, S., Singh, S., Mishra, R.K., Sharma, S., Dubey, N.K., Tripathi, D.K., Chauhan, D.K., 2016. Current trends of engineered nanoparticles (ENPs) in sustainable agriculture: an overview. J. Environ. Anal. Toxicol. 6, 397. http://dx.doi.org/10.4172/2161-0525.1000397.

Arumugam, A., Karthikeyan, C., Hameed, A.S.H., Gopinath, K., Gowri, S., Karthika, V., 2015. Synthesis of cerium oxide nanoparticles using *Gloriosa superba* L. leaf extract and their structural, optical and antibacterial properties. Mater. Sci. Eng. 49, 408–415.

Asli, S., Neumann, P.M., 2009. Colloidal suspensions of clay or titanium dioxide nanoparticles can inhibit leaf growth and transpiration via physical effects on root water transport. Plant Cell Environ. 32, 577–584.

Baker, C., Pradhan, A., Pakstis, L., Pochan, D.J., Shah, S.I., 2005. Synthesis and antibacterial properties of silver nanoparticles. J. Nanosci. Nanotechnol. 5, 244–249.

Balamurugan, M., Saravanan, S., Soga, T., 2014. Synthesis of iron oxide nanoparticles by using Eucalyptus globulus plant extract. E J. Surf. Sci. Nanotechnol. 12, 363–367.

Boonyanitipong, P., Kositsup, B., Kumar, P., Baruah, S., Dutta, J., 2011. Toxicity of ZnO and TiO_2 nanoparticles on germinating rice seed *Oryza sativa* L. Int. J. Biosci. Biochem. Bioinform. 1, 282.

Bowen, P., Menzies, J., Ehret, D., Samuel, L., Glass, A.D.M., 1992. Soluble silicon sprays inhibit powdery development in grape leaves. J. Am. Soc. Hortic. Sci. 117, 906–912.

Castiglione, M.R., Giorgetti, L., Geri, C., Cremonini, R., 2011. The effects of nano-TiO_2 on seed germination, development and mitosis of root tip cells of *Vicia narbonensis* L. and *Zea mays* L. J. Nanopart. Res. 13, 2443–2449.

Chandran, S.P., Chaudhary, M., Pasricha, R., Ahmad, A., Sastry, M., 2006. Synthesis of gold nanotriangles and silver nanoparticles using *Aloe vera* plant extract. Biotechnol. Prog. 22, 577–583.

Chao, S.H.L., Choi, H.S., 2005. Method for Providing Enhanced Photosynthesis. Korea Research Institute of Chemical Technology, Jeonju, South Korea. Bulletin 10 pp.

Corral-Diaz, B., Peralta-Videa, J.R., Alvarez-Parrilla, E., Rodrigo-García, J., Morales, M.I., Osuna-Avila, P., Niu, G., Hernandez-Viezcas, J.A., Gardea-Torresdey, J.L., 2014. Cerium oxide nanoparticles alter the antioxidant capacity but do not impact tuber ionome in *Raphanus sativus* (L). Plant Physiol. Biochem. 84, 277–285.

Cox, A., Venkatachalam, P., Sahi, S., Sharma, N., 2017. Reprint of: silver and titanium dioxide nanoparticle toxicity in plants: a review of current research. Plant Physiol. Biochem. 110, 33–49.

Curtis, A., Wilkinson, C., 2001. Nanotechniques and applications in biotechnology. Trends Biotechnol. 19 (3), 97–101.

Dahl, J.A., Maddux, B.L.S., Hutchison, J.E., 2007. Toward greener nanosynthesis. Chem. Rev. 107, 2228–2269.

Das, R.K., Gogoi, N., Bora, U., 2011. Green synthesis of gold nanoparticles using *Nyctanthes arbortristis* flower extract. Bioprocess Biosyst. Eng. 34, 615–619.

Din, M.I., Rehan, R., 2016. Synthesis, characterization, and applications of copper nanoparticles. Anal. Lett. http://dx.doi.org/10.1080/00032719.2016.1172081.

Du, W., Sun, Y., Ji, R., Zhu, J., Wu, J., Guo, H., 2011. TiO_2 and ZnO nanoparticles negatively affect wheat growth and soil enzyme activities in agricultural soil. J. Environ. Monit. 13, 822–828.

Du, W., Tan, W., Peralta-Videa, J.R., Gardea-Torresdey, J.L., Ji, R., Yin, Y., Guo, H., 2017. Interaction of metal oxide nanoparticles with higher terrestrial plants: physiological and biochemical aspects. Plant Physiol. Biochem. 110, 210–225.

Dubey, S., Lahtinen, M., Silanpaa, M., 2010. Green synthesis and characterization of silver and gold nanoparticles using leaf extract of *Rosa rugosa*. Colloids Surf. A 364 (1–3), 34–41.

Dutta, D., Mukherjee, R., Patra, M., Banik, M., Dasgupta, R., Mukherjee, M., Basu, T., 2016. Green synthesized cerium oxide nanoparticle: a prospective drug against oxidative harm. Colloids Surf. B Biointerfaces 1 (147), 45–53.

Feizi, H., Moghaddam, P.R., Shahtahmassebi, N., Fotovat, A., 2012. Impact of bulk and nanosized titanium dioxide (TiO_2) on wheat seed germination and seedling growth. Biol. Trace Elem. Res. 146, 101–106.

Feynman, R.P., 1960. There's plenty of room at the bottom. Eng. Sci. 23, 22–36.

Ghosh, M., Bandyopadhyay, M., Mukherjee, A., 2010. Genotoxicity of titanium dioxide (TiO_2) nanoparticles at two trophic levels: plant and human lymphocytes. Chemosphere 81 (10), 1253–1262.

Gogos, A., Knauer, K., Bucheli, T.D., 2012. Nanomaterials in plant protection and fertilization: current state, foreseen applications, and research priorities. J. Agric. Food Chem. 60 (39), 9781–9792.

Gopinath, Mubarak, D., Priyadarshini, S., Priyadharsshini, N.M., Thajuddin, N., Velusamy, P., 2012. Biosynthesis of silver nanoparticles from *Tribulus terrestris* and its antimicrobial activity: a novel biological approach. Colloids Surf. B 96, 69–74.

Gui, X., Deng, Y., Rui, Y., Gao, B., Luo, W., Chen, S., Liu, S., Han, Y., Liu, L., Xing, B., 2015. Response difference of transgenic and conventional rice (*Oryza sativa*) to nanoparticles (γFe_2O_3). Environ. Sci. Pollut. Res. 22, 17716–17723.

Halliwell, B., Gutteridge, J.M., 2015. Free Radicals in Biology and Medicine. Oxford University Press, USA.

He, F., Zhao, D., 2005. Preparation and characterization of a new class of starch-stabilized bimetallic nanoparticles for degradation of chlorinated hydrocarbons in water. Environ. Sci. Technol. 39, 3314–3320.

Honary, S., Barabadi, H., Gharaei–Fathabad, E., Naghibi, F., 2012. Green synthesis of copper oxide nanoparticles using *Penicillium aurantiogriseum*, *Penicillium citrinum* and *Penicillium waksmanii*. Dig. J. Nanomater. Biostruct. 7 (3), 999–1005.

Hong, F.S., Yang, P., Gao, F.Q., Liu, C., Zheng, L., Yang, F., Zhou, J., 2005. Effect of nano-anatase TiO_2 on spectral characterization of photosystem II particles from spinach. Chem. Res. Chin. Univ. 21, 196–200.

Hu, C.M.J., Fang, R.H., Copp, J., Luk, B.T., Zhang, L., 2013. A biometric nanosponge that absorbs pore-forming toxins. Nat. Nanotechnol. 8, 336–340.

Hudlikar, M., Joglekar, S., Dhaygude, M., Kodam, K., 2012. Green synthesis of TiO_2 nanoparticles by using aqueous extract of *Jatropha curcas* L. latex. Mater. Lett. 75, 196–199.

Inbathamizh, L., Ponnu, T.M., Mary, E.J., 2013. *In vitro* evaluation of antioxidant and anticancer potential of *Morinda pubescens* synthesized silver nanoparticles. J. Pharm. Res. 6 (1), 32–38.

Iqbal, P., Preece, J.A., Mendes, P.M., 2012. Nanotechnology: the top down and bottom up approaches. In: Supramolecular Chemistry: From Molecules to Nanomaterials. Wiley Online library. http://dx.doi.org/10.1002/9780470661345.smc195.

Iravani, S., Korbekandi, H., Mirmohammadi, S.V., Zolfaghari, B., 2014. Synthesis of silver nanoparticles: chemical, physical and biological methods. Res. Pharm. Sci. 9 (6), 385–406.

Jiang, J., Oberdorster, G., Biswas, P., 2009. Characterization of size, surface charge, and agglomeration state of nanoparticle dispersions for toxicological studies. J. Nanopart. Res. 11, 77–89.

Jong, W.H.D., Borm, P.J.A., 2008. Drug delivery and nanoparticles: applications and hazards. Int. J. Nanomed. 3 (2), 133–149.

Kannan, N., Subbalaxmi, S., 2011. Green synthesis of silver nanoparticles using *Bacillus subtillus* IA751 and its antimicrobial activity. Res. J. Nanosci. Nanotechnol. 1, 87–94.

Kannan, S.K., Sundrarajan, M., 2014. A Green approach for the synthesis of a cerium oxide nanoparticle: characterization and antibacterial activity. Int J Nanosci 13 (3), 1450018.

Kathad, U., Gajera, H.P., 2014. Synthesis of copper nanoparticles by two different methods and size comparison. Int. J. Pharm. Bio. Sci. 5 (3), 533–540.

Kaur, R., Giordano, C., Gradzielski, M., Mehta, S.K., 2014. Synthesis of highly stable, water-dispersible copper nanoparticles as catalysts for nitrobenzene reduction. Chem. Asian J. 9 (1), 189–198.

Kaushik, U., Joshi, S.C., 2015. Silver nanoparticles: green synthesis, optical properties, antimicrobial activity and its mechanism using *Citrus sinensis*. Asian J. Pharm. Clin. Res. 8 (6), 179–184.

Kelly, K.L., Coronado, E., Zhao, L.L., Schatz, G.C., 2003. The optical properties of metal. Nanoparticles: the influence of size, shape, and dielectric environment. J. Phys. Chem. B 107, 668–677.

Kim, S.J., Nam, Y.S., Rhee, D.M., Park, H.S., Park, W.H., 2007. Macromolecular Nanotechnology: preparation and characterization of antimicrobial polycarbonate nanofibrous membrane. Eur. Polym. J. 43, 3146–3152.

Kirthi, A.V., Rahuman, A.A., Rajakumar, G., Marimuthu, S., Kumar, T.S., Jayaseelan, C., Velayutham, K., 2011. Acaricidal, pediculocidal and larvicidal activity of synthesized ZnO nanoparticles using wet chemical route against blood feeding parasites. Parasitol. Res. http://dx.doi.org/10.1007/s00436-011-2277-8.

Kumar, P.V., Pammi, S.V., Kollu, P., Satyanarayan, K.V., Shameem, U., 2014. Green synthesis and characterization of silver nanoparticles using *Boerhaavia diffusa* plant extract and their antibacterial activity. Ind. Crop. Prod. 52, 562–566.

Lee, H., Lee, G., Jang, N., Yun, J., Song, J., Kim, B., 2011. Biological synthesis of copper nanoparticles using plant extract. Nanotechnology 1 (1), 371–374.

Leela, A., Vivekanandan, M., 2008. Tapping the unexploited plant resources for the synthesis of silver nanoparticles. Afr. J. Biotechnol. 7, 3162–3165.

Li, J., Chang, P.R., Huang, J., Wang, Y., Yuan, H., Ren, H., 2013. Physiological effects of magnetic iron oxide nanoparticles towards watermelon. J. Nanosci. Nanotechnol. 13, 5561–5567.

Li, J., Hu, J., Ma, C., Wang, Y., Wu, C., Huang, J., Xing, B., 2016. Uptake, translocation and physiological effects of magnetic iron oxide (γ-Fe_2O_3) nanoparticles in corn (*Zea mays* L.). Chemosphere 159, 326–334.

Liu, R., Zhang, H., Lal, R., 2016. Effects of stabilized nanoparticles of copper, zinc, manganese, and iron oxides in low concentrations on lettuce (*Lactuca sativa*) seed germination: nanotoxicants or nanonutrients? Water Air Soil Pollut. 227, 42.

Ma, J., Li, R., Qu, G., Liu, H., Yan, B., Xia, T., Liu, Y., Liu, S., 2016. Carbon nanotubes stimulate synovial inflammation by inducing systemic pro-inflammatory cytokines. Nanoscale 8, 18070–18086.

Ma, X., Wang, Q., Rossi, L., Zhang, W., 2015. Cerium oxide nanoparticles and bulk cerium oxide leading to different physiological and biochemical responses in *Brassica rapa*. Environ. Sci. Technol. 50, 6793–6802.

Maier, S.A., Brongersma, M.L., Kik, P.G., Meltzer, S., Requicha, A.A., Atwater, H.A., 2001. Plasmonics—a route to nanoscale optical devices. Adv. Mater. 13 (19), 1501–1505.

Mallik, K., Mandal, M., Pradhan, N., Pal, T., 2001. Seed mediated formation of bimetallic nanoparticles by UV irradiation: a photochemical approach for the preparation of "coreshell" type structures. Nano Lett. 1 (6), 319–322.

Mallikarjuna, K., Dillip, G.R., Narasimha, G., John Sushma, N., Deva Prasad Raju, B., 2012. Phytofabrication and characterization of silver nanoparticles from *Piper betle* broth. Res. J. Nanosci. Nanotechnol. 2, 17–23.

Mandal, S., De, S., 2015. Catalytic and fluorescence studies with copper nanoparticles synthesized in polysorbates of varying hydrophobicity. Colloids Surf. A 467, 233–250.

Marimuthu, S., Rahuman, A.A., Jayaseelan, C., Kirthi, A.V., Santhoshkumar, T., Velayutham, K., Bagavan, A., Kamaraj, C., Elango, G., Iyappan, M., Siva, C., 2013. Acaricidal activity of synthesized titanium dioxide nanoparticles using *Calotropis gigantea* against *Rhipicephalus microplus* and *Haemaphysalis bispinosa*. Asian Pac. J. Trop. Med. 6, 682–688.

Martínez-Fernández, D., Vítková, M., Bernal, M.P., Komárek, M., 2015. Effects of nano-maghemite on trace element accumulation and drought response of *Helianthus annuus* L. in a contaminated mine soil. Water Air Soil Pollut. 226, 101.

Marusenko, Y., Shipp, J., Hamilton, G.A., Morgan, J.L., Keebaugh, M., Hill, H., Dutta, A., Zhuo, X., Upadhyay, N., Hutchings, J., Herckes, P., Anbar, A.D., Shock, E., Hartnett, H.E., 2013. Bioavailability of nanoparticulate hematite to *Arabidopsis thaliana*. Environ. Pollut. 174, 150–156.

Mittal, A.K., Chisti, Y., Banerjee, U.C., 2013. Synthesis of metallic nanoparticles using plant extracts. Biotechnol. Adv. 31, 346–356.

Monica, R.C., Cremonini, R., 2009. Nanoparticles and higher plants. Caryologia 62, 161–165.

Mulvaney, P., 1996. Surface plasmon spectroscopy of nanosized metal particles. Langmuir 12 (3), 788–800.

Niihara, K., Izaki, K., Nakahira, A., 1991. Nano–composite materials based on Si_3N_4 and SiC. Key Eng. Mater. 56 (57), 319–326.

Noruzi, M., Zare, D., Khoshnevisan, K., Davoodi, D., 2011. Rapid green synthesis of gold nanoparticles using *Rosa hybrida* petal extract at room temperature. Spectrochim. Acta Part A 79, 1461–1465.

Pandey, A., Pandey, G., 2013. Nanotechnology for herbal drugs and plant research. RRJPNT 2 (1), 13–16.

Park, H.J., Kim, S.H., Kim, H.J., Choi, S.H., 2006. A new composition of nanosized silica–silver for control of various plant diseases. Plant Pathol. J. 22, 295–302.

Peralta-Videa, J.R., Hernandez-Viezcas, J.A., Zhao, L., Diaz, B.C., Ge, Y., Priester, J.H., Holden, P.A., Gardea-Torresdey, J.L., 2014. Cerium dioxide and zinc oxide nanoparticles alter the nutritional value of soil cultivated soybean plants. Plant Physiol. Biochem. 80, 128–135.

Prasad, A.S., 2016. Iron oxide nanoparticles synthesized by controlled bio-precipitation using leaf extract of garlic vine (*Mansoa alliacea*). Mater. Sci. Semicond. Process. 53, 79–83.

Priya, G.S., Kanneganti, A., Kumar, K.A., Rao, K.V., Bykkam, S., 2014. Biosynthesis of cerium oxide nanoparticles using *Aloe barbadensis* miller gel. Int. J. Sci. Res. Publ. 4, 2250–3153.

Purkait, P.K., Maitra, S., Mandal, S., Bhunia, S., Das, G.C., Chaudhuri, M.G., 2015. Green synthesis of titanium dioxide nanoparticles: a review. Sci. Voyage 1 (4), 1–11.

Rajakumar, G., Rahuman, A.A., Priyamvada, B., Khanna, V.G., Kumar, D.K., Sujin, P.J., 2012. Eclipta prostrate leaf aqueous extract mediated synthesis of titanium dioxide nanoparticles. Mater. Lett. 68, 115–117.

Ramimoghadam, D., Bagheri, S., Abd Hamid, S.B., 2014. Biotemplated synthesis of anatase titanium dioxide nanoparticles via lignocellulosic waste material. BioMed Res. Int. http://dx.doi.org/10.1155/2014/205636.

Rao, K.G., Ashok, C.H., Rao, K.V., Chakra, C.S., Rajendar, V., 2015. Synthesis of TiO_2 nanoparticles from orange fruit waste. Synthesis 2, 1.

Ren, H.X., Liu, L., Liu, C., He, S.Y., Huang, J., Li, J.L., Zhang, Y., Huang, X.J., Gu, N., 2011. Physiological investigation of magnetic iron oxide nanoparticles towards Chinese mung bean. J. Biomed. Nanotechnol. 7, 677–684.

Rezaei, F., Moaveni, P., Mozafari, H., 2015. Effect of different concentrations and time of nano TiO_2 spraying on quantitative and qualitative yield of soybean (*Glycine max* L.) at Shahr-e-Qods, Iran. Biol. Forum 7 (1), 957.

Rico, C.M., Barrios, A.C., Tan, W., Rubenecia, R., Lee, S.C., Varela-Ramirez, A., Peralta-Videa, J.R., Gardea-Torresdey, J.L., 2015a. Physiological and biochemical response of soil-grown barley (*Hordeum vulgare* L.) to cerium oxide nanoparticles. Environ. Sci. Pollut. Res. Int. 22 (14), 10551–10558.

Rico, C.M., Peralta-Videa, J.R., Gardea-Torresdey, J.L., 2015b. Chemistry, biochemistry of nanoparticles, and their role in antioxidant defense system in plants. Nanotechnol. Plant Sci. 1–17 Springer International Publishing.

Roopan, S.M., Bharathi, A., Prabhakarn, A., Rahuman, A.A., Velayutham, K., Rajakumar, G., Padmaja, R.D., Lekshmi, M., Madhumitha, G., 2012. Efficient phyto-synthesis and structural characterization of rutile TiO_2 nanoparticles using *Annona squamosa* peel extract. Spectrochim. Acta Part A 98, 86–90.

Rossi, L., Zhang, W., Lombardini, L., Ma, X., 2016. The impact of cerium oxide nanoparticles on the salt stress responses of *Brassica napus* L. Environ. Pollut. 219, 28–36.

Roy, S., Das, T.K., 2015. Plant mediated green synthesis of silver nanoparticles: a review. Int. J. Plant Biol. Res. 3 (3), 1044–1055.

Ruparelia, J.P., Chatterjee, A.K., Duttagupta, S.P., Mukherji, S., 2008. Strain specificity in antimicrobial activity of silver and copper nanoparticles. Acta Biomater. 4 (3), 707–716.

Salam, H.A., Sivaraj, R., 2014. *Ocimum basilicum* L. var. purpurascens Benth.-LAMIACEAE mediated green synthesis and characterization of titanium dioxide nanoparticles. Adv. Biores. 5 (3).

Sasikala, A., Linga Rao, M., Savithramma, N., Prasad, T.N.V.K.V., 2015. Synthesis of silver nanoparticles from stem bark of *Cochlospermum religiosum* (L.) Alston: an important medicinal plant and evaluation of their antimicrobial efficacy. Appl. Nanosci. 5 (7), 827–835.

Sastry, A.B.S., Aamanchi, R.B.K., Prasad, C.S.R.L., Murty, B.S., 2013. Large–scale green synthesis of Cu nanoparticles. Environ. Chem. Lett. 11 (2), 183–187.

Seeger, E.M., Baun, A., Kästner, M., Trapp, S., 2009. Insignificant acute toxicity of TiO_2 nanoparticles to willow trees. J. Soils Sediments 9, 46–53.
Shankramma, K., Yallappa, S., Shivanna, M.B., Manjanna, J., 2016. Fe_2O_3 magnetic nanoparticles to enhance *S. lycopersicum* (tomato) plant growth and their biomineralization. Appl. Nanosci. 6, 983–990.
Shweta, Tripathi, D.K., Singh, S., Singh, S., Dubey, N.K., Chauhan, D.K., 2016. Impact of nanoparticles on photosynthesis: challenges and opportunities. Mater. Focus 5 (5), 405–411.
Singh, S., Tripathi, D.K., Dubey, N.K., Chauhan, D.K., 2016. Effects of nano-materials on seed germination and seedling growth: striking the slight balance between the concepts and controversies. Mater. Focus 5 (3), 195–201.
Sun, S., Murray, C., Weller, D., Folks, L., Moser, A., 2000. Monodisperse FePt nanoparticles and ferromagnetic FePt nanocrystal superlattices. Science 287 (5460), 1989–1992.
Sundrarajan, M., Gowri, S., 2011. Green synthesis of titanium dioxide nanoparticles by *Nyctanthes arbor-tristis* leaves extract. Chalcogenide Lett. 8, 447–451.
Surendra, T.V., Roopan, S.M., 2016. Photocatalytic and antibacterial properties of phytosynthesized CeO_2 NPs using *Moringa oleifera* peel extract. J. Photochem. Photobiol. B 161, 122–128.
Thakkar, K.N., Mhatre, S.S., Parikh, R.Y., 2010. Biological synthesis of metallic nanoparticles. Nanomedicine 6 (2), 257–262.
Theivasanthi, T., Alagar, M., 2010. X-ray diffraction studies of copper nanopowder. Arch. Phys. Res. 1 (2), 112–117.
Tolochko, N.K., 2009. History of nanotechnology. Nanosci. Nanotechnol. 1–6 EOLSS.
Tran, Q.H., Nguyen, V.Q., Le, A.T., 2013. Silver nanoparticles: synthesis, properties, toxicology, applications and perspectives. Adv. Nat. Sci. 4, 1–21.
Tripathi, D.K., Singh, S., Singh, V.P., Prasad, S.M., Chauhan, D.K., Dubey, N.K., 2016. Silicon nanoparticles more efficiently alleviate arsenate toxicity than silicon in maize cultiver and hybrid differing in arsenate tolerance. Front. Environ. Sci. 4, 46.
Tripathi, D.K., Singh, S., Singh, V.P., Prasad, S.M., Dubey, N.K., Chauhan, D.K., 2017e. Silicon nanoparticles more effectively alleviated UV-B stress than silicon in wheat (*Triticum aestivum*) seedlings. Plant Physiol. Biochem. 110, 70–81.
Tripathi, D.K., Singh, V.P., Prasad, S.M., Chauhan, D.K., Dubey, N.K., 2015. Silicon nanoparticles (SiNp) alleviate chromium (VI) phytotoxicity in *Pisum sativum* (L.) seedlings. Plant Physiol. Biochem. 96, 189–198.
Tripathi, D.K., Mishra, R.K., Singh, S., Singh, S., Vishwakarma, K., Sharma, S., Singh, V.P., Singh, P.K., Prasad, S.M., Dubey, N.K., Pandey, A.C., 2017b. Nitric oxide ameliorates zinc oxide nanoparticles phytotoxicity in wheat seedlings: implication of the ascorbate–glutathione cycle. Front. Plant Sci. 8.
Tripathi, D.K., Singh, S., Singh, S., Pandey, R., Singh, V.P., Sharma, N.C., Prasad, S.M., Dubey, N.K., Chauhan, D.K., 2017a. An overview on manufactured nanoparticles in plants: uptake, translocation, accumulation and phytotoxicity. Plant Physiol. Biochem. 110, 2–12.
Tripathi, D.K., Singh, S., Singh, S., Srivastava, P.K., Singh, V.P., Singh, S., Prasad, S.M., Singh, P.K., Dubey, N.K., Pandey, A.C., Chauhan, D.K., 2017c. Nitric oxide alleviates silver nanoparticles (AgNps)-induced phytotoxicity in *Pisum sativum* seedlings. Plant Physiol. Biochem. 110, 167–177.
Tripathi, D.K., Tripathi, A., Shweta, S.S., Singh, Y., Vishwakarma, K., Yadav, G., Sharma, S., Singh, V.K., Mishra, R.K., Upadhyay, R.G., Dubey, N.K., 2017d. Uptake, accumulation and toxicity of silver nanoparticle in autotrophic plants, and heterotrophic microbes: a concentric review. Front. Microbiol. 8.
Umesh, B.J., Vishwas, A.B., 2013. Green synthesis of silver nanopartcles using *Artocarpus heterophyllus* Lam. Seed extract and its antibacterial activity. Ind. Crops Prod. 46, 132–137.
Vankar, P.S., Bajpai, D., 2010. Preparation of gold nanoparticles from *Mirabilis jalapa* flowers. Indian J. Biochem. Biophys. 47, 157–160.
Varshney, R., Bhadauria, S., Gaur, M.S., Pasricha, R., 2010. Characterization of copper nanoparticles synthesized by a novel microbiological method. JOM 62 (12), 102–104.
Velayutham, K., Rahuman, A.A., Rajakumar, G., Santhoshkumar, T., Marimuthu, S., Jayaseelan, C., Bagavan, A., Kirthi, A.V., Kamaraj, C., Zahir, A.A., Elango, G., 2012. Evaluation of *Catharanthus roseus* leaf extract-mediated biosynthesis of titanium dioxide nanoparticles against *Hippobosca maculata* and *Bovicola ovis*. Parasitol. Res. 111, 2329–2337.
Wang, Q., Ma, X., Zhang, W., Pei, H., Chen, Y., 2012. The impact of cerium oxide nanoparticles on tomato (*Solanum lycopersicum* L.) and its implications for food safety. Metallomics 4, 1105–1112.
Wang, X.M., Gao, F.Q., Ma, L.L., Liu, J., Yin, S.T., Yang, P., Hong, F.S., 2008. Effects of nano-anatase on ribulose-1, 5-bisphosphate carboxylase/oxygenase mRNA expression in spinach. Biol. Trace Elem. Res 126 (13), 280–289.

Wang, T., Lin, J., Chen, Z., Megharaj, M., Naidu, R., 2014. Green synthesized iron nanoparticles by green tea and eucalyptus leaves extracts used for removal of nitrate in aqueous solution. J. Clean. Prod. 83, 413–419.

Wang, Y., He, X., Wang, K., Zhang, X., Tan, W., 2009. Barbated skull cup herb extract-mediated biosynthesis of gold nanoparticles and its primary application in electrochemistry. Colloids Surf. B 73, 75–79.

Wunder, S., Polzer, F., Lu, Y., Mei, Y., Ballauff, M., 2010. Kinetic analysis of catalytic reduction of 4-nitrophenol by metallic nanoparticles immobilized in spherical polyelectrolyte brushes. J. Phys. Chem. C 114 (19), 8814–8820.

Xu, X., Cortie, M.B., 2006. Shape change and color gamut in gold nanorods, dumbbells, and dog bones. Adv. Funct. Mater. 16 (16), 2170–2176.

Yallappa, S., Manjanna, J., Sindhe, M.A., Satyanarayan, N.D., Pramod, S.N., Nagaraja, K., 2013. Microwave assisted rapid synthesis and biological evaluation of stable copper nanoparticles using *T. arjuna* bark extract. Spectrochim. Acta Part A 110, 108–115.

Yang, J., Cao, W., Rui, Y., 2017. Interactions between nanoparticles and plants: phytotoxicity and defense mechanisms. J. Plant Interact. 12, 158–169.

Zheng, L., Hong, F., Lu, S., Liu, C., 2005. Effect of nano-TiO_2 on strength of naturally aged seeds and growth of spinach. Biol. Trace Elem. Res. 83–91.

Zhu, H., Han, J., Xiao, J.Q., Jin, Y., 2008. Uptake, translocation, and accumulation of manufactured iron oxide nanoparticles by pumpkin plants. J. Environ. Monit. 10 (6), 713–717.

Further Reading

Ahmed, K.B.A., Subramanian, S., Sivasubramanian, A., Veerappan, G., Veerappan, A., 2014. Preparation of gold nanoparticles using *Salicornia brachiata* plant extract and evaluation of catalytic and antibacterial activity. Spectrochim. Acta Part A 130, 54–58.

Aromal, S.A., Philip, D., 2012. Green synthesis of gold nanoparticles using *Trigonella foenum–graecum* and its size–dependent catalytic activity. Spectrochim. Acta Part A 97, 1–5.

Balamurugan, M., Saminathan, R., 2014. Characterization of *coriander sativum* mediated silver nanoparticles and evaluation of its antimicrobial and wound healing activity. IOSR JPTE 1 (2), 22–25.

Basavegowda, N., Sobczak-Kupiec, A., Malina, D., Yathirajan, H.S., Keerthi, V.R., Chandrashekar, N., Dinkar, S., Liny, P., 2013. Plant mediated synthesis of gold nanoparticles using fruit extracts of *Ananas comosus* (L.) (pineapple) and evaluation of biological activities. Adv. Mat. Lett. 4 (5), 332–337.

Bielinska, A., Eichman, J., Lee, I., Baker, J., Balogh, L., 2002. Imaging {Au0–PAMAM} gold–dendrimer nanocomposites in cells. J. Nanopart. Res. 4 (5), 395–403.

British Standards Institute, 2007. Terminology for Nanomaterials. PAS 136. British Standards Institute, London. ISBN: 978-0-580-61321-0-16.

Bunghez, I.R., Barbinta Patrascu, M.E., Badea, N.M., Doncea, S.M., Popescu, A., 2012. Antioxidant silver nanoparticles green synthesized using ornamental plants. J. Optoelectron. Adv. Mater. 14, 1016–1022.

Cruz, D., Fale, P.L., Mourato, A., Vaz, P.D., Serralheiro, M.L., Lino, A.R.L., 2010. Preparation and physicochemical characterization of Ag nanoparticles biosynthesized by *Lippia citriodora* (*Lemon verbena*). Colloids Surf. B 81 (1), 67–73.

Daisy, P., Saipriya, K., 2012. Biochemical analysis of *Cassia fistula* aqueous extract and phytochemically synthesized gold nanoparticles as hypoglycemic treatment for diabetes mellitus. Int. J. Nanomed. 7, 1189–1202.

Daniel, M.C., Astruc, D., 2004. Gold nanoparticles: assembly, supramolecular chemistry, quantum–size–related properties, and applications toward biology, catalysis, and nanotechnology. Chem. Rev. 104, 293–346.

Das, R., Nath, S.S., Chakdar, D., Gope, G., Bhattacharjee, R., 2009. Preparation of silver nanoparticles and their characterization. J. Nanotechnol. 5, 1–6.

Das, J., Das, P.M., Velusamy, P., 2013a. *Sesbania grandiflora* leaf extract mediated green synthesis of antibacterial silver nanoparticles against selected human pathogens. Spectrochim. Acta Part A 10, 265–270.

Das, S., Das, J., Samadder, A., Battacharyya, S.S., Das, D., Khuda–Bukhsh, A.R., 2013b. Biosynthesized silver nanoparticles by ethanolic extracts of *Phytolacca decandra, Gelsemium sempervirens, Hydrastis canadensis* and *Thuja occidentalis* induce differential cytotoxicity through G2/M arrest in A375 cells. Colloids Surf. B 101, 325–336.

Devi, G.D., Murugan, K., Selvam, C.P., 2014. Green synthesis of silver nanoparticles using *Euphorbia hirta* (Euphorbiaceae) leaf extract against crop pest of cotton bollworm, *Helicoverpa armigera* (Lepidoptera: Noctuidae). J. Biopestic. 7, 54–66.

Dutta, D., Mukherjee, R., Patra, M., Banik, M., Dasgupta, R., Mukherjee, M., Basu, T., 2016. Green synthesized cerium oxide nanoparticle: a prospective drug against oxidative harm. Colloids Surf. B. 147, 45–53.

Furno, F., Morley, K.S., Wong, B., Sharp, B.L., Arnold, P.L., Howdle, S.M., 2004. Silver nanoparticles and polymeric medical devices: a new approach to prevention of infection. J. Antimicrob. Chemother. 54, 1019–1024.
Galletti, A., Antonetti, M.R.C., Marracci, M., Piccinelli, F., Tellini, B., 2013. Novel microwave–synthesis of Cu nanoparticles in the absence of any stabilizing agent and their antibacterial and antistatic applications. Appl. Surf. Sci. 280, 610–618.
Gardea-Torresdey, J.L., Parsons, J.G., Gomez, E., Peralta–Videa, J., Troiani, H.E., Santiago, P., Jose Yacaman, M., 2002. Formation and growth of Au nanoparticles inside live alfalfa plants. Nano Lett. 2, 397–401.
Gardea-Torresdey, J.L., Parsons, J.G., Gomez, E., Peralta–Videa, J., Troiani, H.E., Santiago, P., Jose Yacaman, M., 2003. Alfalfa spouts: a natural source for the synthesis of silver nanoparticles. Langmuir 19, 1357–1361.
Geethalakshmi, R., Sarada, D.V.L., 2013. Characterization and antimicrobial activity of gold and silver nanoparticles synthesized using saponin isolated from *Trianthema decendra* L. Ind. Crop. Prod. 51, 107–115.
Ghaffari, M.M., Dabanlou, R.H., 2014. Plant mediated green synthesis and antibacterial activity of silver nanoparticles using *Crataegus douglasii* fruit extract. J. Ind. Eng. Chem. 20 (2), 739–744.
Ghosh, S., Patil, S., Ahire, M., Kitture, R., Kale, S., Pardesi, K., Cameotra, S.S., Bellare, J., Dhavale, D.D., Jabgunde, J., Chopade, B.A., 2012. Synthesis of silver nanoparticles using *Dioscorea bulbifera* tuber extract and evaluation of its synergistic potential in combination with antimicrobial agents. Int. J. Nanomed. 7, 483–496.
Gnanadesigan, M., Anand, M., Ravikumar, S., Maruthupandy, M., 2011. Biosynthesis of silver nanoparticles by using mangrove plant extract and their potential mosquito property. Asian Pac. J. Trop. Med. 4, 799–803.
Gogoi, S.K., Gopinath, P., Paul, A., Ramesh, A., Ghosh, S.S., Chattopadhyay, A., 2006. Green fluorescent protein–expressing *Escherichia coli* as a model system for investigating the antimicrobial activities of silver nanoparticles. Langmuir 22, 9322–9328.
Gosh, S., Patil, S., Ahire, M., Kitture, R., Gurav, D.D., Jabgunde, A.M., Kale, S., Pardesi, K., Shinde, V., Bellare, J., Dhavale, D.D., Chopade, B.A., 2012. *Gnidia glauca* flower extract mediated synthesis of gold nanoparticles and evaluation of its chemocatalytic potential. J. Nanobiotechnol. 10 (17). http://dx.doi.org/10.1186/1477-3155-10-17.
Grag, S., Chandra, A., Mazumder, A., Mazumder, R., 2014. Green synthesis of silver nanoparticles using *Arnebia nobilities* root extract and wound healing potential of its hydrogen. Asian J. Pharm. 95–101.
Gupta, N., Singh, H.P., Sharma, R.K., 2010. Single-pot synthesis: plant mediated gold nanoparticles catalyzed reduction of methylene blue in presence of stannous chloride. Colloids Surf. A 367, 102–107.
Hwang, C.G., 2003. Nanotechnology enables a new memory growth model. Proc. IEEE 91 (11), 1765–1771.
Ibrahim, H.M.M., 2015. Green synthesis and characterization of silver nanoparticles using banana peel extract and their antimicrobial activity against representative microorganisms. J. Radiat. Res. Appl. Sci. 8 (3), 265–275.
Jagtap, U.B., Bapat, V.A., 2013. Green synthesis of silver nanoparticles using *Artocarpus heterophyllus* Lam. seed extract and its antibacterial activity. Ind. Crop. Prod. 46, 132–137.
Jain, D., Daima, H.K., Kachwala, S., Kothari, S.L., 2009. Synthesis of plant mediated silver nanoparticles using papaya fruit extract and evaluation of their anti microbial activities. Dig. J. Nanomater. Biostruct. 4 (3), 557–563.
Jayaseelan, C., Rahuman, A.A., Rajakumar, G., Kirthi, A.V., Santoshkumar, T., Marimuthu, S., Bagavan, A., Kamaraj, C., Zahir, A.A., Elango, G., 2011. Synthesis of pediculocidal and larvicidal silver nanoparticles by leaf extract from heartleaf moonseed plant, *Tinospora cordifolia* Miers. 109 (1), 185–194.
Karimi, J., Mohsenzadeh, S., 2015. Synthesis and reactivity in inorganic, metal–organic, and nano–metal chemistry. J. Chem. 45 (6), 895–898.
Kirubha, R., Alagumuthu, G., 2014. Plant mediated synthesis of gold nanoparticles. Int. J. Adv. Sci. Tech. Res. 4 (4), 891–900.
Kitching, M., Ramani, M., Marsili, E., 2015. Fungal biosynthesis of gold nanoparticles: mechanism and scale up. Microb. Biotechnol. 8 (6), 904–917.
Kong, H., Jang, J., 2008a. Antibacterial properties of novel (methyl meth–acrylate) nanofibre containing silver nanoparticles. Langmuir 24 (5), 2051–2056.
Kong, H., Jang, J., 2008b. Synthesis and antimicrobial properties of novel silver/polyrhodanine nanofibres. Biomacromolecules 9 (10), 2677–2681.
Kora, A.J., Sashidhar, R.B., Jayaraman, A., 2012. Aqueous extract of gum olibanum (*Boswellia serrata*): a reductant and stabilizer for the biosynthesis of antibacterial silver nanoparticles. Process Biochem. 47 (10), 1516–1520.
Krishnaraj, C., Jagan, E.G., Rajasekar, S., Selvakumar, P., Kalaichelvan, P.T., Mohan, N., 2010. Colloids Surf. B 76 (1), 50–56.
Kumar, P.R., Kumar, S.S., 2014. Antifungal activity of extracellularly synthesized silver nanoparticles from *Morinda citrifolia* L. IJTRA 2 (4), 108–111.

Kumar, M.S., Krishnamurthy, S., Sung, W.W., Yeoung, Y.S., 2009. *Cinnamon zeylanicum* bark extract and powder mediated green synthesis of nano–crystalline silver particles and its bactericidal activity. Colloids Surf. B 73 (2), 332–338.

Kumar, S.A., Ravi, S., Velmurugan, S., 2013. Green synthesis of silver nanoparticles from *Gloriosa superba* L. leaf extract and their catalytic activity. Spectrochim. Acta Part A 115, 388–392.

Kumara, S.D., Jeganathan, N.S., 2013. Antimicrobial activity of bio synthesis silver nanoparticles prepared from the leaf extract of *Lantana camara*. IRJP 4 (5), 203–207.

Lei, Z., Mingyu, S., Chao, L., Liang, C., Hao, H., Xiao, W., Liu, X.Q., Yang, F., Gao, F.Q., Hong, F.S., 2007. Effects of nanoanatase TiO_2 on photosynthesis of spinach chloroplasts under different light illumination. Biol. Trace Elem. Res. 119, 68–76.

Mingyu, S., Fashui, H., Chao, L., Xiao, W., Xiaoqing, L., Liang, C., Liang, C., Fengqing, G., Fan, Y., 2007. Effects of nano-anatase TiO_2 on absorption, distribution of light, and photoreduction activities of chloroplast membrane of spinach. Biol. Trace Elem. Res. 118, 120–130.

Mirkin, C.A., Taton, T.A., 2000. Semiconductors meet biology. Nature 405, 626–627.

Mishra, A.N., Bhadauria, S., Gaur, M.S., Pasricha, R., Kushwah, B.S., 2010. Synthesis of gold nanoparticles by leaves of zero–calorie sweetener herb (*Stevia rebaudiana*) and their nanoscopic characterization by spectroscopy and microscopy. Int. J. Green Nanotechnol. Phys. Chem. 1, 118–124.

Morones, J.R., Elechiguerra, J.L., Camacho, A., Holt, K., Kouri, J.B., Ramirez, J.T., Yacaman, M.J., 2005. The bactericidal effect of silver nanoparticles. Nanotechnology 16 (10), 2346–2353.

Mubarak, A.D., Thajuddin, N., Jaganathan, K., Gunasekaran, M., 2011. Plant extract mediated synthesis of silver and gold nanoparticles and its antibacterial activity against clinically isolated pathogens. Colloids Surf. B 85 (2), 360–365.

Murugaraj, J., Manoharan, R., Renganathan, A., Davoodbasha, M., Gnanasekar, S., Ganeshan, S., Gnanajothi, K.D., Markandan, M., Kumpati, P., Nooruddin, T., Ganapathi, A., 2013. An investigation on the cytotoxicity and caspase–mediated apoptotic effect of biologically synthesized silver nanoparticles using *Podophyllum hexandrum* on human cervical carcinoma cells. Colloids Surf. B 102, 708–717.

Nagati, V.B., Alwala, J., Koyyati, R., Donda, M.R., Banala, R., 2012. Green synthesis of plant mediated silver nanoparticles using *Withania somnifera* leaf extract and evaluation of their anti microbial activity. IJAR 1 (9), 307–313.

Naik, S., Mark, P.K., Sree vennela, P., Devi, V.K., 2013. Green synthesis of silver nanoparticles using *strawberry* leaf extract (*Arbutus unedo*) and evaluvation of its antimicrobial activity a novel study. Int. J. Nanomater. Biostruct. 7, 47–50.

Nam, J.W., Thaxton, C.S., Mirkin, C.A., 2003. Nanoparticle–based bio–bar codes for the ultrasensitive detection of proteins. Science 301 (5641), 1884–1886.

Nezamdoost, T., Bagherieh-Najjar, M.B., Aghdasi, M., 2014. Biogenic synthesis of stable bioactive silver chloride nanoparticles using *Onosma dichroantha* Boiss. root extract. Mater. Lett. 137, 225–228.

Niraimathi, K.L., Sudha, V., Lavanya, R., Brindha, P., 2012. Biosynthesis of silver nanoparticles using *Alternanthera sessilis* (Linn.) extract and their antimicrobial, antioxidant activities. Colloids Surf. B 88, 34–39.

Otari, S.V., Patil, R.M., Ghosh, S.J., Pawar, S.H., 2014. Green phytosynthesis of silver nanoparticles using aqueous extract of *Manilkara zapota* (L.) seeds and its inhibitory action against *Candida* species. Materials Letters 116, 367–369.

Pal, S.L., Jana, U., Manna, P.K., Mohanta, G.P., Manavalan, R., 2011. Nanoparticle: an overview of preparation and characterization. J. Appl. Pharm. Sci. 1 (6), 228–234.

Pen, S.G., He, L., Nattan, M.J., 2003. Nanoparticles for biosensors. Curr. Opin. Chem. Biol. 7, 609–615.

Philip, D., 2010. Green synthesis of gold and silver nanoparticles using *Hibiscus rosa– sinensis*. Physica E 42, 1417–1424.

Praba, P.S., Vasantha, V.S., Jeyasundari, J., Brightson, Y., Jacob, A., 2015. Synthesis of plant–mediated silver nanoparticles using *Ficus microcarpa* leaf extract and evaluation of their antibacterial activities. Eur. Chem. Bull. 4 (3), 117–120.

Prakash, P., Gnanaprakasam, P., Emmanuel, R., Arokiyaraj, S., Saravanan, M., 2013. Green synthesis of silver nanoparticles from leaf extract of *Mimusops elengi* Linn. For enhanced antibacterial activity against multi drug resistant clinical isolates. Colloids Surf. B 108, 255–259.

Prow, T., Smith, J.N., Grebe, R., Salazar, J.H., Wang, N., Kotov, N., Lutty, G., Leary, J., 2006. Construction, gene delivery, and expression of DNA tethered nanoparticles. Mol. Vis. 12, 606–615.

Ramteke, C., Chakrabarti, T., Sarangi, B.K., Pandey, R.A., 2013. Synthesis of Silver nanoparticles from the aqueous extract of leaves of *Ocimum sanctum* for enhanced antibacterial activity. J. Chem. http://dx.doi.org/10.1155/2013/278925.

Reddy, J., Nagoor, V.D., Rani, M., Rani, S., 2014. Evaluation of antioxidant, antibacterial and cytotoxic effects of green synthesized silver nanoparticles by *Piper longum* fruit. Mater. Sci. Eng. 34, 115–122.

Rico-Oller, B., Boudjemaa, A., Bahruji, H., Kebir, M., Prashar, S., Bachari, K., Fajardo, M., Gómez-Ruiz, S., 2016. Photodegradation of organic pollutants in water and green hydrogen production via methanol photoreforming of doped titanium oxide nanoparticles. Sci. Total Environ. 563, 921–932.

Russell, A.D., Hugo, W.B., 1994. Antimicrobial activity and action of silver nanoparticle. J. Nanosci. Nanotechnol. 31, 351–370.

Sankar, R., Karthik, A., Prabu, A., Karthik, S., Shivashangari, K.S., Ravikumar, V., 2013. *Origanum vulgare* mediated biosynthesis of silver nanoparticles for its antibacterial and anticancer activity. Colloids Surf. B 108, 80–84.

Santhoshkumar, T., Rahuman, A.A., Jayaseelan, C., Rajakumar, G., Marimuthu, S., Kirthi, A.V., Kim, S.K., 2014. Green synthesis of titanium dioxide nanoparticles using *Psidium guajava* extract and its antibacterial and antioxidant properties. Asian Pac. J. Trop. Dis. 7, 968–976.

Satyavani, K., Gurudeeban, S., Ramanathan, T., Balasubramanian, T., 2011. Biomedical potential of silver nanoparticles synthesized from calli cells of *Citrullus colocynthis* (L.) Schrad. J Nanobiotechnol. 9, 43.

Shahverdi, A.R., Fakhimi, A., Shahverdi, H.R., Minaian, S., 2007. Synthesis and effect of silver nanoparticles on the antibacterial activity of different antibiotics against *Staphylococcus aureus* and *Escherichia coli*. Nanomedicine 3 (2), 168–171.

Shameli, K., Ahmad, M.B., Zamanian, A., Sangpour, P., Shabanzadeh, P., Abdollahi, Y., Zargar, M., 2012. Green biosynthesis of silver nanoparticles using *Curcuma longa* tuber powder. Int. J. Nanomed. 7, 5603–5610.

Shankar, S.S., Ahmad, A., Sastry, M., 2003. Geranium leaf assisted biosynthesis of silver nanoparticles. Biotechnol. Prog. 19, 1627–1631.

Shankar, S.S., Rai, A., Ankamwar, B., Singh, A., Ahmad, A., Sastry, M., 2004a. Biological synthesis of triangular gold nanoprisms. Nat. Mater. 3, 482–488.

Shankar, S.S., Rai, A., Ankamwar, B., Singh, A., Ahmad, A., Sastry, M., 2004b. Biosynthesis of silver and gold nanoparticles from extracts of different parts of the geranium plant. Appl. Nanosci. 1, 69–77.

Sharma, N.C., Sahi, S.V., Nath, S., Parsons, J.G., Gardea-Torresdey, J.L., Pal, T., 2007. Synthesis of plant-mediated gold nanoparticles and catalytic role of biomatrix-embedded nanomaterials. Environ. Sci. Technol. 41 (14), 5137–5142.

Silver, S., 2003. Bacterial silver resistance: molecular biology and uses and misuses of silver compounds. FEMS Microbiol. Rev. 27 (2), 341–353.

Singh, C., Sharma, V., Naik, P.K., Khandelwal, V., Singh, H., 2010. A green biogenic approach for synthesis of gold and silver nanoparticles using *Zingiber officinale*. Dig. J. Nanomater. Biostruct. 6, 535–542.

Singh, S., Saikia, J.P., Buragohain, A.K., 2013. A novel 'green' synthesis of colloidal silver nanoparticles (SNP) using *Dillenia indica* fruit extract. Colloids Surf. B 102, 83–85.

Singh, K., Panghal, M., Kadyan, S., Chaudhary, U., Yadav, J.P., 2014. Green silver nanoparticles of *Phyllanthus amarusas* antibacterial agent against multi drug resistant clinical isolates of *Pseudomonas aeruginosa*. J. Nanobiotechnol. 12 (40). http://doi.org/10.1186/s12951-014-0040-x.

Singh, S., Vishwakarma, K., Singh, S., Sharma, S., Dubey, N.K., Singh, V.K., Liu, S., Tripathi, D.K., Chauhan, D.K., 2017. Understanding the plant and nanoparticle interface at transcriptomic and proteomic level: a concentric overview. Plant Gene. http://dx.doi.org/10.1016/j.plgene.2017.03.006.

Somasundaram, K., Santhanalakshmi, J., Vishwanathan, B., Srinivasan, K., 2011. Biosynthesis of silver nanoparticles using *Citrus sinensis* peel extract and its antibacterial activity. Spectrochim. Acta Part A 79 (3), 594–598.

Song, J.Y., Kim, B.S., 2009. Rapid biological synthesis of silver nanoparticles using plant leaf extract. Korean J. Chem. Eng. 25, 808–811.

Srinivasan, B., 2013. Synthesis of silver nanoparticles using herbal extract and its antimicrobial activity. J. Phylogenet. Evol. Biol. 3 (1). http://dx.doi.org/10.4172/2329-9002.1000146.

Sukirthaa, R., Kandula, M.P., Jacob, J.A., Soundarajan, K., Ramar, T., Palani, G., Muthukalingan, K., Shanmugam, A., 2012. Cytotoxic effect of green synthesized silver nanoparticles using *Melia azedarach* against *in vitro* HeLa cell lines and lymphoma mice model. Process Biochem. 47, 273–279.

Sulaiman, G.M., Mohammed, W.H., Marzoog, T.R., Al-Amiery, A.A., Kadhum, A.A., Mohamad, A.B., 2013. Green synthesis, antimicrobial and cytotoxic effects of silver nanoparticles using *Eucalyptus chapmaniana* leaves extract. Asian Pac. J. Trop. Biomed. 3 (1), 58–63.

Suriyakalaa, U., Antony, J.J., Suganya, S., Siva, D., Sukirtha, R., Kamalakkannan, S., Pichiah, P.B.T., Achiraman, S., 2013. Hepatocurative activity of biosynthesized silver nanoparticles fabricated using *Andrographis paniculata*. Colloids Surf. B 102, 189–194.

Thombre, R., Parekh, F., Patil, N., 2014. Green synthesis of silver nanoparticles using seed extract of *Argyreia nervosa*. Int. J. Pharm. Biol. Sci. 5, 119–144.

Tripathy, Raichur, A.M., Chandrasekaran, N., Prathna, T.C., Mukherjee, A., 2010. Process variables in biomimetic synthesis of silver nanoparticles by aqueous extract of *Azadirachta indica* (Neem) leaves. J. Nanopart. Res. 12, 243–246.

Usha, C., Gladys Angelin Rachel, D., 2004. Biogenic synthesis of silver nanoparticles by *Acacia nilotia* and their antibacterial activity. Int. J. Sci. Res. 3 (6), 2277–2279.

Velmurugan, P., Choa, M., Lim, S.S., Seo, S.K., 2015. Phytosynthesis of silver nanoparticles by *Prunus yedoensis*leaf extract and their antimicrobial activity. Mater Lett. 138, 272–275.

Vilchis-Nestor, A.R., Sánchez-Mendieta, V., Camacho-López, M.A., Gómez-Espinosa, R.M., Camacho-López, M.A., Arenas-Alatorre, J.A., 2008. Solventless synthesis and optical properties of Au and Ag nanoparticles using *Camiellia sinensis* extract. Mater Lett. 62, 3103–3105.

Vivek, R., Thangan, R., Muthuchelian, K., Gunasekaran, P., Kaveri, K., Kannan, S., 2012. Green biosynthesis of silver nanoparticles from *Annona squamosa* leaf extract and its *in vitro* cytotoxic effect on MCF–7 cells. Process Biochem. 47, 2405–2410.

Wang, Z., Fang, C., Mallavarapu, M., 2015. Characterization of iron–polyphenol complex nanoparticles synthesized by sage (*Salvia officinalis*) leaves. Environ. Technol. Innov. 4, 92–97.

Xuming, W., Fengqing, G., Linglan, M., Jie, L., Sitao, Y., Ping, Y., Fashui, H., 2008. Effects of nano-anatase on ribulose-1, 5-bisphosphate carboxylase/oxygenase mRNA expression in spinach. Biol. Trace Elem. Res. 126, 280.

Yatesen, A.K., Qu, H., Manbachi, A., Butt, H., Dokmeci, M.R., Hinestroza, J.P., Skorobogatiy, M., Khademhosseini, A., Yun, S.H., 2016. Nanotechnology in textiles. ACS Nano 10 (3), 3042–3068.

Zahir, A.A., Rahuman, A.A., 2012. Evaluation of different extracts and synthesised silver nanoparticles from leaves of *Euphorbia prostrata* against *Haemaphysalis bispinosa* and *Hippobosca maculata*. Vet. Parasitol. 187, 511–520.

Zhang, P., Ma, Y., Zhang, Z., He, X., Li, Y., Zhang, J., Heng, L., Zhao, Y., 2015. Species-specific toxicity of ceria nanoparticles to *Lactuca* plants. Nanotoxicology 9, 1–8.

CHAPTER

9

Gold Nanomaterials to Plants: Impact of Bioavailability, Particle Size, and Surface Coating

Nitin Kumar[1], *Pranav Tripathi*[2], *Seema Nara*[2]

[1]Centre for Medical Diagnostic and Research, Motilal Nehru National Institute of Technology, Allahabad, India; [2]Motilal Nehru National Institute of Technology, Allahabad, India

9.1 INTRODUCTION

The natural existence of nanomaterials has been well established by many reports but engineered nanomaterials are also of significant interest because of their controlled manufacture and design. Nanomaterials can be readily discovered in everyday products such as stain-resistant clothing, cosmetics, tires, sunscreens, electronics, and sporting goods along with medicines, diagnostic instruments, imaging, and drug delivery technology (Salisu et al. 2014; Al-Halafi, 2014). Designing of nanomaterials takes place at the molecular level to extract the exact advantageous effects of size and novel characteristics that are very different from their bulky counterpart. Two major reasons are attributed to the distinctive properties of nanostructures, namely, new quantum effects and increased relative surface area (Safiuddin et al., 2014; Zhou et al., 2015). Conventional structures differ from nanostructures in terms of surface area-to-volume ratio, which exacerbates chemical reactivity and strength of material. With respect to quantum effects, new optical, electrical, and magnetic behaviors are revealed in nanostructures (Rajapaksha et al., 2015; Auffan et al., 2009).

Increasing use of engineered nanomaterials in rapidly increasing products such as nanofertilizers, industrial effluents or paints, delivery vehicles for bioactive molecules, field monitoring sensors, monitoring of seed health, etc. (Fig. 9.1) is paving the way for their entry into wastewater, sludge, soils, and ultimately into plants (Grieger et al., 2009; Zhang et al. 2015; Liu and Lal, 2015). Besides normal exposure through soil, there can also be direct exposure, which cannot be neglected. In recent years, many advantageous and detrimental effects of nanomaterials have been reported with respect to crop growth. Such effects are based on source, type, and sizes of nanomaterials along with plant species and duration of exposure to

Nanomaterials in Plants, Algae, and Microorganisms
http://dx.doi.org/10.1016/B978-0-12-811487-2.00009-8

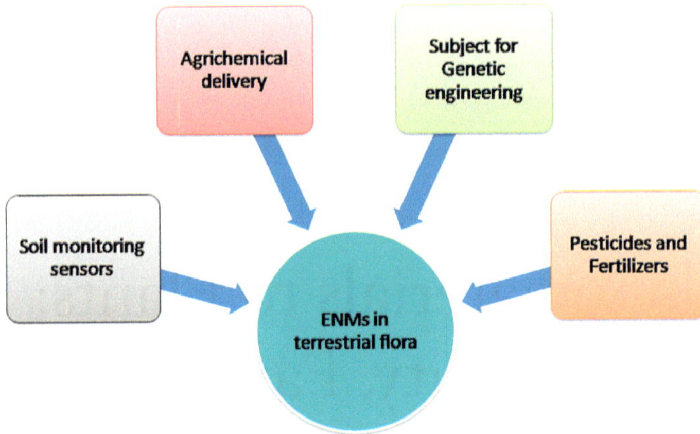

FIGURE 9.1 Applications of nanomaterials for terrestrial plants. *ENMs*, engineered nanomaterials.

plants (Roghayyeh et al., 2010; Rico et al., 2014; Bandyopadhyay et al., 2015; Lalau et al., 2015; An et al., 2008; Tripathi et al., 2015, 2016, 2017a). Studies demonstrating the beneficial effects of nanoparticles (NPs) on plants have been reported, such as increment in the ascorbate and chlorophyll content in the leaves of asparagus in response to silver NPs and iron increment in biomass of soybean plant upon treatment with iron NPs (Roghayyeh et al., 2010; An et al., 2008). In a report, increment in root and shoot length and dry weight of maize seedlings is observed after treatment with silica NPs (Suriyaprabha et al., 2012). In another report, titanium dioxide NPs help to decrease hydrogen peroxide, electrolyte leakage, and malondialdehyde in *Cicer arietinum* plants (Mohammadi et al., 2014). Although nanotechnology presents a plethora of applications for plant science or agriculture, well-defined studies are needed to understand plant–NP interaction because the presence/accumulation of NPs from various origins in soil in unknown concentrations may pose a threat to the ecosystem (Anjum et al., 2013; Keller et al., 2013). To know the realistic effects of nanomaterials on plants, it is therefore important to have an in-depth understanding of the bioavailability, uptake, translocation, and localization of NPs to plant systems. In addition, toxicity associated with these NPs, which can damage internal cellular structures or normal cell functioning, is also a major concern that needs to be addressed (Bergeson, 2010; Kah et al., 2013; Dutschk et al., 2014; Singh et al., 2016, 2017; Shweta et al., 2016; Tripathi et al., 2017a,b,c,d,e).

Of the various nanomaterials studied with plant systems, gold nanomaterials are popularly used NPs because of their easy synthesis, surface modifications, various possible sizes and shapes [spherical NPs, nanorods, nanourchins, nanoshells, and nanocages (Fig. 9.2)], stability, and relatively low toxicity. Among all the gold nanostructures synthesized to date, NPs and nanorods fascinate researchers the most. Because of exclusive surface optical, electrical, and chemical properties, gold nanomaterials have immense applications in the field of bioimaging and phototherapy (Perrault and Chan, 2010), biosensors, and gene/drug delivery. Localization of surface plasmon resonance (Oldenburg et al., 1999) enables near-infrared (NIR) absorption, hence efficient photothermal properties are available for cancer theranostics. Gold nanomaterials have also been seen as sensing probes for biomolecular detection through surface-enhanced Raman scattering. Gold NPs, because of their small size, make

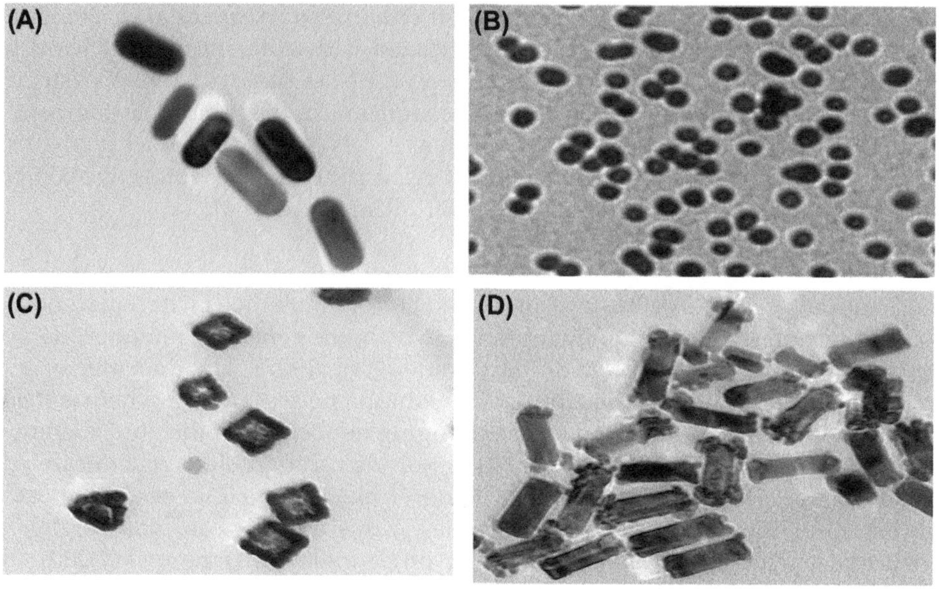

FIGURE 9.2 Gold nanostructures of varied shapes: (A) nanorods, (B) nanospheres, (C) nanocages, and (D) bimetallic *dotted* gold nanorods.

their way into cells by endocytosis or diffusion, enabling them to be used as an ideal agent for gene/drug delivery.

Gold NPs are transferred from natural waters and soils to invertebrates and secondary consumers. Hence their bioavailability and toxicity studies are essential to assess possible risks and impact on ecosystems. Therefore this chapter intends to discuss the interaction of gold NPs with terrestrial plants, particularly focusing on their effect on bioavailability, uptake, and toxicity. The chapter discusses various possible mechanisms involved in the uptake of NPs from soil. It also discusses the effect of physicochemical characteristics of NPs such as their size or surface coating or charge on the bioavailability of these NPs to plants. This chapter intends to present a holistic picture of gold NP interaction with plants.

9.1.1 Gold Nanomaterials: Types and Properties

9.1.1.1 Gold Nanospheres

Gold nanospheres (also known as colloidal gold) can be synthesized by reducing hydrogen tetraaurochloric acid ($HAuCl_4$) solution with various reducing agents in controlled conditions. Preferably, citrate is used as a reducing agent for the production of gold nanospheres in aqueous media, which confers negative charge on their surface and maintains stability of NP solution through electrostatic repulsive forces (Turkevich et al., 1951). The diameter of the nanospheres is varied by changing the citrate:gold ratio. Alternatively, sodium borohydride-mediated reduction of $HAuCl_4$ is used popularly for NP synthesis in organic media (Christopher et al., 2014). These NPs have better control over size and shape but since the synthesis takes place in organic media, phase transfer (organic to aqueous) is required for using them with biomolecules (Leff et al., 1996; Weare et al., 2000; Hiramatsu and Osterloh, 2004). Gold NPs can be functionalized with different types of short ligands having different surface charges to stabilize them, for example, cysteamine, cysteine, thioglycolic acid, glutaraldehyde, etc. Such modifications make them more biocompatible and stable. Introduction of biomolecules such as proteins, DNA, and antibodies on their surface allows them to be used for localization studies or specific delivery of bioactive components to the target (Garcia et al., 1999; Manna et al., 2001; Kim et al., 2004).

9.1.1.2 Gold Nanorods

Gold nanorods can be synthesized using the template method, wherein deposition of gold occurs in pores of template membranes, which are generally nanoporous polycarbonate or alumina (Martin, 1994; Van der Zande et al., 1997). The diameter of the gold nanorods can be calculated by the template membrane pore diameter, whereas the length of gold nanorods can be modulated by the changing deposition of the gold content on the pores of the membrane. Formation of gold nanorods through electrochemical synthesis has also been reported (Yu et al., 1997; Chang et al., 1999). However, the most popular synthesis method is seed-mediated synthesis; this method can make higher aspect ratio nanorods, which cannot be done using other methods (Jana et al., 2001; Busbee et al., 2003). In this approach, generally, gold nanoseeds are synthesized by reduction of $HAuCl_4$ by a strong reducing agent ($NaBH_4$). The seeds then act as nucleation centers for nanorod synthesis. These seeds are then added to another solution called "growth solution," which contains $HAuCl_4$, a weak reducing agent such as ascorbic acid, and hexadecyltrimethylammonium bromide (CTAB). Aspect ratios of the gold nanorods can be modulated by the amount of gold nanoseeds with respect to the gold precursor. The surface charge of these gold nanorods is positive because of CTAB capping, which can be easily modified according to the need.

9.1.1.3 Gold Nanoshells

Variation in the dimensions and composition of layers can be employed in fabrication of shell-like nanostructures termed gold nanoshells, and the characteristic surface plasmon resonance (SPR) peaks range from the visible to the NIR region (Oldenburg et al., 1999). SPR peaks of gold nanoshell can be tuned by changing shell width and core size ratio. Silica or polymer bead-coated shells of inconsistent thickness show SPR peak in the NIR region (Caruso et al., 2001; Oldenburg et al., 1998). Gold nanoshells are serene nanostructures with a dielectric core superficially covered by gold shells of low thickness. The small size of nanomaterials enables them to penetrate into cells through endocytosis or diffusion and act as a method for efficient gene or drug delivery.

9.1.1.4 Gold Nanocages

Synthesis through a galvanic replacement reaction can give rise to cage-like nanostructures called gold nanocages. These nanocages can also be synthesized using a template structure of hexamethylenetetramine molecules (Chen et al., 2006). In the galvanic replacement reaction, evenly distributed silver nanostructures act as seed and subsequent incorporation of gold atoms generates hollow interior gold nanocages (Chen et al., 2005, 2006). The molar ratio of silver and $HAuCl_4$ is employed to generate controlled synthesis of nanocages with high precision. As per the previously mentioned reports, gold nanostructures are used in conjugation with various other metallic combinations in form of physical allure.

9.1.2 Gold Nanostructures: Applications in Plants

Fertilizers play a key role in efficient plant growth and development when used in a controlled manner. However, the bioavailability of fertilizers is negatively affected because of leaching, decomposition, hydrolysis, and degradation by photolysis. Emerging nanotechnology has been of great use in minimizing nutrient losses toward increments in crop yield in the form of nanofertilizers or nanoencapsulated nutrients. Nanofertilizers or nanoencapsulated nutrients are supposed to have characteristics beneficial to crops, such as sustained release of nutrients and enhancement of target activity (DeRosa et al., 2010; Nair and Chung, 2014; Giraldo et al., 2014). As per the report of Galbraith (2007), engineered NPs are capable of traversing in plant cells and are also able to ship DNA and chemical entities in cells. However, in comparison to other types of nanomaterials, gold nanomaterials are less studied. Gold NPs have demonstrated enhanced seed germination in lettuce, cucumber, *Brassica juncea*, *Boswellia ovalifoliolata*, and *Gloriosa superba* (Arora et al., 2012; Gopinath et al., 2014). Gold NPs are also reported to improve leaf number, plant height, leaf area, plant height, and chlorophyll content leading to increased crop yield (Savithramma et al., 2012; Barrena et al., 2009; Arora et al., 2012; Gopinath et al., 2014). Christou et al. (1988) introduced a gene named neomycin phosphotransferase II gene in soybean genome using DNA-coated gold particles establishing a potent effect in gene delivery. Kumar (2009) revealed that gold NPs play a significant role in enhancement of seed germination and antioxidant system in *Arabidopsis thaliana* along with alteration in levels of microRNAs expression regulating certain physiological, metabolic, and morphological processes in plants. As per the report of Govorov and Carmeli (2007), NPs efficiently encourage photosynthetic systems. Through these reports an outstanding hypothesis has revealed that gold and silver NPs form a system that may produce 10-fold higher excited electrons because of SPR and fast separation of electrons and holes. Such a successfully established hypothesis may lead to the designing of novel light-harvesting systems. So we can say that nanotechnology possesses an enormous potential to develop new tools for augmentation of existing functions in plants leading to beneficial functions (Cossins, 2014).

9.2 UPTAKE AND TRANSLOCATION OF NANOSTRUCTURES IN PLANTS

The occurrence of engineered nanoparticles (ENPs) in soil can be attributed to the budding commercial market of ENP-based products, which is drastically increasing day by day (Park et al., 2008; Alexis et al., 2009; Zhao et al., 2012a,b). The primary sinks of ENPs in the environment are oceans and soils, for example, the soil obtains these NPs through sludge application (Shi et al., 2014; Zhu et al., 2008; Zhao et al., 2012a,b). Throughout the globe, ENPs are mainly released into the soil from landfills where approximately 63%–91% of ENPs end up. They can also enter the soil via direct release from industries, by intentionally subjecting the soil to agrochemicals manufactured from NPs for the purpose of remediation, by the use of wastewater for irrigation purposes, and from rain that transfers ENPs from the atmosphere to the soil (Slomberg and Schoenfisch, 2012; Zhao et al., 2013; Li et al., 2014). The source of ENPs in soil and water is a major concern because ENPs can be introduced into the food web as contaminants. These NPs are taken up by plants through various routes and are subsequently translocated to different parts of plants. This section discusses in detail the routes of

NP uptake in plants. Furthermore, translocation and transmission of nanomaterials through plants is presented in detail.

9.2.1 Bioavailability and Uptake

9.2.1.1 Through Roots

Limited studies involving the bioavailability of ENPs for plant root uptake from soil have been reported. These studies show that variables such as soil type, soil biota, and its chemistry may play a key role in ENP uptake through roots. The first report of evidence of ENP uptake by plant roots is presented by Zhu et al. (2008), which demonstrates the uptake of Fe_3O_4 ENPs by pumpkin plants in hydroponics and shows that Fe_3O_4 ENPs are taken up by plants in neither soil nor sand. Since then, the uptake of a wide range of ENPs by an array of plant species in hydroponic exposures has been reported. At the beginning of work in this field, the study of Lin and Xing (2008) showed the accumulation of ZnO ENPs in the protoplast of endodermal cells but translocation of these ENPs to the shoots and leaves is not seen. This is speculated because of the aggregation of ENPs induced by high ionic strength of the nutrient media that probably led to decreased bioavailability of nanomaterials to plants. Such studies addressing the bioavailability of nanomaterials to plants have been reviewed as well (Miralles et al., 2012; Rico et al., 2011). Although multiple studies are performed for examining the plant root uptake of ENPs, many important aspects such as ENP characterization data, ENP dissolution, controls employed, and spatially resolved analysis proving the actual uptake have not been adequately reported. In addition, a plethora of combinations of ENPs coupling with plant species illustrated conflicting data and has limited our understanding of the bioavailability of ENPs for real-time scenarios (Lin and Xing, 2008; Kouhi et al., 2014; Wang et al., 2012; Zhang et al., 2012a,b; Hernandez-Viezcas et al., 2013; Peng et al., 2015; Wang et al., 2012).

Deficient standard ENP exposure protocols are likely the reason for these observed conflicting results since different studies used differing protocols for various analytical techniques including exposure times and media. For instance, most studies have been conducted in hydroponic solutions but a few studies involve the use of solid agar-like media to suspend ENPs. One such study demonstrates the comparison of uptake of ENPs in different states of the media, where 2.8nm TiO_2 ENPs are taken up in the liquid media but not in the solid media. Similarly, another study compared exposure times and it was found that the exposure time can vary from hours (78–79h) to a few months. The effect of these variations has not been investigated systematically (Lin and Xing, 2008). Varying physiochemical properties of different plant species that include hydraulic conductivity, cell wall pore size, and root physiology also influence the bioaccumulation of ENPs. To sustain this hypothesis, uptake studies of NPs have been carried out on different plant species, which shows different effects. For instance, species-based accumulation of magnetite ENPs is observed in *Vigna radiata* (mung bean) but not in *Phaseolus lunatus* (lima bean) (Zhu et al., 2008). Likewise, analysis of tobacco biomass shows accumulated gold NPs in the range of 2.2–53.5mg/kg, while for wheat biomass no significant uptake is observed (Marschner, 1995). Adequate research has not been done and has limited our understanding of how the difference in plant species impacts the uptake of ENPs. However, certain differences such as monocots and dicots releasing differing chemicals into the rhizosphere could also contribute in the modification of ENP stability and its bioavailability.

Other studies pertaining to mycorrhizae-induced root uptake of nanomaterials have shown somewhat negative results, despite the fact that mycorrhizae have always been a source of attraction for botanists and biotechnologists since their inception because mycorrhizae share a symbiotic relationship with host plants through hyphae in exchange of carbohydrates and inorganic ions. Mycorrhizal hyphae may aid in the entry of ENPs by acting as an entry site for root penetration (Judy and Bertsch, 2014) and the exact mechanism dealing with this uptake has been a subject of investigation (Judy et al., 2015; Judy and Bertsch, 2014; Watts-Williams et al., 2014; Feng et al., 2013; Whiteside et al., 2009). Watts-Williams et al. have reported that mycorrhizae colonization in tomatoes grown in a sand/soil mixture of 4:1 remains unaffected by exposure to ZnO NPs. Concurrently, it is also reported that uptake of Zn is also the same for mycorrhizal and nonmycorrhizal plants (Watts-Williams et al., 2014). In 2016, Judy et al. presented a study to demystify the mechanism of gold nanomaterial bioaccumulation in two genotypes of *Solanum lycopersicum* (tomato), namely, progenitor 76R and its mutant genotype, which shows reduced mycorrhizal colonization. It is observed that there is minute or no role for arbuscular mycorrhizal fungi colonization in the trafficking of gold nanomaterials (Judy et al., 2015; Watts-Williams et al., 2014; Feng et al., 2013; Whiteside et al., 2009).

Uptake of NPs through roots also seems to vary with the type or nature of NPs, for example, Antisari et al. (2014) demonstrated that tomato plants exposed to metal NPs such as Ag, Ni, and Co show their bioaccumulation in both shoots and roots, whereas in the case of metal oxide NPs, uptake is seen only at the roots of the tomato plant. ENPs also tend to aggregate under the influence of environmental variables, which may be common or different among plant species, such as release of exudates from plants (mugineic acid, phenolic compounds) in the exposure media or ionic strength of the media or even the intrinsic properties of ENPs themselves such as particle size (Lin and Xing, 2008). Such induced aggregation of ENPs is speculated to affect their bioavailability to plants. Studies show that NPs in an aggregated state are less bioavailable to plants (Marschner, 1995). However, a study showing the dependence of gold ENP aggregation on bioavailability in wheat and tobacco plants also justified this speculation (Judy et al., 2012). Further investigations on the state of aggregation of ENPs prior to and after exposure to plant growth medium would shed light on the bioavailability of ENPs to plants.

It is also reported that uptake by roots causes transformation of chemicals to different forms, sometimes even to crystalline forms (Chandran et al., 2006). Various studies have shifted toward the capability of plants to convert gold ions to crystalline nanostructures. Some plants such as *Medicago sativa* (alfalfa) when grown on $AuCl_4$-enriched medium can take up ionic gold and transform it into atomic gold NPs, which can be seen accumulated in different plant parts (Gardea-Torresdey et al., 2000). NPs synthesized by these plants are said to be biogenic (Bali and Harris, 2010). Environmental adaption of plants to adverse conditions also affects their bioaccumulation capability, which may reach up to 3500 mg/kg (0.35%) dry weight, as seen in plants growing on mine waste with respect to gold NPs. *M. sativa* grown on mine waste rich in gold ions shows both the formation and bioaccumulation of gold NPs (Gardea-Torresdey et al., 2000). Plants grown on soil, which is rich in both gold and silver ions, show bioaccumulation of nanoalloys of Au and Ag. Copper shows no effect when placed with Au and Ag in the soil. However, the interesting fact is that both Cu and Ag have decreased the size of gold NPs, and also limited the extent of reduction of gold. Another interesting fact in this report is that high concentrations of gold in soil resulted in small NPs, and when gold is kept in soil for longer durations, large particles in micrometer scale are obtained. The mechanism for biogenic synthesis of gold NPs in *Sesbania* has been hypothesized by Sharma et al. It is proposed that roots of plants sequester gold from surrounding medium because of a strong affinity between carboxylic acid groups present on the cell wall of roots and Au(III). The trapped gold then transports through roots to other conducting tissues and aerial plant parts.

Transformation has been reported with NPs other than gold NPs, for example, after exposure of cucumber plants to CeO_2 NPs hydroponically, it is observed that a small quantity of these CeO_2 NPs changed to $CePO_4$ at the roots, whereas it is present in the form of cerium carboxylates in shoots. Even in lettuce roots, Cui et al. (2014) observed that small amounts of Ce(IV) are transformed to Ce(III). Hernandez-Viezcas et al. (2013) demonstrated the formation of an oxygen-bound zinc complex that is similar to zinc citrate when soybean grains are exposed to ZnO NPs (Hernandez-Viezcas et al., 2013; Thiruvengadam et al., 2014; Moon et al., 2014; Xiang et al., 2015).

It is also found that uptake of NPs by plants sometimes becomes highly size selective as ENPs need to passively cross an intact cell via the cell wall pores in roots. Various techniques have been utilized to estimate the diameter of these cell wall pores. Solute exclusion experiments were carried out by Carpita (1979) to estimate the diameter of root hair cell wall pores of *Raphanus sativus*, which is 3.5–3.8 nm, while the diameter of pores for palisade parenchyma cells of the leaves of *Xanthium strumarium* was found to be 4.5–5.2 nm.Adani et al. (2011) demonstrated that the diameter of the majority of the cell wall pores is less than 8 nm but a few of them have a diameter as large as 50 nm and are characterized as mesopores. Studies discussing the restrictions imposed by the cell wall pore size on uptake of nanomaterials of a particular size by plants are presented later in the chapter.

9.2.1.2 Through Foliar Uptake

There exist pathways, other than root uptake, for the bioaccumulation of ENPs. Because of rainfall, wind, and/or mechanical disruption, soil particles resuspend on the surface of the leaves and thus penetrate the terrestrial food web. Similarly, foliar uptake of ENPs takes place in the plants once they are on leaf surfaces. The ENP enters the plant vasculature by two mechanisms, namely, cuticular diffusion and transversion through stomata (Lin et al., 2009; Punshon et al., 2003; Schönherr, 2000). Cuticular diffusion can further occur by two pathways: hydrophilic pathway and lipophilic pathway. Polar molecules enter via the hydrophilic pathway using aqueous pores, while nonpolar molecules utilize the lipophilic pathway. Therefore the hydrophobicity or hydrophilicity of ENPs, which in turn is also governed by its surface coating, must be considered to ensure entry of ENPs into the vasculature through these pathways. The mechanism of cuticular diffusion is highly size selective because of the restriction imposed by size of the cuticular pores. The average size of these pores in *Coffea arabica* and *Populus* x *Canadensis* is 4 and 4.8 nm, respectively. In other plants, cuticular pores of sizes as small as 0.6 nm are also seen. Hence the difference in the size of these pores among different plants makes the assessment of this mechanism highly cumbersome (Schönherr, 2006; Schreiber, 2005). The entry of ENPs into plant vascular systems by traversing the stomata mode is not very well studied and is still unclear. Reports showed the uptake of fluorescent polystyrene NPs of 43 nm in size through transverse stomata and suggested that a pore size greater than 40 nm is required for uptake through this pathway. Further studies stated that physiological factors have a low impact on this mechanism, while environmental and circumstantial factors play greater roles (Lin et al., 2009; Punshon et al., 2003; Schönherr, 2000, 2006; Schreiber, 2005; Eichert 2008a,b; Popp, 2005).

9.2.1.3 *Through Endocytosis*

An alternate and significant transport pathway for ENPs is through endocytosis. With suitable surface chemistry, an ENP is capable of inducing endocytosis by activation of membrane receptors. Liu et al. (2009) have performed an experiment to investigate the uptake of ENPs by endocytosis in which *Nicotiana tabacum* L. bright yellow (BY-2) cells are exposed to single-walled carbon nanotubes labeled with fluorescein at 4 and 26°C. Through confocal fluorescence microscopy, it is observed that the uptake of ENPs at 26°C occurs efficiently, while at 4°C they are not taken up. Despite the increase in incubation time at 4°C, the uptake does not take place. This result may correspond to the fact that plant endocytosis drastically reduces at lower temperatures. However, at relatively higher temperatures, endocytosis proves to be an important mechanism for the uptake of ENPs by plants (Liu et al., 2009).

9.2.2 Transmission and Translocation

ENPs that move from the soil to the plant roots face various barriers imposed by the plant root physiology. ENPs apoplastically enter the plants by adsorbing to the root surfaces and passing through the root epidermis and root cortex. The cell walls of the endodermal cells present in the interior of the root cortex contain casparian bands, which are hydrophobic layers of suberin and lignin. ENPs move into the stele symplastically either by crossing the cell wall or plasma membrane of the endodermal or exodermal root, or cross the root cell wall at the exodermis or the exterior of the endodermis. To be transported to the shoot system and the leaves, the ENPs must be able to move without hindrance in the xylem after entering the symplast (Fig. 9.3) (Hose et al., 2001). However, multiple factors such as plant species, growth conditions, geometry and size of the ENPs, surface chemistry of the ENPs, and cultivars impact the translocation of NPs in plants. The metal and metal oxide NPs translocate to the aerial parts of the plant by accumulating at the roots. The amount of NPs required for this translocation depends on the species of the plant, exposure concentration, and the size and type of NPs along with their accumulation (Le et al., 2014; Rico et al., 2015; Zhou et al., 2011; Zhang et al., 2011; Larue et al., 2012a,b,c; Gorczyca et al., 2015; Song et al., 2013). Cifuentes et al. (2010) showed that in plants such as pea (*Pisum sativum* L.), sunflower (*Helianthus annuus* L.), tomato (*S. lycopersicum* L.), and wheat (*Triticum aestivum*), magnetic carbon-coated NPs entered the plants by penetrating the roots. The NPs then translocate through vasculature to reach the aerial parts via transpiration in the xylem. NPs that are applied via foliar spray are taken up by leaves and translocated to other parts of the plant. CaO NPs of 69.9 nm reach leaves and stems via phloem in groundnut plants as compared to bulk sources of sprayed Ca (calcium oxide and calcium nitrate). In two independent studies by Hong et al. (2014) and Larue et al. (2014), it has been shown that CeO_2 NPs in cucumber and TiO_2 NPs in lettuce are internalized and seen in different parts of the crop after their foliar application. Studies on rice and CuO NPs have shown that NPs entered the root stele via lateral roots and are then translocated to the leaves. Although these studies show that NPs are accumulated at the roots and then

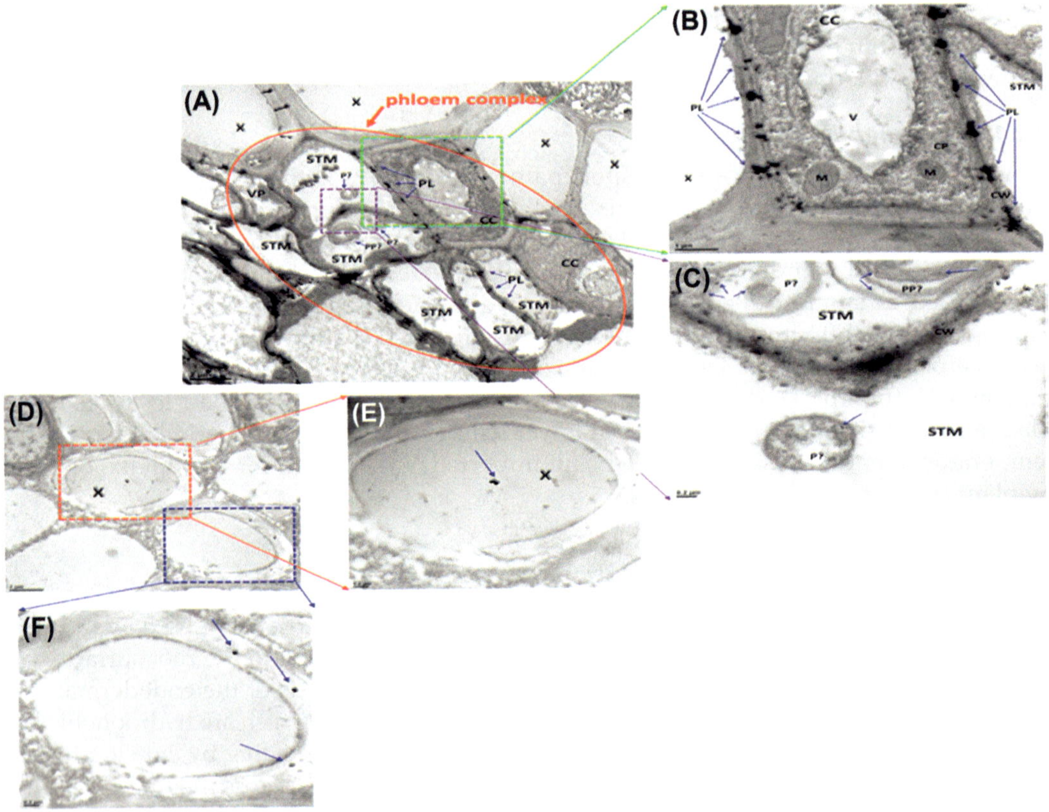

FIGURE 9.3 Gold nanoparticles (NPs) in roots (A–C) and leaves (D–F) of poplars exposed to 15 nm gold NPs on day 6. *CC*, companion cell; *CP*, cytoplasm; *CW*, cell wall; *M*, mitochondria; *P*, plastid; *PL*, plasmodesmata; *PP*, P-protein; *STM*, sieve tube member; *V*, vacuole; *VP*, vascular parenchyma cell; *X*, xylem. *Reprinted (adapted) with permission from Zhai, G., Walters, K.S., Peate, D.W., Alvarez, P.J., Schnoor, J.L., 2014. Transport of gold nanoparticles through plasmodesmata and precipitation of gold ions in woody poplar. Environ. Sci. Technol. Lett. 1, 146–151. Copyright (2014) American Chemical Society.*

translocated to aerial parts of plants there are still studies that show opposite results. An et al. (2008) demonstrated that increase in concentration of CeO_2 NPs showed increased accumulation of Ce at the roots. However, the concentration remains constant in leaves, grains, and hull, suggesting that translocation of Ce has not taken place. Wang et al. (2013) have reported that when maize is exposed to CuO NPs of size 20–40 nm, the xylem vessels serve to translocate the NPs from roots to parts of the shoot and the reverse translocation from the shoots to the roots occurs through phloem. Further studies on ZnO NPs by Wang et al. (2013) have shown that no upward translocation took place in cowpea when grown in both soil and media. In ryegrass, it is seen that ZnO NPs are accumulated at the roots and are also observed in the endodermis and stele of the roots, specifically in the apoplast and the protoplast (Cui et al., 2014; Foltête et al., 2011; Vannini et al., 2013).

Graphite-coated iron ENPs are injected into the pith cavity of the leaf of pumpkin plants to study the transport of ENPs in the xylem. It is found that ENPs of ~46 nm entered the xylem and could be seen at a distance from the injection site. This suggested that, perhaps, ENPs larger than 46nm cannot be transported through the xylem (Corredor et al., 2009). Other reports suggested the role of phloem for the loading and transport process of ENPs. One such study is performed by Wang et al. (2012) in which 20–40nm CuO ENPs are exposed to corn plants and their translocation is analyzed in the plant vascular system. These CuO ENPs are found in the xylem sap, and to confirm that these ENPs are indeed the ones used for treatment, electron diffraction is used. Furthermore, to analyze root exposures and phloem loading and transport, plant roots are exposed to a suspension of ENPs or placed in deionized water. The separate studies show that ENPs are present in the root. The work of Lin et al. (2009) served to bolster these claims where they show the translocation of ENPs to the embryo through phloem and demonstrate that it is the only possible pathway for the translocation to occur (Lin et al., 2009).

9.3 EFFECT OF GOLD NANOSTRUCTURES ON PLANTS

Although a range of ENP compositions have been studied to examine the effects of ENPs on plants, still few data are available on how the properties of gold nanomaterials affect plants. The bioavailability and fate of action of gold nanomaterials on plants is influenced by various factors inherent to NPs and plant species as shown in Fig. 9.4.

9.3.1 Impact of Surface Coating

The bioavailability of ENPs to plants is influenced by the surface chemistry of the ENP. Plant species consist of cation exchange sites in the cell wall of roots, wherein a positively charged ENP adsorbs on to the carboxylic group of the acids (galacturonic and glucuronic) present in the cell wall pores while apoplastically moving into the root (Lee et al., 2008; Zhu et al., 2008; Judy et al., 2011). The amount of these sites differs on the cell surfaces of plants and usually this number decreases with the decrease in pH. Similarly, anion exchange sites are also present because of the presence of organic cations found in the cell wall matrix. Together, these sites assist the transfer of nutrient ions across the plasma membrane. Both the anion exchange sites and the cation exchange sites are present at the bioaccumulation sites for this purpose (Larue et al., 2012a,b,c; Wang et al., 2012; Clarkson, 1988; McLaughlin, 2002; Allan and Jarrell, 1989). There is a possibility that these exchange sites could further facilitate the transport of ENPs but their role is quite unclear at the moment. To elucidate this, studies have been performed in rice, radish, pumpkin, and perennial ryegrass to examine the uptake of gold ENPs. It is observed that the roots take up positively charged ENPs at higher concentrations and translocate them to the shoot of the plants. However, the uptake is not seen for negatively charged NPs (Baluska et al., 2005; Arvizo et al., 2010). The specific impact of the intrinsic properties of NPs on their uptake via endocytosis is a compelling area for future research because of insufficient information available regarding this nondestructive uptake pathway. As mentioned previously, endocytosis can be stimulated by the activation of specific membrane receptors, which can be achieved through modification of the ENP's surface chemistry. In accordance with the foregoing studies (Haynes, 1980; Baker and Hall, 1975; Zhu et al., 2012), Nadine et al. (2015) studied the effect of citrate-capped gold NP uptake in barley (Fig. 9.5). Barley seeds are cultivated with gold NPs for 2weeks before transferring them to NP-free nutrient media for a further 3weeks. Transmission

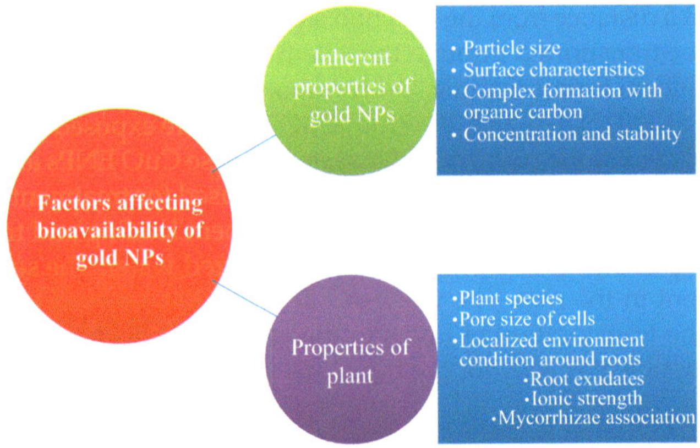

FIGURE 9.4 Factors affecting bioavailability of gold nanoparticles (NPs) to terrestrial plants.

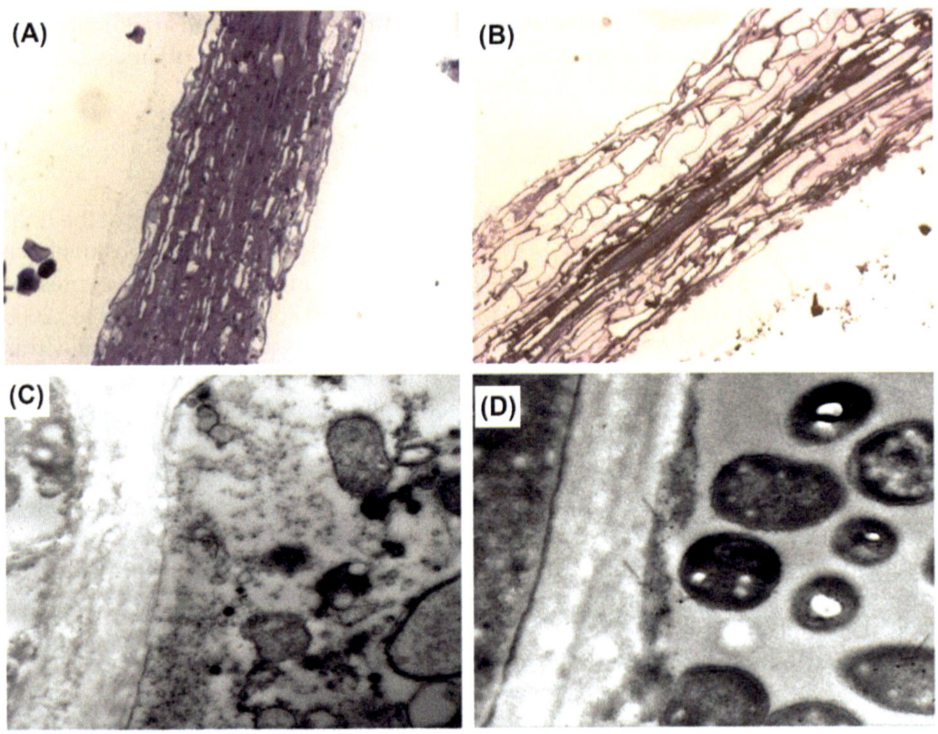

FIGURE 9.5 Transmission electron microscopy (TEM) investigations were performed to localize gold nanoparticles (NPs) and effects in root tips of barley plant. Light microscopic observation (magnification, 10 times) of longitudinal sections of barley primary root tips of (A) control plants and (B) plants exposed to 10 μg/mL gold NPs; TEM images of root cross-sections of (C) control plants and (D) plants exposed to 10 μg/mL gold NPs. Bacteria (*asterisk*) and gold NPs (*arrows*). *IM*, Inner matrix; *OM*, outer matrix; *TW*, thick wall. *Reprinted (adapted) with permission from Nadine, S.F., Walther, P., Leopold, K., 2015. Uptake, effects, and regeneration of barley plants exposed to gold nanoparticles. Environ. Sci. Pollut. Res. 22, 8549–8558. Copyright (2015) Springer.*

electron microscopy (TEM) and X-ray fluorescence analysis of leaves, roots, and other parts of plants show that bioaccumulation of gold NPs in roots is very high as compared to other plant parts. Zhu et al. (2012) used small gold NPs with different surface coatings to investigate their effect on uptake and biodistribution in terrestrial plants such as rice, ryegrass, pumpkin, and radish. The study shows that positively charged gold NPs are more prone to uptake by roots than negatively charged coated gold NPs. These are then translocated to plant shoots and leaves through plasmodesmata. Roots of ryegrass and radish accumulate more gold NPs than pumpkin and rice (Zhu et al., 2012). A study on *Brassica* sp. and *Medicago* sp. has shown that there is an increase in gold NP uptake with increase in exposure duration and substrate. Both of these species are considered gold NP hyperaccumulators. This accumulation is seen primarily in the vacuoles and nucleus of cells of plant roots (Bali and Harris, 2010). Few data suggest that nanomaterials can enter plant cells by binding to carrier protein, through endocytosis, ion channels, and aquaporins. Before transferring to plants, metal NPs also form complexes with root exudates and metal transporters (Tani and Barrington, 2005; Kurepa et al., 2010; Maine et al., 2001; Feretti et al., 2007).

Lia et al. (2016) studied the surface coating effect on plants by testing the effect of cysteine, cysteamine, and thioglycolic acid-coated gold NPs on rice (*Oryza sativa* L.) and tomato (*S. lycopersicum*). The study contradicts Zhu et al. (2012) and Koelmel et al. (2013) in terms of charge effect of gold NPs on root uptake by plants. It is demonstrated that negatively charged gold NPs are more prone to root uptake compared to positively charged NPs. The absorption of gold NPs has involved both clathrin-dependent and -independent mechanisms. Cysteine-coated gold NPs help in endocytosis of gold NPs into the plants. Gold NPs in their atomic form are accumulated in plants and disrupt their nutrient transport system. There are studies that show that gold NPs cause fluorescence quenching of chlorophylls, which is caused by increased electron movement from excited chlorophyll molecules to gold NPs.

9.3.2 Impact of Size

Adequate information is not available as to how the intrinsic properties of ENPs impact the bioavailability of ENPs for plant root uptake. However, properties such as particle size and electrophoretic mobility of ENPs have been hypothesized to influence their uptake by plants. Particle size is considered to be a major factor because of the assumption that ENPs primarily enter the cell through cell wall pores. Investigations of this assumption led to mixed results. By using scanning X-ray fluorescence microscopy (μXRF) mapping and TEM, Sabo-Attwood et al. (2011) demonstrated the uptake of smaller citrate-coated gold NPs (3.5 nm) in *N. tabacum*, while 18 nm gold NPs are not taken up. It shows pores have a size exclusion limit of 18 nm. However, another study on *N. tabacum* demonstrated the uptake of gold NPs of 10–50 nm by bulk elemental and spatially resolved laser ablation inductively coupled plasma mass spectrometry analysis and proved that there is no significant dependence of uptake on the particle size (Fig. 9.6) (Sabo-Attwood et al., 2011; Judy et al., 2012). The inadequacy of quantitative data in the former work can be the reason for contradicting reports from the two studies. In one such study, wheat plants are exposed to TiO_2 NPs ranging from 14 to 655 nm. Above 140 nm, uptake is not observed and ENPs above 36 nm do not translocate to aerial parts of the leaves (Larue et al., 2012a,b,c). The differential uptake of ENPs with different diameters, hence, is not only governed by the size of pores in cell walls but is also influenced by ENP aggregation. The presence of gold NPs of sizes ranging from 0.5 to 100 nm is also reported inside the plant parts where small, spherical gold NPs of 2–19 nm size, however, do not affect the germination of barley seeds but cause darkening of roots, yellowing of leaves, and decreased biomass (Judy et al., 2012).

Glenn et al. (2012) used three experimental plants, namely, *Azolla caroliniana*, *Egeria densa*, and *Myriophyllum simulans*, to study the bioavailability for gold nanomaterials. All these subjects are exposed to gold NPs of varying diameters of 4, 18, and 30 nm. It is seen that uptake of NPs is proportionately influenced by roots, organic carbon content, and NP size. Roots and NP size directly affect the NP uptake in a species-dependent manner, whereas organic carbon content indirectly affects the uptake by inducing nanoparticle aggregation. Nanoparticle–organic carbon complex formation is also shown to be particle size dependent, for example, 4 and 18 nm NPs show complex formation resulting in low NP absorption but 30 nm particles are not observed to complex with carbon. So it can be stated that both biotic and abiotic factors are responsible for forecasting bioavailability of nanomaterials to plants.

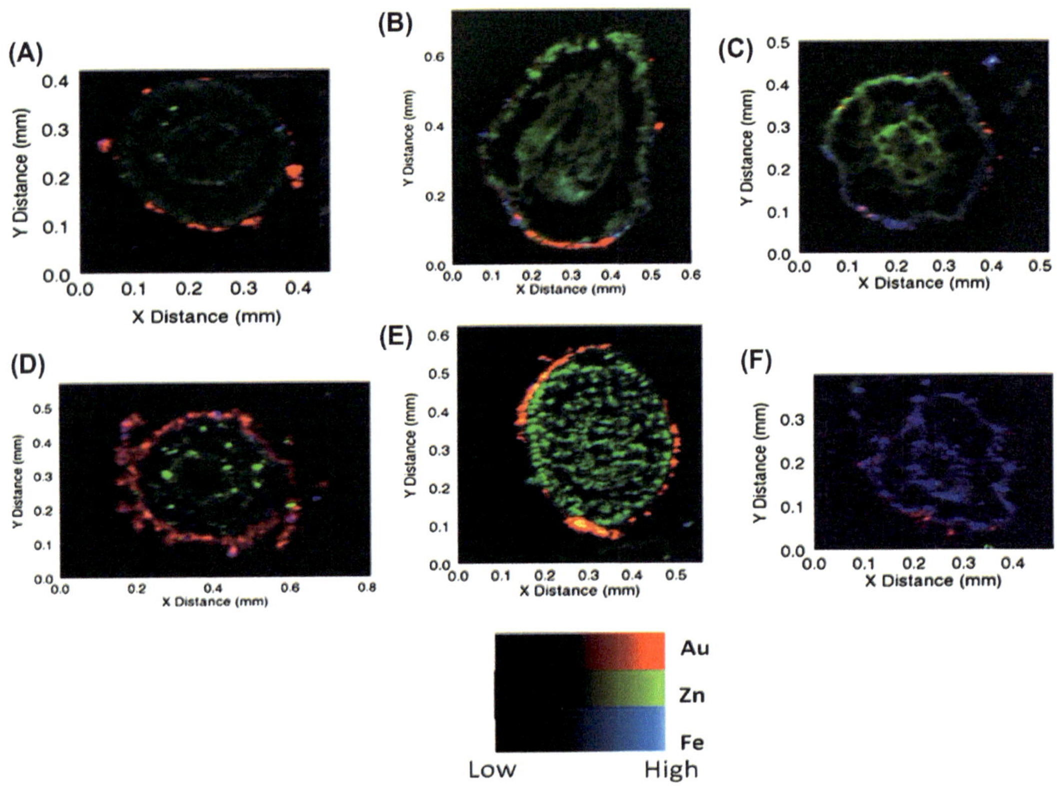

FIGURE 9.6 Scanning X-ray fluorescence microscopy (μXRF) image of fluorescence from the L-α1 edge of Au, depicted in red, the Zn K-β1 edge, depicted in green, and the K-α1 edge of Fe, depicted in blue, for wheat plants exposed to (A) 10-nm, (B) 30-nm, and (C) 50-nm citrate-coated manufactured Au nanomaterials (C-MNMS), and (D) 10-nm, (E) 30-nm, and (F) 50-nm tannate-coated Au MNMs (T-MNMs). *Reprinted (adapted) with permission from Judy, J.D., Unrine, J.M., Rao, W., Wirick, S., Bertsch, P.M., 2012. Bioavailability of gold nanomaterials to plants: importance of particle size and surface coating. Environ. Sci. Technol. 46 (15), 8467–8474. http://dx.doi.org/10.1021/es3019397. Copyright (2012) American Chemical Society.*

N. tabacum and *T. aestivum* were exposed to 10–50 nm diameter gold nanomaterials for 3 or 7 days. Using the laser ablation technique and μXRF, distribution of gold nanomaterials in plant tissue is seen. It is observed that tobacco responds to treatment while wheat plant is not affected by any treatment. This study also illustrates that particle size is an important parameter for discussing plant bioavailability of gold NPs; variable results are found for 10 and 50 nm particles. Bioaccumulation of 50 nm tannate-capped nanomaterials is considerably lower than that of 10 and 30 nm particles in tobacco. A controversial finding is thus seen here, because previously it was stated that nanomaterials pass through cell wall pores and cells reject particles larger than 20 nm (Sabo-Attwood et al., 2011). As it is a well-established fact that cell wall pores are in the order of 5–20 nm (Carpita, 1979), mechanism suggesting the mighty bypass is not well understood. These results point toward the fact that there are undiscovered mechanisms that lead to differential bioavailability of nanomaterials to plants and further their bioaccumulation, which is also dependent on plant species and independent of particle size (Judy et al., 2012).

9.4 TOXICITY ASSESSMENT OF GOLD NANOMATERIALS ON PLANTS

In terrestrial plants, ENPs may pose a threat to crop yield and food safety. Bioavailability and toxicity of ENPs is not very well understood because of inadequate data regarding dissolution during exposure and unacceptable ENP concentrations. This section presents the scenario of toxicity and its mechanism of ENPs to plants.

9.4.1 Toxicity

Crops such as wheat, maize, rice, and soybean have been shown to be adversely affected by metal NPs by retardation of their growth and physiology. To add to this, metal and metal oxide NPs have been shown to be highly toxic crops in comparison to bulk metal particles. NPs have the ability to be transported and translocated to different parts of the plant after being taken up. However, seed germination is decreased in a number of plants. The other toxic effects induced by NPs include decrease in biomass, growth, and grain yield. Additionally, many physiology-related disorders such as reduction in photosynthesis and gas exchange are also observed. ENPs also induce oxidative stress by decreasing the activity of antioxidant enzymes resulting in cytotoxicity and genotoxicity. Studies such as Hawthorne et al. (2012) have used Au, Si, and Cu ENPs on zucchini plants and investigated the effects of transpiration and biomass production. Cu and Si seem to have a negative effect on transpiration and biomass production because a decrease is observed, while Au ENPs do not show any adverse effect on plants. There is a sharp increase in toxicity when humic acid is added. In another such study, Shah and Belozerova (2009) exposed a range of ENPs to lettuce plants in two series of experiments. In the first series, ENPs are placed in the soil and lettuce seeds are planted after 15 days. In the second series, seeds are planted and then ENP treatment is given. Au, Si, Cu, 3-aminopropyl, and Pd ENPs within a matrix of aluminum hydroxide are used. On day zero exposure, the assay endpoints are not affected by any of these ENP treatments. In the tests of 15 days' incubation, low concentrations of Pd and Au exert a positive effect on the endpoint, while high concentrations of Cu and Si have the same effect. Bradfield et al. in 2016 revealed effects in the context of application of ZnO, CuO, and CeO_2 NPs upon experimentation with sweet potato. It is reported that Ce nanoparticles are accumulated in peel and flesh of the sweet potato tubers. In the context of ZnO NPs, about 70% of the NP content is retained in the tubers.

Many other studies have suggested the role of NPs in impairing cell division stages in root tips and also decreasing the mitotic index. It is also demonstrated that NPs modify the expression of certain genes that affect root growth. Other than these mechanisms of direct induction of toxicity, NPs also indirectly affect plants by altering soil microbial population, injuring plant roots, modifying the growth medium, and aiding plants in taking up cocontaminants. Hence NPs are both a boon and a bane to the crops and may enter the food web and cause unknown repercussions to all the organisms involved. To further understand the effects of metal and metal oxide NPs on crops, studies should be carried out to understand the mechanism of toxicity.

9.4.2 Mechanism

Toxicity as a result of ENP exposure could be the result of the action of one or a combination of several mechanisms. Traditionally, the acute ecotoxicity of some metals has been described by models such as the Biotic Ligand Model (BLM).57 The BLM describes metal toxicity as the accumulation of a critical concentration of free metal ion–biotic ligand complexes, where the biotic ligand is a generic site of action for metal ion binding (e.g., sodium and calcium channel proteins in a fish gill; carboxylic ligands associated with membrane surfaces), in the presence of competing ions and organic matter.57 Some ENPs, such as Ag, Cu, and ZnO, may exert toxic effects via dissolution and release of free metal ions within the soil pore water or the organism. Insoluble ENPs, such as Au, CeO_2, and TiO_2, are unlikely to induce toxicity as a result of free ion activity under most environmentally relevant exposure scenarios. However, other than the release of metal ions, insoluble ENPs also exert various other effects that may render them toxic. For example, the production of reactive oxygen species such as oxyradicals in an organism damage cellular components such as proteins, lipids, and DNA under the influence of UV and visible light, in addition to the presence of transition metals. Insoluble ENPs also mechanically disrupt the membranes and cell walls to induce toxicity. The fate, potential toxicity, bioavailability, and transport mechanism of these ENPs are influenced by a number of factors and hence testing all possible combinations of these factors is a tedious process. Particle BRM is a conceptual construct wherein toxicity is defined based on the interactions of free metal ions, pristine ENPs, as well as ENPs with a biotic receptor such as a cell membrane or membrane-bound macromolecule. ENPs pose a major risk to humans and terrestrial ecosystems based on their extent of bioavailability and toxicity to plants, which constitute the base of several food webs.

9.5 CONCLUSION AND FUTURE PROSPECTS

To summarize, NPs have an impact on the supply and growth of various important agricultural crops. The uptake of the NPs by plants depends on the physicochemical characteristics of NPs such as their size, type, surface coating, and the nature of interacting plant species. The chapter discussed the mechanisms of NP uptake and its translocation in terrestrial plants. Studies on gold NP interactions with terrestrial plants were discussed, particularly highlighting the impact of NP size and surface coating on their bioavailability. Table 9.1 summarizes

TABLE 9.1 Summary of Studies on Bioavailability of Gold Nanoparticles to Terrestrial Plants

Nanoparticles/Capping/Size	Plant Species	Uptake/Translocation	References
Gold NPs	*Arabidopsis thaliana*	Uptake and translocation	Kumar (2009)
	Tobacco	Uptake	Marschner (1995)
	Wheat	No uptake	
Gold NPs: DNA coated	Soybean		Christou et al. (1988)
Gold NPs: citrate capped	Barley	Uptake	Nadine et al. (2015)
Gold NPs: cysteamine capped	*Oryza sativa* L.	Uptake	Zhu et al. (2012)
Gold NPs: cysteine and cysteamine capped	*O. sativa* L.	No uptake	Lia et al. (2016)
Gold NPs: citrate capped 3.5 nm	*Nicotiana tabacum*	Uptake	Sabo-Attwood et al. (2011)
Gold NPs: citrate capped 18 nm	*N. tabacum*	No uptake	
Gold NPs: citrate capped 10–50 nm	*N. tabacum*	Uptake	Sabo-Attwood et al. (2011)
Gold NPs: 4–18 nm	*Azolla caroliniana, Egeria densa* and *Myriophyllum simulans*	Uptake	Glenn et al. (2012)
Gold NPs: 30 nm		No uptake	
TiO_2	*Spinacia oleracea*		Ma et al. (2008)
TiO_2: 14–36 nm	Wheat	Uptake and translocation	Larue et al. (2012a,b,c)
TiO_2: 36–140 nm	Wheat	Uptake	
TiO_2: 140–655 nm	Wheat	No uptake	
Fe_3O_4	Pumpkin	Uptake	Zhu et al. (2008)
	Vigna radiata	Uptake	
	Phaseolus lunatus	No uptake	
Fe_3O_4: carbon coated	*Pisum sativum* L.	Uptake and translocation	Cifuentes et al. (2010)
	Helianthus annuus L.	Uptake and translocation	
	Solanum lycopersicum L.	Uptake and translocation	
ZnO	Cowpea	Uptake	Wang et al. (2013)
Single-walled carbon nanotubes: fluorescein coated	*N. tabacum*	Uptake via endocytosis	Liu et al. (2009)
CeO_2	Cucumber	Uptake and translocation	Hong et al. (2014)
CuO	Maize	Uptake and translocation	Wang et al. (2013)

NPs, nanoparticles.

these studies on gold NPs and their interactions with plants. Toxicity of NPs to plants and their proposed mechanism was also described in the chapter.

These NPs primarily enter the plants through root uptake or foliar uptake. Uptake by plants of metal NPs such as gold NPs does not have energy-free transport through cell pores. Other than nanomaterial intrinsic properties, soil components as well as root exudates affect its transport into plants. However, much research, mostly at the molecular level, still needs to be done for a proper understanding of plant–NP interactions. There is relatively scarce literature on the understanding of the bioavailability of gold NPs in terrestrial plants. Further investigations are necessary to characterize the adsorption and uptake of NPs in different plant species in differing environmental conditions. In addition, experiments using soil with appropriate environmental conditions are yet to be performed on a real-time basis. NPs are also raising concern because of their ability to penetrate different parts of the plants through which they find a way to enter the food chain resulting in biomagnification of engineered nanomaterials. Therefore it has become necessary to have a mechanism to counter this that drives the need for further studies for understanding the interaction between nanomaterials and terrestrial flora.

Acknowledgment

The authors are grateful to the Department of Biotechnology, Motilal Nehru National Institute of Technology, Allahabad, and the Council for Scientific and Industrial Research (CSIR), India, for providing required facilities and support to accomplish this work.

References

Adani, F., Papa, G., Schievano, A., Cardinale, G., D'Imporzano, G., Tambone, F., 2011. Nanoscale structure of the cell wall protecting cellulose from enzyme attack. Environ. Sci. Technol. 45, 1107–1113.

Al-Halafi, A.M., 2014. Nanocarriers of nanotechnology in retinal diseases. Saudi J.Ophthalmol. 28, 204–309.

Alexis, D.O., Tyronne, M., Joseph, C., Indy, H., Barbara, H.H., 2009. Nanotoxicology: characterizing the scientific literature, 2000–2007. J. Nanopart. Res. 11, 251–257.

Allan, D.L., Jarrell, W.M., 1989. Proton and copper adsorption to maize and soybean root cell walls. Plant Physiol. 89, 823–832.

An, J., Zhang, M., Wang, S., Tang, J., 2008. Physical, chemical and microbiological changes in stored green asparagus spears as affected by coating of silver nanoparticles-PVP. LWT Food Sci. Technol. 41, 1100–1107.

Anjum, N.A., Gill, S.S., Duarte, A.C., Pereira, E., Ahmad, I., 2013. Silver nanoparticles in soil–plant systems. J. Nanopart. Res. 15, 1–26.

Antisari, L.V., Carbone, S., Gatti, A., Vianello, G., Nannipieri, P., 2014. Uptake and translocation of metals and nutrients in tomato grown in soil polluted with metal oxide (CeO_2, Fe_3O_4, SnO_2, TiO_2) or metallic (Ag, Co Ni) engineered nanoparticles. Environ. Sci. Pollut. Res. http://dx.doi.org/10.1007/s11356-014-3509-0.

Arora, S., Sharma, P., Kumar, S., Nayan, R., Khanna, P.K., Zaidi, M.G.H., 2012. Gold-nanoparticle induced enhancement in growth and seed yield of *Brassica juncea*. Plant Growth Regul. 66, 303–310.

Arvizo, R.R., Miranda, O.R., Thompson, M.A., Pabelick, C.M., Bhattacharya, R., Roberston, J.D., Rotello, V.M., Prakash, Y.S., Mukherjee, P., 2010. Effect of nanoparticle surface charge at the plasma membrane and beyond. Nano Lett. 10, 2543–2548.

Auffan, M., Rose, J., Bottero, J.Y., Lowry, G.V., Jolivet, J.P., Wiesner, M.R., 2009. Towards a definition of inorganic nanoparticles from an environmental, health and safety perspective. Nat. Nanotechnol. 4, 634–641.

Baker, D.A., Hall, J.L., 1975. In: Baker, D.A., Hall, J.L. (Eds.), Ion Transport in Plant Cells and Tissues. North Holland Publishing Co.

Bali, R., Harris, A.T., 2010. Biogenic synthesis of Au nanoparticles using vascular plants. Ind. Eng. Chem. Res. 49, 12762–12772.

Baluska, F., et al., 2005. In: Samaj, J., et al. (Ed.), Plant Endocytosis. Springer-Verlag.
Bandyopadhyay, S., Plascencia-Villa, G., Mukherjee, A., Rico, C.M., José-Yacamán, M., Peralta-Videa, J.R., Gardea-Torresdey, J.L., 2015. Comparative phytotoxicity of ZnO NPs bulk ZnO, and ionic zinc onto the alfalfa plants symbiotically associated with *Sinorhizobium meliloti* in soil. Sci. Total Environ. 515, 60–69.
Barrena, R., Casals, E., Colon, J., Font, X., Sanchez, A., Puntes, V., 2009. Evaluation of the ecotoxicity of model nanoparticles. Chemosphere 75, 850–857.
Bergeson, L.L., 2010. Nanosilver: US EPA's pesticide office considers how best to proceed. Environ. Qual. Manag. 19, 79–85.
Busbee, B.D., Obare, S.O., Murphy, C.J., 2003. An improved synthesis of high-aspect-ratio gold nanorods. Adv. Mater. 15, 414–416.
Carpita, N., 1979. Determination of the pore size of cell walls of living plant cells. Science 205 (4411), 1144–1147.
Caruso, F., Spasova, M., Salgueirino-Maceira, V., et al., 2001. Multilayer assemblies of silica-encapsulated gold nanoparticles on decomposable colloid templates. Adv. Mater. 13, 1090–1094.
Chandran, S.P., Chauhary, M., Pasricha, R., Ahmad, A., Sastry, M., 2006. Synthesis of gold nanoparticles and silver nanoparticles using *Aloe vera* plant extract. Biotechnol. Prog. 22, 577–583.
Chang, S.S., Shih, C.W., Chen, C.D., et al., 1999. The shape transition of gold nanorods. Langmuir 15, 701–709.
Chen, J., Saeki, F., Wiley, B.J., et al., 2005. Gold nanocages: bioconjugation and their potential use as optical imaging contrast agents. Nano Lett. 5, 473–477 (PubMed: 15755097).
Chen, J., McLellan, J.M., Siekkinen, A., et al., 2006. Facile synthesis of gold-silver nanocages with controllable pores on the surface. J Am Chem Soc 128, 14776–14777 (PubMed: 17105266).
Christopher, J.W., Tronndorf, R., Alicia, S.E., Maria, L.A., et al., 2014. Seed-mediated growth of gold nanorods: limits of length to diameter ratio control. J. Nanomater, pp 7.
Christou, P., McCabe, D.E., Swain, W.F., 1988. Stable transformation of soybean callus by DNAcoated gold particles. Plant. Physiol. 87, 671–674.
Cifuentes, Z., Custardoy, L., de la Fuente, J.M., Marquina, C., Ibarra, M.R., Rubiales, D., Pérez-de-Luque, A., 2010. Absorption and translocation to the aerial part of magnetic carbon-coated nanoparticles through the root of different crop plants. J. Nanobiotechnol. 8, 1–8.
Clarkson, 1988. In: Baker, D.A., Hall, J.L. (Eds.), Solute Transport in Plant Cells and Tissues. Longman, pp. 251–304.
Corredor, E., Testillano, P.S., Coronado, M.-J., González-Melendi, P., Fernández-Pacheco, R., Marquina, C., et al., 2009. Nanoparticle penetration and transport in living pumpkin plants: *in situ* subcellular identification. BMC Plant Biol. 9 (45).
Cossins, D., March 16, 2014. Next generation: nanoparticles augment plant functions. The incorporation of synthetic nanoparticles into plants can enhance photosynthesis and transform leaves into biochemical sensors. Sci. News Opin. http://www.the-scientist.com/?articles.view/articleNo/39440/title/Next-Generation–Nanoparticles-Augment-Plant-Functions/.
Cui, D., Zhang, P., Ma, Y., He, X., Li, Y., Zhang, J., et al., 2014. Effect of cerium oxide nanoparticles on asparagus lettuce cultured in an agar medium. Environ. Sci. Nano 1, 459–465.
DeRosa, M.C., Monreal, C., Schnitzer, M., Walsh, R., Sultan, Y., 2010. Nanotechnology in fertilizers. Nat. Nanotechnol. 5, 91.
Dutschk, V., Karapantsios, T., Liggieri, L., McMillan, N., Miller, R., Starov, V.M., 2014. Smart and green interfaces: from single bubbles/drops to industrial environmental and biomedical applications. Adv. Colloid Interface Sci. 209, 109–126.
Eichert, T., 2008a. Size exclusion limits and lateral heterogeneity of the stomatal foliar uptake pathway for aqueous solutes and water suspended nanoparticles. Physiol. Plant. 134, 151–160.
Eichert, T., 2008b. Equivalent pore radii of hydrophilic foliar uptake routes in stomatous and astomatous leaf surfaces-further evidence for a stomatal pathway. Physiol. Plant. 132, 491–502.
Feng, Y., Cui, X., He, S., Dong, G., Chen, M., Wang, J., Lin, X., 2013. The role of metal nanoparticles in influencing arbuscular mycorrhizal fungi effects on plant growth. Environ. Sci. Technol. 47, 9496–9504.
Feretti, D., Zerbini, I., Zani, C., Ceretti, E., Moretti, M., Monarca, S., 2007. *Allium cepa* chromosome aberration and micronucleus tests applied to study genotoxicity of extracts from pesticide-treated vegetables and grapes. Food Addit. Contam. 24 (6), 561–572.
Foltête, A.S., Masfaraud, J.F., Bigorgne, E., Nahmani, J., Chaurand, P., Botta, C., et al., 2011. Environmental impact of sunscreen nanomaterials: ecotoxicity and genotoxicity of altered TiO_2 nanocomposites on *Vicia faba*. Environ. Pollut. 159, 2515–2522.

Galbraith, D.W., 2007. Nanobiotechnology: silica breaks through in plants. Nat. Nanotechnol. 2, 272–273.
Garcia, M.E., Baker, L.A., Crooks, R.M., 1999. Preparation and characterization of dendrimer-gold colloid nanocomposites. Anal. Chem. 71, 256–258.
Gardea-Torresdey, J.L., Tiemann, K.J., Gamez, G., 2000. Reduction and accumulation of gold (III) by *Medicago sativa* alfalfa biomass: X-ray absorption spectroscopy, pH, and temperature dependence. Environ. Sci. Technol. 34, 4392–4396.
Giraldo, J.P., Landry, M.P., Faltermeier, S.M., McNicholas, T.P., Iverson, N.M., Boghossian, A.A., Reuel, N.F., Hilmer, A.J., Sen, F., Brew, J.A., Strano, M.S., 2014. Plant nanobionics approach to augment photosynthesis and biochemical sensing. Nat. Mater. 13, 400–408.
Glenn, J.B., White, S.A., Klaine, S.J., 2012. Interactions of gold nanoparticles with aquatic macrophytes are size and species dependent. Environ. Toxicol. Chem. 31, 194–201.
Gopinath, K., Gowri, S., Karthika, V., Arumugam, A., 2014. Green synthesis of gold nanoparticles from fruit extract of *Terminalia arjuna*, for the enhanced seed germination activity of *Gloriosa superba*. J. Nanostruct. Chem. 4, 1–11.
Gorczyca, A., Pociecha, E., Kasprowicz, M., Niemiec, M., 2015. Effect of nanosilver in wheat seedlings and *Fusarium culmorum* culture systems. Eur. J. Plant Pathol. 142 (2), 251–261.
Govorov, A.O., Carmeli, I., 2007. Hybrid structures composed of photosynthetic system and metal nanoparticles: plasmon enhancement effect. Nano Lett. 7, 620–625.
Grieger, K.D., Hansen, S.F., Baun, A., 2009. The known unknowns of nanomaterials: describing and characterizing uncertainty within environmental health and safety risks. Nanotoxicology 3, 222–233.
Hawthorne, J., Musante, C., Sinha, S.K., White, J.C., 2012. Accumulation and phytotoxicity of engineered nanoparticles to *Cucurbita pepo*. Int. J. Phytoremediation 14, 429–442.
Haynes, R.J., 1980. Ion exchange properties of roots and ionic interactions with the root apoplasm: their role ion accumulation by plants. Bot. Rev. 46 (1), 75–94.
Hernandez-Viezcas, J.A., Castillo-Michel, H., Andrews, J.C., et al., 2013. In situ synchrotron X-ray fluorescence mapping and speciation of CeO_2 and ZnO nanoparticles in soil cultivated soybean *Glycine max*. ACS Nano 7, 1415–1423.
Hiramatsu, H., Osterloh, F.E., 2004. A simple large-scale synthesis of nearly monodisperse gold and silver nanoparticles with adjustable sizes and with exchangeable surfactants. Chem. Mater. 16, 2509–2511.
Hong, J., Peralta-Videa, J.R., Rico, C., et al., 2014. Evidence of translocation and physiological impacts of foliar applied CeO_2 nanoparticles on cucumber *Cucumis sativus* plants. Environ. Sci. Technol. 48, 4376–4385.
Hose, E., Clarkson, D.T., Steudle, E., Schreiber, L., Hartung, W., 2001. The exodermis: a variable apoplastic barrier. J. Exp. Bot. 52 (365), 2245–2264.
Jana, N.R., Gearheart, L., Murphy, C.J., 2001. Wet chemical synthesis of high aspect ratio cylindrical gold nanorods. J. Phys. Chem. B 105, 4065–4067.
Judy, J.D., Bertsch, P.M., 2014. Bioavailability, toxicity and fate of manufactured nanomaterials in terrestrial ecosystems. In: Sparks, D. (Ed.). Sparks, D. (Ed.), Advances in Agronomy, 123. Elsevier, Amsterdam, Netherlands, pp. 1–64.
Judy, J.D., Unrine, J.M., Bertsch, P.M., 2011. Evidence for biomagnification of gold nanoparticles within a terrestrial food chain. Environ. Sci. Technol. 45 (2), 776–781.
Judy, J.D., Unrine, J.M., Rao, W., Wirick, S., Bertsch, P.M., 2012. Bioavailability of gold nanomaterials to plants: importance of particle size and surface coating. Environ. Sci. Technol. 46 (15), 8467–8474. http://dx.doi.org/10.1021/es3019397.
Judy, J.D., Kirby, J.K., Creamer, C., McLaughlin, M.J., Fiebiger, C., Wright, C., Cavagnaro, T., Bertsch, P.M., 2015. Effects of silver sulfide nanomaterials on mycorrhizal colonization of tomato plants and soil microbial communities in biosolid-amended soil. Environ. Pollut. 206, 256–263.
Kah, M., Beulke, S., Tiede, K., Hofmann, T., 2013. Nanopesticides: state of knowledge environmental fate, and exposure modeling. Crit. Rev. Environ. Sci. Technol. 43, 1823–1867.
Keller, A., McFerran, S., Lazareva, A., Suh, S., 2013. Global life cycle releases of engineered nanomaterials. J. Nanopart. Res. 15, 1–17.
Kim, Y.G., Oh, S.K., Crooks, R.M., 2004. Preparation and characterization of 1–2 nm dendrimer-encapsulated gold nanoparticles having very narrow size distributions. Chem. Mater. 16, 167–172.
Koelmel, J., Leland, T., Wang, H., Amarasiriwardena, D., Xing, B., 2013. Investigation of gold nanoparticles uptake and their tissue level distribution in rice plants by laser ablation-inductively coupled-mass spectrometry. Environ. Pollut. 174, 222–228.

Kouhi, S.M.M., Lahouti, M., Ganjeali, A., Entezari, M.H., 2014. Comparative phytotoxicity of ZnO nanoparticles ZnO microparticles, and Zn^{2+} on rapeseed *Brassica napus* L.: investigating a wide range of concentrations. Toxicol. Environ. Chem. 96, 861–868.

Kumar, C.S.S.R. (Ed.), 2009. Nanomaterials for the Life Sciences Mixed Metal Nanomaterials, 3. Wiley-VCH Verlag GmbH & Co. KGaA, Weinheim, pp. 1–44.

Kurepa, J., Paunesku, T., Vogt, S., Arora, H., Rabatic, B.M., Lu, J., Wanzer, M.B., Woloschak, G.E., Smalle, J.A., 2010. Uptake and distribution of ultrasmall anatase TiO_2 alizarin red S nanoconjugates in arabidopsis thaliana. Nano Lett. 10 (7), 2296–2302. http://dx.doi.org/10.1021/nl903518f.

Lalau, C.M., Mohedano, R.D.A., Schmidt, E.C., et al., 2015. Toxicological effects of copper oxide nanoparticles on the growth rate photosynthetic pigment content, and cell morphology of the duckweed *Landoltia punctata*. Protoplasma 252, 221–229.

Larue, C., Laurette, J., Herlin-Boime, N., 2012a. Accumulation, translocation and impact of TiO_2 nanoparticles in wheat *Triticum aestivum* spp.: influence of diameter and crystal phase. Sci. Total Environ. 431, 197–208.

Larue, C., Pinault, M., Czarny, B., Georgin, D., Jaillarde, D., Bendiab, N., Mayn, M., Taran, F., Dive, V., Carrière, M., 2012b. Quantitative evaluation of multi-walled carbon nanotube uptake in wheat and rapeseed. J. Hazard. Mater. http://dx.doi.org/10.1016/j.jhazmat.2012.05.033.

Larue, C., Veronesi, G., Flank, A.M., Surble, S., Herlin-Boime, N., Carrière, M., 2012c. Comparative uptake and impact of TiO_2 nanoparticles in wheat and rapeseed. J. Toxicol. Environ. Health. Part A 75, 722–734.

Larue, C., Castillo-Michel, H., Sobanska, S., Trcera, N., 2014. Fate of pristine TiO_2 nanoparticles and aged paint-containing TiO_2 nanoparticles in lettuce crop after foliar exposure. J. Hazard. Mater. 273, 17–26.

Oldenburg, S.J., Averitt, R.D., Westcott, S.L., et al., 1998. Nanoengineering of optical resonances. Chem. Phys. Lett. 288, 243–247.

Le, V.N., Rui, Y., Gui, X., Li, X., Liu, S., Han, Y., 2014. Uptake, transport, distribution and Bio-effects of SiO_2 nanoparticles in Bt-transgenic cotton. J. Nanobiotechnol. 12, 50.

Lee, W., An, Y.J., Yoon, H., Kweon, H.S., 2008. Toxicity and bioavailability of copper nanoparticles to the terrestrial plants mung bean (*Phaseolus radiatus*) and wheat (*Triticum aestivum*): plant agar test for water-insoluble nanoparticles. Environ. Toxicol. Chem. 27 (9), 1915–1921.

Leff, D.V., Brandt, L., Heath, J.R., 1996. Synthesis and characterization of hydrophobic, organically soluble gold nanocrystals functionalized with primary amines. Langmuir 12, 4723–4730.

Li, X., Gui, X., Rui, Y., Ji, W., Yu, Z., Peng, S., 2014. Bt-transgenic cotton is more sensitive to CeO_2 nanoparticles than its parental non-transgenic cotton. J. Hazard. Mater. 274, 173–180.

Lia, H., Yea, X., Guob, X., Genga, Z., Wanga, G., 2016. Effects of surface ligands on the uptake and transport of gold nanoparticles in rice and tomato. J. Hazard. Mater. 314, 188–196.

Lin, D., Xing, B., 2008. Root uptake and phytotoxicity of ZnO nanoparticles. Environ. Sci. Technol. 42, 5580–5585.

Lin, S., Reppert, J., Hu, Q., Hudson, J.S., Reid, M.L., Ratnikova, T.A., Rao, A.M., Luo, H., Ke, P.C., 2009. Uptake, translocation, and transmission of carbon nanomaterials in rice plants. Small 5 (10), 1128–1132.

Liu, R., Lal, R., 2015. Potentials of engineered nanoparticles as fertilizers for increasing agronomic productions. Sci. Total Environ. 514, 131–139.

Liu, Q., Chen, B., Wang, Q., Shi, X., Xiao, Z., Lin, J., Fang, X., 2009. Carbon nanotubes as molecular transporters for walled plant cells. Nano Lett. 9 (3), 1007–1010.

Maine, M.A., Duarte, M.V., Sune, N.L., 2001. Cadmium uptake by floating macrophytes. Water Res. 35 (11), 2629–2634.

Manna, A., Imae, T., Aoi, K., et al., 2001. Synthesis of dendrimer-passivated noble metal nanoparticles in a polar medium: comparison of size between silver and gold particles. Chem. Mater. 13, 1674–1681.

Marschner, H., 1995. Mineral Nutrition of Higher Plants. Academic Press, San Diego, CA.

Martin, C.R., 1994. Nanomaterials: a membrane-based synthetic approach. Science 266, 1961–1966 (PubMed: 17836514).

McLaughlin, M., 2002. In: Allen, H. (Ed.), Bioavailability of Metals in Terrestrial Ecosystems. Society of Environmental Toxicology and Chemistry (SETAC), pp. 39–68.

Miralles, P., Church, T.L., Harris, A.T., 2012. Toxicity, uptake, and translocation of engineered nanomaterials in vascular plants. Environ. Sci. Technol. 46 (17), 9224–9239.

Mohammadi, R., Maali-Amiri, R., Mantri, N.L., 2014. Effect of TiO_2 nanoparticles on oxidative damage and antioxidant defense systems in chickpea seedlings during cold stress. Rus. J. Plant Physiol. 61, 768–775.

Moon, Y.S., Park, E.S., Kim, T.O., Lee, H.S., Lee, S.E., 2014. SELDI-TOF MS-based discovery of a biomarker in *Cucumis sativus* seeds exposed to CuO nanoparticles. Environ. Toxicol. Pharmacol. 38, 922–931.

Nadine, S.F., Walther, P., Leopold, K., 2015. Uptake, effects, and regeneration of barley plants exposed to gold nanoparticles. Environ. Sci. Pollut. Res. 22, 8549–8558.
Nair, P.M.G., Chung, I.M., 2014. A mechanistic study on the toxic effect of copper oxide nanoparticles in soybean (*Glycine max* L.) root development and lignification of root cells. Biol. Trace Elem. Res. 162, 342–352. https://www.ncbi.nlm.nih.gov/pubmed/25190470.
Oldenburg, S.J., Westcott, S.L., Averitt, R.D., et al., 1999. Surface enhanced Raman scattering in the near infrared using metal nanoshell substrates. J. Chem. Phys. 111, 4729–4735.
Park, B., Donaldson, K., Duffin, R., Tran, L., Kelly, F., Mudway, I., et al., 2008. Hazard and risk assessment of a nanoparticulate cerium oxide-based diesel fuel additive-a case study. Inhal. Toxicol. 20, 547–566.
Peng, C., Duan, D., Xu, C., Chen, Y., Sun, L., Zhang, H., et al., 2015. Translocation and biotransformation of CuO nanoparticles in rice (*Oryza sativa* L.) plants. Environ. Pollut. 197, 99–107.
Perrault, S.D., Chan, W.C.W., 2010. In vivo assembly of nanoparticle components to improve targeted cancer imaging. Proc. Natl. Acad. Sci. 107 (25), 11194–11199.
Popp, C., et al., 2005. Characterization of hydrophilic and lipophilic pathways of *Hedera helix* L. cuticular membranes: permeation of water and uncharged organic compounds. J. Exp. Bot. 56 (421), 2797–2806.
Punshon, T., Gaines, K.G., Bertsch, P.M., Burger, J., 2003. Bioavailability of uranium and nickel to vegetation in a contaminated riparian ecosystem. Environ. Toxicol. Chem. 22 (5), 1146–1154.
Rajapaksha, A.U., Ahmad, M., Vithanage, M., Kim, K.R., Chang, J.Y., Lee, S.S., Ok, Y.S., 2015. The role of biochar, natural iron oxides, and nanomaterials as soil amendments for immobilizing metals in shooting range soil. Environ. Geochem. Health. http://org/10.1007/s10653-015-9694-z.
Rico, C.M., Majumdar, S., Duarte-Gardea, M., Peralta-Videa, J.R., Gardea-Torresdey, J.L., 2011. Interaction of nanoparticles with edible plants and their possible implications in the food chain. J. Agric. Food Chem. 59 (8), 3485–3498.
Rico, C.M., Lee, S.C., Rubenecia, R., Mukherjee, A., Hong, J., Peralta-Videa, J.R., Gardea-Torresdey, J.L., 2014. Cerium oxide nanoparticles impact yield and modify nutritional parameters in wheat (*Triticum aestivum* L.). J. Agric. Food Chem. 62, 9669–9675.
Rico, C.M., Barrios, A.C., Tan, W., Rubenecia, R., Lee, S.C., Varela-Ramirez, A., 2015. Physiological and biochemical response of soil-grown barley (*Hordeumvulgare* L.) to cerium oxide nanoparticles. Environ. Sci. Pollut. Res. http://dx.doi.org/10.1007/s11356-015-4243-y.
Roghayyeh, S.M.S., Mehdi, T.S., Rauf, S.S., 2010. Effects of nano-iron oxide particles on agronomic traits of soybean. Not. Sci. Biol. 2, 112–113.
Sabo-Attwood, T., Unrine, J.M., Stone, J.W., Murphy, C.J., Ghoshroy, S., Blom, D., Bertsch, P.M., Newman, L.A., 2011. Uptake, distribution and toxicity of gold nanoparticles in tobacco (*Nicotiana xanthi*) seedlings. Nanotoxicology 6, 353–360.
Safiuddin, M., Gonzalez, M., Cao, J.W., Tighe, S.L., 2014. State of-the-art report on use of nano materials in concrete. Int. J. Pavement Eng. 15, 940–949.
Salisu, I.B., Abubakar, A.S., Sa'id, M.A., 2014. A novel biosynthesis, characterization and antimicrobial activity of silver nanoparticles using leaves extract of *Aloe vera* plant. Int. J. Sci. Res. 3 (6), 311–314 ISSN (Online): 2319-7064.
Savithramma, N., Ankanna, S., Bhumi, G., 2012. Effect of nanoparticles on seed germination and seedling growth of *Boswellia ovalifoliolata* an endemic and endangered medicinal tree taxon. Nano Vis. 2, 61–68.
Schönherr, J., 2000. Calcium chloride penetrates plant cuticles via aqueous pores. Planta 212 (1), 112–118.
Schönherr, J., 2006. Characterization of aqueous pores in plant cuticles and permeation of ionic solutes. J. Exp. Bot. 57 (11), 2471–2491.
Schreiber, L., 2005. Polar paths of diffusion across plant cuticles: new evidence for an old hypothesis. Ann. Bot. 95, 1069–1073.
Shah, V., Belozerova, I., 2009. Influence of metal nanoparticles on the soil microbial community and germination of lettuce seeds. Water Air Soil Pollut. 197, 143–148.
Shi, J., Peng, C., Yang, Y., Yang, J., et al., 2014. Phytotoxicity and accumulation of copper oxide nanoparticles to the Cu-tolerant plant *Elsholtzia splendens*. Nanotoxicology 8, 179–188.
Shweta, Tripathi, D.K., Singh, S., Singh, S., Dubey, N.K., Chauhan, D.K., 2016. Impact of nanoparticles on photosynthesis: challenges and opportunities. Mater. Focus 5 (5), 405–411.
Singh, S., Tripathi, D.K., Dubey, N.K., Chauhan, D.K., 2016. Effects of nano-materials on seed germination and seedling growth: striking the slight balance between the concepts and controversies. Mater. Focus 5 (3), 195–201.

Singh, S., Vishwakarma, K., Singh, S., Sharma, S., Dubey, N.K., Singh, V.K., Liu, S., Tripathi, D.K., Chauhan, D.K., 2017. Understanding the plant and nanoparticle interface at transcriptomic and proteomic level: a concentric overview. Plant Gene. http://dx.doi.org/10.1016/j.plgene.2017.03.006.

Slomberg, S.L., Schoenfisch, M.H., 2012. Silica nanoparticle phytotoxicity to *Arabidopsis thaliana*. Environ. Sci. Technol. 46, 10247–10254.

Song, U., Jun, H., Waldman, B., Roh, J., Kim, Y., Yi, J., Lee, E.J., 2013. Functional analyses of nanoparticle toxicity: a comparative study of the effects of TiO_2 and Ag on tomatoes *Lycopersicon esculentum*. Ecotoxicol. Environ. Saf. 93, 60–67.

Suriyaprabha, R., Karunakaran, G., Yuvakkumar, R., Prabu, P., Rajendran, V., Kannan, N., 2012. Growth and physiological responses of maize (*Zea mays* L.) to porous silica nanoparticles in soil. J. Nanopart. Res. 14, 1–14.

Tani, F.H., Barrington, S., 2005. Zinc and copper uptake by plants under two transpiration rates. Part I. Wheat (*Triticumaestivum* L.). Environ. Pollut. 138 (3), 538–547.

Thiruvengadam, M., Gurunathan, S., Chung, I.M., 2014. Physiological, metabolic, and transcriptional effects of biologically-synthesized silver nanoparticles in turnip (*Brassica rapa* ssp. *rapa* L.). Protoplasma. http://dx.doi.org/10.1007/s00709-014-0738-5.

Tripathi, D.K., Singh, V.P., Prasad, S.M., Chauhan, D.K., Dubey, N.K., 2015. Silicon nanoparticles (SiNp) alleviate chromium (VI) phytotoxicity in *Pisum sativum* (L.) seedlings. Plant Physiol. Biochem. 96, 189–198.

Tripathi, D.K., Singh, S., Singh, V.P., Prasad, S.M., Chauhan, D.K., Dubey, N.K., 2016. Silicon nanoparticles more efficiently alleviate arsenate toxicity than silicon in maize cultiver and hybrid differing in arsenate tolerance. Front. Environ. Sci. 4, 46.

Tripathi, D.K., Singh, S., Singh, S., Pandey, R., Singh, V.P., Sharma, N.C., Prasad, S.M., Dubey, N.K., Chauhan, D.K., 2017a. An overview on manufactured nanoparticles in plants: uptake, translocation, accumulation and phytotoxicity. Plant Physiol. Biochem. 110, 2–12.

Tripathi, D.K., Mishra, R.K., Singh, S., Singh, S., Vishwakarma, K., Sharma, S., Singh, V.P., Singh, P.K., Prasad, S.M., Dubey, N.K., Pandey, A.C., 2017b. Nitric oxide ameliorates zinc oxide nanoparticles phytotoxicity in wheat seedlings: implication of the ascorbate–glutathione cycle. Front. Plant Sci. 8.

Tripathi, D.K., Singh, S., Singh, S., Srivastava, P.K., Singh, V.P., Singh, S., Prasad, S.M., Singh, P.K., Dubey, N.K., Pandey, A.C., Chauhan, D.K., 2017c. Nitric oxide alleviates silver nanoparticles (AgNps)-induced phytotoxicity in *Pisum sativum* seedlings. Plant Physiol. Biochem. 110, 167–177.

Tripathi, D.K., Tripathi, A., Shweta, S.S., Singh, Y., Vishwakarma, K., Yadav, G., Sharma, S., Singh, V.K., Mishra, R.K., Upadhyay, R.G., Dubey, N.K., 2017d. Uptake, accumulation and toxicity of silver nanoparticle in autotrophic plants, and heterotrophic microbes: a concentric review. Front. Microbiol. 8.

Tripathi, D.K., Singh, S., Singh, V.P., Prasad, S.M., Dubey, N.K., Chauhan, D.K., 2017e. Silicon nanoparticles more effectively alleviated UV-B stress than silicon in wheat (*Triticum aestivum*) seedlings. Plant Physiol. Biochem. 110, 70–81.

Turkevich, J., Stevenson, P.C., Hillier, J., 1951. A study of the nucleation and growth processes in the synthesis of colloidal gold. Discuss. Faraday Soc. 11, 55–75.

Van der Zande, B.M.I., Boehmer, M.R., Fokkink, L.G.J., et al., 1997. Aqueous gold sols and rod-shaped particles. J. Phys. Chem. B 101, 852–854.

Vannini, C., Domingo, G., Onelli, E., Prinsi, B., Marsoni, M., Espen, L., Bracale, M., 2013. Morphological and proteomic responses of *Eruca sativa* exposed to silver nanoparticles or silver nitrate. PLoS One 8, e68752.

Wang, Q., Ma, X., Zhang, W., Pei, H., Chen, Y., 2012. The impact of cerium oxide nanoparticles on tomato *Solanum lycopersicum* L. and its implications for food safety. Metallomics 4, 1105–1112.

Wang, P., Menzies, N.W., Lombi, E., McKenna, B.A., Johannessen, B., Glover, C.J., et al., 2013. Fate of ZnO nanoparticles in soils and cowpea *Vigna unguiculata*. Environ. Sci. Technol. 47, 13822–13830.

Watts-Williams, S.J., Turney, T.W., Patti, A.F., Cavagnaro, T.R., 2014. Uptake of zinc and phosphorus by plants is affected by zinc fertiliser material and arbuscular mycorrhizas. Plant Soil 376, 165–175.

Weare, W.W., Reed, S.M., Warner, M.G., et al., 2000. Improved synthesis of small (d_{CORE} ~ 1.5 nm) phosphine-stabilized gold nanoparticles. J. Am. Chem. Soc. 122, 12890–12891.

Whiteside, M.D., Treseder, K.K., Atsatt, P.R., 2009. The brighter side of soils: quantum dots track organic nitrogen through fungi and plants. Ecology 90 (1), 100–108.

Xiang, L., Zhao, H.M., Li, Y.W., Huang, X.P., Wu, X.L., Zhai, T., et al., 2015. Effects of the size and morphology of zinc oxide nanoparticles on the germination of Chinese cabbage seeds. Environ. Sci. Pollut. Res. http://dx.doi.org/10.1007/s11356-015-4172-9.

Yu, Y.Y., Chang, S.S., Lee, C.L., et al., 1997. Gold nanorods: electrochemical synthesis and optical properties. J. Phys. Chem. B 101, 6661–6664.

Zhai, G., Walters, K.S., Peate, D.W., Alvarez, P.J., Schnoor, J.L., 2014. Transport of gold nanoparticles through plasmodesmata and precipitation of gold ions in woody poplar. Environ. Sci. Technol. Lett. 1, 146–151.
Zhang, Z., He, X., Zhang, H., Ma, Y., Zhang, P., Ding, Y., Zhao, Y., 2011. Uptake and distribution of ceria nanoparticles in cucumber plants. Metallomics 3, 816–822.
Zhang, P., Ma, Y., Zhang, Z., He, X., Guo, Z., Tai, R., et al., 2012a. Comparative toxicity of nanoparticulate/bulk Yb_2O_3 and $YbCl_3$ to cucumber *Cucumis sativus*. Environ. Sci. Technol. 46, 1834–1841.
Zhang, P., Ma, Y., Zhang, Z., He, X., Zhang, J., Guo, Z., 2012b. Biotransformation of ceria nanoparticles in cucumber plants. ACS Nano 6, 9943–9950.
Zhang, R., Zhang, H., Tu, C., Hu, X., Li, L., Luo, Y., Christie, P., 2015. Phytotoxicity of ZnO nanoparticles and the released Zn (II) ion to corn (*Zea mays* L.) and cucumber (*Cucumis sativus* L.) during germination. Environ. Sci. Pollut. Res. http://dx.doi.org/10.1007/s11356-015-4325-x.
Zhao, L., Peng, B., Hernandez-Viezcas, J.A., Rico, C., Sun, Y., Peralta-Videa, J.R., et al., 2012a. Stress response and tolerance of *Zea mays* to CeO_2 nanoparticles: cross talk among H_2O_2, heat shock protein, and lipid peroxidation. ACS Nano 6, 9615–9622.
Zhao, L., Peralta-Videa, J.R., Ren, M., Varela-Ramirez, A., Li, C., Hernandez-Viezcas, J.A., et al., 2012b. Transport of Zn in a sandy loam soil treated with ZnO NPs and uptake by corn plants: electron microprobe and confocal microscopy studies. Chem. Eng. J. 184, 1–8.
Zhao, L., Hernandez-Viezcas, J.A., Peralta-Videa, J.R., Bandyopadhyay, S., Peng, B., Munoz, B., 2013. ZnO nanoparticle fate in soil and zinc bioaccumulation in corn plants *Zea mays* influenced by alginate. Environ. Sci. 15, 260–266.
Zhou, D., Jin, S., Li, L., Wang, Y., Weng, N., 2011. Quantifying the adsorption and uptake of CuO nanoparticles by wheat root based on chemical extractions. J. Environ. Sci. 23 (11), 1852–1857.
Zhou, X., Torabi, M., Lu, J., Shen, R., Zhang, K., 2015. Nanostructured energetic composites: synthesis ignition/combustion modeling, and applications. ACS Appl. Mater. Interfaces 6, 3058–3074.
Zhu, H., Han, J., Xiao, J., Jin, Y., 2008. Uptake, translocation, and accumulation of manufactured iron oxide nanoparticles by pumpkin plants. J. Environ. Monit. 10, 713–717.
Zhu, Z., Wang, H., Yan, B., Zheng, H., Jiang, Y., Miranda, O.R., Rotello, V.M., Xing, B., Vachet, R.W., 2012. Effect of surface charge on the uptake and distribution of gold nanoparticles in four plant species. Environ. Sci. Technol. 46 (22), 12391–12398.

Further Reading

Anshup, A., Venkataraman, J.S., Subramaniam, C., et al., 2005. Growth of gold nanoparticles in human cells. Langmuir 21, 11562–11567 (PubMed: 16316080).
Atha, D.H., Wang, H., Petersen, E.J., Cleveland, D., Holbrook, R.D., Jaruga, P., et al., 2012. Copper oxide nanoparticle mediated DNA damage in terrestrial plant models. Environ. Sci. Technol. 46, 1819–1827.
Birbaium, K., Brogioli, R., Schellenberg, M., Stark, W., Gunther, D., Limbach, L., 2010. No evidence for cerium dioxide nanoparticle translocation in maize plants. Environ. Sci. Technol. 44, 8718–8723.
Brust, M., Walker, M., Bethell, D., et al., 1994. Synthesis of thiol-derivatized gold nanoparticles in a two-phase liquid-liquid system. J. Chem. Soc. Chem. Commun. 801–802.
Cao, Y.C., Jin, R., Mirkin, C.A., 2002. Nanoparticles with Raman spectroscopic fingerprints for DNA and RNA detection. Science 297, 1536–1540 (PubMed: 12202825).
Chithrani, B.D., Ghazani, A.A., Chan, W.C.W., 2006. Determining the size and shape dependence of gold nanoparticle uptake into mammalian cells. Nano Lett. 6 (4).
Deepa, M., Sudhakar, P., Nagamadhuri, K.V., Reddy, K.B., Krishna, T.G., Prasad, T.N., 2014. First evidence on phloem transport of nanoscale calcium oxide in groundnut using solution culture technique. Appl. Nanosci. 5, 545–551.
Dimkpa, C.O., McLean, J.E., Latta, D.E., Manangon, E., Britt, D.W., Johnson, W.P., et al., 2012. CuO and ZnO nanoparticles: phytotoxicity, metal speciation, and induction of oxidative stress in sand-grown wheat. J. Nanopart. Res. 14, 1–15. http://dx.doi.org/10.1007/s11051-012-1125-9.
Dimkpa, C.O., Hansen, T., Stewart, J., McLean, J.E., Britt, D.W., Anderson, A.J., 2014. ZnO nanoparticles and root colonization by a beneficial pseudomonad influence essential metal responses in bean (*Phaseolus vulgaris*). Nanotoxicology. http://dx.doi.org/10.3109/17435390.2014.900583.
Doshi, R., Braida, W., Christodoulatos, C., Wazne, M., O'Connor, G., 2008. Nano-aluminum: transport through sand columns and environmental effects on plants and soil communities. Environ. Res. 106, 296–303.

Du, W., Sun, Y., Ji, R., Zhu, J., Wu, J., Guo, H., 2011. TiO_2 and ZnO nanoparticles negatively affect wheat growth and soil enzyme activities in agricultural soil. J. Environ. Monit. 13, 822–828.

Ge, Y., Priester, J.H., Van De Werfhorst, L.C., Walker, S.L., et al., 2014. Soybean plants modify metal oxide nanoparticle effects on soil bacterial communities. Environ. Sci. Technol. 48, 13489–13496.

Kim, B., Park, C.S., Murayama, M., Hochella, M.F., 2010. Discovery and characterization of silver sulfide nanoparticles in finial sewage sludge products. Environ. Sci. Technol. 44, 7509–7514.

Kiser, M.A., Westerhoff, P., Benn, T., Wang, Y., Perez-Rivera, J., Hristovski, K., 2009. Titanium nanomaterial removal and release from wastewater treatment plants. Environ. Sci. Technol. 43, 6757–6763.

Klaine, S.J., Alvarez, P.J., Batley, G.E., Fernandes, T.F., Handy, R.D., Lyon, D.Y., Mahendra, S., McLaughlin, M.J., Lead, J.R., 2008. Nanomaterials in the environment: behavior, fate, bioavailability and effects. Environ. Toxicol. Chem. 27 (9), 1825–1851.

Kumari, M., Mukherjee, A., Chandrasekaran, N., 2009. Genotoxicity of silver nanoparticles in *Allium cepa*. Sci. Total Environ. 407, 5243–5246.

Kumari, M., Khan, S.S., Pakrashi, S., Mukherjee, A., Chandrasekaran, N., 2011. Cytogenetic and genotoxic effects of zinc oxide nanoparticles on root cells of *Allium cepa*. J. Hazard. Mater. 190, 613–621.

Ma, X., Quah, B., 2016. Effects of surface charge on the fate and phytotoxicity of gold nanoparticles to *Phaseolus vulgaris*. J. Food Chem. Nanotechnol. 2 (1).

Ma, Y., He, X., Zhang, P., Zhang, Z., 2011. Phytotoxicity and biotransformation of La_2O_3 nanoparticles in a terrestrial plant cucumber (*Cucumis sativus*). Nanotoxicology 5, 743–753.

Mahmoodzadeh, H., Aghili, R., Nabavi, M., 2013. Physiological effects of TiO_2 nanoparticles on wheat (*Triticum aestivum*). Tech. J. Eng. Appl. Sci. 3, 1365–1370.

Meyer, J., Lord, C.A., Yang, X.Y., Turner, D.A., Badireddy, A.R., Marinako, S.M., Chilkotib, A., Wiesner, M.R., Auffan, M., 2010. Intracellular uptake and associated toxicity of silver nanoparticles in *Caenorhabditis elegans*. Aquat. Toxicol. 100 (2), 140–150.

Ojea-Jiménez, I., López, X., Arbiol, J., Puntes, V., 2012. Citrate-coated gold nanoparticles as smart scavengers for mercury (II) removal from polluted waters. ACS Nano. 6 (3), 2253–2260.

Parsons, J.G., Lopez, M.L., Gonzalez, C.M., Peralta-Videa, J.R., Gardea-Torresdey, J.L., 2010. Toxicity and biotransformation of uncoated and coated nickel hydroxide nanoparticles on mesquite plants. Environ. Toxicol. Chem. 29, 1146–1154.

Peterson, C., Enstone, D., 1990. A survey of angiosperm species to detect hypodermal Casparian bands. I. Roots with a uniseriate hypodermis and epidermis. Bot. J. Linn. Soc. 103, 93–112.

Peterson, C., Enstone, D., 1996. Functions of passage cells in the endodermis and exodermis of roots. Physiol. Plant. 97, 592–598.

Servin, A.D., Castillo-Michel, H., Hernandez-Viezcas, J.A., Diaz, B.C., Peralta-Videa, J.R., Gardea-Torresdey, J.L., 2012. Synchrotron micro-XRF and micro-XANES confirmation of the uptake and translocation of TiO_2 nanoparticles in cucumber (*Cucumis sativus*) plants. Environ. Sci. Technol. 46, 7637–7643.

Shaw, A.K., Ghosh, S., Kalaji, H.M., Bosa, K., Brestic, M., Zivcak, M., Hossain, Z., 2014. Nano-CuO stress induced modulation of antioxidative defense and photosynthetic performance of Syrian barley (*Hordeum vulgare* L.). Environ. Exp. Bot. 102, 37–47.

Shoults-Wilson, A., Zhurbich, O.I., McNear, D.H., Tsyusko, O.V., Bertsch, P.M., Unrine, J.M., 2011. Evidence for avoidance of Ag nanoparticles by earthworms (*Eisenia fetida*). Ecotoxicology 20, 385–396.

Thuesombat, P., Hannongbua, S., Akasit, S., Chadchawan, S., 2014. Effect of silver nanoparticles on rice (*Oryza sativa* L cv. KDML 105) seed germination and seedling growth. Ecotoxicol. Environ. Saf. 104, 302–309.

Toro Di, D.M., Allenn, H.E., Berghman, H.L., Meyer, J.S., Paquin, P.R., Santore, R.C., 2001. Biotic ligand model of the acute toxicity of metals. 1. Technical basis. Environ. Toxicol. Chem. 20 (10), 2383–2396.

Torresdey, J.L., 2011. Interaction of nanoparticles with edible plants and their possible implications in the food chain. J. Agric. Food Chem. 59 (8), 3485–3498.

Unrine, J.M., Bertsch, P.M., Hunyadi, S.E., 2008. In: Grassian, V.H. (Ed.), Nanoscience and Nanotechnology: Environmental and Health Impacts. John Wiley and Sons, Inc, pp. 345–366 (Chapter 14).

Vannini, C., Domingo, G., Onelli, E., Mattia, F.D., Bruni, I., Marsoni, M., Bracale, M., 2014. Phytotoxic and genotoxic effects of silver nanoparticles exposure on germinating wheat seedlings. J. Plant Physiol. 171, 1142–1148.

Wang, H., Kou, X., Pei, Z., Xiao, J.Q., Shan, X., Xing, B., 2011. Physiological effects of magnetite (Fe_3O_4) nanoparticles on perennial ryegrass (*Lolium perenne* L.) and pumpkin (*Cucurbita mixta*) plants. Nanotoxicology 5, 30–42.

Wang, Z., Xie, X., Zhao, J., Liu, X., Feng, W., White, J.C., Xing, B., 2012. Xylem- and phloem-based transport of CuO nanoparticles in maize (*Zea mays* L.). Environ. Sci. Technol. 46, 4434–4441.

Wang, S., Liu, H., Zhang, Y., Xin, H., 2015. The effect of CuO NPs on ROS and the cell cycle gene expression in roots of rice. Environ. Toxicol. Chem. http://dx.doi.org/10.1002/etc.2826.

Wilson, M., 2001. Interactions between ultrafine particles and transition metals in vivo and in vitro. Toxicol. Appl. Pharmocol. 184 (3), 172–179.

Yamakoshi, Y., 2003. Active oxygen species generated from photoexcited fullerene (C_{60}) as potential medicines: O_2^{-} versus $^1O_2$. J. Am. Chem. Soc. 125, 12803–12809.

Yin, L., Cheng, Y., Espinasse, B., Colman, B.P., Auffan, M., Wiesner, M., Rose, J., Liu, J., Bernhardt, E., 2011. More than the ions: the effect of silver nanoparticles on *Lolium multiflorum*. Environ. Sci. Technol. 45, 2360–2367.

Yuan, J.J., Schmid, A., Armes, S.P., et al., 2006. Facile synthesis of highly biocompatible poly(2-(methacryloyloxy) ethyl phosphorylcho-line)-coated gold nanoparticles in aqueous solution. Langmuir 22, 11022–11027 (PubMed: 17154580).

Zhang, L., Fang, M., 2010. Nanomaterials in pollution trace detection and environmental improvement. Nano Today 5, 128–142.

CHAPTER

10

Responses of Plants to Iron Oxide Nanoparticles

Muhammad Zia-ur-Rehman[1], *Asif Naeem*[2], *Hinnan Khalid*[1], *Muhammad Rizwan*[3], *Shafaqat Ali*[3], *Muhammad Azhar*[1]

[1]University of Agriculture, Faisalabad, Pakistan; [2]Pakistan Atomic Energy Commission, Faisalabad, Pakistan; [3]Government College University, Faisalabad, Pakistan

10.1 INTRODUCTION

Nanotechnology is a swiftly emerging industry that has a proficient role in all sectors of life including agriculture. A great diversity of nanomaterial impacts on plants and agriculture has been well documented, which demonstrates that the positive and negative impacts depend on their different properties (Peralta-Videa et al., 2014, 2016; Tripathi et al., 2015, 2016, 2017a,b,c,d,e; Cox et al., 2016; Shweta et al., 2016; Singh et al., 2016, 2017; Tiwari et al., 2017). Furthermore, it is reported that nanotechnology has elevated its investment for various commodities from $4 to $224 billion within the short period from 2005 to 2009 (Engeman et al., 2012; Becker, 2013), while iron oxide nanoparticles are the supporting pillars of this industry. The reason behind this is the resourceful use of iron oxide nanoparticles in every field and technology (Fu et al., 2014). It is said that the estimated production of nanoparticles for various purposes will rise to 58,000 tons by 2020 and a major portion of this will be iron oxide nanoparticles (Maynard et al., 2006). The efficiency of bulk iron oxides in providing iron to plants has increased manyfold by their conversion to nanosized particles (1–100 nm size). Thus small sizes of iron oxide nanoparticles (1–100 nm) along with other vital properties distinguish the rank of iron oxide nanoparticles from other nanoparticles (Ramimoghadam et al., 2014), and they are making their way in the development of agriculture as well as other industries (Naseri et al., 2015). This is because of the unique properties iron oxide nanoparticles, such as quantum confinement, high surface energy, increased surface area-to-volume ratio, and a number of catalytic properties (Sun et al., 2015). Thus iron oxide nanoparticle technology has opened new opportunities for the production of cost-effective and environmentally friendly fertilizers. They have more potential to increase crop yield and production

http://dx.doi.org/10.1016/B978-0-12-811487-2.00010-4

because of augmented nutrient use efficiency by slow release of nutrients, less salt stress to roots by lowering the accumulation of soluble salts in the root zone, cost minimization, and reduction in nutrient loss by restricting their conversion from uptake form to discarded form, thus controlling environmental pollution (Zareii et al., 2014; Huang et al., 2015). Atar et al. (2015) has also pointed out the beneficial role of iron oxide nanoparticles for the stimulation of certain enzymes such as catalase, cytochrome oxidase, and peroxidase. Kaviani et al. (2014) have reported that iron oxide nanoparticles play a supportive role in enhancing the nutrient stress tolerance in plants by provision of cofactors that are the constituents of many antioxidants.

In conjunction with all this, iron oxide nanoparticles also have a promising role for the removal of heavy metals/metalloids such as arsenic, mercury, lead, and cadmium because they are reinforced with the properties of ligand sorption by forming the inner sphere complexes (Li et al., 2014a; Wang et al., 2014; Yoon et al., 2014). Moreover, the high reduction potential of iron oxide nanoparticles enables them to be efficient for the removal of organic pollutants such as atrazine and chlorinated compounds from water (Li et al., 2015; Zhao et al., 2015). With all these blessings in the form of iron oxide nanoparticles, it is obligatory to detail the mechanism of iron oxide nanoparticle-coupled plant growth and development.

10.2 COMPOSITION AND CHARACTERIZATION OF IRON OXIDE NANOPARTICLES

Innovation in the properties of iron oxide nanoparticles has led to their amalgamation in myriad profitable applications. The great thirst for iron oxide nanoparticles in every field of life has accelerated investigations about the types of iron oxide nanoparticles and concluded that there are eight promising types (Kandpal et al., 2014). Based on the uniqueness of properties, magnetite (Fe_3O_4), hematite (α-Fe_3O_4), and maghemite (γ-Fe_3O_4) nanoparticles have gained more importance than other iron oxide nanoparticles (Wei et al., 2015). A brief illustration of the structure and composition of these iron oxide nanoparticles follows.

10.2.1 Magnetite

Magnetite has 32 O^{-2} ions with a face-centered cubic spinal structure and compact packaging toward the center. It has both divalent and trivalent iron stoichiometry. Divalent iron has the ability to reinstate any other divalent metal (Co, Pb, Mn, Cd) (Wei et al., 2015). Thus it has great potential to reduce the contamination coupled with its benefactor role in the growth and development of plants.

10.2.2 Hematite

Hematite is the most stable form of iron oxide nanoparticles and has a wide range of applications. It is an n-type semiconductor that has a very low cost but very high resistance to corrosion. Because hematite can be used as the initiative for the synthesis of magnetite and maghemite, it has great importance in scientific research and technology (Kolen'ko et al., 2014).

10.2.3 Maghemite

Maghemite has a cubic structure caused by the close packing of 32 O^{-2} ions. It is an n-type semiconductor and considered as a completely oxidized magnetite. X-ray diffraction shows that maghemite has a similar crystalline structure to magnetite (Grau-Crespo et al., 2010).

10.3 SYNTHESIS OF IRON OXIDE NANOPARTICLES

There are a number of physical and chemical methods for the production of iron oxide nanoparticles but the most important and commonly used are discussed here.

10.3.1 Surface Modifications With Organic and Inorganic Materials

During the preparation of iron oxide nanoparticles in colloidal form, the most important consideration is stability management. Magnetizable liquids can be formed by the use of colloidal suspensions of various iron oxide nanoparticles (Fe_3O_4, α-Fe_3O_4, γ-Fe_3O_4). They remain in fluid form and have widespread applications because of their ability to interact with magnetic fields; they also have a fluidity property (Kim et al., 2015). However, nonexistence of surface modifications leads to hydrophobic interactions of iron oxide nanoparticles that result in the formation of large-sized nanoparticles by agglutination or huddle formation. This huddle or cluster of particles then shows strong dipole–dipole interaction, which elevates the attraction between the particles to a great extent (Choi et al., 2014). This strong attraction between the particles has negative impacts on the working ability of the iron oxide nanoparticles and surface modification looks indispensable in this situation.

Surface modifications with the use of certain organic and inorganic materials provide safety against this critical issue. These surfactants have a surface-adhering ability on iron oxide nanoparticles because of their substrate specificity phenomenon (Sakulkhu et al., 2015). There is a long list of organic and inorganic materials that can be used for surface modifications of iron oxide nanoparticles (Maurizi et al., 2014). Organic or polymeric material may be natural or synthetic based upon their nature of production. Polyvinyl alcohol, polyethylene-*co*-vinyl acetate, polyethylene glycol, and polyvinylpyrrolidone are the synthetic polymers that can be used for surface modifications (Ramimoghadam et al., 2014), while pullulan, gelatin, chitosan, and dextran are natural polymers (Kango et al., 2013). Sodium carboxymethylcellulose, sodium oleate, and dodecylamine have an elevated dispersive ability in aqueous medium (Layek et al., 2013).

An iron oxide nanoparticle at the inner sphere in the form of a core along with inorganic material (gold, aluminum, silica, and gadolinium) at the outer surface has been devised as a dexterous technique to enhance the working ability of iron oxide nanoparticles (Gupta and Gupta, 2005). These materials reduce the cluster formation ability of particles and have a supportive role for the proper functioning of iron oxide nanoparticles. This modification moves a step forward when various ionic capping ligands are used along with these inorganic materials (Garofalo et al., 2014). These surface modifications have their significant role in increased plant growth and development because iron oxide nanoparticles can easily diffuse within the different plant parts and excite various growth-stimulating enzymes. Moreover, use of silica for surface modification of iron oxide nanoparticles offers the benefits of silicon as well as iron and can raise the production potential manyfold (Mulvaney et al., 2000; Santra et al., 2001; Tartaj et al., 2002).

10.3.2 Biosynthesis of Iron Oxide Nanoparticles

The biosynthesis of iron oxide nanoparticles has proved to be an economical and eco-friendly substitute for the physical and chemical methods of iron oxide nanoparticle synthesis (Narayanan et al., 2012). In this technique, plant extracts are used that are a cheap, simple, and environmentally friendly material (Mittal et al., 2013; Roy et al., 2013). Moreover, the presence of reducing sugars, amino acids, antioxidants, and nitrogenous bases in the plant extract play a supportive role in efficient activity of iron oxide nanoparticles (Iravani et al., 2014, 2015). Another appreciable aspect of this technique is the nontoxic behavior of biosynthetically formed iron oxide nanoparticles, which is the basic threat during the preparation of metal oxide nanoparticles by some conventional chemical methods (Prakash et al., 2013).

Various plants such as hibiscus (*Hibiscus rosa sinensis*), coffee (*Coffea arabica*), sorghum (*Sorghum* spp.), black tea (*Camellia sinensis*), and countless fruit extracts can be used for the promising production of iron oxide nanoparticles (Iravani, 2011; Prakash et al., 2013). Based on size and morphology (nanorods, spheres, and platelets), iron oxide nanoparticles of different types can be synthesized using the extract of black tea (*C. sinensis*) (Shahwan et al., 2011). Sorghum is helpful for the production of a 40–50nm range of iron oxide nanoparticles, which also contain phenolic compounds (Njagi et al., 2011). Thus by the biosynthesis technique, we can obtain nontoxic and more proficient nanoparticles of iron oxides.

10.3.3 Emulsification of Iron Oxide Nanoparticles

Emulsification of iron oxide nanoparticles leads to the conversion of simple, stable, and isotropic mixtures of oil and water along with iron oxides into the discrete form of very small-sized nanoparticles called a microemulsion. Surfactants have the ability to produce a single layer at the interface of oil and water, while having their hydrophilic head in the aqueous solution and hydrophobic tail in the oil phase (Wei et al., 2015). Metal salts and other constituents form the aqueous phase, while olefins along with various hydrocarbons show the potential of forming the oil phase. On the basis of synthesis protocol, there are two common types of emulsion: (1) direct emulsification (oil dispersed in water) and (2) reverse emulsification (water dispersed in oil). Both these emulsifications provide the foundation for the production of various types of iron oxide nanoparticles (Wongwailikhit and Horwongsakul, 2011; Ladj et al., 2013; Sun et al., 2013). Because of all these blessings, emulsification of iron oxide nanoparticles requires due consideration for its widespread and beneficial use in the synthesis of iron oxide nanoparticles to improve their efficiency and activity.

10.3.4 Sol–Gel Reactions

A sol–gel reaction is a typical wet-chemical method that has extensive applications in material science and thus is helpful for the production of iron oxide nanoparticles. Fabrication of metal oxides such as iron oxide is done using this method for improving the efficiency of nanoparticles. Colloidal solution acts as the primer for the production of integrated nanoparticles of iron oxide (Wei et al., 2015). Stable dispersion of colloidal particles or polymers plays the role of sol in this phenomenon, while a gel is comprised of unremitting three-dimensional networks that cover the liquid phase completely. Agglomerations or aggregations of colloidal or subcolloidal particles are responsible for the formation of a network. Iron salts such as acetates, chlorides, and nitrates along with iron alkoxides after various polycondensation and hydrolysis reactions act as the primers for the production of iron oxide nanoparticles (Mittal and Mishra, 2014).

Continuous heat is required for the development of the crystalline structure of iron oxide nanoparticles after passing through various reactions, and at last their various types can be synthesized by using the sol–gel reaction method (Lemine et al., 2012). Sizes and shapes of iron oxide nanoparticles can be controlled by knowing the nature and mechanism of precursors used in this technique. Wu et al. (2008) reported the production of nanorods of iron oxide nanoparticles by controlling the sol–gel reactions. Being an efficient and advanced technique for the production of iron oxide nanoparticles, sol–gel reaction is receiving significant importance in research and education.

10.3.5 Sonolysis

The sonolysis or sonochemical technique is consistent with the production of iron oxide nanoparticles by using the physical impacts of ultrasound that are being produced by acoustic cavitation. Instead of using high pressure, temperature, and lengthy reactions, sonolysis provides an alternative way for iron oxide nanoparticle production (Xu et al., 2013). Because of the ultrasound irradiations, unique cavities or bubbles are formed by sporadic and compressive acoustic waves. The presence of acoustic waves makes the oscillations of cavities or bubbles and thus helps them to store or gather the ultrasonic energy that is then used for the production of various sizes and shapes of iron oxide nanoparticles under ambient circumstances (Hasany et al., 2012). Novel properties of iron oxide nanoparticles formed by sonolysis, such as uniform mixing or application of iron oxide nanoparticles along with a reduction in crystalline growth, has revolutionized the nanotechnology industry and earned eminence as the efficient procedure for the production of iron oxide nanoparticles (Wei et al., 2015).

10.4 APPLICATION METHODS OF IRON OXIDE NANOPARTICLES

There are numerous application methods of iron oxide nanoparticles. Studies have witnessed the application of iron oxide nanoparticles in rooting medium (nutrient solution/sand/soil) as well as foliar spray, e.g., fertilizers. Zhu et al. (2008) reported that using iron oxide nanoparticles by adopting the nutrient solution method remained very beneficial for the growth and development of pumpkin and lima beans. The possible reason for the high efficiency of this method could be the avoidance of the adsorption process that takes place in the case of the soil application method. Similarly, using the suspension method for iron oxide nanoparticle application has been found to increase the germination rate by up to 41% (Feizi et al., 2012). With all these sanctions, the foliar application method has its own benefits and feasibilities where other methods of application fail. A positive correlation was obtained between growth of poinsettia and foliar spray of iron oxide nanoparticles at initiation and 30 days after germination (Kaviani et al., 2014). Depending upon the need and requirements, appropriate application methods can be chosen and outstanding production on a sustainable basis can be achieved.

10.5 UPTAKE, ABSORBANCE, TRANSFER, AND ACCUMULATION MECHANISM OF IRON OXIDE NANOPARTICLES

Anything taken up by plants is accumulated in different tissues and may lead to toxicological effects if not in degradable form. Thus to get the maximum benefits from iron oxide nanoparticles rather than reimbursement impacts, it becomes necessary to discuss the uptake, transfer, and accumulation mechanism of these particles. Several mechanisms for the absorption and uptake of nanoparticles have been proposed. It is reported that metal oxide nanoparticles like iron oxide nanoparticles, having high surface area, are able to bind with the carrier proteins or organic chemicals and be absorbed by the plant tissues (Xu et al., 2011). A second school of thought states that metal (iron) oxide nanoparticles are absorbed by various plants because of the presence of ion transporters in the plasma membrane (Rico et al., 2011). A hydroponic study revealed that iron oxide nanoparticles have the potential to be translocated through the tissues of pumpkin plants soon after their uptake from nutrient solution (Hong et al., 2005). It was believed that iron oxide nanoparticles mostly accumulate near the roots and into the leaf tissues of pumpkin and lima bean plants grown in liquid culture (Zhu et al., 2008). However, a microscopic study of the same experiment also depicted the presence of iron oxide nanoparticles into the xylem vessels, a vital water-transporting channel of plants. In addition, it was also observed that there was less detection of uptake, transfer, and accumulation of iron oxide nanoparticles into the lima beans as compared to the pumpkin in the same experiment, which shows that uptake, transfer, and accumulation of iron oxide nanoparticles varies with crop types and species (Zhu et al., 2008; Van Aken, 2015).

Accumulation of iron oxide particles occurs in various plant tissues, while the extent of accumulation varies with plant species, particle size, and the most important exposure, time. Barhoumi et al. (2015) reported that the concentration of iron oxide nanoparticles increased 10-fold in *Lemna gibba* plants when they were exposed to 400 μg/mL of iron oxide nanoparticle suspension for 7 days as compared to the initial stage (very first day) of exposure. Another study exhibited that iron oxide nanoparticles were accumulated in the pith cavity of the *L. gibba* plant after just 24 h of application. Moreover, a detailed and technical study revealed that iron oxide nanoparticles were also observed in the epidermis of the stem, near trichomes, on the outer side of the cell wall, and in the cytoplasm of pumpkin plants (Corredor et al., 2010). Results from this experiment after a 48-h study depict the transfer of iron oxide nanoparticles toward the interior of stem cells such as parenchyma cells.

Lee et al. (2010) used *Arabidopsis* in his research and treated it with various concentrations of iron oxide nanoparticles. Although his results show the phytotoxic impact of iron oxide nanoparticles on root elongation, in the same era Corredor et al. (2010) achieved high yield and growth of his test crop when he applied iron oxide nanoparticles for the evaluation of plant response to the high-density nanoparticles. Thus with the present study we can conclude that because the uptake, translocation, and accumulation mechanism varies with the plant species, it is necessary to digest the detailed information about the nature of crops along with their absorbing/uptake behavior of iron oxide nanoparticles before their application. Moreover, the application method and amount of iron oxide nanoparticles are also critical factors for provision of the techniques that play a supportive role in lowering the negative impacts of the utilization of iron oxide nanoparticles and raising the beneficial impacts by endorsing the favorable traits into the plants (Wang et al., 2011).

10.6 IRON OXIDE NANOPARTICLES AND PLANT GROWTH

Nanotechnology, especially the use of iron oxide nanoparticles, has revolutionized agriculture and crop production. Water use efficiency enhancement, efficient products of fertilizers, herbicides, and pesticides are some of the smart tools of iron oxide nanoparticles in the agricultural sector (Scott and Chen, 2012), which have induced the soaring growth of roots and shoots, modified plant metabolism including the photosynthetic process, and thus elevated yield quality and quantity. In the following sections, all of these processes and modifications are discussed.

10.6.1 Privileged Growth and Yield of Plants

Iron oxide nanoparticles play a key role in efficient growth and improved yield of plants, starting right from seed germination to localization of its reserve stores. Seed germination rate is enhanced by the application of iron oxide nanoparticles that provokes shoot growth as well as root growth (Wang et al., 2011; Liu et al., 2016). Moreover, all the metabolic processes of plants are stimulated by the addition of iron oxide nanoparticles and all these activities are directly related to bumper yield production. Various studies reveal that iron oxide nanoparticles have the potential to escalate the fresh as well as dry biomass of various crops including rice, wheat, tomato, peanut, soybean, and spinach (Chittaranjan et al., 2013).

The germination rate of cucumber and lettuce was enhanced manyfold by the application of iron oxide nanoparticles that were directly proportional to improved quality and quantity of yield of both these crops (Antisari et al., 2015). El-Temsah and Joner (2012) found similar results for barley (*Hordeum vulgare*), flax (*Linum usitatissimum*), and ryegrass (*Lolium perenne*) along with root and shoot length of these plants by iron oxide nanoparticles while growing in both liquid suspension and soil. Ren et al. (2011) reported that addition of iron oxide nanoparticles to the Chinese mungbean can reveal promotional effects on seed germination up to 78% and root activity up to 50.3%.

Iron is one of the essential nutrients for plant growth and is the cofactor of various enzymes that speed up plant growth and development processes (Feizi et al., 2013). Therefore augmented seedling, root, and shoot growth was obtained by the use of iron oxide nanoparticles in appropriate concentrations (10–40 mg/L) for mungbean, rice, cucumber, and wheat (Sarvendra Kumar et al., 2015). Similar results showing increased growth and development of seedlings, root, and shoot of *Arabidopsis thaliana* because of the stimulation of cell wall loosening were also reported by Kim et al. (2014). Studies revealed that the effect of iron oxide nanoparticles is more obvious for seedlings, shoots, and growth because of the easy uptake and transfer mechanism of these particles via root cells. Thus researchers found a supportive role of iron oxide nanoparticles for the improvement of growth pattern because of the growth-stimulating effect of iron when applied within a specific range of concentrations (Bombin et al., 2015).

10.6.2 Improvement in Physiological Mechanism of Plants

With other sanctions of iron oxide nanoparticles for plant growth, its beneficial effects for improvement of plants' physiological processes are also unlimited. Ren et al. (2011) reported that iron oxide nanoparticles have the potential to support various physiological processes of plants such as stimulated activity of catalase, superoxide dismutase, and peroxidase along with escalated photosynthetic, respiration, and stomatal activity. Photosynthesis is a crucial process for the sustainability of plant growth and development. Iron is the basal component of the chlorophyll molecule and thus the addition of iron oxide nanoparticles eliminates iron deficiency and enhances chlorophyll *a* as well as chlorophyll *b* content in soybean (Mohammad et al., 2013). Studies also revealed that iron oxide nanoparticles improve the biochemical reactions of thylakoid membranes and chloroplast that are helpful for improving photosynthetic activity (Barhoumi et al., 2015). Various enzymatic processes that escalate chlorophyll activity also require iron, and application of iron oxide nanoparticles provides a solution for this need (Liu and Lal, 2015; Ronen, 2016). Therefore it is clear that application of iron oxide nanoparticles enhances photosynthetic activity, while stimulated photosynthetic rate increases plant growth and yield (Delfani et al., 2014; Montalvo et al., 2015). Iron oxide nanoparticles have the potential to intrigue the intrinsic enzymes of the plants and these enzymes are the supporters of proficient activity of all the physiological processes.

Provision of iron oxide nanoparticles raises the soluble content of iron in the soil solution and, being an essential plant nutrient, iron acts as the cofactor for various physiological as well as metabolic processes. The soluble fraction of iron increases because of the rapid release of soluble ions of iron from iron oxide nanoparticles because of their small size and large surface area (Kandpal et al., 2014). This results in higher physiological traits that result in the higher yield of various crops.

10.6.3 Tolerance Against Abiotic Stresses

As sessile creatures, plants have no option but to hide or escape from adverse environmental circumstances such as salinity, drought, water logging, irradiation, extreme temperature, etc. These conditions induce production of reactive oxygen species (ROS) such as the hydroxyl radical, hydrogen peroxide, the hydroperoxy radical, superoxide radical, and singlet oxygen (Khan et al., 2017). Large ROS accumulation deteriorates nucleic acid, proteins, and lipid membranes, stimulates genotoxicity and cytotoxicity (Yadav et al., 2014), and suppresses plant growth (Begum and Fugetsu, 2012). In addition to enhancing plant growth and yield, nanoparticles also have a pronounced role in plant protection against various types of abiotic stresses by minimizing antioxidative enzymatic activities and scavenging ROS (Rico et al., 2013). The large surface area and small size of iron nanoparticles offer binding sites for toxic materials to adsorb and thereby decrease metal availability and toxicity (Worms et al., 2012). Iron nanoparticles have been reported to improve photosynthetic activities by suppressing osmotic and oxidative stress (Haghighi and Pessarakli, 2013; Qi et al., 2013; Siddiqui et al., 2014). Furthermore, iron nanoparticles have defending effects on photosynthesis and parameters against salinity (Torabian et al., 2016), drought (Zaimenko et al., 2014), and heat (Qi et al., 2013).

10.6.3.1 *Drought Stress*

Drought stress is the result of anthropogenic climate alteration, which inhibits crop production and worldwide distribution. Among various techniques adopted to combat drought-stimulated damage in plants, nanoparticle use is proving to be promising (Aghdam et al., 2016). Iron is an essential component of different enzymes and its scarcity leads to chlorosis (Bameri et al., 2013). Foliar supply of iron oxide nanoparticles decreases the harmful effects caused by drought stress and stimulates the yield parameters of plants (Zareii et al., 2014). Martínez-Fernández et al. (2015) observed that nanostructured maghemite considerably opposes drought stress symptoms in *Helianthus annuus*. Similarly, in *Carthamus tinctorius* L. iron nanoparticles reduce the effects of drought and increase yield (Zareii et al., 2014).

10.6.3.2 *Salt Stress*

Salinity is primarily caused by a large accumulation of NaCl, which results in ionic toxicity and osmotic stress in plants. Ionic stress leads to a lower ratio of K^+/Na^+ and osmotic stress decreases the water and nutrient uptake ability of plants (Khan et al., 2017). Large amounts of ROS so produced severely damage biological molecules leading to electrolyte leakage (Ismail et al., 2014). Investigations show that nanoparticles play a vital role in stimulating the growth and yield of crops under salt stress (Khan et al., 2017). Foliar application of Fe_2O_3-based nanoparticles proved to be best in alleviating salinity stress by significantly reducing Na^+ and Cl contents along with increasing plant nutritional status (Zn, Fe, Mn, Mg, K, N, P), protein, proline, carbohydrates, carotenoid total chlorophyll contents, and antioxidant enzymes and nonenzymes (Soliman et al., 2015); in addition, nano-Si particles regulate salt stress gene expressions (Almutairi, 2016).

10.6.3.3 *Temperature Stress*

Nanomaterials have the ability to alleviate negative effects of cold stress by lowering electrolyte leakage and membrane damage (Mohammadi et al., 2013). Nanoparticles increase Chl- and Rubisco-binding protein genes (Hasanpour et al., 2015), superoxide dismutase, catalase, and ascorbate peroxidase activities (Mohammadi et al., 2014), maintain stability of carotenoid and Chl contents, and increase cold stress tolerance (Hasanpour et al., 2015). Application of nanoparticles as a chilling treatment improves biochemical, physiological, and growth attributes of plants under cold-stressed conditions (Haghighi et al., 2014; Kohan-Baghkheirati and Geisler-Lee, 2015).

10.6.4 Endorsement of Agronomic Traits

Iron present in Earth's crust is insoluble and very few forms are available for plant use, but when seeds are exposed to iron oxide nanoparticles, it significantly enhances their growing potential (Hassan et al., 2013), and at tested concentrations of Fe_3O_4 nanoparticles, they do not impart toxicological impacts on plants. Siva and Benita (2016) showed that iron oxide nanoparticles have great potential to improve all the agronomic traits of plants such as plant height, number of leaves, number of tillers, and protein content. Application of iron oxide nanoparticles increased the protein contents as well as gave a 20% higher grain yield of wheat crop (Mitra et al., 2015). Addition of iron oxide nanoparticles stimulated carotenoid content and plant height including shoot and root length much more as compared to chelated iron ethylenediaminetetraacetic acid compounds (Siva and Benita, 2016). Treatment of nanoferric oxide (at 20 ppm) to watermelon has a great capability to scavenge oxygen radicals, lower malondialdehyde contents, and significantly increase chlorophyll contents, which result in healthier cells and increased plant agronomic parameters (Mohammad et al., 2013). In addition to this, foliar application of these nanoparticles at the vegetative growth stage may maximize plant development and growth, alongside significantly enhancing chlorophyll contents, leaf area, plant density, and plant height (Rawashdeh and Sala, 2014).

10.6.5 Substitutes of Typical Fertilizers

Nanotechnology has opened many new opportunities in every aspect of life including agriculture to increase nutrient use efficiency and lessen the cost of environmental protection techniques (Li et al., 2014a,b; Bindraban et al., 2015; Rui et al., 2016). This technology eliminates environmental toxins via the formation of greater value-added products (Garda-Torredey et al., 2002). One of these products is nanofertilizers to replace conventional iron fertilizers, and these have great significance in agriculture. The superiority of nanofertilizers over traditional fertilizers includes: (1) reduced leaching losses of nutrients from root zone and consequently decreased environmental pollution; (2) restricted conversion of usable forms of nutrients to unusable/unwanted forms; (3) limited toxic accumulation of soluble salts in root zone, thus preventing plant damage; (4) cost reduction; and (5) enhancement of plant nutrient uptake with the characteristics of slow release of nutrients (Gui et al., 2015). Nanofertilizers propose controlled release to synchronize the flux of nutrients over time with plant uptake, thus enhancing nutrient availability and decreasing extravagant interactions with air and soil that cause nutrient loss from the agroecosystem. Production of nanobased fertilizers with higher solubility, lowered persistence, stable dispersal, and better target specificity is ensuing at a prompt pace (Rai et al., 2012).

Iron content is usually high in soil but a greater portion is fixed to soil and present as an insoluble form (Fe^{3+}) (Mimmo et al., 2014; Bindraban et al., 2015), thereby soils are deficient in the plant available form of iron (Fe^{2+}) (Ye et al., 2015). Hence it is very important to enhance utilization effectiveness of iron fertilizers. Iron-based nanofertilizers contain a unique complex, having 9% of water-soluble Fe (Kaviani et al., 2016). A large concentration of Fe_2O_3 nanoparticles adheres to soil particles and thus electrons accumulate largely on the edge of soil particles; this reduces loss of nutrients and improves their efficiency as an iron fertilizer (Shuai and Yost, 2010).

Iron oxide nanoparticles when applied as fertilizers have the potential to lower nutrient-deficient stress (Gui et al., 2015). These fertilizers increase chlorophyll contents, which enhance photosynthetic activities and plant growth (Vasconcelos and Grusak, 2014). Nanoferric fertilizer treatment reduces abscisic acid (ABA) concentration in plants (Lucini and Bernardo, 2015), and less ABA content stimulates plant growth and delays senescence (Davière and

Achard, 2016). Iron oxide nanoparticle-based fertilizers can be an ideal replacement for conventional iron fertilizers, though further investigations are still required to evaluate their potential risk to food security and the environment.

10.6.6 Environmental Impacts

Innovative size-dependent physicochemical characteristics of iron oxide nanoparticles have made them potentially applicable in a wide range of fields including biosensors, magnetic storage media, ferrofluids, catalysts, and environmental media (Guo et al., 2012). Among the various nanoparticles used for environmental remediation purposes, iron nanomaterials (nanoparticles) are frequently applied (Peeters et al., 2016). Nanosized zero valent iron (nZVI) particles are commonly used for wastewater and contaminated soil remediation (O'Carroll et al., 2013). They have also been used for wastewater remediation (Mueller et al., 2012; Fu et al., 2014), and enhancement of anaerobic digestion has also attracted increased attention. Because of their reduced size these nZVI particles have greater reactivity with a wide range of contaminants, comprising inorganic pollutants (heavy metals, phosphate, nitrate) and organic pollutants (aromatic hydrocarbons, polycyclics, azo dyes, pharmaceutical wastes, halogenated compounds) (Mueller et al., 2012; Tosco et al., 2014). nZVI is cost-effective, nontoxic, abundant, easy to synthesize, and the process of reduction requires little maintenance. In addition, its broad availability poses noticeable removal efficacy for various contaminants (Fu et al., 2014).

During the remediation process, the important thing is to effectively disperse nanoparticles. Among various methods of dispersion, shaking (Peeters et al., 2015), mixing (Calderon and Fullana, 2015), or usage of various dispersing elements is applied. Tetramethylammonium hydroxide works as a surfactant and forms an electrostatic repulsion layer around Fe_3O_4 nanoparticles and nZVI, which reveals more effective dispersion (Andrade et al., 2012; Tang and Lo, 2013). nZVI effectively (99.9%) removes heavy metals (Cu, Ni, Cd, Cr, Zn) within just 3h of treatment of wastewater. But with the aging of nZVI particles, metals (except for Cr) redissolve from iron corrosion precipitate (Calderon and Fullana, 2015). Degradation of organic pollutant, azo dyes, is significantly increased in the presence of iron nanomotors acting as reducing agents. This type of environmental remediation will revolutionize the systems that carry out environmental remediation (Teo et al., 2011). However, there is still a need for future research to use these particles for environmental implications in a safe manner.

10.7 CONTROVERSIES ABOUT THE PHYTOTOXICITY OF IRON OXIDE NANOPARTICLES

Contrary to the beneficial uses of iron oxide nanoparticles, researchers have also found conflicting results about their impacts: they inhibit plant growth because of toxicity. Bombin et al. (2015) showed that application of iron oxide nanoparticles hinders seed germination because of the high availability of iron and unavailability of other essential nutrients. The reason behind this may be the unavailability of essential nutrients to the seed embryo because of elevated competition among plant nutrients and their uptake mechanisms. Similarly, Shukla et al. (2015) have reported negative effects of iron oxide nanoparticles because of the stunted permeability potential of plasma membrane. Mostly, phytotoxicological effects of iron oxide nanoparticles were observed when plants were exposed for long durations, having high concentration and reduced seed germination, seedling, root, and shoot growth, and other metabolic process that are directly proportional to yield production (Lee et al., 2010; García et al., 2011; Bombin et al., 2015).

Along with inhibitory effects on plant growth and development there are also concerns about the environmental issues related to extensive use of iron oxide nanoparticles in every field of life. The corrosive character of iron dominates when it comes in contact with various entities in the form of small size with large surface area (Palimi et al., 2015). Moreover, air pollution, because of increases in particulate matter concentration, also emerges with unpleasant impacts on plants as well as on humans and animals. Therefore there is a dire need for further research and more critic investigation about the results of iron oxide nanoparticle application in any field of life. There is also a need for the modification and improvement in application methods of iron oxide nanoparticles.

References

Aghdam, M.T.B., Mohammadi, H., Ghorbanpour, M., 2016. Effects of nanoparticulate anatase titanium dioxide on physiological and biochemical performance of *Linum usitatissimum* (Linaceae) under well-watered and drought stress conditions. Braz. J. Bot. 39 (1), 139–146.

Almutairi, Z.M., 2016. Effect of nano-silicon application on the expression of salt tolerance genes in germinating tomato (*Solanum lycopersicum* L.) seedlings under salt stress. Plant Omics J. 9, 106–114.

Andrade, Â.L., Fabris, J.D., Ardisson, J.D., Valente, M.A., Ferreira, J.M., 2012. Effect of tetramethylammonium hydroxide on nucleation, surface modification and growth of magnetic nanoparticles. J. Nanomater. 2012, 15.

Antisari, L.V., Carbone, S., Gatti, A., Vianello, G., Nannipieri, P., 2015. Uptake and translocation of metals and nutrients in tomato grown in soil polluted with metal oxide (CeO_2, Fe_3O_4, SnO_2, TiO_2) or metallic (Ag, Co, Ni) engineered nanoparticles. Environ. Sci. Pollut. Res. 22 (3), 1841–1853.

Atar, N., Eren, T., Yola, M.L., Karimi-Maleh, H., Demirdögen, B., 2015. Magnetic iron oxide and iron oxide@gold nanoparticle anchored nitrogen and sulfur-functionalized reduced graphene oxide electrocatalyst for methanol oxidation. RSC Adv. 5 (33), 26402–26409.

Bameri, M., Abdolshahi, R., Mohammadi-Nejad, G., Yousefi, K., Tabatabaie, S.M., 2013. Effect of different microelement treatment on wheat (*Triticum aestivum*) growth and yield. Int. Res. J. Appl. Basic Sci. 3 (1), 219–223.

Barhoumi, L., Oukarroum, A., Taher, L.B., Smiri, L.S., Abdelmelek, H., Dewez, D., 2015. Effects of superparamagnetic iron oxide nanoparticles on photosynthesis and growth of the aquatic plant *Lemna gibba*. Arch. Environ. Contam. Toxicol. 68 (3), 510–520.

Becker, S., 2013. Nanotechnology in the marketplace: how the nanotechnology industry views risk. J. Nanopart. Res. 15, 1–13.

Begum, P., Fugetsu, B., 2012. Phytotoxicity of multi-walled carbon nanotubes on red spinach (*Amaranthus tricolor* L.) and the role of ascorbic acid as an antioxidant. J. Hazard. Mater. 243, 212–222.

Bindraban, P.S., Dimkpa, C., Nagarajan, L., Roy, A., Rabbinge, R., 2015. Revisiting fertilisers and fertilisation strategies for improved nutrient uptake by plants. Biol. Fertil. Soils 51, 897–911. http://dx.doi.org/10.1007/s00374-015-1039-7.

Bombin, S., LeFebvre, M., Sherwood, J., Xu, Y., Bao, Y., Ramonell, K.M., 2015. Developmental and reproductive effects of iron oxide nanoparticles in *Arabidopsis thaliana*. Int. J. Mol. Sci. 16 (10), 24174–24193.

Calderon, B., Fullana, A., 2015. Heavy metal release due to aging effect during zero valent iron nanoparticles remediation. Water Res. 83, 1–9.

Chittaranjan, K., Kole, P., Randunu, K.M., Choudhary, P., Podila, R., Ke, P.C., Rao, A.M., Marcus, K., 2013. Nanobiotechnology can boost crop production and quality: first evidence from increased plant biomass, fruit yield and phytomedicine content in bitter melon (*Momordica charantia*). BMC Biotechnol. 13, 37.

Choi, Y.S., Yoon, H.Y., Lee, J.S., Wu, J.H., Kim, Y.K., 2014. Synthesis and magnetic properties of size-tunable $Mn_xFe_{3-x}O_4$ ferrite nanoclusters. J. Appl. Phys. 115 (17), 17B517.

Corredor, E., Risueno, M.C., Testillano, P.S., 2010. Carbon iron magnetic nanoparticles for agronomic use in plants promising but still a long way to go. Plant Signal. Behav. 5, 1295–1297.
Cox, A., Venkatachalam, P., Sahi, S., Sharma, N., 2016. Silver and titanium dioxide nanoparticle toxicity in plants: a review of current research. Plant Physiol. Biochem. 107, 147–163.
Davière, J.M., Achard, P., 2016. A pivotal role of DELLAs in regulating multiple hormone signals. Mol. Plant 9 (1), 10–20.
Delfani, M., Baradarn Firouzabadi, M., Farrokhi, N., Makarian, H., 2014. Some physiological responses of black-eyed pea to iron and magnesium nanofertilizers. Commun. Soil Sci. Plant Anal. 45 (4), 530–540.
El-Temsah, Y.S., Joner, E.J., 2012. Impact of Fe and Ag nanoparticles on seed germination and differences in bioavailability during exposure in aqueous suspension and soil. Environ. Toxicol. 27 (1), 42–49.
Engeman, C.D., Baumgartner, L., Carr, B.M., Fish, A.M., Meyerhofer, J.D., Satterfield, T.A., Holden, P.A., Harthorn, B.H., 2012. Governance implications of nanomaterials companies' inconsistent risk perceptions and safety practices. J. Nanopart. Res. 14, 1–12.
Feizi, H., Rezvani Moghaddam, P., Shahtahmassebi, N., Fotovat, A., 2012. Impact of bulk and nanosized titanium dioxide (TiO_2, ZnO, Ag, CNT, fullerenes) for different regions. Environ. Sci. Technol. 43, 9216–9222.
Feizi, H., Moghaddam, P.R., Shahtahmassebi, N., Fotovat, A., 2013. Assessment of concentrations of nano and bulk iron oxide particles on early growth of wheat (*Triticum aestivum* L.).
Fu, F., Dionysiou, D.D., Liu, H., 2014. The use of zero-valent iron for groundwater remediation and wastewater treatment: a review. J. Hazard. Mater. 267, 194–205.
García, A., Espinosa, R., Delgado, L., Casals, E., González, E., Puntes, V., Barata, C., Font, X., Sánchez, A., 2011. Acute toxicity of cerium oxide, titanium oxide and iron oxide nanoparticles using standardized tests. Desalination 269 (1), 136–141.
Garda-Torredey, J.L., et al., 2002. Formation and growth of Au nanoparticles inside live alfalfa plants. Nano Lett. 2 (4), 397–401.
Garofalo, A., Parat, A., Bordeianu, C., Ghobril, C., Kueny-Stotz, M., Walter, A., Jouhannaud, J., Begin-Colin, S., Felder-Flesch, D., 2014. Efficient synthesis of small-sized phosphonated dendrons: potential organic coatings of iron oxide nanoparticles. New J. Chem. 38 (11), 5226–5239.
Grau-Crespo, R., Al-Baitai, A.Y., Saadoune, I., De Leeuw, N.H., 2010. Vacancy ordering and electronic structure of γ-Fe_2O_3 (maghemite): a theoretical investigation. J. Phys. Condens. Matter 22 (25), 255401.
Gui, X., He, X., Ma, Y.H., Zhang, P., Li, Y., Ding, Y., et al., 2015. Quantifying the distribution of ceria nanoparticles in cucumber roots: the influence of labeling. RSC Adv. 5, 4554–4560. http://dx.doi.org/10.1039/c4ra13915a.
Guo, Y., Sun, X., Liu, Y., Wang, W., Qiu, H., Gao, J., 2012. One pot preparation of reduced graphene oxide (RGO) or Au (Ag) nanoparticle-RGO hybrids using chitosan as a reducing and stabilizing agent and their use in methanol electrooxidation. Carbon 50 (7), 2513–2523.
Gupta, A.K., Gupta, M., 2005. Synthesis and surface engineering of iron oxide nanoparticles for biomedical applications. Biomaterials 26, 3995–4021.
Haghighi, M., Pessarakli, M., 2013. Influence of silicon and nano-silicon on salinity tolerance of cherry tomatoes (*Solanum lycopersicum* L.) at early growth stage. Sci. Hortic. 161, 111–117.
Haghighi, M., Abolghasemi, R., Teixeira da Silva, J.A., 2014. Low and high temperature stress affect the growth characteristics of tomato in hydroponic culture with Se and nano-Se amendment. Sci. Hortic. 178, 231–240.
Hasanpour, H., Maali-Amiri, R., Zeinali, H., 2015. Effect of TiO_2 nanoparticles on metabolic limitations to photosynthesis under cold in chickpea. Russ. J. Plant Physiol. 62, 779–787.
Hasany, S.F., Ahmed, I., Rajan, J., Rehman, A., 2012. Systematic review of the preparation techniques of iron oxide magnetic nanoparticles. Nanosci. Nanotechnol. 2 (6), 148–158.
Hassan, F., Parviz, R.M., Nasser, S., Amir, F., 2013. Assessment of concentrations of nano and bulk iron oxide particles on early growth of wheat (*Triticum aestivum* L.). Annu. Rev. Res. Biol. 3 (4), 752–761.
Hong, T.K., Yang, H.S., Choi, C.J., 2005. Study of the enhanced thermal conductivity of Fe nanofluids. J. Appl. Phys. 97 (6), 064311.
Huang, Q., Cao, M., Ai, Z., Zhang, L., 2015. Reactive oxygen species dependent degradation pathway of 4-chlorophenol with Fe@Fe_2O_3 core–shell nanowires. Appl. Catal. B Environ. 162, 319–326.
Iravani, S., 2011. Green synthesis of metal nanoparticles using plants. Green Chem. 13 (10), 2638–2650.
Iravani, S., Korbekandi, H., Mirmohammadi, S.V., Mekanik, H., 2014. Plants in nanoparticle synthesis. Rev. Adv. Sci. Eng. 3 (3), 261–274.

Iravani, S., Korbekandi, H., Zolfaghari, B., 2015. Phytosynthesis of nanoparticles. In: Nanotechnology and Plant Sciences. Springer International Publishing, pp. 203–258.
Ismail, A., Takeda, S., Nick, P., 2014. Life and death under salt stress: same players, different timing? J. Exp. Bot. 65, 2963–2979.
Kandpal, N.D., Sah, N., Loshali, R., Joshi, R., Prasad, J., 2014. Co-precipitation method of synthesis and characterization of iron oxide nanoparticles. J. Sci. Ind. Res. 73, 87–90.
Kango, S., Kalia, S., Celli, A., Njuguna, J., Habibi, Y., Kumar, R., 2013. Surface modification of inorganic nanoparticles for development of organic–inorganic nanocomposites—a review. Prog. Polymer Sci. 38 (8), 1232–1261.
Kaviani, B., Negahdar, N., Ghaziani, M.V.F., 2014. The effect of iron nano-chelate and cycocel on some morphological and physiological characteristics, proliferation and enhancing the quality of *Euphorbia pulcherrima* Willd. Sci. Pap. Ser. B Hortic. 58, 337–342.
Kaviani, B., Ghaziani, M.V.F., Negahdar, N., 2016. The effect of iron nano-chelate fertilizer and cycocel (CCC) on some quantity and quality characters of *Euphorbia pulcherrima* Willd. J. Med. Bioeng. 5 (1).
Khan, M.N., Mobin, M., Abbas, Z.K., AlMutairi, K.A., Siddiqui, Z.H., 2017. Role of nanomaterials in plants under challenging environments. Plant Physiol. Biochem. 110, 194–209.
Kim, J.H., Lee, Y., Kim, E.J., Gu, S., Sohn, E.J., Seo, Y.S., An, H.J., Chang, Y.S., 2014. Exposure of iron nanoparticles to *Arabidopsis thaliana* enhances root elongation by triggering cell wall loosening. Environ. Sci. Technol. 48 (6), 3477–3485.
Kim, Y.Y., Hwang, B., Song, S., Ree, B.J., Kim, Y., Cho, S.Y., Heo, K., Kwon, Y.K., Ree, M., 2015. Well-defined hollow nanochanneled-silica nanospheres prepared with the aid of sacrificial copolymer nanospheres and surfactant nanocylinders. Nanoscale 7 (35), 14774–14785.
Kohan-Baghkheirati, E., Geisler-Lee, J., 2015. Gene expression, protein function and pathways of *Arabidopsis thaliana* responding to silver nanoparticles in comparison to silver ions, cold, salt, drought and heat. Nanomaterials 5, 436–467.
Kolen'ko, Y.V., Bañobre-López, M., Rodríguez-Abreu, C., Carbó-Argibay, E., Sailsman, A., Piñeiro-Redondo, Y., Cerqueira, M.F., Petrovykh, D.Y., Kovnir, K., Lebedev, O.I., Rivas, J., 2014. Large-scale synthesis of colloidal Fe_3O_4 nanoparticles exhibiting high heating efficiency in magnetic hyperthermia. J. Phys. Chem. C 118 (16), 8691–8701.
Ladj, R., Bitar, A., Eissa, M.M., Fessi, H., Mugnier, Y., Le Dantec, R., Elaissari, A., 2013. Polymer encapsulation of inorganic nanoparticles for biomedical applications. Int. J. Pharm. 458 (1), 230–241.
Layek, R.K., Kundu, A., Nandi, A.K., 2013. High-performance nanocomposites of sodium carboxymethylcellulose and graphene oxide. Macromol. Mater. Eng. 298 (11), 1166–1175.
Lee, C.W., Mahendra, S., Zodrow, K., Li, D., Tsai, Y.C., Braam, J., Alvarez, P.J., 2010. Developmental phytotoxicity of metal oxide nanoparticles to *Arabidopsis thaliana*. Environ. Toxicol. Chem. 29 (3), 669–675.
Lemine, O.M., Omri, K., Zhang, B., El Mir, L., Sajieddine, M., Alyamani, A., Bououdina, M., 2012. Sol–gel synthesis of 8 nm magnetite (Fe_3O_4) nanoparticles and their magnetic properties. Superlattices Microstruct. 52 (4), 793–799.
Li, G., Gao, S., Zhang, G., Zhang, X., 2014a. Enhanced adsorption of phosphate from aqueous solution by nanostructured iron(III)–copper(II) binary oxides. Chem. Eng. J. 235, 124–131.
Li, X., Gui, X., Rui, Y., Ji, W., Yu, Z., Peng, S., 2014b. Bt-transgenic cotton is more sensitive to CeO_2 nanoparticles than its parental non-transgenic cotton. J. Hazard. Mater. 274, 173–180.
Li, R., Jin, X., Megharaj, M., Naidu, R., Chen, Z., 2015. Heterogeneous Fenton oxidation of 2,4-dichlorophenol using iron-based nanoparticles and persulfate system. Chem. Eng. J. 264, 587–594.
Liu, R., Lal, R., 2015. Potentials of engineered nanoparticles as fertilizers for increasing agronomic productions. Sci. Total Environ. 514, 131–139.
Liu, R., Zhang, H., Lal, R., 2016. Effects of stabilized nanoparticles of copper, zinc, manganese, and iron oxides in low concentrations on lettuce (*Lactuca sativa*) seed germination: nanotoxicants or nanonutrients? Water Air Soil Pollut. 227 (1), 1–14.
Lucini, L., Bernardo, L., 2015. Comparison of proteome response to saline and zinc stress in lettuce. Front. Plant Sci. 6.
Martínez-Fernández, D., Vítková, M., Bernal, M.P., Komárek, M., 2015. Effects of nano-maghemite on trace element accumulation and drought response of *Helianthus annuus* L. in a contaminated mine soil. Water Air Soil Pollut. 226, 101. http://dx.doi.org/10.1007/s11270-015-2365-y.
Maurizi, L., Sakulkhu, U., Gramoun, A., Vallee, J.P., Hofmann, H., 2014. A fast and reproducible method to quantify magnetic nanoparticle biodistribution. Analyst 139 (5), 1184–1191.
Maynard, A.D., Aitken, R.J., Butz, T., Colvin, V., Donaldson, K., Oberdörster, G., Philbert, M.A., Ryan, J., Seaton, A., Stone, V., Tinkle, S.S., 2006. Safe handling of nanotechnology. Nature 444 (7117), 267–269.

Mimmo, T., Del Buono, D., Terzano, R., Tomasi, N., Vigani, G., Crecchio, C., Pinton, R., Zocchi, G., Cesco, S., 2014. Rhizospheric organic compounds in the soil–microorganism–plant system: their role in iron availability. Eur. J. Soil Sci. 65 (5), 629–642.

Mitra, B., Payam, M., Behzad, S., 2015. The effect of iron nanoparticles spraying time and concentration on wheat. Biol. Forum Int. J. 7 (1), 679–683.

Mittal, H., Mishra, S.B., 2014. Gum ghatti and Fe_3O_4 magnetic nanoparticles based nanocomposites for the effective adsorption of rhodamine B. Carbohydr. Polym. 101, 1255–1264.

Mittal, A.K., Chisti, Y., Banerjee, U.C., 2013. Synthesis of metallic nanoparticles using plant extracts. Biotechnol. Adv. 31 (2), 346–356.

Mohammad, H.G., Mohammad, J.M., Mohammad, R.D., Pieter, S., Morteza, M., 2013. Effects of magnetite nanoparticles on soybean chlorophyll. Environ. Sci. Technol. 47, 10645–10652.

Mohammadi, R., Maali-Amiri, R., Abbasi, A., 2013. Effect of TiO_2 nanoparticles on chickpea response to cold stress. Biol. Trace Elem. Res. 152, 403–410.

Mohammadi, R., MaaliAmiri, R., Mantri, N., 2014. Effect of TiO_2 nanoparticles on oxidative damage and antioxidant defense systems in chickpea seedlings during cold stress. Russ. J. Plant Physiol. 61, 768–775.

Montalvo, D., McLaughlin, M.J., Degryse, F., 2015. Efficacy of hydroxyapatite nanoparticles as phosphorus fertilizer in andisols and oxisols. Soil Sci. Soc. Am. J. 79 (2), 551–558.

Mueller, N.C., Braun, J., Bruns, J., Černík, M., Rissing, P., Rickerby, D., Nowack, B., 2012. Application of nanoscale zero valent iron (NZVI) for groundwater remediation in Europe. Environ. Sci. Pollut. Res. 19 (2), 550–558.

Mulvaney, P., Liz-Marzán, L.M., Giersig, M., Ung, T., 2000. Silica encapsulation of quantum dots and metal clusters. J. Mater. Chem. 2000 (10), 1259–1270.

Narayanan, T.N., Liu, Z., Lakshmy, P.R., Gao, W., Nagaoka, Y., Kumar, D.S., Lou, J., Vajtai, R., Ajayan, P.M., 2012. Synthesis of reduced graphene oxide–Fe_3O_4 multifunctional freestanding membranes and their temperature dependent electronic transport properties. Carbon 50 (3), 1338–1345.

Naseri, M.G., Kamari, H.M., Dehzangi, A., Kamalianfar, A., Saion, E.B., 2015. Fabrication of a novel chromium-iron oxide ($Cr_2Fe_6O_{12}$) nanoparticles by thermal treatment method. J. Magn. Magn. Mater. 389, 113–119.

Njagi, E.C., Huang, H., Stafford, L., Genuino, H., Galindo, H.M., Collins, J.B., Hoag, G.E., Suib, S.L., 2011. Biosynthesis of iron and silver nanoparticles at room temperature using aqueous sorghum bran extracts. Langmuir 27, 264–271.

O'Carroll, D., Sleep, B., Krol, M., Boparai, H., Kocur, C., 2013. Nanoscale zero valent iron and bimetallic particles for contaminated site remediation. Adv. Water Res. 51, 104–122.

Palimi, M.J., Rostami, M., Mahdavian, M., Ramezanzadeh, B., 2015. A study on the corrosion inhibition properties of silane-modified Fe_2O_3 nanoparticle on mild steel and its effect on the anticorrosion properties of the polyurethane coating. J. Coat. Technol. Res. 12 (2), 277–292.

Peeters, K., Lespes, G., Milačič, R., Ščančar, J., 2015. Adsorption and degradation processes of tributyltin and trimethyltin in landfill leachates treated with iron nanoparticles. Environ. Res. 142, 511–521.

Peeters, K., Lespes, G., Zuliani, T., Ščančar, J., Milačič, R., 2016. The fate of iron nanoparticles in environmental waters treated with nanoscale zero-valent iron, FeO NPs and Fe_3O_4 NPs. Water Res. 94, 315–327.

Peralta-Videa, J.R., Hernandez-Viezcas, J.A., Zhao, L., Diaz, B.C., Ge, Y., Priester, J.H., Holden, P.A., Gardea-Torresdey, J.L., 2014. Cerium dioxide and zinc oxide nanoparticles alter the nutritional value of soil cultivated soybean plants. Plant Physiol. Biochem. 80, 128–135.

Peralta-Videa, J.R., Huang, Y., Parsons, J.G., Zhao, L., Lopez-Moreno, L., Hernandez-Viezcas, J.A., Gardea-Torresdey, J.L., 2016. Plant-based green synthesis of metallic nanoparticles: scientific curiosity or a realistic alternative to chemical synthesis? Nanotechnol. Environ. Eng. 1 (1), 4.

Prakash, S., Chakrabarty, T., Singh, A.K., Shahi, V.K., 2013. Polymer thin films embedded with metal nanoparticles for electrochemical biosensors applications. Biosens. Bioelectron. 41, 43–53.

Qi, M., Liu, Y., Li, T., 2013. Nano-TiO_2 improve the photosynthesis of tomato leaves under mild heat stress. Biol. Trace Elem. Res. 156 (1–3), 323–328.

Rai, M.K., Deshmukh, S.D., Ingle, A.P., Gade, A.K., 2012. Silver nanoparticles: the powerful nanoweapon against multidrug-resistant bacteria. J. Appl. Microbiol. 112 (5), 841–852.

Ramimoghadam, D., Bagheri, S., Hamid, S.B.A., 2014. Progress in electrochemical synthesis of magnetic iron oxide nanoparticles. J. Magn. Magn. Mater. 368, 207–229.

Rawashdeh, H., Sala, F., 2014. Influence of iron foliar fertilization on some growth and physiological parameters of wheat at two growth stages. Sci. Pap. Ser. A Agron. 57, 306–309.

Ren, H.X., Liu, L., Liu, C., He, S.Y., Huang, J., Li, J.L., Zhang, Y., Huang, X.J., Gu, N., 2011. Physiological investigation of magnetic iron oxide nanoparticles towards Chinese mung bean. J. Biomed. Nanotechnol. 7 (5), 677–684.

Rico, C.M., Majumdar, S., Duarte-Gardea, M., Peralta-Videa, J.R., Gardea-Torresdey, J.L., 2011. Interaction of nanoparticles with edible plants and their possible implications in the food chain. J. Agric. Food Chem. 59, 3485–3498.

Rico, C.M., Morales, M.I., McCreary, R., Castillo-Michel, H., Barrios, A.C., Hong, J., Tafoya, A., Lee, W.Y., Varela-Ramirez, A., Peralta-Videa, J.R., Gardea-Torresdey, J.L., 2013. Cerium oxide nanoparticles modify the antioxidative stress enzyme activities and macromolecule composition in rice seedlings. Environ. Sci. Technol. 47 (24), 14110–14118.

Ronen, E., 2016. Micro-elements in agriculture. Pract. Hydroponics Greenh. (164), 35.

Roy, N., Gaur, A., Jain, A., Bhattacharya, S., Rani, V., 2013. Green synthesis of silver nanoparticles: an approach to overcome toxicity. Environ. Toxicol. Pharmacol. 36 (3), 807–812.

Rui, M., Ma, C., Hao, Y., Guo, J., Rui, Y., Tang, X., Zhao, Q., Fan, X., Zhang, Z., Hou, T., Zhu, S., 2016. Iron oxide nanoparticles as a potential iron fertilizer for peanut (*Arachis hypogaea*). Front. Plant Sci. 7, 815.

Sakulkhu, U., Mahmoudi, M., Maurizi, L., Coullerez, G., Hofmann-Amtenbrink, M., Vries, M., Motazacker, M., Rezaee, F., Hofmann, H., 2015. Significance of surface charge and shell material of superparamagnetic iron oxide nanoparticle (SPION) based core/shell nanoparticles on the composition of the protein corona. Biomater. Sci. 3 (2), 265–278.

Santra, S., Tapec, R., Theodoropoulou, N., Dobson, J., Hebard, A., Tan, W., 2001. Synthesis and characterization of silica-coated iron oxide nanoparticles in microemulsion: the effect of non-ionic surfactants. Langmuir 17, 2900–2906.

Sarvendra-Kumar, P.A., Datta, S.C., Rosin, K.G., Purakayastha, T.J., 2015. Phytotoxicity of nanoparticles to seed germination of plants. Int. J. Adv. Res. 3, 854–865.

Scott, N., Chen, H., 2012. Nanoscale science and engineering for agriculture and food systems. Ind. Biotechnol. 8 (6), 340–343.

Shahwan, T., Sirriah, S.A., Nairat, M., Boyac, E., Eroglu, A.E., Scott, T.B., Hallam, K.R., 2011. Green synthesis of iron nanoparticles and their application as a fenton-like catalyst for the degradation of aqueous cationic and anionic dyes. Chem. Eng. J. 72, 258–266.

Shuai, X., Yost, R.S., 2010. Kinetics of ion-pair formation on variable-charge minerals using the frequency domain method. Soil Sci. Soc. Am. J. 74 (5), 1568–1576.

Shukla, S., Jadaun, A., Arora, V., Sinha, R.K., Biyani, N., Jain, V.K., 2015. In vitro toxicity assessment of chitosan oligosaccharide coated iron oxide nanoparticles. Toxicol. Rep. 2, 27–39.

Shweta, Tripathi, D.K., Singh, S., Singh, S., Dubey, N.K., Chauhan, D.K., 2016. Impact of nanoparticles on photosynthesis: challenges and opportunities. Mater. Focus 5 (5), 405–411.

Siddiqui, M.H., Al-Whaibi, M.H., Faisal, M., Al Sahli, A.A., 2014. Nano-silicon dioxide mitigates the adverse effects of salt stress on *Cucurbita pepo* L. Environ. Toxicol. Chem. 33 (11), 2429–2437.

Singh, S., Tripathi, D.K., Dubey, N.K., Chauhan, D.K., 2016. Effects of nano-materials on seed germination and seedling growth: striking the slight balance between the concepts and controversies. Mater. Focus 5 (3), 195–201.

Singh, S., Vishwakarma, K., Singh, S., Sharma, S., Dubey, N.K., Singh, V.K., Liu, S., Tripathi, D.K., Chauhan, D.K., 2017. Understanding the plant and nanoparticle interface at transcriptomic and proteomic level: a concentric overview. Plant Gene. http://dx.doi.org/10.1016/j.plgene.2017.03.006.

Siva, G.V., Benita, L.F.J., 2016. Iron oxide nanoparticles promotes agronomic traits of ginger (*Zingiber officinale* Rosc.). Int. J. Adv. Res. Biol. Sci. 3 (3), 230–237.

Soliman, A.S., El-feky, S.A., Darwish, E., 2015. Alleviation of salt stress on *Moringa peregrina* using foliar application of nanofertilizers. J. Hortic. For. 7, 36–47.

Sun, C., Zhou, R., Jianan, E., Sun, J., Ren, H., 2015. Magnetic CuO@Fe_3O_4 nanocomposite as a highly active heterogeneous catalyst of persulfate for 2,4-dichlorophenol degradation in aqueous solution. RSC Adv. 5 (70), 57058–57066.

Sun, L., Wu, W., Zhang, S., Liu, Y., Xiao, X., Ren, F., Cai, G., Jiang, C., 2013. Spindle-like α-Fe_2O_3 embedded with TiO_2 nanocrystalline: ion implantation preparation and enhanced magnetic properties. J. Nanosci. Nanotechnol. 13 (8), 5428–5433.

Tang, S.C., Lo, I.M., 2013. Magnetic nanoparticles: essential factors for sustainable environmental applications. Water Res. 47 (8), 2613–2632.

Tartaj, P., González-Carreño, T., Serna, C.J., 2002. Synthesis of nano-magnets dispersed in colloidal silica cages with applications in chemical separation. Langmuir 18, 4556–4558.

Teo, W.E., Inai, R., Ramakrishna, S., 2011. Technological advances in electrospinning of nanofibers. Sci. Technol. Adv. Mater 12, 013002.

Tiwari, M., Sharma, N.C., Fleischmann, P., Burbage, J., Venkatachalam, P., Sahi, S.V., 2017. Nanotitania exposure causes alterations in physiological, nutritional and stress responses in tomato (*Solanum lycopersicum*). Front. Plant Sci. 8.

Torabian, S., Zahedi, M., Khoshgoftar, A.H., 2016. Effects of foliar spray of two kinds of zinc oxide on the growth and ion concentration of sunflower cultivars under salt stress. J. Plant Nutr. 39 (2), 172–180.

Tosco, T., Papini, M.P., Viggi, C.C., Sethi, R., 2014. Nanoscale zerovalent iron particles for groundwater remediation: a review. J. Clean. Prod. 77, 10–21.

Tripathi, D.K., Singh, V.P., Prasad, S.M., Chauhan, D.K., Dubey, N.K., 2015. Silicon nanoparticles (SiNp) alleviate chromium(VI) phytotoxicity in *Pisum sativum* (L.) seedlings. Plant Physiol. Biochem. 96, 189–198.

Tripathi, D.K., Singh, S., Singh, V.P., Prasad, S.M., Chauhan, D.K., Dubey, N.K., 2016. Silicon nanoparticles more efficiently alleviate arsenate toxicity than silicon in maize cultiver and hybrid differing in arsenate tolerance. Front. Environ. Sci. 4, 46.

Tripathi, D.K., Singh, S., Singh, S., Pandey, R., Singh, V.P., Sharma, N.C., Prasad, S.M., Dubey, N.K., Chauhan, D.K., 2017a. An overview on manufactured nanoparticles in plants: uptake, translocation, accumulation and phytotoxicity. Plant Physiol. Biochem. 110, 2–12.

Tripathi, D.K., Mishra, R.K., Singh, S., Singh, S., Vishwakarma, K., Sharma, S., Singh, V.P., Singh, P.K., Prasad, S.M., Dubey, N.K., Pandey, A.C., 2017b. Nitric oxide ameliorates zinc oxide nanoparticles phytotoxicity in wheat seedlings: implication of the ascorbate–glutathione cycle. Front. Plant Sci. 8.

Tripathi, D.K., Singh, S., Singh, S., Srivastava, P.K., Singh, V.P., Singh, S., Prasad, S.M., Singh, P.K., Dubey, N.K., Pandey, A.C., Chauhan, D.K., 2017c. Nitric oxide alleviates silver nanoparticles (AgNps)-induced phytotoxicity in *Pisum sativum* seedlings. Plant Physiol. Biochem. 110, 167–177.

Tripathi, D.K., Tripathi, A., Shweta, S.S., Singh, Y., Vishwakarma, K., Yadav, G., Sharma, S., Singh, V.K., Mishra, R.K., Upadhyay, R.G., Dubey, N.K., 2017d. Uptake, accumulation and toxicity of silver nanoparticle in autotrophic plants, and heterotrophic microbes: a concentric review. Front. Microbiol. 8.

Tripathi, D.K., Singh, S., Singh, V.P., Prasad, S.M., Dubey, N.K., Chauhan, D.K., 2017e. Silicon nanoparticles more effectively alleviated UV-B stress than silicon in wheat (*Triticum aestivum*) seedlings. Plant Physiol. Biochem. 110, 70–81.

Van Aken, B., 2015. Gene expression changes in plants and microorganisms exposed to nanomaterials. Curr. Opin. Biotechnol. 33, 206–219.

Vasconcelos, M.W., Grusak, M.A., 2014. Morpho-physiological parameters affecting iron deficiency chlorosis in soybean (*Glycine max* L.). Plant Soil 374 (1–2), 161–172.

Wang, H., Kou, X., Pei, Z., Xiao, J.Q., Shan, X., Xing, B., 2011. Physiological effects of magnetite (Fe_3O_4) nanoparticles on perennial ryegrass (*Lolium perenne* L.) and pumpkin (*Cucurbita mixta*) plants. Nanotoxicology 5 (1), 30–42.

Wang, C., Luo, H., Zhang, Z., Wu, Y., Zhang, J., Chen, S., 2014. Removal of As(III) and As(V) from aqueous solutions using nanoscale zero valent iron-reduced graphite oxide modified composites. J. Hazard. Mater. 268, 124–131.

Wei, W., Wu, Z., Yu, T., Jiang, C., Kim, W.S., 2015. Recent progress on magnetic iron oxide nanoparticles: synthesis, surface functional strategies and biomedical applications. Sci. Technol. Adv. Mater. 16, 43.

Wongwailikhit, K., Horwongsakul, S., 2011. The preparation of iron(III) oxide nanoparticles using W/O microemulsion. Mater. Lett. 65 (17), 2820–2822.

Worms, I.A., Boltzman, J., Garcia, M., Slaveykova, V.I., 2012. Cell-wall-dependent effect of carboxyl-CdSe/ZnS quantum dots on lead and copper availability to green microalgae. Environ. Pollut. 167, 27–33.

Wu, W., He, Q., Jiang, C., 2008. Magnetic iron oxide nanoparticles: synthesis and surface functionalization strategies. Nanoscale Res. Lett. 3 (11), 397.

Xu, Y., Qin, Y., Palchoudhury, S., Bao, Y., 2011. Water-soluble iron oxide nanoparticles with high stability and selective surface functionality. Langmuir 27 (14), 8990–8997.

Xu, H., Zeiger, B.W., Suslick, K.S., 2013. Sonochemical synthesis of nanomaterials. Chem. Soc. Rev. 42 (7), 2555–2567.

Yadav, M., Rhee, K.Y., Park, S.J., Hui, D., 2014. Mechanical properties of Fe_3O_4/GO/chitosan composites. Compos. B Eng. 66, 89–96.

Ye, L., Li, L., Wang, L., Wang, S., Li, S., Du, J., Zhang, S., Shou, H., 2015. MPK3/MPK6 are involved in iron deficiency-induced ethylene production in *Arabidopsis*. Front. Plant Sci. 6.

Yoon, S.Y., Lee, C.G., Park, J.A., Kim, J.H., Kim, S.B., Lee, S.H., Choi, J.W., 2014. Kinetic, equilibrium and thermodynamic studies for phosphate adsorption to magnetic iron oxide nanoparticles. Chem. Eng. J. 236, 341–347.

Zaimenko, N.V., Didyk, N.P., Dzyuba, O.I., Zakrasov, O.V., Rositska, N.V., Viter, A.V., 2014. Enhancement of drought resistance in wheat and corn by nanoparticles of natural mineral analcite. Ecol. Balk. 6 (1).

Zareii, F.D., Roozbahani, A., Hosnamidi, A., 2014. Evaluation the effect of water stress and foliar application of Fe nanoparticles on yield, yield components and oil percentage of safflower (*Carthamus tinctorious* L.). Int. J. Adv. Biol. Biomed. Res. 2, 1150–1159.

Zhao, Y.S., Sun, C., Sun, J.Q., Zhou, R., 2015. Kinetic modeling and efficiency of sulfate radical-based oxidation to remove p-nitroaniline from wastewater by persulfate/Fe_3O_4 nanoparticles process. Sep. Purif. Technol. 142, 182–188.

Zhu, H., Han, J., Xiao, J.Q., Jin, Y., 2008. Uptake, translocation, and accumulation of manufactured iron oxide nanoparticles by pumpkin plants. Environ. Monit. 10, 685–784.

CHAPTER

11

Effects of Rare Earth Oxide Nanoparticles on Plants

Hamaad R. Ahmad[1], *Muhammad Zia-ur-Rehman*[1], *Muhammad I. Sohail*[1,2], *Muhammad Anwar ul Haq*[1], *Hinnan Khalid*[1], *Muhammad A. Ayub*[1], *Gohar Ishaq*[1]

[1]University of Agriculture Faisalabad, Faisalabad, Pakistan; [2]University of Florida, Institute of Food and Agricultural Sciences, Indian River Research and Education Center, Fort Pierce, FL, United States

11.1 INTRODUCTION

Nanotechnology has revolutionized many facets of modern society since the early 2000s through extensive application in all fields of life (Deng et al., 2014). Nanoparticles (NPs) are the building blocks of nanotechnology. Nanomaterials possess at least two external or internal dimensions at the nanoscale of 1–100 nm (Lövestam et al., 2010; Remédios et al., 2012). The nanoscale of NPs gives them vital properties such as high surface area per unit volume, more biochemical reactivity, faster transport across the biological membranes, and novel optical, electrical, quantum, and magnetic properties as compared to their bulk ions or materials. Engineered nanoparticles (ENPs) are intentionally produced using different materials (Bandyo-padhyay et al., 2012; Roco, 2004; Aziz et al., 2015; Prasad et al., 2016; Tripathi et al., 2017a). Among listed nanomaterials, metal oxide NPs are more widely produced for commercial application and they show diverse behavior when interacted with plants (Tripathi et al., 2015, 2017a,b,c,d,e; Bradfield et al., 2017; Aziz et al., 2016; Singh et al., 2016; Shweta et al., 2016).

Usually, people are unaware of the blessing and benefits of rare earth elements (REEs) but they are becoming a big part of today's modern lifestyle (Chaturvedi et al., 2014). These REEs are a collection of 17 elements found in the periodic table, which includes 15 lanthanide elements including yttrium (Yt) and scandium (Sc). These REEs are further grouped based on atomic masses. The light REEs include cerium (Ce), lanthanum (La), neodymium (Nd), praseodymium (Pr), promethium (Pm), europium (Eu), and samarium (Sm), and the

Nanomaterials in Plants, Algae, and Microorganisms
http://dx.doi.org/10.1016/B978-0-12-811487-2.00011-6

heavy REEs gadolinium (Gd), erbium (Er), dysprosium (Dy), thulium (Tm), ytterbium (Yb), lutetium (Lu), and terbium (Tb). The oxidation states of REEs vary between +2 and +4 (d'Aquino et al., 2009). REEs are lithophile, meaning they are enriched in the earth's crust, but in living systems they are found in traces. In fact, most REEs exist in the earth's crust at higher concentration, i.e., 100–200 mg/kg (Chaturvedi et al., 2014; Liang et al., 2005). Rare earth oxide nanoparticles (REONPs) are not only naturally abundant but are also manufactured as ENPs on a large scale for diverse applications. Cerium and yttrium oxide NPs are produced at a very large scale and are listed in the top 10 most producing NPs (Keller and Lazareva, 2014). The estimated production of cerium oxide NPs ($nCeO_2$) is around 1000 tons/year globally (Piccinno et al., 2012) and this depiction makes the proximal role of REONPs clear in every aspect of life.

In agriculture, REONPs have created a revolution by increasing crop growth and productivity (Kah and Hofmann, 2014; Campos et al., 2014; Pereira et al., 2014; Grillo et al., 2014). REONPs are at the top of agriculturally beneficial NPs (Servin and White, 2016; Servin et al., 2015) and thus have gained much importance for testing their proficient aspects coupled with economic analysis by the Organization for Economic Cooperation and Development (Servin and White, 2016). Many studies predict that REONPs have potential to interact with crop plants, perhaps for more than one generation (Schwabe et al., 2015). In China, 50–100 million tons of REE oxides are being used for agriculture annually (Liang et al., 2005; Xiong et al., 2001). REE fertilizer use is escalating in many developed countries: Australia, the Philippines, Korea, Switzerland, and Japan (Wang et al., 2008; Redling, 2006). Cerium oxide NPs are widely experimented on as a plant growth promoter; however, the heavy REONPs are being tested less for agricultural production but this does not indicate that heavy REONPs such as Cs, Yb, etc. cannot be used. Plant physiologists are considering cerium as a potentially beneficial element for plants (Pilon-Smits et al., 2009).

Plants are the major pillar of ecosystem sustainability and food security and they are more exposed to REONPs (Cui et al., 2014). REONPs are taken up by plant roots and reach seeds and/or fruits (Cui et al., 2014). The entry of REONPs in agriculture may change the profile of soil–plant systems (Du et al., 2017). There is an urgent need to address possible comprehensive benefits associated with the use of REONPs for agriculture. Rare earth oxides in the form of REONPs are trending in many countries as fertilizer additives for plant growth promotion and stress relief (Barrios et al., 2017; Servin et al., 2015; Gonzalez et al., 2014), and concerns about their entry into the environment and food chain have also been raised (Chen et al., 2016).

REONPs have the ability to improve plant growth and development by modifying growth parameters and physiological processes. REONPs stimulate the antioxidant system, increase absorption and application of mineral nutrients, enhance electron transport frequency in the photosynthesis process, and improve seed germination rate, which has a direct link with crop growth and production (de Oliveira et al., 2015; Giraldo et al., 2014; Ramos et al., 2016). A surge in research has diverted the understanding of the mechanisms of REONP interaction with plants (Khot et al., 2012), but confusion regarding their mechanisms and potential benefits for plants is ongoing (Khot et al., 2012); however, the effects of bulk REE oxides on plant growth and quality through soil and/or foliar application have been well confirmed (d'Aquino et al., 2009; Xu et al., 2002; Zhang and Shan, 2001; Liu et al., 2012; Pang et al., 2002). Along with the positive impacts of REONPs, adverse properties of REONPs on plants have also been testified, and investigators have considered their effects on dissimilar parameters of plant progress, focusing on their use in agronomics (Arruda et al., 2015).

11.2 GEOLOGICAL OCCURRENCE AND SOURCES OF REONPs

11.2.1 Occurrence

REONPs have long been acknowledged as valuable because of their exclusive physicochemical properties. The natural existence of REONPs is strongly reliant on geological situations; in a few locations, they are found in adequate quantities. From an agriculture point of view, rare earth materials have also been of long-standing interest as challenges for additional scientific study. REONPs are mostly found primarily in carbonates (Strambeanu et al., 2015). Igneous rocks contain more than 50% carbonate crystals, mainly calcite. These rocks are the most obvious place to look for rare earth oxides because carbonates always contain rare earth oxide, which constitutes about 3500 ppm of these rocks (Buzea et al., 2007). REONPs are found in large concentrations in some alkaline igneous rocks as well. These rocks constitute about 0.5% of all igneous rocks. They are categorized by their large amounts of alkali metals. There are several secondary deposits of REONPs. These REE oxides are produced by the weathering of primary minerals and their succeeding concentration by chemical and physical means. There are also some REONP mineral concentrations of granite pegmatites and hydrothermal vein systems (Shi et al., 2001). Studies have found increased concentrations of rare earth metals in the deep-sea mud in the oceans.

The major reserves of REONPs are found in the United States including Mountain Pass; in Russia; in China including Bayan Obo; and in Australia including Mount Weld. Significant reserves are also present in Vietnam, India, Thailand, Malaysia, South Africa, Indonesia, Canada, Greenland, and Brazil (Wagner et al., 2014). China contains the highest known amount of REONPs. It is an iron-rich deposit accommodated in dolomite, the origin of which is still a subject of discussion. The geological occurrence of REE oxide metals is of great importance because it affects the cost of mining REONPs. Maoniuping is at present the world's major REE manufacturer, mainly because REONP minerals are easy to concentrate by gravitational approaches (Hassellöv, 2012). Before Bayan Obo (China) came online, the biggest source of REONPs was a carbonatite sheet in the United States at Mountain Pass in California. Mining at Mountain Pass stopped in 2002, but has now restarted and is returning to full production. Mountain Pass is a significant contributor to global supply (Geological Society 2011).

11.2.2 Sources of REONPs

11.2.2.1 Natural Sources

NPs are abundant in nature, and are formed because of volcanic eruptions, photochemical reactions, erosion, forest fires, and by animals and plants. Air pollution is often linked with anthropogenic activities such as industry, automobiles, charcoal burning, and forest burning, which can produce huge quantities of nanomaterials. The biggest single source of environmental REONPs is dust storms. Almost 50% of atmospheric NPs are mineral, originating from deserts (Shi et al., 2005). In extraterrestrial space, NPs exist widely. Grass fires and forest fires have long been a part of earth's natural history. Ash and smoke can be spread by fires and lead to a proliferation of particulate matter including NPs (Sapkota et al., 2005). Volcanos are the biggest contributor to natural sources of NPs. When the volcanic ash reaches the upper stratosphere and troposphere it can spread worldwide for years without boundaries

(Taylor, 2002). A large amount of REONP materials are released from oceans and other water bodies. These sprays are made by water evaporation and when wave-formed water droplets are ejected into the air. Precipitation is also a source of NPs (Buseck and Posfai, 1999). Some organisms such as bacteria and viruses are also a source of NPs. These organisms are smaller than a few microns, e.g., bacteria 30 nm to 700 μm and viruses from 10 to 400 nm. However, it should be made clear that there is a difference between these organisms and NPs. These organisms are short lived and oxygen dependent. In comparison, nanomaterials are typically inorganic and need no supply of oxygen (Ahmad et al., 2005).

11.2.2.2 Anthropogenic Sources

Human beings have produced rare earth oxides for centuries, because they are the consequences of food cooking and simple combustion, welding, chemical manufacturing, and combustion in airplane engines and vehicles (Joner et al., 2008). REONPs are produced by engineering work used to make tires, cosmetics, clothing, and food additives. Vehicle exhausts are the main source of atmospheric NPs in city areas. The size of most of the particles from automobile exhausts is in the range of 20–60 nm for gasoline engines and 20–130 nm for diesel engines. NPs are normally spherical in shape (Westerdahl et al., 2005). Human activities generate a significant amount of REONPs. These REONPs are produced by domestic activities such as smoking, cooking, and burning (Sioutas et al., 2005). Smoking is also a common source of rare earth oxides. Tobacco smoke is composed of nanomaterials with sizes extending from 10 to 700 nm. Tobacco smoke has a very complicated composition, with approximately 100,000 chemical compounds and components (Ning et al., 2006). Building demolition and the cosmetics industry are also sources of NPs. The use of nanomaterials in cosmetics is not new. Many personal and cosmetic care products include nanomaterials. Silver NPs are used as antifungal and antibacterial agents in many applications. Coatings of NPs are extensively used for improving fabrics to create dyes and wrinkle-free properties (Shi et al., 2001). The major source of REONPs in agricultural soil is biosolids (sewage, sludge, composts) and raw sewage effluents.

11.3 CHARACTERIZATION, TYPES, AND SYNTHESIS OF REONPs

11.3.1 Characterization of REONPs

REONPs have been manufactured in different shapes and sizes depending upon the method of preparation. They are characterized through X-ray diffraction (XRD), electron microscopy, Fourier transform infrared spectroscopy, selected area electron diffraction, along with particle size analysis (Arruda et al., 2015). XRD pattern and electron microscope imagery are mostly used to make a report about the dimensions and shape of prepared NPs. Advanced techniques for the characterization of REONPs are given in Fig. 11.1.

11.3.2 Types of REONPs

Engineered REONPs are being produced on a commercial as well as laboratory scale. A list of REONPs is given next.

Sr. No	Name of REONP	Chemical Formulae
1	Cerium oxide nanoparticles	CeO/CeO_2
2	Dysprosium oxide nanoparticles	Dy_2O_3
3	Erbium oxide nanoparticles	Er_2O_3
4	Europium oxide nanoparticles	Eu_2O_3
5	Gadolinium oxide nanoparticles	Gd_2O_3
6	Holmium oxide nanoparticles	Ho_2O_3
7	Lanthanum oxide nanoparticles	La_2O_3
8	Lutetium oxide nanoparticles	Lu_2O_3
9	Neodymium oxide nanoparticles	Nd_2O_3
10	Praseodymium oxide nanoparticles	Pr_6O_{11}
11	Samarium oxide nanoparticles	Sm_2O_3
12	Scandium oxide nanoparticles	Sc_2O_3
13	Terbium oxide nanoparticles	Tb_4O_7
14	Thulium oxide nanoparticles	Tm_2O_3
15	Ytterbium oxide nanoparticles	Yb_2O_3
16	Yttrium oxide nanoparticles	Y_2O_3
17	Lanthanum oxide nanoparticles	La_2O_3

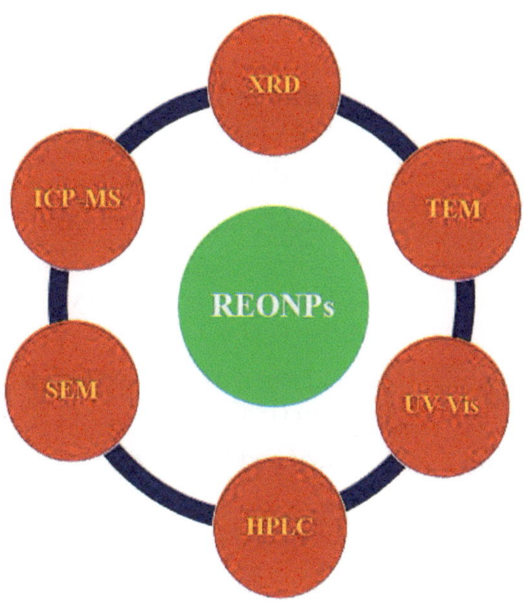

FIGURE 11.1 Different techniques employed for the characterization of rare earth oxide nanoparticles (REONPs) in the environment. *HPLC*, Highperformance liquid chromatography; *ICP-MS*, inductively coupled plasma mass spectrometry; *SEM*, scanning electron microscopy; *TEM*, transmission electron microscopy; *UV-vis*, UV-visible spectrophotometry; *XRD*, X-ray diffraction.

11.3.3 Methods to Prepare REONPs

A wide range of laboratory and commercial production techniques have been used since 1971 for REONP synthesis. A few well-known techniques with references are listed next.

Sr. No.	Mechanism	REONPs	References
1	Calcination of the metal salt	Praseodymium oxide NPs	Borchert et al. (2008)
2	Pechini method	Praseodymium oxide NPs	Borchert et al. (2008)
3	Citrate method	Praseodymium oxide NPs	Borchert et al. (2008)
4	Combustion method	Dysprosium oxide NPs	Zelati et al. (2014a,b)
5	Gas condensation	Yttrium oxide NPs	Gunter and Kumpmann (1992) and Skandan et al. (1991)
		Erbium oxide NPs	Hong et al. (1998), Tissue (1998) and Eilers and Tissue (1995)
		Cerium oxide NPs	Tschope et al. (1995) and Tschöpe and Ying, (1994a,b)
6	Chemical vapor deposition	Yttrium oxide NPs	Sharma et al. (1998) and Tomaszewski et al. (1997)
		Cerium oxide NPs	Hakuta et al. (1998), Chengyun et al. (1996), Hirano and Kato (1996, 1999) and Zhou and Rahaman (1993)
7	Precipitation method	Scandium oxide NPs	Li et al. (2003)
		Yttrium oxide NPs	Sordelet and Akinc (1998), Chen and Chen (1996), Kobayashi, 1992 and Ciftcioglu et al. (1987)
		Lanthanum oxide NPs	Xie et al. (1992)
		Cerium oxide NPs	Chen and Chen (1993) and Xie et al. (1992)
		Samarium oxide NPs	Gao et al. (2002) and Matijević and Hsu (1987)
		Europium oxide NPs	Ma and Yan (1995) and Matijević and Hsu (1987)
		Gadolinium oxide NPs, terbium oxide NPs	Matijević and Hsu (1987)
		Praseodymium oxide NPs	Zinatloo-Ajabshir et al. (2015)

—Cont'd

Sr. No.	Mechanism	REONPs	References
8	Hydrothermal and solvothermal methods	Yttrium oxide NPs	Sharma et al. (1998) and Tomaszewski et al. (1997)
		Cerium oxide NPs	Hirano and Kato (1996, 1999), Hakuta et al. (1998), Chengyun et al. (1996) and Zhou and Rahaman (1993)
		Neodymium oxide NPs	Zawadzki and Kępiński (2004)
9	Sol–gel method	Scandium oxide NPs, dysprosium oxide NPs, erbium oxide NPs, ytterbium oxide NPs	Grosso and Sermon (2000)
		Lanthanum oxide NPs, cerium oxide NPs, gadolinium oxide NPs	Imoto et al. (1988)
		Dysprosium oxide NPs, erbium oxide NPs, ytterbium oxide NPs	Mazdiyasni and Brown (1971)
		Yttrium oxide NPs	Abdulghani and Al-Ogedy (2015)
		Praseodymium oxide NPs	Borchert et al. (2008)
		Ytterbium oxide NPs	Dong et al. (2001)
		Scandium oxide NPs	Poirot et al. (2010)
10	Emulsion and microemulsion method	Yttrium oxide NPs	Lee et al. (2000)
		Cerium oxide NPs	He et al. (2003), Wu et al. (2001) and Masui et al. (1997)
		Neodymium oxide NPs	Que et al. (2001a)
		Erbium oxide NPs	Que et al. (2001b,c)
11	Ultrasound and microwave irradiation method	Yttrium oxide NPs, lanthanum oxide NPs	Wang et al. (2002) and Yin et al. (2002)
		Cesium oxide NPs, samarium oxide NPs	Wang et al. (2002) and Liao et al. (2001)
		Europium oxide NPs, erbium oxide NPs	Pol et al. (2002)
12	Electrospinning	Ytterbium oxide NPs	Henriques et al. (2015)
13	Thermalization of [La(acacen) (NO_3) (H_2O)]	Lanthanum oxide NPs	Saravani and Jehali (2015)

Continued

—Cont'd

Sr. No.	Mechanism	REONPs	References
BIOLOGICAL METHODS			
14	Mycosynthesis method	Cerium oxide NPs	Gopinath et al. (2015), Khan and Ahmad (2013) and Munusamy et al. (2013)
		Gadolinium oxide NPs	Khan et al. (2014)
15	Solution combustion method using fruit extracts	Cerium oxide NPs	Yadav et al. (2016)
16	Microbe (*Lactobacillus* sp.)-mediated biosynthesis	Gadolinium oxide NPs	Jha et al. (2010)
17	Biosynthesis using plants	Cerium oxide NPs	Priya et al. (2014)
		Yttrium oxide NPs	Kannan and Sundrarajan (2015)
		Lanthanum oxide NPs	Chatterjee et al. (2016)

NPs, nanoparticles.

11.4 APPLICATION OF REONPs IN SOIL

11.4.1 Solubility of REONPs in Soil

Soil is a heterogeneous mixture comprised of a variety of constituents contributing to its properties. The solubility of REONPs in soil is highly pH dependent. Arai et al. (2013) used a batch dissolution method and reported that CeO_2 NPs were only soluble in soil at a pH lower than 5 and thus bioavailability of CeO_2 NPs is also controlled by physicochemical properties of soil as it is for all other metals (Ernst, 1996). Soil influences the fractionation of applied REONPs and this is how it controls the solubility of applied REONPs; aggregation of REONPs decreases their solubility in soil solution. Cornelis et al. (2011) reported that dissolution of CeO_2 NPs in soil is close to the detectable limit at pH 7–9, while at pH 4 its dissolution rate is 3.1%. In CeO_2 NPs cerium has oxidation state +4 and it changes its oxidation state to +3 when it comes in contact with soil solution. In oxidizing conditions, Ce is present in +4 and forms cerium oxide, which remains undissolved. Keller et al. (2010) stated that the combined effect of pH, ionic strength, organic matter, and different characters of liquid media can cause aggregation or stabilization of CeO_2 in naturally occurring water matrices. So, application of such water-containing REONPs can also alter its fate in soil solution. Dissolved soil organic matter can stabilize REONPs (Yang et al., 2009; Cornelis et al., 2014), thus making them more bioavailable (Majumdar et al., 2015b). Cao et al. (2001) reported that the pH and redox potential of soil controls release of rare earth metals from soil.

11.4.2 Interaction of REONPs With Inorganic Components of Soil

Inorganic components of soil include clay and minerals of Fe and Al. In acidic soil, pH is lower than 7, thus soil is positively charged. Soil REDOX potential and pH play a vital role in the dissolution of REONPs; phosphate and/or organic molecules such as citrate present in soil can adsorb onto REONPs making them negatively charged (Cornelis et al., 2011). REONPs, when in contact with soluble organic matter, attain a negative charge, thus they become attached with clay sides as well as on Fe/Al minerals. Clay and Fe/Al minerals contribute toward positive charge on soil colloids making sites for specific adsorption of REONPs, thereby increasing bioavailability and decreasing leaching.

11.4.3 Effect of REONPs on Soil Microorganisms

REONPs have been reported to have negative effects of the microbial community when tested in lab conditions. Balusamy et al. (2012) reported that lanthanum oxide NPs had negative effects on the growth of *Staphylococcus aureus* (Gram-positive bacteria), *Escherichia coli*, and *Pseudomonas aeruginosa* (Gram-negative bacteria) in lab conditions, which are responsible for different diseases in humans. Similarly, Anaya et al. (2016) reported that dysprosium oxide NPs can cause damage to the cell membrane of *E. coli* bacterium.

Contrary to this, CeO_2 NPs have been reported to have a significant effect in increasing the metabolic quotient (qCO_2) of soil (Antisari et al., 2013), which is an indicator of soil bacterial action (Schimel and Schaeffer, 2012), while microbial biomass, which is a subtle indicator of soil health (Brookes, 1995), remained unaffected. On dissolution, REONPs can provide rare earth metals, which in low concentration play a role in promoting soil microbial diversity, while in higher concentration destroy it (Jinxia et al., 2010). CeO_2 NPs have no effect on the microbial community in soil without crops, while they can alter the microbial community with crops because of the effect on root exudates and plant growth (Ge et al., 2014). CeO_2 NPs also have a bacteriostatic effect on *Sinorhizobium meliloti*, which is a nitrogen-fixing bacterium that lives in symbiotic association with alfalfa (Bandyopadhyay et al., 2012). Dahle and Arai (2014) reported that CeO_2 NPs were less toxic to the soil denitrification process than Ce^{+3} concentration in soil.

11.4.4 Interaction With Other Mineral Elements

REONPs can release rare earth metal ions in soil, which encounters an anionic community of soil solution and can be precipitated in many forms. The most important anion in this aspect is phosphate. Phosphate is a very important nutrient is soil systems and is exposed to many challenges. Rui et al. (2015) reported that the presence of phosphate in a system leads to the development of needle-like $CePO_4$ crystals outside the epidermis of cucumber seedlings. NPs can undergo many changes in natural conditions (Lowry et al., 2012) making it difficult to study their individual behavior. Phosphate is responsible for the speciation of rare earth metal release in soil. A number of studies (Zhang et al., 2011a,b, 2012a,b; Ma et al., 2010) have reported the formation of $REPO_4$ when phosphate encounters REE oxides.

11.5 DYNAMICS OF REONPs IN SOILS AND PLANTS

11.5.1 Uptake of REONPs

REONPs are a source of rare earth metals in soil and thus release metal ions in soil solution upon dissolution, or REONPs themselves are absorbed by plant roots. There are various mechanisms for the uptake of REONPs by plants under dynamic conditions of the soil matrix. López-Moreno et al. (2010) reported that CeO_2 NPs are taken up by soya bean roots in hydroponic medium, and in cucumber they can also be transported to plant shoots (Zhang et al., 2011a,b). However, some studies (Ma et al., 2015; Schwabe et al., 2015; Rui et al., 2015) reported that CeO_2 releases Ce^{+3} ions, which are taken up and accumulated by plants. Schwabe et al. (2015) reported that CeO_2 NPs were taken up by plants in both CeO_2 NPs and Ce^{+3} ions, which again transformed to CeO_2 inside plant parts. Moreover, the bioavailability of the dissolution of CeO_2 was increased in the presence of plant roots rather than in the absence of plants. Birbaum et al. (2010) reported that foliar applied CeO_2 was taken up by corn leaves but not translocated to new leaves, meaning that it showed immobile behavior inside plants. Singh et al. (2016) reported that CeO_2 NPs can be taken up by plant roots. Hernandez-Viezcas et al. (2013) also reported that CeO_2 NPs can be taken up by plants. They can translocate to aerial parts and Ce can change its oxidation state. Uptake of rare earth ions coming from REONPs increases with increase in soil-applied REONP concentration (López-Moreno et al., 2010). A schematic pathway for REONPs is given in Fig. 11.2.

11.5.2 Transport of REONPs in Plant Parts

Plants intermingle intensely in their polluted environment with harmful substances, including REONPs, and therefore become susceptible to their potential effects. REONPs have a large number of applications because of their specific physical and chemical properties and multidisciplinary applications, and these exposures to plants make them a priority for study (Peng et al., 2012; Anjum et al., 2013).

11.5.2.1 REONPs–Plant Root Interaction and Its Regulation

One of the main routes of REONP entry into plant systems is via leaves during injection, leaf spray, and atmospheric contact (Deng et al., 2014). On the other hand, roots are the primary plant structure close to soils potentially contaminated with REONPs, and the main entry point of REONPs for other plant tissues (Anjum et al., 2015). Procedures such as accumulation, translocation, and sorption of REONPs can be controlled by plant types and by the stability and chemical composition of REONPs (Rico et al., 2011). Furthermore, differential REONP absorption and adsorption is also possible in the roots. Plants are used to produce various REONPs to establish their benefits in the plants, and to confirm the important procedure of REONP delivery in other organs of plants from the roots (Kumari et al., 2015).

11.5.2.2 Processes at Root Level

Because of high specific surface and surface reactivity of the roots, REONPs can be easily adsorbed on the general physical boundary. Certain ENPs can control the interactions

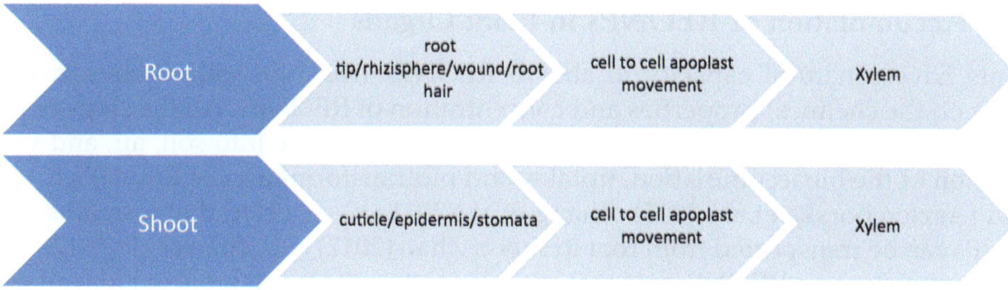

FIGURE 11.2 Pathway of uptake of rare earth oxide nanoparticles (REONPs) from soil to plant through root. *Authors.*

of REONPs with the roots and their accumulation in the epidermis (Zhao, 2012; Zhao et al., 2012). Efforts must be made to reflect both bulk particles and ions in exposure experiments, and assess their levels in the tissues not in direct contact with exposure media (Deng et al., 2014). If the roots in the nutrient solution remain directly in contact with REONPs, their localization and absorption are possible mainly in the epidermis and exodermis (Peng et al., 2012). REONPs can enter the root epidermis and then the cortex, and finally pass to the endodermis (Zhao, 2012; Zhao et al., 2012). Plants use the xylem vessels for the movement of REONPs from the root to the aerial parts of the plants (Cifuentes et al., 2010).

11.5.2.3 Processes at Cellular Level

Before entering the cell and their related intracellular parts, REONPs must intermingle with the plant cell wall, which has a network of cellulose, lignin, and hemicellulose microfibrils that is further infused by pectin in the size range from 5 to 20 nm. The entry of individual and smaller particles into the symplastic and apoplastic flow is possible after diffusion of these REONPs through pores. Alternatively, collaboration of these REONPs with aquaporins, carrier proteins, organic chemicals, and ion channels can also facilitate their entry into the plant cells via the symplastic route. Here the important thing is that the symplastic pathway is the highly regulated pathway for REONP transportation into major crops (Rico et al., 2011). The role of plasmodesmata, especially in the range of 20–50 nm diameters, has been emphasized in the effective transport of REONPs in the plants (Larue et al., 2012).

11.5.2.4 Transport of Root REONPs to Leaves and Edible Plant Parts

The human food chain can be polluted by REONPs harbored in plants that can be donated primarily by REONP accumulation and transportation within the plant parts and species. Mature leaves were claimed as a main site of REONPs transported to shoots. This is because the distance from root to mature leaves is much shorter than the younger leaves, and the exposure time for the mature leaves is also much longer (Anjum et al., 2015). The absorption patterns of water and nutrients within the root facilitate the translocation of root-harbored REONPs to aboveground parts of the plants (Schwabe et al., 2013). This facilitates the movement of REONPs from roots to leaves and other edible parts of the plants (Antisari et al., 2015).

11.5.3 Accumulation of REONPs in Plant Organs

Plants have a natural capacity to absorb REONPs from the soil. Their accumulation depends on the chemical properties and concentration of REONPs. All the steps of production, use, and disposal of REONPs may lead to their discharge into soil, air, and water, so exploration of the bioaccumulation, uptake, and biotransformation of REONPs is urgently needed (Asztemborska et al., 2015). Zhang et al. (2011a,b) reported that a small amount of CeO_2 NPs can be transported from root to shoot. Zhao (2012) and Zhao et al. (2012) reported that surface coating on REONPs and soil organic matter play a vital role in bioavailability of REONPs to plants. From different kinds of REONPs, CeO_2 NPs have significantly different effects on plant communities (Ma et al., 2010). Wang et al. (2012a,b) also reported accumulation of CeO_2 NPs in tomato fruit grown in CeO_2-spiked soils. At times, REONPs can enter plant roots but cannot be transported to shoots, as Chen et al. (2016) reported regarding yttrium oxide NPs.

11.6 EFFECT OF REONPs ON PLANT GROWTH

Researchers are working hard to get a better understanding of the effects of different types of REONPs on plant growth and development. Various research and studies revealed that the effects yielded by REONPs are reliant on the type of REONPs, species of plants, and growing medium (i.e., soil or hydroponics medium) (Shah et al., 2012), and these exerted effects are unpredictable and inconsistent in different studies because of different plant organs/tissues, exposure time, elemental form, genetic variability of the same plant species, and kind of treatment (chronic/acute). Therefore many subjects, experimental studies, and challenges concerning the biological consequence of REONPs are unanswered (Arruda et al., 2015). Several studies confirm that REONPs have varying effects on plants (Remédios et al., 2012). Usually, REONPs have a beneficial effect on plant growth at small concentrations, although adverse effects have also been documented at elevated concentrations, yet the mechanism is not very clear. These effects have been elucidated by the "hormesis effect," which is a dose–response spectacle categorized by low-dose stimulus and high-dose inhibition (Du et al., 2017; Ma et al., 2010; Diatloff et al., 2008; Calabrese and Baldwin, 2002). The possible effects of REONPs are summarized in a schematic diagram given in Fig. 11.3. The beneficial effects of REONPs on plant physiology and development are discussed later.

11.6.1 Effect on Seed Germination

Plants growing under natural environments are constantly exposed to both biotic and abiotic stresses at certain stages of their life cycle (Khan et al., 2017). Seed germination is the first stage of a plant's life cycle when exposed to environmental stresses, and successful establishment of seedlings leads to a plant with an optimum capacity for growth and productivity (Khan et al., 2017). REONPs affect the physical as well as biochemical processes of plants at both genotypic and phenotypic levels (Tumburu et al., 2017). Like other NPs, REONPs could sequester nutrients on their surfaces and thus serve as a nutrient

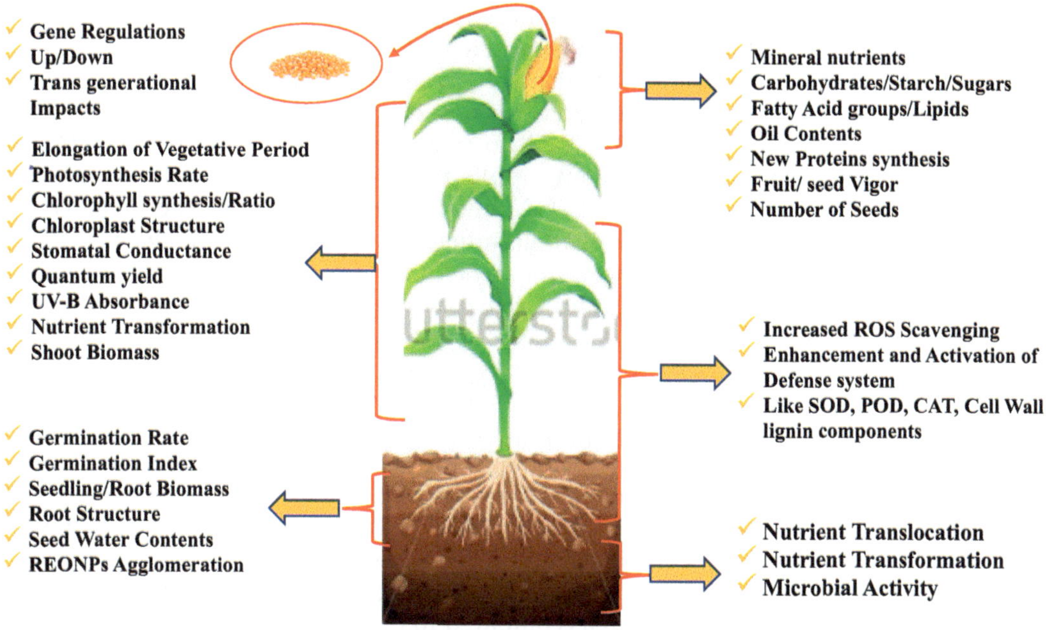

FIGURE 11.3 Schematic diagram of possible effects of rare earth oxide nanoparticles (REONPs) on plant growth. *CAT*, Catalase; *POD*, peroxidase; *ROS*, reactive oxygen species; *SOD*, superoxide dismutase. *Authors.*

stock for growing seeds. However, increased germination rate is the result of penetration of REONPs in the seed coupled with increased seed water uptake (Khodakovskaya et al., 2011). Seed priming with REONPs either alone or a mixture of REONPs at low dose also enhances the moisture contents of sprouting seeds (Peng et al., 2016); this enhancement in water uptake is supposed to be the primary mechanism for the promotion of seed germination. Plant exposure to REONPs causes activation of genes responsible for water channel protein and for better cell growth by regulating the cell cycle. These proximal effects of REONPs exhibit improved seed germination and growth of the plants (Khodakovskaya et al., 2012, 2011). Successful execution of seed germination depends upon the mobilization of stored food by the enzymes amylase and protease for the survival of young plants until they can make their own food by photosynthesis (Khan et al., 2017). The nanooxides of REEs have the ability to alter seed germination in both positive and negative ways depending upon their concentration, seed coat thickness, and growth medium, because REONPs interact with seeds differently in soil and soil-less cultures (Corral-Diaz et al., 2014). Among plants, lettuce is the most sensitive plant to higher doses of REONPs (Ma et al., 2010).

Mild treatment of REONPS enhances seed germination but higher concentrations could adversely affect the germination potential (Corral Diaz et al., 2014) by affecting the fatty acid beta-oxidation that encourages seed dormancy (Pracharoenwattana et al., 2010) or surges the mobilization of the lipid bank in glyoxysomes, which yields significant amounts of reactive oxygen species (ROS) (El-Maarouf-Bouteau and Bailly, 2008), therefore changing seedlings' growth. Chaturvedi et al. (2014) found 100% germination of *Zea mays*, *Vigna radiata*, and *Vigna Mungo* in Hoagland solution containing lanthanum oxide NPs at 50 μmol concentration. Radish seeds can easily tolerate cerium oxide NPs at 500–2000 mg/kg soil treatment and promotes root elongation (Corral Diaz et al., 2014; Ma et al., 2010). López-Moreno et al. (2010) affirmed an enhancement in elongation of cucumber roots and alfalfa stems in plants applied with CeO_2 at 500 mg/L. A high concentration of REONPs (including nano-CeO_2, nano-Gd_2O_3, nano-La_2O_3, and nano-Yb_2O_3) even at 2000 mg/L does not affect the seed germination of rape (*Brassica napus*), *Triticum aestivum* (wheat), *Lactuca sativa* (lettuce), *Brassica oleracea* (cabbage), tomato (*Lycopersicon esculentum*), *Raphanus sativus* (radish), and *Cucumis sativus* (cucumber). Most REONPS at higher concentrations except cerium oxide NPs do not affect root elongation (Ma et al., 2010). REONP application not only promotes the germination of agronomic crops but also enhances the germination of woody plants (Gomez-Garay et al., 2014). It is reported that CeO_2 NPs show better impact on the germination percentage of *Medicago arborea* L. as compared to the bulk cerium oxide application. Some researchers have suggested the use of doping, coating, and upconversion techniques for REONPs with other NPs, metals, or metalloids for plant use. A low concentration (10 μg/mL) of citric acid-coated rare earth upconversion nanophosphors containing a mixed rare earth ion solution (composed of 78 mol% Y, 20 mol% Yb, 1.6 mol% Er, and 0.4 mol% Tm) significantly accelerated the germination and stem growth of mung bean as compared to pure water and citric acid solution priming (Rico et al., 2014).

11.6.2 Enhanced Mineral Nutrient Availability in Soil to Plants

The uptake of mineral nutrients by plant roots is controlled by the accessibility of elements, the capability of the plant to accrue the nutrients, and the antagonism among nutrients (Kabata and Pendias, 2001). The availability of nutrients is affected by a combination of various soil- and plant-related factors (Peralta-Videa et al., 2014). Absorption of sufficient quantities of vital nutrients, largely at the first growth phases, is essential for plant development and growth (Corral Diaz et al., 2014). REONPs modify plant gene expression and related biological pathways, interfering with nutrients and water uptake, finally affecting plant development (Du et al., 2015, 2017). Nano-CeO_2 and bulk CeO_2 significantly enhance the uptake of phosphorus and other nutrients in vegetables and crops (Hong et al., 2016; Morales et al., 2013). Many studies have suggested that REONPs can stimulate plants to absorb, transfer, and assimilate nutrients (Barrios et al., 2015; Chaturvedi et al., 2014). REONP application at the seedling phase changes mineral nutrient concentrations in developed plants, which affects plant health, yield, and quality of produce. Cerium oxide NP application at medium dose in soil enhances the uptake of nitrate (NO_3) and other nutrients from soil to plants (Cui et al., 2014; Gui et al., 2015; Peralta-Videa and Gardea-Torresdey, 2015). Agglomeration of REONPs at the root surfaces alters nutrient availability and uptake. The interaction of REONPs with soil organic matter contents plays an important role in the availability of mineral nutrients to plants, and the dissolution of REONPs in growth media is promoted at acidic pH (Schwabe et al., 2013). On the contrary, some researchers claim that soil organic matter does not affect the REONPs–plant interaction in relation to nutrient uptake and yield productivity (Majumdar et al., 2015b).

Barrios et al. (2015) found that uncoated cerium oxide NPs at any concentration had no significant change in nutrient uptake of tomato plants, but bulk CeO_2 at 62.5 mg/kg increased Zn accumulation in stems. CeO_2 NPs coated with citric acid increase mineral metallic cations in roots and leaves of plants. Trujillo-Reyes et al., (2014) found that citric acid-coated CeO_2 increased the uptake of Al in radish by 93%. Nanoceria reduces Ni uptake from soil medium both at low and high concentrations, while B and Mo uptake increases at medium concentration and P uptake increases at low concentration in radish plants (Corral Diaz et al., 2014). Barrios et al. (2017) found that concentrations of Al, B, Cu, Fe, Mn, Ca, and Mg showed a statistically significant increase compared to controls. Fu et al. (2014) concluded that yttrium increased the Ca, Mg, Mn, Fe, and Mo in the leaves of *Nymphoides peltata* (an aquatic plant used as animal feed). Cerium oxide NPs increased plant height and biomass of transgenic crops, especially of *Bt*-cotton, and a lower application of REONPs also promoted the shoot uptake of base cations (such as Ca, Mg, K); in addition to these elements, root Zn uptake is also promoted (Li et al., 2014). REONPs are commonly known as stable in biological or environmental systems and persist unaltered after being taken up by plant roots. However, REONPs can be transformed into REEs-phosphates and REEs-carboxylates in the plants growing in hydroponic solutions due to action of organic acids and other reducing substances present in the media (Cui et al., 2014). The increased uptake of phosphorus results in an REONP interaction with phosphate ions in cultured media and in plant xylem. The levels of P, K, Ca, Mg, and Mn in plant organs were significantly enhanced in soya bean at low treatment of lanthanum (de Oliveira et al., 2015).

11.6.3 Effect of REONPs on Root Growth and Development

Growth and yield parameters, such as the number of leaves, length of plant organs, size of the plant, and biomass accrual, are symbols of plant healthiness. REE elements stimulate the differentiation and growth of the root (Gomez-Garay et al., 2014). Studies confirm that the response of plants to REONPs varied with the growth stages of plants (Du et al., 2015; Majumdar et al., 2015b). Plants that initially exhibited toxicity of REONPs might recover after some time (Rico et al., 2011). Despite the contrasting results reported by researchers, many studies have proved the growth stimulatory effects of REONPs on different crops (Corral-Diaz et al., 2014). Both, the beneficial and adverse effects of REEs application on plants have been observed but the response of plants toward the REEs exposure depends on concentration of REEs applied and other environmental factors (Cui et al., 2014; Gui et al., 2015). It is clear from the literature that among the plant organs, roots are more sensitive to REONP exposure than other parts of plants because roots directly interact with these materials. A lot of evidence has been collected indicating that REEs exert beneficial effects on the growth and development of plants when they are applied at low concentrations, while high concentrations of REEs are damaging for plants, although the precise mechanism is still unclear; this trend is known as the "hormesis effect" (Cui et al., 2014; Tyler, 2004). Plant treatment with 100 mg/kg significantly accelerates growth, while a minor concentration (50 mg/kg) has no effect on growth. However, higher concentrations significantly decrease the dry weights of roots and shoots (Cui et al., 2014).

Besides plant-related factors, the interaction of $nCeO_2$ with plants depends upon the organic matter content of growth media and doping and surface coating of REONPs, if any. Zhao et al. (2012) found higher Ce concentration in roots of *Zea Mays* grown in organic matter rich soil where pristine coated Ce-oxide NPs were applied but translocation of Ce-oxide

NPs to aerial parts was found lower. López-Moreno et al. (2010) found that $nCeO_2$ at 500 mg/L increased root and shoot measurement in cucumber and alfalfa. Similarly, López-Moreno et al. (2010) and Morales et al. (2013) described that application of $nCeO_2$ amplified root elongation in cucumber, corn, and cilantro plants grown with 125 mg/kg $nCeO_2$. Application of cerium oxide NPs at a rate of 100 mg/kg enhanced the vegetative growth rate of butter head lettuce (Gui et al., 2015). Rico et al. (2014) reported that under greenhouse conditions, $nCeO_2$ increased wheat shoot biomass and plant height. A high dose of $nCeO_2$ promoted stem biomass, while low doses enhanced root biomass in kidney bean and soya bean (Majumdar et al., 2015a; López Moreno et al., 2010). Ma et al. (2010) found growth-stimulating impacts of four oxide NPs (Gd_2O_3, CeO_2, La_2O_3, and Yb_2O_3) on wheat, radish, lettuce, tomato, cabbage, rape, and cucumber species. Similarly, Lin and Xing (2007) testified that the growth of root depended on the type of REONPs and their concentration (Khot et al., 2012). Nano-CeO_2 and bulk CeO_2 can also significantly promote root growth of woody plant species below 400 mg/L concentration. The hypocotyl shoot growth, leaf number, and biomass of *M. arborea* L. are significantly promoted in nanoceria compared to bulk ceria (Gomez-Garay et al., 2014).

Barrios et al. (2015) concluded from his study that at 500 mg/kg, uncoated and coated REONPs increased stem length by 9% and 13%, respectively. A study by Trujillo-Reyes et al., (2014) showed that citric acid-coated $nCeO_2$ increases root biomass. Von Tucher and Schmidhalter (2005) also reported positive effects of lanthanum on shoot development of spinach. Study results of Chaturvedi et al. (2014) showed that 7 days of growth suggested a positive effect of lanthanum treatments on the growth of *Z. mays*, *V. radiata*, and *Vigna Moong*. A continuous increase in fresh and dry weight of all the selected plants indicated the stimulating effect of lanthanum on root and shoot biomass of *V. radiata*, *Z. mays*, and *Vigna Mungo*. Until 2014, yield increments for 50 species of plants including cereals, fruits, sugar, industrial and vegetable crops, and for 20 species of large plants (forest trees) and meadow grasses in response to REE oxides and REONPs have been reported (Chaturvedi et al., 2014). Application of 100 mg/kg of CeO_2 NPs enhanced lettuce growth (Cui et al., 2014). Majumdar et al. (2015a) reported that root and shoot length of kidney bean (*Phaseolus vulgaris* var. red hawk) was not affected with a high dose (2000 mg/kg) of cerium oxide and nanoceria application, but prolonged exposure resulted in cerium biomagnification in kidney beans. The results of Wang et al. (2012a,b) demonstrated that CeO_2 NPs at low concentration (≤ 10 mg/L) increased growth and tomato plant yield. The beneficial effects of bulk REEs other than REE oxides on plants are attributed to their multioxidation states (Yuan et al., 2001).

11.6.4 Impact of REONPs on Physiological and Biochemical Parameters of Plants

NPs are the entities that directly react with cellular metabolites of plant cells and have the ability to alter biochemical efficiency. Plant cells can react to the existence of a high concentration of NPs by altering their subcellular bodies (Cui et al., 2014; Corredor et al., 2010). REONPs have a significant impact on plant physiological processes at the genetic level (Tumburu et al., 2017) and these changes are very complex (Cui et al., 2014).

11.6.4.1 *Effect on Photosynthesis*

The process of photosynthesis is vital for plants to complete their life cycle, and any change in the rate of this energy-yielding process is a result of environmental stress on plants, which represents the plants' adaptability to that change (Niu and Rodriguez, 2008). Photosynthesis is a multiphase process linked with many other physiological processes of plants. Photosynthesis is highly susceptible to changes in light (Shen et al., 2010a,b), and continuous exposure of plants to UV-B radiation causes excessive accumulation of ROS in chloroplasts, which damages photosystem II, decreases electron transfer rate and thylakoid membrane stability (Khan et al., 2017; Eva, 1999), and damages the DNA and even the structure of chloroplasts (Hideg et al., 2013; Wang et al., 2012a,b). Hence REONPs are widely manufactured and utilized in industrial sectors as sunscreen blockers to absorb UV radiations. Thus the damaging effects of UV-B radiation (280–315 nm) and other non-photosynthetically active and nonionizing radiations on photosynthesis are mitigated by the application of REONPs, which improve the rate of photosynthesis by limiting oxidative stress and enhancing chlorophyll biosynthesis, Rubisco activity, light absorbance, transport and transformation of light energy (Sicard et al., 2011; Hong et al., 2005a,b; Gao et al., 2006; Mackerness et al., 2001), and absorbance of UV radiation without scattering the useful visible one (Sicard et al., 2011). At low concentration, lanthanum augmented the overall photosynthetic rate and stomatal conductance in rice plants (Wang et al., 2014). Application of lanthanum also minimized the damage produced by greater UV-B radiation in the seedlings of soya beans (Yang et al., 2015). The increase in photosynthetic rate and total chlorophyll contents in soya bean were attributed to application of lanthanum treatment (de Oliveira et al., 2015).

11.6.4.2 *Oxidative Activity*

Those REEs exhibiting multioxidation states like Ce and Eu possess mixed valence states that give them biological antioxidant capacity because of oxygen vacancies at the surface (Korsvik et al., 2007), and this antioxidant activity of REONPs is efficient at physiological pH (Perez et al., 2008). The ability of these REONPs to act as an antioxidant relies on their ability to reversibly switch from one oxidation state to another (the detailed mechanism for dismutation of ROS is given in Fig. 11.4). This property of REONPs has made them a necessity in engineering, medical, and plant physiological systems. During the plant life cycle, the environmental stresses imposed on plants usually yield ROS, which are detrimental to plant physiology, and to combat these stresses it is necessary to remove these ROS. The production and scavenging of ROS are closely related to photorespiration pathways, photosynthetic apparatus, and mitochondrial respiration (Deng et al., 2014). Oxidative tension produced by ROS molecules is alleviated by plants via numerous mechanisms. Phenols, stress enzymes, and flavonoids are all part of plants' resistance mechanisms (Corral Diaz et al., 2014).

Thus the unique properties of REONPs allow them to quench ROS (Niu et al., 2011), making REONPs in plants an integral part of the defense mechanisms against stresses (Xia et al., 2008). However, if REONPs are bound to any anion molecule at the cellular surface in plants, these complexes experience redox cycling, which may reduce the ROS scavenging activity of REONPs (Xue et al., 2012). Nanoceria yields an intrinsic oxidase activity at low

$$O_2^{\cdot-} + Ce^{III} + 2H^+ \longrightarrow Ce^{IV} + H_2O_2$$

$$O_2^{\cdot-} + Ce^{IV} \longrightarrow O_2 + Ce^{III}$$

$$Ce^{IV} + H_2O_2 \longrightarrow Ce^{III} + H_2O + 0.5O_2$$

FIGURE 11.4 Mechanism for the ability of rare earth oxide nanoparticles (REONPs) to reduce the reactive oxygen species (ROS) concentration in plants.

pH because it can rapidly oxidize a succession of organic compounds short of any oxidizing agent, e.g., H_2O_2 (Asati et al., 2009). The oxidative ability of REONPs is pH and size dependent (Asati et al., 2009). Besides the ability to quench ROS, REONPs such as CeO_2 have the capacity to scavenge reactive nitrogen species in biological systems (Dowding et al., 2012). Rico et al. (2013) found that 62.5 and 125 mg/L of $nCeO_2$ reduced H_2O_2 generation in both roots and shoots of rice seedlings. The application of ceria oxide NP to tuber-producing plants enhances phenol production in leaves compared to tubers (Perez and Lule, 2004; Corral Diaz et al., 2014).

11.6.4.3 *Effect of Enzyme Activity*

Environmental, biological, or chemical stresses are responsible for the activation of plant defense mechanism activation. Plants are equipped with several enzymes such as catalase (CAT) and ascorbate peroxidase (APOX), etc. to cope with these stresses by scavenging and eliminating the ROS generated, e.g., H_2O_2 (Barrios et al., 2015). REONPs such as nanoceria affect the activities of stress enzymes and proteins in plants (Barrios et al., 2015). Zhao et al. (2012) affirmed that CAT and APOX activities were augmented up to day 15 in shoots and seedlings of corn exposed to $nCeO_2$ at 800 mg/kg soil. Rico et al. (2013) claimed a reduction in CAT activities, although they did note an increase in APOX activity in roots of rice exposed to 500 mg $nCeO_2$/kg soil. Majumdar et al. (2014) reported a decrease in APOX in kidney bean leaves exposed for 15 days to 250 and 500 mg $nCeO_2$/kg. Barrios et al. (2015) found that coated ceria increased CAT activity significantly at the rate of 500 mg/kg, while APOX activity was unaltered at a lower dose of uncoated ceria. It is possible that at a concentration of 500 mg/kg, the plant could cope with stress (Barrios et al., 2015), while in bulk CeO_2, Ce occurs as Ce(III)/Ce(IV), and has been known to have superoxide dismutase (SOD) and peroxidase scavenging activity (Rico et al., 2015). Heat shock protein (HSP-70), APOX, and CAT activities were found to enhance the protection of maize plants from oxidative stress (Zhao, 2012).

11.6.4.4 *Chlorophyll Contents*

Chlorophyll is the basic unit of plant energy systems during the process of photosynthesis. Besides light quality, chlorophyll production and activity are influenced by mineral nutrition and chemical metabolites produced in the plant system. REONPs alter chlorophyll production at the genetic level (Tumburu et al., 2017). Barrios et al. (2015) found that bulk cerium oxide had increased chlorophyll contents in tomato plants compared to nanoceria oxide. Zhao et al. (2015, 2014) described that $nCeO_2$ did not change leaf net gas exchange, transpiration rate, stomata conductance, photosynthetic rate, and total chlorophyll content in cucumber and corn. These results contrast with previous studies that noted that NPs procure a negative effect on chlorophyll content (Majumdar et al., 2014; Perreault et al., 2014; Rico et al., 2013; Sadiq et al., 2011). This corroborates with evidence that chlorophyll response to NPs varies with several factors, including growth medium, the environment, and plant species. Nanoceria at low concentration yields high chlorophyll contents in woody plants such as *M. arborea* L. as compared to high application of bulk cerium oxide, but photosynthetic efficiency is not affected by nanoceria (Gomez-Garay et al., 2014). Application of REONPs at low rates through soil medium significantly increases chlorophyll and protein production in butter head lettuce (Gui et al., 2015), and application of RuO_2 NPs increases the total chlorophyll concentration in broccoli at lower dose, but the protein content decreases because of high consumption of sugars in the defense mechanism at higher concentration (Hussain et al., 2015).

11.6.5 Impact of REONPs on Plant Productivity

Metabolomics, proteomics, genomics, and transcriptomics approaches are currently being implemented for a comprehensive understanding of the interactions of REONPs at the molecular level for plant productivity controlling mechanisms such as phytohormone balance, mineral assimilation, cell differentiation, and synthesis of stored reserves in the endosperm (Majumdar et al., 2015a,b). Studies proved that metal oxide NP exposure attributed to variation in plant uptake of nutrient/water and developing processes, resulting in a difference in final biomass and yield. They may also affect other nutritional components influencing fruit/seed flavor and nutritional value, including carbohydrates such as sugar, starch, or fibers, or proteins causing effects on the growth of second-generation seedlings (Du et al., 2017; Wang et al., 2013b). Application of REONPs enhances the vegetative growth period of many plant species (Yoon et al., 2014) and this claim was verified in barley plants by Marchiol et al. (2016) and this fact per se may not be negative. In fact, a longer vegetative phase may promote higher biomass and grain yield because plants can produce more photosynthetically active leaves and therefore more photosynthates are produced.

Many studies suggest that $nCeO_2$ has positive effects in terms of growth and yield at low concentrations (10–125 ppm) (Majumdar et al., 2014; Rico et al., 2013). Application of $nCeO_2$ increases the yield of wheat (Rico et al., 2014) and tomato (*Solanum lycopersicum* L.) (Wang et al., 2012a,b) in greenhouse experiments. Barrios et al., (2015) reported that application of uncoated or coated Ce-oxide NPs on tomato, yielded the ripen fruits after 151 days of germination while the application of bulk Ce-Oxide had no significant effect on the ripening and fruit production. A maximum number of 18 fruits were harvested from cerium oxide NP-uncoated treatment applied at 62.5 mg/kg as compared to six fruits harvested from respective control treatment, and only 27% of tomatoes from cerium oxide NP-uncoated treatment applied at 62.5 mg/kg remained unripened at the end of the life cycle of the tomato plant. Majumdar et al. (2015b) reported that the number of pods per plant and number of seeds per pot of kidney bean were significantly increased under full life cycle exposure at 0–500 mg/kg nanoceria in both high and low organic matter-containing soils. The application of REONPs not only has beneficial effects for fruit-bearing crops but also influences the productivity of root crops. Bradfield et al. (2017) observed a significant increase in the number of tubers per plant, tuber mass, tuber diameter, tuber length, and tuber water contents of sweet potato because of the application of Ce oxide NPs at a rate of 100–1000 mg/kg. Nanoceria up to 500 mg/kg has significant growth and yield promoting effects on wheat crops (Rico et al., 2014). Root and shoot biomass of soya bean was found to be significantly increased because of lanthanum treatment.

11.6.6 Seed Quality Enhancement

The balance of nutrient contents of seeds/grains is fundamental to food security. REONPs have the ability to enhance the nutritional value (Du et al., 2015; Gao et al., 2013), phenolic contents, protein levels (Krystofova et al., 2012; Patra et al., 2013), and carbohydrate contents in grains/seeds of plants. REONPs such as $nCeO_2$ may impact gene expression for protein synthesis during grain development in cereals crops such as wheat, soya bean, and kidney bean (Du et al., 2015; Majumdar et al., 2015b). Carbohydrate metabolism and amino acid synthesis in seeds are also promoted by the application of REONPs in agricultural crops (Rico et al., 2014). The size and weight of some vegetable crops are also enhanced by the application of REONPs (Wang et al., 2012a,b). Uptake and translocation of REONPs in reproductive parts also promotes the synthesis of new groups of proteins, which are usually absent in nontreated plants (Majumdar et al., 2015b). Application of nanoceria also promotes starch synthesis in tomato fruits (Barrios et al., 2017).

A hormetic trend is dose opposite response which means that there are beneficial effects at lower dose while negative effects occur at higher dose of application and Rico et al. (2014) also observed a significant translocation-promoting effect of nanoceria for K, P, and Fe in wheat crops. REONPs react with ionic components of xylem and thus influence the transport of mineral nutrient, especially phosphate ions, which are more reactive to REONPs because REONPs (La_2O_3, Yb_2O_3, and $nCeO_2$) can transform into $REPO_4$ in plants (Rui et al., 2015; Zhang et al., 2011a, 2012a,b; Ma et al., 2011). Application of $nCeO_2$ at 250 mg/kg enhanced P, K, Ca, Mg, S, Fe, Zn, Cu, Al, methionine, aspartic acid, threonine, tyrosine, arginine, and linolenic acid contents in grains (Rico et al., 2015). Many studies confirm that REONPs do not promote the grain accumulation of Cu, Mo, and Mn. REONPs affect fruit and seed quality differently depending upon their route of exposure. When the route of exposure is the foliage, differences in particle size are less significant compared to root-based exposure. This is because exposure through roots is modulated by media components such as root exudates, microorganisms, and soil pH that are absent in the aboveground plant parts and that affect REONPs and bulk counterparts in different ways (Hong et al., 2016). Further investigations are needed to understand the effects of REONPs on fruit firmness through foliar application (Hong et al., 2016). The presowing treatment of REONPs to seed its application at the seedling stage may change the mineral content in mature plants, affecting plant health, yield, and fruit quality (Hong et al., 2016). Soil organic matter contents modulate the impact of REONPs on plants, and the starch content of kidney bean seeds produced in highly organic matter containing soil is increased by the application of nanoceria (Majumdar et al., 2015a). The protein fractions in rice grains were also shown to have a differential response as a function of amylose contents of rice varieties exposed to 500 mg/kg of nano-CeO_2 (Rico et al., 2013). Addition of nano-$nCeO_2$ at 10 mg/L in an organic soil improved tomato fruit production, although the morphology remained unaltered (Wang et al., 2013a,b). Increased growth, soluble sugar content, and vitamin C levels in Chinese cabbage were found as a result of lanthanum treatment (Ma et al., 2014).

11.7 CONTROVERSIES ABOUT THE USE OF REONPs

The US National Nanotechnology Initiative defined REONPs as a material that has at least one measurement in the 1–100nm range. Because of their specific chemical and physical properties, REONPs interact with different components in the cell in certain ways to maximize and induce desired chemical and physiological responses (da Silva et al., 2011). There are significant applications of nanomaterials in biotechnology and biology, and this technology offers new methods for biologists and chemists (Whitesides, 2003). There is a need to investigate the recent trend in nanotechnology whereby nanomaterials interact with biological systems, which is known as nanobiointeraction. These NPs are exposed to cells, animals, plants, or different tissues, and a biological outcome such as toxicity is measured. The benefits of this technology are extensively publicized; study and discussion of the controversial effects should also be promoted to gain proper knowledge of this technology. Considering the danger in the early periods of a technology, the costs of identifying significant environmental and health influences after a technology has widely diffused can be avoided. Nowadays, this nanotechnology is being scrutinized, and research regarding the harmful effects of NPs has increased gradually over the last few years; in fact, it has become a priority in public, private, and government sectors all over the world (Roco, 2004; Helland et al., 2006; Siegrist et al., 2007).

11.7.1 Health Issues Related to the Use of REONPs

The health effects of the rare earth NPs are attracting increasing concern of governments and the public worldwide. It is the most important controversial issue regarding the use of NPs and their negative effects on the health.

11.7.2 Effects of Lungs

In two different publications by Warheit (2004) and Lam et al. (2003), it has been confirmed that nanotubes are toxic. These authors confirmed the pulmonary effects of single-walled carbon nanotubes in vivo after intratracheal instillation in both mice and rats. Both groups reported interstitial inflammation and granuloma formation, and the results showed that if carbon nanotubes reach the lungs, then they will be much more lethal than quartz and carbon black (Agarwal et al., 2013).

11.7.3 Effects on Intestines

The intestinal tract is the most significant gateway for the entry of macromolecules into the body. Charged particles such as carboxylated NPs show poor oral bioavailability through mucus entrapment and electrostatic repulsion. It has been shown in a double-blind randomized study that a low-particle diet (low in exogenous and calcium microparticles) lessens the indications of Crohn's disease (Lomer et al., 2002).

11.7.4 Effects on the Central Nervous System

The effects on neuronal functions of the nervous system involving neurons and NPs have been studied (Oberdörster et al., 2005). The study was carried out by using metal NPs such as Cu, Ag, and Mn on PC12 brain cells to determine the potential of neurotoxicity (Wang et al., 2009). It is estimated that transportation of metal NPs through the blood–brain barrier (BBB) is possible either by carrier-mediated endocytosis or by passive diffusion because the small-sized particles have better mobility (Hoet et al., 2004). For example, AgNPs can enter via the BBB and store in diverse regions of the brain; this may be helpful for drug delivery but it may also carry a risk for the patient (Sarin et al., 2008).

11.7.5 Agricultural Issues Related to the Use of REONPs

Nanotechnology has already been used in many sectors such as agriculture, food, construction, electronics, and cosmetics. However, the implications of this nanotechnology in the food and agriculture sector can be far reaching (Handford et al., 2014). One of the main controversies in the use of rare earth metals is their toxic effects on soil and plants. From an agriculture point of view, NPs have many limitations. Once in the soil ecosystem, NPs may be engaged in vegetation or contaminate water. NPs tend to mount up on the root surfaces of plants, often inhibiting root growth and decreasing hydraulic conductivity. They have been proven to meaningfully reduce germination rates in tomato, corn, and cucumber. Shoot elongation was also disturbed by the presence of NPs in excessive amounts (Dahle and Arai., 2015).

A major global problem is to increase world food production with restricted resources. ENPs of 1–100 nm may have altered chemical and physical properties compared to the naturally occurring NPs. The potential risks and controversies of nanotechnology need to be identified (Husen and Siddiqi, 2014). A large number of rare earth oxide applications to agriculture are at the research and development stage. Possibly the greatest NP-related risks to soil health comes from two sources of contamination. The first source is migration to the soil with the help of contaminated water from nano-coated industrial and consumer goods in landfills. The second source is from treated sewage water applied at least yearly to at least 60–70 million acres of US agricultural lands. The interface of ENPs and soil health is a major concern. Agriculturalists are advised to check the health of their soils at least yearly (Suppan, 2014).

Although REONPs have vast prospects in the agriculture and food sector, there are increasing issues concerning health and safety. Some reviews indicated that the integration of NPs into the food chain may present a totally new array of risks for consumers (Chaudhry and Castle, 2011). The main risk of consumer exposure to nanomaterials is through food packing materials and consumption of edible coatings. Clay NPs and their capability to move out of the film packaging into food under diverse conditions should be subjected to risk assessment. Because of greater surface area, clay is identified as very reactive. This is why concerns have been raised regarding the high reactivity of NPs with foods, which could lead to more poisonous forms of clay NPs during use or production (Handford et al., 2014).

11.7.6 Environmental Issues Related to the Use of REONPs

In the early phases of any technology, risk is a significant issue to consider. Slowly recognized environmental and health risks have terminated many new technologies, e.g., asbestos and chlorofluorocarbons. Natural NPs are a main constituent of our ecosystem, but the activities of new nanoproducts and manufactured NPs in the environment are mostly unknown. In the language of the environmental sciences, ambient particulate matter is characterized in three distributions: (1) particles having a size less than 0.1 μm in diameter called ultrafine particles, (2) particles having a size between 0.1 and 2.5 μm in diameter called mode particles, and (3) coarse mode particles having a diameter larger than 2.5 μm (Buzea et al., 2007).

Small amounts of many metals comprising magnesium, sodium, potassium, copper, and iron are essential for appropriate functioning of biological systems. However, at higher doses, metals can have harmful effects on metabolic process of living organisms. The three particle classes mentioned earlier have specific sources, chemical compositions, and lifetime in the atmosphere. The smaller-sized NPs are transported over greater distances and have longer lifetimes in the atmosphere. This is why they have more opportunities to pollute the air and contaminate the environment (Wenlong et al., 2008). The larger-sized NPs, which settle quicker because of gravity, have a short lifespan in the atmosphere. Atmospheric measurements show that nanomaterials make up a small percentage of mass concentration related to microparticles. In fact, the number of NPs is meaningfully larger than microparticles (Buzea et al., 2007). Because of the polluted environment, humans must face many health effects. When inhaled, elements interact with the normal functions of the body, which can produce many cardiovascular and pulmonary diseases. Generally, NPs seem to be more poisonous than microparticles because of their capability to penetrate into living cells, move within the body, and influence the tasks of the main organs (Buzea et al., 2007). Because of these controversies, NP applications are under observation.

11.7.7 Socioeconomic Issues Related to the Use of REONPs

The rapid appearance of nanotechnology applications in customer products has elevated several societal and ethical concerns in both developing and developed countries, ranging from their effects on animals and human health to the environment. There are many sociopsychological issues that influence the societal response and therefore it is significant to recognize these factors among different shareholder groups for the case of nanotechnology and its success. Expert views should be matched with scientific approaches. To identify these issues, many studies have been conducted. These studies show that there are differences as well as similarities among the experts and the public, e.g., communities were found to be concerned more with ethical issues and the experts focused more on practical factors regarding nanomaterials. Mainly, it appears that community opinion concerning different nanomaterial applications is mostly positive. The exception is those areas where community concern is focused on pesticides. Different agricultural food applications and food packaging are considered more acceptable and environmental and medical applications of this technique are most likely to garner positive communal response (Parisi et al., 2014).

11.8 PROSPECTS OF REONPs

11.8.1 Prospects of REONPs in Food and Agriculture

The current global population stands at nearly six billion with 50% living in Asia. In developing countries, a large proportion of the population faces daily food shortages. The prospect of nanotechnology to revolutionize the food packing and food processing industry has been well publicized. Nanotechnology is used to store and process food so that it can travel all around the globe. In both developed and developing countries, nanotechnology has been defined as the new emerging technology to secure the food market share (Mihranyan et al., 2012). More than 400 organizations around the globe are active in NP research and this number is estimated to rise to more than 1000 within the coming 10 years. Agriculture is the forte of most emerging countries, with more than 65% of people depending on it for their livelihoods. In the future, nanotechnology will have the potential to modernize the agriculture sector with new tools for rapid disease detection, molecular treatment of disease, and improving the capacity of plants to take up nutrients. Smart delivery systems and smart sensors will help agriculture combat viruses and other crop pathogens. Soon, it is estimated that nanostructured catalysts will be accessible, which will upsurge the effectiveness of herbicides and pesticides, permitting lower doses to be used (Joseph and Morrison., 2006).

11.8.2 Wastewater Treatment

An increasing number of countries are facing drinking water problems, and circumstances are very harsh especially in developing countries. Water is the basic need of all living material. In nations such as India, more than 80% of infections are caused by bacterial contamination of drinking water. Now, several procedures are used to clean the water for drinking (Tiwari et al., 2008). Research is proceeding to use NPs in water purification for safe consumption. NPs are predicted to play an important role in water purification because of their extraordinary reactivity due to the large surface-to-volume ratio. There are some issues related to the environmental fate, toxicity, and transport of NPs, but there is no doubt that nanotechnology is better than other technologies. Functional materials for water treatment are divided into four classes of nanoscale: (1) metal containing nanoparticles; (2) dendrimers; (3) carbonaceous nanomaterials; and (4) zeolites. These materials have a large range of physicochemical properties that make them attractive as separation and responsive media for water decontamination (Tiwari et al., 2008).

11.8.3 Potential for Cancer and Other Diseases

Nanotechnology has become a major focus of biomedical study in recent years. With the passage of time its importance in medical science is increasing. Its application includes luminescent biomarkers, drug delivery systems, and tissue engineering. Especially, some specific NPs have shown their specificity for diseases (Nichols and Bae, 2013). For example, CeO NPs, which contain a cerium core bounded by an oxygen matrix, have exposed several applications especially in the medication of cancer. Studies have shown that CeO NPs have a distinctive cytotoxicity toward cancer cells, antiinvasive abilities, and the ability to sensitize cancer cells to radiation-induced cell loss, while defending the neighboring normal tissues. Because of ROS accumulation, several other diseases are categorized with CeO NPs. In diabetes, increased ROS is responsible for removing free radicles and this is the main cause of the disease. CeO NP treatment could return hepatic ROS levels to normal, as well as reduce plasma triglycerides and increase plasma (Wason and Zhao., 2013).

The specific properties of NPs encourage confidence in their application in an extensive range of fields from environmental sciences to medical applications. In the United States, Japan, and China, nanotechnology has become a top research priority. In the field of electronics, nanotechnology is used in microelectronics. Many of the current microelectronics applications are already at the nanoscale (Thompson and Parthasarathy., 2006). The use of nanocrystalline resources can significantly improve the resolution of TV screens by reducing pixel size. To record information on a spinning disc, such as computer hard discs, some utilities have established their ability to magnetize the small area. Tapes and discs comprising ENPs can store large amounts of information. Nowadays, nanomaterials are used in high-energy density batteries to improve their capacity for charging. Because of increased reactivity and high surface area, NPs could be engaged as sensors for detecting parameters such as capacitance and thermal conductivity (Liu et al., 2006).

In transportation, nanomaterials are used in car tires and fenders. Carbon black NPs act as a filler in the polymer matrix. Because of their light weight these nanomaterials are used to make car exteriors and are twice as resistant to scratches. Very thin coatings of nanomaterials are used in a vast range of applications. Titanium oxide NPs are used in self-cleaning windows. In the presence of sunlight and water, the washing of the glass will be easier. Nanomaterials are also used in scratch-resistant materials. In mechanical engineering, cutting tools are constructed of nanomaterials. These are much harder than any other cutting tools in the market (Buzea et al., 2007).

11.9 SUMMARY/CONCLUSIONS

The novel era of nanotechnology has exposed all environments to NPs. REONPs have wide applications in agriculture with a focus on increasing productivity in the agricultural sector. Their unique characteristics and high reactivity have differentiated REONPs among ENPs, and REONPs play a critical role in agricultural sustainability and food security by providing enhanced performance over traditional bulk materials. The broad range of applications in agriculture includes nanofertilizers to increase plant growth and yield, nanopesticides for the control of crop pests and diseases, and nanosensors for soil quality and plant health monitoring. Over the past decade, patents and products incorporating nanomaterials for agricultural practices (e.g., nanopesticides and nanofertilizers) have been developed. It has been previously reported that in 2011 over 3000 patent applications dealing with nanopesticides and nanofertilizers were submitted (Kah et al., 2013).

The collective goal of these approaches is to enhance the efficiency and sustainability of agricultural practices by requiring less input and generating less waste than conventional products and approaches.

Unfortunately, application of REONPs in agriculture is still controversial regarding their fate and unknown effects on humans through food chain contamination. Moreover, comprehensive knowledge of detailed mechanisms describing their dynamics in the soil–plant system and their interaction with plants at the subcellular level is also lacking. With a lack of experimental standardization and the divergent responses, even within similar plant species, it is challenging to conclude what the effects of REONPs are in plants. All these parameters are barriers for the generalization of the beneficial impacts of REONPs on plants. Regarding the reported literature, the effects depend on the concentration of the REONPs, exposure media, and plant species. Current knowledge suggests that more studies are needed to fully understand the physiological and biochemical effects of REONPs in plants. It is suggested that proteomic, genomic, and metabolomic studies along with field trials will be helpful for those researchers aiming to better understand the mechanisms involved in the interactions of REONPs with plants.

References

Abdulghani, A.J., Al-Ogedy, W.M., 2015. Preparation and characterization of yttrium oxide nanoparticles at different calcination temperatures from yttrium hydroxide prepared by hydrothermal and hydrothermal microwave methods. Iraqi J. Sci. 56, 1572–1587.

Agarwal, M., Murugan, M.S., Sharma, A., Rai, R., Kamboj, A., Sharma, H., Roy, S.K., 2013. Nanoparticles and its toxic effects: a review. Int. J. Curr. Microbiol. Appl. Sci. 2 (7682), 34–39.

Ahmad, A., Senapati, S., Khan, M.I., Kumar, R., Sastry, M., 2005. Extra-/intracellular biosynthesis of gold nanoparticles by an alkalotolerant fungus, *Trichothecium* sp. J. Biomed. Nanotechnol. 1 (1), 47–53.

Anaya, N.M., Solomon, F., Oyanedel-Craver, V., 2016. Effects of dysprosium oxide nanoparticles on *Escherichia coli*. Environ. Sci. 3 (1), 67–73.

Anjum, N.A., Gill, S.S., Duarte, A.C., Pereira, E., Ahmad, I., 2013. Silver nanoparticles in soil–plant systems. J. NanoPart. Res. 15 (9), 1896.

Anjum, N.A., Sofo, A., Scopa, A., Roychoudhury, A., Gill, S.S., Iqbal, M., Lukatkin, A.S., Pereira, E., Duarte, A.C., Ahmad, I., 2015. Lipids and proteins—major targets of oxidative modifications in abiotic stressed plants. Environ. Sci. Pollut. Res. 22 (6), 4099–4121.

Antisari, L.V., Carbone, S., Gatti, A., Vianello, G., Nannipieri, P., 2013. Toxicity of metal oxide (CeO_2, Fe_3O_4, SnO_2) engineered nanoparticles on soil microbial biomass and their distribution in soil. Soil Biol. Biochem. 60, 87–94.

Antisari, L.V., Carbone, S., Gatti, A., Vianello, G., Nannipieri, P., 2015. Uptake and translocation of metals and nutrients in tomato grown in soil polluted with metal oxide (CeO_2, Fe_3O_4, SnO_2, TiO_2) or metallic (Ag, Co, Ni) engineered nanoparticles. Environ. Sci. Pollut. Res. 22 (3), 1841–1853.

Arai, N., Yasuoka, K., Zeng, X.C., 2013. A vesicle cell under collision with a Janus or homogeneous nanoparticle: translocation dynamics and late-stage morphology. Nanoscale 5 (19), 9089–9100.

Arruda, S.C.C., Silva, A.L.D., Galazzi, R.M., Azevedo, R.A., Arruda, M.A.Z., 2015. Nanoparticles applied to plant science: a review. Talanta 131, 693–705.

Asati, A., Santra, S., Kaittanis, C., Nath, S., Perez, J.M., 2009. Oxidase-like activity of polymer-coated cerium oxide nanoparticles. Angew. Chem. Int. Ed. 48 (13), 2308–2312.

Asztemborska, M., Steborowski, R., Kowalska, J., Bystrzejewska-Piotrowska, G., 2015. Accumulation of platinum nanoparticles by *Sinapis alba* and *Lepidium sativum* plants. Water Air Soil Pollut. 226 (4), 1–7.

Aziz, N., Faraz, M., Pandey, R., Shakir, M., Fatma, T., Varma, A., et al., 2015. Facile algae-derived route to biogenic silver nanoparticles: synthesis, antibacterial, and photocatalytic properties. Langmuir 31, 11605–11612.

Aziz, N., Pandey, R., Barman, I., Prasad, R., 2016. Leveraging the attributes of *Mucor hiemalis*-derived silver nanoparticles for a synergistic broad-spectrum antimicrobial platform. Front. Microbiol. 7, 1984.

Balusamy, B., Kandhasamy, Y.G., Senthamizhan, A., Chandrasekaran, G., Subramanian, M.S., Kumaravel, T.S., 2012. Characterization and bacterial toxicity of lanthanum oxide bulk and nanoparticles. J. Rare Earths 30 (12), 1298–1302.

Bandyopadhyay, S., Peralta-Videa, J.R., Plascencia-Villa, G., José-Yacamán, M., Gardea-Torresdey, J.L., 2012. Comparative toxicity assessment of CeO_2 and ZnO nanoparticles towards *Sinorhizobium meliloti*, a symbiotic alfalfa associated bacterium: use of advanced microscopic and spectroscopic techniques. J. Hazard. Mater. 241, 379–386.

Barrios, A.C., Rico, C.M., Trujillo-Reyes, J., Medina-Velo, I.A., Peralta-Videa, J.R., Gardea-Torresdey, J.L., 2015. Effects of uncoated and citric acid coated cerium oxide nanoparticles, bulk cerium oxide, cerium acetate, and citric acid on to-mato plants. Sci. Total Environ. http://dx.doi.org/10.1016/j.scitotenv.2015.11.143.

Barrios, A.C., Medina-Velo, I.A., Zuverza-Mena, N., Dominguez, O.E., Peralta-Videa, J.R., Gardea-Torresdey, J.L., 2017. Nutritional quality assessment of tomato fruits after exposure to uncoated and citric acid coated cerium oxide nanoparticles, bulk cerium oxide, cerium acetate and citric acid. Plant Physiol. Biochem. 110, 100–107.

Birbaum, K., Brogioli, R., Schellenberg, M., Martinoia, E., Stark, W.J., Günther, D., Limbach, L.K., 2010. No evidence for cerium dioxide nanoparticle translocation in maize plants. Environ. Sci. Technol. 44 (22), 8718–8723.

Borchert, Y., Sonström, P., Wilhelm, M., Borchert, H., Bäumer, M., 2008. Nanostructured praseodymium oxide: preparation, structure, and catalytic properties. J. Phys. Chem. C 112 (8), 3054–3063.

Bradfield, S.J., Kumar, P., White, J.C., Ebbs, S.D., 2017. Zinc, copper, or cerium accumulation from metal oxide nanoparticles or ions in sweet potato: yield effects and projected dietary intake from consumption. Plant Physiol. Biochem. 110, 128–137.

Brookes, P.C., 1995. The use of microbial parameters in monitoring soil pollution by heavy metals. Biol. Fertil. Soils 19 (4), 269–279.

Buseck, P.R., Posfai, M., 1999. Airborne minerals and related aerosol particles: effects on climate and the environment. Proc. Natl. Acad. Sci. 96 (7), 3372–3379.

Buzea, C., Pacheco, I.I., Robbie, K., 2007. Nanomaterials and nanoparticles: sources and toxicity. Biointerphases 2 (4), MR17–MR71.

Calabrese, E.J., Baldwin, L.A., 2002. Applications of hormesis in toxicology, risk assessment and chemotherapeutics. Trends Pharmacol. Sci. 23 (7), 331–337.

Campos, E.V.R., de Oliveira, J.L., Fraceto, L.F., 2014. Applications of controlled release systems for fungicides, herbicides, acaricides, nutrients, and plant growth hormones: a review. Adv. Sci. Eng. Med. 6, 373–387.

Cao, X., Chen, Y., Wang, X., Deng, X., 2001. Effects of redox potential and pH value on the release of rare earth elements from soil. Chemosphere 44 (4), 655–661.

Chatterjee, A., Nishanthini, D., Sandhiya, N., Abraham, J., 2016. Biosynthesis of titanium dioxide nanoparticles using *Vigna radiata*. Asian J. Pharm. Clin. Res. 28, 85–88.

Chaturvedi, N., Gannavarapu, R., Dhal, N.K., 2014. Effect of lanthanum on the growth and physiological activities of *Zea mays*, *Vigna radiata* and *Vigna mungo*. Int. J. Environ. Sci. 4 (5), 653.

Chaudhry, Q., Castle, L., 2011. Food applications of nanotechnologies: an overview of opportunities and challenges for developing countries. Trends Food Sci. Technol. 22 (11), 595–603.

Chen, P.L., Chen, I.W., 1993. Reactive cerium (IV) oxide powders by the homogeneous precipitation method. J. Am. Ceram. Soc. 76 (6), 1577–1583.

Chen, P.L., Chen, I.W., 1996. Sintering of fine oxide powders: I, microstructural evolution. J. Am. Ceram. Soc. 79 (12), 3129–3141.

Chen, Y., Sanchez, C., Yue, Y., Almeida, M., González, J.M., Parkinson, D.Y., Liang, H., 2016. Observation of yttrium oxide nanoparticles in cabbage (*Brassica oleracea*) through dual energy K-edge subtraction imaging. J. Nanobiotechnol. 14 (1), 1.

Chengyun, W., Yitai, Q., Xie, Y., Changsui, W., Yang, L., Guiwen, Z., 1996. A novel method to prepare nanocrystalline (7 nm) ceria. Mater. Sci. Eng. 39 (3), 160–162.

Ciftcioglu, M., Akinc, M., Burkhart, L., 1987. Effect of agglomerate strength on sintered density for yttria powders containing agglomerates of monosize spheres. J. Am. Ceram. Soc. 70 (11).

Cifuentes, Z., Custardoy, L., de la Fuente, J.M., Marquina, C., Ibarra, M.R., Rubiales, D., Pérez-de-Luque, A., 2010. Absorption and translocation to the aerial part of magnetic carbon-coated nanoparticles through the root of different crop plants. J. Nanobiotechnol. 8 (1), 1.

Cornelis, G., Ryan, B., McLaughlin, M.J., Kirby, J.K., Beak, D., Chittleborough, D., 2011. Solubility and batch retention of CeO_2 nanoparticles in soils. Environ. Sci. Technol. 45 (7), 2777–2782.

Cornelis, G., Hund-Rinke, K., Kuhlbusch, T., 2014. Fate and bioavailability of engineered nanoparticles in soils: a review. Crit. Rev. Environ. Sci. Technol. 44 (24), 2720–2764.

Corral-Diaz, B., Peralta-Videa, J.R., Alvarez-Parrilla, E., Rodrigo-García, J., Morales, M.I., Osuna-Avila, P., Niu, G., Hernandez-Viezcas, J.A., Gardea-Torresdey, J.L., 2014. Cerium oxide nanoparticles alter the antioxidant capacity but do not impact tuber ionome in *Raphanus sativus* (L). Plant Physiol. Biochem. 84, 277–285.

Corredor, E., Testillano, P.S., Coronado, M.J., González-Melendi, P., Fernández-Pacheco, R., Marquina, C., Ibarra, M.R., de la Fuente, J.M., Rubiales, D., Pérez-de-Luque, A., Risueño, M.C., 2010. Nanoparticle penetration and transport in living pumpkin plants: in situ subcellular identification. BMC Plant Biol. 9 (1), 1.

Cui, D., Zhang, P., Ma, Y., He, X., Li, Y., Zhang, J., Zhao, Y., Zhang, Z., 2014. Effect of cerium oxide nanoparticles on asparagus lettuce cultured in an agar medium. Environ. Sci. 1 (5), 459–465.

d'Aquino, L., De Pinto, M.C., Nardi, L., Morgana, M., Tommasi, F., 2009. Effect of some light rare earth elements on seed germination, seedling growth and antioxidant metabolism in *Triticum durum*. Chemosphere 75 (7), 900–905.

da Silva, B.F., Pérez, S., Gardinalli, P., Singhal, R.K., Mozeto, A.A., Barceló, D., 2011. Analytical chemistry of metallic nanoparticles in natural environments. TrAC 30 (3), 528–540.

Dahle, J.T., Arai, Y., 2014. Effects of Ce (III) and CeO_2 nanoparticles on soil-denitrification kinetics. Arch. Environ. Contam. Toxicol. 67 (4), 474–482.

Dahle, J.T., Arai, Y., 2015. Environmental geochemistry of cerium: applications and toxicology of cerium oxide nanoparticles. Int. J. Environ. Res. Public Health 12 (2), 1253–1278.

de Oliveira, C., Ramos, S.J., Siqueira, J.O., Faquin, V., de Castro, E.M., Amaral, D.C., Techio, V.H., Coelho, L.C., e Silva, P.H., Schnug, E., Guilherme, L.R., 2015. Bioaccumulation and effects of lanthanum on growth and mitotic index in soybean plants. Ecotoxicol. Environ. Saf. 122, 136–144.

Deng, Y.Q., White, J.C., Xing, B.S., 2014. Interactions between engineered nanomaterials and agricultural crops: implications for food safety. J. Zhejiang Univ. Sci. A 15 (8), 552–572.

Diatloff, E., Smith, F.W., Asher, C.J., 2008. Effects of lanthanum and cerium on the growth and mineral nutrition of corn and mungbean. Ann. Bot. (Lond.) 101 (7), 971–982.

Dong, X.T., Liu, G.X., Sun, J., Li, J.L., Liu, J.H., Hong, G.G., 2001. Preparation and properties of nanocrystalline Yb2O3.

Dowding, J.M., Dosani, T., Kumar, A., Seal, S., Self, W.T., 2012. Cerium oxide nanoparticles scavenge nitric oxide radical (·NO). Chem. Commun. 48 (40), 4896–4898.

Du, W., Gardea-Torresdey, J.L., Ji, R., Yin, Y., Zhu, J., Peralta-Videa, J.R., Guo, H., 2015. Physiological and biochemical changes imposed by CeO_2 nanoparticles on wheat: a life cycle field study. Environ. Sci. Technol. 49 (19), 11884–11893.

Du, W., Tan, W., Peralta-Videa, J.R., Gardea-Torresdey, J.L., Ji, R., Yin, Y., Guo, H., 2017. Interaction of metal oxide nanoparticles with higher terrestrial plants: physiological and biochemical aspects. Plant Physiol. Biochem. 110, 210–225

Eilers, H., Tissue, B.M., 1995. Synthesis of nanophase ZnO, Eu_2O_3, and ZrO_2 by gas-phase condensation with cw-CO_2 laser heating. Mater. Lett. 24 (4), 261–265.

El-Maarouf-Bouteau, H., Bailly, C., 2008. Oxidative signaling in seed germination and dormancy. Plant Signal. Behav. 3, 175–182.

Ernst, W.H.O., 1996. Bioavailability of heavy metals and decontamination of soils by plants. Appl. Geochem. 11 (1), 163–167.

Eva, H., 1999. Utilizing new adamantly spin traps in studying UV-B-induced oxidative damage of photosystem II. J. Photochem. Photobiol. B 48, 174–179.

Fu, Y., Li, F., Xu, T., Cai, S., Chu, W., Qiu, H., Sha, S., Cheng, G., Xu, Q., 2014. Bioaccumulation, subcellular, and molecular localization and damage to physiology and ultrastructure in *Nymphoides peltata* (Gmel.) O. Kuntze exposed to yttrium. Environ. Sci. Pollut. Res. 21 (4), 2935–2942.

Gao, J., Zhao, Y., Yang, W., Tian, J., Guan, F., Ma, Y., Hou, J., Kang, J., Wang, Y., 2002. Preparation of samarium oxide nanoparticles and its catalytic activity on the esterification. Mater. Chem. Phys. 77 (1), 65–69.

Gao, F.Q., Hong, F.S., Liu, C., Zheng, L., Su, M.Y., Wu, X., Yang, F., Wu, C., Yang, P., 2006. Mechanism of nanoanatase TiO_2 on promoting photosynthetic carbon reaction of spinach: inducing complex of Rubisco-Rubisco activase. Biol. Trace Elem. Res. 11, 239–254.

Gao, J., Xu, G., Qian, H., Liu, P., Zhao, P., Hu, Y., 2013. Effects of nano-TiO_2 on photosynthetic characteristics of *Ulmus elongata* seedlings. Environ. Pollut. 176, 63–70.

Ge, Y., Priester, J.H., Van De Werfhorst, L.C., Walker, S.L., Nisbet, R.M., An, Y.J., Schimel, J.P., Gardea-Torresdey, J.L., Holden, P.A., 2014. Soybean plants modify metal oxide nanoparticle effects on soil bacterial communities. Environ. Sci. Technol. 48 (22), 13489–13496.

Giraldo, J.P., Landry, M.P., Faltermeier, S.M., McNicholas, T.P., Iverson, N.M., Boghossian, A.A., Reuel, N.F., Hilmer, A.J., Sen, F., Brew, J.A., Strano, M.S., 2014. Plant nanobionics approach to augment photosynthesis and biochemical sensing. Nat. Mater. 13 (4), 400–408.

Gomez-Garay, A., Pintos, B., Manzanera, J.A., Lobo, C., Villalobos, N., Martín, L., 2014. Uptake of CeO_2 nanoparticles and its effect on growth of *Medicago arborea* in vitro plantlets. Biol. Trace Elem. Res. 161 (1), 143–150.

Gonzalez, V., Vignati, D.A.L., Leyval, C., Giamberini, L., 2014. Environmental fate and ecotoxicity of lanthanides: are they a uniform group beyond chemistry? Environ. Int. 71, 148–157.

Gopinath, K., Karthika, V., Sundaravadivelan, C., Gowri, S., Arumugam, A., 2015. Mycogenesis of cerium oxide nanoparticles using *Aspergillus niger* culture filtrate and their applications for antibacterial and larvicidal activities. J. Nanostruct. 5 (3), 295–303.

Grillo, R., Rosa, A.H., Fraceto, L.F., 2014. Engineered nanoparticles and organic matter: a review of the state-of-the-art. Chemosphere 119, 608–619.

Grosso, D., Sermon, P.A., 2000. Scandium oxide nanoparticles produced from sol–gel chemistry. J. Mater. Chem. 10 (2), 359–363.

Gui, X., Zhang, Z., Liu, S., Ma, Y., Zhang, P., He, X., Li, Y., Zhang, J., Li, H., Rui, Y., Liu, L., 2015. Fate and phytotoxicity of CeO_2 nanoparticles on lettuce cultured in the potting soil environment. PLoS One 10 (8), 134–161.

Gunter, B., Kumpmann, A., 1992. Ultrafine oxide powders prepared by inert gas evaporation. Nanostruct. Mater. 1 (1), 27–30.

Hakuta, Y., Onai, S., Terayama, H., Adschiri, T., Arai, K., 1998. Production of ultra-fine ceria particles by hydrothermal synthesis under supercritical conditions. J. Mater. Sci. Lett. 17 (14), 1211–1213.

Handford, C.E., Dean, M., Henchion, M., Spence, M., Elliott, C.T., Campbell, K., 2014. Implications of nanotechnology for the agri-food industry: opportunities, benefits and risks. Trends Food Sci. Technol. 40 (2), 226–241.

Hassellöv, M., 2012. Occurrence, identification, fate and behavior of engineered nanoparticles and nanoscale pollutants in marine systems. In: Proc BSIJ.

He, M.L., Ranz, D., Rambeck, W.A., 2003. Study on the performance enhancing effects of rare earth elements in growing and fattening pigs. J. Anim. Physiol. Anim. Nutr. 85, 263–270.

Helland, A., Kastenholz, H., Thidell, A., Arnfalk, P., Deppert, K., 2006. Nanoparticulate materials and regulatory policy in Europe: an analysis of stakeholder perspectives. J. Nanopart. Res. 8 (5), 709–719.

Henriques, M.S., Ferreira, A.C., Cruz, A., Ferreira, L.M., Branco, J.B., Brázda, P., Jurek, K., Stora, T., Gonçalves, A.P., 2015. Preparation of Yb_2O_3 submicron-and nano-materials via electrospinning. Ceram. Int. 41 (9), 10795–10802.

Hernandez-Viezcas, J.A., Castillo-Michel, H., Andrews, J.C., Cotte, M., Rico, C., Peralta-Videa, J.R., Ge, Y., Priester, J.H., Holden, P.A., Gardea-Torresdey, J.L., 2013. In situ synchrotron X-ray fluorescence mapping and speciation of CeO_2 and ZnO nanoparticles in soil cultivated soybean (*Glycine max*). ACS Nano 7 (2), 1415–1423.

Hideg, E., Jansen, M.A., Strid, A., 2013. UV-B exposure, ROS, and stress: inseparable companions or loosely linked associates? Trends Plant Sci. 18, 107–115.

Hirano, M., Kato, E., 1996. The hydrothermal synthesis of ultrafine cenum (w) oxide powders. J. Mater. Sci. Lett. 15 (14), 1249–1250.

Hirano, M., Kato, E., 1999. Hydrothermal synthesis of nanocrystalline cerium (IV) oxide powders. J. Am. Ceram. Soc. 82 (3), 786–788.

Hoet, P.H., Brüske-Hohlfeld, I., Salata, O.V., 2004. Nanoparticles–known and unknown health risks. J. Nanobiotechnol. 2 (1), 12.

Hong, D.S., Meltzer, R.S., Bihari, B., Williams, D.K., Tissue, B.M., 1998. Spectral hole burning in crystalline Eu_2O_3 and Y_2O_3: Eu^{3+} nanoparticles. J. Lumin. 76, 234–237.

Hong, F.S., Yang, F., Liu, C., Gao, Q., Wan, Z.G., Gu, F.G., Wu, C., Ma, Z.N., Zhou, J., Yang, P., 2005a. Influences of nano-TiO_2 on the chloroplast aging of spinach under light. Biol. Trace Elem. Res. 104 (3), 249–260.

Hong, F.S., Zhou, J., Liu, C., Yang, F., Wu, C., Zheng, L., Yang, P., 2005b. Effect of nano-TiO_2 on photochemical reaction of chloroplasts of spinach. Biol. Trace Elem. Res. 105 (1–3), 269–279.

Hong, J., Wang, L., Sun, Y., Zhao, L., Niu, G., Tan, W., Rico, C.M., Peralta-Videa, J.R., Gardea-Torresdey, J.L., 2016. Foliar applied nanoscale and microscale CeO_2 and CuO alter cucumber (*Cucumis sativus*) fruit quality. Sci. Total Environ. 563, 904–911.

Husen, A., Siddiqi, K.S., 2014. Phytosynthesis of nanoparticles: concept, controversy and application. Nanoscale Res. Lett. 9 (1), 1–24.

Hussain, I., Singh, A., Singh, H., Singh, S.C., Singh, N.B., 2015. Physiological response of broccoli exposed to RuO_2 nanoparticle. Trop. Plant Res. 2 (3), 246–252.

Imoto, F., Nanataki, T., Kaneko, S., 1988. Preparation and characterisation of fine rare-earth oxide powders through alkoxide processes. Ceram. Trans. 1, 204–210.

Jha, A.K., Prasad, K., Kulkarni, A.R., 2010. Synthesis of Gd_2O_3 nanoparticles using *Lactobacillus* sp.: a novel green approach. Int. J. Green Nanotechnol. 2 (2), P31–P38.

Jinxia, L.I., Mei, H.O.N.G., Xiuqin, Y.I.N., Jiliang, L.I.U., 2010. Effects of the accumulation of the rare earth elements on soil macrofauna community. J. Rare Earths 28 (6), 957–964.

Joner, E.J., Hartnik, T., Amundsen, C.E., 2008. Environmental Fate and Ecotoxicity of Engineered Nanoparticles Final Report. TA, 2304.2007.

Joseph, T., Morrison, M., 2006. Nanotechnology in Agriculture and Food Nanoforum Report. Institute of nanotechnology.

Kabata, A., Pendias, H., 2001. Trace Elements in Soils and Plants. CRC, Washington.

Kah, M., Hofmann, T., 2014. Nanopesticide research: current trends and future priorities. Environ. Int. 63, 224–235.

Kah, M., Beulke, S., Tiede, K., Hofmann, T., 2013. Nanopesticides: state of knowledge, environmental fate, and exposure modeling. Crit. Rev. Environ. Sci. Technol. 43, 1823–1867.

Kannan, S.K., Sundrarajan, M., 2015. Biosynthesis of yttrium oxide nanoparticles using *Acalypha indica* leaf extract. Bull. Mater. Sci. 38 (4), 945–950.

Keller, A.A., Lazareva, A., 2014. Predicted releases of engineered nanomaterials: from global to regional to local. Environ. Sci. Technol. Lett. 1, 65–70.

Keller, A.A., Wang, H., Zhou, D., Lenihan, H.S., Cherr, G., Cardinale, B.J., Miller, R., Ji, Z., 2010. Stability and aggregation of metal oxide nanoparticles in natural aqueous matrices. Environ. Sci. Technol. 44 (6), 1962–1967.

Khan, S.A., Ahmad, A., 2013. Fungus mediated synthesis of biomedically important cerium oxide nanoparticles. Mater. Res. Bull. 48 (10), 4134–4138.

Khan, S.A., Gambhir, S., Ahmad, A., 2014. Extracellular biosynthesis of gadolinium oxide (Gd_2O_3) nanoparticles, their biodistribution and bioconjugation with the chemically modified anticancer drug taxol. Beilstein. J. Nanotechnol. 5 (1), 249–257.

Khan, M.N., Mobin, M., Abbas, Z.K., AlMutairi, K.A., Siddiqui, Z.H., 2017. Role of nanomaterials in plants under challenging environments. Plant Physiol. Biochem. 110, 194–209.

Khodakovskaya, M.V., de Silva, K., Nedosekin, D.A., Dervishi, E., Biris, A.S., Shashkov, E.V., Galanzha, E.I., Zharov, V.P., 2011. Complex genetic, photothermal, and photoacoustic analysis of nanoparticle-plant interactions. Proc. Natl. Acad. Sci. 108 (3), 1028–1033.

Khodakovskaya, M.V., Silva, K., De Biris, A.S., Dervishi, E., Villagarcia, H., 2012. Carbon nanotubes induce growth enhancement of tobacco cells. ACS Nano 6, 2128–2135.

Khot, L.R., Sankaran, S., Maja, J.M., Ehsani, R., Schuster, E.W., 2012. Applications of nanomaterials in agricultural production and crop protection: a review. Crop Protect 35, 64–70.

Kobayashi, M., 1992. Preparation of ultrafine precursor particles for Y_2O_3 from YCl_3/ethylenediaminetetracetic acid/urea aqueous solution. J. Mater. Sci. Lett. 11 (11), 767–768.

Korsvik, C., Patil, S., Seal, S., Self, W.T., 2007. Superoxide dismutase mimetic properties exhibited by vacancy engineered ceria nanoparticles. Chem. Commun. (10), 1056–1058.

Krystofova, O., Sochor, J., Zitka, O., Babula, P., Kudrle, V., Adam, V., Kizek, R., 2012. Effect of magnetic nanoparticles on tobacco BY-2 cell suspension culture. Inter. J. Environ. Res. Public Health 10 (1), 47–71.

Kumari, M.M., Jacob, J., Philip, D., 2015. Green synthesis and applications of Au–Ag bimetallic nanoparticles. Spectrochim. Acta Part A 137, 185–192.

Lam, D.H., Watkins, E.R., Hayward, P., Bright, J., Wright, K., Kerr, N., Parr-Davis, G., Sham, P., 2003. A randomized controlled study of cognitive therapy for relapse prevention for bipolar affective disorder: outcome of the first year. Arch. Gen. Psychiatry 60 (2), 145–152.

Larue, C., Laurette, J., Herlin-Boime, N., Khodja, H., Fayard, B., Flank, A.M., Brisset, F., Carriere, M., 2012. Accumulation, translocation and impact of TiO_2 nanoparticles in wheat (*Triticum aestivum* spp.): influence of diameter and crystal phase. Sci. Total Environ. 431, 197–208.

Lee, M.H., Oh, S.G., Yi, S.C., 2000. Preparation of Eu-doped Y_2O_3 luminescent nanoparticles in nonionic reverse microemulsions. Adv. Colloid Interface Sci. 226 (1), 65–70.

Li, J.G., Ikegami, T., Mori, T., Yajima, Y., 2003. Monodispersed Sc_2O_3 precursor particles via homogeneous precipitation: synthesis, thermal decomposition, and the effects of supporting anions on powder properties. J. Mater. Res. 18 (05), 1149–1156.

Li, R., Ji, Z., Chang, C.H., Xia, T., 2014. Surface interactions with compartmentalized cellular phosphates explain rare earth oxide nanoparticle hazard and provide opportunities for safer design. ACS Nano 8 (2). http://dx.doi.org/10.1021/nn406166n.

Liang, T., Zhang, S., Wang, L., Kung, H., Wang, Y., Hu, A., Ding, S., 2005. Environ-mental biogeochemical behaviors of rare earth elements in soil-plant systems. Environ. Geochem. Health 27, 301–311.

Liao, X.H., Zhu, J.M., Zhu, J.J., Xu, J.Z., Chen, H.Y., 2001. Preparation of monodispersed nanocrystalline CeO_2 powders by microwave irradiation. Chem. Commun. (10), 937–938.

Lin, D., Xing, B., 2007. Phytotoxicity of nanoparticles: inhibition of seed germination and root growth. Environ. Pollut. 150, 243–250.

Liu, J., Levine, A.L., Mattoon, J.S., Yamaguchi, M., Lee, R.J., Pan, X., Rosol, T.J., 2006. Nanoparticles as image enhancing agents for ultrasonography. Phys. Med. Biol. 51 (9), 2179.

Liu, D., Wang, X., Lin, Y., Chen, Z., Xu, H., Wang, L., 2012. The effects of cerium on the growth and some antioxidant metabolisms in rice seedlings. Environ. Sci. Pollut. Res. 19 (8), 3282–3291.

Lomer, M.C., Thompson, R.P., Powell, J.J., 2002. Fine and ultrafine particles of the diet: influence on the mucosal immune response and association with Crohn's disease. Proc. Nutr. Soc. 61 (1), 123–130.

Lövestam, G., Rauscher, H., Roebben, G., Klüttgen, B.S., Gibson, N., Putaud, J.P., Stamm, H., 2010. Considerations on a Definition of Nanomaterial for Regulatory Purposes. Joint Research Centre (JRC), pp. 80004–80011. Reference Reports.

Lowry, G.V., Gregory, K.B., Apte, S.C., Lead, J.R., 2012. Transformations of nanomaterials in the environment. Environ. Sci. Technol. 46, 6893–6899.

López-Moreno, M.L., de la Rosa, G., Hernández-Viezcas, J.A., Peralta-Videa, J.R., Gardea-Torresdey, J.L., 2010. X-ray absorption spectroscopy (XAS) corroboration of the uptake and storage of CeO_2 nanoparticles and assessment of their differential toxicity in four edible plant species. J. Agric. Food Chem. 58 (6), 3689–3693.

Ma, J., Yan, Q.J., 1995. Preparation of spherical, monodispersed nanometer Eu_2O_3 particles. Chin. Chem. Lett. 6 (10), 915–918.

Ma, Y.H., Kuang, L.L., He, X., Bai, W., Ding, Y.Y., Zhang, Z.Y., Zhao, Y.L., Chai, Z.F., 2010. Effects of rare earth oxide nanoparticles on root elongation of plants. Chemosphere 78 (3), 273–279.

Ma, Y., He, X., Zhang, P., Zhang, Z., Guo, Z., Tai, R., Xu, Z., Zhang, L., Ding, Y., Zhao, Y., Chai, Z., 2011. Phytotoxicity and biotransformation of La_2O_3 nanoparticles in a terrestrial plant cucumber (*Cucumis sativus*). Nanotoxicology 5 (4), 743–753.

Ma, J.J., Ren, Y.J., Yan, L.Y., 2014. Effects of spray application of lanthanum and cerium on yield and quality of Chinese cabbage (Brassica chinensis L) based on different seasons. Biol. Trace Elem. Res. 160 (3), 427–432.

Ma, X., Wang, Q., Rossi, L., Zhang, W., 2015. Cerium oxide nanoparticles and bulk cerium oxide leading to different physiological and biochemical responses in Brassica rapa. Environ. Sci. Technol. 50 (13), 6793–6802.

Mackerness, S.A.H., John, C.F., Jordan, B., Thomas, B., 2001. Early signalling components in ultraviolet-B responses: distinct roles for different reactive oxygen species and nitric oxide. FEBS Lett. 489, 237–242.

Majumdar, S., Peralta-Videa, J.R., Bandyopadhyay, S., Castillo-Michel, H., Hernandez-Viezcas, J.A., Sahi, S., Gardea-Torresdey, J.L., 2014. Exposure of cerium oxide nanoparticles to kidney bean shows disturbance in the plant defense mechanisms. J. Hazard. Mater. 278, 279–287.

Majumdar, S., Almeida, I.C., Arigi, E.A., Choi, H., VerBerkmoes, N.C., Trujillo-Reyes, J., Flores-Margez, J.P., White, J.C., Peralta-Videa, J.R., Gardea-Torresdey, J.L., 2015a. Environmental effects of nanoceria on seed production of common bean (*Phaseolus vulgaris*): a proteomic analysis. Environ. Sci. Technol. 49 (22), 13283–13293.

Majumdar, S., Trujillo-Reyes, J., Hernandez-Viezcas, J.A., White, J.C., Peralta-Videa, J.R., Gardea-Torresdey, J.L., 2015b. Cerium biomagnification in a terrestrial food chain: influence of particle size and growth stage. Environ. Sci. Technol. 50 (13), 6782–6793.

Marchiol, L., Mattiello, A., Pošćić, F., Fellet, G., Zavalloni, C., Carlino, E., Musetti, R., 2016. Changes in physiological and agronomical parameters of barley (*Hordeum vulgare*) exposed to cerium and titanium dioxide nanoparticles. Int. J. Environ. Res. Public Health 13 (3), 332.

Masui, T., Fujiwara, K., Machida, K.I., Adachi, G.Y., Sakata, T., Mori, H., 1997. Trujillomicelles, Characterization of cerium (IV) oxide ultrafine particles prepared using reversed micelles. Chem. Mater. 9 (10), 2197–2204.

Matijević, E., Hsu, W.P., 1987. Preparation and properties of monodispersed colloidal particles of lanthanide compounds: I. Gadolinium, europium, terbium, samarium, and cerium (III). J. Colloid Interface Sci. 118 (2), 506–523.

Mazdiyasni, K.S., Brown, L.M., 1971. Influence of dynamic calcination on crystallite growth of submicron rare-earth oxides. J. Am. Ceram. Soc. 54 (10), 479–483.

Mihranyan, A., Ferraz, N., Strømme, M., 2012. Current status and future prospects of nanotechnology in cosmetics. Prog. Mater. Sci. 57 (5), 875–910.

Morales, M.I., Rico, C.M., Hernandez-Viezcas, J.A., Nunez, J.E., Barrios, A.C., Tafoya, A., Flores-Marges, J.P., Peralta-Videa, J.R., Gardea-Torresdey, J.L., 2013. Toxicity assessment of cerium oxide nanoparticles in cilantro (*Coriandrum sativum* L.) plants grown in organic soil. J. Agric. Food Chem. 61 (26), 6224–6230.

Munusamy, S., Bhakyaraj, K., Vijayalakshmi, L., Stephen, A., Narayanan, V., 2013. Synthesis and characterization of cerium oxide nanoparticles using *Curvularia lunata* and their antibacterial properties. Int. J. Innov. Res. Sci. Eng. Technol. 2347–3207 Online.

Nichols, J.W., Bae, Y.H., 2013. Nanotechnology for cancer treatment: possibilities and limitations. In: Cancer Targeted Drug Delivery. Springer, New York, pp. 37–56.

Ning, Z., Cheung, C.S., Fu, J., Liu, M.A., Schnell, M.A., 2006. Experimental study of environmental tobacco smoke particles under actual indoor environment. Sci. Total Environ. 367 (2), 822–830.

Niu, G., Rodriguez, D.S., 2008. Growth and physiological response to drought stress in four oleander clones. J. Am. Soc. Hort. Sci. 133, 188–196.

Niu, J., Wang, K., Kolattukudy, P.E., 2011. Cerium oxide nanoparticles inhibits oxidative stress and nuclear factor-kB activation in H9c2 cardiomyocytes exposed to cigarette smoke extract. J. Pharmacol. Exp. Ther. 338, 53–61.

Oberdörster, G., Oberdörster, E., Oberdörster, J., 2005. Nanotoxicology: an emerging discipline evolving from studies of ultrafine particles. Environ. Health Perspect. 823–839.

Pang, X., Li, D., Peng, A., 2002. Application of rare-earth elements in the agriculture of China and its environmental behavior in soil. Environ. Sci. Pollut. Res. 9 (2), 143–148.

Parisi, C., Vigani, M., Cerezo, E.R., 2014. In: Proceedings of a Workshop on "Nanotechnology for the Agricultural Sector: From Research to the Field"(No. JRC89736). Inst. for Pros. and Tech. Studies, Joint Res. Centre.

Patra, K.C., Wang, Q., Bhaskar, P.T., Miller, L., Wang, Z., Wheaton, W., Chandel, N., Laakso, M., Muller, W.J., Allen, E.L., Jha, A.K., 2013. Hexokinase 2 is required for tumor initiation and maintenance and its systemic deletion is therapeutic in mouse models of cancer. Cancer Cell 24 (2), 213–228.

Peng, J., Sun, Y., Liu, Q., Yang, Y., Zhou, J., Feng, W., Zhang, X., Li, F., 2012. Upconversion nanoparticles dramatically promote plant growth without toxicity. Nano Res. 5 (11), 770–782.

Peng, H., Liu, Y., Peng, W., Zhang, J., Ruan, R., 2016. Green synthesis and stability evaluation of Ag nanoparticles using bamboo hemicellulose. BioResources 11 (1), 385–399.

Peralta-Videa, J.R., Hernandez-Viezcas, J.A., Zhao, L., Diaz, B.C., Ge, Y., Priester, J.H., Holden, P.A., Gardea-Torresdey, J.L., 2014. Cerium dioxide and zinc oxide nanoparticles alter the nutritional value of soil cultivated soybean plants. Plant physiol. Biochem. 80, 128–135.

Peralta-Videa, J.R., Gardea-Torresdey, J.L., 2015. Effects of uncoated and citric acid coated cerium oxide nanoparticles, bulk cerium oxide, cerium acetate, and citric acid on tomato plants.

Pereira, A.E., Grillo, R., Mello, N.F., Rosa, A.H., Fraceto, L.F., 2014. Application of poly (epsilon-caprolactone) nanoparticles containing atrazine herbicide as an alternative technique to control weeds and reduce damage to the environment. J. Hazard. Mater. 268, 207–215.

Perez, G.R.M., Lule, P.R., 2004. *Raphanus sativus* (Radish): their chemistry and biology. Sci. World J. 4, 811–837.

Perez, J.M., Asati, A., Nath, S., Kaittanis, C., 2008. Synthesis of biocompatible dextran-coated nanoceria with pH-dependent antioxidant properties. Small 4, 552–556. http://dx.doi.org/10.1002/smll.200700824.

Perreault, F., Popovic, R., Dewez, D., 2014. Different toxicity mechanisms between bare and polymer-coated copper oxide nanoparticles in *Lemna gibba*. Environ. Pollut. 185, 219–227.

Piccinno, F., Gottschalk, F., Seeger, S., Nowack, B., 2012. Industrial production quantities and uses of ten engineered nanomaterials in Europe and the world. J. Nanopart. Res. 14 (9), 1–11.

Pilon-Smits, E.A.H., Quinn, C.F., Tapken, W., Malagoli, M., Schiavon, M., 2009. Physiological functions of beneficial elements. Curr. Opin. Plant Biol. 12, 267–274.

Poirot, N., Boy, P., Belleville, P., Bianchi, L., 2010. Synthesis of scandium oxide nanoparticles from solution. In: Advances in Science and Technology, vol. 62. Trans Tech Publications, pp. 70–75.

Pol, V.G., Palchik, O., Gedanken, A., Felner, I., 2002. Synthesis of europium oxide nanorods by ultrasound irradiation. J. Phys. Chem. B 106 (38), 9737–9743.

Pracharoenwattana, I., Zhou, W., Smith, S.M., 2010. Fatty acid beta-oxidation in germinating *Arabidopsis* seeds is supported by peroxisomal hydroxypyruvate reductase when malate dehydrogenase is absent. Plant Mol. Biol. 72, 101–109.

Prasad, R., Pandey, R., Barman, I., 2016. Engineering tailored nanoparticles with microbes: quo vadis? Wiley Interdiscip. Rev. 8, 316–330.

Priya, G.S., Kanneganti, A., Kumar, K.A., Rao, K.V., Bykkam, S., 2014. Biosynthesis of Cerium oxide nanoparticles using *Aloe barbadensis* miller gel. Int. J. Sci. Res. Publ. 4, 2250–3153.

Que, W., Buddhudu, S., Zhou, Y., Lam, Y.L., Zhou, J., Chan, Y.C., Kam, C.H., Gan, L.H., Deen, G.R., 2001a. Up-conversion emission in violet from neodymium oxalate and neodymium oxide phosphors obtained by microemulsion technique. Mater. Sci. Eng. 16 (1), 153–156.

Que, W., Zhou, Y., Kam, C.H., Lam, Y.L., Chan, Y.C., Gan, L.H., Deen, G.R., 2001b. Fluorescence characteristics from microemulsion technique derived erbium (III) oxide nanocrystals. Mater. Res. Bull. 36 (5), 889–895.

Que, W., Buddhudu, S., Zhou, Y., Lam, Y.L., Zhou, J., Chan, Y.C., Kam, C.H., Gan, L.H., Deen, G.R., 2001c. Preparation and characterization of erbium oxalate and erbium oxide nanoparticles by microemulsion technique. Mater. Sci. Eng. 16 (1), 51–54.

Ramos, S.J., Dinali, G.S., de Carvalho, T.S., Chaves, L.C., Siqueira, J.O., Guilherme, L.R., 2016. Rare earth elements in raw materials and products of the phosphate fertilizer industry in South America: content, signature, and crystalline phases. J. Geochem. Explor. 168, 177–186.

Redling, K., 2006. Rare Earth Elements in Agriculture with Emphasis on Animal Husbandry (Doctoral Dissertation, Lmu). Ludwig-Maximilians-Universität München.

Remédios, C., Rosário, F., Bastos, V., 2012. Environmental nanoparticles interactions with plants: morphological, physiological, and genotoxic aspects. J. Bot. 2012.

Rico, C.M., Majumdar, S., Duarte-Gardea, M., Peralta-Videa, J.R., Gardea-Torresdey, J.L., 2011. Interaction of nanoparticles with edible plants and their possible implications in the food chain. J. Agric. Food Chem. 59 (8), 3485–3498.

Rico, C.M., Morales, M.I., Barrios, A.C., McCreary, R., Hong, J., Lee, W.Y., Nunez, J., Peralta-Videa, J.R., Gardea-Torresdey, J.L., 2013. Effect of cerium oxide nanoparticles on the quality of rice (*Oryza sativa* L.) grains. J. Agric. Food Chem. 61 (47), 11278–11285.

Rico, C.M., Lee, S.C., Rubenecia, R., Mukherjee, A., Hong, J., PeraltaVidea, J.R., GardeaTorresdey, J.L., 2014. Cerium oxide nanoparticles impact yield and modify nutritional parameters in wheat (*Triticumaestivum* L.). J. Agric. Food Chem. 62 (40), 9669–9675.

Roco, M.C., 2004. Nanoscale science and engineering: unifying and transforming tools. AIChE J. 50 (5), 890–897.

Rui, Y., Zhang, P., Zhang, Y., Ma, Y., He, X., Gui, X., Li, Y., Zhang, J., Zheng, L., Chu, S., Guo, Z., 2015. Transformation of ceria nanoparticles in cucumber plants is influenced by phosphate. Environ. Pollut. 198, 8–14.

Sadiq, I.M., Pakrashi, S., Chandrasekaran, N., Mukherjee, A., 2011. Studies on toxicity of aluminum oxide (Al_2O_3) nanoparticles to microalgae species: Scenedesmus sp. and Chlorella sp. J. Nanopart. Res. 13 (8), 3287–3299.

Sapkota, A., Symons, J.M., Kleissl, J., Wang, L., Parlange, M.B., Ondov, J., Breysse, P.N., Diette, G.B., Eggleston, P.A., Buckley, T.J., 2005. Impact of the 2002 Canadian forest fires on particulate matter air quality in Baltimore City. Environ. Sci. Technol. 39 (1), 24–32.

Saravani, H., Jehali, M., 2015. Synthesis and characterization of lanthanum oxide and lanthanumoxid carbonate nanoparticles from thermalizes of La (acacen)(NO). Orient. J. Chem. 32 (1), 491–498.

Sarin, H., Kanevsky, A.S., Wu, H., Brimacombe, K.R., Fung, S.H., Sousa, A.A., Auh, S., Wilson, C.M., Sharma, K., Aronova, M.A., Leapman, R.D., Griffiths, G.L., Hall, M.D., 2008. Effective transvascular delivery of nanoparticles across the blood–brain tumor barrier into malignant glioma cells. J. Transl. Med. 6, 80.

Schimel, J.P., Schaeffer, S.M., 2012. Microbial control over carbon cycling in soil. The Causes and Consequences of Microbial Community Structure, vol. 155.

Schwabe, F., Schulin, R., Limbach, L.K., Stark, W., Bürge, D., Nowack, B., 2013. Influence of two types of organic matter on interaction of CeO_2 nanoparticles with plants in hydroponic culture. Chemosphere 91 (4), 512–520.

Schwabe, F., Schulin, R., Rupper, P., Rotzetter, A., Stark, W., Nowack, B., 2015. Dissolution and transformation of cerium oxide nanoparticles in plant growth media. J. Nano Res. 16, 1–11.

Servin, A.D., White, J.C., 2016. Nanotechnology in agriculture: next steps for understanding engineered nanoparticle exposure and risk. NanoImpact 1, 9–12.

Servin, A., Elmer, W., Mukherjee, A., De la Torre-Roche, R., Hamdi, H., White, J.C., Bindraban, P., Dimkpa, C., 2015. A review of the use of engineered nanomaterials to suppress plant disease and enhance crop yield. J. Nanopart. Res. 17 (2), 1–21.

Shah, v., Shah, S., Shah, H., Rispoli, F.J., McDonnell, K.T., Workeneh, S., Karakoti, A., Kumar, A., et al., 2012. Antibacterial activity of polymer coated cerium oxide nanoparticles. PLoS One 7, 47827.

Sharma, P.K., Jilavi, M.H., Naraß, R., Schmidt, H., 1998. Seeding effect in hydrothermal synthesis of nanosize yttria. J. Mater. Sci. Lett. 17 (10), 823–825.

Shen, C.X., Zhang, Q.F., Li, J., Bi, F.C., Yao, N., 2010a. Induction of programmed cell death in Arabidopsis and rice by single-wall carbon nanotubes. Am. J. Bot. 97, 1–8.

Shen, X., Zhou, Y., Duan, L., Li, Z., Eneji, A.E., Li, J., 2010b. Silicon effects on photo-synthesis and antioxidant parameters of soybean seedlings under drought and ultraviolet-B radiation. J. Plant Physiol. 167, 1248–1252.

Shi, J.P., Evans, D.E., Khan, A.A., Harrison, R.M., 2001. Sources and concentration of nanoparticles (<10 nm diameter) in the urban atmosphere. Atmos. Environ. 35 (7), 1193–1202.

Shi, Z., Shao, L., Jones, T.P., Lu, S., 2005. Microscopy and mineralogy of airborne particles collected during severe dust storm episodes in Beijing, China. J. Geophysic. Res. Atmos. 2005 110(D1).

Shweta, Tripathi, D.K., Singh, S., Singh, S., Dubey, N.K., Chauhan, D.K., 2016. Impact of nanoparticles on photosynthesis: challenges and opportunities. Mater. Focus 5 (5), 405–411.

Sicard, C., Perullini, M., Spedalieri, C., Coradin, T., Brayner, R., Livage, J., Jobbagy, M., Bilmes, S.A., 2011. CeO_2 nanoparticles for the protection of photosynthetic organisms immobilized in silica gels. Chem. Mater. 23, 1374–1378.

Siegrist, M., Keller, C., Kastenholz, H., Frey, S., Wiek, A., 2007. Laypeople's and experts' perception of nanotechnology hazards. Risk Anal. 27 (1), 59–69.

Singh, S., Tripathi, D.K., Dubey, N.K., Chauhan, D.K., 2016. Effects of nano-materials on seed germination and seedling growth: striking the slight balance between the concepts and controversies. Mater. Focus 5 (3), 195–201.

Sioutas, C., Delfino, R.J., Singh, M., 2005. Exposure assessment for atmospheric ultrafine particles (UFPs) and implications in epidemiologic research. Environ. Health Perspect. 947–955.

Skandan, G., Hahn, H., Parker, J.C., 1991. Nanostructured Y_2O_3: synthesis and relation to microstructure and properties. Scr. Metall. Mater. 25 (10), 2389–2393.

Sordelet, D., Akinc, M., 1988. Preparation of spherical, monosized Y_2O_3 precursor particles. J. Colloid Interface Sci. 122 (1), 47–59.

Strambeanu, N., Demetrovici, L., Dragos, D., 2015. Natural sources of nanoparticles. In: Nanoparticles' Promises and Risks. Springer International Publishing, pp. 9–19.

Suppan, S., 2014. Nanomaterials risk to soil health. Institute for agri. And trade policy that matter, S.S. In: Christoph Lauterwasser (Ed.), Opportunities and Risks of Nanotechnologies, Joint Report of the Allianz Center for Techn ology and the OECD International Futures Programme 2007.

Taylor, David A., 2002. "Dust in the wind." Environmental Health Perspectives 110 (2), A-80.

Thompson, S.E., Parthasarathy, S., 2006. Moore's law: the future of Si microelectronics. Mater. Today 9 (6), 20–25.

Tissue, B.M., 1998. Synthesis and luminescence of lanthanide ions in nanoscale insulating hosts. Chem. Mater. 10 (10), 2837–2845.

Tiwari, D.K., Behari, J., Sen, P., 2008. Application of Nanoparticles in Waste Water Treatment 1.

Tomaszewski, H., Weglarz, H., De Gryse, R., 1997. Crystallization of yttria under hydrothermal conditions. J. Eur. Ceram. Soc. 17 (2), 403–406.

Tripathi, D.K., Singh, V.P., Prasad, S.M., Chauhan, D.K., Dubey, N.K., 2015. Silicon nanoparticles (SiNp) alleviate chromium (VI) phytotoxicity in *Pisum sativum* (L.) seedlings. Plant Physiol. Biochem. 96, 189–198.

Tripathi, D.K., Singh, S., Singh, S., Pandey, R., Singh, V.P., Sharma, N.C., et al., 2017a. An overview on manufactured nanoparticles in plants: uptake, translocation, accumulation and phytotoxicity. Plant Physiol. Biochem. 110, 2–12.

Tripathi, D.K., Tripathi, A., Guar, S., Singh, S., Singh, Y., Vishwakarma, K., et al., 2017b. Uptake, accumulation and toxicity of silver nanoparticle in autotrophic plants, and heterotrophic microbes: a concentric review. Front. Microbiol. 8, 7. http://dx.doi.org/10.3389/fmicb.2017.00007.

Tripathi, D.K., Singh, S., Singh, S., Srivastava, P.K., Singh, V.P., Singh, S., et al., 2017c. Nitric oxide alleviates silver nanoparticles (AgNps)-induced phytotoxicity in *Pisum sativum* seedlings. Plant Physiol. Biochem. 110, 167–177. http://dx.doi.org/10.1016/j.plaphy.2016.06.015.

Tripathi, D.K., Mishra, R.K., Singh, S., Singh, S., Vishwakarma, K., Sharma, S., Singh, V.P., Singh, P.K., Prasad, S.M., Dubey, N.K., Pandey, A.C., 2017d. Nitric oxide ameliorates zinc oxide nanoparticles phytotoxicity in wheat seedlings: implication of the ascorbate–glutathione cycle. Front. Plant Sci. 8.

Tripathi, D.K., Singh, S., Singh, V.P., Prasad, S.M., Dubey, N.K., Chauhan, D.K., 2017e. Silicon nanoparticles more effectively alleviated UV-B stress than silicon in wheat (*Triticum aestivum*) seedlings. Plant Physiol. Biochem. 110, 70–81.

Trujillo-Reyes, J., Majumdar, S., Botez, C.E., Peralta-Videa, J.R., Gardea-Torresdey, J.L., 2014. Exposure studies of core-shell Fe/Fe_3O_4 and Cu/CuO NPs to lettuce (*Lactuca sativa*) plants: are they a potential physiological and nutritional hazard? J. Hazard. Mater. 267, 255–263.

Tschöpe, A., Ying, J.Y., 1994a. Nanocrystalline cerium oxide catalytic materials. In: NATO ASI Ser. E, pp. 781–784.

Tschöpe, A., Ying, J.Y., 1994b. Synthesis of nanostructured catalytic materials using a modified magnetron sputtering technique. Nanostr. Mater. 4 (5), 617–623.

Tschope, A., Liu, W., Flytzani-Stephanopoulos, M., Ying, J.Y., 1995. Redox activity of nonstoichiometric cerium oxide-based nanocrystalline catalysts. J. Catal. 157 (1), 42–50.

Tumburu, L., Andersen, C.P., Rygiewicz, P.T., Reichman, J.R., 2016. Molecular and physiological responses to titanium dioxide and cerium oxide nanoparticles in *Arabidopsis*. Environ. Toxicolo. Chem. 36 (1), 71–82.

Tyler, G., 2004. Rare earth elements in soil and plant systems-A review. Plant Soil 267 (1–2), 191–206.

von Tucher, S., Schmidhalter, U., 2005. Lanthanum uptake from soil and nutrient solution and its effects on plant growth. J. Plant Nutr. Soil Sci. 168 (4), 574–580.

Wagner, S., Gondikas, A., Neubauer, E., Hofmann, T., von der Kammer, F., 2014. Spot the difference: engineered and natural nanoparticles in the environment-release, behavior, and fate. Angew. Chem. Int. Ed. 53 (46), 12398–12419.

Wang, Y., Yin, L., Gedanken, A., 2002. Sonochemical synthesis of mesoporous transition metal and rare earth oxides. Ultrason. Sonochem. 9 (6), 285–290.

Wang, X.M., Gao, F.Q., Ma, L.L., Liu, J., Yin, S.T., Yang, P., Hong, F.S., 2008. Effects of nano-anatase on ribulose-1, 5-bisphosphate carboxylase/oxygenase mRNA expression in spinach. Biol. Trace Elem. Res. 126 (1–3), 280–289.

Wang, J., Rahman, M.F., Duhart, H.M., Newport, G.D., Patterson, T.A., Murdock, R.C., Hussain, S.M., Schlager, J.J., Ali, S.F., 2009. Expression changes of dopaminergic system-related genes in PC12 cells induced by manganese, silver, or copper nanoparticles. Neurotoxicology 30 (6), 926–933.

Wang, Q., Ebbs, S.D., Chen, Y., Ma, X., 2012a. Trans-generational impact of cerium oxide nanoparticles on tomato plants. Metallomics 5, 753–759.

Wang, Q., Ma, X.M., Zhang, W., Pei, H.C., Chen, Y.S., 2012b. The impact of cerium oxide nanoparticles on tomato (*Solanum lycopersicum* L.) and its implications for food safety. Metallomics 4 (10), 1105–1112.

Wang, P., Menzies, N.W., Lombi, E., McKenna, B.A., Johannessen, B., Glover, C.J., Kappen, P., Kopittke, P.M., 2013a. Fate of ZnO nanoparticles in soils and cowpea (*Vigna unguiculata*). Environ. Sci. Technol. 47, 13822–13830.

Wang, W.-N., Tarafdar, J.C., Biswas, P., 2013b. Nanoparticle synthesis and delivery by an aerosol route for watermelon plant foliar uptake. J. Nanopart. Res. 15 (1), 1–13, 1417.

Wang, L., Wang, W., Zhou, Q., Huang, X., 2014. Combined effects of lanthanum (III) chloride and acid rain on photosynthetic parameters in rice. Chemosphere 112, 355–361.

Warheit, D.B., 2004. Nanoparticles: health impacts? Mater. Today 7 (2), 32–35.

Wason, M.S., Zhao, J., 2013. Cerium oxide nanoparticles: potential applications for cancer and other diseases. Am. J. Transl. Res. 5 (2), 126–131.

Westerdahl, D., Fruin, S., Sax, T., Fine, P.M., Sioutas, C., 2005. Mobile platform measurements of ultrafine particles and associated pollutant concentrations on freeways and residential streets in Los Angeles. Atmos. Environ. 39 (20), 3597–3610.

Whitesides, G.M., 2003. The 'right' size in nanobiotechnology. Nat. Biotechnol. 21 (10), 1161–1165.

Wu, Z., Guo, L., Li, H., Benfield, R.E., Yang, Q., Grandjean, D., Li, Q., Zhu, H., 2001. Cerium oxide nanoparticles coated by surfactant sodium bis (2-ethylhexyl) sulphosuccinate (AOT): local atomic structures and x-ray absorption spectroscopic studies. J. Phys. Condens. Matter 13 (22), 5269.

Xia, T., Kovochich, M., Liong, M., Madler, L., Gilbert, B., Shi, H., Yeh, J.I., Zink, J.I., Nel, A., 2008. Comparison of the mechanism of toxicity of zinc oxide and cerium oxide nanoparticles based on dissolution and oxidative stress properties. ACS Nano 2, 2121–2134.

Xie, C.Y., Li, S.Z., Li, J., 1992. Preparation of La~ 2O~ 3 ultrafine particle by homogeneous precipitation. Chin. Chem. Lett. 3, 951.

Xiong, B.K., Zhang, S.R., Guo, B.S., Zheng, W., 2001. Reviews of the use of REEs-based fertilizer in the past of three decades. In: Symposium of the Use of REEs in Agricultural of China. Baotou. PR China (in Chinese).

Xu, X., Zhu, W., Wang, Z., Witkamp, G.J., 2002. Distribution of rare earths and heavy metals in field-grown maize after application of rare earth-containing fertilizers. Sci. Total Environ. 293, 97–105.

Xue, C.H., Chen, J., Yin, W., Jia, S.T., Ma, J.Z., 2012. Superhydrophobic conductive textiles with antibacterial property by coating fibers with silver nanoparticles. Appl. Surf. Sci. 258 (7), 2468–2472.

Yadav, L.R., Manjunath, K., Archana, B., Madhu, C., Naika, H.R., Nagabhushana, H., Kavitha, C., Nagaraju, G., 2016. Fruit juice extract mediated synthesis of CeO_2 nanoparticles for antibacterial and photocatalytic activities. EPJ Plus 131 (5), 1–10.

Yang, K., Lin, D., Xing, B., 2009. Interactions of humic acid with nanosized inorganic oxides. Langmuir 25 (6), 3571–3576.

Yang, Z., Chen, J., Dou, R.Z., Gao, X., Mao, C.B., Wang, L., 2015. Assessment of the phytotoxicity of metal oxide nanoparticles on two crop plants, maize (Zea mays L.) and rice (Oryza sativa L.). Int. J. Environ. Res. Pub. Health 12 (12), 15100–15109.

Yin, L., Wang, Y., Pang, G., Koltypin, Y., Gedanken, A., 2002. Sonochemical synthesis of cerium oxide nanoparticles—effect of additives and quantum size effect. J. Colloid Interface Sci. 246 (1), 78–84.

Yoon, S.J., Kwak, J.I., Lee, W.M., Holden, P.A., An, Y.J., 2014. Zinc oxide nanoparticles delay soybean development: a standard soil microcosm study. Ecotoxicol. Environ. Saf. 100, 131–137.

Yuan, D., Shan, X.-Q., Huai, Q., Wen, B., Zhu, X., 2001. Uptake and distribution of rare earth elements in rice seeds cultured in fertilizer solution of rare earth elements. Chemosphere 43, 327–337 PMID: 11302578.

Zawadzki, M., Kępiński, L., 2004. Synthesis and characterization of neodymium oxide nanoparticles. J. Alloy. Compd. 380 (1), 255–259.

Zelati, A., Amirabadizadeh, A., Hosseini, A., 2014a. A facile approach to synthesize dysprosium oxide nanoparticles. Int. J. Ind. Chem. 5 (3–4), 69–75.

Zelati, A., Amirabadizadeh, A., Kompany, A., Salamati, H., Sonier, J.E., 2014b. Critical current density and intergranular coupling study of the dysprosium oxide nanoparticle added $Bi_{1.6}Pb_{0.4}Sr_2Ca_2Cu_3O_y$ superconductor. J. Supercond. Novel Magn. 27 (10), 2185–2193.

Zhang, S., Shan, X., 2001. Speciation of rare earth elements in soil and accumulation by wheat with rare earth fertilizer application. Environ. Pollut. 112, 395–405.

Zhang, P., Ma, Y.H., Zhang, Z.Y., He, X., Guo, Z., Tai, R.Z., Ding, Y.Y., Zhao, Y.L., Chai, Z.F., 2011a. Comparative toxicity of nanoparticulate bulk Yb_2O_3 and $YbCl_3$ to cucumber (*Cucumis sativus*). Environ. Sci. Technol. 46, 1834–1841.

Zhang, Z., He, X., Zhang, H., Ma, Y., Zhang, P., Ding, Y., Zhao, Y., 2011b. Uptake and distribution of ceria nanoparticles in cucumber plants. Metallomics 3 (8), 816–822.

Zhang, P., Ma, Y., Zhang, Z., He, X., Guo, Z., Tai, R., Ding, Y., Zhao, Y., Chai, Z., 2012a. Comparative toxicity of nanoparticulate/bulk Yb_2O_3 and $YbCl_3$ to cucumber (*Cucumis sativus*). Environ. Sci. Technol. 46 (3), 1834–1841.

Zhang, P., Ma, Y.H., Zhang, Z.Y., He, X., Zhang, J., Guo, Z., Tai, R.Z., Zhao, Y.L., Chai, Z.F., 2012b. Biotransformation of ceria nanoparticles in cucumber plants. ACS Nano 6, 9943–9950.

Zhao, Y.L., 2012. Phytotoxicity and biotransformation of La_2O_3 nanoparticles in a terrestrial plant cucumber (*Cucumis sativus*). Nanotoxicology 5, 743–753.

Zhao, L., Peralta-Videa, J.R., Varela-Ramirez, A., Castillo-Michel, H., Li, C., Zhang, J., Aguilera, R.J., Keller, A.A., Gardea-Torresdey, J.L., 2012. Effect of surface coating and organic matter on the uptake of CeO_2 NPs by corn plants grown in soil: insight into the uptake mechanism. J. Hazard. Mater. 225, 131–138.

Zhao, L., Peralta-Videa, J.R., Peng, B., Bandyopadhyay, S., Corral-Diaz, B., Osuna-Avila, P., Montes, M.O., Keller, A.A., Gardea-Torresdey, J.L., 2014. Alginate modifies the physiological impact of CeO_2 nanoparticles in corn seedlings cultivated in soil. J. Environ. Sci. 26 (2), 382–389.

Zhao, L., Sun, Y., Hernandez-Viezcas, J.A., Hong, J., Majumdar, S., Niu, G., Duarte-Gardea, M., Peralta-Videa, J.R., Gardea-Torresdey, J.L., 2015. Monitoring the environmental effects of CeO_2 and ZnO nanoparticles through the life cycle of corn (Zea mays) plants and in situ μ-XRF mapping of nutrients in kernels. Environ. Sci. Tech. 49 (5), 2921–2928.

Zhou, Y.C., Rahaman, M.N., 1993. Hydrothermal synthesis and sintering of ultrafine CeO_2 powders. J. Mater. Res. 8 (7), 1680–1686.

Zinatloo-Ajabshir, S., Salavati-Niasari, M., Hamadanian, M., 2015. Praseodymium oxide nanostructures: novel solvent-less preparation, characterization and investigation of their optical and photocatalytic properties. RSC Adv. 5 (43), 33792–33800.

Further Reading

Alidoust, D., Isoda, A., 2013. Effect of γFe_2O_3 nanoparticles on photosynthetic characteristic of soybean (*Glycine max* (L.) Merr.): foliar spray versus soil amendment. Acta Physiol. Plant. 35 (12), 3365–3375.

Aslani, F., Bagheri, S., Muhd Julkapli, N., Juraimi, A.S., Hashemi, F.S.G., Baghdadi, A., 2014. Effects of engineered nanomaterials on plants growth: an overview. Sci. World J. 2014.

Dahle, J.T., Livi, K., Arai, Y., 2015. Effects of pH and phosphate on CeO_2 nanoparticle dissolution. Chemosphere 119, 1365–1371.

Hong, J., Peralta-Videa, J.R., Rico, C., 2014. Evidence of translocation and physiological impacts of foliar applied CeO_2 nanoparticles on cucumber (*Cucumis sativus*) plants. Environ. Sci. Technol. 48 (8), 4376–4385. http://dx.doi.org/10.1021/es404931g.

Khodakovskaya, M.V., Kim, B.S., Kim, J.N., Alimohammadi, M., Dervishi, E., Mustafa, T., Cernigla, C.E., 2013. Carbon nanotubes as plant growth regulators: effects on tomato growth, reproductive system, and soil microbial community. Small. 9, 115–123. http://dx.doi.org/10.1002/smll.201201225.

Li, Y., He, J.G., Huang, X.W., 2011. Synthesis and properties of cerium oxide nanoparticles. In: Advanced Materials Research, vol. 299. Trans Tech Publications, pp. 118–121.

Li, X., Gui, X., Rui, Y., Ji, W., Yu, Z., Peng, S., 2014. Bt-transgenic cotton is more sensitive to CeO_2 nanoparticles than its parental non-transgenic cotton. J. Hazard. Mater. 274, 173–180.

Liang, T., Shiming, D., Wenchong, S., Chong, Z., Zhang, C., Haitao, L., 2008. A review of fractionations of rare earth elements in plants. J. Rare Earths 26 (1), 7–15.

Lin, D., Xing, B., 2006. Root uptake and phytotoxicity of ZnO nanoparticles. Environ. Sci. Technol. 42, 5580–5585.

Nowack, B., Bucheli, T.D., 2007. Occurrence, behavior and effects of nanoparticles in the environment. Environ. Pollut. 150 (1), 5–22.

Rizwan, M., Ali, S., Qayyum, M.F., Ok, Y.S., Adrees, M., Ibrahim, M., Zia-ur-Rehman, M., Farid, M., Abbas, F., 2017. Effect of metal and metal oxide nanoparticles on growth and physiology of globally important food crops: A critical review. J. Hazard. Mat. 322, 2–16.

Schulte, P.A., Salamanca-Buentello, F., 2007. Ethical and scientific issues of nanotechnology in the workplace. Cien. Saude. Colet 12 (5), 1319–1332.

Shi-Ming, D.I.N.G., Liang, T., Zhang, C.S., Li-Jun, W.A.N.G., Qin, S.U.N., 2006. Accumulation and fractionation of rare earth elements in a soil-wheat system. Pedosphere 16 (1), 82–90.

Singh, S., Dosani, T., Karakoti, A.S., Kumar, A., Seal, S., Self, W.T., 2011. A phosphate-dependent shift in redox state of cerium oxide nanoparticles and its effects on catalytic properties. Biomaterials 32 (28), 6745–6753.

Singh, S., Srivastava, P., Kapoor, I.P.S., Singh, G., 2013. Preparation, characterization, and catalytic activity of rare earth metal oxide nanoparticles. J. Therm. Anal. Calorim. 111 (2), 1073–1082.

Trujillo, S., Peralta-Videa, J.R., Trujillo-Reyes, J., Sun, Y., Barrios, A.C., Niu, G., Flores-Margez, J.P., Gardea-Torresdey, J.L., 2016. Soil organic matter influences cerium translocation and physiological processes in kidney bean plants exposed to cerium oxide nanoparticles. Sci. Total Environ. 569, 201–211.

Van Gosen, B.S., Verplanck, P.L., Long, K.R., Gambogi, J., Robert, R., Seal, I.I., 2014. The Rare-earth Elements: Vital to Modern Technologies and Lifestyles (No. 2014-3078). US Geological Survey.

Zhang, L., Gu, F.X., Chan, J.M., Wang, A.Z., Langer, R.S., Farokhzad, O.C., 2008. Nanoparticles in medicine: therapeutic applications and developments. Clin. Pharmacol. Ther. 83 (5), 761–769.

CHAPTER

12

Influence of Titanium Dioxide Nanoparticles (nTiO$_2$) on Crop Plants: A Systematic Overview

Alessandro Mattiello[1], Daniel Lizzi[1,2], Luca Marchiol[1]
[1]University of Udine, Udine, Italy; [2]University of Trieste, Trieste, Italy

12.1 INTRODUCTION

From the last few decades, several research groups of all over the world have paid much attention to understanding the responses of nanomaterials and plant interactions because of the increasing availability of these tiny particles in the environment (Cox et al., 2017; Du et al., 2017; Ochoa et al., 2017; Reddy et al., 2016; Tripathi et al., 2017). Studies revealed that some nanoparticles showed positive responses and enhanced plant growth and development by supporting the plant metabolic system (Tripathi et al., 2016; Singh et al., 2016, 2017; Shweta et al., 2016; López-Moreno et al., 2017; Medina-Velo et al., 2017; Venkatachalam et al., 2017). In contrast, several studies demonstrated the adverse impacts of some nanomaterials (Cox et al., 2017; Ochoa et al., 2017; Reddy et al., 2016). Thus it is required to know the impact of each of these nanomaterials on plants separately.

Titanium (Ti) is the second most abundant transition metal, having an average concentration of 6.320 mg/kg in the earth's crust (Hrubý et al., 2002). It has been reported to have beneficial effects on plants including: (1) increase in macronutrient uptake (Hrubý et al., 2002), (2) increase in activity of reductase enzymes (e.g., nitrate reductase, peroxidase, and catalase) (Pais, 1983), and (3) efficient gaseous exchange, overall enhancing the crop yield (Khodakovskaya and Lahiani, 2014).

The future scenario for food security mandates the use of nanotechnology in the agrifood sector (Singh Sekhon, 2014). Although the application of nanomaterials in agriculture is a relatively new idea as compared to other sectors (e.g., industry, energy, aerospace, medicine), they have been considered to be an important tool for enhancing resistance against stress as well as enhancing the crop yield (Servin and White, 2016).

Nanomaterials in Plants, Algae, and Microorganisms
http://dx.doi.org/10.1016/B978-0-12-811487-2.00012-8

Bulk titanium oxide and nanotitanium dioxide ($nTiO_2$) are being used in industry and day-to-day life (Lu et al., 2015). $nTiO_2$ is manufactured worldwide at an average quantity of up to 88,000 t/year, making it one of the most used nanomaterials and consequently the most abundant nanomaterial released to the environment. For this reason, it has now been included in the group of emerging contaminants (Keller et al., 2013). Sun et al. (2014) reported that $nTiO_2$ input in soils in Europe has reached 0.13 μg/kg year and discharge of sewage sludge could enhance accumulation up to 1200 μg/kg year. Gottschalk et al. (2009) demonstrated an annual increase of 89 and 358 μg $nTiO_2$/kg in soil and sediments amended with sludges, respectively. As a consequence, $nTiO_2$ leaches water bodies and is possibly taken up by crops (Gottschalk et al., 2009).

Moreover, future applications of titanium nanomaterials in agriculture will result in an additional input of $nTiO_2$ in soil (Moll et al., 2016). Concerns have been raised about the possible negative effect on soil microbial communities and fungi (Cox et al., 2016; Simonin et al., 2016; Moll et al., 2017) posed by $nTiO_2$.

12.2 INFLUENCE OF nTIO$_2$ ON PLANT GROWTH

Several reports demonstrated that nanomaterials may induce harmful effects on terrestrial and aquatic ecosystems. However, current knowledge does not yet allow us to define the basis for an environmental risk assessment of nanomaterials. For this reason, it is very important to define the effects induced by exposure to $nTiO_2$ on plants, soil organisms, and ecosystem functions (Zuverza-Mena et al., 2016). Ecotoxicological tests have been carried out to understand whether $nTiO_2$ because of its reactivity poses adverse effects on the environment and human health (Semenzin et al., 2015). So far we have understood that $nTiO_2$ toxicity is size dependent and is far higher than bulk TiO_2 both in aquatic and terrestrial environments (Wang et al., 2009; Xiong et al., 2011). The literature related to the impacts of $nTiO_2$ on crops is not sufficient and is still emerging (Cox et al., 2016). Up to now, experimental data demonstrated that $nTiO_2$ could have both beneficial and toxic effects on plants depending upon the experimental conditions (Mukherjee et al., 2016).

In the following section we have provided brief details of the impact of $nTiO_2$ exposure on some cultivated plants. Several works have been performed on nanoparticles, especially the impact of $nTiO_2$ on crop plants, and on the basis of the literature review we have categorized the types of studies into three groups (on the basis of paper published; Fig. 12.1A). Group I includes those studies that surveyed the effects of $nTiO_2$ on seed germination and seedlings with duration comprised of a few days to a few weeks, Group II includes those studies that investigated the effects of $nTiO_2$ on plant growth and physiological parameters on adult plants, while Group III include those studies performed on the whole crop cycle. The largest numbers of studies (six papers) have been performed by Chinese and American research groups, followed by Iranian and Italian (five papers), then by French (four papers), and lastly Indian and the Republic of Korea (three papers) groups. Finally, studies have been reported from Switzerland (two papers), Israel, Thailand, Czech Republic, and Slovak Republic (one paper) (Fig. 12.1B).

The previously presented survey has been organized according to the type of studies and the order of paper publication in Fig. 12.1A. In addition, the experimental conditions and main results are summarized in Tables 12.1A–C, where the papers are ordered according to particle size.

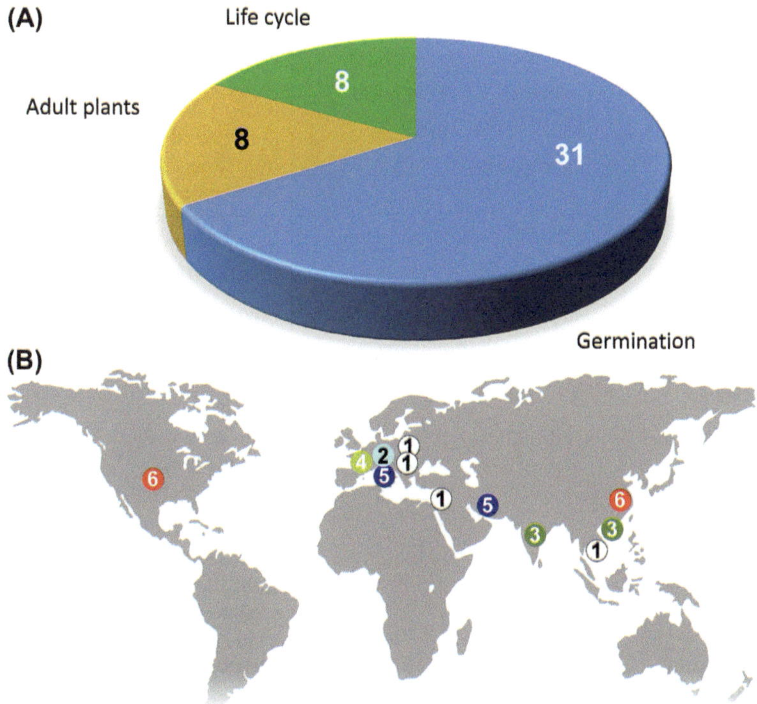

FIGURE 12.1 (A) Number of research papers on nTiO$_2$ effects on crops as classified according to the type of experiment. (B) Geographical distribution of research groups.

12.2.1 Germination and Root Elongation

The advantageous effects of bulk titanium on plants have been known for some time (Carvajal and Alcaraz, 1998). For this reason, considerable interest has developed to clarify the effects of nTiO$_2$ on plant metabolism. In addition to this, the influence of nTiO$_2$ on seed germination and the early stages of vegetative growth of crops have also been studied in recent years. The most important studies in this context are summarized in Tables 12.1A–C. Here we will briefly review these works to summarize both aspects of experimental activities and the main results.

Among several studies on the impact on plants, Zheng et al. (2005) provided theoretical basis and technical details that supported the agricultural applications of nTiO$_2$. It is known that the vigor or strength of seeds decreases as a function of the storage time because of possible accumulation of free radicals. The study by Zheng et al. (2005) provided information about the germination rate and the germination and vigor indexes of naturally aged seeds of *Spinacia oleracea* L. (spinach) treated with nTiO$_2$ and bulk TiO$_2$. Spinach seeds were soaked in nTiO$_2$ and bulk TiO$_2$ and germinated in Petri dishes. The results demonstrated that nTiO$_2$ increases the vigor of aged seeds supporting their germination.

TABLE 12.1A Summary of the Effects of $nTiO_2$ on Germination and Seedling Root Elongation of Cultivated Plants

Species	Particle Size (nm)	Substrate	Treatment	Plant Response	References
Hordeum vulgare	5–10	Agar Hydroponics	0%–20% (m/m)	Root growth inhibition with increasing treatment concentration. Limited nanoparticle transport to aerial fractions	Kořenková et al. (2017)
Vicia faba	<10 (spherical shape) <100 (tetragonal crystals)	Petri dishes	0, 50 mg/L	Phytotoxic and genotoxic effects depending on size and shape of $nTiO_2$	Ruffini Castiglione et al. (2016)
V. faba	Nanocomposites, TiO_2 core 10–50	Hydroponics	0, 5, 25, 50 mg/L	Plant growth and physiological parameters unchanged compared to controls	Foltête et al. (2011)
Triticum aestivum Brassica napus	14–25	Hydroponics	0, 100 mg/L	Increased root elongation in plantlets, seed germination, evapotranspiration, and plant biomass not affected	Larue et al. (2013)
T. aestivum	14–655	Hydroponics	0, 15 mg/L	The smallest nanoparticles promote root elongation in plantlets. Proposed threshold diameters for particle transfer through root epidermis and root-to-shoot transfer	Larue et al. (2012)
T. aestivum	20–50	Sand	0, 250, 500, 1000, 2000 mg/L	Seed germination, shoot–root growth, and cell division unaffected	Ramesh et al. (2014)
T. aestivum	21	Petri dishes	0, 1, 2, 10, 100, 500 mg/L	Shoot and seedling lengths at lower concentrations higher than controls	Feizi et al. (2012)
Foeniculum vulgare	21	Petri dishes	0, 5, 20, 40, 60, 80 mg/L	Enhancement of germination value and mean daily germination	Feizi et al. (2013)

TABLE 12.1B Summary of the Effects of $nTiO_2$ on Germination and Seedling Root Elongation of Cultivated Plants

Species	Particle Size (nm)	Substrate	Treatment	Plant Response	References
Allium cepa *Avena sativa* *Brassica oleracea capitata* *Cucumis sativus* *Daucus carota* *Glycine max* *Lactuca sativa* *Lolium perenne* *Lycopersicum esculentum* *Zea mays*	~23	Petri dishes	0, 250, 500, 1000 µg/mL	Change in germination rate and cotyledon development. Promoted root elongation in cucumber and onion. Hindered root growth in cabbage, corn, lettuce, and oats	Andersen et al. (2016)
Nicotiana tabacum	<25	Petri dishes	0%, 0.1%, 1%, 2.5%, 5%	Germination inhibition, reduction of root elongation, and plant growth	Frazier et al. (2014)
A. cepa	<25	Growth chamber, dark conditions	12.5, 25, 50, 100 mg/mL	Dose-dependent decrease in mitotic index; increase in chromosomal aberrations in root tips	Pakrashi et al. (2014)
Brassica campestris napus *L. sativa* *Phaseolus vulgaris*	27 anatase/rutile = 80:20	Pot	0, 50, 100, 1000, 2500, 5000 mg/L	No differences among the species in germination; root elongation in *L. sativa* negatively affected by the higher treatment	Song et al. (2013a)
Solanum lycopersicum	27	Petri dishes	0, 50, 100, 1000, 2500, 5000 mg/L	No acute toxicity on germination and root elongation	Song et al. (2013b)
C. sativus	27	Hydroponics	0, 50, 250, 500, 1000, 2000, 4000 mg/L	>300% increase in root length. $nTiO_2$ not biotransformed	Servin et al. (2012)
C. sativus	30–50	Petri dishes	0, 250, 500, 2500, 5000 µg/mL	Inhibitory effects on germination and reduction in root growth	Mushtaq (2011)

TABLE 12.1C Summary of the Effects of $nTiO_2$ on Germination and Seedling Root Elongation of Cultivated Plants

Species	Particle Size (nm)	Substrate	Treatment	Plant Response	References
Hordeum vulgare	50	Petri dishes	0, 500, 1000, 2000 mg/L	Effects of random amplified polymorphic DNA banding patterns and reduction of root cell divisions. $nTiO_2$ aggregates within root cells	Mattiello et al. (2015)
Zea mays *Vicia narbonensis*	<100	Petri dishes	0%, 0.2%, 1.0%, 2.0%, 4.0%	Delayed germination. Root elongation affected by higher concentration. Genotoxic effects	Ruffini Castiglione et al. (2011)
Allium cepa *Nicotiana tabacum*	90–110	Sand Garden soil	0, 2, 4, 6, 8, 10 mM	Micronuclei and chromosomal aberrations linked with root growth reduction	Ghosh et al. (2010)
Sinapis alba	98 rutile, 21 anatase Nanowires: ~100 × 10 μm	Petri dishes	0, 10, 100, 1000 mg/L	Size and shape on $nTiO_2$ did not play a crucial role. Metal ions released into cultivation media. Ti accumulated in seedlings and induced phytotoxicity	Landa et al. (2016)
Oryza sativa	252 Å anatase, 330 Å rutile, rutile hydrophilic surfaces 408.9 Å, rutile hydrophobic surfaces 396.1 Å	Hydroponics	0, 10, 1000 mg/L	Anatase particles entered seedling roots through apoplastic route. No $nTiO_2$ root-to-shoot translocation	Cai et al. (2017)
O. sativa	–	Petri dishes	0, 100, 500, 1000 mg/L	$nTiO_2$ did not adversely affected rice seed germination	Boonyanitipong et al. (2011)
Spinacia oleracea	Rutile	Petri dishes	0, 0.25, 0.5, 1.0, 1.5, 2.0, 2.5, 4.0, 6.0‰	$nTiO_2$ increased germination rate, germination index, and vigor index of aged seeds	Zheng et al. (2005)
Petroselinum crispum	–	In vitro	0, 10, 20, 30, 40 mg/mL	Increase in germination %, germination rate, root-and-shoot growth in seedlings, and chlorophyll content	Dehkourdi and Mosavi (2013)

Two classical genotoxic endpoints, i.e., comet assay and the DNA laddering technique, were used by Ghosh et al. (2010) to evaluate the genotoxicity of $nTiO_2$ in *Allium cepa* L. (onion) and *Nicotiana tabacum* L. (tobacco). After germination, onion bulbs were exposed to different $nTiO_2$ solutions. The roots of onion collected after respectively 3, 6, and 24h were used for comet assay test. The plantlets of tobacco at the fourth leaf stage were exposed to the same $nTiO_2$ solutions for 24h and then the comet assay and DNA isolation were performed on leaf tissues collected. The results confirmed the genotoxic potential of $nTiO_2$ on onion and tobacco.

Since the manufactured nanomaterials are dispersed in the environment, studies on the consequences on biological systems are very important. Foltête et al. (2011), using *Vicia faba* (broad bean) as a model plant, carried out an investigation that focused on the effects of $nTiO_2$ coated with aluminum hydroxide and dimethicone films, which are commonly used in sunscreen creams (AFSSET, 2010). After 48h of $nTiO_2$ treatment, plant growth parameters, photosystem II photochemistry, genotoxicity, and phytochelatin concentrations were observed. These results did not show significant differences compared to control plants. However, spectroscopy analysis revealed high Ti concentrations in roots.

Seed germination is a complex physiological process that starts with water imbibition. Boonyanitipong et al. (2011) compared the effects of $nZnO$ and $nTiO_2$ on seed germination and root development in *Oryza sativa* L. (rice). The seeds were soaked in different concentrations of $nZnO$ and $nTiO_2$ for one, two, and three soaking periods, respectively. Thereafter germinated seeds were transferred on filter paper containing 4mL of Milli-Q water or nanoparticle suspensions. After 7days of treatment, seed germination and the number of roots and primary root length were recorded. Seed germination occurred normally under both nanoparticle treatments, probably because of the rice seed coat, which can act as a protector for the embryo. While $nZnO$ negatively affected root development, number of roots, and root length, there was no significant effect on number of roots under $nTiO_2$ treatments, suggesting that $nTiO_2$ did not have much effect on rice root development. In another study by Mushtaq (2011), germinating seeds of *Cucumis sativus* L. (cucumber) were exposed to different concentrations of nFe_3O_4, $nTiO_2$, and carbon nanoparticles. The relative root elongation, relative seed germination, and germination index were calculated. Germination was more inhibited by $nTiO_2$ than nFe_3O_4 and carbon nanoparticles. The germination index for $nTiO_2$ was consistently reduced with increasing the concentration of nanoparticles. All the nanoparticles reduced both the percentage germination and the percentage root elongation in the cucumber seeds.

Another study aimed at verifying the effect of $nTiO_2$ on the germination process of cultivated plants was carried out by Ruffini Castiglione et al. (2011), and in this study *Zea mays* L. (corn) and *Vicia narbonensis* L. (purple broad vetch) seeds were treated with solutions of $nTiO_2$ (up to 4.0% mixture of rutile and anatase). After 24h of exposure, delay in germination of both the species was recorded. Root elongation was affected only at higher doses of $nTiO_2$. To test the potential genotoxicity of $nTiO_2$, mitotic divisions in root tip cells were also analyzed and signals of mitotic and chromosomal abnormalities were evidenced.

In contrast to the foregoing studies, positive effects on germination of *Triticum aestivum* L. (wheat) were reported by Feizi et al. (2012). In particular, it was reported that $nTiO_2$ in a suitable concentration (1–500ppm) promoted seed germination in comparison to bulk TiO_2. The findings by Feizi et al. (2012) on wheat were confirmed by another group of scientists, where Larue et al. (2012) investigated the early effects on germination and root elongation of wheat under $nTiO_2$ (anatase and rutile) treatment with diameters in the range 14–655 nm. This large dimensional interval was necessary to make observations on the effectiveness of barriers present in plant tissues on $nTiO_2$ translocation. The findings demonstrated that $nTiO_2$ did not influence seed germination, whereas it promoted root elongation. Electronic microscope observation showed the presence of $nTiO_2$ both on the walls of cortical parenchyma cells and on the root epidermis. They also observed nanoparticles in cell walls and on both sides of the Casparian band. Subsequently, such nanoclusters were analyzed by in situ X-ray absorption spectroscopy. Clusters of $nTiO_2$ were observed in vascular structures, in the walls of leaf cells, and in vacuoles and chloroplasts as well. Finally, perhaps the most important outcome of this study proposed threshold diameters for $nTiO_2$ translocation. It has been demonstrated that $nTiO_2$ translocation was not allowed for nanostructures having a primary diameter greater than 36 nm.

The impacts of $nTiO_2$ in hydroponically grown *C. sativus* L. (cucumber) were evaluated by Servin et al. (2012). Advanced spectroscopic techniques (micro-X-ray fluorescence and micro-X-ray absorption spectroscopy) were used to define the chemical speciation of Ti and to track $nTiO_2$ within the plant tissue. It was confirmed that $nTiO_2$ stimulated root elongation. Experimental results demonstrated the presence of TiO_2 within the cucumber tissues, suggesting that $nTiO_2$ was not biotransformed.

The impact of $nTiO_2$ on plant growth, possible internalization, and root-to-shoot translocation of *T. aestivum* (wheat) and *Brassica rapa* (rapeseed) was investigated by Larue et al. (2013). In this study, wheat and rapeseed plantlets were exposed to $nTiO_2$ either through root or leaf surface. As for root exposure, $nTiO_2$ suspension was dispensed in a minihydroponic system, whereas for leaf exposure, $nTiO_2$ was administered by immersing one leaf in aqueous $nTiO_2$ suspension. The chemical species of the titanium taken up by plant roots and their distribution in plant tissues were investigated by advanced spectroscopy techniques (μPIXE and μXRF). The results showed that $nTiO_2$ exposure did not affect germination but stimulated root elongation, evapotranspiration, and plant biomass. The results described by Larue et al. (2013) are a complement to those provided in a previous paper by the same research group (Larue et al., 2012).

Song et al. (2013a) investigated the toxicity of $nTiO_2$ on germination of *B. napus* (oilseed rape), *L. sativa* L. (lettuce), and *Phaseolus vulgaris* (kidney bean). The species were selected because they are the recommended crops for the testing of chemicals in the Ecological Effects Test Guidelines provided by the USEPA (1996) and OECD Guidelines (2003). The germination of the studied plants was not influenced by the treatments. The root elongation in *L. sativa* was significantly lowered only at the higher concentration. However, it was interesting to note that seeds treated with the higher $nTiO_2$ (1.64 mg/kg) treatment showed three times lower Ti accumulation than in those treated with an intermediate concentration (0.54 mg/kg).

The potential of nanomaterials in stimulating germination and seedling growth of medicinal plants was also explored by Feizi et al. (2013). The objective of this work was to observe the effects of $nTiO_2$ and bulk TiO_2 on *Foeniculum vulgare* Mill (fennel). After exposure of germinated seeds to different $nTiO_2$ and bulk TiO_2 solutions, typical parameters of germination (germination percentage, vigor index, and mean daily germination) were observed. Also Dehkourdi and Mosavi (2013) investigated the potential of $nTiO_2$ in increasing germination time and percentage of *Petroselinum crispum* L. (parsley), which germinates very slowly. In this case, the percentage of germination, germination rate index, root and shoot length, fresh weight of seedlings, vigor index, and chlorophyll content were evaluated. A positive concentration-dependent effect of $n\mathrm{TiO_2}$ was recorded. At an appropriate concentration of $n\mathrm{TiO_2}$ both seed germination and seedling growth of parsley were stimulated.

A further comparative study on the effects of $n\mathrm{TiO_2}$ and bulk $\mathrm{TiO_2}$ on germination and seedling growth of *T. aestivum* (wheat) was carried out by Ramesh et al. (2014). No significant changes were observed in seed germination and seedling growth under $n\mathrm{TiO_2}$ and bulk $\mathrm{TiO_2}$ exposures compared to controls. Also in this case, no genotoxic effects were observed, while no changes in chlorophyll content and total protein were observed in treated plants. A significant inhibition in the germination rates, root elongation, and growth of *N. tabacum* L. (tobacco) seedlings exposed to $n\mathrm{TiO_2}$ was reported by Frazier et al. (2014). After 3 weeks of treatment, evidence of phytotoxicity and genotoxicity was recorded. A negative dose response on the seedlings' growth was observed. In addition, significant changes in the expression profiles of microRNA was induced by $n\mathrm{TiO_2}$. Taking into account the role played by microRNAs in plant development and tolerance to abiotic stresses suggests a relevant imbalance in plant physiology.

After Ghosh et al. (2010), the genotoxic effects of $n\mathrm{TiO_2}$ on *A. cepa* (onion) were studied by Pakrashi et al. (2014). In particular, the major focus of the study was to establish the genotoxic potential of $n\mathrm{TiO_2}$ and elucidate the possible mechanism of genotoxicity. A series of detailed observations by optical microscopy, fluorescence microscopy, and laser confocal microscopy was performed and results revealed dose-dependent chromosomal aberrations. It was confirmed that $n\mathrm{TiO_2}$ induces genotoxicity in the plant systems even at a lower concentration. Early genotoxic and phytotoxic effects of $n\mathrm{TiO_2}$ in seedlings of *Hordeum vulgare* L. (barley) were studied by Mattiello et al. (2015). In random amplified polymorphic DNA banding patterns as well as at the chromosomal level, genotoxicity was evidenced, whereas at the cellular level $n\mathrm{TiO_2}$ did not influence ROS generation and ATP content. Transmission electron microscopy microanalysis demonstrated the presence of aggregates of $n\mathrm{TiO_2}$ within root cells.

Over the past 2 years, research has shown that toxicity is dependent not only on plant species but also on exposure time, concentrations of $n\mathrm{TiO_2}$, and different physical forms of $n\mathrm{TiO_2}$. A study by Ruffini Castiglione et al. (2016) used *V. faba* L. (broad bean) as model plants to investigate the potential phytotoxicity of a commercial $n\mathrm{TiO_2}$ <100 nm consisting laboratory-made tetragonal particles, and spherical $n\mathrm{TiO_2}$ <10 nm and bulk $\mathrm{TiO_2}$. The treatment (50 mg/L, a low–middle exposure concentration of nano and bulk materials) was applied to seeds up to early seedling development. Another aim of this study was to estimate phytotoxic and genotoxic effects at different levels (i.e., ultrastructural, cytohistological, and physiological). The results demonstrated that $n\mathrm{TiO_2}$ may exert specific actions at different levels of toxicity, depending on their size and shape, and that the bulk counterpart provokes major adverse effects in broad bean roots.

Andersen et al. (2016) suggested previous work inconsistent because it was carried out with a high dose of $n\mathrm{TiO_2}$ that gave variable and inconclusive results. *A. cepa* L. (garlic), *Avena sativa* L. (oat), *Brassica oleracea capitata* L. (cabbage), *C. sativus* L. (zucchini), *Daucus carota* L. (carrot), *Glycine max* L. (soybean), *L. sativa* L. (lettuce), *Lolium perenne* L. (ryegrass), *Lycopersicum esculentum* L. (tomato), and *Z. mays* L. (corn) were exposed to 0–1000 μg/mL of $n\mathrm{TiO_2}$ and the study suggested toxic impact on root growth. However, overall the results confirmed that early root growth is a more effective indicator of $n\mathrm{TiO_2}$ phytotoxicity than seed germination (Andersen et al., 2016).

Works by Kořenková et al. (2017), Landa et al. (2016), and Cai et al. (2017) were also examined. Kořenková et al. (2017) carried out a simple and short germination experiment on *H. vulgare* (barley). Seeds of barley were germinated in two different nTiO$_2$-amended culture media (agar + distilled water and agar + nutrient solution). In both conditions, root growth inhibition effects became visible with increasing concentration of nTiO$_2$. In addition, chlorophyll *a*/*b* as well as seedling biomass did not change in treated plants. Landa et al. (2016) carried out a detailed comparative study on the effects of nAl_2O_3, nCuO, nFe_3O_4, nMnO, nZnO, and nTiO$_2$ on the germination of *Sinapis alba* L. (white mustard). In addition to the nanoparticles, the experimental framework also considered the metal oxides as nanowires and in their bulk form. The germination was not affected by different forms of nTiO$_2$ at tested concentrations. The interaction of nanoparticles and existing contaminants is quite obvious in existing environments, so it is both appropriate and important to study the role of nanoparticles on environmental behavior and fate of coexisting contaminants. The study by Cai et al. (2017) probably opens a new research field. In this study, cultivated plants were exposed to both nanoparticles and additional contaminants, such as inorganic elements and organic chemicals. In hydroponic conditions, Cai et al. (2017) investigated the impact of four different nTiO$_2$ on Pb accumulation in seedlings of *O. sativa* L. (rice). nTiO$_2$ had no adverse effects on seedling growth and development. Because of its small crystallite size (252.2 Å), only one type of nTiO$_2$ (NAnT) entered the rice seedling roots through the apoplastic route. nTiO$_2$ reduced the Pb concentration in rice roots by 80% and by 77%–97% in shoots likely because of the nanoparticles' high sorption potential for Pb in the nutrient solution.

12.2.2 Adult Plants

Apart from germination studies, some studies investigated the effects of nTiO$_2$ on adult plants grown in greenhouses. Although very few studies have been performed regarding adult plants, experiments included both biometric and metabolic variables (Table 12.2). These studies are important because they offer information on the impact on germination and the early stages of plant growth.

It is known that nitrate reductase (NR), glutamate dehydrogenase (GDH), and glutamine synthetase (GS) are the principal enzymes involved in plant nitrogen metabolism. Moreover, in higher plants they promote the synthesis of chlorophyll, amino acids, and proteins. The effects of nTiO$_2$ on the nitrogen metabolism of growing spinach were studied by Yang et al. (2006). Spinach leaves were sprayed with 0.25% nTiO$_2$ solution. After 20–45 days the activities of NR, GDH, GS, and the contents of NO_3–N and $NH_4{}^+$–N were recorded. The results showed that nano-anatase nTiO$_2$ treatment increased the activities of NR, GDH, and GS, accelerated the inorganic nitrogen content to be translated into organic forms, such as protein and chlorophyll, and enhanced the biomass yield of spinach.

A laboratory investigation was conducted by Asli and Neumann (2009) to study whether nTiO$_2$ suspensions supplied to the primary root of *Z. mays* L. (corn) are able to influence root water uptake and leaf transpiration. The results showed that root elongation in corn was not significantly affected after a few days of nTiO$_2$ treatment. Apparently, the plant roots were unaffected, whereas leaf growth and transpiration were negatively influenced.

After germination studies (reported in Table 12.1B), Song et al. (2013a) carried out a pot study with the same plant seedlings exposed to different nTiO$_2$ solutions (1000–5000 mg/L).

TABLE 12.2 Summary of the Effects of $nTiO_2$ Exposure in Adult Plants

Species	Particle Size (nm)	Substrate	Treatment	Plant Response	References
Spinacia oleracea	4–6	Perlite	0.25%	Improved plant growth; increased glutamate dehydrogenase, glutamine synthetase, and glutamic pyruvic transaminase activity	Yang et al. (2006)
Cicer arietinum	7–40	Soil–sand–farmyard manure mixture	0, 2, 5, 10 ppm	Reduction in electrolyte leakage and malondialdehyde content at 5 mg/kg treatment	Mohammadi et al. (2013)
Vigna radiata	10–15	Loamy sand soil	10 ppm (foliar application)	Improvement in root-and-shoot length, root nodules. Chlorophyll content and total soluble leaf protein were increased. Rhizosphere microbial population and soil enzymes were increased	Raliyaa et al. (2015)
Solanum lycopersicum	16	Nutrition pots	0, 0.05, 0.1, 0.2 mg/L	Increase in net photosynthesis, water transpiration, and conductance	Qi et al. (2013)
Brassica campestris napus, Lactuca sativa, Phaseolus vulgaris	27 anatase/rutile = 80:20	Commercial soil	1000, 2500, 5000 mg/L	Species showed high tolerance to $nTiO_2$ indicating less possibility of phytotoxicity	Song et al. (2013a)
Zea mays	30 nm	Hydroponics	0, 300, 1000 mg/L	$nTiO_2$ colloidal suspensions reduced root hydraulic conductivity, leaf expansion, and transpiration	Asli and Neumann (2009)
Trifolium pratense (*Rhizobium trifolii*)	30 nm (80% anatase–20% rutile); 93 nm (100% anatase)	Hydroponics	18, 23 mg/L	No nodule formation in about 20% of treated plants. Symbiosis and red clover growth were impaired	Moll et al. (2016)
S. oleracea	–	Petri dishes	0, 0.25, 0.5, 1.0, 1.5, 2.0, 2.5, 4.0, 6.0‰	Improved formation of chlorophyll and plant growth. Enhancement in Rubisco activity and increase in photosynthetic rate	Zheng et al. (2005)

The total antioxidant capacity and superoxide dismutase (SOD) activity were not influenced by the treatments. According to Wallace et al. (1977), the results demonstrated that although Ti was assimilated it did not have any adverse effects on plant metabolism.

The majority of experiments devoted to elucidate the interactions of nanomaterials with plants were carried out under optimum conditions. On the other hand, to support wide applications of $n\text{TiO}_2$ in agroecosystems their possible role should also be verified in stress conditions. Mohammadi et al. (2013) first reported the effects of $n\text{TiO}_2$ on two genotypes of *Cicer arietinum* L. (chickpea) under cold stress. Cold-sensitive and cold-tolerant lines were grown until they reached five branches and then sprayed twice with different $n\text{TiO}_2$ solutions on fully expanded plant leaves. Plants were subsequently transferred to a chilling growth chamber with a temperature of 4°C. After 2 days, some leaves were sampled and then analyzed for antioxidant enzyme activity. Results showed that $n\text{TiO}_2$ treatments are able to mitigate the cold stress effects.

Heat stress is considered as one of the major abiotic adversities that can affect plant growth and crop yield. According to Kotak et al. (2007), photosystem II, carbon fixation by Rubisco carboxylase, and ATP-generating systems are the primary targets of thermal damage in plants. As for nanomaterials, Lei et al. (2008) showed $n\text{TiO}_2$ can activate protective enzymes, whereas Qi et al. (2013) studied the specific role of $n\text{TiO}_2$ in heat stress in *L. esculentum* (tomato). In this study, tomato plants at the six-leaves stage, previously exposed to moderately high-temperature treatment of 35°C in a phytotron, were treated with 0.05, 0.1, and 0.2 g/L of $n\text{TiO}_2$. The carbon exchange rate, water transpiration, and stomatal conductance were measured. Results showed that all the gas exchange parameters of tomato leaves responded positively after treatments, demonstrating that $n\text{TiO}_2$ plays a positive role under heat stress in tomato leaves.

Raliyaa et al. (2015) synthesized $n\text{TiO}_2$ using the extracellular enzymes of *Aspergillus flavus* TFR 7 and bulk TiO_2. Subsequently, to determine whether $n\text{TiO}_2$ is suitable for plant nutrition, leaves of *Vigna radiata* (mung bean) were sprayed with $n\text{TiO}_2$. The results demonstrated a significant increasing growth under $n\text{TiO}_2$ treatments with subsequent increase in nodulation and root-and-shoot length. In addition, leaves of treated plants showed enhancement in the content of chlorophyll and total soluble proteins.

Soil microorganisms play important roles in nitrogen cycling within ecosystems representing a central ecosystem service (Herridge et al., 2008). The effects of $n\text{TiO}_2$ on N-fixation have already been studied on *Pisum sativum* (pea) by Fan et al. (2014) and *Medicago truncatula* (barrel clover) by Chen et al. (2015). However, it is important to investigate the possible adverse effects on other symbiotic systems. The first report on the impact of $n\text{TiO}_2$ on *Trifolium pratense* (red clover) and *Rhizobium trifolii* was provided by Moll et al. (2016). In this study they observed the effects of two different structures of $n\text{TiO}_2$ (one with 100% anatase and other with 20% rutile–80% anatase) on the growth rate of *R. trifolii* and red clover, and nodule formation by *R. trifolii* on clover roots. The results suggested that the different structure of $n\text{TiO}_2$ determined different effects on plant growth and nitrogen fixation.

12.2.3 Life Cycle Studies

Several studies have demonstrated the root uptake and translocation of nanomaterials to aerial plant tissues. Much of the literature has focused on short-term, high-dose exposure scenarios often conducted in model media such as hydroponics or synthetic soil (Servin and White, 2016). However, none of the previous studies covered the crop life cycle. Therefore information regarding the possible accumulation of nanomaterials in plant tissues other than leaves is quite scarce. To the best of our knowledge, only a few studies investigated the effects of $n\mathrm{TiO_2}$ on the entire life cycle of plants. They are reported in Table 12.3.

Experimental observations on the effects of $n\mathrm{TiO_2}$ on *T. aestivum* L. (wheat) under field conditions were reported by Du et al. (2011). The results demonstrated that $n\mathrm{TiO_2}$ negatively influenced crop growth. In addition, some important observations were made concerning rhizosphere conditions in soil enriched with $n\mathrm{TiO_2}$. In fact, significant changes in soil enzyme activities were inhibited in the presence of the nanoparticles. This raises the suspicion that this material could be toxic for the soil ecosystem.

Since *L. esculentum* L. (tomato) is one of the most widespread vegetable plants, Song et al. (2013b) carried out a comprehensive study aimed at evaluating the toxicity of $n\mathrm{TiO_2}$ in mature

TABLE 12.3 Summary of the Effects of $n\mathrm{TiO_2}$ Exposure Observed in Crop Life Cycle Experiments

Species	Particle Size (nm)	Substrate	Treatment	Plant Response	References
Solanum lycopersicum	27	Standard potting soil	0, 50, 100, 1000, 2500, 5000 mg/L	In adult plants, the highest concentration negatively affected root elongation, chlorophyll contents, and superoxide dismutase activity	Song et al. (2013b)
Hordeum vulgare	<25	Clay soil	0, 500, 1000 mg/kg	Phenological stage delayed; positive dose–response effect on vegetative growth. $n\mathrm{TiO_2}$ was found in the plant leaves	Marchiol et al. (2016)
H. vulgare	<25	Clay soil	0, 500, 1000 mg/kg	Potential beneficial role of $n\mathrm{TiO_2}$ on nutritional parameters of kernels	Pošćić et al. (2016)
Cucumis sativus	27	Sandy loam soil	0, 250, 500, 750 mg/kg	Enhanced metabolic activity in plant leaves. Increased K and P allocation in fruit	Servin et al. (2013)
Triticum aestivum	29–92	Sandy loam soil mixed with sand (50% v/v)	0, 1, 100, 1000 mg/kg	No negative effects on wheat growth and arbuscular mycorrhizal root colonization	Moll et al. (2017)
Ocimum basilicum	50	Commercial soil	0, 125, 250, 500, 750 mg/kg	Coated particles impacted basil nutritional quality (essential elements, starch, and reducing sugars)	Tan et al. (2017)
T. aestivum	<100	Loamy clay soil	~91 mg/kg	Reduced plant biomass. $n\mathrm{TiO_2}$ found mostly stuck on root surface	Du et al. (2011)
Glycine max	–	Agricultural soil	0%, 0.01%, 0.03%, 0.05%	Increased plant height and biomass	Rezaei et al. (2015)

tomato plants grown in greenhouses. Physiological variables and antioxidant activities were observed in the tomato plants. Even though no negative effects of treatments were recorded in the early growth stages of plants (germination and root elongation), nTiO$_2$ had much higher uptake in leaves (6.1%), stem (11.4%), and fruit (2.7%) when exposed to 1000 mg/kg, and leaves (8.4%), stem (11.0%), and fruit (5.9%) when exposed to 5000 mg/kg. On the other hand, a sign of plant stress was evidenced. In fact, SOD activity significantly differed among treatments, suggesting that nTiO$_2$ induces physiological stress on tomato. However, the stress signals were recorded only in plants treated with the highest level of nTiO$_2$ among the tested dose (5000 mg/kg), since this concentration is unlikely to exist under natural conditions. Therefore it can be concluded that in most cases, nTiO$_2$ might not be toxic to plants. A similar approach was used by Servin et al. (2013) where they studied the possible transfer of nTiO$_2$ in *C. sativus* L. (cucumber), which is another important freshly consumed garden vegetable. In nTiO$_2$-treated plants an influence on plant metabolism was observed. In fact, the concentration of catalase enzymes was significantly increased in the leaves of treated plants, whereas ascorbate peroxidase was decreased. Fruits of treated plants showed significant differences in amide, lignin, and carbohydrates compared to control plants. In addition, the concentrations of K and P increased respectively by 35% and 34%. The most important result was obtained by analyzing fruits using synchrotron μ-XRF and μ-XANES. Using this powerful approach, it was demonstrated that nTiO$_2$ translocates to the fruits without biotransformation. In this way nTiO$_2$ could enter the food chain.

To analyze the role of nTiO$_2$ application in stimulating plant growth, Rezaei et al. (2015) studied the effect of foliar application of nTiO$_2$ in *G. max* L. (soybean). Several parameters (plant height, 1000 grain weight, number of pods per plant, pod dry weight, oil percentage, and seed and oil yield of soybean) were investigated. Results showed that parameters such as plant height, grain weight, number of pods per plant, and pod dry weight were positively influenced by nTiO$_2$.

Marchiol et al. (2016) investigated the influence of nTiO$_2$ on the life cycle of *H. vulgare* L. (barley). Seeds of barley were grown in a soil amended with nTiO$_2$ according to the methods proposed by Priester et al. (2012). The duration of the growth stages was delayed by nTiO$_2$ compared to control plants. At physiological maturity, several biometric variables were measured. The plant height as well as the number of tillers and the total leaf area showed a positive dose response. Apart from the positive effects on plant growth, it was reported that nTiO$_2$ clusters in leaf tissues localized inside the chloroplasts were able to cross the biological barriers in plant tissues. This has raised some concerns about the nutritional quality of the barley kernels. In a study by Poščić et al. (2016) it was reported that Ti accumulation in treated plants was not significant. On the other hand, nTiO$_2$ had an impact on composition and nutritional quality of barley kernels in contrasting ways: amylose content was reduced but β-glucans were unaffected. Moreover, the grains showed a higher content of macro- (Na and Ca) and micronutrients (Fe, Mn, and Zn) than controls. In addition, a positive effect of nTiO$_2$ treatment was also observed for several amino acids. The increase in macro-/micronutrient and amino acid concentrations in kernels could be an effect related to the longest vegetative phase.

Tan et al. (2017) carried out a study aimed at determining the biochemical and physiological effects of the surface chemistry of nTiO$_2$ on *Ocimum basilicum* L. (basil). In this study, the physiological and biochemical effects of pristine, hydrophobic, and hydrophilic nTiO$_2$ were evaluated in basil plants cultivated until flowering stage in soil. The hydrophobic and hydrophilic $nTiO_2$ were capped with Al_2O_3 (6 wt%) and encapsulated with dimethicone (2 wt%) and glycerol (1 wt%). Hydrophilic $nTiO_2$ showed a reduction in seed germination and hydrophobic $nTiO_2$ significantly reduced root elongation. In addition, pristine and hydrophobic $nTiO_2$ enhanced oxidative stress in roots and leaves, while hydrophilic $nTiO_2$ enhanced oxidative stress in roots only. Results showed that independent of the surface properties, a small amount of $nTiO_2$ reached the vascular system of plants. Perhaps the most interesting evidence was that the three types of $nTiO_2$ affected the homeostasis of essential elements Ca, P, Mg, Cu, Fe, Mn, Se, and Zn at different levels. Finally, hydrophilic $nTiO_2$ at 500 mg/kg reduced total sugar content, while hydrophobic ones reduced starch, and pristine $nTiO_2$ decreased reducing sugars content.

Previous studies hypothesized that $nTiO_2$ might affect soil organisms differently than the natural occurring bulk titanium dioxide in soils and potentially affect the ecosystem at different trophic levels (Gardea-Torresdey et al., 2014). This is of particular concern since rhizosphere microbial communities perform ecological key functions in terms of plant growth, plant protection, and enhanced nutrient uptake (Burke et al., 2014; Van Der Heijden and Hartmann, 2016). On that basis, Moll et al. (2017) investigated the effect of two different $nTiO_2$ on the diversity and community composition of soil microorganisms and *T. aestivum* L. (wheat) biomass yield. No negative effects on wheat growth and biomass yield were recorded, and no evidence for a dos–response relationship between wheat performance and $nTiO_2$ concentration was found.

12.3 FUTURE RESEARCH

The studies considered in this chapter have answered some questions raised about the influence of $nTiO_2$ on cultivated plants. Most of the studies were concerned with the effects of $nTiO_2$ exposure on germination or phytotoxicity on very young plants, whereas life cycle studies are still lacking, probably because of the long duration and complex organization. So, very little is known about the interactions of $nTiO_2$ with plants in different soils. In addition, very few studies have investigated $nTiO_2$ uptake by plants, which is an important factor providing evidence of adsorption or internalization. There are other research issues of great interest that will need to be explored in the future.

- **Low-dose exposure.** Engineered nanomaterial (ENM) fluxes in environmental compartments were modeled by Holden et al. (2014), who reported that the highest predicted ENM concentration could be found in the range ≤0.001–1 ppm for water and ≤0.001–1000 ppm for biosolids, respectively. However, the existing literature reports experiments that tested relatively short-term exposures, high dose, and insensitive endpoints such as germination, biomass, and pigment production. This demonstrates that there is a paucity of information on actual environmental conditions. Therefore studies need to include exposures at relevant environmental concentrations.
- **Nutritional quality of food.** The existence of ENMs in the agroecosystem exposes crops/edible tissues and/or food quality to potential risks. Some papers have already demonstrated that ENMs are able to interact significantly with both organic

and inorganic constituents in soil. Specific interactions between ENMs and macro-/ microelements could cause imbalances in plant nutrition as well as the synthesis and metabolism of specific biomolecules (Servin and White, 2016).

- **Intergenerational studies.** The influence of ENM exposure across multiple generations is largely unknown. So far, a few published articles have examined the effects of previous exposure on plant performance in the succeeding generation, thus the long-term impacts on seed integrity and food safety across multiple generations and exposure regimes remain completely unexplored (Rico et al., 2017).

12.4 CONCLUSIONS

Although the literature is somewhat contradictory because of differences in experimental conditions, overall $n\text{TiO}_2$ showed a low phytotoxicity, which in turn is influenced by (1) treatment level, (2) particle size, and (3) plant growth stage. However, current studies on plant nanotoxicology are in their early stages and largely incomplete. The most urgent need is to conduct long-term studies under environmentally realistic scenarios and with sensitive endpoints (Servin and White, 2016). Several issues have received attention in the last few years but the current assessment finds that there are significant gaps in the knowledge of the impact of nanoparticles. Therefore relevant amounts of work have to be performed for a better understanding of the balance between the potential of nanotechnologies in agriculture and food production under safe environments.

References

AFSSET, 2010. Évaluation des risquesliés aux nanomatériaux pour la population générale et pour l'environnement, p. 207.

Andersen, C.P., King, G., Plocher, M., Storm, M., Pokhrel, L.R., Johnson, M.G., Rygiewicz, P.T., 2016. Germination and early plant development of ten plant species exposed to titanium dioxide and cerium oxide nanoparticles. Environ. Toxicol. Chem. 35, 2223–2229. http://dx.doi.org/10.1002/etc.3374.

Asli, S., Neumann, M., 2009. Colloidal suspensions of clay or titanium dioxide nanoparticles can inhibit leaf growth and transpiration via physical effects on root water transport. Plant Cell Environ. 32, 577. http://dx.doi.org/10.1111/j.1365-3040.2009.01952.x.

Boonyanitipong, B., Kositsup, B., Kumar, P., Baruah, S., Dutta, J., 2011. Toxicity of ZnO and TiO_2 nanoparticles on germinating rice seed *Oryza sativa* L. Int. J. Biosci. Biochem. Bioinform. 1, 282–285. http://dx.doi.org/10.7763/IJBBB.2011.V1.53.

Burke, D.J., Zhu, S., Pablico-Lansigan, M.P., Hewins, C.R., Samia, A.C.S., 2014. Titanium oxide nanoparticle effects on composition of soil microbial communities and plant performance. Biol. Fertil. Soils 50, 1169–1173. http://dx.doi.org/10.1007/s00374-014-0938-3.

Cai, F., Wu, X., Zhang, H., Shen, X., Zhang, M., Chen, W., Gao, Q., White, J.C., Tao, S., Wang, X., 2017. Impact of TiO_2 nanoparticles on lead uptake and bioaccumulation in rice (*Oryza sativa* L.). NanoImpact 5, 101–108. http://dx.doi.org/10.1016/j.impact.2017.01.006.

Carvajal, M., Alcaraz, C.F., 1998. Why titanium is a beneficial element for plants. J. Plant Nutr. 21, 655–664. http://dx.doi.org/10.1080/01904169809365433.

Chen, C., Unrine, J.M., Judy, J.D., Lewis, R.W., Guo, J., McNear Jr., D.H., Tsyusko, O.V., 2015. Toxicogenomic responses of the model legume Medicago truncatula to aged biosolids containing a mixture of nanomaterials (TiO_2, Ag, and ZnO) from a pilot wastewater treatment plant. Environ. Sci. Technol. 49, 8759–8768. http://dx.doi.org/10.1021/acs.est.5b01211.

Cox, A., Venkatachalam, P., Sahi, S., Sharma, N., 2016. Silver and titanium dioxide nanoparticle toxicity in plants: a review of current research. Plant Physiol. Biochem. 107, 147–163. http://dx.doi.org/10.1016/j.plaphy.2016.05.022.

Cox, A., Venkatachalam, P., Sahi, S., Sharma, N., 2017. Reprint of: silver and titanium dioxide nanoparticle toxicity in plants: a review of current research. Plant Physiol. Biochem. 110, 33–49.

Dehkourdi, E.H., Mosavi, M., 2013. Effect of anatase nanoparticles (TiO_2) on parsley seed germination (*Petroselinum crispum*) in vitro. Biol. Trace Elem. Res. 155, 283–286. http://dx.doi.org/10.1007/s12011-013-9788-3.

Du, W., Tan, W., Peralta-Videa, J.R., Gardea-Torresdey, J.L., Ji, R., Yin, Y., Guo, H., 2017. Interaction of metal oxide nanoparticles with higher terrestrial plants: physiological and biochemical aspects. Plant Physiol. Biochem. 110, 210–225.

Du, W.C., Sun, Y.Y., Ji, R., Zhu, J.G., Wu, J.C., Guo, H.Y., 2011. TiO_2 and ZnO nanoparticles negatively affect wheat growth and soil enzyme activities in agricultural soil. J. Environ. Monit. 13, 822–828. http://dx.doi.org/10.1039/c0em00611d.

Fan, R., Huang, Y.C., Grusak, M.A., Huang, C.P., Sherrier, D.J., 2014. Effects of nano-TiO_2 on the agronomically-relevant *Rhizobium*-legume symbiosis. Sci. Total Environ. 466–467, 503–512. http://dx.doi.org/10.1016/j.scitotenv.2013.07.032.

Feizi, H., Kamali, M., Jafari, L., RezvaniMoghaddam, P., 2013. Phytotoxicity and stimulatory impacts of nanosized and bulk titanium dioxide on fennel (*Foeniculum vulgare* Mill). Chemosphere 91, 506. http://dx.doi.org/10.1016/j.chemosphere.2012.12.012.

Feizi, H., Moghaddam, P.R., Shahtahmassebi, N., Fotovat, A., 2012. Impact of bulk and nano-sized titanium dioxide on wheat seed germination and seedling growth. Biol. Trace Elem. Res. 146, 101. http://dx.doi.org/10.1007/s12011-011-9222-7.

Foltête, A.S., Masfaraud, J.F., Bigorgne, E., Nahmani, J., Chaurand, P., Botta, C., Labille, J., Rose, J., Férard, J.F., Cotelle, S., 2011. Environmental impact of sunscreen nanomaterials: ecotoxicity and genotoxicity of altered TiO_2 nanocomposites on *Vicia faba*. Environ. Pollut. 159, 2515–2522. http://dx.doi.org/10.1016/j.envpol.2011.06.020.

Frazier, T.P., Burklew, C.E., Zhang, B., 2014. Titanium dioxide nanoparticles affect the growth and microRNA expression of tobacco (*Nicotiana tabacum*). Funct. Integr. Genom. 14, 75–83. http://dx.doi.org/10.1007/s10142-013-0341-4.

Gardea-Torresdey, J.L., Rico, C.M., White, J.C., 2014. Trophic transfer, transformation, and impact of engineered nanomaterials in terrestrial environments. Environ. Sci. Technol. 48, 2526–2540. http://dx.doi.org/10.1021/es4050665.

Ghosh, M., Bandyopadhyay, M., Mukherjee, A., 2010. Genotoxicity of titanium dioxide (TiO_2) nanoparticles at two trophic levels: plant and human lymphocytes. Chemosphere 81, 1253–1262. http://dx.doi.org/10.1016/j.chemosphere.2010.09.022.

Gottschalk, F., Sonderer, T., Scholz, R.W., Nowack, B., 2009. Modeled environmental concentrations of engineered nanomaterials (TiO_2), ZnO, Ag, CNT, Fullerenes for different regions. Environ. Sci. Technol. 43, 9216–9222. http://dx.doi.org/10.1021/es9015553.

Herridge, D.F., Peoples, M.B., Boddey, R.M., 2008. Global inputs of biological nitrogen fixation in agricultural systems. Plant Soil 311, 1–18. http://dx.doi.org/10.1007/s11104-008-9668-3.

Holden, P.A., Klaessig, F., Turco, R.T., Priester, J.H., Rico, C.R., Avila-Arias, H., Mortimer, M., Pacpaco, K., Gardea-Torresdey, J.L., 2014. Evaluation of exposure concentrations used in assessing manufactured nanomaterial environmental hazards: are they relevant? Environ. Sci. Technol. 48, 10541–10551. http://dx.doi.org/10.1021/es502440s.

Hrubý, M., Cígler, P., Kuzel, S., 2002. Contribution to understanding the mechanism of titanium action in plants. J. Plant Nutr. 25, 577–598. http://dx.doi.org/10.1081/PLN-120003383.

Keller, A.A., McFerran, S., Lazareva, A., Suh, S., 2013. Global life cycle releases of engineered nanomaterials. J. Nanopart. Res. 6, 1–17. http://dx.doi.org/10.1007/s11051-013-1692-4.

Khodakovskaya, M.V., Lahiani, M.H., 2014. Nanoparticles and plants: from toxicity to activation of growth. In: Handbook of Nanotoxicology, Nanomedicine and Stem Cell Use in Toxicology. Wiley, pp. 121–130. http://dx.doi.org/10.1002/9781118856017.ch7.

Kořenková, L., Šebesta, M., Urík, M., Kolenčík, M., Kratošová, G., Bujdoš, M., Vávra, I., Dobročka, E., 2017. Physiological response of culture media-grown barley (*Hordeum vulgare* L.) to titanium oxide nanoparticles. Acta Agric. Scand. 67, 285–291. http://dx.doi.org/10.1080/09064710.2016.1267255.

Kotak, S., Larkindale, J., Lee, U., Von Koskull-Döring, P., Vierling, E., Scharf, K.D., 2007. Complexity of the heat stress response in plants. Curr. Opin. Plant Biol. 10, 310–316. http://dx.doi.org/10.1016/j.pbi.2007.04.011.

Landa, P., Cyrusova, T., Jerabkova, J., Drabek, O., Vanek, T., Podlipna, R., 2016. Effect of metal oxides on plant germination: phytotoxicity of nanoparticles, bulk materials and metal ions. Water Air Soil Pollut. 227, 448. http://dx.doi.org/10.1007/s11270-016-3156-9.

López-Moreno, M.L., de la Rosa, G., Cruz-Jiménez, G., Castellano, L., Peralta-Videa, J.R., Gardea-Torresdey, J.L., 2017. Effect of ZnO nanoparticles on corn seedlings at different temperatures; X-ray absorption spectroscopy and ICP/OES studies. Microchem. J. 134, 54–61.

Larue, C., Laurette, H., Herlin-Boime, N., Khodja, H., Fayard, B., Flank, A.M., Brisset, F., Carriere, M., 2012. Accumulation, translocation and impact of TiO_2 nanoparticles in wheat (*Triticum aestivum* spp.): influence of diameter and crystal phase. Sci. Total Environ. 431, 197. http://dx.doi.org/10.1016/j.scitotenv.2012.04.073.

Larue, C., Veronesi, G., Flank, A.M., Surble, S., Herlin-Boime, N., Carrière, M., 2013. Comparative uptake and impact of TiO_2 nanoparticles. J. Toxicol. Environ. Health A 75, 722–734. http://dx.doi.org/10.1080/15287394.2012.689800.

Lei, Z., Mingyu, S., Xiao, W., Chao, L., Chunxiang, Q., Liang, C., Hao, H., Xiaoqing, L., Fashui, H., 2008. Antioxidant stress is promoted by nanoanatase in spinach chloroplasts under UV-B radiation. Biol. Trace Elem. Res. 121, 69–79. http://dx.doi.org/10.1007/s12011-007-8028-0.

Lu, P.-J., Huang, S.-C., Chen, Y.-P., Chiueh, L.-C., Shih, D.Y.-C., 2015. Analysis of titanium dioxide and zinc oxide nanoparticles in cosmetics. J. Food Drug Anal. 23, 587–594. http://dx.doi.org/10.1016/j.jfda.2015.02.009.

Marchiol, L., Mattiello, A., Pošćić, F., Fellet, G., Zavalloni, C., Carlino, E., Musetti, R., 2016. Changes in physiological and agronomical parameters of barley (*Hordeum vulgare*) exposed to cerium and titanium dioxide nanoparticles. Int. J. Environ. Res. Public Health 13, 332. http://dx.doi.org/10.3390/ijerph13030332.

Medina-Velo, I.A., Barrios, A.C., Zuverza-Mena, N., Hernandez-Viezcas, J.A., Chang, C.H., Ji, Z., Zink, J.I., Peralta-Videa, J.R., Gardea-Torresdey, J.L., 2017. Comparison of the effects of commercial coated and uncoated ZnO nanomaterials and Zn compounds in kidney bean (*Phaseolus vulgaris*) plants. J. Hazard. Mater. 332, 214–222.

Mattiello, A., Pošćić, F., Musetti, R., Giordano, C., Vischi, M., Filippi, A., Bertolini, A., Marchiol, L., 2015. Evidences of genotoxicity and phytotoxicity in *Hordeum vulgare* exposed to CeO_2 and TiO_2 nanoparticles. Front. Plant Sci. 6, 1043. http://dx.doi.org/10.3389/fpls.2015.01043.

Mohammadi, R., Maali-Amiri, R., Abbas, A., 2013. Effect of TiO_2 nanoparticles on chickpea response to cold stress. Biol. Trace Elem. Res. 152, 403–410. http://dx.doi.org/10.1007/s12011-013-9631-x.

Moll, J., Klingenfuss, F., Widmer, F., Gogos, A., Bucheli, T.D., Hartmann, M., Van Der Heijden, M.G.A., 2017. Effects of titanium dioxide nanoparticles on soil microbial communities and wheat biomass. Soil Biol. Biochem. 111, 85–93. http://dx.doi.org/10.1016/j.soilbio.2017.03.019.

Moll, J., Okupnik, A., Gogos, A., Knauer, K., Bucheli, T.D., Van Der Heijden, M.G.A., Widmer, F., 2016. Effects of titanium dioxide nanoparticles on red clover and its rhizobial symbiont. PLoS One 11, e0155111. http://dx.doi.org/10.1371/journal.pone.0155111.

Mukherjee, A., Majumdar, S., Servin, A.D., Pagano, L., Dhankher, O.P., White, J.C., 2016. Carbon nanomaterials in agriculture: a critical review. Front. Plant Sci. 7, 1073. http://dx.doi.org/10.3389/fpls.2015.00172.

Mushtaq, Y.K., 2011. Effect of nanoscale Fe_3O_4, TiO_2 and carbon particles on cucumber seed germination. J. Environ. Sci. Health A 46, 1732–1735. http://dx.doi.org/10.1080/10934529.2011.633403.

Ochoa, L., Medina-Velo, I.A., Barrios, A.C., Bonilla-Bird, N.J., Hernandez-Viezcas, J.A., Peralta-Videa, J.R., Gardea-Torresdey, J.L., 2017. Modulation of CuO nanoparticles toxicity to green pea (*Pisum sativum Fabaceae*) by the phytohormone indole-3-acetic acid. Sci. Total Environ. 598, 513–524.

OECD, 2003. OECD Guidelines for the Testing of Chemicals: Proposals for Updating Guideline 208-Terrestrial Plant Test: Seedling Emergence and Seedling Growth Test. http://dx.doi.org/10.1016/j.buildenv.2013.04.024.

Pais, I., 1983. The biological importance of titanium. J. Plant Nutr. 6, 3–131. http://dx.doi.org/10.1080/01904168309363075.

Pakrashi, S., Jain, N., Dalai, S., Jayakumar, J., Chandrasekaran, P.T., Raichur, A.M., Chandrasekaran, N., Mukherjee, A., 2014. In vivo genotoxicity assessment of titanium dioxide nanoparticles by *Allium cepa* root tip assay at high exposure concentrations. PLoS One 9, e87789. http://dx.doi.org/10.1371/journal.pone.0087789.

Pošćić, F., Mattiello, A., Fellet, G., Miceli, F., Marchiol, L., 2016. Effects of cerium and titanium oxide nanoparticles in soil on the nutrient composition of barley (*Hordeum vulgare* L.) kernels. Int. J. Environ. Res. Public Health 13, 577. http://dx.doi.org/10.3390/ijerph13060577.

Priester, J.H., Ge, Y., Mielke, R.E., Horst, A.M., Moritz, S.C., Espinosa, K., Gelb, J., Walker, S.L., Nisbet, R.M., An, Y.J., Schimel, J.P., Palmer, R.G., Hernandez-Viezcas, J.A., Zhao, L., Gardea-Torresdey, J.L., Holden, P.A., 2012. Soybean susceptibility to manufactured nanomaterials with evidence for food quality and soil fertility interruption. Proc. Natl. Acad. Sci. U.S.A. 109, 2451–2456. http://dx.doi.org/10.1073/pnas.1205431109.

Qi, M., Liu, Y., Li, T., 2013. Nano-TiO_2 improve the photosynthesis of tomato leaves under mild heat stress. Biol. Trace Elem. Res. 156, 323. http://dx.doi.org/10.1007/s12011-013-9833-2.

Raliyaa, R., Biswasa, P., Tarafdarm, J.C., 2015. TiO_2 nanoparticle biosynthesis and its physiological effect on mung bean (*Vigna radiata* L.). Biotechnol. Rep. 5, 22–26. http://dx.doi.org/10.1016/j.btre.2014.10.009.

Ramesh, M., Palanisamy, K., Babu, K., Sharma, N.K., 2014. Effect of bulk and nano-titanium dioxide and zinc oxide on physio-morphological changes in *Triticum aestivum* Linn. J. Global Biosci. 3, 415–422.

Reddy, P.V.L., Hernández-Viezcas, J.A., Peralta-Videa, J.R., Gardea-Torresdey, J.L., 2016. Lessons learned: are engineered nanomaterials toxic to terrestrial plants? Sci. Total Environ. 568, 470–479. http://dx.doi.org/10.1016/j.scitotenv.2016.06.042.

Rezaei, F., Moaveni, P., Mozafari, H., 2015. Effect of different concentrations and time of nano TiO_2 spraying on quantitative and qualitative yield of soybean (*Glycine max* L.) at Shahr-e-Qods, Iran. Biol. Forum 7, 957–964.

Rico, C.M., Johnson, M.G., Marcus, M.A., Andersen, C.P., 2017. Intergenerational responses of wheat (*Triticum aestivum* L.) to cerium oxide nanoparticles exposure. Environ. Sci. 4, 700. http://dx.doi.org/10.1039/c7en00057j.

Ruffini Castiglione, M., Giorgetti, L., Bellani, L., Muccifora, S., Bottega, S., Spanò, C., 2016. Root responses to different types of TiO_2 nanoparticles and bulk counterpart in plant model system *Vicia faba* L. Environ. Exp. Bot. 130, 11–21. http://dx.doi.org/10.1016/j.envexpbot.2016.05.002.

Ruffini Castiglione, M., Giorgetti, L., Geri, C., Cremonini, R., 2011. The effects of nano-TiO_2 on seed germination, development and mitosis of root tip cells of *Vicia narbonensis* L. and *Zea mays* L. J. Nanopart. Res. 1, 2443–2449. http://dx.doi.org/10.1007/s11051-010-0135-8.

Semenzin, E., Lanzellotto, E., Hristozov, D., Critto, A., Zabeo, A., Giubilato, E., Marcomini, A., 2015. Species sensitivity weighted distribution for ecological risk assessment of engineered nanomaterials: the n-TiO_2 case study. Environ. Toxicol. Chem. 34, 2644–2659. http://dx.doi.org/10.1002/etc.3103.

Servin, A.D., Castillo-Michel, H., Hernández-Viezcas, J.A., Corral Diaz, B., Peralta-Videa, J.R., Gardea-Torresdey, J.L., 2012. Synchrotron micro-XRF and micro-XANES confirmation of the uptake and translocation of TiO_2 nanoparticles in cucumber (*Cucumis sativus*) plants. Environ. Sci. Technol. 46, 7637–7643. http://dx.doi.org/10.1021/es300955b.

Servin, A.D., Morales, M.I., Castillo-Michel, H., Hernández-Viezcas, J.A., Munoz, B., Zhao, L., Nunez, J.E., Peralta-Videa, J.R., Gardea-Torresdey, J.L., 2013. Synchrotron verification of TiO_2 accumulation in cucumber fruit: a possible pathway of TiO_2 nanoparticle transfer from soil into the food chain. Environ. Sci. Technol. 47, 11592–11598. http://dx.doi.org/10.1021/es403368j.

Servin, A.D., White, J.C., 2016. Nanotechnology in agriculture: next steps for understanding engineered nanoparticle exposure and risk. NanoImpact 1, 9–12. http://dx.doi.org/10.1016/j.impact.2015.12.002.

Shweta, Tripathi, D.K., Singh, S., Singh, S., Dubey, N.K., Chauhan, D.K., 2016. Impact of nanoparticles on photosynthesis: challenges and opportunities. Mater. Focus 5 (5), 405–411.

Simonin, M., Richaume, A., Guyonnet, J.P., Dubost, A., Martins, J.M.F., Pommier, T., 2016. Titanium dioxide nanoparticles strongly impact soil microbial function by affecting archaeal nitrifiers. Sci. Rep. 6:33643. http://dx.doi.org/10.1038/srep33643.

Singh Sekhon, B., 2014. Nanotechnology in agri-food production: an overview. Nanotechnol. Sci. Appl. 7, 31–53. http://dx.doi.org/10.2147/NSA.S39406.

Singh, S., Tripathi, D.K., Dubey, N.K., Chauhan, D.K., 2016. Effects of nano-materials on seed germination and seedling growth: striking the slight balance between the concepts and controversies. Mater. Focus 5 (3), 195–201.

Singh, S., Vishwakarma, K., Singh, S., Sharma, S., Dubey, N.K., Singh, V.K., Liu, S., Tripathi, D.K., Chauhan, D.K., 2017. Understanding the plant and nanoparticle interface at transcriptomic and proteomic level: a concentric overview. Plant Gene. http://dx.doi.org/10.1016/j.plgene.2017.03.006.

Song, U., Jun, H., Waldman, B., Roh, J., Kim, Y., Yi, J., Lee, E.J., 2013b. Functional analyses of nanoparticle toxicity: a comparative study of the effects of TiO_2 and Ag on tomatoes (*Lycopersicon esculentum*). Ecotoxicol. Environ. Saf. 93, 60. http://dx.doi.org/10.1016/j.ecoenv.2013.03.033.

Song, U., Shin, M., Lee, G., Roh, J., Kim, Y., Lee, E.J., 2013a. Functional analysis of TiO_2 nanoparticle toxicity in three plant species. Biol. Trace Elem. Res. 155, 93–103. http://dx.doi.org/10.1007/s12011-013-9765-x.

Sun, T.Y., Gottschalk, F., Hungerbühler, K., Nowack, B., 2014. Comprehensive probabilistic modelling of environmental emissions of engineered nanomaterials. Environ. Pollut. 185, 69–76. http://dx.doi.org/10.1016/j.envpol.2013.10.004.

Tan, W., Du, W., Barrios, A.C., Armendariz Jr., R., Zuverza-Mena, N., Ji, Z., Chang, C.H., Zink, J.I., Hernandez-Viezcas, J.A., Peralta-Videa, J.R., Gardea-Torresdey, J.L., 2017. Surface coating changes the physiological and biochemical impacts of nano-TiO_2 in basil (*Ocimum basilicum*) plants. Environ. Pollut. 222, 64–72. http://dx.doi.org/10.1016/j.envpol.2017.01.002.

Tripathi, D.K., Singh, S., Singh, S., Pandey, R., Singh, V.P., Sharma, N.C., Prasad, S.M., Dubey, N.K., Chauhan, D.K., 2017. An overview on manufactured nanoparticles in plants: uptake, translocation, accumulation and phytotoxicity. Plant Physiol. Biochem. 110, 2–12.

Tripathi, D.K., Singh, S., Singh, V.P., Prasad, S.M., Chauhan, D.K., Dubey, N.K., 2016. Silicon nanoparticles more efficiently alleviate arsenate toxicity than silicon in maize cultiver and hybrid differing in arsenate tolerance. Front. Environ. Sci. 4, 46.

United States Environmental Protection Agency (USEPA), 1996. Ecological Effects Test Guidelines (OPPTS 850.4200): Seed Germination/Root Elongation Toxicity Test. http://dx.doi.org/10.1016/j.buildenv.2013.04.024.

Van Der Heijden, M., Hartmann, M., 2016. Networking in the plant microbiome. PLoS Biol. 14, e1002378. http://dx.doi.org/10.1371/journal.pbio.1002378.

Venkatachalam, P., Priyanka, N., Manikandan, K., Ganeshbabu, I., Indiraarulselvi, P., Geetha, N., Muralikrishna, K., Bhattacharya, R.C., Tiwari, M., Sharma, N., Sahi, S.V., 2017. Enhanced plant growth promoting role of phycomolecules coated zinc oxide nanoparticles with P supplementation in cotton (*Gossypium hirsutum* L.). Plant Physiol. Biochem. 110, 118–127.

Wallace, A., Alexander, G.V., Chaudhry, F.M., 1977. Phytotoxicity of cobalt, vanadium, titanium, silver, and chromium. Commun. Soil Sci. Plant Anal. 8, 751–756. http://dx.doi.org/10.1080/00103627709366769.

Wang, H., Wick, R.L., Xing, B., 2009. Toxicity of nanoparticulate and bulk ZnO, Al_2O_3 and TiO_2 to the nematode *Caenorhabditiselegans*. Environ. Pollut. 157, 1171–1177. http://dx.doi.org/10.1016/j.envpol.2008.11.004.

Xiong, D., Fang, T., Yu, L., Sima, X., Zhu, W., 2011. Effects of nano-scale TiO_2, ZnO and their bulk counterparts on zebrafish: acute toxicity, oxidative stress and oxidative damage. Sci. Total Environ. 409, 1444–1452. http://dx.doi.org/10.1016/j.scitotenv.2011.01.015.

Yang, F., Hong, F., You, W., Liu, C., Gao, F., Wu, C., Yang, P., 2006. Influences of nano-anatase TiO_2 on the nitrogen metabolism of growing spinach. Biol. Trace Elem. Res. 110, 179–190. http://dx.doi.org/10.1385/BTER:110:2:179.

Zheng, L., Hong, F., Lu, S., Liu, C., 2005. Effect of nano-TiO_2 on strength of naturally aged seeds and growth of spinach. Biol. Trace Elem. Res. 104, 83. http://dx.doi.org/10.1385/BTER:104:1:083.

Zuverza-Mena, N., Martínez-Fernandez, D., Du, W., Hernandez-Viezcas, J.A., Bonilla-Bird, N., Lopez-Moreno, M.L., Komarek, M., Peralta-Videa, J.R., Gardea-Torresdey, J.L., 2016. Exposure of engineered nanomaterials to plants: insights into the physiological and biochemical responses - a review. Plant Physiol. Biochem. 110, 236–264. http://dx.doi.org/10.1016/j.plaphy.2016.05.037.

CHAPTER

13

Interaction of Copper Oxide Nanoparticles With Plants: Uptake, Accumulation, and Toxicity

Namira Arif[1], *Vashali Yadav*[1], *Swati Singh*[1], *Durgesh K. Tripathi*[1,2,4], *Nawal Kishore Dubey*[2], *Devendra Kumar Chauhan*[4], *Lucia Giorgetti*[3]

[1]**Centre for Medical Diagnostic and Research, Motilal Nehru National Institute of Technology, Allahabad, India;** [2]**Banaras Hindu University, Varanasi, India;** [3]**National Research Council (IBBA-CNR), Pisa, Italy;** [4]**University of Allahabad, Allahabad, India**

13.1 INTRODUCTION

With the advent of the 21st century, industrialization and urbanization are rapidly proceeding worldwide ensuring better life conditions for humans, but, as a result, a number of problems are continuously increasing such as population explosion and environmental pollution. Moreover, the rapidly increasing population with its irrepressible needs has led to many different sources of pollution. Anthropic origin contamination can cause toxicity for natural flora and fauna affecting both ecosystems at various levels, and human health directly or through the food chain. Evident effects on flora and fauna ecosystems are consistent with decreases in plant productivity, poor soil and water quality, loss of biodiversity, and further bioaccumulation of contaminants in higher trophic levels including humans. Therefore to tackle these challenges in today's technological world, continuous research and environmental monitoring are carried out to consider the different kinds of potential toxicant spread by pollution. In this context, nanomaterials (ranging between 1 and 100nm in at least one dimension) (Roco, 2003) have become a hastily rising field of revolutionary science, but somehow they are suspected of being a hindrance rather than boon for both humans and the environment. Great effort has been made to understand the response of the environment and organisms to toxic concentrations of nanoparticles (NPs) because of their commercialization in various fields such the industrial sector, medical research field, in cosmetics, etc. (Oberdörster et al., 2005; Borm et al., 2006); they also have numerous applications in the UN's Millennium Development Goals (UNESCO, 2006; Gavankar et al., 2012).

Nanomaterials in Plants, Algae, and Microorganisms
http://dx.doi.org/10.1016/B978-0-12-811487-2.00013-X

At the beginning of nanotechnology-based research, major pioneering scientific and economic growth areas were provided by nanomaterials that when massively released into the environment could lead to a variety of environmental and human health hazards (Moore, 2006). Since the evolution of earth, NPs have always been present, mainly generated by natural events, but in the last few decades anthropogenic activities have also greatly increased the diffusion of NPs, for instance, in air as ultrafine particles and in water and soil as colloids with an association of different particles (Lead and Wilkinson, 2006). NPs from various anthropogenic activities can accumulate over time in waters and soils causing an unpredictable type of pollution which is difficult to handle and control. Wide application of nanotechnology has urged the need for proper studies regarding NP-induced toxicity. In the last few years, nanotechnological products have being broadly synthesized for their new and favorable physicochemical properties, and have numerous applications in diverse fields, such as electronics, optics, textiles, medical devices, food packaging, water treatment technology, fuel cells, catalysts, biosensors, and agents for environmental remediation (Handy et al., 2008).

Moreover, because the production and discharge of engineered NPs (ENPs) by industries is increasing at a high rate, ENPs can enter and accumulate in the environment through various pathways causing high contamination in different environmental matrices: in the atmosphere; in agricultural soils, especially when amended with sewage sludge from wastewater treatment plants; in drainage water from crop fields, which flows to rivers, lakes, and streams; and from NPs entering aquatic systems (Brar et al., 2010; Oleszczuk et al., 2011). The toxic effects of various ENPs for living beings and their environment have been reported in a number of studies (Blinova et al., 2010, Unrine et al., 2010, Ge et al., 2011, Pachapur et al., 2016) and, despite the scarcity of information related to ENP risks, they are largely used in the composition of many commercial, industrial, and biomedical products. As a consequence, NPs released from ENPs that are used also in household and industrial commodities can contaminate urban waste, accumulate in wastewater treatment plants, and be massively present in wastewater sludge (Andreotti et al., 2015).

Although plenty of information exists concerning the interaction of plants with metals, there is still a need for detailed information about the interaction of metallic NPs with plants because several studies have revealed the diverse impacts of NPs on plants (Ma et al., 2010; Andreotti et al., 2015; Monica and Cremonini, 2009; Castiglione et al., 2016; Tripathi et al., 2015, 2016, 2017a,b,c,d,e; Singh et al., 2016; Shweta et al., 2016; Singh et al., 2017).

Copper (Cu) comes under transition metals in the periodic table, present in D block. Cu, as a periodic element, has atomic number 29 and density 8.96 g/mL. The electronic structure of Cu is Ar (3d10) (4s1), with an oxidation state of +1 and +2, having an atomic weight 63.54. It is an essential earth element for living beings at low concentrations and toxic at higher concentrations. Cu pollution caused by anthropic activities is a result of Cu polishing, paint, mining, plating, and printing operations (Nagajyoti et al., 2010; Ayangbenro and Babalola, 2017). Cu is a vital micronutrient required by plants in very small quantities, playing an important role in maintaining plants' natural metabolism and growth, but its deficiency and excess are proven inhibitors of various physiological functions (Zhao et al., 2010), Cu excess being also genotoxic (Bellani et al., 2014). Cu concentration and availability in soil depend on pH and inorganic matter content (McCauley et al., 2009). Similar to other metal nutrients such as Mn and Zn, Cu is a small element and has positive charges able to strongly bind to the surface of soil particles. At high pH, the bond with soil increases and Cu concentration in soil solution and its availability for plant uptake decreases. On the contrary, at low pH and high organic matter content, Cu solubility and availability for plants increase (McCauley et al., 2009).

Deficiency of Cu in soil also causes many serious problems in plants such as stunted growth, decrease in productivity, delayed flowering, and sterility. Since Cu solubility is dependent on the chemical physical structure of soil, it can be retained in soil for many years and may be considered as a major toxicant to plant cells because it inhibits many physiological and biochemical processes (Ahsan et al., 2007).

Since Cu when present in excess amounts causes toxicity, similarly copper oxide NPs (CuO NPs) have many toxic effects. CuO NPs exhibit a variety of impending physical properties such as high-temperature conductivity, electron correlation effects, and spin dynamics (Cava, 1990; Tranquada et al., 1995). CuO NPs acquire functional photovoltaic and photoconductive properties because their crystal structure has a narrow band gap (Xu et al., 1999). CuO NPs also hold novel features because they can improve fluid viscosity and enhance thermal conductivity and these properties make CuO NPs potentially useful energy-saving materials, thus improving the effect of energy conversion (Namburu et al., 2007). CuO NPs have their applications in different fields including gas sensors (Chowdhuri et al., 2004), catalysis (Jammi et al., 2009), batteries (Zhang et al., 2005), high-temperature superconductors (Dar et al., 2008), solar energy conversion (Yin et al., 2005), and field emission emitters (Dar et al., 2008). CuO NPs are broadly applied in electronics and for coatings on textiles because of their optical, electrical, and antifungal properties (Cioffi et al., 2005; Kim et al., 2011; Song et al., 2015). CuO NPs are also used in biocides (Shi et al., 2011; Perreault et al., 2014). According to a reported study, CuO NPs of 50nm size are considered as a hard material that does not exhibit the same malleability and ductility as bulk Cu (Science Daily, 2014). Engineered CuO NPs possess catalytic, electrical, and optical properties and are being widely used in electronics, ceramics, films, polymers, inks, metallics, and coatings (Yang et al., 2006; Nasibulin et al., 2000; Lee et al., 2008), and also in lubricant oil because they have a good mending effect (Liu et al., 2004; Lee et al., 2008). Because of its ultrafine size, CuO NPs are widely used in solar cells and lithium-ion batteries (Guo et al., 2002, 2009; Sau et al., 2010), polymers/plastics, inks/ceramic pigments, gas sensors, catalysts, and electronics (Li and Kleinstreuer, 2008; Ebrahimnia-Bajestan et al., 2011). Moreover, CuO NPs are being used in antifouling paints for boats; as a consequence, marine paints can become an important source of NP contamination in aquatic systems since they also contain acrylic and styrenic monomers to make a polymeric film used in the coating of CuO NPs (Almeida et al., 2007; Perreault et al., 2012). In particular, CuO NPs are used extensively in antimicrobial formulations and products because of their antimicrobial properties (Gabbay et al., 2006; Borkow and Gabbay, 2009; Abramova et al., 2013). CuO NP synthesis occurs via different methods such as (1) chemical (Zhang et al., 2006; Suramwar et al., 2012), (2) physical (Kim et al., 2006; Blosi et al., 2011), and (3) biological production (Ramanathan et al., 2011; Lee et al., 2013).

Although detailed information on synthesis, characterization, fundamental properties, and applications of CuO nanomaterials is currently present, there are many stones still unturned regarding metallic NP bioavailability, their uptake and translocation by plants, and their toxicity mechanism (Navarro et al., 2008; Andreotti et al., 2015).

The toxicity of CuO NPs to humans and environmental health has been reported from previous studies (Yoon et al., 2007; Griffitt et al., 2007; Aruoja et al., 2009; Gomes et al., 2011).

In general, NPs are potentially more poisonous than their bulk and ionic counterparts because of their enlarged surface area and reactivity, leading to their amplified bioavailability and toxicity (Bhatt and Tripathi, 2011; Scown et al., 2010; Gomes et al., 2011).

NPs and their exploration have now become a need and a challenge for present day researchers, and many studies are going on in this field. This review emphasizes the various aspects related to the complete interaction of CuO NPs with plants, and their uptake, translocation, accumulation, toxicity, and tolerance mechanisms.

13.2 UPTAKE TRANSLOCATION AND ACCUMULATION

CuO NPs can enter into soil via anthropogenic activities and in general soil can represent the primary source of NPs for terrestrial plant uptake. Similar to Cu in its ionic forms, bioavailable CuO NPs first come in direct contact and are absorbed by the plant roots' surface, thereby CuO NPs dissociate from their complex forms and are converted into simpler forms to be absorbed by the plant (Kabata-Pendias and Pendias, 1992; Adrees et al., 2015; Fig. 13.1). CuO NPs can cause toxicity by entering a cell membrane and releasing Cu ions inside the cell (Karlsson et al., 2008; Anjum et al., 2015). The root system plays a key role for the uptake of Cu, and after absorption, Cu ions are transported to the whole plant through vascular tissues, xylem, and phloem (Adrees et al., 2015; Fig. 13.1). Because of the presence of the suberinized

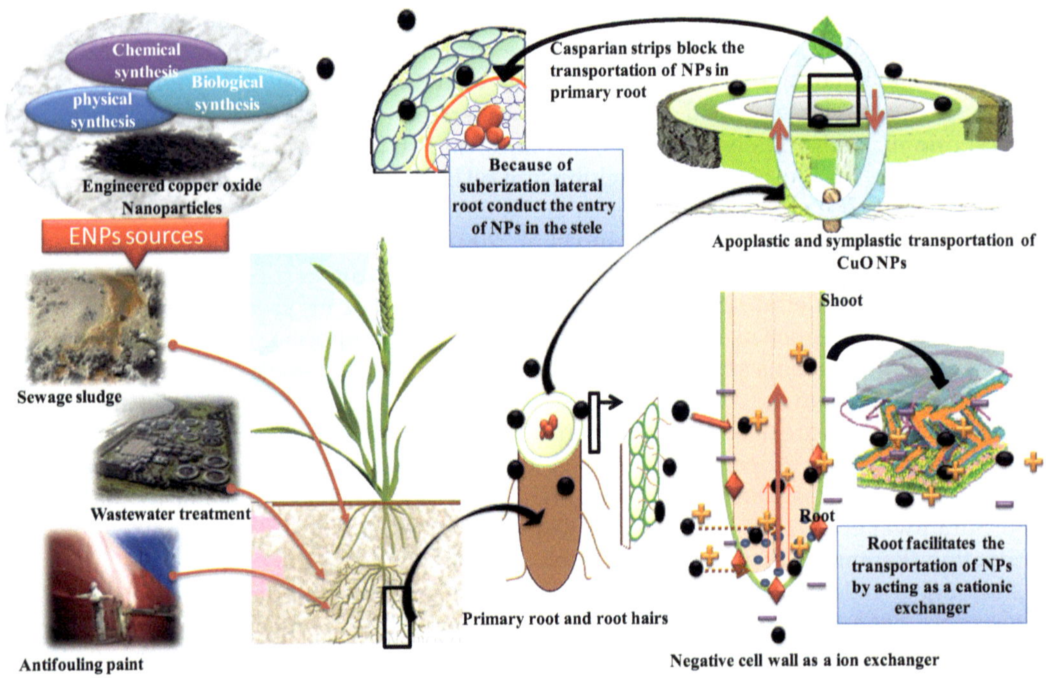

FIGURE 13.1 Sources of copper oxide nanoparticle (CuO NP) environmental contamination by anthropic activities, CuO NP interactions with plant roots, and their transport to the whole plant through vascular tissues, xylem, and phloem.

exodermis and endodermis in the primary roots, uptake of NPs is barred through these primary sites; nevertheless, lateral root development breaks through the cortex favoring apoplastic bypass, facilitating NP entry into xylem through the cortex and central cylinder (Dietz and Herth, 2011; Shaw and Hossain, 2013; Fig. 13.1). The apoplastic region of the root is the principal compartment of metal accumulation in plants (Krzesłowska, 2011). Cu gets adsorbed in cationic form on the root surface interacting with the negative charges of the cell walls which are constituted by a network of cellulose, pectins, and glycoprotein and act as definite ion exchangers (Allan and Jarrell, 1989; Adrees et al., 2015).

CuO NPs are more actively accumulated by plants than Cu ions as demonstrated by Nekrasova et al. (2011). In this research, the comparative effect of Cu ions and CuO NPs was considered in *Elodea densa* plant; at high concentration (5 mg/mL), the biological accumulation coefficients (BACs) for Cu ions was proved to be higher in comparison to CuO NP BACs because of the small radius of Cu and higher penetration capacity, while NPs at higher concentrations hinder their own penetration by being adsorbed on the root cell wall (Nekrasova et al., 2011; Fig. 13.1). When CuO NPs are accumulated in the root and shoot system of plants, they are capable of transferring electrons to molecular oxygen, thus resulting in the formation of reactive superoxide ions and H_2O_2 leading to the formation of the most destructive of reactive oxygen species (ROS)—hydroxyl radicals (Dietz and Herth, 2011; Shaw and Hossain, 2013).

13.3 EFFECT OF CuO NPs ON PLANTS

Plants play a crucial role as ecological receptors and as potential pathways for the transportation and bioaccumulation of NPs at different trophic levels in the ecosystem and food chain (Wang et al., 2012; Monica and Cremonini, 2009; Andreotti et al., 2015). At present, CuO NPs are widely released into the environment through a variety of anthropogenic actions, thus they have become severe hazards to ecological systems, including plants (Shaw and Hossain, 2013). CuO NPs can affect plant development and metabolism and the severity of toxic effects is determined by the size, concentration, and chemistry of the NPs (Dietz and Herth, 2011; Table 13.1). CuO NPs enter from soil through roots and are transported to shoots through xylem cells, and again translocation occurs from shoot to root through phloem (Table 13.1). At increased concentration, CuO NPs inhibit the growth of seedlings and also affect root growth (Wang et al., 2012).

Cu is an indispensable nutrient (essential micronutrient) for the proper metabolism of plants and plays a significant role in photosynthetic reactions, contributes to RNA synthesis by activating several specific enzymes, and improves the performance of photosystems I and II (Leng et al., 2015; Table 13.1). However, numerous reported studies indicate that Cu and CuO NPs are toxic for plants; Cu ions can lead to toxic symptoms once they surpass the physiological tolerance range, for instance; seed germination and plant growth are inhibited by increasing levels of Cu exposition (Sethy and Ghosh, 2013); and the redox activity of Cu at high concentration leads to chlorosis, necrosis, stunted growth, root growth inhibition (Xiong and Wang, 2005; Yruela, 2005; Manceau et al., 2008), and increases ROS generation (Halliwell and Gutteridge, 1999; Manceau et al., 2008). Therefore CuO NPs dispersed as contaminants in the environment can have potential health and toxicity effects on living beings and plants and

TABLE 13.1 Effects of Copper Oxide Nanoparticles (CuO NPs) on Different Plants at Different Concentrations

Studied Plant	Toxic Concentration of CuO NPs	Effect of CuO NPs on Plant	References
Ipomoea batatas	1000 mg/kg	Reduced biomass and yield of sweet potato	Bradfield et al. (2017)
Oryza sativa	1000 mg/L	Decreased rate of germination, reduced root–shoot length and biomass content. Negatively affected photosynthetic rate, transpiration rate, stomatal conductance, and also photosystem II	Da Costa and Sharma, 2016
Bt-transgenic and conventional cotton	1000 mg/L	Hindered growth and development of cotton, decreased uptake of essential nutrient by roots, indole-3-acetic acid, and abscisic acid content	Le Van et al. (2016)
Elsholtzia splendens	1000 mg/L	Deposition of CuO NPs occurred in root and leaf cells. Inhibited seed germination, reduced root and shoot elongation and plant biomass, marginal chlorosis occurred	Shi et al. (2014)
Arabidopsis thaliana	100 mg/L	Decreased total chlorophyll content and plant biomass, reduced root elongation, retarded primary root growth, increased lateral root growth, and also loss of root gravitropism. Reduction in stress-related genes	Nair and Chung. (2014)
Lactuca sativa and *Medicago sativa*	20 mg/L	Decrease in root and shoot length, alteration in nutrient availability and enzymatic activity in both plants	Hong et al. (2016)
L. sativa	20 mg/L	Enhanced accumulation of Cu in roots, decrease in water content, root length, and total biomass. Reduced ascorbic peroxidase (APX) activities. Reduced nutrition quality of lettuce	Trujillo-Reyes et al. (2014)
Brassica juncea	500 mg/L	Significantly decreased shoot–root elongation, lignifications occurred in root and shoot, reduction in photosynthesis, less development in primary and secondary root. Enhanced reactive oxygen species (ROS) production and lipid peroxidation and decreased APX activity	Nair and Chung. (2015)
Vigna radiata	500 mg/L	Reduction in shoot length and biomass and total chlorophyll content, suppressed growth of primary and lateral root. Increased ROS in roots and hydrogen peroxide content and lipid peroxidation also. Roots become much lignified	Nair et al. (2014)
L. sativa	4 ppm	Decline in seed germination, root length elongation. Causes oxidative damage to the plant, thus formation of ROS occurred.	Liu et al. (2016)
O. sativa	100 mg/L	Reduction in root elongation, hindrance in root morphology and ultrastructure, alteration of cell viability and membrane integrity. Oxidative damage caused by ROS production	Peng et al. (2015)
A. thaliana	100 mg/L	CuO NPs ceased the root elongation, reduced the plant biomass, and provoked the development of rosettes of leaves. Plant shows chlorosis and increased wilting. ROS generation also occurred	Yuan et al. (2016)
A. thaliana	20 mg/L	Root cells of *Arabidopsis* were severely damaged. ROS accumulation was induced and as a result 1035 and 623 genes were upregulated and	Tang et al. (2016)

in particular this aspect is largely becoming a great challenge for both the public and scientific researchers (Adhikari et al., 2012; Aruoja et al., 2009; Chang et al., 2012).

Studies have reported that CuO NPs have negative implications in *Oryza sativa* L., influencing growth, morphology, photosynthesis, and antioxidant potential of the plant. Moreover, according to one study, treatment with high concentrations of CuO NPs (1000 mg/L) reduced germination rate and root–shoot length, decreased biomass and increased Cu uptake, decreased the number of thylakoids per granum in chloroplasts, and also resulted in the complete loss of photosystem II photochemical quenching (Da Costa and Sharma, 2016; Table 13.1). Another study on lettuce (*Lactuca sativa*) and alfalfa (*Medicago sativa*) indicated that Cu NP treatment altered the normal physiological parameters of the plants, reducing the size and altering both nutrient content and enzyme activity in both plant species (Hong et al., 2016; Fig. 13.2). Another study on *Pistia stratiotes* L., considering the effects of CuO NP exposure, showed a decrease in ascorbic acid and amino acid content, and reduced concentration of some amino acids (oxoproline, arginine, leucine, ornithine, phenylalanine, proline, serine, and tyrosine) and also some acylcarnitines (Olkhovych et al., 2016). Similarly, CuO NPs also exhibit a toxic impact in different plants such that they decrease the biomass yield in potato (Bradfield et al., 2017), reduce the total biomass in carrot (*Daucus carota*) (Ebbs et al., 2016), diminish the uptake of nutrient, and enhance the expression of the *Bt* toxin protein in cotton leading to inhibited growth and development of transgenic and conventional cottons (Le Van et al., 2016). Although it was thoroughly proven that NPs influenced plant development and

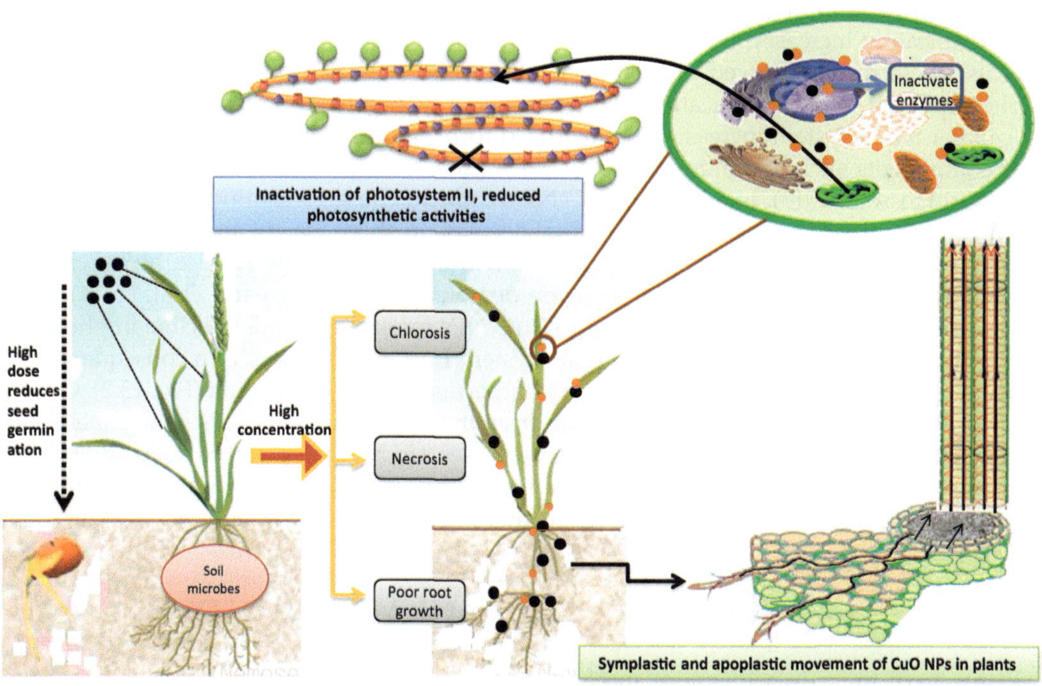

FIGURE 13.2 Interaction of copper oxide nanoparticles (CuO NPs) with different plant tissues (leaves and roots) and their toxic effects at physiological, cellular, and subcellular levels.

metabolism, the extent of their negative or positive impact is influenced by the size, concentration, chemistry, and type of the considered NPs (Dietz and Herth, 2011; Fig. 13.2).

13.4 TOXICITY

Many works showed that as a general rule, CuO NPs have a toxic effect on plants. As reported by Atha et al. (2012), CuO NPs have strong negative effects on perennial ryegrass (*Lolium perenne* L.) root and shoot growth. Shi et al. (2011) also observed inhibition of 50% of growth because of treatment with CuO NPs in duck weeds (*Landoltia punctata*). It is acknowledged that nanomaterials are nowadays being successfully used as nanofertilizers and nanopesticides, but studies on their adverse effect on plants and the environment must be implemented (Selivanov and Zorin, 2001; Raikova et al., 2006). CuO NPs have minimum toxic effects on seed germination; the studies of Adhikari et al. (2012) illustrated the effect of CuO NPs on seed germination on *Glycine max* L. and *Cicer aretinium* L. showing normal seed germination process of the two crops possibly because of the presence of a seed coat that can protect the embryo inside the seed (Adhikari et al., 2012). On the other hand, toxicity symptoms, consisting of a dose-dependent decrease of root elongation, appeared in primary roots (Fig. 13.2).

Toxic effects of CuO NPs were also reported in Syrian barley (*Hordeum vulgare* L.) (Shaw et al., 2014) in which stress-induced modulation of antioxidative defense and decrease in photosynthetic performances were found: antioxidant enzymes of the ascorbate glutathione cycle were differentially modulated under CuO NP treatments as growing concentrations of CuO NPs significantly increased levels of H_2O_2, which can decompose into highly toxic hydroxyl radicals. These results showed toxic impacts of CuO NPs on both physiobiochemical as well as photosynthetic activity indicating that an excess of ROS generation in plants could disturb the cellular redox system increasing the oxidized forms and causing oxidative damage of cellular components such as lipids, proteins, and nucleic acid (Halliwell and Gutteridge, 1989; Shaw and Hossain, 2013). In addition, CuO NP treatment caused a significant effect on barley seedlings growth, where effective and gradual decrease in shoot weight and shoot length were observed with increasing CuO NP concentration within 20 days (Shaw et al., 2014; Fig. 13.2).

Similarly, toxicity effects of CuO NPs were demonstrated by Perreault et al. (2014) in the aquatic plant *Lemna gibba*; plants exposed to CuO NPs for 48 h showed a decline in photosynthetic activity caused by inactivation of photosystem II reaction centers, as a consequence of a decrease of electron transport and an increase of thermal energy dissipation (Fig. 13.2). Cu ions released from NPs are the main factors that drive the toxicity of CuO NPs since their excess (or deficiency) can affect many physiological processes, Cu being a structural element in many regulatory proteins, in photosynthetic electron transport, mitochondrial respiration, oxidative stress responses, cell wall metabolism, and hormone signaling (Marschner, 1996; Yruela, 2005).

13.5 TOLERANCE MECHANISM IN PLANTS

Plants have their own tolerance mechanisms toward many different types of stress that mainly lead to the generation of ROS. ROS are generated as a by-product of normal metabolic pathways during plant growth and development in standard conditions and their physiological presence is restricted to organelles such as chloroplasts, mitochondria, and peroxisomes (Møller et al., 2007; Ma et al., 2015). ROS are eliminated to avoid oxidative damage in plants (Apel and Hirt, 2004) by specific antioxidant enzymes such as superoxide dismutase (SOD), catalase (CAT), or ascorbic peroxidase (APX) (Apel and Hirt, 2004). Generally, oxidative stress stimulates the activities of SOD, CAT, and APX, which fall under different types of antioxidant enzymes in plants (Fenik et al., 1995; Devi and Prasad, 2005). During the early stages of oxidative stress in roots or leaves or in any other part, SOD starts operating and thus converts the superoxide radical in hydrogen peroxide (Nekrasova et al., 2011). CuO NPs are engaged in the Fenton reaction generating ROS, resulting in intensification of lipid peroxidation (Fubini et al., 2007; Nekrasova et al., 2011). There are a number of antioxidant enzymes that help plants to competently detoxify ROS (Mittler, 2002; Visser et al., 2003). Three types of SOD, Fe-SOD, Mn-SOD, and Cu-Zn-SOD, are able to rapidly convert $O_2^{\cdot -}$ to H_2O_2 through a specific process. However, when H_2O_2 is produced in excess, three main additional antioxidant enzyme pathways are activated to detoxify these species. CAT is among one of the common antioxidant enzymes, which helps plants to convert H_2O_2 to H_2O and O_2. APX also carries out a possible reaction that can convert H_2O_2 to H_2O via oxidation of ascorbate into monodehydroascorbate and dehydroascorbate, and these two by-products can be recycled to generate more and more ascorbate (Ma et al., 2015). Besides antioxidant enzymes, nonenzymatic antioxidants, such as flavonoids, phenolic acids, carotenoids, vitamins, etc., are either present inside plant cells or localized higher in the vacuole system. These bioactive substances are both able to scavenge free radicals and chelate metals under stress conditions (Carocho and Ferreira, 2013; Mourato et al., 2012; Gould et al., 2004). Moreover, since the bioaccumulation of CuO NPs was observed (Perreault et al., 2014) and their possible tolerance mechanism in plants noted, further studies would be necessary to evaluate possible contamination of organisms at higher trophic levels.

13.6 CONCLUSION AND FUTURE REMARKS

In this review we have focused on CuO NPs, their practical applications, toxic impacts, and interaction with plants. Because of the escalating demand of nanotechnology and nanomaterials in the present technological era, study of the potential risk of metal-based NPs on the environment has become a high priority argument. Nanomaterials affect not only yield and quality of crops but also the diversity of plants and microorganisms associated with plants in soil systems. The fate of nanomaterials and their effect on the terrestrial environment have a critical knowledge gap, which needs to be filled to facilitate conscious exposure and risk assessment toward NPs in the environment. CuO NPs have several negative impacts on plants such as their increasing concentration, which decreases plant growth and productivity. In the published literature on metal-based CuO NPs and their interaction with plants, results have shown that CuO NPs caused adverse effects on plants at both physiological (such as root length inhibition, biomass decrease, and plant development delay) and cellular (such as chlorophyll synthesis disruption, cell membrane damage, and chromosomal aberration) levels. A better understanding of the effects and implications of metal-based NPs has now become essential and future research will be needed to focus on their proper study and positive and negative impacts before their commercialization in the marketplace and their release in the environment.

References

Abramova, A., Gedanken, A., Popov, V., Ooi, E.H., Mason, T.J., Joyce, E.M., Bayazitov, V., 2013. A sonochemical technology for coating of textiles with antibacterial nanoparticles and equipment for its implementation. Mater. Lett. 96, 121–124.

Adhikari, T., Kundu, S., Biswas, A.K., Tarafdar, J.C., Rao, A.S., 2012. Effect of copper oxide nano particle on seed germination of selected crops. J. Agric. Sci. Technol. 2 (6A), 815.

Adrees, M., Ali, S., Rizwan, M., Ibrahim, M., Abbas, F., Farid, M., Bharwana, S.A., 2015. The effect of excess copper on growth and physiology of important food crops: a review. Environ. Sci. Pollut. Res. 22 (11), 8148–8162.

Ahsan, N., Lee, D.G., Lee, S.H., Kang, K.Y., Lee, J.J., Kim, P.J., Lee, B.H., 2007. Excess copper induced physiological and proteomic changes in germinating rice seeds. Chemosphere 67 (6), 1182–1193.

Allan, D.L., Jarrell, W.M., 1989. Proton and copper adsorption to maize and soybean root cell walls. Plant Physiol. 89 (3), 823–832.

Almeida, E., Diamantino, T.C., de Sousa, O., 2007. Marinepaints: the particular case of antifouling paints. Prog. Org. Coat. 59 (1), 2–20.

Andreotti, F., Mucha, A.P., Caetano, C., Rodrigues, P., Gomes, C.R., Almeida, C.M.R., 2015. Interactions between salt marsh plants and Cu nanoparticles–effects on metal uptake and phytoremediation processes. Ecotoxicol. Environ. Saf. 120, 303–309.

Anjum, N.A., Singh, H.P., Khan, M.I.R., Masood, A., Per, T.S., et al., 2015. Too much is bad-an appraisal of phytotoxicity of elevated plant-beneficial heavy metal ions. Environ. Sci. Pollut. Res. 22, 3361–3382.

Apel, K., Hirt, H., 2004. Reactive oxygen species: metabolism, oxidative stress, and signal transduction. Annu. Rev. Plant Biol. 55, 373–399.

Aruoja, V., Dubourguier, H-C., Kasemets, K., Kahru, A., 2009. Toxicity of nanoparticles of CuO, ZnO and TiO_2 to microalgae *Pseudokirchneriella subcapitata*. Sci. Total. Environ. 407, 1461–1468.

Atha, D.H., Wang, H., Petersen, E.J., Cleveland, D., Holbrook, R.D., Jaruga, P., Dizdaroglu, M., Xing, B., Nelson, B.C., 2012. Copper oxide nanoparticle mediated DNA damage in terrestrial plant models. Environ. Sci. Technol. 46, 1819–1827.

Ayangbenro, A.S., Babalola, O.O., 2017. A new strategy for heavy metal polluted environments: a review of microbial biosorbents. Int. J. Environ. Res. Public Health 14 (1), 94.

Bellani, L.M., Muccifora, S., Giorgetti, L., 2014. Response to copper bromide exposure in *Vicia sativa* L. seeds: analysis of genotoxicity, nucleolar activity and mineral profile. Ecotoxicol. Environ. Saf. 107, 245–250.

Bhatt, I., Tripathi, B.N., 2011. Interaction of engineered nanoparticles with various components of the environment and possible strategies for their risk assessment. Chemosphere 82 (3), 308–317.

Blinova, I., Ivask, A., Heinlaan, M., Mortimer, M., Kahru, A., 2010. Ecotoxicity of nanoparticles of CuO and ZnO in natural water. Environ. Pollut. 158 (1), 41–47.

Blosi, M., Albonetti, S., Dondi, M., Martelli, C., Baldi, G., 2011. Microwave-assisted polyol synthesis of Cu nanoparticles. J. Nanopart. Res. 13 (1), 127–138.

Borkow, G., Gabbay, J., 2009. Copper, an ancient remedy returning to fight microbial, fungal and viral infections. Curr. Chem. Biol. 3 (3), 272–278.

Borm, P.J., Robbins, D., Haubold, S., Kuhlbusch, T., Fissan, H., Donaldson, K., Krutmann, J., 2006. The potential risks of nanomaterials: a review carried out for ECETOC. Part. Fibre Toxicol. 3 (1), 11.

Bradfield, S.J., Kumar, P., White, J.C., Ebbs, S.D., 2017. Zinc, copper, or cerium accumulation from metal oxide nanoparticles or ions in sweet potato: yield effects and projected dietary intake from consumption. Plant Physiol. Biochem. 110, 128–137.

Brar, S.K., Verma, M., Tyagi, R.D., Surampalli, R.Y., 2010. Engineered nanoparticles in wastewater and wastewater sludge–evidence and impacts. Waste Manag. 30 (3), 504–520.

Carocho, M., Ferreira, I.C.F.R., 2013. A review on antioxidants, prooxidants and related controversy: natural and synthetic compounds, screening and analysis methodologies and future perspectives. Food Chem. Toxicol. 51 (0), 15–25.

Castiglione, M.R., Giorgetti, L., Bellani, L., Muccifora, S., Bottega, S., Spanò, C., 2016. Root responses to different types of TiO_2 nanoparticles and bulk counterpart in plant model system *Vicia faba* L. Environ. Exp. Bot. 130, 11–21.

Cava, R.J., 1990. Structural chemistry and the local charge picture of copper oxide superconductors. Science 247, 656–662.

Chang, Y.N., Zhang, M., Xia, L., Zhang, J., Xing, G., 2012. The toxic effects and mechanisms of CuO and ZnO nanoparticles. Materials 5 (12), 2850–2871.

Chowdhuri, A., Gupta, V., Sreenivas, K., Kumar, R., Mozumdar, S., Patanjali, P.K., 2004. Response speed of SnO_2-based H_2S gas sensors with CuO nanoparticles. Appl. Phys. Lett. 84, 1180–1182.
Cioffi, N., Torsi, L., Ditaranto, N., Tantillo, G., Ghibelli, L., Sabbatini, L., 2005. Copper nanoparticles/polymer composites with antifungal and bacteriostatic properties. Chem. Mater. 17, 5255–5262.
Da Costa, M.V.J., Sharma, P.K., 2016. Effect of copper oxide nanoparticles on growth, morphology, photosynthesis, and antioxidant response in *Oryza sativa*. Photosynthetica 54 (1), 110–119.
Dar, M.A., Kim, Y.S., Kim, W.B., Sohn, J.M., Shin, H.S., 2008. Structural and magnetic properties of CuO nanoneedles synthesized by hydrothermal method. Appl. Surf. Sci. 254, 7477–7481.
Devi, S.R., Prasad, M.N.V., 2005. Antioxidant capacity of *Brassica juncea* plants exposed to elevated levels of copper. Russ. J. Plant Physiol. 52 (2), 205–208.
Dietz, K.J., Herth, S., 2011. Plant nanotoxicology. Trends Plant Sci. 16 (11), 582–589.
Ebbs, S.D., Bradfield, S.J., Kumar, P., White, J.C., Musante, C., Ma, X., 2016. Accumulation of zinc, copper, or cerium in carrot (*Daucus carota*) exposed to metal oxide nanoparticles and metal ions. Environ. Sci. 3, 114–126.
Ebrahimnia-Bajestan, E., Niazmand, H., Duangthongsuk, W., Wongwises, S., 2011. Numerical investigation of effective parameters in convective heat transfer of nanofluids flowing under a laminar flow regime. Int. J. Heat Mass Transf. 54 (19), 4376–4388.
Fenik, S.I., Trofimyak, T.B., Blyum, Y.B., 1995. Mechanisms of development of plant tolerance to heavy metals. Uspe khisovremennoĭbiologii 115, 259–275.
Fubini, B., Fenoglio, I., Tomatis, M., 2007. Physicochemical characteristics of nanoparticles that determine potential toxicity. In: Nanotoxicology: Characterization, Dosing, and Health Effects. Informa Healthcare, New York, pp. 59–70.
Gabbay, J., Borkow, G., Mishal, J., Magen, E., Zatcoff, R., Shemer-Avni, Y., 2006. Copper oxide impregnated textiles with potent biocidal activities. J. Ind. Text. 35 (4), 323–335.
Gavankar, S., Suh, S., Keller, A.F., 2012. Life cycle assessment at nanoscale: review and recommendations. Int. J. Life Cycle Assess. 17, 295.
Ge, Y., Schimel, J.P., Holden, P.A., 2011. Evidence for negative effects of TiO_2 and ZnO nanoparticles on soil bacterial communities. Environ. Sci. Technol. 45 (4), 1659–1664.
Gomes, T., Pinheiro, J.P., Cancio, I., Pereira, C.G., Cardoso, C., Bebianno, M.J., 2011. Effects of copper nanoparticles exposure in the mussel *Mytilus galloprovincialis*. Environ. Sci. Technol. 45 (21), 9356–9362.
Gould, K.S., 2004. Nature's Swiss army knife: the diverse protective roles of anthocyanins in leaves. BioMed Res. Int. 2004 (5), 314–320.
Griffitt, R.J., Weil, R., Hyndman, K.A., Denslow, N.D., Powers, K., Taylor, D., Barber, D.S., 2007. Exposure to copper nanoparticles causes gill injury and acute lethality in zebrafish (*Danio rerio*). Environ. Sci. Technol. 41 (23), 8178–8186.
Guo, K., Pan, Q., Wang, L., Fang, S., 2002. Nano-scale copper-coated graphite as anode material for lithium-ion batteries. J. Appl. Electrochem. 32 (6), 679–685.
Guo, Z., Ng, H.W., Yee, G.L., Hahn, H.T., 2009. Differential scanning calorimetry investigation on vinyl ester resin curing process for polymer nanocomposite fabrication. J. Nanosci. Nanotechnol. 9 (5), 3278–3285.
Halliwell, B., Gutteridge, J.M.C., 1989. Free Radicals in Biology and Medicine, second ed. Clarendon Press, Oxford.
Halliwell, B., Gutteridge, J.M., 1999. Free Radicals in Biology and Medicine, vol. 3. Oxford University Press, Oxford.
Handy, R.D., Kamme, F., Lead, J.R., Hasselov, M., Owen, R., Crane, M., 2008. The ecotoxicology and chemistry of manufactured nanoparticles. Ecotoxicology 17, 287–314.
Hong, J., Rico, C.M., Zhao, L., Adeleye, A.S., Keller, A.A., Peralta-Videa, J.R., Gardea-Torresdey, J.L., 2016. Toxic effects of copper-based nanoparticles or compounds to lettuce (*Lactuca sativa*) and alfalfa (*Medicago sativa*). Environ. Sci. 17, 177–185.
https://www.sciencedaily.com/terms/nanoparticle.htm. 2014.
Jammi, S., Sakthivel, S., Rout, L., Mukherjee, T., Mandal, S., Mitra, R., Saha, P., Punniyamurthy, T., 2009. CuO nanoparticles catalyzed C-N, C-O, and C-S cross-coupling reactions: scope and mechanism. J. Org. Chem. 74, 1971–1976.
Kabata-Pendias, A., Pendias, H., 1992. Trace Elements in Soils and Plants, second ed. CRC Press, Boca Raton.
Karlsson, H.L., Cronholm, P., Gustafsson, J., Moller, L., 2008. Copper oxide nanoparticles are highly toxic: a comparison between metal oxide nanoparticles and carbon nanotubes. Chem. Res. Toxicol. 21 (9), 1726–1732.

Kim, Y.H., Lee, D.K., Cha, H.G., Kim, C.W., Kang, Y.C., Kang, Y.S., 2006. Preparation and characterization of the antibacterial Cu nanoparticle formed on the surface of SiO_2 nanoparticles. J. Phys. Chem. B 110 (49), 24923–24928.

Kim, D.-K., Yoo, S.M., Park, T.J., Yoshikawa, H., Tamiya, E., Park, J.Y., Lee, S.Y., 2011. Plasmonic properties of the multispot copper-capped nanoparticle array chip and its application to optical biosensors for pathogen detection of multiplex DNAs. Anal. Chem. 83, 6215–6222.

Krzesłowska, M., 2011. The cell wall in plant cell response to trace metals: polysaccharide remodeling and its role in defense strategy. Acta Physiol. Plant. 33 (1), 35–51.

Le Van, N., Ma, C., Shang, J., Rui, Y., Liu, S., Xing, B., 2016. Effects of CuO nanoparticles on insecticidal activity and phytotoxicity in conventional and transgenic cotton. Chemosphere 144, 661–670.

Lead, J.R., Wilkinson, K.J., 2006. Environmental colloids: knowledge and future developments. In: Wilkinson, K.J., Lead, J.R. (Eds.), Environmental Colloids: Behavior, Structure and Characterization. John Wiley, Chichester, UK, pp. 1–15.

Lee, W.M., An, Y.J., Yoon, H., Kweon, H.S., 2008. Toxicity and bioavailability of copper nanoparticles to the terrestrial plants mung bean (*Phaseolus radiatus*) and wheat (*Triticum aestivum*): plant agar test for water-insoluble nanoparticles. Environ. Toxicol. Chem. 27 (9), 1915–1921.

Lee, H.J., Song, J.Y., Kim, B.S., 2013. Biological synthesis of copper nanoparticles using *Magnolia kobus* leaf extract and their antibacterial activity. J. Chem. Technol. Biotechnol. 88 (11), 1971–1977.

Leng, X., Jia, H., Sun, X., Shangguan, L., Mu, Q., Wang, B., Fang, J., 2015. Comparative transcriptome analysis of grapevine in response to copper stress. Nature Publishing Group Sci. Rep. 5, 17749.

Li, J., Kleinstreuer, C., 2008. Thermal performance of nanofluid flow in microchannels. Int. J. Heat Fluid Flow 29 (4), 1221–1232.

Liu, G., Li, X., Qin, B., Xing, D., Guo, Y., Fan, R., 2004. Investigation of the mending effect and mechanism of copper nanoparticles on a tribologically stressed surface. Tribol. Lett. 17, 961–965.

Liu, R., Zhang, H., Lal, R., 2016. Effects of stabilized nanoparticles of copper, zinc, manganese, and iron oxides in low concentrations on lettuce (*Lactuca sativa*) seed germination: nanotoxicants or nanonutrients? Water Air Soil Pollut. 227 (1), 42.

Ma, X., Geiser-Lee, J., Deng, Y., Kolmakov, A., 2010. Interactions between engineered nanoparticles (ENPs) and plants: phytotoxicity, uptake and accumulation. Sci. Total Environ. 408 (16), 3053–3061.

Ma, C., White, J.C., Dhankher, O.P., Xing, B., 2015. Metal-based nanotoxicity and detoxification pathways in higher plants. Environ. Sci. Pollut. Res. 49 (12), 7109–7122.

Manceau, A., Nagy, K.L., Marcus, M.A., Lanson, M., Geoffroy, N., Jacquet, T., Kirpichtchikova, T., 2008. Formation of metallic copper nanoparticles at the soil–root interface. Environ. Sci. Technol. 42 (5), 1766–1772.

Marschner, H., 1996. Mineral Nutrition of Higher Plants. Academic Press, London.

McCauley, A., Jones, C., Jacobsen, J., 2009. Soil pH and organic matter. Nutr. Manag. Modul. 8, 1–12.

Mittler, R., 2002. Oxidative stress, antioxidants and stress tolerance. Trends Plant Sci. 7 (9), 405–410.

Møller, I.M., Jensen, P.E., Hansson, A., 2007. Oxidative modifications to cellular components in plants. Annu. Rev. Plant Biol. 58, 459–481.

Monica, R.C., Cremonini, R., 2009. Nanoparticles and higher plants. Caryologia 62 (2), 161–165.

Moore, M.N., 2006. Do nanoparticles present ecotoxicological risks for the health of the aquatic environment? Environ. Int. 32, 967–976.

Mourato, M., Reis, R., Martins, M.L.L., 2012. Characterization of plant antioxidative system in response to abiotic stresses: a focus on heavy metal toxicity. Adv. Select. Plant Physiol. Asp. 23–44.

Nagajyoti, P.C., Lee, K.D., Sreekanth, T.V.M., 2010. Heavy metals, occurrence and toxicity for plants: a review. Environ. Chem. Lett. 8 (3), 199–216.

Nair, P.M.G., Chung, I.M., 2014. Impact of copper oxide nanoparticles exposure on *Arabidopsis thaliana* growth, root system development, root lignificaion, and molecular level changes. Environ. Sci. Pollut. Res. Int. 21 (22), 12709–12722.

Nair, P.M.G., Chung, I.M., 2015. Study on the correlation between copper oxide nanoparticles induced growth suppression and enhanced lignification in Indian mustard (*Brassica juncea* L.). Ecotoxicol. Environ. Saf. 113, 302–313.

Nair, P.M.G., Kim, S.H., Chung, I.M., 2014. Copper oxide nanoparticle toxicity in mung bean (*Vigna radiata* L.) seedlings: physiological and molecular level responses of in vitro grown plants. Acta Physiol. Plant. 36 (11), 2947–2958.

Namburu, P.K., Kulkarni, D.P., Misra, D., Das, D.K., 2007. Viscosity of copper oxide nanoparticles dispersed in ethylene glycol and water mixture. Exp. Therm. Fluid Sci. 32 (2), 397–402.

Nasibulin, A.G., Ahonen, P.P., Richard, O., Kauppinen, E.I., 2000. Copper and copper oxide nanoparticle formation by chemical vapor nucleation from copper (II) acetylacetonate. J. Aerosol Sci. 31, 552–553.

Navarro, E., Piccapietra, F., Wagner, B., Marconi, F., Kaegi, R., Odzak, N., Sigg, L., Behra, R., 2008. Toxicity of silver nanoparticles to *Chlamydomonas reinhardtii*. Environ. Sci. Technol. 42, 8959–8964.

Nekrasova, G.F., Ushakova, O.S., Ermakov, A.E., Uimin, M.A., Byzov, I.V., 2011. Effects of copper (II) ions and copper oxide nanoparticles on *Elodea densa* Planch. Russ. J. Ecol. 42 (6), 458–463.

Oberdörster, G., Oberdörster, E., Oberdörster, J., 2005. Nanotoxicology: an emerging discipline evolving from studies of ultrafine particles. Environ. Health Perspect. 823–839.

Oleszczuk, P., Jośko, I., Xing, B., 2011. The toxicity to plants of the sewage sludges containing multiwalled carbon nanotubes. J. Hazard. Mater. 186 (1), 436–442.

Olkhovych, O., Volkogon, M., Taran, N., Batsmanova, L., Kravchenko, I., 2016. The effect of copper and zinc nanoparticles on the growth parameters, contents of ascorbic acid, and qualitative composition of amino acids and acylcarnitines in *Pistia stratiotes* L. (Araceae). Nanoscale Res. Lett. 11, 218.

Pachapur, V.L., Larios, A.D., Cledon, M., Brar, S.K., Verma, M., Surampalli, R.Y., 2016. Behavior and characterization of titanium dioxide and silver nanoparticles in soils. Sci. Total Environ. 563, 933–943.

Peng, C., Duan, D., Xu, C., Chen, Y., Sun, L., Zhang, H., Zhen, X., 2015. Translocation and biotransformation of CuO nanoparticles in rice (*Oryza sativa* L.) plants. Environ. Pollut. 197, 99–107.

Perreault, F., Oukarroum, A., Melegari, S.P., Matias, W.G., Popovic, R., 2012. Polymer coating of copper oxide nanoparticles increases nanoparticles uptake and toxicity in the green alga *Chlamydomonas reinhardtii*. Chemosphere 87, 1388–1394.

Perreault, F., Samadani, M., Dewez, D., 2014. Effect of soluble copper released from copper oxide nanoparticles solubilisation on growth and photosynthetic processes of *Lemna gibba* L. Nanotoxicology 8, 374–382.

Raikova, O.P., Panichkin, L.A., Raikova, N.N., 2006. Studies on the effect of ultrafine metal powders produced by different methods on plant growth and development. Nanotechnologies and information technologies in the 21st century. In: Proceedings of the International Scientific and Practical Conference, pp. 108–111.

Ramanathan, R., Bhargava, S.K., Bansal, V., 2011. Biological synthesis of copper/copper oxide nanoparticles. In: Chemeca 2011: Engineering a Better World: Sydney Hilton Hotel, NSW, Australia, 18–21 September 2011, 1991.

Roco, M.C., 2003. Nanotechnology: convergence with modern biology and medicine. Curr. Opin. Biotechnol. 14, 337–346.

Sau, T.K., Rogach, A.L., Jäckel, F., Klar, T.A., Feldmann, J., 2010. Properties and applications of colloidal nonspherical noble metal nanoparticles. Adv. Mater. 22 (16), 1805–1825.

Scown, T.M., Van Aerle, R., Tyler, C.R., 2010. Review: do engineered nanoparticles pose a significant threat to the aquatic environment? Crit. Rev. Toxicol. 40 (7), 653–670.

Selivanov, V.N., Zorin, E.V., 2001. Sustained action of ultrafine metal powders on seeds of grain crops. Perspect. Mater. 4, 66–69.

Sethy, S.K., Ghosh, S., 2013. Effect of heavy metals on germination of seeds. J. Nat. Sci. Biol. Med. 4 (2), 272–275.

Shaw, A.K., Hossain, Z., 2013. Impact of nano-CuO stress on rice (*Oryza sativa* L.) seedlings. Chemosphere 93 (6), 906–915.

Shaw, A.K., Ghosh, S., Kalaji, H.M., Bosa, K., Brestic, M., Zivcak, M., Hossain, Z., 2014. Nano-CuO stress induced modulation of antioxidative defense and photosynthetic performance of Syrian barley (*Hordeum vulgare* L.). Environ. Exp. Bot. 102, 37–47.

Shi, J., Abid, A.D., Kennedy, I.M., Hristova, K.R., Silk, W.K., 2011. To duckweeds (*Landoltia punctata*), nanoparticulate copper oxide is more inhibitory than the soluble copper in the bulk solution. Environ. Pollut. 159 (5), 1277–1282.

Shi, J., Peng, C., Yang, Y., Yang, J., Zhang, H., Yuan, X., Hu, T., 2014. Phytotoxicity and accumulation of copper oxide nanoparticles to the Cu-tolerant plant *Elsholtzia splendens*. Nanotoxicology 8 (2), 179–188.

Shweta, Tripathi, D.K., Singh, S., Singh, S., Dubey, N.K., Chauhan, D.K., 2016. Impact of nanoparticles on photosynthesis: challenges and opportunities. Mater. Focus 5 (5), 405–411.

Singh, S., Tripathi, D.K., Dubey, N.K., Chauhan, D.K., 2016. Effects of nano-materials on seed germination and seedling growth: striking the slight balance between the concepts and controversies. Mater. Focus 5 (3), 195–201.

Singh, S., Vishwakarma, K., Singh, S., Sharma, S., Dubey, N.K., Singh, V.K., Liu, S., Tripathi, D.K., Chauhan, D.K., 2017. Understanding the plant and nanoparticle interface at transcriptomic and proteomic level: a concentric overview. Plant Gene. http://dx.doi.org/10.1016/j.plgene.2017.03.006.

Song, L., Vijver, M.G., Peijnenburg, W.J., 2015. Comparative toxicity of copper nanoparticles across three Lemnaceae species. Sci. Total Environ. 518, 217–224.

Suramwar, N.V., Thakare, S.R., Khaty, N.T., 2012. Synthesis and catalytic properties of nano CuO prepared by soft chemical method. Int. J. Nano Dimens. 3 (1), 75–80.

Tang, Y., He, R., Zhao, J., Nie, G., Xu, L., Xing, B., 2016. Oxidative stress-induced toxicity of CuO nanoparticles and related toxicogenomic responses in *Arabidopsis thaliana*. Environ. Pollut. 212, 605–614.

Tranquada, J.M., Sternlieb, B.J., Axe, J.D., Nakamura, Y., Uchida, S., 1995. Evidence for stripe correlations of spins and holes in copper oxide superconductors. Nature 375 (6532), 561.

Tripathi, D.K., Singh, V.P., Prasad, S.M., Chauhan, D.K., Dubey, N.K., 2015. Silicon nanoparticles (SiNp) alleviate chromium (VI) phytotoxicity in *Pisum sativum* (L.) seedlings. Plant Physiol. Biochem. 96, 189–198.

Tripathi, D.K., Singh, S., Singh, V.P., Prasad, S.M., Chauhan, D.K., Dubey, N.K., 2016. Silicon nanoparticles more efficiently alleviate arsenate toxicity than silicon in maize cultiver and hybrid differing in arsenate tolerance. Front. Environ. Sci. 4, 46.

Tripathi, D.K., Singh, S., Singh, S., Pandey, R., Singh, V.P., Sharma, N.C., Prasad, S.M., Dubey, N.K., Chauhan, D.K., 2017a. An overview on manufactured nanoparticles in plants: uptake, translocation, accumulation and phytotoxicity. Plant Physiol. Biochem. 110, 2–12.

Tripathi, D.K., Mishra, R.K., Singh, S., Singh, S., Vishwakarma, K., Sharma, S., Singh, V.P., Singh, P.K., Prasad, S.M., Dubey, N.K., Pandey, A.C., 2017b. Nitric oxide ameliorates zinc oxide nanoparticles phytotoxicity in wheat seedlings: implication of the ascorbate–glutathione cycle. Front. Plant Sci. 8.

Tripathi, D.K., Singh, S., Singh, S., Srivastava, P.K., Singh, V.P., Singh, S., Prasad, S.M., Singh, P.K., Dubey, N.K., Pandey, A.C., Chauhan, D.K., 2017c. Nitric oxide alleviates silver nanoparticles (AgNps)-induced phytotoxicity in *Pisum sativum* seedlings. Plant Physiol. Biochem. 110, 167–177.

Tripathi, D.K., Tripathi, A., Shweta, S.S., Singh, Y., Vishwakarma, K., Yadav, G., Sharma, S., Singh, V.K., Mishra, R.K., Upadhyay, R.G., Dubey, N.K., 2017d. Uptake, accumulation and toxicity of silver nanoparticle in autotrophic plants, and heterotrophic microbes: a concentric review. Front. Microbiol. 8.

Tripathi, D.K., Singh, S., Singh, V.P., Prasad, S.M., Dubey, N.K., Chauhan, D.K., 2017e. Silicon nanoparticles more effectively alleviated UV-B stress than silicon in wheat (*Triticum aestivum*) seedlings. Plant Physiol. Biochem. 110, 70–81.

Trujillo-Reyes, J., Majumdar, S., Botez, C.E., Peralta-Videa, J.R., Gardea-Torresdey, J.L., 2014. Exposure studies of core–shell Fe/Fe_3O_4 and Cu/CuO NPs to lettuce (*Lactuca sativa*) plants: are they a potential physiological and nutritional hazard? J. Hazard. Mater. 267, 255–263.

UNESCO, 2006. The Ethics and Politics of Nanotechnology. United Nations Educational, Scientific and Cultural Organization.

Unrine, J.M., Tsyusko, O.V., Hunyadi, S.E., Judy, J.D., Bertsch, P.M., 2010. Effects of particle size on chemical speciation and bioavailability of copper to earthworms (*Eisenia fetida*) exposed to copper nanoparticles. J. Environ. Qual. 39 (6), 1942–1953.

Visser, E.J.W., Voesenek, L.A.C.J., Vartapetian, B.B., Jackson, M.B., 2003. Flooding and plant growth. Ann. Bot. 91 (2), 107–109.

Wang, Z., Xie, X., Zhao, J., Liu, X., Feng, W., White, J.C., Xing, B., 2012. Xylem-and phloem-based transport of CuO nanoparticles in maize (*Zea mays* L.). Environ. Sci. Technol. 46 (8), 4434–4441.

Xiong, Z.T., Wang, H., 2005. Copper toxicity and bioaccumulation in Chinese cabbage (*Brassica pekinensis* Rupr.). Environ. Toxicol. 20 (2), 188–194.

Xu, J.F., Ji, W., Shen, Z.X., Tang, S.H., Ye, X.R., Jia, D.Z., Xin, X.Q., 1999. Preparation and characterization of CuO nanocrystals. J. Solid State Chem. 147 (2), 516–519.

Yang, J.G., Okamoto, T., Ichino, R., Bessho, T., Satake, S., Okido, M., 2006. A simple way for preparing antioxidation nano-copper powders. Chem. Lett. 35 (6), 648–649.

Yin, M., Wu, C.K., Lou, Y., Burda, C., Koberstein, J.T., Zhu, Y., O'Brien, S., 2005. Copper oxide nanocrystals. J. Am. Chem. Soc. 127 (26), 9506–9511.

Yoon, K.Y., Byeon, J.H., Park, J.H., Hwang, J., 2007. Susceptibility constants of *Escherichia coli* and *Bacillus subtilis* to silver and copper nanoparticles. Sci. Total Environ. 373, 572–575.

Yruela, I., 2005. Copper in plants. Braz. J. Plant Physiol. (Londrina) 17 (1), 145–156.

Yuan, J., He, A., Huang, S., Hua, J., Sheng, G.D., 2016. Internalization and phytotoxic effects of CuO nanoparticles in *Arabidopsis thaliana* as revealed by fatty acid profiles. Environ. Sci. Technol. 50 (19), 10437–10447.

Zhang, D.W., Yi, T.H., Chen, C.H., 2005. Cu nanoparticles derived from CuO electrodes in lithium cells. Nanotechnology 16 (10), 2338.

Zhang, V., Liu, J., Peng, Q., Wang, X., Li, Y., 2006. Nearly monodisperse Cu_2O and CuO nanospheres: preparation and applications for sensitive gas sensors. Chem. Mater. 18, 867–871.

Zhao, S., Liu, Q., Qi, Y., Duo, L., 2010. Responses of root growth and protective enzymes to copper stress in turf grass. Acta Biol. Crac. Ser. Bot. 52 (2), 7–11.

CHAPTER

14

Impacts of Cerium Oxide Nanoparticles ($nCeO_2$) on Crop Plants: A Concentric Overview

Daniel Lizzi[1,2], *Alessandro Mattiello*[1], *Luca Marchiol*[1]
[1]University of Udine, Udine, Italy; [2]University of Trieste, Trieste, Italy

14.1 INTRODUCTION

Nanotechnology is a science that is expanding with constant and rapid evolution with the potential to revolutionize everyday life. For this reason, it is considered to be responsible for the beginning of the "Nano Revolution" (Ramsden, 2005). Nanotechnology is one of the EU Key-Enabling Technologies, which promotes the creation of new types of products (Kawazoe and Meech, 2005) and permits the development of new materials for applications in different sciences such as medicine, engineering, agriculture, life sciences, plant science, chemical sciences, marine sciences, geography, and geology (Borm and Berube, 2008; Soni, 2014; Tripathi et al., 2015, 2016, 2017a,b,c,d,e; Chan et al., 2016; Singh et al., 2016, 2017; Shweta et al., 2016; Cox et al., 2017; Du et al., 2017). These new technologies are great tools to address many problems, because of their wide applications in several fields such as energy, medical diagnosis and health care, computer chips and information technology, manufacturing, industry, and agriculture (Servin et al., 2017).

Cerium is a metal that belongs to the lanthanide series in the modern periodic table and is one of the most abundant rare metals in rocks and soil of the earth's crust (Hedrick, 2004). It is available in the form of different mineral species, such as carbonates, oxides, phosphates, and silicates, in particular in cerite, allanite, zircon, and bastnasite (Hammond, 2000; Cervini-Silva et al., 2005). In the last few years, $nCeO_2$ has become very important from the point of view of the nanotechnology global market, because of its useful applications in electronics and optics manufacturing with additional scope in fuel cells and additives, catalytic converters in diesel fuels (Corma et al., 2004), paints, gas sensors, glass, ceramic and metallurgical applications, oxygen pumps, and solar cells (Lin et al., 2006). With an estimated annual global production of 100–1000 tons/year, $nCeO_2$ is among the most widely utilized metal oxide nanoparticle in

Nanomaterials in Plants, Algae, and Microorganisms
http://dx.doi.org/10.1016/B978-0-12-811487-2.00014-1

Europe (Sun et al., 2014; Piccinno et al., 2012). Since the release of $nCeO_2$ in the environment is quite significant (Dahle and Arai, 2015), the Organization for Economic Cooperation and Development included $nCeO_2$ among the nanoparticles being studied and analyzed for risk assessment (OECD, 2008).

However, this will only be possible if we are able to improve the methods for detecting $nCeO_2$ in the environment and to know in depth its effects on living organisms. In fact, at the moment we have contradictory information about the ecotoxicology of $nCeO_2$. Some evidence indicates it is responsible for oxidative stress, while other evidence suggests it is a promoter of antioxidant substances (Siddiqi and Husen, 2017). This poses a challenge in future regulations for the application of $nCeO_2$ and its risk assessment in the environment.

14.2 INFLUENCE OF $nCeO_2$ ON PLANT GROWTH

The remarkable and rapid increase in the use of $nCeO_2$ in many application areas caused the spread of these nanoparticles in different environmental matrices. When $nCeO_2$ reaches the soil, it is absorbed by vegetation, and it could penetrate deeper into the soil layers, contaminating aquatic groundwater, or it could also enter the food chain (López-Moreno et al., 2010a; Hernandez-Viezcas et al., 2013). Nanoparticles generally come into plants through root uptake or gaseous exchange of leaves. $nCeO_2$ could cause several effects on the plant system by reducing germination rates and improving, reducing, or inhibiting radical growth (López-Moreno et al., 2010a) where plants grow more slowly and reduce their leaf coverage or produce less biomass; however, it is also observed that these nanoparticles sometimes do not affect plant physiology and biochemistry (Ma et al., 2010). Responses given by plants depend on parameters such as plant species and particle size, treatment, and concentration of $nCeO_2$.

This chapter mainly discusses the effects of $nCeO_2$ exposure on edible plants in synchrony with studies already published. Studies on germination and seedling stage precede those on adult plants and yield quality. Finally, studies covering the entire crop cycle and multigenerational studies are examined. Tables 14.1–14.4 are summarized reports of the experimental conditions and inferences of studies.

14.2.1 Effect on Germination and Root Elongation

The exponential growth of nanotechnology and its introduction to agriculture has prompted researchers to clarify the relationships between nanomaterials and plants. During early stages of plant development, various parameters can be studied to evaluate the effects of exposition to toxic agents or other abiotic stresses. When germinating seeds are exposed to nanoparticles, different effects could be verified, depending upon species, particle size, and concentration (Fig. 14.1) (Miralles et al., 2012).

Different studies have reported different effects of $nCeO_2$ on crops. López-Moreno et al. (2010a) observed the germination and root elongation of *Medicago sativa* (alfalfa), *Zea mays* (corn), *Cucumis sativus* (cucumber), and *Lycopersicum esculentum* (tomato). Both parameters were influenced by $nCeO_2$. In particular, a significant reduction in germination was recorded in tomato and corn at moderate concentrations. However, low toxicity was reported for alfalfa and cucumber, wherein contrasting evidence was recorded for root

TABLE 14.1 Summary of the Effects of $nCeO_2$ on Germination and Seedling Root Elongation of Cultivated Plants

Plant Species	Particle Size (nm)	Substrate	$nCeO_2$ Treatment	Plant Response	References
Cucumis sativus *Medicago sativa* *Lycopersicum esculentum* *Zea mays*	7	Petri dishes	0, 500, 1000, 2000, 4000 mg/L	Seed germination reduced in corn, tomato, and cucumber at 2000 mg/L. Root elongation promoted in corn and reduced in alfalfa and tomato	López-Moreno et al. (2010a)
Glycine max	7	Petri dishes	0, 500, 1000, 2000, 4000 mg/L	Seed germination reduced in soybean at higher levels of treatment. Evidence of genotoxicity. Root elongation promoted at all concentrations	López-Moreno et al. (2010b)
Allium cepa *Avena sativa* *Brassica oleracea capitata* *C. sativus* *Daucus carota* *G. max* *Lactuca sativa* *Lolium perenne* *L. esculentum* *Z. mays*	23–25	Petri dishes	0, 250, 500, 1000 mg/L	Root elongation: (1) promoted in onion and ryegrass; (2) reduced in lettuce and tomato	Andersen et al. (2016)
Hordeum vulgare	50	Petri dishes	0, 500, 1000, 2000 mg/L	Root elongation reduced at 500 mg/L	Mattiello et al. (2015)

growth as well. At an $nCeO_2$ concentration of 2000 mg/L, root elongation was stimulated in cucumber and delayed in alfalfa and tomato. In the framework of a similarly organized experiment, López-Moreno et al. (2010b) also studied the influence of cubic $nCeO_2$ on seed germination and early growth of soybean plantlets. No statistically significant effects on germination were reported, with the exception of the highest treatment (2000 mg/L). However, root growth increased by 75% in treated plants compared to control. In addition to this study, certain groups observed the potential of $nCeO_2$ genotoxicity. The random amplified polymorphic DNA technique was applied to root cell samples to verify the presence of genetic aberrations following exposure to $nCeO_2$. The results confirmed that higher concentrations of $nCeO_2$ are able to induce genotoxic effects on soybean plants. Seed germination is commonly used as an endpoint in phytotoxicity studies. Andersen et al. (2016)

TABLE 14.2 Summary of the Effects of $nCeO_2$ Exposure in Adult Plants

Plant Species	Particle Size (nm)	Substrate	$nCeO_2$ Treatment	Plant Response	References
Glycine max	8	Organic farm soil	100, 500, 1000 mg/kg	Decrease of plant growth and yield. Shutdown of nitrogen fixation at high $nCeO_2$ concentration	Priester et al. (2012)
G. max	8	Organic farm soil	1000 mg/kg	Edible fractions in treated plants contain CeO_2 compounds	Hernández-Viezcas et al. (2013)
Lycopersicum esculentum	Rods, 8 × 231	Miraclegro Potting soil	Uncoated and coated (citric acid) particles 0, 62.5, 125, 250, 500 mg/kg	Homeostasis of nutrient elements not affected. Coated particles reduced Ce uptake by roots. Unclear effects on fruit production	Barrios et al. (2015)
Lactuca sativa	10–22	Sand	0, 100, 500, 1000, 2000 mg/kg	Reduction in chlorophyll at 1000 and 2000 mg/kg. Decrease in biomass at 1000 and 2000 mg/kg. Altered antioxidant activity. Increase in nitrate-N in shoot	Zhang et al. (2017)
Brassica rapa	10–25	Loamy sand soil + vermiculite (1:1 v/v)	Increasing $nCeO_2$ concentrations during the crop cycle	Plant responses varied with the particle sizes and the growth stages of plants	Ma et al. (2016)
Cucurbita pepo *G. max* *L. sativa* *Zea mays*	25	Sandy loam soil Fine sandy loam soil (biochar amended)	0, 500, 1000, 2000 mg/kg	Biomass varied based on species and cerium concentration. Pigments and lipid peroxidation concentration varied based on plant species and cerium concentration	Servin et al. (2017)
L. esculentum	50–105	Natural soil and peat (1:4 v/v)	20 mg/kg	Changes in uptake of macronutrients and their translocation in plant fractions	Vittori Antisari et al. (2015)

report the appropriateness of this parameter to screen $nCeO_2$ toxicity and the same was evaluated on some important crops. In this study, seeds of *Allium cepa* L. (onion), *Avena sativa* L. (oat), *Brassica oleracea capitata* L. (cabbage), *C. sativus* L. (cucumber), *Daucus carota* L. (carrot), *Glycine max* L. (soybean), *Lactuca sativa* L. (lettuce), *Lolium perenne* L. (perennial ryegrass), *L. esculentum* L. (tomato), and *Z. mays* L. (corn) were exposed to 0–1000 μg/mL of $nCeO_2$. Germination and early plant development were observed. According to López-Moreno et al. (2010a,b), it was verified that $nCeO_2$ does not cause acute toxicity in the early stages of plant development. Moreover, root elongation was more sensitive than germination. It is likely that the coated seed structures were able to protect seed germination,

TABLE 14.3 Summary of the Effects of $nCeO_2$ on Food Quality

Species	Particle Size (nm)	Substrate	$nCeO_2$ Treatment	Plant Response	References
Oryza sativa	Rods, 8 × 231	Earthgro Potting soil	500 mg/kg	Modification of amino acid and fatty acid contents	Rico et al. (2013)
Triticum aestivum	Rods, 8 × 231	Earthgro Potting soil	0, 125, 250, 500 mg/L	Modification in amino acid composition and S and Mn concentration in grains	Rico et al. (2014)
T. aestivum	Rods, 8 × 231	Silt clay soil	0, 100, 400 mg/kg	Morphological modifications of root and leaf cell organelle. Lower chlorophyll concentration and increased catalase/superoxide dismutase at highest treatment. Increase of grain protein content	Du et al. (2015)
Phaseolus vulgaris	Rods, 8 × 67	Low organic matter soil; organic matter-enriched soil	0, 62.5, 125, 250, 500 mg/kg	Increase antioxidant enzyme activities at each concentration. Decrease of Mo and Fe. Increase of fruit fresh weight, B, and starch	Majumdar et al. (2015)
Zea mays	Rods, 8 × 67	Loamy sand soil	0, 400, 800 mg/kg	Undisturbed leaf chlorophyll. Carbon exchange rate not affected. Changes in nutrient allocation in kernels	Zhao et al. (2015)
Hordeum vulgare	Rods, 8 × 231	Miraclegro Potting soil	0, 125, 250, 500 mg/kg	Concentration of Ca, K, Zn, Mg, Cu, Al, Fe, P, and S in grains increased	Rico et al. (2015)
Lycopersicum esculentum	Rods, 8 × 231	Miraclegro Potting soil	0, 62.5, 125, 250, 500 mg/kg	Reduction in fruit dry weight. Decrease in reducing sugars and starch concentration. Reduction in B, Fe, Mn, Ca, and Al concentration	Barrios et al. (2017)
H. vulgare	<25	Clay soil	0, 500, 1000 mg/kg	Decrease in amylose content. Increase in amino acids and crude protein content. Decrease in K and S content	Pošćić et al. (2016)

whereas the $nCeO_2$ exposure of seedlings determined different consequences. In fact, root elongation was stimulated in onion and ryegrass and stunted in lettuce and tomato. A further confirmation of the fact that seed germination is not influenced by the presence of $nCeO_2$ in the culture medium was provided by Mattiello et al. (2015). In their experimental conditions the germination of *Hordeum vulgare* L. (barley) was not affected by $nCeO_2$ even at the highest level (2000 mg/L). Once more, the effects of the treatments were recorded at seedling stage. In particular, in this case seedlings exposed to the milder $nCeO_2$ concentration showed a reduction in root elongation compared to control.

TABLE 14.4 Summary of the Effects of $nCeO_2$ on Plant Multigenerations

Species	Particle Size (nm)	Substrate	$nCeO_2$ Treatment	Plant Response	References
Triticum aestivum	Rods, 8 × 231	Miraclegro Potting soil	0, 125, 500 mg/kg	Decrease of Mn, Ca, K, Mg, and P in kernels of plants treated in S0 and S1 with 125 mg/kg	Rico et al. (2017)
Lycopersicum esculentum	20	Hydroponics	10 mg/L	Reduction in biomass of second-generation plantlets. Decreased water transpiration. Increased reactive oxygen species concentration	Wang et al. (2013)
Brassica rapa	<25	Loamy sand soil + vermiculite (1:1 v/v)	0, 10, 100, 1000 mg/L	Slowdown of vegetative growth in second and third generation. Reduction in seeds per plant and increase in concentration of H_2O_2 in third generation	Ma et al. (2016)

FIGURE 14.1 Example of germination trial of *Hordeum vulgare* L. at different concentrations (0–2000 mg/L) of $nCeO_2$ (Mattiello et al., 2015).

14.2.2 Adult Plant Studies

To determine severity of effect, potential interactions of $nCeO_2$ and crops were investigated not only in the early stages of development but also along the entire life cycle. In this orientation, the first aspect investigated by the researchers was the plants' ability to take up $nCeO_2$. Then, several other traits of plant physiology and metabolism were also studied.

The accumulation study of $nCeO_2$ in plants was first documented by Priester et al. (2012) in *G. max* L. (soybean). Ce concentration in the root tissues was linearly correlated with soil $nCeO_2$ concentration and accumulated into the roots with uptake levels similar to those observed by López-Moreno et al. (2010b). They also reported evidence of great concern from an ecological point of view. Observations made by environmental scanning electron microscope showed that root nodules also acquire $nCeO_2$. Low levels of $nCeO_2$ did not significantly alter nitrogen fixation; however, plant growth was stunted. With medium and high $nCeO_2$ concentrations, nitrogen fixation was severely affected resulting in treated plants with a lower number of leaves compared with controls. In addition, crop yield was affected as well. Although the number of pods per plant and seeds per pod were not affected by the treatment, a statistically significant reduction in pod dry weight was recorded. In another study, Hernandez-Viezcas et al. (2013) investigated the uptake of $nCeO_2$ of soybean plants. In this case the Ce concentration in roots of treated plants was higher than those in controls. However, the results demonstrated a low root-to-shoot translocation rate of the $nCeO_2$ in the aerial plant biomass. And to demonstrate the possibility of direct uptake of $nCeO_2$, they investigated the Ce oxidation state in the plant tissues. The results demonstrate that about 79% and 88% of total Ce detected in root nodules and pods was present as $nCeO_2$.

The behavior in the environment and the ecotoxicology of doped or coated nanoparticles is even less well known than traditional ones. On the other hand, there is a growing use of these next-generation nanomaterials and therefore an increase in their flow to the biota is expected with unpredictable consequences. Taking this scenario into account, Barrios et al. (2015) compared the effects of $nCeO_2$ and citric acid-coated $nCeO_2$ on the growth and uptake of Ce and essential elements in *L. esculentum* (tomato). In addition, physiological as well as chlorophyll contents and the activity of catalase (CAT) and ascorbate peroxidase enzymes were measured in fully developed plants of *L. esculentum* (tomato). The treatments with coated and uncoated $nCeO_2$ did not affect the nutrient element uptake and plant metabolism. The release of ionic Ce was smaller in coated Ce than in uncoated. As a consequence, a lower Ce root uptake was observed. The influence of coated $nCeO_2$ on fruit production was not clearly demonstrated.

In studies by Vittori Antisari et al. (2015), tomato plants were used as a model, and they studied the influence of $nCeO_2$, nFe_3O_4, $nSnO_2$, $nTiO_2$, and metallic nAg, nCo, and nNi on plant growth, the possible uptake of nanoparticles, and the content of macronutrients in different plant fractions. $nCeO_2$ treatment did not affect plant growth; however, a decrease in dry plant biomass was observed. As far as the uptake of macronutrients is concerned, $nCeO_2$ seemed to cause an inhibition of root-to-shoot K translocation. Additionally, $nCeO_2$ did not allow Ca, Mg, and P to be translocated to leaves. According to CarvalhoBertoli et al. (2012), this evidence suggests a competitive inhibition between K and other cations, e.g., metal component of nanomaterials. Cerium did not translocate in the treated tomato, perhaps because of the size of the $nCeO_2$ employed in the experiment (50–105nm).

Since the potentially different impact of $nCeO_2$ and its bulk counterparts to plants is not so clear, Ma et al. (2016) studied the physiological and biochemical adjustments at different growth stages in *Brassica rapa* L. (turnip). The results indicated that bulk CeO_2 enhanced plant biomass, while $nCeO_2$ did not. While the bulk CeO_2 treatment resulted in significantly higher hydrogen peroxide (H_2O_2) in plant tissues at the vegetative stage, $nCeO_2$ led to significantly higher H_2O_2 in plant tissues at the floral stage. Superoxide dismutase (SOD) activity had a growth-stage-dependent response to different sizes of $nCeO_2$. CAT activity was not affected by either size of $nCeO_2$ throughout the turnip life cycle.

Finally, we can consider other studies by Zhang et al. (2017) and Servin et al. (2017), where Zhang et al. (2017) studied the $nCeO_2$ uptake by sand-cultured *L. sativa* L. (romaine lettuce) and demonstrated the dose-dependent accumulation of Ce in plant roots. In this report, electron microscopy observations confirmed the presence of $nCeO_2$ aggregates both in the root surface and within root tissues (in the intercellular space) and within root cells (vacuole). Once the relationship between $nCeO_2$ and plant uptake was confirmed, the authors studied the potential effects on plant metabolism and growth. In particular, the SOD and peroxidase activities and malondialdehyde contents were analyzed in roots and shoots tissues. At higher concentrations (1000–2000 mg/kg) $nCeO_2$ caused a decrease in chlorophyll content and in fresh and dry weight of plant fractions. The present study suggests $nCeO_2$ could induce significant oxidative damage, reduction of chlorophyll content, and decrease of biomass production to romaine lettuce. Moreover, romaine lettuce showed higher sensitivity to $nCeO_2$ in sand media than that in soil media. These results indicate that plants increased the antioxidant enzyme activities as a defense against oxidative stress. $nCeO_2$ at high concentrations induced higher oxidative stresses, which cannot be effectively attenuated by the antioxidant system. According to Rico et al. (2013), these results confirmed that oxidative stress was involved in the toxicity of $nCeO_2$ to romaine lettuce.

The interactions between engineered nanomaterials and soil are still limited. However, soils are known to contain or be amended with organic matter, humic substances, and biochar that can significantly interact with nanomaterials. In particular, experimental evidence by Yi et al. (2015) showed that biochar and $nCeO_2$ tend to aggregate because of the attraction of negatively charged sites on biochar and the positively charged surfaces of $nCeO_2$.

In connection with this, Servin et al. (2017) studied the effect of interactions between $nCeO_2$ and biochar on Ce accumulation, biomass, chlorophyll production, and lipid peroxidation in *Cucurbita pepo* (zucchini), *G. max* (soybean), *L. sativa* (lettuce), and *Z. mays* (corn). The plants were grown in two different soils amended with two different types of biochar. The results showed that the biochar type influenced significantly the interactions between $nCeO_2$ and plants, with relevant consequences on particle fate.

14.2.3 Crop Yield Quality

Since very few investigations reported the possible influence of nanomaterials on crop yield and food quality, these issues are far from being understood. In particular, knowledge of the implications on nanoparticles in the nutritional traits of food crops is still limited. In fact, only a few studies have been devoted to this important topic (Table 14.3). However, observations have dealt with the four most common cereals in the world: rice, wheat, corn, and barley (Fig. 14.2). Basic information about these reports is discussed in Table 14.3.

First evidence of the aforementioned factoid is the study of Rico et al. (2013) discussing the significant impacts of $nCeO_2$ on the nutritive value of rice. Three rice varieties with respectively high, medium, and low amylose content were grown in greenhouse conditions in $nCeO_2$-amended soil. The rice grains were harvested and analyzed with biochemical assays and spectroscopic techniques to study the effects of $nCeO_2$ on Ce accumulation, nutrient content, antioxidant property, and their nutritional quality. Another study carried out by the same research group was focused on *Triticum aestivum* L. (wheat). Also in this case, mature kernels collected from plants grown on a soil amended with $nCeO_2$ were analyzed for macro- and microelements, fatty acids, and amino acid content. Results showed that $nCeO_2$ altered all yield parameters, except for spike length and 100-grain weight. However, most important of all is that $nCeO_2$ modified S and Mn allocation in grains. In addition, $nCeO_2$ influenced the amino acid composition and modified the fatty acid content (linoleic and linolenic acid, respectively, decreased and increased) of the grains. Therefore this study showed clear demarcations of the potential of $nCeO_2$ to modify crop physiology and food quality (Rico et al., 2014).

FIGURE 14.2 Life cycle experiment. Plants of *Hordeum vulgare* L. growing in soil amended with 500 and 1000 mg nCeO$_2$/kg at heading stage (Marchiol et al., 2016).

Another study of the entire life cycle of *T. aestivum* L. (wheat) was carried out by Du et al. (2015) in the field using lysimeters. Evidence of physiological stress (increase of CAT and SOD activities) and changes on root and leaf cell microstructures (chromatin agglomeration in nuclei and size reduction of starch grains in endosperm) were reported. However, there were no effects on the final biomass and yield. Seed quality modifications with regard to size and number of starch granules in endosperm suggest that nCeO$_2$ could have serious implications on the nutritional value of wheat grains.

The peculiarity of previous studies is that they used the same type of nCeO$_2$ provided by the University of California Center for Environmental Implications of Nanotechnology. For the same reason these works are also associated with the study by Majumdar et al. (2015), which examined the effect of nCeO$_2$ exposure on *Phaseolus vulgaris* L. (kidney bean) productivity and seed quality. Since soil organic matter content is supposed to play a significant influence in the environmental fate of engineered nanoparticles by influencing their aggregation and surface charge (Schwabe et al., 2013), in this study two soils, with low and high organic matter content, respectively, were considered. Although the plants did not show toxicity symptoms, both protein and carbohydrate metabolism in nCeO$_2$-treated plants were negatively influenced. Therefore the reports of Majumdar et al. (2015) suggest that nCeO$_2$ may negatively affect the nutritional quality of kidney beans.

Zhao et al. (2015) carried out an elegant study that compared the effects of nCeO$_2$ and nZnO through the life cycle of *Z. mays* L. (corn). During the crop cycle, gas exchanges were monitored to investigate whether corn plants maintained normal photosynthesis rate under treatment. The accumulation of Ce and Zn in plant tissues, as well as the nutrient concentration and distribution in grains, was evaluated by in situ micro-X-ray fluorescence (μ-XRF). The corn yield (−38%) and nutritional quality of grains were influenced. Moreover, μ-XRF mapping showed that in $nCeO_2$-treated plants, Cu, K, Mn, and Zn were mainly localized at the insertion of kernels into cobs.

Although *H. vulgare* L. (barley) is among the most cultivated cereals in the world, very few studies have considered the effects of exposure to engineered nanomaterials on its nutritional properties. To the best of our knowledge the implications of $nCeO_2$ on its nutritional value were studied by Rico et al. (2015) and Pošćić et al. (2016), and so far Rico et al. (2015) studied the agronomic traits and yield components of barley plants grown in greenhouse conditions in $nCeO_2$-amended soil. Barley grains from plants treated with different levels of $nCeO_2$ showed both harmful and beneficial effects. At higher $nCeO_2$ levels, plants did not form grains, whereas at intermediate concentration, grains were formed and presented higher levels of amino acids than controls (Rico et al., 2015). A similar study was carried out by Pošćić et al. (2016). In this case the researchers evaluated the accumulation of Ce and mineral nutrients in kernels and studied some seed quality parameters (β-glucans, amino acid, and crude protein). However, results were in contrast with those by Rico et al. (2015). In general, Pošćić et al. (2016) demonstrated that barley kernels are negatively affected by $nCeO_2$, which did not affect β-glucans but reduced amylose content. Potassium, S, and B were negatively impacted by $nCeO_2$. With regard to amino acids, $nCeO_2$ promoted and increased the synthesis of lysine and proline.

In a previous study, Barrios et al. (2015) compared the effects of uncoated and coated $nCeO_2$ on the metabolism and growth of *L. esculentum* L. (tomato). Barrios et al. (2017) hypothesized that in those conditions, uncoated and coated $nCeO_2$ are able to influence in a different way the physiological and biochemical parameters of tomato fruits. So, the changes in macro- and micronutrient accumulation, starch and sugar content, and the lycopene content in tomato fruits of treated plants were studied. Results showed that fruit macromolecules were mainly affected by coated $nCeO_2$, while nutritional elements were affected by uncoated $nCeO_2$. However, all treatments altered in some way the nutritional quality of tomato fruit.

14.2.4 Multigenerational Studies

Upon analysis of the literature concerning the impact of $nCeO_2$ on plants, it is quite clear that most studies have been conducted in the short term in artificial conditions, far away from studies conducted in realistic conditions. The general conclusion from the studies was that the impact of $nCeO_2$ on plants depends upon plant species, exposure concentration, and particle dimensions.

While such studies have shed light on the basic relationships between nanomaterials and plants, it still remains unknown whether and how nanomaterials will affect plant development under realistic conditions, which will likely involve exposure over a much longer time but at relatively lower concentrations (Servin and White, 2016). The first of these issues was addressed by a number of authors who conducted multigenerational exposure to nanomaterials.

Wang et al. (2013) addressed the issue of the effects on successive generations of plants. The main objective of this paper was to verify (1) whether the treatment of *L. esculentum* L. (tomato) plants with nontoxic levels of $nCeO_2$ for the entire life cycle was able to influence the seed quality, and (2) possible responses in second-generation seedlings. It has been noted that the $nCeO_2$-treated second-generation seedlings were generally smaller and weaker. Interestingly, all treated second-generation seedlings developed excessive root hairs in the presence of $nCeO_2$. Ma et al. (2016) studied the effect of $nCeO_2$ on plant growth and stress metabolism in three generations of *B. rapa* L. (turnip). The results showed that $nCeO_2$-treated plants in the second and third generation displayed slower plant growth, smaller biomass, and produced fewer siliques. In addition, plant tissues in the later generations generally contained higher H_2O_2. The results demonstrated that the impact of $nCeO_2$ varied across generations; in fact, the second and third generation of turnips had higher oxidative stress than the first-generation plants. So, it is evident that only long-term evaluation extending several generations of plant growth is necessary to obtain a realistic understanding of the impact of engineered nanoparticles on crops.

Another report discussing the multigenerational exposure to nanomaterials was published by Rico et al. (2017). This research group reported the relationships between plants and nanomaterials. In this work they performed a soil microcosm study to assess the effects of multigenerational exposure of *T. aestivum* L. (wheat) to $nCeO_2$. Seeds from plants that were exposed to different levels of $nCeO_2$ in the first generation were in turn exposed to $nCeO_2$ to produce second-generation plants. Biometric parameters, elemental, isotopic, and XRF and X-ray absorption near edge structure data were collected on second-generation plants. Results showed that plants treated during the first generation only had reduced accumulation of Ce, Fe, and Mn in their roots. Plants treated in both generations produced grains that had lower Mn, Ca, K, Mg, and P relative to plants treated in the second generation only. Synchrotron XRF data revealed limited transformation of $nCeO_2$ with no evidence of plant uptake or accumulation. The findings demonstrated that first-generation exposure of wheat to $nCeO_2$ affects the physiology and nutrient profile of the second-generation plants. It was supposed that complex physiological processes that alter uptake and metabolism of $nCeO_2$ in wheat are involved. This clearly indicates that there is scope in this area of research with adequate options to generate multidimensional studies regarding plant and nanomaterial interaction over a period of generations.

14.3 CONCLUDING REMARKS

Nanotechnology is a technology that is in constant and rapid evolution with the potential to revolutionize everyday life. For this reason, it is considered responsible for the beginning of the so-called "Nano Revolution" (Ramsden, 2005). One of the consequences of the massive use of such materials is their introduction into both terrestrial and aquatic ecosystems.

In this chapter we reviewed papers reporting data about the consequences of the presence of $nCeO_2$ in the agroecosystem. Cerium uptake seems to be limited at root portions of the treated plants and translocation to the upper part is generally very limited except for some plant species grown in particular substrates enriched with organic matter substances.

Cerium uptake negatively affects root development. Conversely, the effect on the upper part changes based on plant species and the characteristics of growth substrate. In fact, in some experimental works the treated plants show a positive effect on shoot biomass, total leaf area, and height reached, but in other experiments these parameters are negatively affected. In addition, the grains and fruits produced by plants were analyzed and there was a general negative effect in terms of both number and biomass. Cerium uptake modifies the photosynthetic process. In fact, the experimental results showed that nCeO$_2$ determined the decrease the chlorophyll concentration in the majority of the studied crops. At the same time the results showed a positive effect on transpiration rate and stomatal conductance parameters. Another process of plant physiology influenced by nCeO$_2$ is stress metabolism. High nCeO$_2$ concentrations tend to cause an increase in H_2O_2 level in the plants; however, at low doses, the effect is the opposite. In plants having high levels of H_2O_2, electronic leakage and lipid peroxidation increase, as well as the activities of the principal enzymes involved in the antioxidant metabolism.

In conclusion, nCeO$_2$ has in general a negative effect on plants but there is variability related to the plant species, the substrate, and nCeO$_2$ characteristics along with great scope to explore.

References

Andersen, C.P., King, G., Plocher, M., Storm, M., Pokhrel, L.R., Johnson, M.G., Rygiewicz, P.T., 2016. Germination and early plant development of ten plant species exposed to titanium dioxide and cerium oxide nanoparticles. Env. Toxicol. Chem. 35, 2223–2229. http://dx.doi.org/10.1002/etc.3374.

Barrios, A.C., Medina-Velo, I.A., Zuverza-Mena, N., Dominguez, O.D., Peralta-Videa, J.R., Gardea-Torresdey, J.L., 2017. Nutritional quality assessment of tomato fruits after exposure to uncoated and citric acid coated cerium oxide nanoparticles, bulk cerium oxide, cerium acetate and citric acid. Plant Physiol. Biochem. 110, 100–107. http://dx.doi.org/10.1016/j.plaphy.2016.04.017.

Barrios, A.C., Rico, C.R., Trujillo-Reyes, J., Medina-Velo, I.A., Peralta-Videa, J.R., Gardea-Torresdey, J.L., 2015. Effects of uncoated and citric acid coated cerium oxide nanoparticles, bulk cerium oxide, cerium acetate and citric acid on tomato plants. Sci. Tot. Environ. 563–564. 956–64. http://dx.doi.org/10.1016/j.scitotenv.2015.11.143.

Borm, P.J., Berube, D., 2008. A tale of opportunities, uncertainties, and risks. Nano Today 3 (1), 56–59.

Carvalho Bertoli, C.A., Cannata, M.G., Carvalho, R., RibeiroBastos, A.R., Freitas, M.P., Dos Santos, A., 2012. *Lycopersicon esculentum* submitted to Cd-stressful conditions in nutrition solution: nutrient contents and translocation. Ecotoxicol. Environ. Saf. 86, 176–181. http://dx.doi.org/10.1016/j.ecoenv.2012.09.011.

Cervini-Silva, J., Fowle, D.A., Banfield, J., 2005. Biogenic dissolution of a soil cerium-phosphate mineral. Am. J. Sci. 305, 711–726. http://dx.doi.org/10.2475/ajs.305.6-8.711.

Chan, W.W.C., Chhowalla, M., Glotzer, S., Gogotsi, Y., Hafner, J.H., Hammond, P.T., Hersam, M.C., et al., 2016. Nanoscience and Nanotechnology Impacting Diverse Fields of Science, Engineering, and Medicine, pp. 10615–10617.

Corma, A., Chane-Ching, J.Y., Airiau, M., Martinez, C., 2004. Synthesis and catalytic properties of thermally and hydrothermally stable, high surface area SiO_2-CeO_2 mesostructured composite materials and their application for the removal of sulphur compounds from gasoline. J. Catal. 224, 441–448. http://dx.doi.org/10.1016/j.jcat.2004.03.004.

Cox, A., Venkatachalam, P., Sahi, S., Sharma, N., 2017. Reprint of: silver and titanium dioxide nanoparticle toxicity in plants: a review of current research. Plant Physiol. Biochem. 110, 33–49.

Dahle, J.T., Arai, Y., 2015. Environmental geochemistry of cerium: applications and toxicology of cerium oxide nanoparticles. Int. J. Environ. Res. Public Health. 12, 1253–1278. http://dx.doi.org/10.3390/ijerph120201253.

Du, W., Gardea-Torresdey, J.L., Ji, R., Yin, Y., Zhu, J., Peralta-Videa, J.R., Guo, H., 2015. Physiological and biochemical changes imposed by CeO_2 nanoparticles on wheat: a life cycle field study. Environ. Sci. Technol. 49, 11884–11893. http://dx.doi.org/10.1021/acs.est.5b03055.

Du, W., Tan, W., Peralta-Videa, J.R., Gardea-Torresdey, J.L., Ji, R., Yin, Y., Guo, H., 2017. Interaction of metal oxide nanoparticles with higher terrestrial plants: physiological and biochemical aspects. Plant Physiol. Biochem. 110, 210–225.

Hammond, C.R., 2000. The elements. In: Handbook of Chemistry and Physics, 81st edition. CRC Press.

Hedrick, J.B., 2004. Rare Earths. Minerals Yearbook. Vol I. Metals and Minerals. US Geological Survey, US Department of the Interior, Reston, VA, USA.

Hernandez-Viezcas, J.A., Castillo-Michel, H., Cooke Andrews, J., Cotte, M., Rico, C., Peralta-Videa, J.R., Ge, Y., Priester, J.H., Holden, P.A., Gardea-Torresdey, J.L., 2013. In situ synchrotron X-ray fluorescence mapping and speciation of CeO_2 and ZnO nanoparticles in soil cultivated soybean (Glycine max). ACS Nano. 7, 1415–1423. http://dx.doi.org/10.1021/nn305196q.

Kawazoe, Y., Meech, J.A., 2005. Welcome to IPPM'03-nanotechnology: do good things really come in small packages? In: Meech, J., Kawazoe, Y., Kumar, V., Maguire, J.F. (Eds.), Proc. 4th Int. Conf. Intelligent Processing and Manufacturing of Materials. DSEtech Publications, Inc., Lancaster, PA, USA, pp. 3–11.

Lin, W., Huang, Y., Zhou, X., Ma, Y., 2006. Int. J. Toxicol. 25, 451–457. http://dx.doi.org/10.1080/10915810600959543.

List of Manufactured Nanomaterials and List of Endpoints for Phase One of the OECD Testing Programme, 2008. Safety of Manufactured Nanomaterials. No. 6. ENV/JM/MONO/13/REV. Organisation for Economic Co-operation and Development, Paris.

López-Moreno, M.L., De La Rosa, G., Hernandez-Viezcas, J.A., Castillo-Michel, H., Botez, C.E., Peralta-Videa, J.R., Gardea-Torresdey, J.L., 2010b. Evidence of the differential biotransformation and genotoxicity of ZnO and CeO_2 nanoparticles on soybean (*Glycine max*) plants. Environ. Sci. Technol. 44, 7315–7320. http://dx.doi.org/10.1021/es903891g.

López-Moreno, M.L., De La Rosa, G., Hernandez-Viezcas, J.A., Peralta-Videa, J.R., Gardea-Torresdey, J.L., 2010a. X-ray absorption spectroscopy (XAS) corroboration of the uptake and storage of CeO_2 nanoparticles and assessment of their differential toxicity in four edible plant species. J. Agric. Food Chem. 58, 3689–3693. http://dx.doi.org/10.1021/jf904472e.

Ma, X., Wang, Q., Rossi, L., Ebbs, S.D., White, J.C., 2016. Multigenerational exposure to ceriumoxide nanoparticles: physiological and biochemical analysis reveals transmissible changes in rapid cycling *Brassica rapa*. NanoImpact. 1, 46–54. http://dx.doi.org/10.1016/j.impact.2016.04.001.

Ma, Y.H., Kuang, L.L., He, X., Bai, W., Ding, Y.Y., Zhang, Z.Y., Zhao, Y., Chai, Z., 2010. Effects of rare earth oxide nanoparticles on root elongation of plants. Chemosphere. 78, 273–279. http://dx.doi.org/10.1016/j.chemosphere.2009.10.050.

Majumdar, S., Almeida, I.C., Arigi, E.A., Choi, H., VerBerkmoes, N.C., Trujillo-Reyes, J., Flores-Margez, J.P., White, J.C., Peralta-Videa, J.R., Gardea-Torresdey, J.L., 2015. Environmental effects of nanoceria on seed production of common bean (*Phaseolus vulgaris*): a proteomic analysis. Environ. Sci. Technol. 49, 13283–13293. http://dx.doi.org/10.1021/acs.est.5b03452.

Marchiol, L., Mattiello, A., Pošćić, F., Fellet, G., Zavalloni, C., Carlino, E., Musetti, R., 2016. Changes in physiological and agronomical parameters of barley (*Hordeum vulgare*) exposed to cerium and titanium dioxide nanoparticles. Int. J. Env. Res Pub. Health. 13, 332. http://dx.doi.org/10.3390/ijerph13030332.

Mattiello, A., Pošćić, F., Musetti, R., Giordano, C., Vischi, M., Filippi, A., Bertolini, A., Marchiol, L., 2015. Evidences of genotoxicity and phytotoxicity in *Hordeum vulgare* exposed to CeO_2 and TiO_2 nanoparticles. Front. Plant Sci. 6, 1043. http://dx.doi.org/10.3389/fpls.2015.01043.

Miralles, P., Church, T.L., Harris, A.T., 2012. Toxicity, uptake and translocation of engineered nanomaterials in vascular plants. Environ. Sci. Technol. 46, 9224–9239. http://dx.doi.org/10.1021/es202995d.

Piccinno, F., Gottschalk, F., Seeger, S., Nowack, B., 2012. Industrial production quantities and uses of ten engineered nanomaterials in Europe and the world. J. Nanopart. Res. 14, 1109. http://dx.doi.org/10.1007/s11051-012-1109-9.

Pošćićć, F., Mattiello, A., Fellet, G., Miceli, F., Marchiol, L., 2016. Effects of cerium and titanium oxide nanoparticles in soil on the nutrient composition of barley (*Hordeumvulgare* L.) kernels. Int. J. Environ. Res. Public Health. 13, 577. http://dx.doi.org/10.3390/ijerph13060577.

Priester, J.H., Ge, Y., Mielke, R.E., Horst, A.M., Moritz, S.C., Espinosa, K., Gelb, J., Walker, S.L., Nisbet, R.M., An, Y.J., Schimel, J.P., Palmer, R.G., Hernandez-Viezcas, J.A., Zhao, L., Gardea-Torresdey, J.L., Holden, P.A., 2012. Soybean susceptibility to manufactured nanomaterials with evidence for food quality and soil fertility interruption. Proc. Natl. Acad. Sci. U.S.A. 109, 2451–2456. http://dx.doi.org/10.1073/pnas.1205431109.

Ramsden, J.J., 2005. What is nanotechnology? Nanotechnol. Percept. 1, 3–17. http://dx.doi.org/10.4024/N03RA05/01.01.

Rico, C.M., Barrios, A.C., Tan, W., Rubenecia, R., Lee, S.C., Varela-Ramirez, A., Peralta-Videa, J.R., Gardea-Torresdey, J.L., 2015. Physiological and biochemical response of soil-grown barley (*Hordeumvulgare* L.) to cerium oxide nanoparticles. Environ. Sci. Pollut. Res. 22, 10551–10558. http://dx.doi.org/10.1007/s11356-015-4243-y.

Rico, C.M., Johnson, M.G., Matthew, M.A., Andersen, C.P., 2017. Intergenerational responses of wheat (*Triticumaestivum* L.) to cerium oxide nanoparticles exposure. Environ. Sci. 4, 700–711. http://dx.doi.org/10.1039/C7EN00057J.

Rico, C.M., Morales, M.I., Barrios, A.C., McCreary, R., Hong, J., Lee, W.Y., Nunez, J., Peralta-Videa, J.R., Gardea-Torresdey, J.L., 2013. Effect of cerium oxide nanoparticles on the quality of rice (*Oryza sativa* L.) grains. J. Agric. Food Chem. 61, 11278–11285. http://dx.doi.org/10.1021/jf404046v.

Rico, C.R., Lee, S.C., Rubenecia, R., Mukherjee, A., Hong, J., Peralta-Videa, J.R., Gardea-Torresdey, J.L., 2014. Cerium oxide nanoparticles impact yield and modify nutritional parameters in wheat (*Triticumaestivum* L.). J. Agric. Food Chem. 62, 9669–9675. http://dx.doi.org/10.1021/jf503526r.

Schwabe, F., Schulin, R., Limbach, L.K., Stark, W., Bürge, D., Nowack, B., 2013. Influence of two types of organic matter on interaction of CeO_2 nanoparticles with plants in hydroponic culture. Chemosphere. 91, 512–520. http://dx.doi.org/10.1016/j.chemosphere.2012.12.025.

Servin, A.D., De La Torre-Roche, R., Castillo-Michel, H., Pagano, L., Hawthorne, J., Musante, C., Pignatello, J., Uchimiya, M., White, J.C., 2017. Exposure of agricultural crops to nanoparticle CeO_2 in biochar-amended soil. Plant Physiol. Biochem. 110, 147–157. http://dx.doi.org/10.1016/j.plaphy.2016.06.003.

Servin, A.D., White, J.C., 2016. Nanotechnology in agriculture: Next steps for understanding engineered nanoparticle exposure and risk. NanoImpact. 1, 9–12. http://dx.doi.org/ 10.1016/j.impact.2015.12.002.

Shweta, Tripathi, D.K., Singh, S., Singh, S., Dubey, N.K., Chauhan, D.K., 2016. Impact of nanoparticles on photosynthesis: challenges and opportunities. Mater. Focus 5 (5), 405–411.

Siddiqi, K.S., Husen, A., 2017. Plant response to engineered metal oxide nanoparticles. Nanoscale Res. Lett. 12, 92. http://dx.doi.org/10.1186/s11671-017-1861-y.

Singh, S., Tripathi, D.K., Dubey, N.K., Chauhan, D.K., 2016. Effects of nano-materials on seed germination and seedling growth: striking the slight balance between the concepts and controversies. Mater. Focus 5 (3), 195–201.

Singh, S., Vishwakarma, K., Singh, S., Sharma, S., Dubey, N.K., Singh, V.K., Liu, S., Tripathi, D.K., Chauhan, D.K., 2017. Understanding the plant and nanoparticle interface at transcriptomic and proteomic level: a concentric overview. Plant Gene. http://dx.doi.org/10.1016/j.plgene.2017.03.006.

Soni, S. (Ed.), 2014. Handbook of Research on Diverse Applications of Nanotechnology in Biomedicine, Chemistry, and Engineering. IGI Global.

Sun, T.Y., Gottschalk, F., Hungerbuhler, K., Nowack, B., 2014. Comprehensive probabilistic modelling of environmental emissions of engineered nanomaterials. Environ. Pollut. 185, 69–76. http://dx.doi.org/10.1016/j.envpol.2013.10.004.

Tripathi, D.K., Singh, S., Singh, S., Pandey, R., Singh, V.P., Sharma, N.C., Prasad, S.M., Dubey, N.K., Chauhan, D.K., 2017a. An overview on manufactured nanoparticles in plants: uptake, translocation, accumulation and phytotoxicity. Plant Physiol. Biochem. 110, 2–12.

Tripathi, D.K., Mishra, R.K., Singh, S., Singh, S., Vishwakarma, K., Sharma, S., Singh, V.P., Singh, P.K., Prasad, S.M., Dubey, N.K., Pandey, A.C., 2017b. Nitric oxide ameliorates zinc oxide nanoparticles phytotoxicity in wheat seedlings: implication of the ascorbate–glutathione cycle. Front. Plant Sci. 8.

Tripathi, D.K., Singh, S., Singh, S., Srivastava, P.K., Singh, V.P., Singh, S., Prasad, S.M., Singh, P.K., Dubey, N.K., Pandey, A.C., Chauhan, D.K., 2017c. Nitric oxide alleviates silver nanoparticles (AgNps)-induced phytotoxicity in *Pisum sativum* seedlings. Plant Physiol. Biochem. 110, 167–177.

Tripathi, D.K., Tripathi, A., Shweta, S.S., Singh, Y., Vishwakarma, K., Yadav, G., Sharma, S., Singh, V.K., Mishra, R.K., Upadhyay, R.G., Dubey, N.K., 2017d. Uptake, accumulation and toxicity of silver nanoparticle in autotrophic plants, and heterotrophic microbes: a concentric review. Front. Microbiol. 8.

Tripathi, D.K., Singh, S., Singh, V.P., Prasad, S.M., Dubey, N.K., Chauhan, D.K., 2017e. Silicon nanoparticles more effectively alleviated UV-B stress than silicon in wheat (*Triticum aestivum*) seedlings. Plant Physiol. Biochem. 110, 70–81.

Tripathi, D.K., Singh, S., Singh, V.P., Prasad, S.M., Chauhan, D.K., Dubey, N.K., 2016. Silicon nanoparticles more efficiently alleviate arsenate toxicity than silicon in maize cultiver and hybrid differing in arsenate tolerance. Front. Environ. Sci. 4, 46.

Tripathi, D.K., Singh, V.P., Prasad, S.M., Chauhan, D.K., Dubey, N.K., 2015. Silicon nanoparticles (SiNp) alleviate chromium (VI) phytotoxicity in *Pisum sativum* (L.) seedlings. Plant Physiol. Biochem. 96, 189–198.

Vittori Antisari, L., Carbone, S., Gatti, A., Vianello, G., Nannipieri, P., 2015. Uptake and translocation of metals and nutrients in tomato grown in soil polluted with metal oxide (CeO_2, Fe_3O_4, SnO_2, TiO_2)or metallic (Ag, Co, Ni) engineered nanoparticles. Environ. Sci. Pollut. Res. 22, 1841–1853. http://dx.doi.org/10.1007/s11356-014-3509-0.

Wang, Q., Ebbs, S.D., Chenc, Y., Ma, X., 2013. Trans-generational impact of cerium oxide nanoparticles on tomato plants. Metallomics. 5, 753. http://dx.doi.org/10.1039/c3mt00033h.

Yi, P., Pignatello, J.J., Uchimiya, M., White, J.C., 2015. Heteroaggregation of cerium oxide nanoparticles and nanoparticles of pyrolyzed biomass. Environ. Sci. Technol. 49, 13294–13303. http://dx.doi.org/10.1021/acs.est.5b03541.

Zhang, P., Ma, Y., Liu, S., Wang, G., Zhang, J., He, X., Zhang, J., Rui, Y., Zhang, Z., 2017. Phytotoxicity, uptake and transformation of nano-CeO_2 in sand cultured romaine lettuce. Environ. Pollut. 220, 1400–1408. http://dx.doi.org/10.1016/j.envpol.2016.10.094.

Zhao, L., Sun, Y., Hernandez-Viezcas, J.A., Hong, J., Majumdar, S., Niu, G., Duarte-Gardea, M., Peralta-Videa, J.R., Gardea-Torresdey, J.L., 2015. Monitoring the environmental effects of CeO2 and ZnO nanoparticles through the life cycle of Corn (*Zea mays*) plants and in situ μ-XRF mapping of nutrients in kernels. Environ. Sci. Technol. 49, 2921–2928. http://dx.doi.org/10.1021/es5060226.

CHAPTER

15

Plant and Nanoparticle Interface at the Molecular Level: An Integrated Overview

Gausiya Bashri, Parul Parihar, Rachana Singh, Anuradha Patel, Sheo M. Prasad
University of Allahabad, Allahabad, India

15.1 INTRODUCTION

Technological advancement is needed to help feed and shelter the ever-growing world population. Among different technologies, nanotechnology is the one that is growing the fastest and could soon become a trillion-dollar industry (Nel et al., 2006). Because of excess use of nanomaterials they have been discharged into the environment. Nanomaterials are those materials that have sizes between 1 and 100 nm and fall in a transitional zone between individual molecules and corresponding bulk materials, and therefore have very large specific surface area, high surface energy, and quantum confinement (Taylor and Walton, 1993; ASTM, 2006). Because of these unique properties, nanoparticles (NPs) have been widely used in everyday products such as catalysts, semiconductors, cosmetics, drug carriers, etc. (Nel et al., 2006). NPs can be categorized into five types, i.e., carbonaceous NPs, metal oxides, quantum dots, zero-valent metals, and nanopolymers (Christian et al., 2008; You and Reid, 2008). The excessive use of NPs has created a new problem of nanotoxicity (Gottschalk et al., 2015; Tolaymat et al., 2015). NPs mainly enter the soil through agricultural application and atmospheric deposition, then they start to accumulate in the soil. Soil is an important component of the environment and accumulation of NPs in the soil can affect all kinds of organisms positively and negatively. Plants are the producers of the ecosystem and higher concentrations of NPs in the soil can affect the plants (McKee and Filser, 2016; Singh et al., 2017). Accumulation of NPs depends upon their size: small-sized NPs enter plant cells easily through the stoma of the leaf, cross the cell wall, and reach the plasma membrane. NPs can alter the physiology of plants because of their unique properties (Hong et al., 2014). Apart from inhibiting physiological functioning, NPs can stimulate the production of reactive oxygen species (ROS), and when the level of ROS exceeds the

Nanomaterials in Plants, Algae, and Microorganisms
http://dx.doi.org/10.1016/B978-0-12-811487-2.00015-3

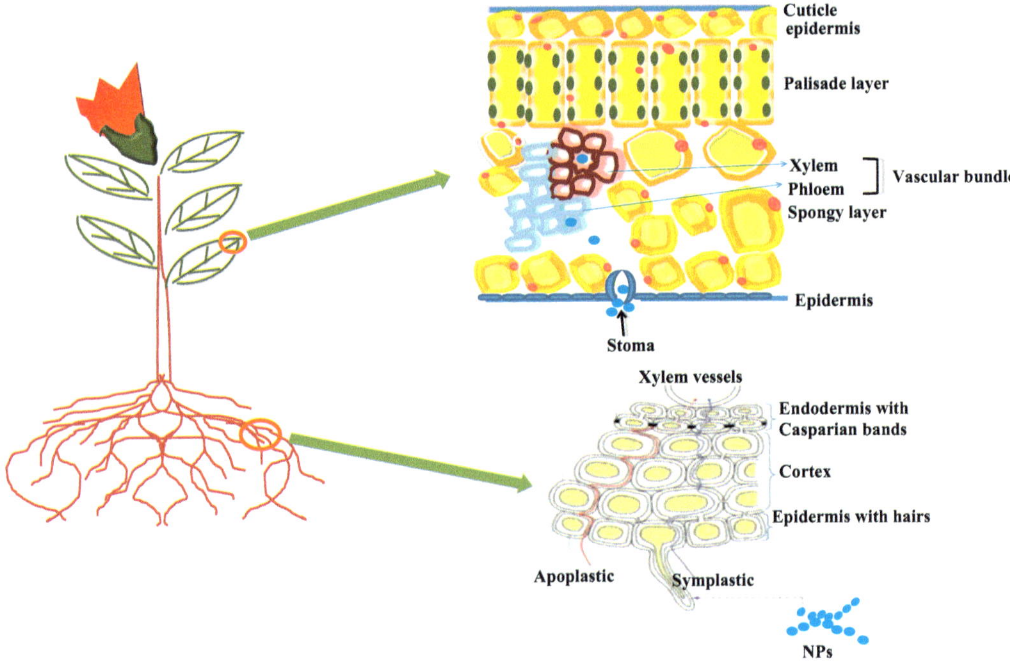

FIGURE 15.1 Showing pathways for uptake and translocation of nanoparticles (NPs) in plants. *Modified from Tripathi, D.K., Singh, S., Singh, S., Pandey, R., Singh, V.P., Sharma, N.C., Prasad, S.M., Dubey, N.K., Chauhan, D.K., 2017a. An overview on manufactured nanoparticles in plants: uptake, translocation, accumulation and phytotoxicity. Plant Physiol. Biochem. 110, 2–12.*

threshold level, a state of "oxidative stress" arises. The subsequent production of large amounts of ROS damages the membrane system by initiating radical reactions and aggravating the formation of more radicals (Montillet et al., 2005; Singh et al., 2017; Tripathi et al., 2017a,b,c,d,e). The formation of ROS radicals directly damages the DNA, proteins, and biomolecules (Sharma et al., 2012). Damage to the membrane caused by radicals initiates oxidation of the aldehydic group and thereby enhances the formation of lipid peroxides (Tanou et al., 2009; Halliwell and Gutteridge, 2015). Studies performed on plants exposed to NP stress suggested that these stimulate ROS production and in turn damage the membrane, causing loss to the plant system (Tolaymat et al., 2015; Yang et al., 2017). Therefore this chapter is aimed to discuss the uptake and translocation mechanisms of NPs, their impacts on the growth and biochemistry as well as positive/or negative modifications at genomic, transcriptomic, and proteomic levels in the plants (Figs. 15.1 and 15.2).

15.2 UPTAKE AND TRANSLOCATION OF NPs IN PLANTS

NPs enter the plant's system either through the stomata or roots, or sometimes they are absorbed through leaf epidermal layers. Entry through the stomata is size specific because it depends on the pore size, therefore the size of NPs is quite small. In the case of roots, NPs enter cells via the symplastic or apoplastic pathway and reach different parts of the plant's system. NPs enter the cell wall and membrane by penetrating the epidermis of the root, from

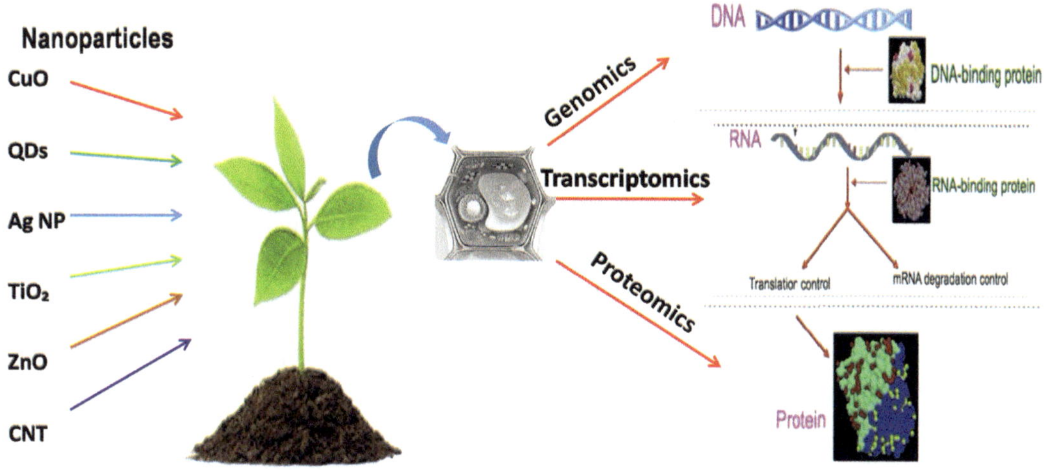

FIGURE 15.2 Showing different nanoparticles (NPs) and approaches to study their toxicity.

where they are translocated to the vascular system symplastically. Movement through the cell wall/membrane is dependent on pore size (Ma et al., 2010; Rico et al., 2011). Apart from the symplastic pathway, NPs also reach the vascular system via the apoplastic pathway. After reaching the stellar system, NPs are translocated into the leaves where they interact with the important components of the chloroplast, thereby disturbing the physiological processes of the plant (Fig. 15.1). The following is a brief discussion on how NPs enter the plant's system.

15.2.1 Leaf

There are two entry points in leaves: one is via the stomata and the other is absorption through the epidermis of the leaf. A study by Eichert et al. (2008) using confocal microscopy in *Vicia faba* showed that entry of NPs through the stomata is size specific; NPs of 43 nm penetrated the cell system, while those of 1.1 μm did not. Another study using mass spectrometry and electronic and confocal microscopy in leaves of *Zea mays* suggested that cerium NPs having size 37 nm entered the plant system but were not translocated to the stem; however, this study could not confirm whether entry was via the leaf epidermis or through the stomata (Birbaum et al., 2010). Likewise, other studies performed with electron and X-ray fluorescence microscopy in *Arabidopsis* seedlings suggested that NPs could penetrate the epidermis and reach the palisade layer by entering via the stomata as well as via absorption by endocytic vessels (Kurepa et al., 2010; Wang et al., 2013). Apart from studies conducted for stomatal entry, some studies were also performed to investigate entry through the epidermis of the leaf. Work performed by Taran et al. (2014), using Fe, Zn, and Mn (nonionic colloidal solution) NPs, showed that these are absorbed by the epidermis of leaves and are thereafter translocated. Another study by Larue et al. (2014) conducted on *Lactuga* using Ag NPs as well as bulk salt (both having hydrodynamic diameter 47.9 nm) showed entry of both forms via absorption through the cuticle as well as through the stomata and thereafter translocation to the vascular system. In addition, the study also suggested that Ag^+ binds with thiol groups and was interconverted into Ag NPs.

15.2.2 Roots

Many researchers studied the maize, spinach, rapeseed, and desert plants (Taran et al., 2014; Zheng et al., 2005; Răcuciu and Creangă, 2009; Stampoulis et al., 2009; De la Rosa et al., 2011; Pokhrel and Dubey, 2013; Kouhi et al., 2014) and showed that NPs could penetrate via the root epidermis and thereafter translocate to stems and leaves (Rico et al., 2011). In a study by Zhu et al. (2008), entry of iron oxide NPs through the epidermis of the root was noticed, and further translocation to shoots and leaves was also seen in pumpkin plants grown in aqueous medium. In another study performed on *Arabidopsis thaliana*, Ti nanoconjugates (size 2.8 nm) were found to penetrate inside root cells and were translocated to vacuoles as well as the nucleus (Kurepa et al., 2010). Some studies have been conducted in soil systems as well, and transmission electron microscopy showed accumulation and adherence of Ti NPs in the cell wall of roots and further translocated particles in the cortex and up to the vacuoles (Du et al., 2011). Moreover, several experiments employing inductively coupled plasma optical emission spectrometry, X-ray fluorescence, and confocal microscopy have been performed and these studies suggested that NPs made their entry via roots and were further translocated to shoots using the apoplastic pathway (Zhao et al., 2012a,b, 2014). Studies conducted in soil systems also suggested that mobility of NPs is affected by organic substances in the soil.

15.3 EFFECTS OF NANOPARTICLES ON PLANTS

NPs have unique properties as mentioned earlier, thats why they can alter the growth, physiology, and biochemistry of the plant negatively or positively on the basis of their concentration. In this section we shall address phytotoxicity induced by NPs on plants. The extensive use of nanotechnology has increased the concentration of NPs in the environment, which affects plant growth and physiology (Shweta et al., 2016; Singh et al., 2016, 2017; Tripathi et al., 2015, 2016, 2017a,b,c,d,e). Lin and Xing (2007) reported inhibition in seed germination and root growth with the treatment of ZnO NPs at a concentration of 2000 mg/L in the six higher plant species (radish, rape, ryegrass, lettuce, corn, and cucumber). Furthermore, Lin and Xing (2008) found decreased biomass of *Lolium perenne* (ryegrass) on treatment with ZnO NPs. Similarly, SiO_2 NPs inhibited the growth of *Bacillus thuringiensis* (*Bt*)-transgenic cotton (Le Van et al., 2014). Shaw and Hossain (2013) found that CuO NPs decreased the germination rate and biomass of rice seedlings, and root length and fresh weights of *Arabidopsis* seedlings. Both Stampoulis et al. (2009) and Wang et al. (2012) found that CuO NPs suppressed root elongation of zucchini and maize. Treatment of CuO NPs inhibited growth and structure of the root in wheat plants (Dimkpa et al., 2012; Tang et al., 2016). The rare earth oxide NPs (CeO_2, La_2O_3, Gd_2O_3, and Yb_2O_3) had a negative effect on plant growth of radish, tomato, rape, lettuce, wheat, cabbage, cucumber, and corn at high concentrations (Ma et al., 2010; López-Moreno et al., 2010). The US Environmental Protection Agency has set an indicator of phytotoxicity to seed germination and root elongation, and there are many reports that confirm nanotoxicity (Stampoulis et al., 2009). Furthermore, several NPs such as multi-walled carbon nanotubes, single-walled carbon nanotubes, ZnO NPs, Ag NPs, and Fe NPs negatively inhibited seed germination, shoot length, biomass, and gene expression of soybean, maize, wheat, and ryegrass (Lin and Xing, 2007, 2008; Dimkpa et al., 2012; El-Temsah and Joner, 2012; Riahi-Madvar et al., 2012; Yan et al., 2013; Ghosh et al., 2015), and many of them are presented in Table 15.1.

TABLE 15.1 Impact of Nanoparticles on Growth and Physiology of Plant

Nanomaterial	Size/Dose	Model System	Response	References
ZnO	2000 mg/L	Radish, rape, ryegrass, lettuce, corn, and cucumber	Inhibition in seed germination and root growth	Lin and Xing (2007)
ZnO	2000 mg/L	Ryegrass	Reduced biomass accumulation	Lin and Xing (2008)
ZnO	44.46 nm; 0, 1, 5, 100, 1000, and 2000 mg/L	*Fagopyrum esculentum*	ZnO at higher concentration increased reactive oxygen species (ROS) generation and reduced glutathione level and catalase activity, which would be a predictive biomarker of nanotoxicity	Lee et al. (2013)
Ag NPs: triangular	47 nm, spherical 8 nm, decahedral (45 nm)	*Arabidopsis thaliana*	Ag NPs increased ROS accumulation	Syu et al. (2014)
ZnO TiO_2	<150 nm; 100 ppm	*A. thaliana*	Upregulated the genes of both abiotic (oxidative, salt, water deprivation) and biotic (wounding and defense to pathogens)	Landa et al. (2012)
Ag NPs or $AgNO_3$	0.1, 1, 10, 20, 100 ppm	*Eruca sativa* (common name: rocket)	Alteration of some proteins related to the endoplasmic reticulum and vacuole indicating these two organelles as targets of Ag NP action	Vannini et al. (2013)
Ag NPs	30, 60 ppm	Rice (*Oryza sativa* L. cv. IR651)	Increased proteins related to oxidative stress, Ca^{2+} regulation signaling, transcription, protein degradation, cell wall synthesis, cell division, and apoptosis	Mirzajani et al. (2014)
Ag NPs	20 nm; 5 ppm	*A. thaliana*	Upregulated genes primarily associated with metals and oxidative stress, while downregulated genes linked to biotic and hormonal stimuli	Kaveh et al. (2013)
Ag, TiO_2 ZnO, QDs	>20 nm; 1–10 ppm	*Chlamydomonas reinhardtii*	Genes associated with photosynthesis were markedly decreased on exposure to TiO_2	Simon et al. (2013)
Nd_2O_3	30–45 nm; 5–100 mg/L	Pumpkin	SOD and POD activity increased, APX and CAT decreased under 100 mg/L Nd_2O_3 NPs and bulk exposure	Chen et al. (2016)
CeO_2	16.5 ± 6.8 nm; 0–2000 mg/L	Romaine lettuce	Oxidative stress. Increased nitrate-N level in shoots and POD was increased but SOD was decreased under 2000 mg/L CeO_2 exposure	Zhang et al. (2017)

15.4 MECHANISM OF PHYTOTOXICITY IN PLANTS GENERATED BY NPs

Phytotoxicity induced by NPs enhances the generation of ROS such as superoxide radical ($O_2^{\%-}$)· hydroxyl radical (•OH), hydrogen peroxide (H_2O_2), and singlet oxygen ($^1O_2$). ROS are produced in the chloroplast and mitochondria by the disruption of electron flow. This disruption in electron flow leaks the electron from the electron transport chain and because of the aerobic environment, oxygen takes this leaked electron and generates the superoxide radical ($O_2^{\%-}$)· $O_2^{\%-}$ converts into H_2O_2 by the action of enzyme superoxide dismutase (SOD). Although ROS at a low level work as a signaling molecule, excess ROS would cause adverse effects called oxidative stress, which emerges when the ROS level exceeds the defense mechanisms and is able to pose a threat to cells by inducing DNA damage, protein oxidation, and lipid peroxidation, which result in electrolyte leakage, membrane damage, and finally cell death (Meriga et al., 2004; Gill and Tuteja, 2010; Sharma et al., 2012). Among all ROS, hydroxyl radical (•OH) is the most reactive, and is generated by Fenton reaction from the H_2O_2 through a transition metal-catalyzed reaction and it has a single unpaired electron, therefore it could interact with all biological molecules and cause subsequent cellular damage such as lipid peroxidation, protein damage, membrane destruction (Foyer et al., 1997; Freinbichler et al., 2011), and ultimately premature cell death. The peroxidation of lipid creates membrane leakage because of damage in the plasma membrane basic structure (Halliwell and Gutteridge, 2015). We can say that NPs may increase the generation of ROS and enhance the rate of lipid peroxidation. Lin and Xing (2008) worked on ZnO NPs in ryegrass, and proposed that ROS formation and lipid peroxidation cause phytotoxicity. Tarasenko et al. (2012) used the ROS-sensitive dye 3,3′-diaminobenzidine to find the accumulation of H_2O_2 in plant roots treated with CeO_2 and La_2O_3 NPs, and showed that the insoluble deep brown-colored product was produced under the treatment of NPs. Likewise, Oukarroum et al. (2013) reported that Ag NPs up to 10mg/mL treatment caused generation of ROS, which decreased the viable cell numbers.

To combat ROS-induced oxidative stress, plants are equipped with an antioxidant system composed of enzymatic [SOD, catalase (CAT), guaiacol peroxidase, ascorbate peroxidase (APX), glutathione reductase (GR), monodehydroascorbate reductase, dehydroascorbate reductase (DHAR), glutathione *S*-transferase (GST), and glutathione peroxidase (GPx)] and nonenzymatic [(ascorbate (ASC), glutathione (GSH), proline, cysteine, and non protein thiols (NP-SHs)] components (Bashri and Prasad, 2016). Yasur and Rani (2013) reported increased ROS production accompanied by enhanced activity of peroxidase (POD) and SOD in castor bean (*Ricinus communis*) seeds treated with Ag NPs. Among the enzymatic antioxidants, SOD is the first line of defense, which dismutases $O_2^{\%-}$ into less toxic H_2O_2. SOD contains three isozymes including iron SOD, manganese SOD, and copper/zinc SOD (Fridovich, 1989). Faisal et al. (2013) observed that the impact of NiO NPs at 0–1000mg/L increased SOD at low-level treatment, while at high level SOD activity was inhibited. Similarly, Ma et al. (2016) also reported an increase in SOD activity at low levels of CeO_2 NP exposure, while at high-level treatment there is no change in enzyme activity. Rajeshwari et al. (2015) also found similar results using 1 and 100mg/L Al_2O_3 NPs exposed on onion root. Furthermore, CAT is also an important enzyme that converts H_2O_2 into O_2 and H_2O and plays a role in ROS detoxification under stress (Garg and Manchanda, 2009). APX also scavenges H_2O_2 but at the expense of ascorbate and plays a central role in regulating the ascorbate–glutathione (ASC–GSH) cycle, which comprises GR, APX, and DHAR that control the excess levels of ROS in plants. Likewise, Wei and Wang (2013) reported increased activity of CAT on treatment with Fe_3O_4 NPs, Co_3O_4 NPs, and CeO_2 NPs, GPX on treatment with Au NPs, MnO_2 NPs, Fe_3O_4 NPs, CuO NPs, Co_3O_4 NPs, and CeO_2 NPs, and SOD on treatment with Pt NPs and CeO_2 NPs. Faisal et al. (2013) reported a dose-dependent increase in SOD and CAT activity in tomato by the treatment of NiO NPs. Furthermore, Majumdar et al. (2014) found that CeO_2 NP exposure disturbed the antioxidant defense system and generated phytotoxicity in *Phaseolus vulgaris*. Ma et al. (2016) found that the activities of CAT, SOD, POD, and APX were significantly ameliorated under CeO_2 NP exposure in *A. thaliana*, while elevations were found only for POD and SOD. Similarly, Chen et al. (2016) studied the impact of Nd_2O_3 NPs on pumpkin, which decreased growth and nutrient levels (K, Ca, Mg, and S) and activity of CAT and APX, and increased the activities of SOD. Similar to enzymatic antioxidants, nonenzymatic antioxidants such as ASC and GSH take part in the detoxification of ROS. Shaw and Hossain (2013) found consistent increases in ASC on treatment with CuO NPs. GSH is another important metabolite in the thiol group, which plays a role in binding metals and pesticides. GSH also helps in the regeneration of ASC through the ASC–GSH cycle (Foyer and Noctor, 2003). Published studies on metal and metal-based NP phytotoxicity suggest that NPs can induce oxidative stress in many plant species through the generation of ROS, which is balanced with the defense system (Dimkpa et al., 2012; Shaw and Hossain, 2013; Cui et al., 2014; Zhang et al., 2015, 2017). There has been rapid progress in the study of NP phytotoxicity in recent years; however, there are many gaps and uncertainties in our knowledge of ROS-dependent damage and its effects on plants. Therefore there is a need to study the complete mechanism of nanotoxicity in plants through enzymatic and nonenzymatic antioxidant defense systems.

15.5 EFFECT OF NPs ON GENOMICS

Genomics deals with changes at gene or DNA level in any organism. It plays an important role in the study of stress physiology for understanding the mechanisms of toxicity through high-throughput methods such as cDNA microarrays or quantitative real-time polymerase chain reaction (Xu et al., 2011). As discussed earlier, NPs negatively affect plants, and Table 15.2 has a brief account of nanotoxicity at the genomics level in plants. Gene expression analyses are performed to test the changes at the genetic level for any morphological and/or physiological pathways and modes of action of any compound (Ankley et al., 2006; Dietz and Herth, 2011). Gene expression analysis helps to identify sensitive or tolerant genes for particular stresses, which can help in the development of transgenic plants for particular stresses (Merrick and Bruno, 2004; Thomas et al., 2011). NPs showed changes in gene expression at very low dose, which may help to investigate the cellular impact of chronic toxicity associated with NPs (Poma and Di Giorgio, 2008; Poynton and Vulpe, 2009). Therefore genomics analysis is very useful for the correlation of any morphological/physiological changes to the genetic level of the plants and also provides the mechanism of toxicity of NPs. Thus genomics analysis of NP-mediated toxicity is associated with disruption of the basic processes of electron transport chain signaling, which ultimately leads to the impairment of the cell cycle of the organism, as shown in Fig. 15.3.

TABLE 15.2 Genome Analysis of Nanoparticle-Mediated Toxicity

Nanomaterial	Size/Dose	Model System	Effect on the Cellular System	References
Ag NPs	60nm	*Vicia faba*	Impairment in mitosis causing chromosomal aberration and micronucleus induction, which caused genotoxicity	Patlolla et al. (2012)
Ag NPs	10nm	*Triticum aestivum*	Enhancement in oxidative stress that resulted in accumulation of oxidized glutathione (GSSG), and metallothionein encoding gene expression was enhanced	Dimkpa et al. (2013)
Ag NPs	20nm	*Arabidopsis*	Enhancement in the expression of genes associated with sulfur assimilation, glutathione *S*-transferase, and glutathione reductase	Nair and Chung (2014a)
Ag NPs	20nm	*Arabidopsis*	Variable expression of genes; some were upregulated, while others were downregulated; downregulation of DNA mismatch repair protein (MSH), while upregulation of proliferating cell nuclear antigen (PCNA)	Nair and Chung (2014b)
Ag NPs	20nm	*Arabidopsis*	Up- and downregulation of genes associated with protein family domain (PFAM) and interpro protein	Kohan-Baghkheirati and Geisler-Lee (2015)
Ag NPs	45–47nm	*Arabidopsis*	Accumulation of Cu/Zn superoxide dismutase (CSD2), cell-division-cycle kinase 2 (CDC2), protochlorophyllide oxidoreductase (POR), and fructose-1, 6-bisphosphate aldolase (FBA). Enhancement in expression of genes associated with indoleacetic acid protein 8 (IAA8), 9-cis-epoxycarotenoid dioxygenase (NCED3), and dehydration-responsive RD22, while decline in expression of aminocyclopropane-1-carboxylic acid (ACC)-derivatives, ACC synthase 7 (ACS7), and ACC oxidase 2 (ACO_2)	Syu et al. (2014)
Ag NPs, ZnO, and TiO_2	–	*Medicago*	Induced shift in expression profiles of genes associated with biological pathways such as nitrogen metabolism, nodulation, metal homeostasis, and stress responses	Chen et al. (2015)
Ag–silica hybrid complex	30nm	*Arabidopsis*	Upregulation of pathogenesis related (PR) genes, which is implicated in systemic acquired resistance (SAR)	Chu et al. (2012)
Al_2O_3	Not specified	*Nicotiana tabacum*	Enhancement in the expression of miRNA involved in mediating stress responses	Burklew et al. (2012)
CuO	Not specified	*Brassica juncea*	Upregulation in genes associated with CuZn superoxide dismutase (CuZnSOD), while no significant change in expression of catalase (CAT) and ascorbate peroxidase (APX) genes	Nair and Chung (2015)

TABLE 15.2 Genome Analysis of Nanoparticle-Mediated Toxicity—cont'd

Nanomaterial	Size/Dose	Model System	Effect on the Cellular System	References
Multiwalled carbon nanotubes (MWCNTs)	20 nm	*Nicotiana tabacum*	Increment in expression of genes associated with aquaporin, cell division, and cell wall extension	Khodakovskaya et al. (2012)
MWCNTs	15–40 nm	*Glycine max, Hordeum vulgare, Zea mays*	Increase in expression of gene encoding water channel proteins	Lahiani et al. (2013)
Single-walled carbon nanotubes (SWCNTs)	1–2 nm	*Arabidopsis, Oryza sativa*	Enhanced cell aggregation and chromatin condensation, negative impact on protoplast leading to oxidative stress	Shen et al. (2010)
SWCNTs	1–2 nm	*Z. mays*	Variable expression of genes, such as genes associated with increase in seminal root;epigenetic modification enzymes were enhanced, while those of root hair were reduced	Yan et al. (2013)
SWCNTs	50–100 nm	*Hordeum vulgare, Z. mays, O. sativa, G. max, Panicum virgatum, Solanum lycopersicum, N. tabacum*	Genes associated with stress responses, cellular responses, and metabolic processes were variably expressed	Lahiani et al. (2015)
TiO_2	25 nm	*Nicotiana tabacum*	Expression profiles of miRNA involved in regulating important genes related to tolerance were affected	Frazier et al. (2014)
TiO_2	Not specified	*Arabidopsis*	Differential expression of genes involved in DNA metabolism, hormone metabolism, tetrapyrrole synthesis, and photosynthesis	Tumburu et al. (2015)
Graphene oxide (GO)	40–50 nm	*Arabidopsis*	Altered expression of genes involved in development of abiotic stress tolerance and induction in oxidative stress	Wang et al. (2014)
CeO_2, In_2O_3	10–30 nm 20–70 nm	*Arabidopsis*	Alteration in expression of major stress-responsive genes, i.e., glutathione biosynthetic gene and sulfur assimilation	Ma et al. (2013)
ZnO	20 nm	*Arabidopsis*	Differential expression of genes, such as genes associated with stress, were upregulated, while those associated with cell organization and biogenesis and DNA/RNA metabolism were downregulated	Landa et al. (2015)
ZnO, CeO_2	7–8 nm	*G. max*	Damage to DNA and mutation, expression of new bands	López-Moreno et al. (2010)

(Continued)

TABLE 15.2 Genome Analysis of Nanoparticle-Mediated Toxicity—cont'd

Nanomaterial	Size/Dose	Model System	Effect on the Cellular System	References
ZnO TiO_2	<100 nm <150 nm	*A. thaliana*	Upregulation in genes associated with response under biotic and abiotic stress and downregulation in genes involved in translation, nucleosome assembly, and processes involved with microtubule and cell organization and biogenesis	Landa et al (2012)
Ag TiO_2	10–80 nm 10–40 nm	*A. thaliana*	Repression in transcriptional factors related to pathogenesis and phosphate starvation	García-Sánchez et al. (2015)
Ag	20 nm	*A. thaliana*	Genes related to response to oxidative stress, namely, vacuolar cation/proton exchanger, superoxide dismutase, cytochrome P450-dependent oxidase, and peroxidase, were upregulated, while those related with pathogenesis and hormonal stimuli were downregulated	Kaveh et al. (2013)
Ag TiO_2 ZnO Quantum dots	20 nm 5 nm 20 nm 6–10 nm	*Chlamydomonas reinhardtii*	Decline in expression of genes associated with photosynthesis and enhancement in transcripts encoding cell wall and flagella components	Simon et al. (2013)

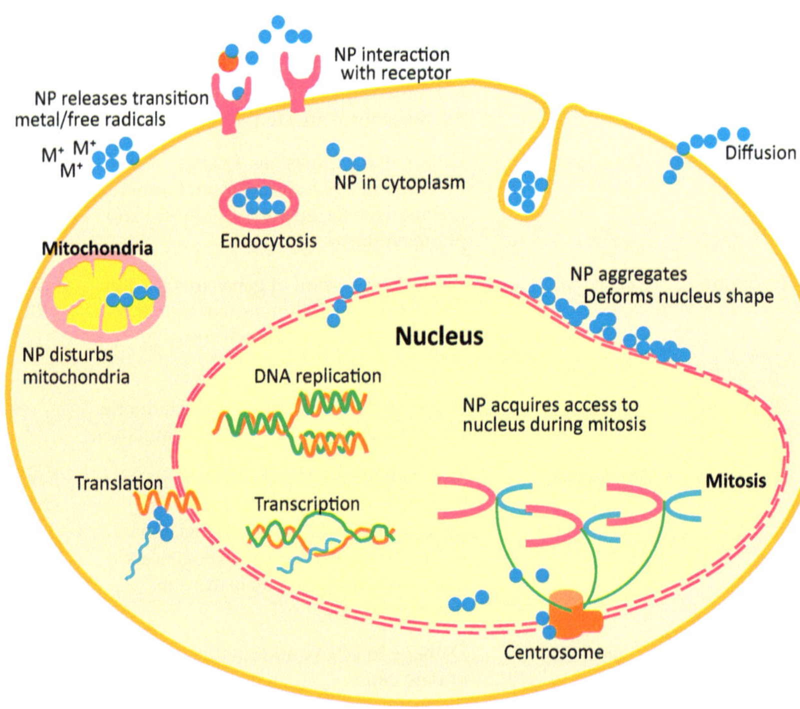

FIGURE 15.3 Cellular uptake and cyto-/genotoxicity mediated by nanoparticles. *Modified from Magdolenova, Z., Collins, A., Kumar, A., Dhawan, A., Ston, V., Dusinska., M., 2013. Mechanisms of genotoxicity. A review of in vitro and in vivo studies with engineered nanoparticles. Nanotoxicology, 1–46. http://dx.doi.org/10.3109/17435390.2013.773464.*

15.6 EFFECT OF NPs ON TRANSCRIPTOMICS

With the advancement in omic approaches the changes occurring at the gene, transcriptional, as well as protein levels could be studied easily and provide a brief mechanism of how a particular stimulus alters expression at every stage. Several studies have reported alteration in the genetic level followed by transcriptional changes. In a study on *A. thaliana*, Landa et al. (2012) reported altered gene expression when exposed to ZnO and TiO_2 NPs; these NPs upregulated 660 and 80 genes, while they downregulated 826 and 74 genes, respectively, and all the genes whose expression was altered were associated with stress responsiveness. In another study on the same system, downregulation in the expression of gene as well as transcription factor was noticed when subjected to NPs. Analysis employing the microarray technique has also suggested expression of 89 and 95 genes that were involved in the defense mechanism, metabolite synthesis (primary as well as secondary), transcription, transportation, etc. in two lines of *Brassica napus* (Li et al., 2009). In a study on Au NPs, the effect on *Arabidopsis* has been studied and results revealed differential expression of transcription factors of families WRKY, MYB, and BHLH, which are involved in the homeostasis of iron and other minerals (Tiwari et al., 2016). Studies have also revealed the downregulation of iron-regulated transporter genes as well as cation transporter genes. Apart from this downregulation in genes, transcripts of water channels were also noticed. Several other transcriptome analyses have been performed with NPs, which are summarized along with their impact on cell systems in Table 15.3. Overall, transcriptome analysis revealed that NP-mediated toxicity is associated with upregulation of genes, and their transcripts are involved in vacuolar cation/proton exchange, antioxidant systems, hormonal responses, pathogen-responsive genes, aquaporin genes, lowering of transcript formation of genes involved in photosynthesis, as well as genes involved in biogenesis and the organization of cells (Fig. 15.4).

15.7 EFFECT OF NPs ON PROTEOMICS

Toxicity assessment caused by toxicants or environmental stressors using proteomics falls under the branch of "toxicoproteomics," and when this toxicity assessment is mainly conducted to characterize NP toxicity, it is called "nanotoxicoproteomics" (Wetmore and Merrick, 2004; Gao et al., 2009). It combines the principles as well as the methods that could be employed to study the mechanisms involved with NP toxicity. It also provides a technique to identify stress-responsive proteins and how their expressions are changed when they interact with abiotic stresses (Barkla et al., 2013). With the help of two-dimensional polyacrylamide gel electrophoresis and mass spectrometry, differentially expressed proteins could be identified. Another emerging method for protein identification and its interaction with other biomolecules is the protein chip or microarray technique. Several studies on NP-induced toxicity in cells with the help of proteomic analysis have been performed, and these studies revealed that NPs affect the expression of important proteins by upregulating or downregulating them. Wigginton et al. (2010) with the help of proteomic approaches studied the cytotoxic effect of Ag NPs in cells of *Escherichia coli* and showed that Ag NPs affected the precursors of envelope proteins that in turn affected the proton motive force. In another study on *Staphylococcus aureus* by Li et al. (2011), Ag NP exposure resulted in variable expression of proteins: some

TABLE 15.3 Transcriptome Analysis of Nanoparticle-Mediated Toxicity

Nanomaterial	Size/Dose (nm)	Model System	Effect on the Cellular System	References
Ag NPs	20	*Arabidopsis*	Upregulation of genes such as vacuolar cation/proton exchanger, superoxide dismutase, cytochrome P450-dependent oxidase, and peroxidase associated with metal and oxidative stress, and downregulation of genes related with pathogenic and hormonal responses such as ARGOS (auxin-regulated gene involved in organs size) and SAR (signaling pathway and systemic acquired resistance)	Kaveh et al. (2013)
Multiwalled carbon nanotubes, single-walled carbon nanotubes, graphene	10–35	*Solanum lycopersicum*	Alteration in gene expression in leaves and roots with upregulation of genes associated with stress such as those associated with pathogen infection and water channel (LeAqp2 gene)	Khodakovskaya et al. (2011)
TiO_2, Ag NP	10–40 10–80	*Arabidopsis*	Decline in expression of genes involved in pathogen infection as well as repression of genes activated during phosphate starvation	García-Sánchez et al. (2015)
Cadmium sulfide quantum dots	5	*Arabidopsis*	Alteration in genes whose transcription was enhanced, thereby providing tolerance to cells, and mutation in gene involved in photosynthesis	Marmiroli et al. (2014)
ZnO TiO_2	<100 <150	*Arabidopsis*	Downregulation in gene involved in organization of cell and biogenesis, upregulation of stress-responsive genes	Landa et al. (2012)

were upregulated five times, while others were downregulated six times. Proteomic analysis on *B. thuringiensis* exposed to Ag NPs stress showed differential expression of proteins as well as disturbance to the proton motive force (Mirzajani et al., 2014). Likewise, proteome analysis of *Oryza sativa* seedlings exposed to Ag NP stress revealed similar results to those obtained in other studies, i.e., up- and downregulation of several proteins involved in stress responses and important growth regulating processes (Hecker et al., 2009). Similar to Ag NPs, other NP toxicities have been studied in several cell systems, e.g., CeO_2 NPs have been shown to reduce maize growth by affecting the expression of stress-responsive proteins such as CAT, APX, and heat shock protein (Zhao et al., 2012a). The impact of other NPs on cell systems is briefly mentioned in Table 15.4. Overall, NP toxicity has been shown to be mediated by destabilization and decline in plasma membrane potential, thereby depleting intracellular ATP levels, enhancing quorum sensing and stress-responsive proteins, expressing proteins related to oxidative stress tolerance, Ca^{2+} signaling and its release, transcriptional proteins,

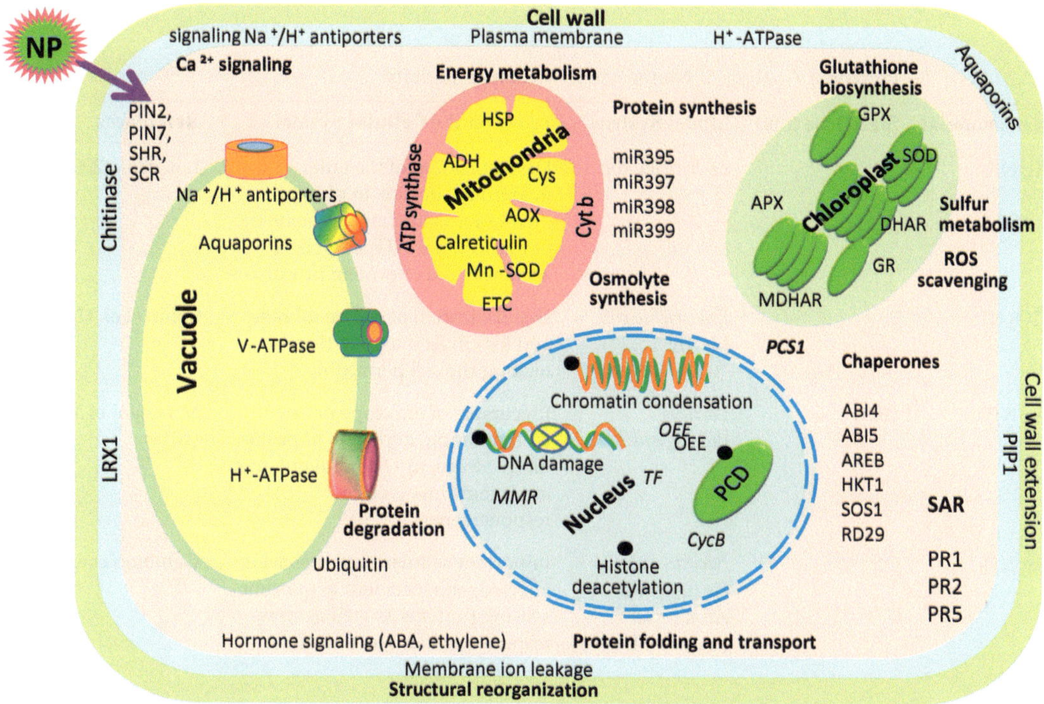

FIGURE 15.4 Transcriptome and proteome modulation by nanoparticles (NPs) in cell systems. *ABI*, ABA-insensitive; *ADH*, alcohol dehydrogenase; *AOX*, alternate oxidase; *APX*, ascorbate peroxidase; *AREB*, ABA-responsive elements-binding proteins; *ARGOS*, auxin-regulated gene involved in organ size; *BiP*, binding immunoglobulin proteins; *CycB*, cyclin B; *CyS*, cysteine synthase; *ETC*, electron transport chain; *GR*, glutathione reductase; *GlyII*, glyoxalase II; *GPX*, glutathione peroxidase; *HKT1*, high-affinity K^+ transporter 1; *HSP*, heat shock protein; *MDHAR*, monodehydroascorbate reductase; *miR*, microRNA; *MMR*, mismatch repair gene; *MT*, metallothionein; *OEE2*, oxygen-evolving complex 2; *PCNA*, proliferating cell nuclear antigen; *PCS1*, phytochelatin synthase 1; *PIN*, peptidyl-prolyl cis-trans isomerase NIMA-interacting; *PIP1*, plasma membrane intrinsic protein 1; *PR*, pathogenesis related; *RbcL*, large subunit of Rubisco; *RD29A*, responsive to desiccation 29A; *ROS*, reactive oxygen species; *SAR*, systemic acquired resistance; *SCR*, scarecrow; *SOD*, superoxide dismutase; *SOS1*, salt overly sensitive 1; *SHR*, short root; *TF*, transcription factors. *Modified from Jha, S., Pudake, R.N., 2016. Molecular mechanism of plant–nanoparticle interactions. In: Kole, et al. (Eds.), Plant Nanotechnology, Springer International Publishing, Switzerland. http://dx.doi.org/10.1007/978-3-319-42154-4_7.*

and also downregulating important proteins involved in growth-regulating processes such as photosystem II functioning, sulfur assimilation, and cell metabolism (Fig. 15.4).

15.8 CONCLUSION AND FUTURE PERSPECTIVES

The rapid growth of nanotechnology has created a new debate on the toxicity and environmental impacts of released NPs. This rise in the level of NPs has increased the attention of many scientists to study the impact of NPs on different organisms. NPs have characteristic properties such as large specific surface area and high surface energy. Because of these properties, NPs

TABLE 15.4 Proteome Analysis of Nanoparticle-Mediated Toxicity

Nanomaterial	Size/Dose (nm)	Model System	Effect on the Cellular System	References
Ag NPs	9.3	*Escherichia coli*	Destabilization of the outer membrane, decline in plasma membrane potential, depletion of intracellular ATP, as well as accumulation of protein precursors	Lok et al. (2006)
TiO_2 NPs	23	*Escherichia coli* MG1655	Depolarization of cell membrane and loss in its integrity, thereby making the cell permeable	Sohm et al. (2015)
Ag NPs	18.34	*Bacillus thuringiensis*	Precursors of envelope protein accumulation, depletion of proton motive force, up-/downregulation in proteins involved in responsiveness	Mirzajani et al. (2014)
Ag NPs	10	*Bacillus subtilis* Cu1065 and *Azotobacter vinelandii* UW136	Important cellular processes were affected, enhancement in quorum sensing proteins as well as stress response proteins, extracellular polysaccharides	Gambino et al. (2015)
Ag NPs	18	*Oryza sativa*	Variable expression of 28 proteins and their expression regulates oxidative stress tolerance, Ca^{2+} release and signaling, transcription, cell division, and apoptosis	Mirzajani et al. (2014)
Ag NPs	10	*Eruca sativa*	Alteration in proteins of endoplasmic reticulum and vacuoles, as well as in the proteins involved in sulfur metabolism	Vannini et al. (2013)
Ag NPs	10	*Triticum aestivum*	Expression of different cellular compartmental proteins in shoots and roots mainly involved in metabolism and cell defense	Vannini et al. (2014)
Ag NPs	15	*Glycine max*	Alteration in 107 differential root proteins that were associated with stress, cell metabolism, and signaling	Mustafa et al (2015)
CeO_2 NPs	10	*Zea mays*	Increased accumulation of H_2O_2 and heat shock protein, and enhancement in activity of ascorbate peroxidase and catalase	Zhao et al. (2012a)
Ag NPs	20	*Chlamydomonas*	Downregulation of oxygen-evolving enhancer protein 2 of photosystem II	Lindgren (2014)

can have positive or negative effects on the basis of their concentration. Plants are essential for the environment and therefore play an important role in energy conversion. The rising concentration of NPs has generated nanotoxicity, which affects plant growth by inhibiting seed germination and root elongation, and ultimately reducing yield. Nanotoxicity is generated by the excessive generation of ROS. To understand the mechanism of nanotoxicity, scientists have used different omics tools such as genomics, trancriptomics, and proteomics. The positive and negative biological effects of NPs are still to be fully understood. There is a further need for more studies on different organisms to test nanotoxicity, and new tools such as next-generation sequencing techniques and posttranslational proteomics are proving to be very useful.

References

American Society for Testing and Materials, 2006. Standard Terminology Relating to Nanotechnology p. E2456–06. ASTM International, West Conshohocken, PA. www.astm.org.

Ankley, G.T., Daston, G.P., Degitz, S.J., 2006. Toxicogenomics in regulatory ecotoxicology. Environ. Sci. Technol. 40, 4055–4065.

Barkla, B.J., Vera-Estrella, R., Pantoja, O., 2013. Progress and challenges for abiotic stress proteomics of crop plants. Proteomics 13, 1801–1815.

Bashri, G., Prasad, S.M., 2016. Exogenous IAA differentially affects growth, oxidative stress and antioxidants system in Cd stressed *Trigonella foenum-graecum* L. seedlings: toxicity alleviation by up-regulation of ascorbate-glutathione cycle. Ecotoxicol. Environ. Saf. 132, 329–338.

Birbaum, K., Brogioli, R., Schellenberg, M., Martinoia, E., Stark, W.J., Günther, D., Limbach, L.K., 2010. No evidence for cerium dioxide nanoparticle translocation in maize plants. Environ. Sci. Technol. 44, 8718–8723.

Burklew, C.E., Ashlock, J., Winfrey, W.B., Zhang, B.H., 2012. Effects of aluminum oxide nanoparticles on the growth, development, and microRNA expression of tobacco (*Nicotiana tabacum*). PLoS One 7.

Chen, C., Unrine, J.M., Judy, J.D., Lewis, R.W., Guo, J., McNear Jr., D.H., Tsyusko, O.V., 2015. Toxicogenomic responses of the model legume *Medicago truncatula* to aged biosolids containing a mixture of nanomaterials (TiO_2, Ag, and ZnO) from a pilot wastewater treatment plant. Environ. Sci. Technol. 49, 8759–8768.

Chen, G., Ma, C., Mukherjee, A., Musante, C., Zhang, J., White, J.C., Dhankher, O.P., Xing, B., 2016. Tannic acid alleviates bulk and nanoparticle Nd_2O_3 toxicity in pumpkin: a physiological and molecular response. Nanotoxicology 10, 1243–1253.

Christian, P., von der Kammer, F., Baalousha, M., Hofmann, T., 2008. Nanoparticles: structure, properties, preparation and behaviour in environmental media. Ecotoxicology 17, 326–343.

Chu, H., Kim, H.J., Kim, J.S., Kim, M.S., Yoon, B.D., Park, H.J., Kim, C.Y., 2012. A nanosized Ag-silica hybrid complex prepared by gamma-irradiation activates the defense response in *Arabidopsis*. Radiat. Phys. Chem. 81, 180–184.

Cui, D., Zhang, P., Ma, Y., He, X., Li, Y., Zhang, J., Zhao, Y., Zhang, Z., 2014. Effect of cerium oxide nanoparticles on asparagus lettuce cultured in an agar medium. Environ. Sci. Nano 1, 459–465.

De la Rosa, G., Lopez-Moreno, M.L., Hernandez-Viescaz, J., Montes, M.O., Peralta-Videa, J.R., Gardea-Torresdey, J.L., 2011. Toxicity and biotransformation of ZnO nanoparticles in the desert plants *Prosopis juliflora-velutina, Salsola tragus* and *Parkinsonia florida*. Int. J. Nanotechnol. 492–506.

Dietz, K.J., Herth, S., 2011. Plant nanotoxicology. Trends. Plant Sci. 16, 582–589.

Dimkpa, C.O., McLean, J.E., Latta, D.E., Manangón, E., Britt, D.W., Johnson, W.P., Boyanov, M.I., Anderson, A.J., 2012. Cuo and ZnO nanoparticles: phytotoxicity, metal speciation, and induction of oxidative stress in sand-grown wheat. J. Nanopart. Res. 14, 1–15.

Dimkpa, C.O., McLean, J.E., Martineau, N., Britt, D.W., Haverkamp, R., Anderson, A.J., 2013. Silver nanoparticles disrupt wheat (*Triticum aestivum* L.) growth in a sand matrix. Environ. Sci. Technol. 47, 1082–1090.

Du, W., Sun, Y., Ji, R., Zhu, J., Wu, J., Guo, H., 2011. TiO_2 and ZnO nanoparticles negatively affect wheat growth and soil enzyme activities in agricultural soil. J. Environ. Monit. 13, 822–828.

Eichert, T., Kurtz, A., Steiner, U., Goldbach, H.E., 2008. Size exclusion limits and lateral heterogeneity of the stomatal foliar uptake pathway for aqueous solutes and water-suspended nanoparticles. Physiol. Plant. 134, 151–160.

El-Temsah, Y.S., Joner, E.J., 2012. Impact of Fe and Ag nanoparticles on seed germination and differences in bioavailability during exposure in aqueous suspension and soil. Environ. Toxicol. 27, 42–49.

Faisal, M., Saquib, Q., Alatar, A.A., Al-Khedhairy, A.A., Hegazy, A.K., Musarrat, J., 2013. Phytotoxic hazards of NiO-nanoparticles in tomato: a study on mechanism of cell death. J. Hazard. Mater. 250–251, 318–332.

Foyer, C.H., Lopez-Delgado, H., Dat, J.F., Scott, I.M., 1997. Hydrogen peroxide- and glutathione-associated mechanisms of acclimatory stress tolerance and signalling. Physiol. Plant 100, 241–254.

Foyer, C.H., Noctor, G., 2003. Redox sensing and signalling associated with reactive oxygen in chloroplasts, peroxisomes and mitochondria. Physiol. Plant 119, 355–364.

Frazier, T.P., Burklew, C.E., Zhang, B.H., 2014. Titanium dioxide nanoparticles affect the growth and microRNA expression of tobacco (*Nicotiana tabacum*). Funct. Integr. Genom. 14, 75–83.

Freinbichler, W., Colivicchi, M.A., Stefanini, C., Bianchi, L., Ballini, C., Misini, B., Weinberger, P., Linert, W., Varešlija, D., Tipton, K.F., Corte, L.D., 2011. Highly reactive oxygen species: detection, formation, and possible functions. Cell. Mol. Life Sci. 68, 2067–2079.

Fridovich, I., 1989. Superoxide dismutases. An adaptation to a paramagnetic gas. J. Biol. Chem. 264, 7761–7764.

Gambino, M., Marzano, V., Villa, F., Vitali, A., Vannini, C., Landini, P., Cappitelli, F., 2015. Effects of sublethal doses of silver nanoparticles on *Bacillus subtilis* planktonic and sessile cells. J. Appl. Microbiol. 118, 1103–1115.

García-Sánchez, S., Bernales, I., Cristobal, S., 2015. Early response to nanoparticles in the *Arabidopsis* transcriptome compromises plant defence and root-hair development through salicylic acid signalling. BMC Genom. 16, 341.

Garg, N., Manchanda, G., 2009. ROS generation in plants: boon or bane? Plant Biosyst. 143, 81–96.

Gao, Y., Holland, R.D., Yu, L.R., 2009. Quantitative proteomics for drug toxicity. Brief Funct. Genomic Proteomic 8, 158–166.

Ghosh, M., Bhadra, S., Adegoke, A., Bandyopadhyay, M., Mukherjee, A., 2015. MWCNT uptake in *Allium cepa* root cells induces cytotoxic and genotoxic responses and results in DNA hyper-methylation. Mutat. Res. Fundam. Mol. Mech. Mutagen. 774, 49–58.

Gill, S.S., Tuteja, N., 2010. Reactive oxygen species and antioxidant machinery in abiotic stress tolerance in crop plants. Plant Physiol. Biochem. 48, 909–930.

Gottschalk, F., Lassen, C., Kjoelholt, J., Christensen, F., Nowack, B., 2015. Modeling flows and concentrations of nine engineered nanomaterials in the Danish environment. Int. J. Environ. Res. Public Health 127, 5581–5602.

Halliwell, B., Gutteridge, J.M., 2015. Free Radicals in Biology and Medicine. Oxford University Press, USA.

Hecker, M., Reder, A., Fuchs, S., Pagels, M., Engelmann, S., 2009. Physiological proteomics and stress/starvation responses in *Bacillus subtilis* and *Staphylococcus aureus*. Res. Microbiol. 160, 245–258.

Hong, J., Peralta-Videa, J.R., Rico, C., Sahi, S., Viveros, M.N., Bartonjo, J., Zhao, L., Gardea-Torresdey, J.L., 2014. Evidence of translocation and physiological impacts of foliar applied CeO_2 nanoparticles on cucumber (*Cucumis sativus*) plants. Environ. Sci. Technol. 48, 4376–4385.

Jha, S., Pudake, R.N., 2016. Molecular mechanism of plant–nanoparticle interactions. In: Kole, et al. (Ed.), Plant Nanotechnology. Springer International Publishing, Switzerland. http://dx.doi.org/10.1007/978-3-319-42154-4_7.

Kaveh, R., Li, Y.S., Ranjbar, S., Tehrani, R., Brueck, C.L., van Aken, B., 2013. Changes in *Arabidopsis thaliana* gene expression in response to silver nanoparticles and silver ions. Environ. Sci. Technol. 47, 10637–10644.

Khodakovskaya, M.V., de Silva, K., Nedosekin, D.A., Dervishi, E., Biris, A.S., Shashkov, E.V., Galanzha, E.I., Zharov, V.P., 2011. Complex genetic, photothermal, and photoacoustic analysis of nanoparticle-plant interactions. Proc. Natl. Acad. Sci. U.S.A. 108, 1028–1033.

Khodakovskaya, M.V., de Silva, K., Biris, A.S., Dervishi, E., Villagarcia, H., 2012. Carbon nanotubes induce growth enhancement of tobacco cells. ACS Nano 6, 2128–2135.

Kohan-Baghkheirati, E., Geisler-Lee, J., 2015. Gene expression, protein function and pathways of *Arabidopsis thaliana* responding to silver nanoparticles in comparison to silver ions, cold, salt, drought, and heat. Nanomaterials 5, 436–467.

Kouhi, S.M., Lahouti, M., Ganjeali, A., Entezari, M.H., 2014. Comparative phytotoxicity of ZnO nanoparticles, ZnO microparticles, and Zn^{2+} on rapeseed (*Brassica napus* L.): investigating a wide range of concentrations. Toxicol. Environ. Chem. 96, 861–868.

Kurepa, J., Paunesku, T., Vogt, S., Arora, H., Rabatic, B.M., Lu, J., Wanzer, M.B., Woloschak, G.E., Smalle, J.A., 2010. Uptake and distribution of ultra-small anatase TiO_2 alizarin red S nanoconjugates in *Arabidopsis thaliana*. Nano Lett. 10, 2296–2302.

Lahiani, M.H., Dervishi, E., Chen, J., Nima, Z., Gaume, A., Biris, A.S., Khodakovskaya, M.V., 2013. Impact of carbon nanotube exposure to seeds of valuable crops. ACS Appl. Mater. Interfaces 5, 7965–7973.

Lahiani, M.H., Chen, J., Irin, F., Puretzky, A.A., Green, M.J., Khodakovskaya, M.V., 2015. Interaction of carbon nanohorns with plants: uptake and biological effects. Carbon 81, 607–619.

Landa, P., Vankova, R., Andrlova, J., Hodek, J., Marsik, P., Storchova, H., White, J.C., Vanek, T., 2012. Nanoparticle-specific changes in *Arabidopsis thaliana* gene expression after exposure to ZnO, TiO_2, and fullerene soot. J. Hazard. Mater. 241, 55–62.

Landa, P., Prerostova, S., Petrova, S., Knirsch, V., Vankova, R., Vanek, T., 2015. The transcriptomic response of *Arabidopsis thaliana* to Zinc Oxide: a comparison of the impact of nanoparticle, bulk, and ionic zinc. Environ. Sci. Technol. 49, 14537–14545.

Larue, C., Castillo-Michel, H., Sobanska, S., Bureau, S., Barthès, V., Ouerdane, L., Carrière, M., Sarret, G., 2014. Foliar exposure of the crop *Lactuca sativa* to silver nanoparticles: evidence for internalization and changes in Ag speciation. J. Hazard. Mater. 261, 98–106.

Le Van, N., Rui, Y., Gui, X., Li, X., Liu, S., Han, Y., 2014. Uptake, transport, distribution and bio-effects of SiO_2 nanoparticles in Bt-transgenic cotton. J. Nanobiotechnol. 12, 50.

Lee, S., Kim, S., Kim, S., Lee, I., 2013. Assessment of phytotoxicity of ZnO NPs on a medicinal plant, *Fagopyrum esculentum*. Environ. Sci. Pollut. Res. 20, 848–854.

Li, X., Gao, P., Gjetvaj, B., Westcott, N., Gruber, M.Y., 2009. Analysis of the metabolome and transcriptome of *Brassica carinata* seedlings after lithium chloride exposure. Plant Sci. 177, 68–80.

Li, W.R., Xie, X.B., Shi, Q.S., Duan, S.S., Ouyang, Y.S., Chen, Y.B., 2011. Antibacterial effect of silver nanoparticles on *Staphylococcus aureus*. Biometals 24, 135–141.

Lin, D., Xing, B., 2007. Phytotoxicity of nanoparticles: inhibition of seed germination and root growth. Environ. Pollut. 150, 243–250.

Lin, D., Xing, B., 2008. Root uptake and phytotoxicity of ZnO nanoparticles. Environ. Sci. Technol. 42, 5580–5585.

Lindgren, A.L., 2014. The Effects of Silver Nitrate and Silver Nanoparticles on Chlamydomonas Reinhardtii: A Proteomic Approach. Degree Project, Department of Biology and Environmental Sciences, University of Gothenburg, Germany.

Lok, C.N., Ho, C.M., Chen, R., He, Q.Y., Yu, W.Y., Sun, H., Tam, P.K., Chiu, J.F., Che, C.M., 2006. Proteomic analysis of the mode of antibacterial action of silver nanoparticles. J. Proteome Res. 5, 916–924.

López-Moreno, M., de la Rosa, G., Hernandez-Viezcas, J., Castillo-Michel, H., Botez, C., Peralta-Videa, J., Gardea-Torresdey, J., 2010. Evidence of the differential biotransformation and genotoxicity of ZnO and CeO_2 nanoparticles on soybean (*Glycine max*) plants. Environ. Sci. Technol. 44, 7315–7320.

Ma, Y., Kuang, L., He, X., Bai, W., Ding, Y., Zhang, Z., Zhao, Y., Chai, Z., 2010. Effects of rare earth oxide nanoparticles on root elongation of plants. Chemosphere 78, 273–279.

Ma, C.X., Chhikara, S., Xing, B.S., Musante, C., White, J.C., Dhankher, O.P., 2013. Physiological and molecular response of *Arabidopsis thaliana* (L.) to nanoparticle cerium and indium oxide exposure. ACS Sustain. Chem. Eng. 1, 768–778.

Ma, C., Liu, H., Guo, H., Musante, C., Coskun, S.H., Nelson, B.C., White, J.C., Xing, B., Dhankher, O.P., 2016. Defense mechanisms and nutrient displacement in *Arabidopsis thaliana* upon exposure to CeO_2 and In_2O_3 nanoparticles. Environ. Sci. Nano 3, 1369–1379.

Magdolenova, Z., Collins, A., Kumar, A., Dhawan, A., Ston, V., Dusinska, M., 2013. Mechanisms of genotoxicity. A review of in vitro and in vivo studies with engineered nanoparticles. Nanotoxicology 1–46. http://dx.doi.org/10.3109/17435390.2013.773464.

Majumdar, S., Peralta-Videa, J.R., Bandyopadhyay, S., Castillo-Michel, H., Hernandez-Viezcas, J.A., Sahi, S., Gardea-Torresdey, J.L., 2014. Exposure of cerium oxide nanoparticles to kidney bean shows disturbance in the plant defense mechanisms. J. Hazard. Mater. 278, 279–287.

Marmiroli, M., Pagano, L., Sardaro, M.L.S., Villani, M., Marmiroli, N., 2014. Genome-wide approach in *Arabidopsis thaliana* to assess the toxicity of cadmium sulfide quantum dots. Environ. Sci. Technol. 48, 5902–5909.

McKee, M.S., Filser, J., 2016. Impacts of metal-based engineered nanomaterials on soil communities. Environ. Sci. Nano 3, 506–533.

Meriga, B., Reddy, B.K., Rao, K.R., Reddy, L.A., Kishor, P.B.K., 2004. Aluminium-induced production of oxygen radicals, lipid peroxidation and DNA damage in seedlings of rice (*Oryza sativa*). J. Plant Physiol. 161, 63–68.

Merrick, B.A., Bruno, M.E., 2004. Genomic and proteomic profiling for biomarkers and signature profiles of toxicity. Curr. Opin. Mol. Ther. 6, 600–607.

Mirzajani, F., Askari, H., Hamzelou, S., Schober, Y., Rompp, A., Ghassempour, A., Spengler, B., 2014. Proteomics study of silver nanoparticles toxicity on *Bacillus thuringiensis*. Ecotoxicol. Environ. Saf. 100, 122–130.

Montillet, J.-L., Chamnongpol, S., Rustérucci, C., Dat, J., Van de Cotte, B., Agnel, J.-P., Battesti, C., Inzé, D., Van Breusegem, F., Triantaphylidès, C., 2005. Fatty acid hydroperoxides and H_2O_2 in the execution of hypersensitive cell death in tobacco leaves. Plant Physiol. 138, 1516–1526.

Mustafa, G., Sakata, K., Hossain, Z., Komatsu, S., 2015. Proteomic study on the effects of silver nanoparticles on soybean under flooding stress. J. Proteome 122, 100–118.

Nair, P.M.G., Chung, I.M., 2014a. Assessment of silver nanoparticle-induced physiological and molecular changes in *Arabidopsis thaliana*. Environ. Sci. Pollut. Res. 21, 8858–8869.

Nair, P.M.G., Chung, I.M., 2014b. Cell cycle and mismatch repair genes as potential biomarkers in *Arabidopsis thaliana* seedlings exposed to silver nanoparticles. Bull. Environ. Contam. Toxicol. 92, 719–725.

Nair, P.M., Chung, I.M., 2015. Study on the correlation between copper oxide nanoparticles induced growth suppression and enhanced lignification in Indian mustard (*Brassica juncea* L.). Ecotoxicol. Environ. Saf. 113, 302–313.

Nel, A., Xia, T., Mädler, L., Li, N., 2006. Toxic potential of materials at the nanolevel. Science 311, 622–627.

Oukarroum, A., Barhoumi, L., Pirastru, L., Dewez, D., 2013. Silver nanoparticle toxicity effect on growth and cellular viability of the aquatic plant *Lemna gibba*. Environ. Toxicol. Chem. 32, 902–907.

Patlolla, A.K., Berry, A., May, L., Tchounwou, P.B., 2012. Genotoxicity of silver nanoparticles in *Vicia faba*: a pilot study on the environmental monitoring of nanoparticles. Int. J. Environ. Res. Public Health 9, 1649–1662.

Pokhrel, L.R., Dubey, B., 2013. Evaluation of developmental responses of two crop plants exposed to silver and zinc oxide nanoparticles. Sci. Total Environ. 452–453, 321–332.

Poma, A., Di Giorgio, M.L., 2008. Toxicogenomics to improve comprehension of the mechanisms underlying responses of in vitro and in vivo systems to nanomaterials: a review. Curr. Genom. 9, 571–585.

Poynton, H., Vulpe, C., 2009. Ecotoxicogenomics: emerging technologies for emerging contaminants. J. Am. Water Resour. Assoc. 45, 83–96.

Răcuciu, M., Creangă, D.E., 2009. Cytogenetical changes induced by β-cyclodextrin coated nanoparticles in plant seeds. Rom. J. Phys. 54, 125–131.

Rajeshwari, A., Kavitha, S., Alex, S.A., Kumar, D., Mukherjee, A., Chandrasekaran, N., Mukherjee, A., 2015. Cytotoxicity of aluminum oxide nanoparticles on *Allium cepa* root tip – effects of oxidative stress generation and biouptake. Environ. Sci. Pollut. Res. 22, 11057–11066.

Riahi-Madvar, A., Rezaee, F., Jalali, V., 2012. Effects of alumina nanoparticles on morphological properties and antioxidant system of *Triticum aestivum*. Iran. J. Plant Physiol. 3, 595–603.

Rico, C.M., Majumdar, S., Duarte-Gardea, M., Peralta-Videa, J.R., Gardea-Torresdey, J.L., 2011. Interaction of nanoparticles with edible plants and their possible implications in the food chain. J. Agric. Food Chem. 59, 3485–3498.

Sharma, P., Jha, A.B., Dubey, R.S., Pessarakli, M., 2012. Reactive oxygen species, oxidative damage, and antioxidative defense mechanism in plants under stressful conditions. J. Bot. 1–26.

Shaw, A.K., Hossain, Z., 2013. Impact of nano-CuO stress on rice (*Oryza sativa* L.) seedlings. Chemosphere 93, 906–915.

Shen, C.X., Zhang, Q.F., Li, J., Bi, F.C., Yao, N., 2010. Induction of programmed cell death in *Arabidopsis* and rice by single-wall carbon nanotubes. Am. J. Bot. 97, 1602–1609.

Shweta, Tripathi, D.K., Singh, S., Singh, S., Dubey, N.K., Chauhan, D.K., 2016. Impact of nanoparticles on photosynthesis: challenges and opportunities. Mater. Focus 5 (5), 405–411.

Simon, D.F., Domingos, R.F., Hauser, C., Hutchins, C.M., Zerges, W., Wilkinson, K.J., 2013. Transcriptome sequencing (RNA-seq) analysis of the effects of metal nanoparticle exposure on the transcriptome of *Chlamydomonas reinhardtii*. Appl. Environ. Microbiol. 79, 4774–4785.

Singh, S., Tripathi, D.K., Dubey, N.K., Chauhan, D.K., 2016. Effects of nano-materials on seed germination and seedling growth: striking the slight balance between the concepts and controversies. Mater. Focus 5 (3), 195–201.

Singh, S., Vishwakarma, K., Singh, S., Sharma, S., Dubey, N.K., Singh, V.K., Liu, S., Tripathi, D.K., Chauhan, D.K., 2017. Understanding the plant and nanoparticle interface at transcriptomic and proteomic level: a concentric overview. Plant Gene. http://dx.doi.org/10.1016/j.plgene.2017.03.006.

Sohm, B., Immel, F., Bauda, P., Pagnout, C., 2015. Insight into the primary mode of action of TiO_2 nanoparticles on *Escherichia coli* in the dark. Proteomics 15, 98–113.

Stampoulis, D., Sinha, S.K., White, J.C., 2009. White assay-dependent phytotoxicity of nanoparticles to plants. Environ. Sci. Technol. 43, 9473–9479.

Syu, Y.Y., Hung, J.H., Chen, J.C., Chuang, H.W., 2014. Impact of size and shape of silver nanoparticles on *Arabidopsis* plant growth and gene expression. Plant Physiol. Biochem. 83, 57–64.

Tang, Y., He, R., Zhao, J., Nie, G., Xu, L., Xing, B., 2016. Oxidative stress- induced toxicity of CuO nanoparticles and related toxicogenomic responses in *Arabidopsis thaliana*. Environ. Pollut. 212, 605–614.

Tanou, G., Molassiotis, A., Diamantidis, G., 2009. Induction of reactive oxygen species and necrotic death-like destruction in strawberry leaves by salinity. Environ. Exp. Bot. 65, 270–281.

Taran, N., Batsmanova, L., Konotop, Y., Okanenko, A., 2014. Redistribution of elements of metals in plant tissues under treatment by non-ionic colloidal solution of biogenic metal nanoparticles. Nanoscale Res. Lett. 9, 354–357.

Tarasenko, V., Garnik, E.Y., Shmakov, V., Konstantinov, Y.M., 2012. Modified alternative oxidase expression results in different reactive oxygen species contents in *Arabidopsis* cell culture but not in whole plants. Biol. Plant 56, 635–640.

Taylor, R., Walton, D.R.M., 1993. The chemistry of fullerenes. Nature 363, 685–693.
Thomas, C.R., George, S., Horst, A.M., Ji, Z.X., Miller, R.J., Peralta-Videa, J.R., Xia, T.A., Pokhrel, S., Madler, L., Gardea-Torresdey, J.L., et al., 2011. Nanomaterials in the environment: from materials to high-throughput screening to organisms. ACS Nano 5, 13–20.
Tiwari, M., Krishnamurthy, S., Shukla, D., Kiiskila, J., Jain, A., Datta, R., Sharma, N., Sahi, S.V., 2016. Comparative transcriptome and proteome analysis to reveal the biosynthesis of gold nanoparticles in *Arabidopsis*. Sci. Rep. 6.
Tolaymat, T., El Badawy, A., Sequeira, R., Genaidy, A., 2015. A system-of- systems approach as a broad and integrated paradigm for sustainable engineered nanomaterials. Sci. Total Environ. 511, 595–607.
Tripathi, D.K., Singh, V.P., Prasad, S.M., Chauhan, D.K., Dubey, N.K., 2015. Silicon nanoparticles (SiNp) alleviate chromium (VI) phytotoxicity in *Pisum sativum* (L.) seedlings. Plant Physiol. Biochem. 96, 189–198.
Tripathi, D.K., Singh, S., Singh, V.P., Prasad, S.M., Chauhan, D.K., Dubey, N.K., 2016. Silicon nanoparticles more efficiently alleviate arsenate toxicity than silicon in maize cultiver and hybrid differing in arsenate tolerance. Front. Environ. Sci. 4, 46.
Tripathi, D.K., Singh, S., Singh, S., Pandey, R., Singh, V.P., Sharma, N.C., Prasad, S.M., Dubey, N.K., Chauhan, D.K., 2017a. An overview on manufactured nanoparticles in plants: uptake, translocation, accumulation and phytotoxicity. Plant Physiol. Biochem. 110, 2–12.
Tripathi, D.K., Mishra, R.K., Singh, S., Singh, S., Vishwakarma, K., Sharma, S., Singh, V.P., Singh, P.K., Prasad, S.M., Dubey, N.K., Pandey, A.C., 2017b. Nitric oxide ameliorates zinc oxide nanoparticles phytotoxicity in wheat seedlings: implication of the ascorbate–glutathione cycle. Front. Plant Sci. 8.
Tripathi, D.K., Singh, S., Singh, S., Srivastava, P.K., Singh, V.P., Singh, S., Prasad, S.M., Singh, P.K., Dubey, N.K., Pandey, A.C., Chauhan, D.K., 2017c. Nitric oxide alleviates silver nanoparticles (AgNps)-induced phytotoxicity in *Pisum sativum* seedlings. Plant Physiol. Biochem. 110, 167–177.
Tripathi, D.K., Tripathi, A., Shweta, S.S., Singh, S., Singh, Y., Vishwakarma, K., et al., 2017d. Uptake, accumulation and toxicity of silver nanoparticle in autotrophic plants, and heterotrophic microbes: a concentric review. Front. Microbiol. 8 (07). http://dx.doi.org/10.3389/fmicb.2017.00007.
Tripathi, D.K., Singh, S., Singh, V.P., Prasad, S.M., Dubey, N.K., Chauhan, D.K., 2017e. Silicon nanoparticles more effectively alleviated UV-B stress than silicon in wheat (*Triticum aestivum*) seedlings. Plant Physiol. Biochem. 110, 70–81.
Tumburu, L., Andersen, C.P., Rygiewicz, P.T., Reichman, J.R., 2015. Phenotypic and genomic responses to titanium dioxide and cerium oxide nanoparticles in *Arabidopsis germinants*. Environ. Toxicol. Chem. 34, 70–83.
Vannini, C., Domingo, G., Onelli, E., Prinsi, B., Marsoni, M., Espen, L., Bracale, M., 2013. Morphological and proteomic responses of *Eruca sativa* exposed to silver nanoparticles or silver nitrate. PLoS One 8, e68752.
Vannini, C., Domingo, G., Onelli, E., De Mattia, F., Bruni, I., Marsoni, M., Bracale, M., 2014. Phytotoxic and genotoxic effects of silver nanoparticles exposure on germinating wheat seedlings. J. Plant Physiol. 171, 1142–1148.
Wang, Z., Xie, X., Zhao, J., Liu, X., Feng, W., White, J.C., Xing, B., 2012. Xylem- and phloem-based transport of CuO nanoparticles in maize (*Zea mays* L.). Environ. Sci. Technol. 46, 4434–4441.
Wang, W.-N., Tarafdar, J.C., Biswas, P., 2013. Nanoparticle synthesis and delivery by an aerosol route for watermelon plant foliar uptake. J. Nanopart. Res. 15. http://dx.doi.org/10.1007/s11051-013-1417-8.
Wang, Q., Zhao, S., Zhao, Y., Rui, Q., Wang, D., 2014. Toxicity and translocation of graphene oxide in *Arabidopsis* plants under stress conditions. RSC Adv. 4, 60891–60901.
Wei, H., Wang, E., 2013. Nanomaterials with enzyme-like characteristics (nanozymes): next-generation artificial enzymes. Chem. Soc. Rev. 42, 6060–6093.
Wetmore, B.A., Merrick, B.A., 2004. Invited review: toxicoproteomics: proteomics applied to toxicology and pathology. Toxicol. Pathol. 32, 619–642.
Wigginton, N.S., de Titta, A., Piccapietra, F., Dobias, J., Nesatyy, V.J., Suter, M.J., et al., 2010. Binding of silver nanoparticles to bacterial proteins depends on surface modifications and inhibits enzymatic activity. Environ. Sci. Technol. 44, 2163–2168.
Xu, L.M., Takemura, T., Xu, M.S., Hanagata, N., 2011. Toxicity of silver nanoparticles as assessed by global gene expression analysis. Mater. Express 1, 74–79.
Yan, S., Zhao, L., Li, H., Zhang, Q., Tan, J., Huang, M., He, S., Li, L., 2013. Single- walled carbon nanotubes selectively influence maize root tissue development accompanied by the change in the related gene expression. J. Hazard. Mater. 246–247, 110–118.
Yang, J., Cao, W., Rui, Y., 2017. Interactions between nanoparticles and plants: phytotoxicity and defense mechanisms. J. Plant Interact. 12 (1), 158–169.
Yasur, J., Rani, P.U., 2013. Environmental effects of nanosilver: impact on castor seed germination, seedling growth, and plant physiology. Environ. Sci. Pollut. Res. 20, 8636–8648.

You, J., Reid, J.R., 2008. Manufactured nanoparticles: an overview of their chemistry, interactions and their potential environmental implications. Sci. Total Environ. 400, 396–414.

Zhang, P., Ma, Y., Zhang, Z., He, X., Li, Y., Zhang, J., Zheng, L., Zhao, Y., 2015. Species-specific toxicity of ceria nanoparticles to *Lactuca* plants. Nanotoxicology 9, 1–8.

Zhang, P., Ma, Y., Liu, S., Wang, G., Zhang, J., He, X., Zhang, J., Rui, Y., Zhang, Z., 2017. Phytotoxicity, uptake and transformation of nano-CeO_2 in sand cultured romaine lettuce. Environ. Pollut. 220, 1400–1408.

Zhao, L., Peng, B., Hernandez-Viezcas, J.A., Rico, C., Sun, Y., Peralta-Videa, J.R., Tang, X., Niu, G., Jin, L., Varela-Ramirez, A., Zhang, J.Y., Gardea-Torresdey, J.L., 2012a. Stress response and tolerance of *Zea mays* to CeO_2 nanoparticles: cross talk among H_2O_2, heat shock protein, and lipid peroxidation. ACS Nano 6, 9615–9622.

Zhao, L., Peralta-Videa, J.R., Varela-Ramirez, A., Castillo-Michel, H., Li, C., Zhang, J., Aguilera, R.J., Keller, A.A., Gardea-Torresdey, J.L., 2012b. Effect of surface coating and organic matter on the uptake of CeO_2–NPs by corn plants grown in soil: insight into the uptake mechanism. J. Hazard. Mater. 225–226, 131–138.

Zhao, L., Peralta-Videa, J.R., Peng, B., Bandyopadhyay, S., Corral-Diaz, B., Osuna- vila, P., Montes, M.O., Keller, A.A., Gardea-Torresdey, J.L., 2014. Alginate modifies the physiological impact of CeO_2 nanoparticles in corn seedlings cultivated in soil. J. Environ. Sci. 26, 382–389.

Zheng, L., Hong, F., Lu, S., Liu, C., 2005. Effect of nano-TiO_2 on strength of naturally aged seeds and growth of spinach. Biol. Trace Elem. Res. 104, 83–91.

Zhu, H., Han, J., Xiao, J.Q., Jin, Y., 2008. Uptake, translocation, and accumulation of manufactured iron oxide by pumpkin plants. J. Environ. Monit 10, 713–717.

Further Reading

Ma, Y., Zhang, P., Zhang, Z., He, X., Li, Y., Zhang, J., Zheng, L., Chu, S., Yang, K., Zhao, Y., Chai, Z., 2015. Origin of the different phytotoxicity and biotransformation of cerium and lanthanum oxide nanoparticles in cucumber. Nanotoxicology 9, 262–270.Zhu, H., Han, J., Xiao, J.Q., Jin, Y., 2008. Uptake, translocation, and accumulation of manufactured iron oxide by pumpkin plants. J. Environ. Monit. 10, 713–717.

CHAPTER

16

Nanotechnology in Crop Protection

Sunita Ojha, Deepika Singh, Arghya Sett, Hasnahana Chetia, Debajyoti Kabiraj, Utpal Bora

Indian Institute of Technology Guwahati, Guwahati, India

16.1 INTRODUCTION

Nanoparticles (NPs) are particles in a nanorange structure with unique optical, magnetic, electrical, and thermal properties (Luo et al., 2015). Compared to their bulk materials they have more strength with superior conductivity and reactivity (Annadhasan et al., 2014). The large number of surface atoms alters the surface-related properties of the particle in their nanosize. These surface atoms make these NPs very reactive for use as catalysts. In addition, high surface energy and spatial confinement properties of the NPs aid in quantum effects. Quantum confinement alters the energy band structure and charge carrier density, which in turn alters the optical and electronic properties (Buzea et al., 2007; Roduner, 2006).

Nanotechnology utilizing these nanomaterials is a multifaceted technology that has been applied in material science, electronics, energy sectors, biotechnology, medicine, and many other sectors. In agriculture, it has gained momentum over time with an abundance of public funding and government policies to combat food crises. Numerous scientific reports and patents filed in this field suggest the advancement of nanotechnology in crop protection and disease management (Parisi et al., 2015). In farming, nanointervention improves yield by increasing the efficiency of nutrient uptake and protects crops from pests through nanoformulations of pesticides. Development of new-generation pesticides with better carrier systems and decontamination of water and soil using nanotechnology also helps in crop protection. Moreover, bionanotechnology helps in better understanding host–parasite interaction at the molecular level. Preservation and packaging of food are some of the other applications of nanotechnology. It is a boon to the concept of precision farming, which aims for targeted delivery, firm attachment, and controlled release of the active material. Encapsulation or entrapment of agrochemicals using polymers and dendrimers by surface ionic attachment helps in their controlled release and improves solubility and stability. Because of their antimicrobial activity, nanoparticles of metals such as silver, zinc oxide, and titanium oxide, and nonmetals such as silica and sulfur can be used to eradicate bactericide- and fungicide-resistant pathogens (Chowdappa and Gowda, 2013; Aziz et al., 2015, 2016).

Nanomaterials in Plants, Algae, and Microorganisms
http://dx.doi.org/10.1016/B978-0-12-811487-2.00016-5

The uptake, translocation, and accumulation of nanomaterials depend on their size, concentration, suspension medium, and also the plants that can accept and reject their absorption (Tripathi et al., 2017a,b). Nanomaterials such as carbon nanomaterials are absorbed through roots or by making pores in seeds and are then translocated to the aerial parts (Rico et al., 2011; Husen and Siddiqi, 2014). Interaction of organic materials in the suspension medium with carbon nanomaterials increases their hydrophilicity and easy transport into the plant system. Their transportation takes place through capillary action and accumulates when there is a blockage or the size of the capillary is small. Fullerene has been found to be genetically transferred to the next generation (Husen and Siddiqi, 2014). Nanomaterials have a mixed effect on plants such as altering the metabolism and increasing biomass and grain yield at one side and phytotoxicity on the other side (Aslani et al., 2014; Singh et al., 2016; Tripathi et al., 2016, 2017c). Their bioavailability and impact on the environment need to be assessed. Sensors based on nanomaterials are used as diagnostic tools to detect plant infection, nutrient and moisture content, pesticide residual, temperature, and soil condition. This information can be helpful for periodic application of nutrients, pesticides, and herbicides, timely plantation, and harvesting of crops.

Here we briefly discuss various nanobased materials and their utility in plant growth, disease management, soil and water improvement, and as diagnostic tools and delivery vehicles for pesticides and genetic material.

16.2 NANOTECHNOLOGY AND PLANT GROWTH

Applications of nanomaterials have been studied to examine their negative and positive effects on plants. Effects of carbon-based nanomaterials such as fullerene, fullerol, graphene, carbon nanotubes (CNTs), and metal and metal oxide NPs have also been studied on plant growth (Aslani et al., 2014; Tripathi et al., 2015; Shweta et al., 2016). Research in nanomaterials in plant growth demonstrated an increase in tobacco cell culture growth rate by 55%–64% using multiwalled carbon nanotubes (MWCNTs) (Khodakovskaya et al., 2012). CNTs as growth regulators upregulate the genes involved in water transport, cell division, and cell wall formation such as aquaporin NtPIP1, CycB, and NtLRX1 and help in seed germination. MWCNTs can easily penetrate through the hard seed coat and enhance water uptake thereby increasing germination rate. Studies have shown a 20% increase in tomato seed germination rate as compared to control (Chowdappa and Gowda, 2013). Therefore there is scope for functionalizing MWCNT to deliver desired molecules and promote growth and resistance.

Zinc is a cofactor for phytase and phosphatase enzymes, which hydrolyze the ester bond between phosphorus and calcium/aluminum/iron. Free phosphorous groups thus formed are available for plants to absorb them from soil. Both the enzymes are of plant or microbial origin, which curtails the addition of phosphorus. Reports suggest that zinc oxide NPs also have the same effect as bulk ZnO, and plants (mungbean) treated with ZnO NPs display higher stem height, root length and area, and nodule, biomass, and grain yield (Raliya et al., 2016). However, in another case, ZnO NPs showed toxic impacts in wheat seedlings and caused a significant alteration in the physiological, morphological, and biochemical traits of plants (Tripathi et al., 2017d). Despite huge advances in the utility of nanomaterials for pest control, their effects are contradictory (Husen and Siddiqui, 2014). Various roles of nanotechnology are shown in Fig. 16.1.

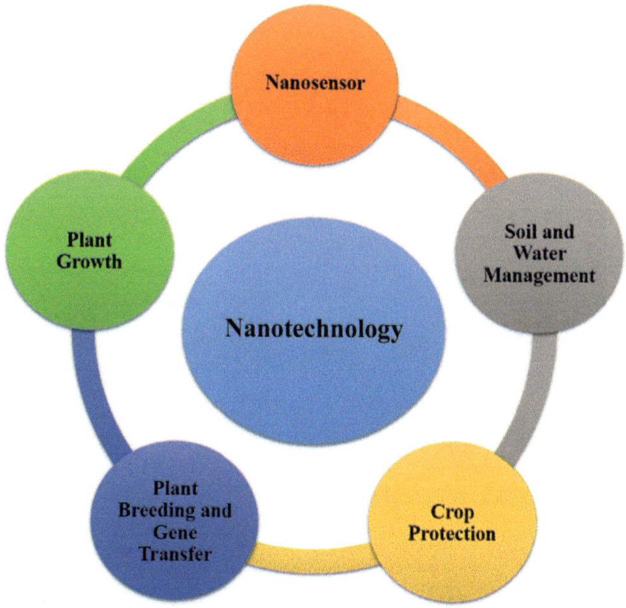

FIGURE 16.1 Various applications of nanotechnology in crop improvement.

16.3 NANOTECHNOLOGY IN CROP PROTECTION

Crops are attacked by a wide variety of attackers including weeds (monocots, dicots, and parasitic weeds), pests (insets, mites, rodents, birds, and mammals), and phytopathogens (bacteria, fungi, and viruses). These attackers can be classified as stand reducers, photosynthetic rate reducers, leaf senescence accelerators, light stealers, assimilate sappers, and tissue consumers (Kitherian, 2017). For example, *Helicoverpa armigera* feeds on more than 150 crops throughout the world including tomato, chickpea, pigeon pea, etc. (Vinutha et al., 2013). The fruitfly (*Bactrocera dorsalis* Hendel) is a polyphagus pest that feeds on various vegetable crops and fruits such as guava, apple, mango, orange, and banana. Insects also act as vectors for various plant diseases causing serious crop damage. Plant bacterial pathogens *Pseudomonas syringae, Ralstonia solanacearum, Agrobacterium tumefaciens, Xanthomonas oryzae* pv. *oryzae, Xanthomonas campestris, Xanthomonas axonopodis* pathovars, *Erwinia amylovora, Xyllela fastidiosa, Dickeya*, and *Pectobacterium carotovorum* are important (Mansfield et al., 2012). Fungal pathogens such as *Aspergillus, Blumeria graminis, Botrytis cinerea, Cladosporium, Colletotrichum* sp., *Fusarium graminearum, Fusarium oxysporum, Magnaporthe oryzae, Melampsora lini, Mycosphaerella graminicola, Penicillium, Puccinia* sp., *Phytophthora, Pythium, Rhizopus*, sp., *Rhizoctonia solani, Sclerotinia sclerotiorum*, and *Ustilago maydis* cause damage equally (Dean et al., 2012). Crop rotation, selection of the right variety of crop, time of sowing, and mixed cropping are some of the traditional ways of crop protection. Later, chemical pesticides were brought in to eradicate pests and pathogens (Glass and Thurston, 2014). However, most of the pesticides used for crop protection are lost in air or as runoff or spray drift, which results in off-target deposition. Only less than 0.1% of the applied pesticides reach the target; the rest

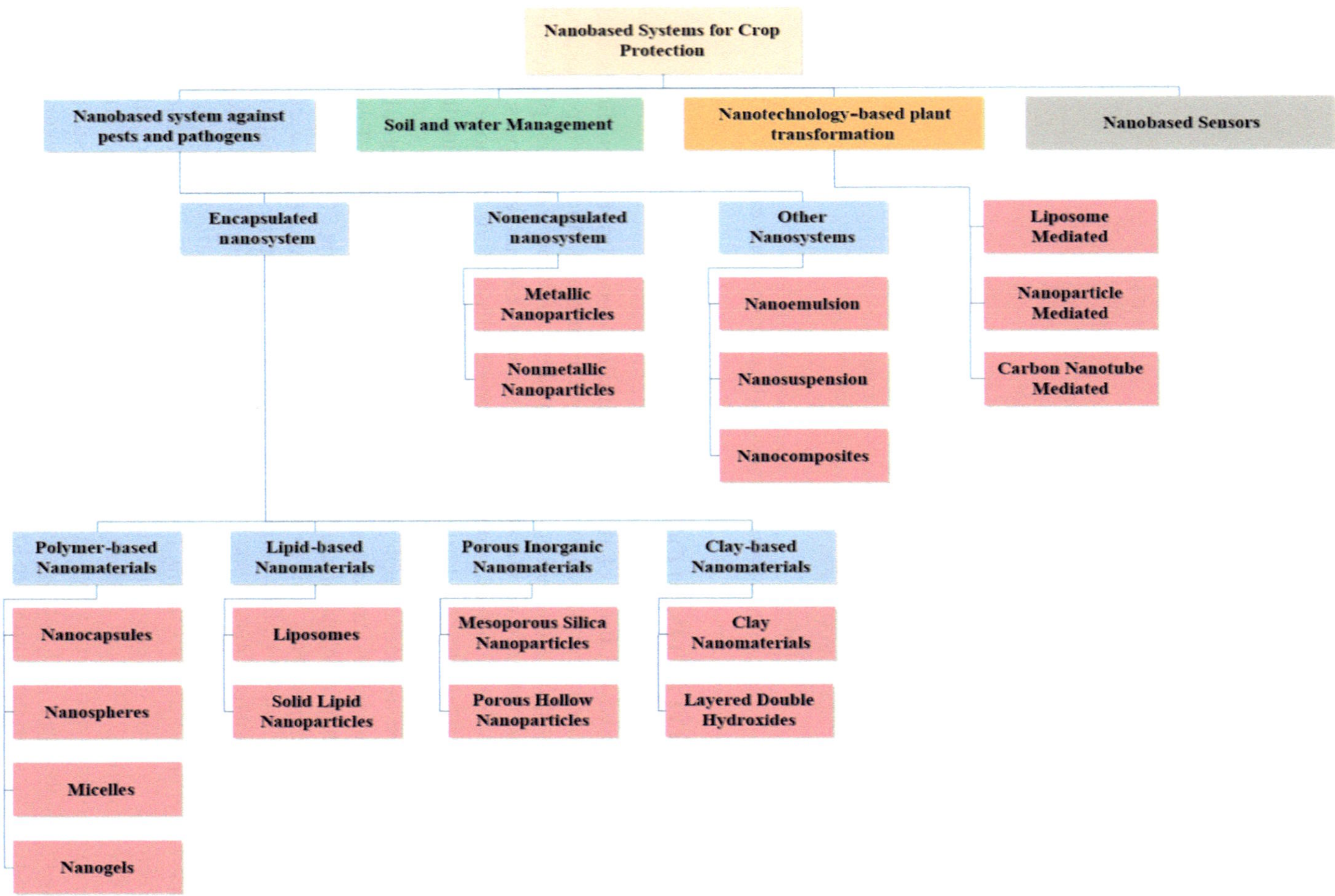

FIGURE 16.2 Different nanobased systems for crop protection. *Adapted from Nuruzzaman, M., Rahman, M.M., Liu, Y., Naidu, R., 2016. Nanoencapsulation, nano-guard for pesticides: a new window for safe application. J. Agric. Food Chem. 64 (7), 1447–1483.*

are photodegraded or pollute the environment (Chowdappa and Gowda, 2013). Utility of nanomaterials is an advanced and alternative method to overcome the foregoing limitations. Various nano-based systems involved in crop protection are systematically represented in Fig. 16.2. The advantages of using nanomaterials in agriculture are:

- Lessening the amount of agrochemicals being used for pest control;
- Target-specific delivery of the biocide;
- Controlled release of active ingredients;
- Nanoformulation of the insoluble chemicals to increase their solubility;
- Increased stability of pesticides by protecting them from washaway, leaching, evaporation, photodegradation, hydrolysis, and microbial degradation;
- Bioaccumulation of harmful chemicals and environmental health protection as a result of fewer pesticides.

Pesticides are encapsulated into various forms to achieve the foregoing advantages. The encapsulating material that forms the shell around the active ingredient controls the physical, chemical, and release property of the nanopesticides thus formed. For encapsulation, the active ingredients are loaded inside or trapped in the polymeric material forming a nanosized structure. However, there is controversy as to whether 1–1000 nm is the nanosize of a colloidal system for pesticide delivery or if it should be 1–100 nm, which is the actual range of nano. In this chapter, we have considered 1–1000 nm as the nanorange (Nuruzzaman et al., 2016). The various nanobased systems for crop protection are discussed in detail in this chapter.

16.3.1 Encapsulated Nanosystems for Crop Protection

16.3.1.1 Biopolymers

Biopolymers are derived from biological sources such as plants, microbes, animals, agriculture wastes, and fossils or synthesized chemically from biological monomer units such as sugars, amino acids, natural fats and oils, and nucleotides (Sanchez-Vazquez et al., 2013; Perlatti et al., 2013). These biopolymers include cellulose, starch, chitin, chitosan proteins, peptides, DNA, and RNA. As compared to synthetic polymers, biopolymers are structurally well defined (lignocelluloses being an exception) and depend on their primary structure constituted by sequential repetition of monomer units. They are readily available and can be produced in large quantities. They are a sustainable source of nontoxic, biodegradable, eco-friendly, and low-cost polymeric materials. Their biocompatibility has been explored in agriculture (Perlatti et al., 2013).

16.3.1.2 Synthetic Polymer

Amphiphilic block copolymers have drawn researchers' interest for the effective delivery of pesticides because they have both hydrophilic and hydrophobic parts. Block polymers are synthesized using two or more different polymer chains that are linked together. Biblock (AB), triblock (ABA), multiblock (A_nB_n), heteroarmstar (AB_3), Y structure (A_2B), and miktoarm (ABC) are the types of copolymer structures depending upon the number of blocks used (Ruzette and Leibler, 2005). Examples of synthetic copolymers are poly(D, L-lactide) and poly(ε-caprolactone) (Kothamasu et al., 2012).

16.3.1.3 *Polymeric-Based Nanomaterials for Crop Protection*

Polymers such as natural biopolymers and synthetic polymers are used to encapsulate the active ingredients for pesticide delivery (Nuruzzaman et al., 2016). These encapsulating materials form various kinds of structures such as nanocapsule, nanosphere, micelle, and nanogel (Fig. 16.3) for the delivery of active ingredients, which are briefly described next.

16.3.1.3.1 NANOCAPSULE

Nanocapsules consist of an inner solid or liquid core containing active material or pesticides and a protective shell of polymeric material. They may contain a liquid core surrounded by polymeric membrane, polymeric matrix covered by a polymeric membrane, or active substance in molecular dispersion (Mora-Huertas et al., 2010). The materials used for nanocapsule formation are listed in Table 16.1. There are various methods of forming nanocapsules such as nanoprecipitation, emulsification–diffusion, emulsification–coacervation, double emulsification, polymer coating, and layer-by-layer methods. In addition, other modified methods include melt dispersion, emulsion polymerization, interfacial polymerization, interfacial deposition method, solvent displacement technique, emulsion evaporation, etc. Of all these methods, interfacial deposition of preformed polymer is commonly used. In this method, drug is mixed in a water-miscible solvent with or without surfactant and a similar solvent-miscible oil is added afterward. The mixture is then dispersed in an aqueous phase containing hydrophilic surfactant, which upon agitation forms nanocapsules (Letchford and Burt, 2007; Nuruzzaman et al., 2016). In a study, azidobenzaldehyde and amphiphillic biocopolymer were mixed with carboxymethyl chitosan. Nanocapsules with aqueous core were formed when polymers were self-assembled in aqueous media followed by photocrosslinking. Methomyl dispersed in aqueous medium of pH 4.0 was then loaded into the nanocapsules after sonication. At pH 6.0, methomyl was released with a half released time ($t_{1/2}$) of 36.3–69.5 h. The insecticidal activity of the methomyl loaded showed improved insecticidal activity against armyworm larvae (Sun et al., 2014).

16.3.1.3.2 NANOSPHERE

The active ingredients are uniformly embedded in a polymeric matrix as a nanosphere. The synthesis process of a nanosphere is similar to that of nanocapsules; however, their polymerization is crucial (Letchford and Burt, 2007). For polymeric nanosphere formation,

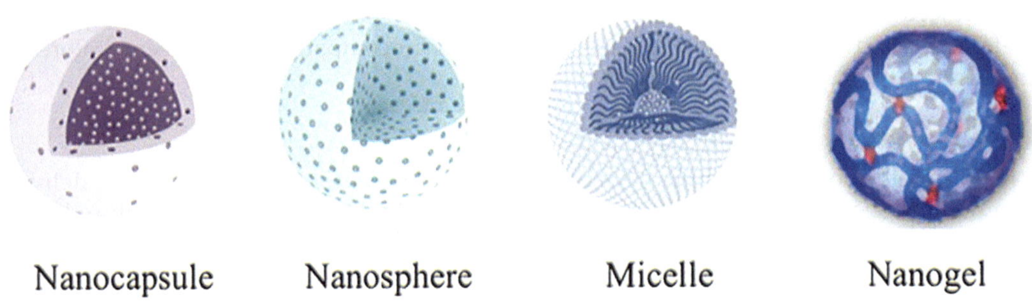

FIGURE 16.3 Various nanoforms of polymeric-encapsulated pesticides. *Reprinted with permission from Nuruzzaman, M., Rahman, M.M., Liu, Y., Naidu, R., 2016. Nanoencapsulation, nano-guard for pesticides: a new window for safe application. J. Agric. Food Chem. 64 (7), 1447–1483. Copyright © 2016, American Chemical Society.*

TABLE 16.1 List of Polymer-Based Nanomaterials Used for Crop Protection

Nanostructure	Active Ingredient	Bipolymer	Effect/Target Phytopathogen	References
Microsphere	Diazinon, trifluralin, and alachlor	Intelimer polymer	Differential release in response to temperature, prolonged activity of pesticides, reduced crop phytotoxicity	Greene et al. (1992)
	Carbaryl	Gelucire 54/02	Slow release of active material, reduced vertical mobility of the pesticide into the soil	Quaglia et al. (2001)
	Acetamiprid	D, L-Polylactide poly(ε-caprolactone)	Slow release of active material	Takei et al. (2008)
Nanosphere	Active ingredient	Poly(ε-caprolactone)	Enhanced penetration in the plant because of its small size	Boehm et al. (2003)
Microcapsule	Sulprofos	Hydrolyzed polyvinyl alcohol	*Heliothis virescens* (tobacco budworm)	Latheef et al. (1993)
	Cyfluthrin	Not mentioned	*Sitophilus oryzae*	Arthur (1999)
	Bacillus thuringiensis Berliner	Lignin with pregelatinized corn flour	*Ostrinia nubilalis*, *Trichoplusia ni*	Tamez-Guerra et al. (2000b)
	Lavandin oil	Polyethylene glycol (PEG)	Increased stability of the active ingredient	Varona et al. (2010)
	Pyriproxyfen	Polyurethane	Quick liberation of the active ingredient	Tsuda et al. (2012)
Nanocapsule	Tebuconazole and chlorothalonil	Polyvinylpyridine and polyvinylpyridine-*co*-styrene	Controlled release of the active ingredient	Liu et al. (2001)
	Carbaryl	Sodium alginate, sodium alginate/gelatin, sodium alginate/sodium carboxymethyl cellulose blend bead linked by glutaraldehyde	Carbaryl release in gelatin and carboxymethyl cellulose is higher	Işiklan (2006)
	Garlic essential oil	PEG	Insecticidal activity against adult *Tribolium castaneum*	Yang et al. (2009a,b)
	Piperonyl butoxide and deltamethrin	High-density and low-density polyethylene	For bed nets against mosquitoes	Frandsen et al. (2015)

(Continued)

TABLE 16.1 List of Polymer-Based Nanomaterials Used for Crop Protection—cont'd

Nanostructure	Active Ingredient	Bipolymer	Effect/Target Phytopathogen	References
	Essential oil of *Cuminum cyminum*	Poly(urea-formaldehyde)	Highly effective against *T. castaneum*	Negahban et al. (2012)
	Imidacloprid	Lignin–PEG matrix	Decreases the amount of applied insecticide	Flores-Céspedes et al. (2012)
	β-Cyfluthrin	PEG	*Callosobruchus maculatus*	Loha et al. (2012)
	Imidacloprid	Sodium alginate	More effective than nonencapsulated pesticide	Kumar et al. (2014)
	Pyridalyl		Effective against *Helicoverpa armigera* larvae	Saini et al. (2014)
	Fipronil in oil	Silica shell	Sustained release of the insecticide	Wibowo et al. (2014)
	Methomyl	Bicopolymer of azidobenzaldehyde and carboxymethyl chitosan	Increased insecticidal activity against armyworm	Sun et al. (2014)
Microgranules	Nuclear polyhedrosis	Pregelatinized corn flour and potassium lignate	*T. ni*	Tamez-Guerra et al. (2000a, 2002)
	Acetamiprid	Tapioca starch, urea, and sodium borate	Controlled release and provides stability to the pesticide	Cao et al. (2005)
Microcrystals	Imidacloprid	Chitosan and sodium alginate	Prolonged release and increased stability of insecticide	Guan et al. (2008)
Spray-dried nanocapsules	*Cydia pomonella* L. granulovirus	Lignin	Virus solar protection	Arthurs et al. (2006)
Nanomicelle	Carbofuran	Copolymer of PEG with dimethyl 5-hydroxyisophthalate	Controlled release of insecticide	Shakil et al. (2010)
	Imidacloprid	Copolymer of PEG and aliphatic diacids	Controlled release of insecticide	Adak et al. (2012)
	Thiamethoxam	Copolymer of PEG and aliphatic and aromatic diacids	Controlled release formulation	Sarkar et al. (2012)
	Diethylphenyl-acetamide	PEG	Increased stability and bioefficacy of poorly soluble diethylphenyl-acetamide	Balaji et al. (2015)

TABLE 16.1 List of Polymer-Based Nanomaterials Used for Crop Protection—cont'd

Nanostructure	Active Ingredient	Bipolymer	Effect/Target Phytopathogen	References
	Carbendazim	PEG-based functionalized amphiphilic copolymers	Controlled release formulations to increase bioefficacy against *Rhizoctonia solani*	Koli et al. (2015)
	Rotenone	*N*-(Octadecanol-1-glycidyl ether)-*O*-sulfate chitosan,	Increased solubility of rotenone,	Lao et al. (2010)
		N,N-dimethylhexadecyl carboxymethyl chitosan	controlled release of rotenone	Kamari et al. (2016)
	Dichlorophenoxyacetic acid	PEG-conjugated *o*-nitrobenzyl group grafted	Controlled release of herbicide	Ding et al. (2016)
Nanogel	*Lippia sidoides* oil	Chitosan and cashew gum	Controlled release and more effective larvicidal activity	Ashton Acton (2013)
	Pheromone (methyl eugenol)	All-trans tri(*p*-phenylenevinylene) bis-aldoxime	Entrapment of pests	Bhagat et al. (2013)
	Cuminum cyminum essential oil	Myristic acid–chitosan	Improved efficacy of cumin oil against *Sitophilus granarius*, *Tribolium confusum*	Ziaee et al. (2014)

poly(ε-caprolactone) and an active ingredient were dissolved in acetone and injected into aqueous medium containing surfactant with moderate stirring. The size, dispersity, and encapsulation efficiency of the nanospheres differ with the type of surfactant being used. Boehm et al. (2000) reported that among various surfactants used in the study, the highest encapsulation efficiency was obtained with the surfactant "Monotanox 80." The active ingredient release was controlled by the surrounding medium. Various agrochemicals loaded in nanospheres are listed in Table 16.1.

Nanospheres of 135 nm size containing 3.5% insecticide were obtained by nanoprecipitation with Eudragit S100 (Boehm et al., 2003). The effect was studied when used on aphid-infested cotton plant with control experiments. The study revealed that speed of action and controlled release were not as good as the reference compound. However, the nanospheres showed better penetration because of their small size (Boehm et al., 2003).

16.3.1.3.3 MICELLE

Micelles are used to deliver water-insoluble agrochemicals. Amphiphillic block copolymers self-assemble themselves in an aqueous medium to form micelles. In the case of water-soluble copolymers such as polyethylene glycol (PEG) and polyvinylpyrrolidone, two methods can be employed for micelle formation. First, copolymers are dissolved in an aqueous medium above critical micelle concentration and pesticides are added, which form the core of the micelle. Second, copolymers and pesticide dissolved in a volatile solvent are cast to form a thin film. Then, warm water is added with continuous agitation to obtain pesticide-loaded micelle. For water-insoluble copolymers, the dialysis method or oil-in-water emulsion method is commonly used. In the dialysis method, water is added to a water-miscible solvent containing polymer and pesticide. Conversely, the solvent containing polymer and pesticide can be added to water in a stirring condition and after dialysis with water, micelle can be recovered. In the oil-in-water emulsion method, water-insoluble polymer and pesticide are mixed in an organic solvent and added to water with or without surfactant at stirring condition. Evaporation of the organic solvent then results in micelle formation (Letchford and Burt, 2007). In a study, PEG-based amphiphillic polymers were used to form the main chains of copolymer using dimethyl 5-hydroxyisophthalate as linker in an enzymatic reaction. These polymers and carbofuran were mixed in dichloromethane and the solvent was evaporated after shaking. Water was added to the residue and mixed well. After filtration and freeze drying, nanomicelles encapsulating carbofuran were obtained (Shakil et al., 2010).

R. solani, a fungal pathogen, causes various plant diseases such as root collar rot, damping off, and wire stem and sheath blight in rice plants resulting in major loss of yield. Carbendazim, a broad-spectrum benzimidazole fungicide, is used to control *R. solani* fungus. A nanoformulation of carbendazim has been developed in the laboratory using PEG and the release kinetics have been studied by Koli et al. (2015). The release of carbendazim can be optimized by optimizing molecular weight of the PEG so that single application of the formulation can control disease for the whole lifespan of crops (Koli et al., 2015).

16.3.1.3.4 NANOGELS

Nanogels are formed of hydrophilic polymers crosslinked by van der Waal's forces or covalent bonds. They can be prepared by various methods described in detail by Sultana et al. (2013). Crosslinking of polymers can be done by UV light as used in photolithography and the nanogels can be molded into shapes using a quartz template. Similar to photolithography, the micromodeling method involves suspension of cells in the hydrogel precursor solution in the presence of a photoinitiator such as polyethylene (glycol) diacrylate (Sultana et al., 2013). The resultant solution was poured onto a hydrophilic pattern division multiple access pattern and photocrosslinked. Microgels of polymers such as chitosan and hyaluronan can be obtained by water-in-oil heterogeneous emulsion, aqueous homogeneous gelation, spray drying, and chemical crosslinking (Sultana et al., 2013).

The water-in-oil emulsion method involves emulsification of aqueous solution of water-soluble polymers in oil phase in the presence or absence of surfactant with vigorous stirring. Crosslinking of aqueous droplets is performed by adding crosslinkers such as adipic dihydrazide. The microgel particles are suspended in organic solvents, centrifuged, and washed with organic solvents. In the membrane emulsification technique, a membrane is used through which solution (to be dispersed phase) is passed. The size of microgels depends on the pores of the membrane and the flowing phase inside the membrane system. These emulsions can be water-in-oil, oil-in-water, oil-in-water-in-oil, and water-in-oil-in-water. In the chemical process of crosslinking, nanogels are formed by precipitation polymerization, emulsion polymerization, dispersion polymerization, and radical polymerization. Nanogels can also be prepared by grinding down the macrogels and sieving for the particular size. The active ingredients are loaded by covalent conjugation, physical entrapment, or self-assembly methods. Sultana et al. (2013) reported preparation of nanogel made up of chitosan and myristic acid, which were then crosslinked by amide linkage using 1-ethyl-3-(3-dimethyl-aminopropyl)carbodiimide (EDC). *Cuminum cyminum* essential oil was added to the nanogel and sonicated resulting in oil-loaded nanogels. Fumigant toxicity of the oil-loaded nanogel against *Sitophilus granarius* has been found to show prolonged effect compared to only oil (Ziaee et al., 2014).

B. dorsalis is a polyphagus pest that infects vegetables and a variety of fruits such as guava, banana, mango, orange, etc. (Bhagat et al., 2013). Several methods have been adopted to control this pest such as destruction of infested fruits, cold and heat treatment, bait sprays, insecticides, etc. (Bhagat et al., 2013). However, some of these methods are ineffective and some cause damage to off-target species. Therefore nowadays pheromones are used to attract insects because of their species specificity and nontoxicity toward other species. Hydrogels that were used to deliver these pheromones were inefficient because of their changing nature with changing climatic conditions (Bhagat et al., 2013). Researchers tried to immobilize methyl eugenol (pheromone) in nanogels prepared from gelator I to trap insects (Bhagat et al., 2013). This results in high residual activity, increased shelf-life, and excellent efficacy of the pheromones. The pheromone evaporates slowly in ambient conditions and attracts insects. This nanogel-based insect trap has been tested in the field and the results are promising (Bhagat et al., 2013).

16.3.1.4 *Lipid-Based Nanomaterials for Crop Protection*

Lipid-based nanomaterials such as nanoliposomes and solid lipid NPs are useful because they can readily target the insect cuticle and deliver hydrophilic, hydrophobic, and lipophilic active ingredients (Nuruzzaman et al., 2016). Nanoliposomes are made up of an inner aqueous core surrounded by a lipid bilayer mostly composed of phospholipids. Other molecules such as sterols are also incorporated in the bilayer membrane ensuring membrane stability because of increased fluidity. When energy in the form of sonication and high-pressure homogenization is applied, vesicles are formed. Methods such as extrusion, microfluidization, detergent removal, and reversed phase evaporation are also used for vesicle formation (Taylor et al., 2005; Mozafari, 2005, 2010). On the other hand, in the absence of applied energy, membranes such as flattened structures are formed. Nanoliposomes are stabilized by lyophilization, spray drying, freeze drying, and supercritical fluid technology. Solid lipid NPs are superior in terms of stability, nontoxicity, loading capacity, and target-specific release properties. Their preparation requires melted lipid molecules and active ingredients at the same temperature and dispersed in aqueous medium containing surfactant using a high-pressure homogenizer. Recrystallization of solid lipid NPs results in active ingredients trapped in the NPs. For temperature-labile active ingredients, temperature is lowered after adding the ingredients to the melted lipid. Then the solid lipid microparticles are ground and dispersed in cold aqueous medium containing surfactant homogenized with high pressure. The active material can be present as a homogeneous matrix, concentrated in the core, or dispersed in the shell of the solid lipid NP. Deltamethrin is a common pesticide that is degraded under UV light. Chitosan-coated, deltamethrin-loaded, beeswax solid lipid NPs were prepared using the hot homogenization method and sonication, which protected the pesticide from photolysis

(Nguyen et al., 2012). *Spodoptera litura* is a polyphagus insect that infects 150 kinds of plants. Etofenprox is an insecticide that is used to control *S. litura* infestation in Korea. However, there is a need for frequent spraying of the insecticides because of decreased residual activity. Therefore nanosized controlled release formulations of these insecticides were prepared where lecithin forms the liposome and chitosan forms the coating. The release time of etofenprox was controlled by varying the concentration of chitosan and using chitosan of different molecular weight. This encapsulated etofenprox has a dual mechanism of action: one is contact poison and the other is digestive poison. The insect either directly comes in contact with the encapsulated pesticides or swallows it. In the latter case, the pesticide is released in the midgut of the insect because of the alkaline environment. Controlled release formulations of pesticides enhance the effect of the pesticides by protecting them from various environmental conditions (Hwang et al., 2011). Various effects of lipid-based nanomaterials on crop protection are given in Table 16.2.

16.3.1.5 Inorganic Porous Nanomaterials for Crop Protection

Porous silica NPs are formed when active ingredients are adsorbed inside the NPs without forming any linkage. Surfactant forms the basic skeleton on which silica oxides are deposited and condensed. The removal of the surfactants results in the synthesis of these NPs. These NPs are nontoxic and biocompatible with large surface area and tunable pore size and volume. They offer a variety of surface functionalizations, which can be used for controlled release of pesticides against plant pathogens (Lodha et al., 2012). Mesoporous silica NPs and hollow silica NPs are two major forms of inorganic porous NPs. The functionalized mesoporous silica NPs are absorbed in the insect cuticular membrane lipids, damaging their protective waxy layer and causing dehydration. Wang et al. (2016) reported that loading of avermectin pesticide in the inner core of mesoporous silica NPs resulted in their protection

TABLE 16.2 List of Lipid-Based Nanomaterials for Crop Protection and Their Effects

Nanostructure	Active Ingredient	Polymer	Effect	References
Nanoliposomes	Etofenprox	Chitosan-coated lecithin-forming liposome	Controlled release of insecticides	Hwang et al. (2011)
	Alpha-cypermethrin			Bang et al. (2011)
Solid lipid nanoparticle	*Artemisia arborescens* L. essential oil	Glycerol behenate as lipid and Poloxamer 188 or Miranol Ultra C32 as surfactants	Reduced evaporation of the essential oil	Lai et al. (2006)
	Deltamethrin	Beeswax solid lipid nanoparticle coated with chitosan	Deltamethrin is protected against photodegradation	Nguyen et al. (2012)
	Carbendazim, tebuconazole	Myritol	Reduced toxicity and controlled release of fungicides	Campos et al. (2015)
	Geranium essential oil	Stearic acid	Increased efficiency against *Phthorimaea operculella*	Adel et al. (2015)

against photodegradation and slow release. In this study, mesoporous silica NPs were prepared in a step-by-step process. Calcium carbonate was dissolved in distilled water and surfactant (hexadecyltrimethyl ammonium bromide) was added. After heating at 353 K, sodium silicate was added with continuous stirring. Two hours later, the suspension was filtered and vacuum dried at 373 K. The calcium carbonate shell was removed after calcination by HCl treatment. The NPs were collected after filtration and vacuum dried. Avermectin was then loaded into the prepared mesoporous silica NPs by supercritical carbon dioxide technology. Acetone was added to the mixture and kept in high pressure adsorption apparatus, which after vacuum drying formed mesoporous silica NP-loaded avermectin (Wang et al., 2016). The UV shielding properties of the avermectin-loaded porous hollow silica NPs were evaluated along with the effect of temperature and pH of the medium. Because of the entrapment of the avermectin, photostability of the pesticide increased. However, UV shielding properties of entrapped avermectin declined with increased intensity of UV, pH, and temperature (Wang et al., 2016). Similarly, Cao et al. (2016) reported successful loading of pyraclostrobin into mesoporous silica NPs, which resulted in sustained release of the pesticide. The capping of the mesoporous silica NPs with water-soluble chitosan resulted in increased pesticide loading efficiency and faster release rate than bare mesoporous silica NPs. The fungicidal activity of the encapsulated pesticide against *Phomopsis asparagi* (Sacc.) was studied and it was reported that half doses of pyraclostrobin-loaded capped mesoporous silica NPs possessed the same fungicidal activity (Cao et al., 2016). The effects of inorganic porous nanomaterials on crop protection are described in detail in Table 16.3.

16.3.1.6 Clay-Based Nanomaterials for Crop Protection

Generally, aluminum silicates or hydrous silicates or phyllosilicates constitute clay materials. Tetrahedral silicate or octahedral aluminum are two types of sheets and depending on the sheet arrangement, they can be classified as 1:1 (kaolinite), 2:1 (montmorillonite), or 2:1:1 (chlorite). The thickness of the sheets is in the nanometer range and therefore regarded as nanomaterials irrespective of their length, which is in millimeters. Moreover, the layers are interactive because of van der Waal's forces, hydrogen bonding, and electrostatic force between interlayer cations. The active ingredients can move into the interlayer spaces, which is called intercalation. The clays are modified to facilitate the desired intercalation of active ingredients. An ion exchange reaction between hydrophilic clay and organic cation results in widened interlayer spaces and changes in surface properties (Suresh et al., 2010). Polymers and nanoclays can be nanocomposited to form various types of polymer–nanoclay nanocomposites, as discussed later. The diffusion of the loaded active ingredient is slow because of less porosity of the nanoclay and therefore results in obstructed diffusion. Chevillard et al. (2012) prepared a nanoformulation of pesticide (ethofumesate) by combining wheat gluten and montmorillonites using a bis-vis extrusion process. The release of the pesticide could be affected by the interaction between pesticide and montmorillonites, and the interaction between gluten and montmorillonites. The slow release of pesticide was governed by the nature of clays, matrix, and pesticide. The release of hydrophobic ethofumesate can be slowed down in the presence of hydrophobic clays. Apart from clay materials, clay-based nanomaterials include layered double hydroxides (LDHs), which are made up of minerals or synthetic materials with metal hydroxides constituting positively charged brucite-type layers (Table 16.3). LDHs are synthesized by direct synthesis methods such as coprecipitation,

TABLE 16.3 List of Inorganic Porous Nanomaterials and Clay-Based Nanomaterials for Crop Protection and Their Effect

Nanostructure	Active Ingredient	Polymeric Material	Effect	References
Porous nanoparticles	Validamycin	Porous hollow silica nanoparticles	Sustained release of pesticides	Liu et al. (2006)
	Abamectin	Porous silica nanoparticles	Controlled release, photostability of pesticides, and increased water dispersity of nanoparticles	Wang et al. (2014)
Porous hollow nanoparticles	Avermectin	Porous hollow silica nanoparticles	Controlled release of pesticide	Li et al. (2007)
Mesoporous nanoparticles	Pyoluteorin	Mesoporous silica nanoparticles	Slow release and prolonged antifungal effects	Chen et al. (2010)
	Imidacloprid	Mesoporous silica nanoparticles	Adsorption and controlled release of imidacloprid against termites	Popat et al. (2012)
Mesoporous nanoparticles	Fabricated on stainless steel fiber	Mesoporous TiO_2 nanoparticles	Solid phase microextraction of organochlorine pesticides	Liu et al. (2015)
	Pyraclostrobin	Quaternized chitosan-capped mesoporous silica nanoparticles	Reduced the amount of pesticide application retaining its efficacy	Cao et al. (2016)
	Large surface area-to-volume ratio, active surface sites, and open channel pores	Mesoporous alumina sphere nanoparticles	High antifungal efficacy *Fusarium oxysporium*	Shenashen et al. (2017)
	Metalaxyl	Tetraethylorthosilicate, cetyltrimethylammonium bromide	Storage and controlled release	Wanyika (2013)
Layered double hydroxides (LDHs)	Beta-cypermethrin in sulfonated hydroxyethyl-β-cyclodextrin	ZnAl-LDH	Controlled release of pesticide at specific pH	Zhang et al. (2016)

salt oxide, hydrothermal synthesis, induced hydrolysis, the sol–gel method, or by indirect synthesis involving ion exchange in solution, ion exchange in acidic medium, and regeneration methods. The detailed procedures of the methods are described by Benício et al. (2015). Pesticides can be intercalated through an ion exchange process for their controlled release (Nalawade et al., 2009). In a study 2,4-dichlorophenoxyacetate was ion exchanged with zinc ions of zinc–aluminum LDHs. $ZnCl_2$ and $AlCl_3$ aqueous solution was mixed in a reactor at a fixed ratio and NaOH was added. After stirring for 34 h at room temperature, precipitate was filtered, washed, and dried at 30°C. However, at pH 7, exchange reactions were carried out and the total exchange was observed when the dichlorophenoxyacetic acid (2,4-D)/[Zn—Al—Cl] ratio was 11 (Lakraimi et al., 2000).

16.3.2 Nonencapsulated Nanopesticide for Crop Protection

A nonpolymeric system of pesticide delivery includes use of metallic, nonmetallic NPs and their combination as hybrids against plant pests and pathogens (Chowdappa and Gowda, 2013). They exhibit unique physical, chemical, optical, and biological properties (Swathy, 2014). Metallic NPs including silver, silver oxide, gold, zinc oxide, titanium oxide, magnesium oxide, and copper oxide NPs exhibit antimicrobial activity (Dizaj et al., 2014). These NPs are explored for their bactericidal, insecticidal, and fungicidal activity against phytopathogens alone or in combination with other metallic NPs. Metal NPs inhibit or delay the growth of many pathogen species because of their diverse mode of inhibition. Therefore these NPs can be used as new antimicrobial agents and as an alternative to synthetic fungicide (Chowdappa and Gowda, 2013). Metallic NPs can be synthesized via two major approaches: a top-down approach or a bottom-up approach. In the top-down approach, a bulk metal is deconstructed to NPs using mechanical or chemical techniques, but they suffer from surface imperfections. In the bottom-up approach, NPs are synthesized from atoms, molecules, or clusters and are homogeneous in chemical composition. The bottom-up approach can be performed mechanically, chemically, and biologically. Among all, the biological approach involving synthesis of NPs using biological components is energy saving, cost-effective, and eco-friendly (Thakkar et al., 2010; Keat et al., 2015; Ahmed et al., 2016). Green synthesis is one of the bottom-up approaches which uses phyto-constituents (phenolics, flavonoids, tannins, etc.), enzymes, proteins, sugars, and polysaccharides for the synthesis of metallic nanoparticles. The process involves reduction of metal salts, capping, and stabilization. Apart from metallic NPs, other materials such as silica-based, C-based, sulfur-based, and aluminum-based NPs are also used in crop improvement (Suriyaprabha et al., 2014; Suleiman et al., 2013; Wang et al., 2009; Servin et al., 2015). Various methods of metal NP synthesis are depicted in Fig. 16.4.

16.3.2.1 Silver Nanoparticles

Silver nanoparticles (AgNPs) can be synthesized biologically from $AgNO_3$ using plant extract or microorganisms for bioreduction of Ag^{++} to Ag^0 (Ahmed et al., 2016; Srikar et al., 2016; Prasad et al., 2016). AgNPs can be used as nanopesticides because these are toxic to microbes but nontoxic to animals. They have a broad spectrum of antibacterial activity and are also potent against spore-forming fungi. Because of their high surface area-to-volume ratio, they easily come in contact with microorganisms and induce gaps in the plasma membrane because of its interaction with sulfhydryl groups of proteins, finally causing membrane damage. After penetrating into the bacterial cell, interaction with DNA causes cell cycle halt at the G2/M phase, which leads to apoptosis. The antimicrobial activity depends on the size, shape, and surface of the AgNPs (Keat et al., 2015). Reports suggested that AgNPs are toxic against a broad range of plant pathogens. AgNPs damage *Raffaelea* sp. fungal hyphae by interfering with their nutrient uptake and hence inhibiting their growth and conidia formation.

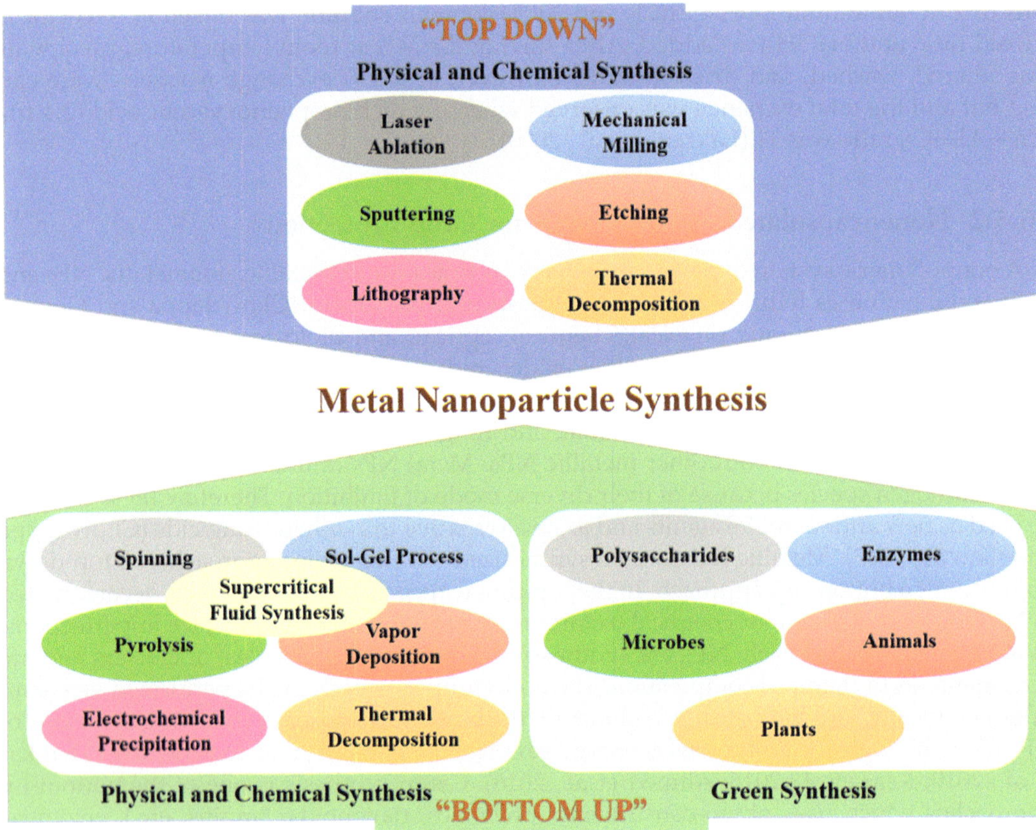

FIGURE 16.4 Methods of metallic nanoparticle synthesis. *Modified from Keat, C.L., Aziz, A., Eid, A.M., Elmarzugi, N.A., 2015. Biosynthesis of nanoparticles and silver nanoparticles. Bioresour. Bioprocess. 2, 47. http://dx.doi.org/10.1186/s40643-015-0076-2.*

They obstruct microbial growth by inhibiting spore germination, which makes them a potent nanopesticide (Alghuthaymi et al., 2015). As a potent antimicrobial, AgNPs can be used for plant pathogen management. The list of phytopathogens tested for AgNP efficacy can be found in Table 16.4.

16.3.2.2 ZnO Nanoparticles

Synthesis of ZnO NPs has been described in detail by Kołodziejczak-Radzimska and Jesionowski (2014) and Mishra et al. (2014a,b). ZnO NPs impose a photocatalysis, photooxidizing effect on biological species because of their enhanced surface reactivity (Sirelkhatim et al., 2015). Antibacterial activity of ZnO NPs is caused by membrane damage caused by direct or electrostatic interaction between NPs and the bacterial cell surface. The NPs are internalized by the bacterial cells, which cause reactive oxygen species (ROS) release. ZnO NPs under UV illumination cause phototoxicity in aqueous solution because of the production of H_2O_2 and superoxide ions (O^{2-}). Photostimulation causes desorption of oxygen molecules from the surface. The H_2O_2, O_2^-, and OH* generated by UV light

TABLE 16.4 List of Nanoparticles and Their Antimicrobial Activity Against Phytopathogens

Nanoparticles	Activity Against Phytopathogenic Microorganisms	References
Silver nanoparticles	*Bipolaris sorokiniana* and *Magnaporthe grisea*	Jo et al. (2009)
	Oak wilt pathogen *Raffaelea* sp.	Kim et al. (2009)
	Colletotrichum gloeosporioides	Aguilar-Méndez et al. (2011)
	Colletotrichum sp., pepper anthracnose	Lamsal et al. (2011a)
	Powdery mildews	Lamsal et al. (2011b)
	Alternaria alternata, Sclerotinia sclerotiorum, Macrophomina phaseolina, Rhizoctonia solani, Botrytis cinerea, and *Curvularia lunata*	Krishnaraj et al. (2012)
	A. alternata, Alternaria brassicicola, Alternaria solani, B. cinerea, Cladosporium cucumerinum, Corynespora cassiicola, Cylindrocarpon destructans, Didymella bryoniae, Fusarium oxysporum f. sp. *cucumerinum, F. oxysporum* f. sp. *lycopersici, F. oxysporum, Fusarium solani, Glomerella alternata, Monosporascus cannonballus, Pythium aphanidermatum, Pythium spinosum, Stemphylium lycopersici*	Kim et al. (2012)
	Xanthomonas campestris pv. *malvacearum*	Rajesh et al. (2012)
	Colletotrichum coccodes, Monilinia sp., *Pyricularia* sp.	Lee et al. (2013)
	Citrobacter freundii, Erwinia cacticida	Paulkumar et al. (2014)
	Bipolaris sorokiniana	Mishra et al. (2014a,b)
Ag–silica hybrid complex	Activates expression of defense-related genes (system-acquired resistance marker genes)	Chu et al. (2012)
Nanosized silver and silica	*Pythium ultimum, M. grisea, B. cinerea, R. solani, C. gloeosporioides*	Park et al. (2006)
Silver–chitosan nanocomposite	*B. cinerea* Pers.	Moussa et al. (2013)
Colloidal silver nanoparticles	*R. solani, S. sclerotiorum, Sarracenia minor*	Min et al. (2009)
ZnO nanoparticle	*B. cinerea* and *Penicillium expansum*	He et al. (2011)
	Fusarium graminearum	Dimkpa et al. (2013)
Sulfur nanoparticles	*F. solani* and *Venturia inaequalis*	Rao and Paria (2013)
	Drosophila melanogaster	Araj et al. (2015)
Surface-modified sulfur nanoparticles	*Aspergillus niger, F. oxysporum*	Choudhury et al. (2011)
Silver/copper nanoparticles, copper nanoparticles	*A. alternata, B. cinerea*	Ouda (2014)

(Continued)

TABLE 16.4 List of Nanoparticles and Their Antimicrobial Activity Against Phytopathogens—cont'd

Nanoparticles	Activity Against Phytopathogenic Microorganisms	References
Copper nanoparticles	*Xanthomonas axonopodis* pv. *punicae*	Mondal and Mani (2012)
	Phoma distructiva, Curvularia lunata, A. alternata, F. oxysporum	Kanhed et al. (2014)
	Fusarium sp.	Bramhanwade et al. (2016)
	Xanthomonas oryzae pv. *oryzae*	Kala et al. (2016)
Magnesium oxide nanoparticles	Spores of *Rhizopus stolonifer, A. alternata, F. oxysporum, Mucor plumbeus*	Wani and Shah (2012)
	F. oxysporum f. sp. *lycopersici*	Parizi et al. (2014)
	Ralstonia solanacearum Systemic resistance by activating salicylic acid-, jasmonate-, and ethylene-signaling pathway	Imada et al. (2015)
$CaCO_3$	*Aonidiella aurantii* and *Bactrocera dorsalis*	Hua et al. (2015)

damages bacteria by causing damage to enzymes, proteins, and DNA. NPs, because of their small size, can easily pass through the bacterial cell membrane, especially Gram-negative bacteria. ZnO NP morphology changes with pH, temperature, solvents, and precursor. The rod-shaped structure with (111) and (100) facets has higher antibacterial activity compared to spherical ZnO NPs, which have only a (100) facet. The (111) facet has higher atomic density than the (100) facet resulting in more oxygen vacancies, which in turn generates more ROS and more antimicrobial activity. Moreover, the rod-shaped and flower-shaped ZnO NPs are more easily internalized than spherical ones. When the surface morphology of ZnO NPs is not uniform, it causes abrasive injury to the cell membrane through its edges and corners. The size of ZnO NPs also affects their antibacterial activity. Less the size of ZnO NPs, more the dissolution of ZnO NPs to Zn^{+2} and therefore more is the toxicity. Modifying surfaces by annealing oxygen on ZnO surfaces also increases its antibacterial activity. The following reaction occurs when ZnO is suspended in aqueous medium with bacteria:

$$ZnO + h\upsilon \rightarrow e^- + h^+, \tag{16.1}$$

$$h^+ + H_2O \rightarrow {}^{\cdot}OH + H^+, \tag{16.2}$$

$$e^- + O_2 \rightarrow {}^{\cdot}O_2^-, \tag{16.3}$$

$${}^{\cdot}O_2 + H^+ \rightarrow H_2O_2{}^{\cdot}, \tag{16.4}$$

$$HO_2{}^{\cdot} + H^+ + e^- \rightarrow H_2O_2 \tag{16.5}$$

The peroxides generated pass through the bacterial membrane more easily than superoxides and hydroxyl radicals because of their negative charges and cause damage to the bacteria (Sirelkhatim et al., 2015). ZnO NP treatment also induces increased expression of stress-related genes such as oxidative stress genes (*ahpC* and *katA*) and the general stress response gene (*dnaK*). This confirms increased stress levels because of ZnO NP treatment (Xie et al., 2011).

16.3.2.3 Silica Nanoparticles

Silica NP synthesis was pioneered by Stober and thus the process is known as Stober's process. However, many other methods also are also used to synthesize silica NPs (Rao et al., 2005). Silica in its nanoform is toxic to microorganisms because of its large surface area affecting adhesion, differentiation, and proliferation. Silica when present as a nanocomposite with Ag and Au is more effective against pathogens (Dizaj et al., 2014). However, these NPs are more effective in promoting the growth and development of crop plants and also have significant ability to alleviate adverse impacts of UV-B radiation and heavy metal stress (Tripathi et al., 2015, 2016, 2017e).

16.3.2.4 Copper Nanoparticles

CuNPs can be synthesized physically, chemically, and biologically (Camacho-Flores et al., 2015; Khodashenas and Ghorbani, 2014). Because of the high surface area-to-volume ratio, these NPs are more reactive. Inside the bacterial cells, they bind to nucleic acids and cause intermolecular crosslinking. They also bind with the sulfhydryl group and/or carboxyl group of amino acid and damage the protein so that the biological processes are hampered. The US Environmental Protection Agency has also approved the use of CuNPs as an antimicrobial agent (Shobha et al., 2014).

16.3.2.5 TiO_2 Nanoparticles

TiO_2 NPs can be synthesized physically and chemically as described by Chen and Mao (2007) or through green synthesis (Kanti et al., 2015). The antimicrobial activity of TiO_2 NPs is attributed to their crystal structure. The mode of action basically depends on the ROS generation because of photocatalytic activity of TiO_2 NPs similar to that of ZnO. The ROS cause site-specific damage in DNA and proteins of the pathogen (Ravishankar and Jamuna, 2011). A report showed that TiO_2 NPs induce larvicidal activity in *Aedes aegypti* (Durairaj et al., 2014). These metallic NPs cause disturbances in the cellular biochemistry resulting in reduced viability of the phytopathogens (Kala et al., 2016).

16.3.2.6 MgO Nanoparticles

MgO NPs can be synthesized by the sol–gel method, precipitation method, and green synthesis method (Agrawal et al., 2015; Sushma et al., 2016; Meenakshi et al., 2012). The antibacterial activity of MgO NPs depends on their size and concentration. However, studies suggest that MgO NPs are more reactive against Gram-positive bacteria than Gram-negative bacteria. This can be explained because of the presence of thin layers of lipid A, lipopolysaccharide, and peptidoglycan on the Gram-positive cell membrane, whereas the Gram-negative bacterial cell membrane has a thick layer of peptidoglycan. Similar to AgNPs, MgO NPs cause ROS formation inside the cell. These oxygen species cause cell wall damage in the bacteria. Also the electrostatic interaction between bacterial cell membrane and MgO NPs causes membrane damage (Tang and Lv, 2014). MgO NPs result in an increase in pH of the medium caused by hydration of the metal NPs (Dizaj et al., 2014).

16.3.3 Other Nano-Based Systems for Crop Protection

Pesticides can also be nanoformulated as microemulsions, nanoemulsions, nanosuspensions, or nanocomposites. Two immiscible liquids such as oil and water when mixed together in the presence of surfactants, which act as a stabilizer for interfacial film, form microdroplets (Paul and Moulik, 2001). Based on the concentration of three factors (oil, water, and surfactants), different structures can be formed such as micelles, reverse micelles, water-in-oil microemulsions, oil-in-water microemulsions, and bicontinuous microemulsions. Microemulsion-based delivery systems are considered to deliver natural pesticides and essential oils. A microemulsion has various advantages such as stability, permeability, solubility, and low viscosity. However, it requires large amounts of surfactants and retains low amounts of pesticides, which is its major disadvantage. A nanoemulsion requires the same ingredients as a microemulsion: two immiscible liquids, surfactants, and cosurfactant, with additional high-energy inputs in the form of a high-pressure homogenizer, high shear stirrer, and microfluidizer to obtain kinetic stability (Patil and Waghmare, 2013). Nanoemulsions can be prepared by various methods such as high-pressure homogenization, microfluidization, spontaneous emulsification, low-energy emulsification, solvent evaporation, and the hydrogel method (Baboota et al., 2007). They are used to deliver hydrophobic pesticides such as neem oil, clove oil, and eucalyptus oil. In both micro- and nanoemulsion-based processes, the size may increase because of Ostwald ripening. However, with increase in surfactant concentration, Ostwald ripening is inhibited. A nanosuspension is the colloidal system in which encapsulated pesticides or NPs are suspended in a liquid. Nanopowder formed from the solvent evaporation of a microemulsion or nanoemulsion containing pesticides when redispersed in water forms a nanosuspension (Nuruzzaman et al., 2016). The roles of these systems in pesticide delivery are depicted in Table 16.5.

16.3.3.1 Nanocomposites

Nanocomposites are the systems in which at least one component must be in the nanorange. Depending on the matrix materials, they can be classified into ceramic matrix composites (Al_2O_3/SiO_2, SiO_2/Ni, Al_2O_3/TiO_2, Al_2O_3/SiC, Al_2O_3/CNT), metal matrix composites (Fe—Cr/Al_2O_3, Ni/Al_2O_3, Co/Cr, Fe/MgO, Al/CNT), and polymer matrix (polymer/layered silicates, polymer/CNT, polymer/LDHs, polyester/TiO_2, thermoplastic/thermoset) composites (Camargo et al., 2009). Nanocomposites are prepared using nanofillers as skeleton and other materials covering it. Nanofillers can be classified into layered silicates (e.g., clays), spherical (silica), and acicular (CNTs) depending on the geometry and aspect ratio (Sahithya and Das, 2015). Metal matrix nanocomposites consist of an alloy metal on which nanosized materials are reinforced. A polymer/CNT nanocomposite has been explored because of its multifunctional structures and properties (Table 16.6). Pesticides such as zineb and mancozeb that were encapsulated into the MWCNT-grafted poly(citric acid) showed a marked effect on *Alternaria alternata* compared to bulk material (Sarlak et al., 2014). Radian and Mishael (2008) found that among two polymer/clay-based nanocomposites—polydiallyldimethylammonium chloride (PDADMAC)–montmorillonite and PDADMAC–kaolinite—the former showed higher electrostatic interaction and subsequent slow release of the pesticide imazapyr but still maintained its activity. In another study, Mg/Al LDH was intercalated with anionic herbicide 2,4-D + 2-methyl-4-chlorophenoxyacetic acid by three

TABLE 16.5 List of Nanoformulations and Their Effects

Structure	Active Ingredient	Ingredient	Effects	References
Microemulsion	Nicotine carboxylate	Capric acid, lauric acid, palmitic acid, stearic acid, oleic acid	Increased encapsulation efficiency obtained for capric acid	Casanova et al. (2005)
	Chlorpyrifos	Ethyl acetate, agricultural emulsifier no. 600, natural carboxylate, glycol, water	Environmental protection of chlorpyrifos	Feng and Yaqian (2011)
	Lavandin essential oil	*n*-Octenyl succinic anhydride-modified starches, water	Stable microemulsions were produced	Varona et al. (2009)
Evaporation of oil-in-water microemulsions to nanoparticles in powder form	Novaluron	Isopropanol with isobutanol, surfactants (sodium dodecyl sulfate, benzene sulfonate, alkyl naphthalene sulfonate condensate sodium salt), water	Toxicity of nanoparticles of novaluron resembled that of the commercial formulation	Elek et al. (2010)
	Permethrin	*n*-Butyl acetate, *sec*-butyl alcohol, soybean phosphatidylcholine, ammonium glycyrrhizinate, sucrose, water	Increased pesticidal life and improved specificity	Kumar et al. (2013)
Nanoemulsion	β-Cypermethrin	Poly(oxyethylene) lauryl ether, ethylene oxide and methyl decanoate, water	Reduced pesticide application	Wang et al. (2007)
	Triazophos	Water, polyoxyethylene ether,/noctyl-2-pyrrolidone	Improved stability of the pesticide	Song et al. (2009)
	Neem oil	Tween 20, water	Effective larvicidal agent against *Culex quinquefasciatus*	Anjali et al. (2012)
	Glyphosate isopropylamine	Esterified vegetable oils (short-chain fatty acid methyl esters, long-chain fatty acid methyl esters, carbohydrate-derived alkylpolyglucoside), surfactants (short-chain alkylpolyglucosides, long-chain alkylpolyglucosides), water	Enhanced bioactivity against *Asystasia gangetica*, *Diodia ocimifolia*, *Paspalum conjugatum*	Lim et al. (2013)
	Ethanolic extract of *Manilkara subsericea* and its triterpenes	Octyldodecyl myristate, sorbitan, monooleate/ polysorbate 80, water	Enhanced water solubility of the natural extract	Fernandes et al. (2014)
	Clove oil	Polyethylene glycol sorbitan monooleate, sorbitan monooleate, water	Stable clove oil nanoemulsion formation	Shahavi et al. (2015)
	Eucalyptus oil	Tween 20, water	Larvicidal activity against *Pectinophora gossypiella* and *Earias insulana*	Moustafa et al. (2015)

(Continued)

TABLE 16.5 List of Nanoformulations and Their Effects—cont'd

Structure	Active Ingredient	Ingredient	Effects	References
Nanosuspension	Bifenthrin	Stabilizer (copolymer of polyacrylic acid and polybutylacrylate), water	Increased stability	Liu et al. (2008c)
	Carbofuran	Polyvinylpyrrolidone, Altox 4913, Miglyol, ethylene glycol, AT-120, water	Reduction in dose of pesticide, increased solubility of the pesticide	Chin et al. (2011)
	Lambda-cyhalothrin	Maleic rosin-polyoxypropylene-polyoxyethylene ether sulfonate, 1-dodecanesulfonic acid sodium salt (sodium dodecyl sulfate), water	Perspective in plant protection for improved bioavailability	Pan et al. (2015)

TABLE 16.6 List of Nanocomposites and Their Effects

Nanostructure	Components	Effect	References
Nanocomposite	DNA-directed silver nanoparticles on graphene oxide	Enhanced antibacterial activity against *Xanthomonas perforans*, improved stability, strengthened adsorption properties	Ocsoy et al. (2013)
	Multiwalled carbon nanotube-grafted poly(citric acid)	Pesticide is encapsulated in the polycitric shell	Sarlak et al. (2014)
	Polyvinyl alcohol/alginate–montmorillonite nanocomposite	Controlled release-encapsulated NeemAzal	Rashidzadeh et al. (2014)
	Arizona montmorillonite with the cationic polymer hexadimethrine	Potential as adsorbent for pesticide	Gámiz et al. (2015)
	Chitosan/carbon nanotube nanocomposite	Controlled delivery of Azimphos	Bibi et al. (2016)
	Alginate/exfoliated montmorillonite nanocomposite	Controlled release of acetamiprid	Yan et al. (2016)

methods: direct synthesis, regeneration, and ion exchange method (Cardoso et al., 2006). Carbonate and hydroxyl ions replaced the herbicide from the complex resulting in its slow release (Cardoso et al., 2006). Furthermore, Chowdappa et al. (2014) reported that chitosan–AgNP nanocomposite exhibited antifungal activity against *Colletotrichum gloeosporioides* and anthracnose.

16.3.4 Mode of Entry of Nanomaterials in Plants and Controlled Release of Agrochemicals

The plant cell wall is the primary barrier for any foreign particle to enter into the plant cell. It acts as a sieve with a specific pore size allowing particles of 40–50nm to pass through. NPs are also internalized by the plasma membrane through endocytosis (Pérez-de-Luque, 2017). They can move across the cell membrane with the help of embedded transport carrier proteins or through ion channels. Once inside the cytoplasm, they bind to the cell organelles or proteins and interfere with the cellular processes (Jia et al., 2005). When present on the leaf surface, they move inside through stomata openings and block gaseous exchange resulting in altered physiological functions (Nair et al., 2010).

The materials loaded in the nanomaterials can be crop protection agents (insecticides, herbicides, fungicides), biochemicals (antibiotics, RNA-based vaccines for phytopathogens), semiochemicals (repellants, pheromones, allomones), etc. (Peteu et al., 2010). The efficiency of nanomaterials carrying agrochemicals depends on the controlled release of the agrochemical at the target site in its active form (Peteu et al., 2010). The released chemical should remain inactive at nontarget sites. The controlled release of the biocide also depends on the interaction of the active material and the carrier molecule. Strong interaction between the biocide results in its slow release and vice versa (Peteu et al., 2010).

The use of smart or intelligent polymers is anticipated for the controlled release of agrochemicals. Some of them are responsive to various stimuli such as ultrasound, light of a particular wavelength, plant-derived enzymes, etc., which are depicted in Fig. 16.5 (Peteu et al., 2010).

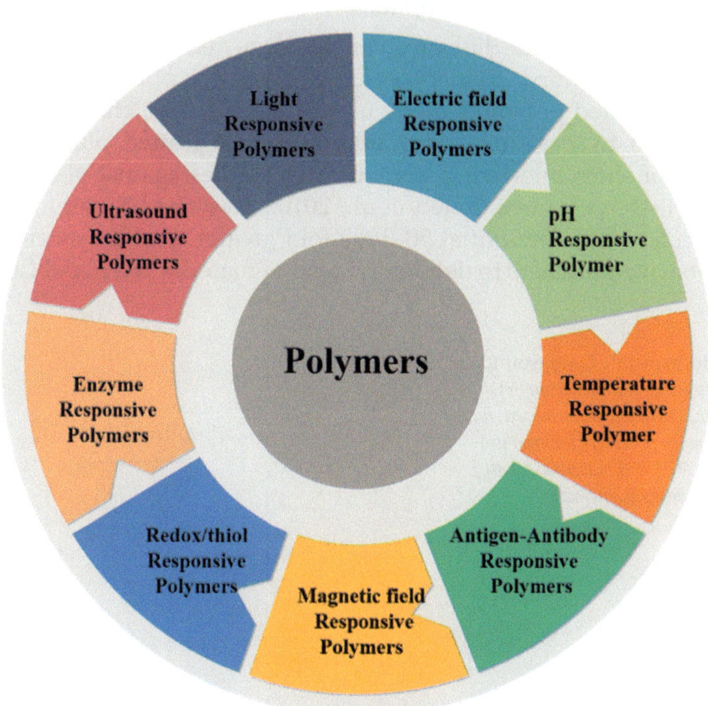

FIGURE 16.5 Various stimuli responsive polymers.

16.3.4.1 Ultrasound Responsive Polymers

The application of ultrasound energy generates waves of both high and low energy, which induce the opening of pores in polymers and the release of chemicals present inside them. The mechanism of pesticide release from ultrasound responsive polymer is depicted in Fig. 16.6. These types of polymer can be biodegradable such as polyglycolides, polylactides, and bis(*p*-carboxyphenoxy)alkane anhydride with sebacic acid, or nonbiodegradable such as ethylene-vinyl acetate copolymers, poly(lactide-*co*-glycolide) microspheres, poly(hydroxyethyl methacrylate) hydrogels, etc. (Peteu et al., 2010)

16.3.4.2 Light Responsive Polymers

These photoresponsive polymers release the chemicals in response to a certain wavelength of light. These polymers contain photoreactive groups such as azobenzene, triphenylmethane, and spiropyran groups. These groups can be entrapped, crosslinked, or introduced as side chains of polymers. Upon irradiation, azobenzene groups undergo trans-to-cis isomerization. The two forms have different geometries, polarities, and electrical properties, therefore properties of the polymers are photocontrolled (Nicoletta et al., 2012). Similarly, triphenylmethane groups change from nonionic form to cationic form on illumination, which causes changes in hydrophobicity and permeability of membrane polymers containing these groups. Spiropyran transforms the closed ring form to the open ring form and hence changes the properties of the polymer containing these groups (Nicoletta et al., 2012).

16.3.4.3 Redox/Thiol Responsive Polymers

Polymers containing disulfide linkages can undergo redox reaction in the presence of reducing agents or thiol groups (Kamada et al., 2010). These polymers can be micelle, dendrimer, or drug-conjugated polymers made up of PEG, dextran, chitosan, grapheme oxide, silica NPs crosslinked by cystamine, dithiothreitol, thiol-grafted pluronic, etc. (Wen and Li, 2014). Delivery of short interfering RNA-based vaccines against plant viruses could be achieved through these polymers (Peteu et al., 2010). Cysteine, glutathione, and reductive amino acid molecules when present at 50–1000-fold greater than the exterior concentration allow redox responsive polymers to deliver the loaded material (Wen and Li, 2014).

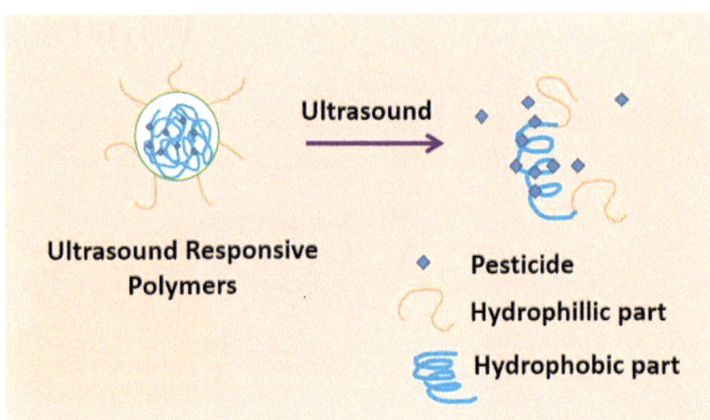

FIGURE 16.6 Mechanism of ultrasound-mediated pesticide release. *Adapted from Peteu, S.F., Oancea, F., Sicuia, O.A., Constantinescu, F., Dinu, S., 2010. Responsive polymers for crop protection. Polymers 2 (3), 229–251; License: https://creativecommons.org/licenses/by-nc-sa/3.0/.*

16.3.4.4 *Magnetic Field Responsive Polymers*

This type of delivery system involves magnetic NP-loaded polymer, which is supposed to release the active material in response to a magnetic field. The release of cargo depends on the polymer in which these NPs are embedded (Thévenot et al., 2013). The polymer could be pH-sensitive or thermosensitive hydrogels so that the chemical is released in response to their respective stimulus (Peteu et al., 2010).

16.3.4.5 *Enzyme Responsive Polymers*

Enzyme responsive polymers can undergo disintegration, reversible assembly and disassembly, molecular transition, gelation, and gel–sol transition. Examples are chitosan-, albumin-, dextran-, heparin-, alginate-, and hyaluronic acid-entrapped polymers, which are substrates for specific enzymes (Hu et al., 2012). Because of the enzymatic reactions, the polymers swell or collapse resulting in the release of the active ingredient (Hu et al., 2012). These polymers are attractive for their specificity, accuracy, reproducibility, and reduced response time (Hu et al., 2012).

16.3.4.6 *Antigen–Antibody Responsive Polymers*

These types of polymer are made up of hydrogels grafted with antigen and its antibody. The interaction between antigen and antibody ensures crosslinking of the hydrogel (Miyata et al., 1999). In the presence of free antigens, the crosslinks between antigen and antibody diminish and the hydrogel swells resulting in leakage of the loaded chemical (Miyata et al., 1999).

16.3.4.7 *Electric Field Responsive Polymers*

Electric field responsive hydrogels respond to the applied electric field caused by the presence of polyelectrolytes in them (Peteu et al., 2010). The ions move collectively toward the cathode or anode according to the charge they carry. This results in reversible shrinkage and swelling of the hydrogels. The conducting polymers can be used in nanobiosensors or to deliver agrochemicals. An example of such polymers is the interpenetrating polymer network hydrogel of (9-phenanthryl)methyl methacrylate and chitosan (Peteu et al., 2010).

16.3.4.8 *pH and Temperature Responsive Polymers*

pH of the environment controls intermolecular interaction of the polymer. Change in pH alters their ionic interactions and hydrogen bonding resulting in altered conformation, configuration, solubility, and volume (Dai et al., 2008). Because of the altered ionic interaction, hydrogen bonding, and hydrophobic interaction, self-reorganization occurs (Dai et al., 2008). Similarly, variation in temperature causes change in the solvation state of temperature responsive polymers and hydrogels (Bawa et al., 2009). These polymers commonly have ethyl, methyl, or propyl groups. At a critical solution temperature, such polymers and solution exhibit discontinuous change in their phase. Lower critical solution temperature (LCST) is a temperature below which partially soluble polymers are in a miscible state and upper critical solution temperature is a temperature above which the polymers are in a miscible state (Bawa et al., 2009). An example of a temperature responsive polymer is poly(*N*-isopropylacrylamide) whose LCST is 32°C in aqueous solution (Constantin et al., 2011).

16.4 NANOTECHNOLOGY IN SOIL AND WATER MANAGEMENT

The growth and healthy sustenance of crops is directly related to the nutrients they absorb from the soil. Apart from crop protection against diseases, proper management of soil is necessary to ensure the availability of nutritional elements to the plants (Robertson, 1987). Concentration of the nutrients should also be maintained in such a manner that the plant faces neither deficiency because of nutrient shortage nor toxicity arising from the presence of higher concentrations (Ongley, 1996). Soil management is gaining importance globally because the need to upscale food production has been gradually mounting with expansion of the world population. Maintaining soil sustainability is crucial to ensure food security, which requires soil nutrient management and conservation. Similarly, population growth and climate change have given rise to the need to maintain water quality for production of healthy drinking water and reduction of water pollution (Ongley, 1996). Unmonitored use of fertilizers, weedicides, etc. to spike growth has also skyrocketed with the freshwater demands. Agricultural wastewater deposits itself on the surface of groundwater, thereby contributing toward water pollution. Again, when the same contaminated water is reused for irrigation, crops suffer. In the field of agriculture, remediation and recycling are of keen interest because of the growing polluted nature of irrigation sources. These problems in soil and water management need urgent solutions to prevent further damage.

Nanotechnology can be utilized to address these problems using nanoscale materials for soil nutrient/water retention and even dissemination in agriculture, and has been named as one of the six "Key Enabling Technologies" for sustainable competitiveness and growth by the European Commission (https://ec.europa.eu/growth/industry/key-enabling-technologies_en). CNTs, nanoclays or zeolites, nanoscale membranes, etc. can be applied for the purification of water. For example, a novel technology using nanoscale Fe particles has been found to be promising for the treatment of polluted water at the site of origin itself. This nanomaterial with zero valency has great affinity for fertilizers/pesticides, organic solvents, heavy metals, etc. (Mueller et al., 2012). Another good example is the usage of TiO_2 NP-coated filters used for the photocatalytic degradation of agro-based chemicals in polluted water (McMurray et al., 2006). Similarly, soil remediation can be achieved by the use of nanosized metals that are embedded into pollutant-adsorbing resins or beads, which increase adsorption specificity and stability. Again, nanomaterials are used for soil enhancement by retention of applied agrochemicals and their sustained release to the crops. An example of such nanomaterial is zeolites. In a study performed using natural zeolites, it was observed that they are "effective as slow-releasing nitrogen fertilizer" because their application resulted in slowing of the nitrification process in the soil (Torma et al., 2014). Nanoclay–polymer composites from soil clay are also being tested for the controlled release of nutrients into soil (Sarkar et al., 2014).

16.5 NANOTECHNOLOGY IN PLANT BREEDING AND GENETIC TRANSFORMATION

Apart from the foregoing methods, crop protection can also be achieved by strengthening plants' defense systems, by integrating resistance genes, and by manipulating their existing defense mechanisms to attain enhanced protection. In the plant biotransformation process, various forms of NPs such as CNTs, quantum dots (QDs), liposomes, and lipofectins have been used as gene carriers (Chen et al., 2011). A nanotechnology-based gene transformation system seems to be very effective, time saving, and protects DNA during the transformation process compared to conventional processes (Rafsanjani et al., 2012). Physical biotransformation processes such as *Agrobacterium*-mediated gene transfer suffer from instability of genes in the environment, deletion of genes because of high-energy impact, and poor penetration and reproducibility (Rafsanjani et al., 2012). However, a nanotechnology-armed conventional gene transfer system reduces the probability of energy-induced DNA damage related to gene mutation (Rafsanjani et al., 2012). Various nanotechnology-armed gene transfer systems are discussed in this chapter.

16.5.1 Liposome-Mediated Gene Transfer

Liposomes are typically vesicular and colloidal nanostructures forming one or more lipid bilayers surrounding equal numbers of aqueous compartments. Because of the presence of aqueous core and lipophilic vesicle structure, the plasmid/gene is expected to be encapsulated in the aqueous environment. A novel liposomal gene transfer technique has been successfully tried and adopted in gene delivery (Deshayes et al., 1985).

Liposomes offer the following advantages over conventional methods:

- Stability in the process of gene transfer;
- Reduction in the deletion in the DNA when used with physical techniques;
- Surface modification of liposomes (positively charged) to accelerate penetration within a cell;
- Higher degree of reproducibility;
- Applicability to various cell types;
- Lack of cellular toxicity.

In this technique, the mechanism of DNA entry through the protoplast is facilitated by an endocytosis process, which may involve the following steps:

- Adhesion of liposomes to the protoplast surface;
- Fusion of liposomes at the site of adhesion;
- Endosome formation within the cytoplasm; and
- Release of plasmids inside the cell.

Liposome-mediated gene transfer has been successfully used to deliver DNA into the protoplasts of a number of plant species (e.g., tobacco, petunia, carrot, etc.) (Deshayes et al., 1985).

16.5.2 Modified or Neoliposomal Technique

Wiesman et al. (2007) synthesized novel cationic amphiphilic compounds from vernonia oil (a natural epoxidized triglyceride) and further utilized the same in the preparation of liposomes that were able to encapsulate DNA (Wiesman et al., 2007). The developed liposomes have a positive charge on the vesicle surface and excellent stability because of charge equilibrium provided by an amphiphilic compound. The liposome was characterized for surface morphology and vesicular size by various microscopy techniques such as transmission electron microscopy, scanning electron microscopy, atomic-force microscopy (AFM), and confocal laser systems, which showed the spherical shape with diameters between 40 and 110 nm (Wiesman et al., 2007). Liposomes were capable of encapsulating a packed 5.2-kb plasmid (pJD328). They interacted and crossed the cuticle membrane in intact form. In addition, the delivery rate of encapsulated radiolabeled 2,4-D obtained with these vesicles was greater as compared to liposomes prepared from dipalmitoyl phosphatidyl choline and the control (unencapsulated 2,4-D). These observations clearly indicate the significance of positive charge over the liposomal surface introduced by amphiphilic compounds used in the preparation.

16.5.3 Nanoparticle-Mediated Gene Transfer

In plant biotechnology, NPs in the form of liposomes and other nanostructures as gene carriers in the biotransformation process protect DNA from ultrasound and improve transformation efficiency (Yu-qin et al., 2012). Liu et al. (2008a,b,c) used a green fluorescence protein-tagged NP to demonstrate its effectiveness and reported that the NP alone enters the cytoplasm as seen in green cytoplasmic content. In addition, different NP-based delivery carriers such as polymeric NPs, metallic NPs, and QDs have also been successfully adopted as gene carriers (Akhtar and Benter, 2017; De Rosa et al., 2009; Truong et al., 2011).

Scientists have developed several NP-mediated strategies for successful gene transfer. They successfully developed surface-functionalized silica NPs akin to honeycomb mesoporous form with a pore size of 3 nm that can transport DNA into isolated plant cells and intact leaves (Rai et al., 2015). They also developed capped and uncapped AuNPs, and reported that chemicals leached out from the NP and triggered gene expression in the plant under controlled release conditions (Rai et al., 2015). It is reported that fluorescent-labeled starch NPs were used as a plant transgenic vehicle that can traverse through cell walls, cell membranes, and nuclear membranes by forming pores with the help of ultrasound waves (Rai et al., 2015). Thus nanobiomaterial can efficiently solve the problem of traversing exterior genes into the plant cell wall.

16.5.4 Carbon Nanotube-Mediated Gene Transfer

CNTs have extraordinary unique features, which have attracted scientists' attention. These nanomaterials have become increasingly popular in various fields simply because of their tiny size and unique optical, electric, and magnetic properties when used alone or with other metals (Akhter et al., 2011). It has been reported that rice cell suspension (*Oryza sativa* L.) cultured with MWCNTs increased ROS and decreased cell viability (Tan et al., 2009). In the case of CNT exposure on *Arabidopsis* cell culture, it seems that the CNTs induced hypersensitive signals that resulted in defense responses in the cells causing cell death (Lin et al., 2009).

16.5.5 Lipofectin

Lipofectins are cationic liposomes capable of gene transfer in the plant cell line. Sawahel and other researchers have reported that lipofectin effectively transfuses the mesophyll protoplast of *Nicotiana tabacum* and *Nicotiana plumbaginifolia* (Sawahel, 2002; Mohapatra et al., 2011). The transformation efficiency was more in the case of PEG-mediated transformation than electroporation (Meijer et al., 1991). The advantage with the use of lipofectin in biotransformation is similar to that of liposome with additional advantages of stably transformed plants with or without additional application of electroporation (Mohapatra et al., 2011).

16.5.6 Other Novel Nanomaterial-Based Gene Transfer Processes

Based on the earlier studies of liposomal and NP-based gene transfer, the efficiency of the novel nonviral gene-protected biotransformation method can be improved through several modifications. It is a well-known fact that protoplast and plasma membrane have negative surface charge and their surface potential varies from −10 to −30 mV (Reid et al., 2002). Therefore adding positive charge to the surface can help in increasing the adhesive property of the encapsulated DNA. This idea was successfully tried earlier by some investigators (Wiesman et al., 2007). Straylamine, triethylamine, and natural biodegradable polymers such as chitosan can easily functionalize liposomal or nanomaterial surfaces with a positive charge (Li et al., 2011).

These scientific methodologies may pave the way for further development of efficient plant transformation. Patents and patent applications such as US patent applications 20110065092 A1 (gene-encapsulated nonviral particle) and US6534484 (gene-encapsulated liposome for biotransformation) indicate the importance of this novel technology in biotransformation (Rafsanjani et al., 2012). Combining nanotechnology-based transformation with conventional methods of transformation will improve the efficiency and productivity of the process with minimal transgene silencing. Though the complete mechanism of the nanotechnology-mediated transformation process is not clear, endocytosis as well as transcellular and paracellular transport might be involved in nanocarrier transport (Rafsanjani et al., 2012).

16.6 NANO-BASED DIAGNOSTIC SENSORS

Besides nanomaterials, nanosensors/nanobiosensors are some of the main application tools of nanotechnology that can be helpful in the field of agriculture including crop protection. As compared to analytical sensors or biosensors, nanosensors have great potential in crop protection with several advanced characteristics such as small size, portability, sensitivity, accuracy, reproducibility, stability, etc. (Liu et al., 2008a,b,c; Rai and Ingle, 2012). These are the analytical devices with dimensions not more than 100 nm and can be fabricated into various forms for monitoring physicochemical changes related to crop improvement. Some of these properties include soil quality and crop health analysis, easy biochemical sensing and control, water management, pesticide and nutrient delivery, etc. (Ghormade et al., 2011).

For agriculturists, precise or accurate farming has always been a major target for crop productivity. Nanosensor systems, because of their high sensitivity and specificity, are being employed in the field of agriculture for monitoring pests, pathogens, and pesticides for crop growth management (Cioffi et al., 2004; Rai and Ingle, 2012). The utility of nanosensors can aid in avoiding the overdose of agricultural chemicals by reducing inputs of herbicides, fertilizers, and pesticides, thereby increasing crop growth and yield. This will further reduce agricultural wastes and thus pollution (FAO and WHO, 2013) (Fraceto et al., 2016). Studies have reported the use of nanosensors as diagnostic tools for the detection of pesticides, pests, and pathogens in crops/plants (Table 16.7). Development of nanosensors using chemically modified AFM

TABLE 16.7 List of Nanosensors for Detection of Pesticides/Herbicides/Pathogens

Pesticide/Herbicide	Sensing Material	Detection Limit	Sensor Type/ Detection Type	References
2,4,5-Trichlorophenoxy acetic acid (systemic herbicide)	Poly-*o*-toluidine/zirconium (IV) phosphate nanocomposite	1 mM	Electrochemical	Khan and Akhtar (2011)
Fenitrothion in water	Nano-TiO_2/nafion composite	0.13 mM	Electrochemical	Kumaravel and Chandrasekaran (2011)
Melamine in milk	18-Crown-6 ether functionalized AuNPs	6 ppb	Optical	Kuang et al. (2011)
Aflatoxin (AFM1) in milk	Magnetic nanoparticles coated with antibody (anti-AFM1)	0.01 ppb	Electrochemical	Paniel et al. (2010)
Organochlorine and organophosphorus pesticides in cabbage	Amino-functionalized nanocomposite with tetraethylene pentamine	0.29 mg/kg	Chromatography	Zhao et al. (2011)
Dimethoate	Iron oxide (30 nm), zirconium oxide (31.5 nm) nanoparticles	5.6×10^{-4} ng/mL	Amperometric	Gan et al. (2010)
Methyl parathion in water	Nano-ZrO_2/graphite/paraffin	2 ng/mL	Electrochemical	Parham and Rahbar (2010)
Bacillus cereus in diverse food samples	Anti-*B. cereus* antibodies functionalized with polyaniline nanowires	10^1 and 10^6 CFU/mL	Conductometric	Pal et al. (2008)
Methyl parathion in vegetables (cabbage, spinach, lettuce)	Nano-Au/nafion composite	10^{-7} M	Electrochemical	Kang et al. (2010)
Methyl parathion, chlorpyrifos	Nanosized polyaniline matrix with single-walled carbon nanotube, single-stranded DNA, and enzyme	1 pM	Electrochemical	Viswanathan et al. (2009)
Fenamithion and acetamiprid in water	Cd-tellurium quantum dots (QDs) with *p*-sulfonatocalix[4]arene	0.12 and 0.34 nM	Luminescence	Qu et al. (2009)
Herbicide metsulfuron-methyl, an acetolactate synthase (ALS) inhibitor	Atomic force microscopy tip functionalized with the ALS enzyme	Adhesion force 250%	–	Da Silva et al. (2013)
Parathion in water	Nano-ZrO_2/Au composite	3 ng/mL	Electrochemical	Wang and Li (2008)
Dichlorvos and paraoxon in drinking water	Acetylcholinesterase and pyranine immobilized on nanosized liposomes	1010 M	Optical	Vamvakaki and Chaniotakis (2007)

Parathion in vegetables	Nano-TiO_2 on glassy carbon electrode	108 M	Electrochemical	Li et al. (2006)
Pirimicarb in vegetables	Molecular-imprinted nanopolymers (methacrylic acid with carboxyl functional groups)	8×10^6 M	Piezoelectric	Sun and Fung (2006)
Imazaquin, metsulfuron-methyl and atrazine (herbicide detection)	Chemically modified AFM tips	Not mentioned	Piezoelectric	Deda et al. (2013)
2,4-Dichlorophenoxyacetic acid (2,4-D) (herbicide)	Cadmium telluride QDs	2,4-D up to 250 pg/L	–	Vinayaka et al. (2009)
2,4-D (herbicide)	Anti-2,4-D antibody labeled with AuNPs	3 ng/mL	Chemiluminescence	Boro et al. (2011)
Atrazine (herbicide)	Core–shell nanostructure	1.8 μM	Molecularly imprinted polymers fluorescent chemosensor	Liu et al. (2011)
Dichlorodiphenyltrichloroethane (pesticide)	Bovine serum albumin was covalently immobilized on the gold-coated side of the cantilever	Not mentioned	Nanomechanical/ immunochemical	Alvarez et al. (2003)
Atrazine	Cantilevers (nanoscale) functionalized with highly specific and site-directed antiatrazine antibodies	1 ppt sensitivity	Nanomechanical	Suri et al. (2008)
Insecticide cartap (nereistoxin derivative)	NaYF4:Yb,Ho/Au nanocomposite	High sensitivity (10 ppb)	Chemosensor	Wang et al. (2013)
OP neurotoxin	AuNPs attached via amino or sulfhydryl groups to the organophosphate hydrolase (OPH)	Not mentioned	Optical	Simonian et al. (2005)
Paraoxon	(CdSe)ZnS core–shell QDs and an OPH bioconjugate	10^{-8} M	Optical	Ji et al. (2005)
OP (paraoxon)	Localized surface plasmon resonance of AuNPs conjugated with acetylcholinesterase (AChE)	0.234 ppb	Optical	Lin et al. (2006)
OP compounds	CNTs–AChE/CHO	AChE/CHO	Electrochemical	Lin et al. (2004)

tips and microcantilevers has been widely utilized by researchers for detecting various analytes, proteins, pesticides, microorganisms, etc. Alvarez et al. (2003) for the first time reported the detection of the pesticide dichlorodiphenyltrichloroethane, which has been widely used for insect pest control using a nanomechanical biosensor. This biosensor worked on the basis of immunochemical assay between the pesticide and bovine serum albumin that was covalently immobilized on the gold-coated surface of the cantilever and real-time monitoring of nanoscale cantilever deflection (Alvarez et al., 2003). A nanomechanical biosensor comprises receptor molecules immobilized on a microcantilever surface without scanning the probes and results in the bending or deflection of the cantilever as a nanomechanical response (Tamayo et al., 2001). Suri et al. (2008) reported the label-free ultrasensitive detection of atrazine, an herbicide known to contaminate drinking water and many agricultural products, based on the nanomechanical biosensor. The detection of this herbicide was also performed by Liu et al. (2011) through a core–shell nanostructured molecular imprinting fluorescent chemosensor with high sensitivity. Pesticides such as carbofuran and triazophos could be detected using a one-step AuNP-based portable strip assay developed by Guo et al. (2009). This assay has the potential to monitor the pesticides in a short period of time without requiring any equipment. Da Silva et al. (2013) developed a nanobiosensor that was based on the functionalization of AFM tips with acetolactate synthase (ALS) enzyme for the detection of metsulfuron-methyl (an ALS inhibitor) in wheat fields. They found ~250% greater adhesion force as compared to the unfunctionalized AFM tips. Again, Deda et al. (2013) reported the detection of herbicides such as imazaquin and atrazine along with metsulfuron-methyl at low concentrations using nanosensors based on the functionalized AFM tips. The detection was based on the specific interactions between the herbicide (target) and biomolecules (enzyme/antibody) with an enhanced adhesion force. Another study employed chemiluminescence immunoassay for the detection of the herbicide 2,4-D using AuNPs as catalysts (Boro et al., 2011).

Organophosphate (OP) pesticides are widely used for agricultural pesticide-contaminated food and water. Several nanosensors have been developed and showed their potential for monitoring OP compounds including paraoxon, parathion, and coumaphos (Liu et al., 2008a,b,c). A study reported the detection of OP pesticides using NP-based optical sensors based on fluorescence intensity via an AuNP–OP hydrolase conjugate (Simonian et al., 2005). In another study, Lin et al. (2006) monitored OP pesticides using a nanosensor based on localized surface plasmon resonance of AuNPs covalently linked with acetylcholinesterase. Ji et al. (2005) detected paraoxon using a nanosensor based on electrostatic interactions between the (CdSe)ZnS core–shell QDs and OP hydrolase bioconjugate with a detection limit of 10^{-8} M. Furthermore, a new disposable and highly sensitive electrochemical biosensor was developed by Lin et al. (2004) for detecting OP pesticides based on carbodiimide chemistry of EDC (coupling agent) and CNTs (catalysts) with a low detection limit of 0.05 μM. Thus the nanosensors have great potential for monitoring OP pesticides and can be employed in the future.

For crop protection against diseases, diagnosis is the prerequisite, which is now possible with nanobiosensors. Nanobased sensors such as NPs, carbon-based materials, or QDs have the potential to diagnose and quantify various pathogens and contaminants, e.g., bacteria, fungi, viruses, toxins, and biohazardous compounds in agricultural crops with low detection limits (Yao et al., 2009; Etefagh et al., 2013). Nanochips have been reported to detect single nucleotide variations in pathogens such as bacteria and viruses (Lopez et al., 2009). Similarly, *Xanthomonas axonopodis* causing bacterial spot disease has been detected using antibody conjugated with fluorescence silica NPs in peppers and tomatoes (Yao et al., 2009). Singh et al.

(2010) reported that Karnal bunt disease caused by *Tilletia indica* could be detected using a nanogold-based immunosensor in wheat. Etefagh et al. (2013) reported the synthesis of CuO NPs and developed nanostructural layer biosensors to detect the fungi *Aspergillus niger* with high sensitivity. More studies need to be performed for utilizing nanosensors in the detection of various disease-causing pathogens and treatment of diseases in crops. Apart from crop protection, they can also be utilized as packaging materials in protecting post-harvest products. For instance, a nanocomposite film (with silicate NPs) called Durethan produced by Bayer Polymers prevents food spoilage by reducing oxygen and other gases as well as preserving moisture (Sharon et al., 2010). Similarly, other NPs can be employed in the preservation of crops as well as the post-harvest products. In addition, nanosensors can efficiently and timely monitor the suitable environmental conditions required for efficient crop improvement such as crop monitoring, detecting nutrients and pollutants in soil, water purification, etc. (Wang et al., 2010). Also they can prove to be helpful in measuring the soil conditions required for optimum crop growth, such as pH, temperature, humidity, and toxins in crops and soil. Such nanosensors, which can detect plant physiological changes, level of soil nutrients, metabolites, etc., have been reported by some researchers (Chen and Yada, 2011). For instance, Wang et al. (2010) developed an electrochemical nanosensor using the plant stimuli to detect salicylic acid levels in oil seeds and the fungi *S. sclerotiorum*. This sensor utilizes a gold electrode with copper NPs for its development. Environmental pollution, which can affect crop production, yield as well as quality, can be detected using nanosmart dust, which involves tiny wireless sensors as well as transponders.

Automation of irrigation systems for better crop productivity is another area where the use of nanosensors can be applied. They can be employed to increase the efficiency of water usage, particularly in the case of water shortage. Thus they can be used in sustainable irrigation management by estimating soil water tension in real time through coupling with autonomous irrigation controllers (de Medeiros et al., 2001). Detection of genetically modified organisms on agricultural crops is now possible using nanosensors, which showed promising results. These sensors are known as genosensors, which are based on DNA detection technology through hybridization reaction. These sensors can be applied to monitor the presence and level of genetically modified ingredients in food products (Feng et al., 2008; Yang et al., 2008; Manzanares-Palenzuela et al., 2015). Several sensors that have been developed to use Au, Pt, CNTs, nanopolyaniline chitosan, etc. for the detection of genetically modified food components are listed in Table 16.8.

Based on the studies, we conclude that nanosensors can be widely employed for real-time monitoring of soil conditions, crop diseases, and crop productivity. Despite the applications of nanotechnology in crop improvement, the utility of nanosensors is limited. Nanosensors should be incorporated in this field to detect the adverse conditions related to crop growth. These nanosensors and rapid monitoring systems can prove to be highly beneficial in the near future for sustainable agriculture and precision farming practices.

16.7 LIMITATION OF NANOMATERIALS

Despite several benefits of engineered nanomaterials, a main concern is their harmful impact on crop health (Lowry and Casman, 2009; Alshatwi et al., 2013). Thus it is very important to distinguish the toxicity of NPs, which typically depends on the chemical composition, structure, and size of these particles (Nowack and Bucheli, 2007). In the presence

TABLE 16.8 List of Nanosensors for the Detection of Genetically Modified Ingredients

Target	Electrode Type	DNA Immobilization Strategy	Detection Method	Detection Limit (M)	References
PAT	GCE	Adsorption onto nanogold/PDC	EIS (label-free)	$2.4 \cdot 10^{-11}$	Yang et al. (2007)
PAT	GCE	Adsorption onto nanogold/nano-PANI chitosan	EIS (label-free)	$3.1 \cdot 10^{-13}$	Feng et al. (2008)
PAT	Au	Potential-controlled adsorption onto aSiO_2-PATP	EIS (label free)	$1.5 \cdot 10^{-12}$	Ma et al. (2008)
PAT	GCE	Potential-controlled adsorption onto ZrO_2/nanogold	DPV with MB as indicator	$3.1 \cdot 10^{-11}$	Zhang et al. (2008)
PAT	GCE	Adsorption onto nano-PANI-ZrO_2/tyrosine	EIS (label free)	$2.68 \cdot 10^{-14}$	Yang et al. (2012a,b)
PAT and NOS amplicons	GCE	Adsorption onto ZrO_2/SWCNTs/PDC	EIS (label free)	$1.38 \cdot 10^{-12}$ (PAT) not reported for amplicons	Yang et al. (2007)
PAT and NOS amplicons	GCE	PDDA/PDC-SWCNTs films	DPV with MB as indicator	$2.6 \cdot 10^{-12}$ (PAT) not reported for amplicons	Yang et al. (2008)
PAT and NOS amplicons	CPE	Potential-controlled adsorption onto polylysine/SWCNTs	EIS (label free)	$3.1 \cdot 10^{-13}$ (PAT) not reported for amplicons	Jiang et al. (2008)
PAT and NOS amplicons	CPE	Adsorption onto PANI-MWCNTs/chitosan	EIS (label free)	$2.7 \cdot 10^{-14}$ (PAT) not reported for amplicons	Yang et al. (2009a,b)
PAT and NOS amplicons	CPE	Potential-controlled adsorption onto nanogold-CNT/nano-PANI	EIS (label free)	$5.6 \cdot 10^{-13}$ (PAT) not reported for amplicons	Zhou et al. (2009)
PEP gene	GCE	Adsorption onto nano-(Au–Pt)-polytyramine	EIS (label free)	$3.6 \cdot 10^{-13}$	Yang et al. (2012a,b)
35S	GCE	Adsorption onto Pt NPs	SWV with [Co(phen)3]+3 as indicator	$1 \cdot 10^{-9}$ (synthetic)	Wang et al. (2008)
35S	GCE	SAMs onto Au electrodes	DPASV using PbS NPs onto GCE	$4.38 \cdot 10^{-12}$	Sun et al. (2008)
35S	Au-interdigitated microelectrodes	Electrocopolymerization with PPy/MWCNTs	EIS (label free)	Not reported	Lien et al. (2010)
35S	SPCE	Covalent attachment to succinimide-functionalized acrylic microspheres onto the AuNPs/SPCE	DPV with anthraquinone-2-sulfonic acid monohydrate sodium salt as indicator	$7.79 \cdot 10^{-16}$	Ulianas et al. (2014)

AuNP, gold nanoparticle; *CNT*, carbon nanotube; *CPE*, carbon-paste electrode; *DPASV*, differential pulse anodic solid voltammetry; *DPV*, differential pulse voltammetry; *EIS*, electrochemical impedance spectroscopy; *GCE*, glassy carbon electrode; *MB*, methylene blue; *MWCNT*, multi-walled carbon nanotube; *NOS*, nopaline synthase; *PANI*, polyaniline; *PAT*, phosphinothricin acetyltransferase gene; *PATP*, p-aminothiophenol; *PbS*, lead sulfide; *PDC*, 2,6-pyridinedicarboxylic acid; *PDDA*, poly(diallyldimethyl ammonium chloride); *PEP*, phosphoenolpyruvate carboxylase promoter; *PPy*, polypyrrole; *SPCE*, screen-printed carbon electrode; *SWCNT*, single-walled CNT; *SWV*, square of particular growing environments, plants possibly absorb essential and nonessential components, which at a higher concentration than the threshold might cause toxicity. The Joint FAO/WHO Meeting on Pesticide Residues has issued guidance on field trials prior to approving pesticides. However, the guidance does not contain specific points with regard to pesticides containing nanocomponents. Some nanopesticides have been approved along with nonnanopesticides using the same process and there is no specific parameter used to identify nanovariants (http://www.fao.org/fileadmin/templates/agns/pdf/topics/Nano_Report_Final_20120625.pdf). Use of less pesticide will contribute to a lower risk of residues, thus ensuring safety. By properly addressing the risk assessment of nanotechnologies in the food and agriculture sectors for the purposes of safety, engineered nanomaterials need to be regulated before and after coming to market (Takeuchi et al., 2014, Food safety at FAO-www.fao.org/food/food-safety-quality/home-page/en/).

16.8 CONCLUSION

Application of nanomaterials in the area of agriculture has proved to be a boon in crop protection. Nanobased delivery of agrochemicals in the form of nanocapsules, nanospheres, micelles, nanocomposites, etc. is now possible with several advantages over the conventional method of crop protection. The effect of encapsulated pesticides lasts for a long time because of their controlled release and protection against photocatalysis, hydrolysis, and/or other degrading mechanisms. Considering the benefits of NPs over bactericides and fungicides, their potential as nanopesticides is remarkable. Real-time monitoring of biophysicochemical properties related to crop improvement can now be achieved with high sensitivity through nanobased sensors. Gene transfer using nanotechnology is also attracting many researchers to efficient plant transformation. However, application of nanomaterials can be phytotoxic at a certain concentration, which should be assessed prior to their field application. Evaluation of the toxic effects of these nanomaterials on animals and the environment with their bioavailability is thus essential. Before releasing nanobased plant protectants to the market, they should undergo proper regulatory assessment to measure their efficiency and toxicity.

Acknowledgments

The authors thank the Department of Biotechnology, Govt. of India, for supporting the research and development activities through the Institutional Biotech Hub Project (Project No. BT/04/NE/2009). We also thank IITG and MHRD for financial support in the form of a fellowship.

References

Adak, T., Kumar, J., Shakil, N.A., Walia, S., 2012. Development of controlled release formulations of imidacloprid employing novel nano-ranged amphiphilic polymers. J. Environ. Sci. Health B 47 (3), 217–225.

Adel, M.M., Atwa, W.A., Hassan, M.L., Salem, N.Y., Farghaly, D.S., Ibrahim, S.S., 2015. Biological activity and field persistence of *Pelargonium graveolens* (Geraniales: *Geraniaceae*) loaded solid lipid nanoparticles (SLNs) on *Phthorimaea operculella* (Zeller) (PTM) (*Lepidoptera*: *Gelechiidae*). Int. J. Sci. Res. 4 (11), 514–520.

Agrawal, R.M., Charpe, S.D., Raghuwanshi, F.C., Lamdhade, G.T., 2015. Synthesis and characterization of magnesium oxide nanoparticles with 1: 1 molar ratio via liquid-phase method. Int. J. Appl. Innov. Eng. Manag. 4 (2), 141–145.

Aguilar-Méndez, M.A., San Martín-Martínez, E., Ortega-Arroyo, L., Cobián-Portillo, G., Sánchez-Espíndola, E., 2011. Synthesis and characterization of silver nanoparticles: effect on phytopathogen *Colletotrichum gloesporioides*. J. Nanopart. Res. 13 (6), 2525–2532.

Ahmed, S., Ahmad, M., Swami, B.L., Ikram, S., 2016. A review on plants extract mediated synthesis of silver nanoparticles for antimicrobial applications: a green expertise. J. Adv. Res. 7 (1), 17–28.

Akhtar, S., Benter, I.F., 2007. Nonviral delivery of synthetic siRNAs in vivo. J. Clin. Invest. 117 (12), 3623–3632.

Akhter, S., Zaki Ahmad, M., Singh, A., Ahmad, I., Rahman, M., Anwar, M., Kumar Jain, G., Jalees Ahmad, F., Krishen Khar, R., 2011. Cancer targeted metallic nanoparticle: targeting overview, recent advancement and toxicity concern. Curr. Pharm. Des. 17 (18), 1834–1850.

Alghuthaymi, M.A., Almoammar, H., Rai, M., Said-Galiev, E., Abd-Elsalam, K.A., 2015. Myconanoparticles: synthesis and their role in phytopathogens management. Biotechnol. Biotechnol. Equip. 29 (2), 221–236.

Alshatwi, A.A., Subbarayan, P.V., Ramesh, E., Al-Hazzani, A.A., Alsaif, M.A., Alwarthan, A.A., 2013. Aluminium oxide nanoparticles induce mitochondrial-mediated oxidative stress and alter the expression of antioxidant enzymes in human mesenchymal stem cells. Food Addit. Contam. A 30 (1), 1–10.

Alvarez, M., Calle, A., Tamayo, J., Lechuga, L.M., Abad, A., Montoya, A., 2003. Development of nanomechanical biosensors for detection of the pesticide DDT. Biosens. Bioelectron. 18 (5), 649–653.

Anjali, C.H., Sharma, Y., Mukherjee, A., Chandrasekaran, N., 2012. Neem oil (*Azadirachta indica*) nanoemulsion-a potent larvicidal agent against *Culex quinquefasciatus*. Pest Manag. Sci. 68 (2), 158–163.

Annadhasan, M., Muthukumarasamyvel, T., Sankar Babu, V.R., Rajendiran, N., 2014. Green synthesized silver and gold nanoparticles for colorimetric detection of Hg^{2+}, Pb^{2+}, and Mn^{2+} in aqueous medium. ACS Sustain. Chem. Eng. 2 (4), 887–896.

Araj, S.E.A., Salem, N.M., Ghabeish, I.H., Awwad, A.M., 2015. Toxicity of nanoparticles against *Drosophila melanogaster* (*Diptera*: *Drosophilidae*). J. Nanomater. 2015 (5).

Arthur, F.H., 1999. Evaluation of an encapsulated formulation of cyfluthrin to control *Sitophilus oryzae* (L.) on stored wheat. J. Stored Prod. Res. 35 (2), 159–166.

Arthurs, S.P., Lacey, L.A., Behle, R.W., 2006. Evaluation of spray-dried lignin-based formulations and adjuvants as solar protectants for the granulovirus of the codling moth, *Cydia pomonella* (L). J. Invertebr. Pathol. 93 (2), 88–95.

Ashton Acton, Q., 2013. Issues in Biochemistry and Biophysics Research: 2013 ed.

Aslani, F., Bagheri, S., Muhd Julkapli, N., Juraimi, A.S., Hashemi, F.S.G., Baghdadi, A., 2014. Effects of engineered nanomaterials on plants growth: an overview. Sci. World J. 2014, 1–28.

Aziz, N., Faraz, M., Pandey, R., Shakir, M., Fatma, T., Varma, A., et al., 2015. Facile algae-derived route to biogenic silver nanoparticles: synthesis, antibacterial, and photocatalytic properties. Langmuir 31, 11605–11612.

Aziz, N., Pandey, R., Barman, I., Prasad, R., 2016. Leveraging the attributes of *Mucor hiemalis*-derived silver nanoparticles for a synergistic broad-spectrum antimicrobial platform. Front. Microbiol. 7, 1984. http://dx.doi.org/10.3389/fmicb.2016.01984.

Baboota, S., Shakeel, F., Ahuja, A., Ali, J., Shafiq, S., 2007. Design, development and evaluation of novel nanoemulsion formulations for transdermal potential of celecoxib. Acta Pharm. 57 (3), 315–332.

Balaji, A.P.B., Mishra, P., Kumar, R.S., Mukherjee, A., Chandrasekaran, N., 2015. Nanoformulation of poly (ethylene glycol) polymerized organic insect repellent by PIT emulsification method and its application for Japanese encephalitis vector control. Colloids Surf. B Biointerfaces 128, 370–378.

Bang, S.H., Hwang, I.C., Yu, Y.M., Kwon, H.R., Kim, D.H., Park, H.J., 2011. Influence of chitosan coating on the liposomal surface on physicochemical properties and the release profile of nanocarrier systems. J. Microencapsul. 28 (7), 595–604.

Bawa, P., Pillay, V., Choonara, Y.E., Du Toit, L.C., 2009. Stimuli-responsive polymers and their applications in drug delivery. Biomed. Mater. 4 (2), 022001.

Benício, L.P.F., Silva, R.A., Lopes, J.A., Eulálio, D., Santos, R.M.M.D., Aquino, L.A.D., Vergütz, L., Novais, R.F., Costa, L.M.D., Pinto, F.G., Tronto, J., 2015. Layered double hydroxides: nanomaterials for applications in agriculture. Revista Brasileira de Ciência do Solo 39 (1), 1–13.

Bhagat, D., Samanta, S.K., Bhattacharya, S., 2013. Efficient management of fruit pests by pheromone nanogels. Sci. Rep. 3.

Bibi, S., Nawaz, M., Yasin, T., Riaz, M., 2016. Chitosan/CNTs nanocomposite as green carrier material for pesticides controlled release. J. Polymer Res. 23 (8), 154.

Boehm, A.L.L.R., Zerrouk, R., Fessi, H., 2000. Poly epsilon-caprolactone nanoparticles containing a poorly soluble pesticide: formulation and stability study. J. Microencapsul. 17 (2), 195–205.

Boehm, A.L., Martinon, I., Zerrouk, R., Rump, E., Fessi, H., 2003. Nanoprecipitation technique for the encapsulation of agrochemical active ingredients. J. Microencapsul. 20 (4), 433–441.

Boro, R.C., Kaushal, J., Nangia, Y., Wangoo, N., Bhasin, A., Suri, C.R., 2011. Gold nanoparticles catalyzed chemiluminescence immunoassay for detection of herbicide 2, 4-dichlorophenoxyacetic acid. Analyst 136 (10), 2125–2130.

Bramhanwade, K., Shende, S., Bonde, S., Gade, A., Rai, M., 2016. Fungicidal activity of Cu nanoparticles against *Fusarium* causing crop diseases. Environ. Chem. Lett. 1–7.

Buzea, C., Pacheco, I.I., Robbie, K., 2007. Nanomaterials and nanoparticles: sources and toxicity. Biointerphases 2 (4), MR17–MR71.

Camacho-Flores, B.A., Martínez-Álvarez, O., Arenas-Arrocena, M.C., Garcia-Contreras, R., Argueta-Figueroa, L., de la Fuente-Hernández, J., Acosta-Torres, L.S., 2015. Copper: synthesis techniques in nanoscale and powerful application as an antimicrobial agent. J. Nanomater. 2015, 1–11.

Camargo, P.H.C., Satyanarayana, K.G., Wypych, F., 2009. Nanocomposites: synthesis, structure, properties and new application opportunities. Mater. Res. 12 (1), 1–39.

Campos, E.V.R., de Oliveira, J.L., da Silva, C.M.G., Pascoli, M., Pasquoto, T., Lima, R., Abhilash, P.C., Fraceto, L.F., 2015. Polymeric and solid lipid nanoparticles for sustained release of carbendazim and tebuconazole in agricultural applications. Sci. Rep. 5, 13809.

Cao, Y., Huang, L., Chen, J., Liang, J., Long, S., Lu, Y., 2005. Development of a controlled release formulation based on a starch matrix system. Int. J. Pharm. 298 (1), 108–116.

Cao, L., Zhang, H., Cao, C., Zhang, J., Li, F., Huang, Q., 2016. Quaternized chitosan-capped mesoporous silica nanoparticles as nanocarriers for controlled pesticide release. Nanomaterials 6 (7), 126.

Cardoso, L.P., Celis, R., Cornejo, J., Valim, J.B., 2006. Layered double hydroxides as supports for the slow release of acid herbicides. J. Agric. Food Chem. 54 (16), 5968–5975.

Casanova, H., Araque, P., Ortiz, C., 2005. Nicotine carboxylate insecticide emulsions: effect of the fatty acid chain length. J. Agric. Food Chem. 53 (26), 9949–9953.

Chen, X., Mao, S.S., 2007. Titanium dioxide nanomaterials: synthesis, properties, modifications, and applications. Chem. Rev. 107 (7), 2891–2959.

Chen, H., Yada, R., 2011. Nanotechnologies in agriculture: new tools for sustainable development. Trends Food Sci. Technol. 22 (11), 585–594.

Chen, J., Wang, W., Xu, Y., Zhang, X., 2010. Slow-release formulation of a new biological pesticide, pyoluteorin, with mesoporous silica. J. Agric. Food Chem. 59 (1), 307–311.

Chen, Z.Y., Liang, K., Qiu, R.X., Luo, L.P., 2011. Ultrasound-and liposome microbubble-mediated targeted gene transfer to cardiomyocytes in vivo accompanied by polyethylenimine. J. Ultrasound Med. 30 (9), 1247–1258.

Chevillard, A., Angellier-Coussy, H., Guillard, V., Gontard, N., Gastaldi, E., 2012. Controlling pesticide release via structuring agropolymer and nanoclays based materials. J. Hazard. Mater. 205, 32–39.

Chin, C.P., Wu, H.S., Wang, S.S., 2011. New approach to pesticide delivery using nanosuspensions: research and applications. Ind. Eng. Chem. Res. 50 (12), 7637–7643.

Choudhury, S.R., Ghosh, M., Mandal, A., Chakravorty, D., Pal, M., Pradhan, S., Goswami, A., 2011. Surface-modified sulfur nanoparticles: an effective antifungal agent against *Aspergillus Niger* and *Fusarium oxysporum*. Appl. Microbiol. Biotechnol. 90 (2), 733–743.

Chowdappa, P., Gowda, S., 2013. Nanotechnology in crop protection: status and scope. Pest Manag. Hortic. Ecosyst. 19 (2), 131–151.

Chowdappa, P., Gowda, S., Chethana, C.S., Madhura, S., 2014. Antifungal activity of chitosan-silver nanoparticle composite against *Colletotrichum gloeosporioides* associated with mango anthracnose. Afr. J. Microbiol. Res. 8 (17), 1803–1812.

Chu, H., Kim, H.J., Kim, J.S., Kim, M.S., Yoon, B.D., Park, H.J., Kim, C.Y., 2012. A nanosized Ag–silica hybrid complex prepared by γ-irradiation activates the defense response in *Arabidopsis*. Radiat. Phys. Chem. 81 (2), 180–184.

Cioffi, N., Torsi, L., Ditaranto, N., Sabbatini, L., Zambonin, P.G., Tantillo, G., Ghibelli, L., D'Alessio, M., Bleve-Zacheo, T., Traversa, E., 2004. Antifungal activity of polymer-based copper nanocomposite coatings. Appl. Phys. Lett. 85 (12), 2417–2419.

Constantin, M., Cristea, M., Ascenzi, P., Fundueanu, G., 2011. Lower critical solution temperature versus volume phase transition temperature in thermoresponsive drug delivery systems. Express Polymer Lett. 5 (10), 839–848.

Da Silva, A.C., Deda, D.K., da Róz, A.L., Prado, R.A., Carvalho, C.C., Viviani, V., Leite, F.L., 2013. Nanobiosensors based on chemically modified AFM probes: a useful tool for metsulfuron-methyl detection. Sensors 13 (2), 1477–1489.

Dai, S., Ravi, P., Tam, K.C., 2008. pH-responsive polymers: synthesis, properties and applications. Soft Matter 4 (3), 435–449.
De Medeiros, G.A., Arruda, F.B., Sakai, E., Fujiwara, M., 2001. The influence of crop canopy on evapotranspiration and crop coefficient of beans (*Phaseolus vulgaris* L.). Agric. Water Manag. 49 (3), 211–224.
De Rosa, G., La Rotonda, M.I., 2009. Nano and microtechnologies for the delivery of oligonucleotides with gene silencing properties. Molecules 14 (8), 2801–2823.
Dean, R., Van Kan, J.A., Pretorius, Z.A., Hammond-Kosack, K.E., Di Pietro, A., Spanu, P.D., Rudd, J.J., Dickman, M., Kahmann, R., Ellis, J., Foster, G.D., 2012. The Top 10 fungal pathogens in molecular plant pathology. Mol. Plant Pathol. 13 (4), 414–430.
Deda, D.K., Pereira, B., Bueno, C.C., Silva, A.N.D., Ribeiro, G.A., Amarante, A.M., Franca, E.F., Leite, F.L., 2013. The use of functionalized AFM tips as molecular sensors in the detection of pesticides. Mater. Res. 16 (3), 683–687.
Deshayes, A., Herrera-Estrella, L., Caboche, M., 1985. Liposome-mediated transformation of tobacco mesophyll protoplasts by an *Escherichia coli* plasmid. EMBO J. 4 (11), 2731.
Dimkpa, C.O., McLean, J.E., Britt, D.W., Anderson, A.J., 2013. Antifungal activity of ZnO nanoparticles and their interactive effect with a biocontrol bacterium on growth antagonism of the plant pathogen *Fusarium graminearum*. Biometals 26 (6), 913–924.
Ding, K., Shi, L., Zhang, L., Zeng, T., Yin, Y., Yi, Y., 2016. Synthesis of photoresponsive polymeric propesticide micelles based on PEG for the controlled release of a herbicide. Polymer Chem. 7 (4), 899–904.
Dizaj, S.M., Lotfipour, F., Barzegar-Jalali, M., Zarrintan, M.H., Adibkia, K., 2014. Antimicrobial activity of the metals and metal oxide nanoparticles. Mater. Sci. Eng. C 44, 278–284.
Durairaj, B., Xavier, T., Muthu, S., 2014. Fungal generated titanium dioxide Nanoparticles: a potent mosquito (*Aedes aegypti*) larvicidal agent. Sch. Acad. J. Biosci. (SAJB) 2 (9), 651–658.
Elek, N., Hoffman, R., Raviv, U., Resh, R., Ishaaya, I., Magdassi, S., 2010. Novaluron nanoparticles: formation and potential use in controlling agricultural insect pests. Colloids Surf. A Physicochem. Eng. Asp. 372 (1), 66–72.
Etefagh, R., Azhir, E., Shahtahmasebi, N., 2013. Synthesis of CuO nanoparticles and fabrication of nanostructural layer biosensors for detecting Aspergillus niger fungi. *Scientia Iranica* 20 (3), 1055–1058.
Europa. https://ec.europa.eu/growth/industry/key-enabling-technologies_en.
FAO. http://www.fao.org/fileadmin/templates/agns/pdf/topics/Nano_Report_Final_20120625.pdf.
Feng, Z., Yaqian, W., 2011. Formula optimization design of pesticide microemulsion. In: Pesticide–Formulations, Effects, Fate, pp. 63–68.
Feng, Y., Yang, T., Zhang, W., Jiang, C., Jiao, K., 2008. Enhanced sensitivity for deoxyribonucleic acid electrochemical impedance sensor: gold nanoparticle/polyaniline nanotube membranes. Anal. Chim. Acta 616 (2), 144–151.
Fernandes, C.P., de Almeida, F.B., Silveira, A.N., Gonzalez, M.S., Mello, C.B., Feder, D., Apolinário, R., Santos, M.G., Carvalho, J.C.T., Tietbohl, L.A.C., Rocha, L., 2014. Development of an insecticidal nanoemulsion with *Manilkara subsericea* (Sapotaceae) extract. J. Nanobiotechnol. 12 (1), 1.
Flores-Céspedes, F., Figueredo-Flores, C.I., Daza-Fernández, I., Vidal-Peña, F., VillafraFlores-Céspedesnca-Sánchez, M., Fernández-Pérez, M., 2012. Preparation and characterization of imidacloprid lignin–polyethylene glycol matrices coated with ethylcellulose. J. Agric. Food Chem. 60 (4), 1042–1051.
Fraceto, L.F., Grillo, R., de Medeiros, G.A., Scognamiglio, V., Rea, G., Bartolucci, C., 2016. Nanotechnology in Agriculture: which innovation potential does it have? Front. Environ. Sci. 4, 20.
Frandsen, M.V., Pedersen, M.S., Zellweger, M., Gouin, S., Roorda, S.D., Phan, T.Q.C., April 2, 2015. Insecticidal Polymer Matrix Comprising PBO and DM United States Patent US 20150093427 A1.
Gámiz, B., Hermosín, M.C., Cornejo, J., Celis, R., 2015. Hexadimethrine-montmorillonite nanocomposite: characterization and application as a pesticide adsorbent. Appl. Surf. Sci. 332, 606–613.
Gan, N., Yang, X., Xie, D., Wu, Y., Wen, W., 2010. A disposable organophosphorus pesticides enzyme biosensor based on magnetic composite nano-particles modified screen printed carbon electrode. Sensors 10 (1), 625–638.
Ghormade, V., Deshpande, M.V., Paknikar, K.M., 2011. Perspectives for nano-biotechnology enabled protection and nutrition of plants. Biotechnol. Adv. 29 (6), 792–803.
Glass, E.H., Thurston, H.D., 2014. Traditional and modern crop protection in perspective. Bioscience 28 (2), 109–115.
Greene, L.C., Meyers, P.A., Springer, J.T., Banks, P.A., 1992. Biological evaluation of pesticides released from temperature-responsive microcapsules. J. Agric. Food Chem. 40 (11), 2274–2278.
Guan, H., Chi, D., Yu, J., Li, X., 2008. A novel photodegradable insecticide: preparation, characterization and properties evaluation of nano-Imidacloprid. Pestic. Biochem. Physiol. 92 (2), 83–91.

Guo, Y.R., Liu, S.Y., Gui, W.J., Zhu, G.N., 2009. Gold immunochromatographic assay for simultaneous detection of carbofuran and triazophos in water samples. Anal. Biochem. 389 (1), 32–39.

He, L., Liu, Y., Mustapha, A., Lin, M., 2011. Antifungal activity of zinc oxide nanoparticles against *Botrytis cinerea* and *Penicillium expansum*. Microbiol. Res. 166 (3), 207–215.

Hu, J., Zhang, G., Liu, S., 2012. Enzyme-responsive polymeric assemblies, nanoparticles and hydrogels. Chem. Soc. Rev. 41 (18), 5933–5949.

Hua, K.H., Wang, H.C., Chung, R.S., Hsu, J.C., 2015. Calcium carbonate nanoparticles can enhance plant nutrition and insect pest tolerance. J. Pest. Sci. 40 (4), 208–213.

Husen, A., Siddiqi, K.S., 2014. Carbon and fullerene nanomaterials in plant system. J. Nanobiotechnol. 12 (1), 1.

Hwang, I.C., Kim, T.H., Bang, S.H., Kim, K.S., Kwon, H.R., Seo, M.J., Youn, Y.N., Park, H.J., Yasunaga-Aokt, C., Yu, Y.M., 2011. Insecticidal effect of controlled release formulations of etofenprox based on nano-bio technique. J. Fac. Agric. Kyushu Univ. 56 (1), 33–40.

Imada, K., Sakai, S., Kajihara, H., Tanaka, S., Ito, S., 2015. Magnesium oxide nanoparticles induce systemic resistance in tomato against bacterial wilt disease. Plant Pathol. 551–560.

Işiklan, N., 2006. Controlled release of insecticide carbaryl from sodium alginate, sodium alginate/gelatin, and sodium alginate/sodium carboxymethyl cellulose blend beads crosslinked with glutaraldehyde. J. Appl. Polymer Sci. 99 (4), 1310–1319.

Ji, X., Zheng, J., Xu, J., Rastogi, V.K., Cheng, T.C., DeFrank, J.J., Leblanc, R.M., 2005. (CdSe) ZnS quantum dots and organophosphorus hydrolase bioconjugate as biosensors for detection of paraoxon. J. Phys. Chem. B 109 (9), 3793–3799.

Jia, G., Wang, H., Yan, L., Wang, X., Pei, R., Yan, T., Zhao, Y., Guo, X., 2005. Cytotoxicity of carbon nanomaterials: single-wall nanotube, multi-wall nanotube, and fullerene. Environ. Sci. Technol. 39 (5), 1378–1383.

Jiang, C., Yang, T., Jiao, K., Gao, H., 2008. A DNA electrochemical sensor with poly-l-lysine/single-walled carbon nanotubes films and its application for the highly sensitive EIS detection of PAT gene fragment and PCR amplification of NOS gene. Electrochim. Acta 53 (6), 2917–2924.

Jo, Y.K., Kim, B.H., Jung, G., 2009. Antifungal activity of silver ions and nanoparticles on phytopathogenic fungi. Plant Dis. 93 (10), 1037–1043.

Kala, A., Soosairaj, S., Mathiyazhagan, S., Raja, P., 2016. Green synthesis of copper bionanoparticles to control the bacterial leaf blight disease of rice. Curr. Sci. 110 (10).

Kamada, J., Koynov, K., Corten, C., Juhari, A., Yoon, J.A., Urban, M.W., Balazs, A.C., Matyjaszewski, K., 2010. Redox responsive behavior of thiol/disulfide-functionalized star polymers synthesized via atom transfer radical polymerization. Macromolecules 43 (9), 4133–4139.

Kamari, A., Aljafree, N.F.A., Yusoff, S.N.M., 2016. *N, N*-dimethylhexadecyl carboxymethyl chitosan as a potential carrier agent for rotenone. Int. J. Biol. Macromol. 88, 263–272.

Kang, T.F., Wang, F., Lu, L.P., Zhang, Y., Liu, T.S., 2010. Methyl parathion sensors based on gold nanoparticles and nafion film modified glassy carbon electrodes. Sens. Actuators B Chem. 145 (1), 104–109.

Kanhed, P., Birla, S., Gaikwad, S., Gade, A., Seabra, A.B., Rubilar, O., Duran, N., Rai, M., 2014. In vitro antifungal efficacy of copper nanoparticles against selected crop pathogenic fungi. Mater. Lett. 115, 13–17.

Kanti, P., Maitra, S., Mandal, S., Bhunia, S., Das, G.C., Chaudhury, M.G., 2015. Green synthesis of titanium dioxide nanoparticles: a review. Sci. Voyag. 1 (4), 28–38.

Keat, C.L., Aziz, A., Eid, A.M., Elmarzugi, N.A., 2015. Biosynthesis of nanoparticles and silver nanoparticles. Bioresour. Bioprocess. 2, 47. http://dx.doi.org/10.1186/s40643-015-0076-2.

Khan, A.A., Akhtar, T., 2011. Adsorption thermodynamics studies of 2, 4, 5-trichlorophenoxy acetic acid on poly-o-toluidine Zr (IV) phosphate, a nano-composite used as pesticide sensitive membrane electrode. Desalination 272 (1), 259–264.

Khodakovskaya, M.V., de Silva, K., Biris, A.S., Dervishi, E., Villagarcia, H., 2012. Carbon nanotubes induce growth enhancement of tobacco cells. ACS Nano 6 (3), 2128–2135.

Khodashenas, B., Ghorbani, H.R., 2014. Synthesis of copper nanoparticles: an overview of the various methods. Korean J. Chem. Eng. 31 (7), 1105–1109.

Kim, S.W., Kim, K.S., Lamsal, K., Kim, Y.J., Kim, S.B., Jung, M., Sim, S.J., Kim, H.S., Chang, S.J., Kim, J.K., Lee, Y.S., 2009. An in vitro study of the antifungal effect of silver nanoparticles on oak wilt pathogen *Raffaelea* sp. J. Microbiol. Biotechnol. 19 (8), 760–764.

Kim, S.W., Jung, J.H., Lamsal, K., Kim, Y.S., Min, J.S., Lee, Y.S., 2012. Antifungal effects of silver nanoparticles (AgNPs) against various plant pathogenic fungi. Mycobiology 40 (1), 53–58.

Kitherian, S., 2017. Review article nano and bio-nanoparticles for insect control. Res. J. Nanosci. Nanotechnol. 7 (1), 1–9.

Koli, P., Singh, B.B., Shakil, N.A., Kumar, J., Kamil, D., 2015. Development of controlled release nanoformulations of carbendazim employing amphiphilic polymers and their bioefficacy evaluation against *Rhizoctonia solani*. J. Environ. Sci. Health B 50 (9), 674–681.

Kołodziejczak-Radzimska, A., Jesionowski, T., 2014. Zinc oxide—from synthesis to application: a review. Materials 7 (4), 2833–2881.

Kothamasu, P., Kanumur, H., Ravur, N., Maddu, C., Parasuramrajam, R., Thangavel, S., 2012. Nanocapsules: the weapons for novel drug delivery systems. Bioimpacts 2 (2), 71–81.

Krishnaraj, C., Ramachandran, R., Mohan, K., Kalaichelvan, P.T., 2012. Optimization for rapid synthesis of silver nanoparticles and its effect on phytopathogenic fungi. Spectrochim. Acta A Mol. Biomol. Spectroscopy 93, 95–99.

Kuang, H., Chen, W., Yan, W., Xu, L., Zhu, Y., Liu, L., Chu, H., Peng, C., Wang, L., Kotov, N.A., Xu, C., 2011. Crown ether assembly of gold nanoparticles: melamine sensor. Biosens. Bioelectron. 26 (5), 2032–2037.

Kumar, R.S., Shiny, P.J., Anjali, C.H., Jerobin, J., Goshen, K.M., Magdassi, S., Chandrasekaran, N., 2013. Distinctive effects of nano-sized permethrin in the environment. Environ. Sci. Pollut. Res. 20 (4), 2593–2602.

Kumar, S., Bhanjana, G., Sharma, A., Sidhu, M.C., Dilbaghi, N., 2014. Synthesis, characterization and on field evaluation of pesticide loaded sodium alginate nanoparticles. Carbohydr. Polym. 101, 1061–1067.

Kumaravel, A., Chandrasekaran, M., 2011. A biocompatible nano TiO_2/nafion composite modified glassy carbon electrode for the detection of fenitrothion. J. Electroanal. Chem. 650 (2), 163–170.

Lai, F., Wissing, S.A., Müller, R.H., Fadda, A.M., 2006. *Artemisia arborescens* L essential oil-loaded solid lipid nanoparticles for potential agricultural application: preparation and characterization. AAPS PharmSciTech 7 (1), E10–E18.

Lakraimi, M., Legrouri, A., Barroug, A., De Roy, A., Besse, J.P., 2000. Preparation of a new stable hybrid material by chloride–2, 4-dichlorophenoxyacetate ion exchange into the zinc–aluminium–chloride layered double hydroxide. J. Mater. Chem. 10 (4), 1007–1011.

Lamsal, K., Kim, S.W., Jung, J.H., Kim, Y.S., Kim, K.S., Lee, Y.S., 2011a. Application of silver nanoparticles for the control of *Colletotrichum* species in vitro and pepper anthracnose disease in field. Mycobiology 39 (3), 194–199.

Lamsal, K., Kim, S.W., Jung, J.H., Kim, Y.S., Kim, K.S., Lee, Y.S., 2011b. Inhibition effects of silver nanoparticles against powdery mildews on cucumber and pumpkin. Mycobiology 39 (1), 26–32.

Lao, S.B., Zhang, Z.X., Xu, H.H., Jiang, G.B., 2010. Novel amphiphilic chitosan derivatives: synthesis, characterization and micellar solubilization of rotenone. Carbohydr. Polym. 82 (4), 1136–1142.

Latheef, M.A., Dailey, O.D., Franz, E., 1993. Efficacy of polymeric controlled release formulations of sulprofos against tobacco budworm, *Heliothis virescens* (*Lepidoptera*: *Noctuidae*) on cotton. In: Pesticide Formulations and Application Systems, vol. 13. ASTM International.

Lee, K.J., Park, S.H., Govarthanan, M., Hwang, P.H., Seo, Y.S., Cho, M., Lee, W.H., Lee, J.Y., Kamala-Kannan, S., Oh, B.T., 2013. Synthesis of silver nanoparticles using cow milk and their antifungal activity against phytopathogens. Mater. Lett. 105, 128–131.

Letchford, K., Burt, H., 2007. A review of the formation and classification of amphiphilic block copolymer nanoparticulate structures: micelles, nanospheres, nanocapsules and polymersomes. Eur. J. Pharm. Biopharm. 65 (3), 259–269.

Li, C., Wang, C., Wang, C., Hu, S., 2006. Development of a parathion sensor based on molecularly imprinted nano-TiO_2 self-assembled film electrode. Sens. Actuators B Chem. 117 (1), 166–171.

Li, Z.Z., Chen, J.F., Liu, F., Liu, A.Q., Wang, Q., Sun, H.Y., Wen, L.X., 2007. Study of UV-shielding properties of novel porous hollow silica nanoparticle carriers for avermectin. Pest Manag. Sci. 63 (3), 241–246.

Li, C., Guo, T., Zhou, D., Hu, Y., Zhou, H., Wang, S., Chen, J., Zhang, Z., 2011. A novel glutathione modified chitosan conjugate for efficient gene delivery. J. Controll. Release 154 (2), 177–188.

Lien, T.T.N., Dai Lam, T., An, V.T.H., Hoang, T.V., Quang, D.T., Khieu, D.Q., Tsukahara, T., Lee, Y.H., Kim, J.S., 2010. Multi-wall carbon nanotubes (MWCNTs)-doped polypyrrole DNA biosensor for label-free detection of genetically modified organisms by QCM and EIS. Talanta 80 (3), 1164–1169.

Lim, C.J., Basri, M., Omar, D., Abdul Rahman, M.B., Salleh, A.B., Rahman, R.A., Zaliha, R.N., 2013. Green nanoemulsion-laden glyphosate isopropylamine formulation in suppressing creeping foxglove (*A. gangetica*), slender button weed (*D. ocimifolia*) and buffalo grass (*P. conjugatum*). Pest Manag. Sci. 69 (1), 104–111.

Lin, Y., Lu, F., Wang, J., 2004. Disposable carbon nanotube modified screen-printed biosensor for amperometric detection of organophosphorus pesticides and nerve agents. Electroanalysis 16 (1–2), 145–149.

Lin, T.J., Huang, K.T., Liu, C.Y., 2006. Determination of organophosphorous pesticides by a novel biosensor based on localized surface plasmon resonance. Biosens. Bioelectron. 22 (4), 513–518.
Lin, C., Fugetsu, B., Su, Y., Watari, F., 2009. Studies on toxicity of multi-walled carbon nanotubes on *Arabidopsis* T87 suspension cells. J. Hazard. Mater. 170 (2), 578–583.
Liu, Y., Yan, L., Heiden, P., Laks, P., 2001. Use of nanoparticles for controlled release of biocides in solid wood. J. Appl. Polymer Sci. 79 (3), 458–465.
Liu, F., Wen, L.X., Li, Z.Z., Yu, W., Sun, H.Y., Chen, J.F., 2006. Porous hollow silica nanoparticles as controlled delivery system for water-soluble pesticide. Mater. Res. Bull. 41 (12), 2268–2275.
Liu, J., Wang, F.H., Wang, L.L., Xiao, S.Y., Tong, C.Y., Tang, D.Y., Liu, X.M., 2008a. Preparation of fluorescence starch-nanoparticle and its application as plant transgenic vehicle. J. Cent. South Univ. Technol. 15, 768–773.
Liu, S., Yuan, L., Yue, X., Zheng, Z., Tang, Z., 2008b. Recent advances in nanosensors for organophosphate pesticide detection. Adv. Powder Technol. 19 (5), 419–441.
Liu, Y., Tong, Z., Prud'homme, R.K., 2008c. Stabilized polymeric nanoparticles for controlled and efficient release of bifenthrin. Pest Manag. Sci. 64 (8), 808–812.
Liu, R., Guan, G., Wang, S., Zhang, Z., 2011. Core-shell nanostructured molecular imprinting fluorescent chemosensor for selective detection of atrazine herbicide. Analyst 136 (1), 184–190.
Liu, S., Xie, L., Zheng, J., Jiang, R., Zhu, F., Luan, T., Ouyang, G., 2015. Mesoporous TiO_2 nanoparticles for highly sensitive solid-phase microextraction of organochlorine pesticides. Anal. Chim. Acta 878, 109–117.
Lodha, A., Lodha, M., Patel, A., Chaudhuri, J., Dalal, J., Edwards, M., Douroumis, D., 2012. Synthesis of mesoporous silica nanoparticles and drug loading of poorly water soluble drug cyclosporin A. J. Pharm. Bioallied Sci. 4 (Suppl. 1), S92.
Loha, K.M., Shakil, N.A., Kumar, J., Singh, M.K., Srivastava, C., 2012. Bio-efficacy evaluation of nanoformulations of β-cyfluthrin against *Callosobruchus maculatus* (*Coleoptera*: *Bruchidae*). J. Environ. Sci. Health B 47 (7), 687–691.
López, M.M., Llop, P., Olmos, A., Marco-Noales, E., Cambra, M., Bertolini, E., 2009. Are molecular tools solving the challenges posed by detection of plant pathogenic bacteria and viruses? Curr. Issues Mol. Biol. 11 (1), 13.
Lowry, G.V., Casman, E.A., 2009. Nanomaterial transport, transformation, and fate in the environment. In: Linkov, I., Steevens, J. (Eds.), Nanomaterials: Risks Benefits. Springer Netherlands, Dordrecht, pp. 125–137. http://dx.doi.org/10.1007/978-1-4020-9491-0_9.
Luo, C.H., Shanmugam, V., Yeh, C.S., 2015. Nanoparticle biosynthesis using unicellular and subcellular supports. NPG Asia Mater. 7.
Ma, Y., Jiao, K., Yang, T., Sun, D., 2008. Sensitive PAT gene sequence detection by nano-SiO_2/*p*-aminothiophenol self-assembled films DNA electrochemical biosensor based on impedance measurement. Sens. Actuators B Chem. 131 (2), 565–571.
Mansfield, J., Genin, S., Magori, S., Citovsky, V., Sriariyanum, M., Ronald, P., Dow, M.A.X., Verdier, V., Beer, S.V., Machado, M.A., Toth, I.A.N., 2012. Top 10 plant pathogenic bacteria in molecular plant pathology. Mol. Plant Pathol. 13 (6), 614–629.
Manzanares-Palenzuela, C.L., Martín-Fernández, B., López, M.S.P., López-Ruiz, B., 2015. Electrochemical genosensors as innovative tools for detection of genetically modified organisms. Trends Anal. Chem. 66, 19–31.
McMurray, T.A., Dunlop, P.S.M., Byrne, J.A., 2006. The photocatalytic degradation of atrazine on nanoparticulate TiO_2 films. J. Photochem. Photobiol. A Chem. 182 (1), 43–51.
Meenakshi, S.D., Rajarajan, M., Rajendran, S., Kennedy, Z.R., Brindha, G., 2012. Synthesis and characterization of magnesium oxide nanoparticles. Elixir Int. J. 50 (9), 10618–10620.
Meijer, E.G.M., Schilperoort, R.A., Rueb, S., Os-Ruygrok, P.V., Hensgens, L.A.M., 1991. Transgenic rice cell lines and plants: expression of transferred chimeric genes. Plant Mol. Biol. 16 (5), 807–820.
Min, J.S., Kim, K.S., Kim, S.W., Jung, J.H., Lamsal, K., Kim, S.B., Jung, M.Y., Lee, Y.S., 2009. Effects of colloidal silver nanoparticles on sclerotium-forming phytopathogenic fungi. Plant Pathol. J. 25 (4), 376–380.
Mishra, S., Singh, B.R., Singh, A., Keswani, C., Naqvi, A.H., Singh, H.B., 2014a. Biofabricated silver nanoparticles act as a strong fungicide against *Bipolaris sorokiniana* causing spot blotch disease in wheat. PLoS One 9 (5), e97881.
Mishra, V., Sharma, R., Jasuja, N.D., Gupta, D.K., 2014b. A review on green synthesis of nanoparticles and evaluation of antimicrobial activity. Int. J. Green Herb. Chem. 3 (1), 81–94.
Miyata, T., Asami, N., Uragami, T., 1999. A reversibly antigen-responsive hydrogel. Nature 399 (6738), 766–769.
Mohapatra, D.P., Gassara, F., Brar, S.K., 2011. Nanoparticles—production and role in biotransformation. J. Nanosci. Nanotechnol. 11 (2), 899–918.

Mondal, K.K., Mani, C., 2012. Investigation of the antibacterial properties of nanocopper against *Xanthomonas axonopodispv. punicae*, the incitant of pomegranate bacterial blight. Ann. Microbiol. 62 (2), 889–893.

Mora-Huertas, C.E., Fessi, H., Elaissari, A., 2010. Polymer-based nanocapsules for drug delivery. Int. J. Pharm. 385 (1), 113–142.

Moussa, S.H., Tayel, A.A., Alsohim, A.S., Abdallah, R.R., 2013. Botryticidal activity of nanosized silver-chitosan composite and its application for the control of gray mold in strawberry. J. Food Sci. 78 (10), M1589–M1594.

Moustafa, H.Z., Mohamad, T.G.M., Torkey, H., 2015. Effect of formulated nanoemulsion of *Eucalyptus* oil on the cotton Bollworms. J. Biol. Chem. Res. 32 (2), 478–484.

Mozafari, M.R., 2005. Liposomes: an overview of manufacturing techniques. Cell. Mol. Biol. Lett. 10 (4), 711.

Mozafari, M.R., 2010. Nanoliposomes: preparation and analysis. In: Liposomes: Methods and Protocols, Volume 1: Pharmaceutical Nanocarriers, pp. 29–50.

Mueller, N.C., Braun, J., Bruns, J., Černík, M., Rissing, P., Rickerby, D., Nowack, B., 2012. Application of nanoscale zero valent iron (NZVI) for groundwater remediation in Europe. Environ. Sci. Pollut. Res. 19 (2), 550–558.

Nair, R., Varghese, S.H., Nair, B.G., Maekawa, T., Yoshida, Y., Kumar, D.S., 2010. Nanoparticulate material delivery to plants. Plant Sci. 179 (3), 154–163.

Nalawade, P., Aware, B., Kadam, V.J., Hirlekar, R.S., 2009. Layered double hydroxides: a review. J. Sci. Ind. Res 68 (4), 267–272.

Negahban, M., Moharramipour, S., Zandi, M., Hashemi, S.A., Ziayee, F., 2012. Nano-insecticidal activity of essential oil from *Cuminum cyminum* on *Tribolium castaneum*. In: Proc. 9th. Int. Conf. On Controlled Atmosphere and Fumigation in Stored Products, Antalya, Turkey, pp. 63–68.

Nguyen, H.M., Hwang, I.C., Park, J.W., Park, H.J., 2012. Photoprotection for deltamethrin using chitosan-coated beeswax solid lipid nanoparticles. Pest Manag. Sci. 68 (7), 1062–1068.

Nicoletta, F.P., Cupelli, D., Formoso, P., De Filpo, G., Colella, V., Gugliuzza, A., 2012. Light responsive polymer membranes: a review. Membranes 2 (1), 134–197.

Nowack, B., Bucheli, T.D., 2007. Occurrence, behavior and effects of nanoparticles in the environment. Environ. Pollut. 150 (1), 5–22.

Nuruzzaman, M., Rahman, M.M., Liu, Y., Naidu, R., 2016. Nanoencapsulation, nano-guard for pesticides: a new window for safe application. J. Agric. Food Chem. 64 (7), 1447–1483.

Ocsoy, I., Paret, M.L., Ocsoy, M.A., Kunwar, S., Chen, T., You, M., Tan, W., 2013. Nanotechnology in plant disease management: DNA-directed silver nanoparticles on graphene oxide as an antibacterial against *Xanthomonas perforans*. ACS Nano 7 (10), 8972–8980.

Ongley, E.D., 1996. Control of Water Pollution from Agriculture (No. 55). Food & Agriculture Org.

Ouda, S.M., 2014. Antifungal activity of silver and copper nanoparticles on two plant pathogens, *Alternaria alternata* and *Botrytis cinerea*. Res. J. Microbiol. 9 (1), 34–42.

Pal, S., Ying, W., Alocilja, E.C., Downes, F.P., 2008. Sensitivity and specificity performance of a direct-charge transfer biosensor for detecting *Bacillus cereus* in selected food matrices. Biosyst. Eng. 99 (4), 461–468.

Pan, Z., Cui, B., Zeng, Z., Feng, L., Liu, G., Cui, H., Pan, H., 2015. Lambda-Cyhalothrin nanosuspension prepared by the melt emulsification-high pressure homogenization method. J. Nanomater. 16 (1), 263.

Paniel, N., Radoi, A., Marty, J.L., 2010. Development of an electrochemical biosensor for the detection of aflatoxin M1 in milk. Sensors 10 (10), 9439–9448.

Parham, H., Rahbar, N., 2010. Square wave voltammetric determination of methyl parathion using ZrO 2-nanoparticles modified carbon paste electrode. J. Hazard. Mater. 177 (1), 1077–1084.

Parisi, C., Vigani, M., Rodríguez-Cerezo, E., 2015. Agricultural Nanotechnologies: what are the current possibilities? Nano Today 10 (2), 124–127.

Parizi, M.A., Moradpour, Y., Roostaei, A., Khani, M., Negahdari, M., Rahimi, G., 2014. Evaluation of the antifungal effect of magnesium oxide nanoparticles on *Fusarium oxysporum* F. Sp. *lycopersici*, pathogenic agent of tomato. Eur. J. Exp. Biol. 4 (3), 151–156.

Park, H.J., Kim, S.H., Kim, H.J., Choi, S.H., 2006. A new composition of nanosized silica-silver for control of various plant diseases. Plant Pathol. J. 22 (3), 295–302.

Patil, H., Waghmare, J., 2013. Nanoemulsion: Current state and perspectives. Research J. Topical and Cosmetic Sci. 4 (1), 32–40.

Paul, B.K., Moulik, S.P., 2001. Uses and applications of microemulsions. Curr. Sci. (Bangalore) 80 (8), 990–1001.

Paulkumar, K., Gnanajobitha, G., Vanaja, M., Rajeshkumar, S., Malarkodi, C., Pandian, K., Annadurai, G., 2014. *Piper nigrum* leaf and stem assisted green synthesis of silver nanoparticles and evaluation of its antibacterial activity against agricultural plant pathogens. Sci. World J. 2014, 1–9.

Pérez-de-Luque, A., 2017. Interaction of nanomaterials with plants: what do we need for real applications in agriculture? Front. Environ. Sci. 5, 12.

Perlatti, B., Fernandes, J.B., Maria Fátima das Graças Fernandes da Silva, Forim, M.R., de Souza Bergo, P.L., 2013. Polymeric Nanoparticle-Based Insecticides: A Controlled Release Purpose for Agrochemicals. INTECH Open Access Publisher.

Peteu, S.F., Oancea, F., Sicuia, O.A., Constantinescu, F., Dinu, S., 2010. Responsive polymers for crop protection. Polymers 2 (3), 229–251.

Popat, A., Liu, J., Hu, Q., Kennedy, M., Peters, B., Lu, G.Q.M., Qiao, S.Z., 2012. Adsorption and release of biocides with mesoporous silica nanoparticles. Nanoscale 4 (3), 970–975.

Prasad, R., Pandey, R., Barman, I., 2016. Engineering tailored nanoparticles with microbes: quo vadis? Wiley Interdiscip. Rev. Nanomed. Nanobiotechnol 8, 316–330.

Qu, F., Zhou, X., Xu, J., Li, H., Xie, G., 2009. Luminescence switching of CdTe quantum dots in presence of *p*-sulfonatocalix[4]arene to detect pesticides in aqueous solution. Talanta 78 (4), 1359–1363.

Quaglia, F., Barbato, F., De Rosa, G., Granata, E., Miro, A., La Rotonda, M.I., 2001. Reduction of the environmental impact of pesticides: waxy microspheres encapsulating the insecticide carbaryl. J. Agric. Food Chem. 49 (10), 4808–4812.

Radian, A., Mishael, Y.G., 2008. Characterizing and designing polycation– clay nanocomposites as a basis for imazapyr controlled release formulations. Environ. Sci. Technol. 42 (5), 1511–1516.

Rafsanjani, M.S.O., Alvari, A., Samim, M., Amin Hejazi, M., Abdin, M.Z., 2012. Application of novel nanotechnology strategies in plant biotransformation: a contemporary overview. Recent Pat. Biotechnol. 6 (1), 69–79.

Rai, M., Ingle, A., 2012. Role of nanotechnology in agriculture with special reference to management of insect pests. Appl. Microbiol. Biotechnol. 94 (2), 287–293.

Rai, M., Ribeiro, C., Mattoso, L., Duran, N., 2015. Nanotechnologies in Food and Agriculture. Springer. 347 pages.

Rajesh, S., Raja, D.P., Rathi, J.M., Sahayaraj, K., 2012. Biosynthesis of silver nanoparticles using *Ulva fasciata* (Delile) ethyl acetate extract and its activity against *Xanthomonas campestris* pv. *malvacearum*. J. Biopestic. 5 (119), 119–128.

Raliya, R., Tarafdar, J.C., Biswas, P., 2016. Enhancing the mobilization of native phosphorus in the mung bean rhizosphere using ZnO nanoparticles synthesized by soil fungi. J. Agric. Food Chem. 64 (16), 3111–3118.

Rao, K.J., Paria, S., 2013. Use of sulfur nanoparticles as a green pesticide on *Fusarium solani* and *Venturia inaequalis* phytopathogens. RSC Adv. 3 (26), 10471–10478.

Rao, K.S., El-Hami, K., Kodaki, T., Matsushige, K., Makino, K., 2005. A novel method for synthesis of silica nanoparticles. J. Colloid Interface Sci. 289 (1), 125–131.

Rashidzadeh, A., Olad, A., Salari, D., Jalil Hejazi, M., 2014. On the encapsulation of natural pesticide using polyvinyl alcohol/alginate–montmorillonite nanocomposite for controlled release application. Polymer Eng. Sci. 54 (12), 2707–2714.

Ravishankar Rai, V., Jamuna Bai, A., 2011. Nanoparticles and their potential application as antimicrobials. In: Science Against Microbial Pathogens, Communicating Current Research and Technological Advances. Formatex, Badajoz, pp. 197–209.

Reid, W., Zhang, Q., Sekimoto, H., 2002. Influence of membrane surface charge on nutrient uptake by plants. Dev. Plant Soil Sci. 92, 198–199.

Rico, C.M., Majumdar, S., Duarte-Gardea, M., Peralta-Videa, J.R., Gardea-Torresdey, J.L., 2011. Interaction of nanoparticles with edible plants and their possible implications in the food chain. J. Agric. Food Chem. 59 (8), 3485–3498.

Robertson, G.A., 1987. Soil Management for Sustainable Agriculture. Government of Western Australia, Perth, 1–138. ISSN 0729-3135.

Roduner, E., 2006. Size matters: why nanomaterials are different. Chem. Soc. Rev. 35 (7), 583–592.

Ruzette, A.V., Leibler, L., 2005. Block copolymers in tomorrow's plastics. Nature Mater. 4 (1), 19–31.

Sahithya, K., Das, N., 2015. Remediation of pesticides using nanomaterials: an overview. Int. J. ChemTech Res. 8 (8), 86–91.

Saini, P., Gopal, M., Kumar, R., Srivastava, C., 2014. Development of pyridalyl nanocapsule suspension for efficient management of tomato fruit and shoot borer (*Helicoverpa armigera*). J. Environ. Sci. Health B 49 (5), 344–351.

Sanchez-Vazquez, S.A., Hailes, H.C., Evans, J.R.G., 2013. Hydrophobic polymers from food waste: resources and synthesis. Polym. Rev. 53 (4), 627–694.

Sarkar, D.J., Kumar, J., Shakil, N.A., Walia, S., 2012. Release kinetics of controlled release formulations of thiamethoxam employing nano-ranged amphiphilic PEG and diacid based block polymers in soil. J. Environ. Sci. Health A 47 (11), 1701–1712.

Sarkar, S., Datta, S.C., Biswas, D.R., 2014. Synthesis and characterization of nanoclay–polymer composites from soil clay with respect to their water-holding capacities and nutrient-release behavior. J. Appl. Polymer Sci. 131 (6).

Sarlak, N., Taherifar, A., Salehi, F., 2014. Synthesis of nanopesticides by encapsulating pesticide nanoparticles using functionalized carbon nanotubes and application of new nanocomposite for plant disease treatment. J. Agric. Food Chem. 62 (21), 4833–4838.

Sawahel, W.A., 2002. The production of transgenic potato plants expressing human alpha-interferon using lipofectin-mediated transformation. Cell. Mol. Biol. Lett. 7 (1), 19–30.

Servin, A., Elmer, W., Mukherjee, A., De la Torre-Roche, R., Hamdi, H., White, J.C., Bindraban, P., Dimkpa, C., 2015. A review of the use of engineered nanomaterials to suppress plant disease and enhance crop yield. J. Nanopart. Res. 17 (2), 92.

Shahavi, M.H., Hosseini, M., Jahanshahi, M., Meyer, R.L., Darzi, G.N., 2015. Evaluation of critical parameters for preparation of stable clove oil nanoemulsion. Arab. J. Chem., 1–6.

Shakil, N.A., Singh, M.K., Pandey, A., Kumar, J., Pankaj, Parmar, V.S., Singh, M.K., Pandey, R.P., Watterson, A.C., 2010. Development of poly (ethylene glycol) based amphiphilic copolymers for controlled release delivery of carbofuran. J. Macromol. Sci. A Pure Appl. Chem. 47 (3), 241–247.

Sharon, M., Choudhary, A.K., Kumar, R., 2010. Nanotechnology in agricultural diseases and food safety. J. Phytol. 2 (4).

Shenashen, M., Derbalah, A., Hamza, A., Mohamed, A., El Safty, S., 2017. Antifungal activity of fabricated mesoporous alumina nanoparticles against root rot disease of tomato caused by *Fusarium oxysporium*. Pest Manag. Sci. 73 (6), 1121–1126.

Shobha, G., Moses, V., Ananda, S., 2014. Biological synthesis of copper nanoparticles and its impact-a review. Int. J. Pharm. Sci. Invent. 3 (8), 28–38.

Shweta, Tripathi, D.K., Singh, S., Singh, S., Dubey, N.K., Chauhan, D.K., 2016. Impact of nanoparticles on photosynthesis: challenges and opportunities. Mater. Focus 5 (5), 405–411.

Simonian, A.L., Good, T.A., Wang, S.S., Wild, J.R., 2005. Nanoparticle-based optical biosensors for the direct detection of organophosphate chemical warfare agents and pesticides. Anal. Chim. Acta 534 (1), 69–77.

Singh, S., Singh, M., Agrawal, V.V., Kumar, A., 2010. An attempt to develop surface plasmon resonance based immunosensor for Karnal bunt (*Tilletia indica*) diagnosis based on the experience of nano-gold based lateral flow immuno-dipstick test. Thin Solid Films 519 (3), 1156–1159.

Singh, S., Tripathi, D.K., Dubey, N.K., Chauhan, D.K., 2016. Effects of nano-materials on seed germination and seedling growth: striking the slight balance between the concepts and controversies. Mater. Focus 5 (3), 195–201.

Sirelkhatim, A., Mahmud, S., Seeni, A., Kaus, N.H.M., Ann, L.C., Bakhori, S.K.M., Mohamad, D., 2015. Review on zinc oxide nanoparticles: antibacterial activity and toxicity mechanism. Nano-Micro Lett. 7 (3), 219–242.

Song, S., Liu, X., Jiang, J., Qian, Y., Zhang, N., Wu, Q., 2009. Stability of triazophos in self-nanoemulsifying pesticide delivery system. Colloids Surf. A Physicochem. Eng. Asp. 350 (1), 57–62.

Srikar, S.K., Giri, D.D., Pal, D.B., Mishra, P.K., Upadhyay, S.N., 2016. Green synthesis of silver nanoparticles: a review. Green Sustain. Chem. 6 (01), 34–56.

Suleiman, M., Al Ali, A., Hussein, A., Hammouti, B., Hadda, T.B., Warad, I., 2013. Sulfur nanoparticles: synthesis, characterizations and their applications. J. Mater. Environ. Sci 4, 1029–1033.

Sultana, F., Imran-Ul-Haque, M., Arafat, M., Sharmin, S., 2013. An overview of nanogel drug delivery system. J. Appl. Pharm. Sci. 3, 95–105.

Sun, H., Fung, Y., 2006. Piezoelectric quartz crystal sensor for rapid analysis of pirimicarb residues using molecularly imprinted polymers as recognition elements. Anal. Chim. Acta 576 (1), 67–76.

Sun, W., Zhong, J., Qin, P., Jiao, K., 2008. Electrochemical biosensor for the detection of cauliflower mosaic virus 35 S gene sequences using lead sulfide nanoparticles as oligonucleotide labels. Anal. Biochem. 377 (2), 115–119.

Sun, C., Shu, K., Wang, W., Ye, Z., Liu, T., Gao, Y., Zheng, H., He, G., Yin, Y., 2014. Encapsulation and controlled release of hydrophilic pesticide in shell cross-linked nanocapsules containing aqueous core. Int. J. Pharm. 463 (1), 108–114.

Suresh, R., Borkar, S.N., Sawant, V.A., Shende, V.S., Dimble, S.K., 2010. Nanoclay drug delivery system. Int. J. Pharm. Sci. Nanotechnol. 3, 901–905.

Suri, C.R., Kaur, J., Gandhi, S., Shekhawat, G.S., 2008. Label-free ultra-sensitive detection of atrazine based on nanomechanics. Nanotechnology 19 (23), 235502.

Suriyaprabha, R., Karunakaran, G., Kavitha, K., Yuvakkumar, R., Rajendran, V., Kannan, N., 2014. Application of silica nanoparticles in maize to enhance fungal resistance. IET Nanobiotechnol. 8 (3), 133–137.

Sushma, N.J., Prathyusha, D., Swathi, G., Madhavi, T., Raju, B.D.P., Mallikarjuna, K., Kim, H.S., 2016. Facile approach to synthesize magnesium oxide nanoparticles by using *Clitoria ternatea*-characterization and in vitro antioxidant studies. Appl. Nanosci. 6 (3), 437–444.

Swathy, B., 2014. A review on metallic silver nanoparticles. IOSR J. Pharm. 4, 38–44.

Takei, T., Yoshida, M., Hatate, Y., Shiomori, K., Kiyoyama, S., 2008. Preparation of polylactide/poly (ε-caprolactone) microspheres enclosing acetamiprid and evaluation of release behavior. Polymer Bull. 61 (3), 391–397.

Takeuchi, M.T., Kojima, M., Luetzow, M., 2014. State of the art on the initiatives and activities relevant to risk assessment and risk management of nanotechnologies in the food and agriculture sectors. Food Res. Int. 64, 976–981.

Tamayo, J., Humphris, A.D.L., Malloy, A.M., Miles, M.J., 2001. Chemical sensors and biosensors in liquid environment based on microcantilevers with amplified quality factor. Ultramicroscopy 86 (1), 167–173.

Tamez-Guerra, P., McGuire, M.R., Behle, R.W., Hamm, J.J., Sumner, H.R., Shasha, B.S., 2000a. Sunlight persistence and rainfastness of spray-dried formulations of baculovirus isolated from *Anagrapha falcifera* (*Lepidoptera: Noctuidae*). J. Econ. Entomol. 93 (2), 210–218.

Tamez-Guerra, P., McGuire, M.R., Behle, R.W., Shasha, B.S., Wong, L.J.G., 2000b. Assessment of microencapsulated formulations for improved residual activity of *Bacillus thuringiensis*. J. Econ. Entomol. 93 (2), 219–225.

Tamez-Guerra, P., McGuire, M.R., Behle, R.W., Shasha, B.S., Pingel, R.L., 2002. Storage stability of *Anagrapha falcifera* nucleopolyhedrovirus in spray-dried formulations. J. Invert. Pathol. 79 (1), 7–16.

Tan, X.M., Lin, C., Fugetsu, B., 2009. Studies on toxicity of multi-walled carbon nanotubes on suspension rice cells. Carbon 47 (15), 3479–3487.

Tang, Z.X., Lv, B.F., 2014. MgO nanoparticles as antibacterial agent: preparation and activity. Braz. J. Chem. Eng. 31 (3), 591–601.

Taylor, T.M., Weiss, J., Davidson, P.M., Bruce, B.D., 2005. Liposomal nanocapsules in food science and agriculture. Crit. Rev. Food Sci. Nutr. 45 (7–8), 587–605.

Thakkar, K.N., Mhatre, S.S., Parikh, R.Y., 2010. Biological synthesis of metallic nanoparticles. Nanomed. Nanotechnol. Biol. Med. 6 (2), 257–262.

Thévenot, J., Oliveira, H., Sandre, O., Lecommandoux, S., 2013. Magnetic responsive polymer composite materials. Chem. Soc. Rev. 42 (17), 7099–7116.

Torma, S., Vilcek, J., Adamisin, P., Huttmanova, E., Hronec, O., 2014. Influence of natural zeolite on nitrogen dynamics in soil. Turk. J. Agric. For. 38 (5), 739–744.

Tripathi, D.K., Singh, V.P., Prasad, S.M., Chauhan, D.K., Dubey, N.K., 2015. Silicon nanoparticles (SiNp) alleviate chromium (VI) phytotoxicity in *Pisum sativum* (L.) seedlings. Plant Physiol. Biochem. 96, 189–198.

Tripathi, D.K., Singh, S., Singh, V.P., Prasad, S.M., Chauhan, D.K., Dubey, N.K., 2016. Silicon nanoparticles more efficiently alleviate arsenate toxicity than silicon in maize cultivar and hybrid differing in arsenate tolerance. Front. Environ. Sci. 4, 46.

Tripathi, D.K., Singh, S., Singh, S., Pandey, R., Singh, V.P., Sharma, N.C., et al., 2017a. An overview on manufactured nanoparticles in plants: uptake, translocation, accumulation and phytotoxicity. Plant Physiol. Biochem. 110, 2–12.

Tripathi, D.K., Tripathi, A., Guar, S., Singh, S., Singh, Y., Vishwakarma, K., et al., 2017b. Uptake, accumulation and toxicity of silver nanoparticle in autotrophic plants, and heterotrophic microbes: a concentric review. Front. Microbiol. 8, 7.

Tripathi, D.K., Singh, S., Singh, S., Srivastava, P.K., Singh, V.P., Singh, S., et al., 2017c. Nitric oxide alleviates silver nanoparticles (AgNps)-induced phytotoxicity in Pisum sativum seedlings. Plant Physiol. Biochem. 110, 167–177.

Tripathi, D.K., Mishra, R.K., Singh, S., Singh, S., Vishwakarma, K., Sharma, S., Singh, V.P., Singh, P.K., Prasad, S.M., Dubey, N.K., Pandey, A.C., 2017d. Nitric oxide ameliorates zinc oxide nanoparticles phytotoxicity in wheat seedlings: implication of the ascorbate–glutathione cycle. Front. Plant Sci. 8.

Tripathi, D.K., Singh, S., Singh, V.P., Prasad, S.M., Dubey, N.K., Chauhan, D.K., 2017e. Silicon nanoparticles more effectively alleviated UV-B stress than silicon in wheat (*Triticum aestivum*) seedlings. Plant Physiol. Biochem. 110, 70–81.

Truong, N.P., Jia, Z., Burgess, M., Payne, L., McMillan, N.A., Monteiro, M.J., 2011. Self-catalyzed degradable cationic polymer for release of DNA. Biomacromolecules 12 (10), 3540–3548.

Tsuda, N., Ohtsubo, T., Fuji, M., 2012. Preparation of self-bursting microcapsules by interfacial polymerization. Adv. Powder Technol. 23 (6), 724–730.

Ulianas, A., Heng, L.Y., Ahmad, M., Lau, H.Y., Ishak, Z., Ling, T.L., 2014. A regenerable screen-printed DNA biosensor based on acrylic microsphere–gold nanoparticle composite for genetically modified soybean determination. Sens. Actuators B Chem. 190, 694–701.

Vamvakaki, V., Chaniotakis, N.A., 2007. Pesticide detection with a liposome-based nano-biosensor. Biosens. Bioelectron. 22 (12), 2848–2853.

Varona, S., Martín, Á., Cocero, M.J., 2009. Formulation of a natural biocide based on lavandin essential oil by emulsification using modified starches. Chem. Eng. Process. Process Intensif. 48 (6), 1121–1128.

Varona, S., Kareth, S., Martín, Á., Cocero, M.J., 2010. Formulation of lavandin essential oil with biopolymers by PGSS for application as biocide in ecological agriculture. J. Supercrit. Fluids 54 (3), 369–377.

Vinayaka, A.C., Basheer, S., Thakur, M.S., 2009. Bioconjugation of CdTe quantum dot for the detection of 2, 4-dichlorophenoxyacetic acid by competitive fluoroimmunoassay based biosensor. Biosens. Bioelectron. 24 (6), 1615–1620.

Vinutha, J.S., Bhagat, D., Bakthavatsalam, N., 2013. Nanotechnology in the management of polyphagous pest Helicoverpa armigera. J. Acad. Ind. Res. 1 (10), 606–608.

Viswanathan, S., Radecka, H., Radecki, J., 2009. Electrochemical biosensor for pesticides based on acetylcholinesterase immobilized on polyaniline deposited on vertically assembled carbon nanotubes wrapped with ssDNA. Biosens. Bioelectron. 24 (9), 2772–2777.

Wang, M., Li, Z., 2008. Nano-composite ZrO_2/Au film electrode for voltammetric detection of parathion. Sens. Actuators B Chem. 133 (2), 607–612.

Wang, L., Li, X., Zhang, G., Dong, J., Eastoe, J., 2007. Oil-in-water nanoemulsions for pesticide formulations. J. Colloid Interface Sci. 314 (1), 230–235.

Wang, H., Wick, R.L., Xing, B., 2009. Toxicity of nanoparticulate and bulk ZnO, Al_2O_3 and TiO_2 to the nematode *Caenorhabditis elegans*. Environ. Pollut. 157 (4), 1171–1177.

Wang, Z., Wei, F., Liu, S.Y., Xu, Q., Huang, J.Y., Dong, X.Y., Yu, J.H., Yang, Q., Zhao, Y.D., Chen, H., 2010. Electrocatalytic oxidation of phytohormone salicylic acid at copper nanoparticles-modified gold electrode and its detection in oilseed rape infected with fungal pathogen *Sclerotinia sclerotiorum*. Talanta 80 (3), 1277–1281.

Wang, Z., Wu, L., Shen, B., Jiang, Z., 2013. Highly sensitive and selective cartap nanosensor based on luminescence resonance energy transfer between $NaYF_4$: Yb, Ho nanocrystals and gold nanoparticles. Talanta 114, 124–130.

Wang, Y., Cui, H., Sun, C., Zhao, X., Cui, B., 2014. Construction and evaluation of controlled-release delivery system of Abamectin using porous silica nanoparticles as carriers. Nanoscale Res. Lett. 9, 655.

Wang, P., Lombi, E., Zhao, F.J., Kopittke, P.M., 2016. Nanotechnology: a new opportunity in plant sciences. Trends Plant Sci. 21 (8), 699–712.

Wang, M.-Q., Du, X.-Y., Sun, Q., Jiang, X.C., 2008. DNA biosensor prepared by electrodeposited Pt-nanoparticles for the detection of specific deoxyribonucleic acid sequence in genetically modified soybean. Chin. J. Anal. Chem. 36 (7), 890–894.

Wani, A.H., Shah, M.A., 2012. A unique and profound effect of MgO and ZnO nanoparticles on some plant pathogenic fungi. J. Appl. Pharm. Sci. 2 (3), 40–44.

Wanyika, H., 2013. Sustained release of fungicide metalaxyl by mesoporous silica nanospheres. J. Nanopart. Res. 15 (8), 1–9.

Wen, H., Li, Y., 2014. Redox sensitive nanoparticles with disulfide bond linked sheddable shell for intracellular drug delivery. Med. Chem. 4 (11), 748–755.

Wibowo, D., Zhao, C.X., Peters, B.C., Middelberg, A.P., 2014. Sustained release of fipronil insecticide in vitro and in vivo from biocompatible silica nanocapsules. J. Agric. Food Chem. 62 (52), 12504–12511.

Wiesman, Z., Dom, N.B., Sharvit, E., Grinberg, S., Linder, C., Heldman, E., Zaccai, M., 2007. Novel cationic vesicle platform derived from vernonia oil for efficient delivery of DNA through plant cuticle membranes. J. Biotechnol. 130 (1), 85–94.

Xie, Y., He, Y., Irwin, P.L., Jin, T., Shi, X., 2011. Antibacterial activity and mechanism of action of zinc oxide nanoparticles against *Campylobacter jejuni*. Appl. Environ. Microbiol. 77 (7), 2325–2331.

Yan, H., Chen, X., Feng, Y., Xiang, F., Li, J., Shi, Z., Wang, X., Lin, Q., 2016. Modification of montmorillonite by ball-milling method for immobilization and delivery of acetamiprid based on alginate/exfoliated montmorillonite nanocomposite. Polymer Bull. 73 (4), 1185–1206.

Yang, J., Jiao, K., Yang, T., 2007. A DNA electrochemical sensor prepared by electrodepositing zirconia on composite films of single-walled carbon nanotubes and poly (2, 6-pyridinedicarboxylic acid), and its application to detection of the PAT gene fragment. Anal. Bioanal. Chem. 389 (3), 913–921.

Yang, T., Zhang, W., Du, M., Jiao, K., 2008. A PDDA/poly (2, 6-pyridinedicarboxylic acid)-CNTs composite film DNA electrochemical sensor and its application for the detection of specific sequences related to PAT gene and NOS gene. Talanta 75 (4), 987–994.

Yang, F.L., Li, X.G., Zhu, F., Lei, C.L., 2009a. Structural characterization of nanoparticles loaded with garlic essential oil and their insecticidal activity against *Tribolium castaneum* (Herbst) (*Coleoptera*: *Tenebrionidae*). J. Agric. Food Chem. 57 (21), 10156–10162.
Yang, T., Zhou, N., Zhang, Y., Zhang, W., Jiao, K., Li, G., 2009b. Synergistically improved sensitivity for the detection of specific DNA sequences using polyaniline nanofibers and multi-walled carbon nanotubes composites. Biosens. Bioelectron. 24 (7), 2165–2170.
Yang, J., Wang, X., Shi, H., 2012a. An electrochemical DNA biosensor for highly sensitive detection of phosphinothricin acetyltransferase gene sequence based on polyaniline-(mesoporous nanozirconia)/poly-tyrosine film. Sens. Actuators B Chem. 162 (1), 178–183.
Yang, T., Zhou, N., Li, Q., Guan, Q., Zhang, W., Jiao, K., 2012b. Highly sensitive electrochemical impedance sensing of PEP gene based on integrated Au–Pt alloy nanoparticles and polytyramine. Colloids Surf. B Biointerfaces 97, 150–154.
Yao, K.S., Li, S.J., Tzeng, K.C., Cheng, T.C., Chang, C.Y., Chiu, C.Y., Liao, C.Y., Hsu, J.J., Lin, Z.P., 2009. Fluorescence silica nanoprobe as a biomarker for rapid detection of plant pathogens. Adv. Mater. Res. 79, 513–516.
Yu-qin, F., Lu-hua, L., Pi-wu, W., Jing, Q., Yong-ping, F., Hui, W., Jing-ran, S., 2012. Delivering DNA into plant cell by gene carriers of ZnS nanoparticles. Chem. Res. Chin. Univ. 28 (4), 672–676.
Zhang, W., Yang, T., Jiang, C., Jiao, K., 2008. DNA hybridization and phosphinothricin acetyltransferase gene sequence detection based on zirconia/nanogold film modified electrode. Appl. Surf. Sci. 254 (15), 4750–4756.
Zhang, X., Liu, J., Hou, W., Tong, J., Ren, L., Sun, G., Sun, Y., 2016. Preparation and properties of pesticide/cyclodextrin complex intercalated into ZnAl-layered double hydroxide. Ind. Eng. Chem. Res. 55 (6), 1550–1558.
Zhao, Y.G., Shen, H.Y., Shi, J.W., Chen, X.H., Jin, M.C., 2011. Preparation and characterization of amino functionalized nano-composite material and its application for multi-residue analysis of pesticides in cabbage by gas chromatography–triple quadrupole mass spectrometry. J. Chromatogr. A 1218 (33), 5568–5580.
Zhou, N., Yang, T., Jiang, C., Du, M., Jiao, K., 2009. Highly sensitive electrochemical impedance spectroscopic detection of DNA hybridization based on Au nano–CNT/PAN nano films. Talanta 77 (3), 1021–1026.
Ziaee, M., Moharramipour, S., Mohsenifar, A., 2014. MA-chitosan nanogel loaded with *Cuminum cyminum* essential oil for efficient management of two stored product beetle pests. J. Pest Sci. 87 (4), 691–699.

Further Reading

Cui, B., Feng, L., Pan, Z., Yu, M., Zeng, Z., Sun, C., Zhao, X., Wang, Y., Cui, H., 2015. Evaluation of stability and biological activity of solid nanodispersion of lambda-cyhalothrin. PLoS One 10 (8), e0135953.
Kuo-Hsun, H., Hsiang-Chuan, W., Ren-Shih, C., Ju-Chun, H., 2015. Calcium carbonate nanoparticles can enhance plant nutrition and insect pest tolerance. J. Pestic. Sci. 40 (4).
Li, K.-E., Chang, Z.-Y., Shen, C.-X., Yao, N., 2015. Toxicity of Nanomaterials to plants. In: Nanotechnology and Plant Sciences. Springer International Publishing, Cham, pp. 101–123.
Liu, G., Lin, Y., 2005. Electrochemical sensor for organophosphate pesticides and nerve agents using zirconia nanoparticles as selective sorbents. Anal. Chem. 77 (18), 5894–5901.
Liu, J., Liu, X.M., Xiao, S.Y., Tong, C.Y., Tang, D.Y., Zhao, L.J., 2005. Bioconjugated nanoparticle for DNA protection from ultrasound damage. Anal. Sci. 21 (3), 193–195.

CHAPTER

17

Impact of Nanoparticles on Oxidative Stress and Responsive Antioxidative Defense in Plants

Vinay Kumar[1], *Mansi Sharma*[1], *Tushar Khare*[1], *Shabir H. Wani*[2]

[1]Modern College of Arts, Science and Commerce, Savitribai Phule Pune University, Ganeshkhind, Pune, India; [2]Mountain Research Centre for Field Crops, Khudwani, Sher-e-Kashmir University of Agricultural Sciences and Technology, Anantnag, India

17.1 INTRODUCTION

Nanotechnology has witnessed incredible growth in recent years and nanoparticles have received great attention because of their remarkable and multidimensional potentials. These nanoscale (1–100 nm in at least one dimension) materials have gained unprecedented attention because of their vast array of applications in agriculture, the environment, and health (Hossain et al., 2015; Mapara et al., 2015; Patil et al., 2016). Nanoparticles have noteworthy applications in cosmetics, biomedicines, electronic devices, biosensors, herbicides, and food additives, and include silver, titanium oxide, zinc oxide, and cerium oxide nanoparticles (Ma et al., 2015).

The division of bulk materials into nanoscale pieces gives them unique physical and chemical properties (Jefferson, 2000). Besides their strikingly small sizes, these particles are known for their high surface area-to-volume ratio along with unusual surface properties (Ghosh et al., 2016). However, these properties of nanoparticles make them highly reactive and enable them to easily penetrate cells causing possible nanotoxicity to microorganisms, plants, as well as animals. The ever-increasing synthesis and applications of nanoparticles have raised the risks of environmental exposure (Ghosh et al., 2016). There are therefore global concerns for potential phytotoxic effects caused by unspecified release of nanoparticles into the environment (Hossain et al., 2015). Agricultural regions face higher risk of exposure, particularly

Nanomaterials in Plants, Algae, and Microorganisms
http://dx.doi.org/10.1016/B978-0-12-811487-2.00017-7

to engineered nanoparticles, of which thousands of tons are released in these environments (Keller and Lazareva, 2014).

As a primary producer converting direct solar energy into organic matter, plants play a crucial role in the functioning and maintenance of ecosystems (McKee and Filser, 2016). At the same time, engineered nanoparticles may be released into different ecosystems including water, air, and soil, either through their direct applications or accidently through indirect modes such as landfills and waste materials (reviewed by Tripathi et al., 2017a). Nanoparticles released in the environment may interact with plants before being taken up by them , and reports have indicated that both carbon- as well as metal-based engineered nanoparticles cause toxicities at higher concentrations by altering morphoanatomical, physiological, and genetic constituents of plants (Rico et al., 2015; Shweta et al., 2016; Tripathi et al., 2015, 2016a,b, 2017a,b,c,d,e; Singh et al., 2017). A few reports have also concentrated on the negative impacts of engineered nanoparticles on the seed quality of different plants (Singh et al., 2016) such as wheat (Rico et al., 2014) and common bean (Majumdar et al., 2015). Hence the plant system serves as a means of transportation of various nanomaterials to different ecosystems (Rico et al., 2011). However, an in-depth understanding of the interactions between plant systems and engineered nanoparticles is necessary to evaluate their toxic effects (Gardea-Torresdey et al., 2014).

The interaction between various nanoparticles and the plant ecosystem may be caused by the application of nanoparticles in different agricultural chemicals (fertilizers and pesticides), leakage of nanoparticles from landfills, application of nanoparticles for remediation procedures, atmospheric fallouts, as well as their accidental release in the environment (Rizwan et al., 2017). The overall interaction is the general reason for alteration of overall plant metabolism, which ultimately leads to a reduction in yield and quality of important crops. The possible reasons for plant– nanoparticle interaction are briefly summarized in Fig. 17.1. Several reports have confirmed that nanoparticles induce negative impacts on plants via generating reactive oxygen species (ROS, the main determinants of oxidative damage) in plants (Ma et al., 2015; Rico et al., 2015; Siddiqui et al., 2015). These ROS, such as singlet oxygen, superoxide radical, hydrogen peroxide, and hydroxyl radical, cause serious damage to lipid membranes and other essential macromolecules including proteins and nucleic acids, which ultimately results in cell death in plants (Khare et al., 2014; Wani et al., 2016; Kumar et al., 2017). However, ROS also hold significance for their ability to work as signaling molecules and their importance in regulating development and stress responses (Ismail et al., 2014).

An unfortunate consequence of stress conditions faced by plants is the excess generation of ROS (Khare et al., 2014; Kumar and Khare, 2014), and therefore overproduction of ROS in plant cells after their exposure to nanoparticles indicates their negative impacts on plants. There has been a recent debate over oxidative damage in plants as a consequence of nanoparticle exposure and the responsive antioxidant machinery deployed by plants to mitigate this damage, though greater insights at the molecular level are needed and should be a matter for future research. We present herein a discussion about various engineered nanoparticles available in the environment, their interactions with plants, and their roles in inducing oxidative stress followed by responsive antioxidant machinery activated by the stressed plants as an adaptive mechanism.

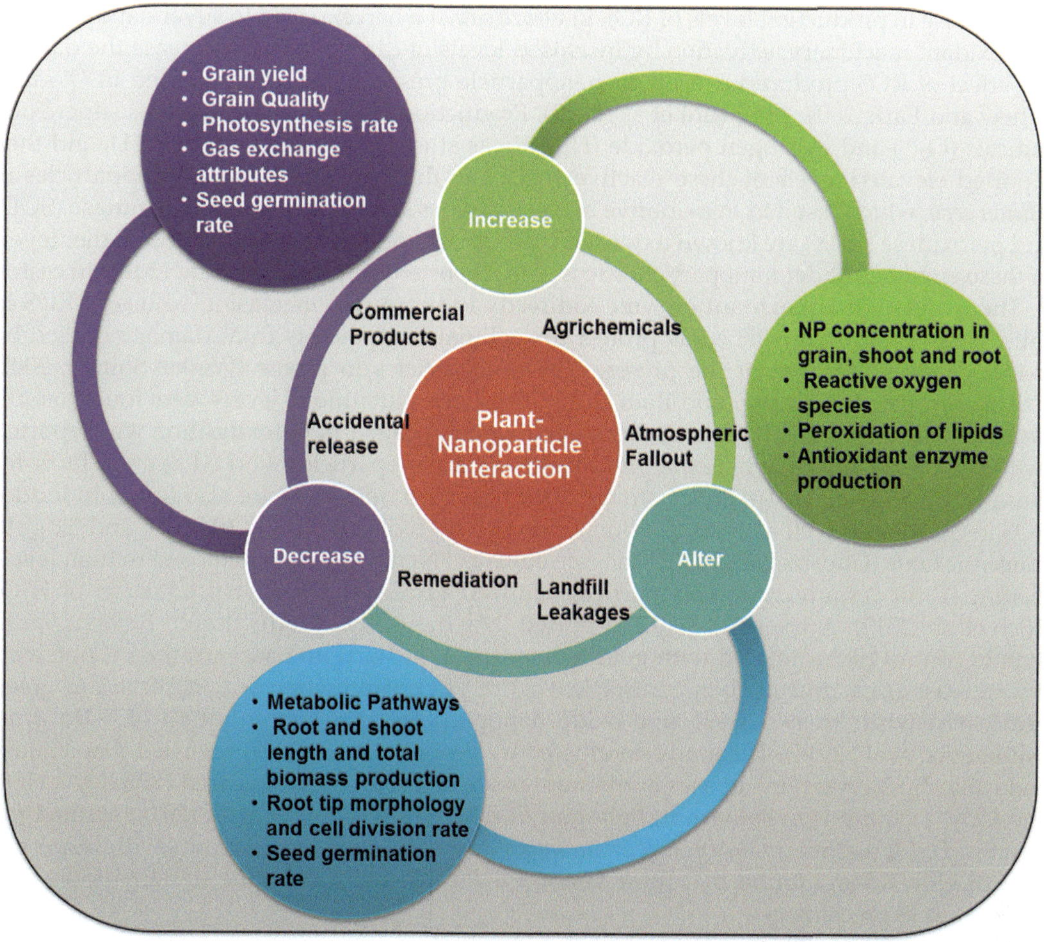

FIGURE 17.1 The possible sources for entry of nanoparticles (NPs) in the environment causing the plant–NP interaction and their effect on the general physiological and biochemical status of the affected plant.

17.2 NANOPARTICLE-INDUCED OXIDATIVE STRESS IN PLANTS: GENERATION OF ROS

Commonly used nanoparticles mainly include silver (Ag), gold (Au), cerium oxide (CeO), titanium dioxide (TiO_2), zinc oxide (ZnO), and copper (Cu) nanoparticles, whereas manganese (Mn), iron (Fe), and copper oxide (CuO) nanoparticles are gaining in popularity (Rico et al., 2015). One of the well-established ways to investigate whether a certain plant is experiencing the presence or absence of a compound as a stress inducer is by detecting the ROS generated as well as activated antioxidant machinery. Also changes in antioxidant defense responses by plants have been taken into consideration as biomarkers for ecotoxicological analysis in the presence of nanoparticles (Anjum et al., 2013). Mirzajani et al. (2013) reported the increase in production levels of ROS in *Oryza sativa* when exposed to silver nanoparticles. Antioxidant machinery activation by increased levels of carotenoids to decrease the damaging effect of ROS produced caused by nanoparticle presence has been reported in *O. sativa* (Chew and Park, 2004; Mirzajani et al., 2013). Production of specific ROS such as superoxide radicals ($O_2^{\cdot-}$) and hydrogen peroxide (H_2O_2) was studied by Panda et al. (2011), and they reported elevated levels of these reactive species in the presence of silver nanoparticles in *Allium cepa*, which resulted in oxidative burst. Enzymes such as superoxide dismutase (SOD) and peroxidase (POX) are known to work in cells to reduce oxidative stress, and higher levels of these enzymes under nanoparticle stress were reported by Yasur and Rani (2013) in castor.

The increase in antioxidant enzyme indirectly indicates the increasing values of ROS in cells. Phenols and phenolic acids protect the cellular components from damage caused by oxidative stress because of the presence of nanoparticles in plants (Winkel-Shirley, 2002; Krishnaraj et al., 2012; Yasur and Rani, 2013). Oxidized glutathione levels were found amplified and also activation of the metallothionein gene engaged in detoxification was reported by Dimkpa et al. (2013), where the presence of silver nanoparticles worked as an inducer for these elevated levels in *Triticum aestivum*. Zinc oxide nanoparticles are also found to induce ROS generation as well as H_2O_2 production in *A. cepa* roots at varying (0.2, 0.4, and 0.8 g/L) concentrations (Ghosh et al., 2016). The rice cultivar "Neptune" when exposed to high levels (500 mg/L) of cerium oxide nanoparticles showed three times higher production of H_2O_2 (Rico et al., 2013). Arora et al. (2012) reported 29% more ROS production in *Brassica juncea* than in control plants treated with gold nanoparticles at 100 ppm concentration. A noteworthy increase (61% greater than control leaves) of H_2O_2 production was observed in green pea leaves when treated with zinc oxide nanoparticles at a concentration of 500 mg/kg (Mukherjee et al., 2014). The study conducted by Fenoglio et al. (2009) revealed that anatase and rutile, both crystalline phases of titanium oxide nanoparticles, increased ROS ($O_2^{\cdot-}$, $HO^{\cdot}$, and $CO_2^{\cdot-}$) synthesis in spinach in light and dark conditions. Wang et al. (2011) studied the effect of Fe_3O_4 nanoparticles on *Cucurbita mixta* and *Lolium perenne* L. plants with respect to ROS production and found that these nanoparticles affected membrane stability because of elevated ROS production.

Reactive nitrogen species, H_2O_2, and ROS production were induced in *Spirodela punctata* when exposed to silver and zinc nanoparticles (Thwala et al., 2013). ROS production was also observed in ryegrass in the presence of zinc oxide nanoparticles (Lin and Xing, 2007). ROS accumulated in the presence of carbon nanotubes in the root tips of *Arabidopsis thaliana* seedlings (Liu et al., 2010) and in cell cultures of *Nicotiana tabacum* (Liu et al., 2013). Various reports suggest that ROS are produced in the presence of different nanoparticles such as nCeO$_2$, nFe$_3$O$_4$, nCoFe$_2$O$_4$, and nTiO$_2$ (Mingyu et al., 2007; Ursache-Oprisan et al., 2011; Rico et al., 2013). Responses to zinc oxide nanoparticle exposure are related to their morphology, and higher exposure to polar faces results in the production of ROS (Li et al., 2008; Mclaren et al., 2009). Kumari et al. (2011) reported production of ROS in *A. cepa* because of zinc oxide nanoparticles as well as nanoparticle-released zinc ions. Similarly, Ghodake et al. (2011) and Lee et al. (2013) discovered that zinc oxide nanoparticles were responsible for oxidative stress in *A. cepa* and *Fagopyrum esculentum*, respectively. Concentration-dependent ROS generation in tomato roots and protoplasts isolated from roots because of nickel oxide nanoparticles was studied by Faisal et al. (2013), but it is not clear whether this effect was caused by nanoparticles or released ions as both were detected in root cells. Similar studies report that oxidative stress is caused by either copper oxide nanoparticles or released copper ions (Shi et al., 2011; Lee et al., 2013; Nair and Chung, 2014, 2015). However, based on experiments it was found that copper oxide nanoparticles are solubilized inside plant cells, which in turn results in the release of copper ions, therefore causing oxidative damage by redox reaction between Cu^{2+} and Cu^{+} (Shi et al., 2011). Lipid peroxidation was credited to ROS production by Fenton reaction because of the presence of polyvalent copper oxide nanoparticles (Fubini et al., 2007).

In *B. juncea* the induction of H_2O_2 production was found after silver nanoparticle treatment for 7 days (Sharma et al., 2012). A concentration-dependent increase in H_2O_2 levels in cabbage, tomato, and red spinach leaves was observed by Begum et al. (2011). Tan et al. (2009) studied the effects of sonicated multiwalled carbon nanotubes in rice suspension cells at a concentration of 20 mg/L and observed an increase in ROS concentration with time.

17.3 OXIDATIVE DAMAGE CAUSED BY GENERATED ROS

Nanoparticle-induced ROS production in plants has been reported that leads to membrane damage and ultimately cell death (Shen et al., 2010). Though studies relating nanoparticle stress and ROS generation have been reported, the after effects of the produced ROS on DNA and proteins are less documented (Zhao et al., 2012). A study conducted by Yang and Watts (2005) found inhibition of root elongation in five different plant species, *Zea mays, Cucumis sativus, Glycine max, Brassica oleracea,* and *Daucus carota,* caused by nano-Al_2O_3 suspension at a concentration of 2000 μg/mL. The negative impacts of oxidative stress on DNA, proteins, and lipids are well known. Atha et al. (2012) reported DNA damage and growth inhibition in *Raphanus sativus, Lolium perenne,* and *Lolium rigidum* caused by copper oxide nanoparticles. Atomic force microscope imaging has revealed the interaction of titanium oxide nanoparticles with DNA, and also the cause of DNA damage in *A. cepa* and *N. tabacum* (Ghosh et al., 2010). Zinc oxide nanoparticle-mediated membrane damage was investigated by Mukherjee et al. (2014) in *Pisum sativum,* and the authors recorded relatively higher lipid peroxidation in plants treated with nanoparticles than their control (nontreated) counterparts. In the same study the overproduction of H_2O_2 under zinc oxide nanoparticles was reported in *P. sativum.*

The extent of lipid peroxidation of *A. cepa* roots was found to increase with an increase in zinc oxide nanoparticles, as reported by Kumari et al. (2011). DNA damage caused by cerium oxide nanoparticles in *G. max* was detected by random amplified polymorphic DNA analysis. Treatment of soybean plant with 2000 μg/mL gave four new bands, whereas 4000 μg/mL gave three new bands (López-Moreno et al., 2010). Abnormal growth of root hairs was seen in *Solanum lycopersicum* seedlings after treating them with high concentrations of titanium oxide and Fe_3O_4 nanoparticles as compared to control, but no morphological changes were seen in shoots (Giordani et al., 2012). Majumdar et al. (2014) studied the effect of cerium oxide nanoparticles on kidney beans. Malondialdehyde (MDA) content was used as an indicator of lipid peroxidation. They studied the effect of nanoparticles over a concentration of 0–500 mg/L in roots, leaves, and stem. No substantial change in MDA content was seen as compared to control and stem and root, but a considerable decrease in MDA content in leaves compared to control was seen in 500 mg/L-treated plants over a period of 15 days. The study was conducted by Mirzajani et al. (2013) on rice plant using silver nanoparticles. Treatment was given at a concentration of 30–60 mg/L, and they found that the higher concentrations of nanoparticles negatively affected plants in terms of cell wall and shoot growth. Reduced chlorophyll content was found in *A. thaliana* when treated with iron nanoparticles by Marusenko et al. (2013).

Faisal et al. (2013) studied the effects of nickel oxide nanoparticles in tomato plants. The induction of oxidative stress was observed in roots and the higher formation of oxidative stress in protoplasts was detected by flow cytometry. Increased reactive stress was correlated with elevated levels of enzymes such as catalase (CAT) and SOD. The NiO-NP induced DNA damage was assessed using comet assay, and number of apoptotic cells was increased with the concentration of NPs, as exhibited by the typical tail pattern observed during the assay (Faisal et al., 2013).

The study conducted by Dimkpa et al. (2012) on evaluating the possible effects of copper oxide and zinc oxide nanoparticles revealed an increase in the extent of lipid peroxidation and reduction in chlorophyll content, indicators of oxidative stress (Dimkpa et al., 2012). Lipid peroxidation caused by copper oxide nanoparticles was found to be increased to around fourfold and about a twofold increase was recorded because of zinc oxide nanoparticles compared to their respective controls in wheat. On the other hand, a considerable decrease in chlorophyll content in the presence of both of these nanoparticles was found. On similar lines, the effect of cerium oxide nanoparticles was studied in *Z. mays* by Zhao et al. (2012) with 5–10 times higher H_2O_2 levels in nanoparticle-treated roots and shoots over control tissues; however, surprisingly, no significant levels of lipid peroxidation were reported.

The increase in MDA content indicating lipid peroxidation as well as elevated antioxidant enzyme activity was observed in *O. sativa* seedlings after exposure to copper oxide nanoparticles (Shaw and Hossain, 2013). On similar lines, an experiment performed by Shaw et al. (2014) studied the effect of copper oxide nanoparticles on *Hordeum vulgare* and found that root and shoot length was affected and the photosynthetic performance index was also reduced. Cerium oxide nanoparticles were found to damage membrane resulting in lipid peroxidation and cell membrane damage as well as inhibition of root elongation in asparagus lettuce (Cui et al., 2014).

17.4 ACTIVATION OF ANTIOXIDANT MACHINERY IN RESPONSE TO NANOPARTICLE EXPOSURE

In response to oxidative damage, plants have developed a defense system that involves triggering antioxidative defense mechanisms with enzymatic as well as nonenzymatic components. These enzymes and low molecular weight antioxidants scavenge or detoxify the ROS-generated plants exposed to nanoparticles. These enzymes include SOD, CAT, peroxidases (POXes) including guaiacol peroxidase (GPOX), ascorbate peroxidase (APOX), glutathione reductase, and dehydroascorbate reductase. Similarly, ascorbates and thiols (glutathione) are the common low molecular weight antioxidants. Different forms of POX and CAT quench ROS as well as the peroxyl radicals, whereas SOD scavenges the superoxide radicals (Khare et al., 2014). The overall interactions of engineered nanoparticles in plant systems through roots and/or leaves and plants' enzymatic response and ROS scavenging are depicted in Fig. 17.2.

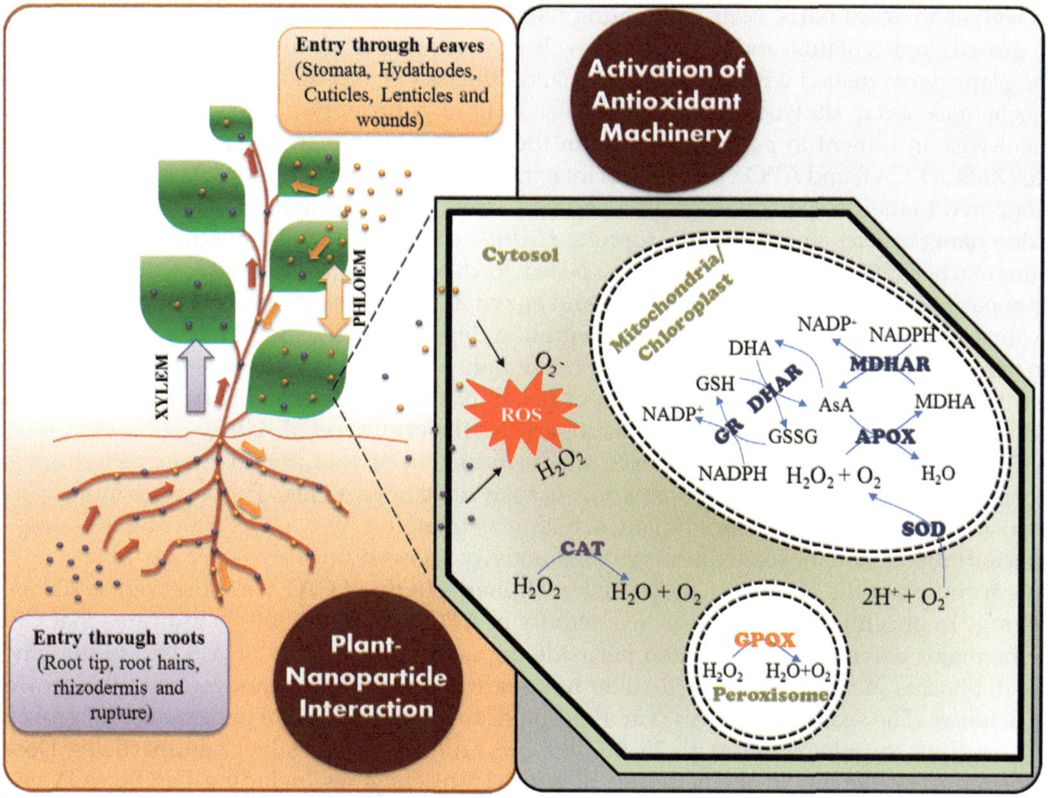

FIGURE 17.2 Entry of nanoparticles (NPs) through roots and leaves, their interaction with plant systems leading to generation of reactive oxygen species (ROS), and activation of responsive antioxidant machinery. $^1O_2$, singlet oxygen; *APOX*, ascorbate peroxidase; *AsA*, ascorbic acid; *CAT*, catalase; *DHA*, dehydroascorbate; *DHAR*, dehydroascorbate reductase; *GPOX*, guaiacol peroxidase; *GR*, glutathione reductase; *GSSG*, oxidized glutathione; H_2O_2, hydrogen peroxide; *MDAR*, monodehydroascorbate; *NADPH*, nicotinamide adenine dinucleotide phosphate; $O_2^{\cdot -}$, superoxide radical; $OH^{\cdot}$, hydroxyl radicals.

Dimkpa et al. (2012) examined the extent of phytotoxicity level generated in sand-grown wheat plants in response to the treatment of nano-oxides of copper and zinc. The study incorporated the sand-based growth system where the autoclaved sand was amended with 500 mg/kg of either of the nanoparticles. CAT and POX analysis was performed from root samples. Compared to the control plants, copper nanoparticle-treated plants showed improved CAT and POX activities, while such increase was absent in the case of zinc nanoparticle treatment. The impact of magnetite iron oxide nanoparticles on wheat development was studied by Iannone et al. (2016) for evaluation of oxidative damage. In this work, physiological effects and possible cell internalization of citric acid-coated iron oxide nanoparticles (5, 10, 15, and 20 mg/L) on wheat plants grown in a hydroponic system were assessed (Iannone et al., 2016). In general, nanoparticle treatment was reported to be responsible for the increment in enzymatic activities, but the degree of increase was dependent on the concentration of applied nanoparticles. Increased enzymatic activities, namely, CAT, APOX, GPOX, and SOD, were observed in roots as well as in aerial parts. Nair and Chung (2014) conducted a study to decipher the changes in growth, redox status, and expression levels of stress-related genes in *Cicer arietinum*, when the plants were treated with copper oxide nanoparticles (50, 100, 200, 300, 400, and 500 mg/L). Biochemical assay showed a gradual increase in lipid peroxidation levels along with a treatment-wise increment in peroxidase level in the plants. Similarly, mRNA expression levels of Cu/ZnSOD, CAT, and APOX showed an increment in copper oxide nanoparticle-treated plants compared to their respective controls. Oxidative stress responses against exposure to titanium oxide nanoparticles in aquatic macrophyte *Hydrilla verticillata* were studied by Okupnik and Pflugmacher (2016). The plants were exposed to different concentrations of titanium oxide nanoparticles (0.01, 0.1, 1, and 10 mg/L), and enzymatic activities including CAT and GR were evaluated. The study indicated the concentration-dependent increase in activities of both the enzymes. Cytotoxic effects of aluminum oxide nanoparticles showed dose-dependent increment in SOD activities in *A. cepa* when root tips of plants were exposed to 0.01, 1, and 100 μg/mL concentration of aluminum oxide nanoparticles (Rajeshwari et al., 2015).

Riahi-Madvar et al. (2012) proposed the introduction of oxidative damage as a mode of persuading toxicity in plant systems via uptake of nanoparticles. This was demonstrated by using *T. aestivum* as a model plant, which was treated with 50, 200, 500, and 1000 mg/L concentrations of nanoscale alumina. SOD activity showed an increase of up to 500 mg/L concentration of the alumina nanoparticles. Higher activity of CAT was observed at 200 and 500 mg/L concentrations, whereas low activity of APOX was seen, which indicates that CAT is the major scavenger of hydrogen peroxide regulating the activity of APOX. On the other hand, Sharma et al. (2012) studied silver nanoparticle-mediated enhancement of the growth of *B. juncea*. The seeds of *B. juncea* (var. Pusa Jaikisan) were allowed to germinate and grow in MS medium complemented with 25, 50, 100, 200, and 400 ppm of silver nanoparticles. Dose-dependent increments in the activities of antioxidant enzymes including GPOX, POX, and CAT were observed in the 7-day-old seedlings, with highest enzymatic activities observed at the highest concentration of silver nanoparticles (400 ppm).

Rao and Shekhawat (2016) particularized the consequences of oxidative stress-causing copper oxide and titanium dioxide nanoparticles on *B. juncea*. The plants were treated with different amounts of the mentioned nanoparticles (200, 500, 1000, and 1500 mg/L). A treatment-dependent increment in APOX activity was observed in roots and shoots of the plant under the influence of both the nanomaterials; however, the extent of increase in APOX was more in the copper oxide nanoparticle treatment compared to titanium oxide nanoparticles. Similarly, copper oxide nanoparticles showed dosage-related higher CAT activity, whereas no significant change was observed in the titanium oxide nanoparticle treatment. SOD enzyme showed similar activity to that shown by APOX and CAT. Fe_3O_4 nanoparticles have shown a reduction in the uptake and accumulation of heavy metals such as lead, zinc, cadmium, and copper, and mitigation of the toxic effect caused by these heavy metals by incrementing the activities of antioxidant enzymes in wheat seedlings (Konate et al., 2017). The wheat seedlings showed reduction in SOD and POX activities under the influence of low concentrations of heavy metals. However, addition of Fe_3O_4 nanoparticles (2000 mg/L) proved to be responsible for initiation of protective mechanisms to reduce oxidative stress, which can be correlated to the increased SOD and POX activities in the wheat seedlings with reduction in the peroxidation of lipids (MDA content). Zinc nanoparticles synthesized by the green synthesis approach were tested to check their potency for induction of oxidative stress and to study the activation of antioxidant machinery in grass pea (*Lathyrus sativus*) by Panda et al. (2017).

The germinating seedlings were treated with a range of different concentrations of green-synthesized zinc nanoparticles (10, 20, 40, 80, and 100 mg/L) by dipping the sprouting roots in the medium containing the nanoparticles. Oxidative damage was documented by noticing the dose-dependent increment in SOD, GPX, and APOX activities, whereas there was a decline in CAT activity.

Apart from the fact that nanoparticles induce antioxidant machinery in plants, Wei and Wang (2013) reviewed the antioxidant capability of nanoparticles and mechanisms that mimic the activity of natural enzymes. The authors reviewed the potentials of various NPs for their enzyme-like activities for instance CAT-like activity by CeO_2, Fe_3O_4, and Co_3O_4 NPs; POX-like activities by CeO_2, Fe_3O_4, Co_3O_4, MnO, CuO, and Au NPs; SOD-like activities by CeO2 and Pt NPs, and presented the NPs as potential candidates for developing artificial nanozymes (Wei and Wang, 2013). Still it is very challenging to identify and correlate these activities with nanoparticle exposure to whole individual plants. Though many nanophytotoxicity investigations indicated the possible role of nanoparticles in enzyme activation, no evidence is currently available that could correlate with the chemical properties of nanoparticles. Some studies also showed unequal and random effects of nanoparticles on enzyme activities. Lei et al. (2008) reported higher SOD, CAT, APOX, and GPOX activities in spinach when the plants were treated with titanium oxide nanoparticles. On a similar line, Song et al. (2012) showed an increase in SOD, CAT, and GPOX activities in *Lemna minor*. However, contrary results were also reported by a few authors, for instance Foltete et al. (2011), where decreased GR and APOX activities were observed in *V. faba* and GPOX, SOD, CAT, in *L. minor* (Song et al., 2012), but decreased GR and APOX activities in *V. faba*. Thus the role of nanoparticle chemical characteristics on the variation of antioxidant defense systems in plants is still a less explored area. A summarized list of different types of nanoparticles and their impacts on induction/activation of antioxidative enzymes in different plant species is given in Table 17.1.

TABLE 17.1 Effect of Types of Nanoparticles (NPs) on Induction/Activation of Antioxidative Enzymes in Different Plant Species

NPs	Plant	Responsive Enzymes	References
CuO	*Triticum aestivum*	CAT↑, POX↑	Dimkpa et al. (2012)
ZnO	*T. aestivum*	CAT=, POX=	Dimkpa et al. (2012)
Fe_3O_4	*T. aestivum*	CAT↑, APOX↑, GPOX↑, SOD↑	Iannone et al. (2016)
CuO	*Cicer arietinum*	POX↑	Nair and Chung (2015)
TiO_2	*Hydrilla verticillata*	CAT↑, GR↑	Okupnik and Pflugmacher (2016)
Al_2O_3	*Allium cepa*	SOD↑	Rajeshwari et al. (2015)
TiO_2	*Spinacia oleracea*	SOD↑, CAT↑, APOX↑, GPOX↑	Lei et al. (2008)
TiO_2	*Lemna minor*	SOD↑, CAT↑, GPOX↑	Song et al. (2012)
TiO_2	*Vicia faba*	GR↓, APOX↓	Foltete et al. (2011)

APOX, ascorbate peroxidase; *CAT*, catalase; *GPOX*, guaiacol peroxidase; *GR*, glutathione reductase; *POX*, peroxidase; *SOD*, superoxide dismutase.

17.5 CONCLUSION AND FUTURE OUTLOOK

Because of unusual properties and the wide spectrum of applications, use of nanoscale materials or nanoparticles, particularly engineered nanoparticles, has seen unprecedented production and demand in recent years. Though contemporary reports confirm the successful use of nanoparticles in improving the efficiency of fertilizers/growth enhancers and pesticides, there are also reports of negative impacts of nanoparticles on plant growth and development. There has been a recent debate on assessing the impacts of nanoparticles on inducing negative impacts mainly through oxidative stress; consequently, several reports have been published establishing such roles. Many recently published reports have shown that nanoparticles impart oxidative stress and cellular damage in plant systems and in response, plants activate their antioxidant defense. However, greater in-depth understanding of these phenomena at the molecular level is needed to target and mitigate these negative impacts, and further research in this direction is advocated.

Acknowledgments

VK gratefully acknowledges the financial support from the Science and Engineering Research Board, Government of India, in the form of a Young Scientist Project [SERB-OYS; grant number SR/FT/LS-93/2011]. The authors acknowledge the support from the Department of Science and Technology, Government of India (DST-FIST Program).

References

Anjum, N.A., Gill, S.S., Duarte, A.C., Pereira, E., Ahmad, I., 2013. Silver nanoparticles in soil-plant systems. J. Nanopart. Res. 15, 1896.

Arora, S., Sharma, P., Kumar, S., Nayan, R., Khanna, P., Zaidi, M., 2012. Gold-nanoparticle induced enhancement in growth and seed yield of *Brassica juncea*. Plant Growth Regul. 66, 303–310.

Atha, D.H., Wang, H., Petersen, E.J., Cleveland, D., Holbrook, R.D., Jaruga, P., Dizdaroglu, M., Xing, B., Nelson, B.C., 2012. Copper oxide nanoparticle mediated DNA damage in terrestrial plant models. Environ. Sci. Technol. 46, 1819–1827.

Begum, P., Ikhtiari, R., Fugetsu, B., 2011. Graphene phytotoxicity in the seedling stage of cabbage, tomato, red spinach, and lettuce. Carbon 49, 3907–3919.

Chew, B.P., Park, J.S., 2004. Carotenoid action on the immune response. J. Nutr. 134, 257S–261S.

Cui, D., Zhang, P., Ma, Y., He, X., Li, Y., Zhang, J., Zhao, Y., Zhang, Z., 2014. Effect of cerium oxide nanoparticles on asparagus lettuce cultured in an agar medium. Environ. Sci. 1, 459–465.

Dimkpa, C.O., McLean, J.E., Latta, D.E., Manangón, E., Britt, D.W., Johnson, W.P., Boyanov, B.I., Anderson, A.J., 2012. CuO and ZnO nanoparticles: phytotoxicity, metal speciation, and induction of oxidative stress in sand-grown wheat. J. Nanopart. Res. 14, 1125.

Dimkpa, C.O., McLean, J.E., Martineau, N., Britt, D.W., Haverkamp, R., Anderson, A.J., 2013. Silver nanoparticles disrupt wheat (*Triticum aestivum* L.) growth in a sand matrix. Environ. Sci. Technol. 47, 1082–1090.

Faisal, M., Saquib, Q., Alatar, A.A., Al-Khedhairy, A.A., Hegazy, A.K., Musarrat, J., 2013. Phyotoxic hazards of NiO-nanoparticles in tomato: a study on mechanism of cell death. J. Hazard. Mater. 250–251, 318–332.

Fenoglio, I., Greco, G., Livraghi, S., Fubini, B., 2009. Non-UV-induced radical reactions at the surface of TiO_2 nanoparticles that may trigger toxic responses. Chem. Eur. J. 15, 4614–4621.

Foltete, A.S., Masfaraud, J.F., Bigorgne, E., et al., 2011. Environmental impact of sunscreen nanomaterials: ecotoxicity and gentoxicity of altered TiO_2 nanocomposites on *Vicia faba*. Environ. Pollut. 159, 2515–2522.

Fubini, B., Fenoglio, I., Tomatis, M., 2007. Physicochemical characteristics of nanoparticles that determine potential toxicity. In: Monteiro-Riviere, N.A., Tran, C.L. (Eds.), Nanotoxicology: Characterization, Dosing, and Health Effects. Informa Healthcare, New York, pp. 59–70.

Gardea-Torresdey, J.L., Rico, C.M., White, J.C., 2014. Trophic transfer, transformation, and impact of engineered nanomaterials in terrestrial environments. Environ. Sci. Technol. 48, 2526–2540. http://dx.doi.org/10.1021/es4050665.

Ghodake, G., Seo, Y.D., Lee, D.S., 2011. Hazardous phytotoxic nature of cobalt and zinc oxide nanoparticles assessed using *Allium cepa*. J. Hazard. Mater. 186, 952–955.

Ghosh, M., Bandopadhyay, M., Mukherjee, A., 2010. Genotoxicity of titanium oxide (TiO_2) nanoparticles at two tropic levels: plant and human lymphocytes. Chemosphere 81, 1253–1262.

Ghosh, M., Jana, A., Sinha, S., Jothiramajayam, M., Nag, A., Chakraborty, A., Mukherjee, A., Mukherjee, A., 2016. The effect of ZnO nanoparticles in plants: cytotoxicity, genotoxicity, deregulation of antioxidant defenses, and cell-cycle arrest. Mutat. Res. 807, 25–32.

Giordani, T., Fabrizi, A., Guidi, L., Natali, L., Giunti, G., Ravasi, F., Cavallini, A., Pardossi, A., 2012. Response of tomato plants exposed to treatment with nanoparticles. Environ. Qual. 8, 27–38.

Hossain, Z., Mustafa, G., Komatsu, S., 2015. Plant responses to nanoparticle stress. Int. J. Mol. Sci. 16, 26644–26653.

Iannone, M.F., Groppa, M.D., de Sousa, M.E., van Raap, M.B.F., Benavides, M.P., 2016. Impact of magnetite iron oxide nanoparticles on wheat (*Triticum aestivum* L.) development: evaluation of oxidative damage. Environ. Exp. Bot. 131, 77–88.

Ismail, A., Takeda, S., Nick, P.R., 2014. Life and death under salt stress: same players, different timing? J. Exp. Bot. http://dx.doi.org/10.1093/jxb/eru159.

Jefferson, D.A., 2000. The surface activity of ultrafine particles. Philos. Trans. R. Soc. Lond. A 358, 2683–2692.

Keller, A.A., Lazareva, A., 2014. Predicted releases of engineered nanomaterials: from global to regional to local. Environ. Sci. Technol. Lett. 1, 65–70.

Khare, T., Kumar, V., Kishor, P.B.K., 2014. Na^+ and Cl^- ions show additive effects under NaCl stress on induction of oxidative stress and the responsive antioxidative defense in rice. Protoplasma 252, 1149–1165. http://dx.doi.org/10.1007/s00709-014-0749-2.

Konate, A., He, X., Zhang, Z., Ma, Y., Zhang, P., Alugongo, G.M., Rui, Y., 2017. Magnetic (Fe_3O_4) nanoparticles reduce heavy metals uptake and mitigate their toxicity in wheat seedling. Sustainability 9, 790. http://dx.doi.org/10.3390/su9050790.

Krishnaraj, C., Jagan, G., Ramachandran, R., Abirami, S.M., Mohan, N., Kalaichelvan, P.T., 2012. Effect of biologically synthesized silver nanoparticles on *Bacopa monnieri* L. Wettst. plant growth metabolism. Process Biochem. 47, 651–658.

Kumar, V., Khare, T., 2014. Individual and additive effects of Na^+ and Cl^- ions on rice under salinity stress. Arch. Agron. Soil Sci. http://dx.doi.org/10.1080/03650340.2014.936400.

Kumar, V., Shriram, V., Hoque, T.S., Hasan, M.M., Burritt, D.J., Hossain, M.A., 2017. Glycinebetaine-mediated abiotic oxidative-stress tolerance in plants: physiological and biochemical mechanisms. In: Sarwat, M., et al. (Ed.), Stress Signaling in Plants: Genomics and Proteomics Perspective, vol. 2. Springer International, Switzerland, pp. 111–133. http://dx.doi.org/10.1007/978-3-319-42183-4-5.

Kumari, M., Khan, S.S., Pakrashi, S., Mukherjee, A., Chandrasekaran, N., 2011. Cytogenetic and genotoxic effects of zinc oxide nanoparticles on root cells of *Allium cepa*. J. Hazard. Mater. 190, 613–621.

Lee, S., Chung, H., Kim, S., Lee, I., 2013. The genotoxic effect of ZnO and CuO nanoparticles on early growth of buckwheat, *Fagopyrum esculentum*. Water Air Soil Pollut. 224, 1668–1678.

Lei, Z., Mingyu, S., Xiao, W., et al., 2008. Antioxidant stress is promoted by nano-anatase in spinach chloroplasts under UV-B radiation. Biol. Trace Elem. Res. 121, 69–79.

Li, G.R., Hu, T., Pan, G.L., Yan, T.Y., Gao, X.P., Zhu, H.Y., 2008. Morphology-function relationship of ZnO: polar planes, oxygen vacancies, and activity. J. Phys. Chem. C 112, 11859–11864. http://dx.doi.org/10.1021/jp8038626.

Lin, D., Xing, B., 2007. Phytotoxicity of nanoparticles: inhibition of seed germination and root growth. Environ. Pollut. 150, 243–250.

Liu, Q., Zhang, X., Zhao, Y., et al., 2013. Fullerene-induced increase of glycosyl residue on living plant cell wall. Environ. Sci. Technol. 47, 7490–7498.

Liu, Q., Zhao, Y., Wan, Y., et al., 2010. Study of the inhibitory effect of water-soluble fullerenes on plant growth at the cellular level. ACS Nano 4, 5743–5748.

López-Moreno, M.L., de la Rosa, G., Hernández-Viezcas, J.A., Castillo-Michel, H., Botez, C.E., Peralta-Videa, J.R., Gardea-Torresdey, J.L., 2010. Evidence of the differential biotransformation and genotoxicity of ZnO and CeO_2 nanoparticles on soybean (*Glycine max*) plants. Environ. Sci. Technol. 44 (19), 7315–7320.

Ma, C., White, J.C., Dhankher, O.P., Xing, B., 2015. Metal-based nanotoxicity and detoxification pathways in higher plants. Environ. Sci. Technol. 49, 7109–7122.

Majumdar, S., Almeida, I.C., Arigi, E.A., Choi, H., VerBerkmoes, N.C., Trujillo-Reyes, J., Flores-Margez, J.P., White, J.C., Peralta-Videa, J.R., Gardea-Torresdey, J.L., 2015. Environmental effects of nanoceria on seed production of common bean (*Phaseolus vulgaris*): a proteomic analysis. Environ. Sci. Technol. 49, 13283–13293.

Majumdar, S., Peralta-Videa, J.R., Bandopadhyay, S., Castillo-Michel, H., Hernandez-Viezcas, J.-A., Sahi, S., Gardea-Torresdey, J.L., 2014. Exposure of cerium oxide nanoparticles to kidney bean shows disturbance in the plant defence mechanisms. J. Hazard. Mater. 278, 279–287.

Mapara, N., Sharma, M., Shriram, V., Bharadwaj, R., Mohite, K.C., Kumar, V., 2015. Antimicrobial potentials of *Helicteres isora* silver nanoparticles against extensively drug-resistant (XDR) clinical isolates of *Pseudomonas aeruginosa*. Appl. Microbiol. Biotechnol. 99, 10655–10667.

Marusenko, Y., Shipp, J., Hamilton, G.A., Morgan, J.L.L., Keebaugh, M., Hill, H., Dutta, A., Zhuo, X., Upadhyay, N., Hutchings, J., 2013. Bioavailability of nanoparticles hematite to *Arabidopsis thaliana*. Environ. Pollut. 174, 150–156.

McKee, M.S., Filser, J., 2016. Impacts of metal-based engineered nanomaterials on soil communities. Environ. Sci. 3, 506–533.

Mclaren, A., Valdes-Solis, T., Li, G., Tsang, S.C., 2009. Shape and size effects of ZnO nanocrystals on photocatalytic activity. J. Am. Chem. Soc. 131, 12540–12541.

Mingyu, S., Xiao, W., Chao, L., Chunxiang, Q., Xiaoqing, L., Kiang, C., Hao, H., Fashui, H., 2007. Promotion of energy transfer and oxygen evolution in spinach photosystem II by nano-anatase TiO_2. Biol. Trace Elem. Res. 119, 183–192.

Mirzajani, F., Askari, H., Hamzelou, S., Farzaneh, M., Ghassempour, A., 2013. Effect of silver nanoparticles on *Oryza sativa* L. and its rhizosphere bacteria. Ecotoxicol. Environ. Saf. 88, 48–54.

Mukherjee, A., Peralta-Videa, J.R., Bandopadhyay, S., Rico, C.M., Zhao, L., Gardea-Torresdey, L., 2014. Physiological effects of nanoparticulate ZnO in green peas (*Pisum sativum* L.) cultivated in soil. Metallomics 6, 132–138.

Nair, P.M.G., Chung, I.M., 2014. Impact of copper oxide nanoparticles exposure on *Arabidopsis thaliana* growth, root system development, root lignification, and molecular level changes. Environ. Sci. Pollut. Res. 21, 12709–12722. http://dx.doi.org/10.1007/s11356-014-3210-3.

Nair, P.M.G., Chung, I.M., 2015. Changes in the growth, redox status and expression of oxidative stress related genes in chickpea (*Cicer arietinum* L.) in response to copper oxide nanoparticle exposure. J. Plant Growth Regul. 34, 350–361.

Okupnik, A., Pflugmacher, S., 2016. Oxidative stress response of the aquatic macrophyte *Hydrilla verticillata* exposed to TiO_2 nanoparticles. Environ. Toxicol. Chem. 35, 2859–2866. http://dx.doi.org/10.1002/etc.3469.

Panda, K.K., Achary, V.M.M., Krishnaveni, R., Padhi, B.K., Sarangi, S.N., Sahu, S.N., Panda, B.B., 2011. In vitro biosynthesis and genotoxicity bioassay of silver nanoparticles using plants. Toxicol. Vitro 25, 1097–1105.

Panda, K.K., Golari, D., Venugopal, A., Achary, V.M.M., Phaomei, G., Parinandi, N.L., Sahu, H.K., Panda, B.B., 2017. Green synthesized zinc oxide (ZnO) nanoparticles induce oxidative stress and DNA damage in *Lathyrus sativus* L. root bioassay system. Antioxidants 6, 35. http://dx.doi.org/10.3390/antiox6020035.

Patil, S.S., Shedbalkar, U.U., Truskewycz, A., Chopade, B.A., Ball, A.S., 2016. Nanoparticles for environmental clean-up: a review of potential risks and emerging solutions. Environ. Technol. Innov. 5, 10–21.

Rajeshwari, A., Kavitha, S., Alex, S.A., Kumar, D., Mukherjee, A., Chandrasekaran, N., Mukherjee, A., 2015. Cytotoxicity of aluminum oxide nanoparticles on *Allium cepa* root tip-effects of oxidative stress generation and biouptake. Environ. Sci. Pollut. Res. 22, 11057–11066.

Rao, S., Shekhawat, G.S., 2016. Phytotoxicity and oxidative stress perspective of two selected nanoparticles in *Brassica juncea*. 3 Biotech 6, 244.

Riahi-Madvar, A., Rezaee, F., Jalili, V., 2012. Effects of alumina nanoparticles on morphological properties and antioxidant system of *Triticum aestivum*. Iran. J. Plant Physiol. 3, 595–603.

Rico, C.M., Lee, S.C., Rubenecia, R., Mukherjee, A., Hong, J., Peralta-Videa, J.R., Gardea-Torresdey, J.L., 2014. Cerium oxide nanoparticles impact yield and modify nutritional parameters in wheat (*Triticum aestivum* L.). J. Agric. Food Chem. 62, 9669–9675.

Rico, C.M., Majumdar, S., Duarte-Gardea, M., Peralta-Videa, J.R., Gardea-Torresdey, J.L., 2011. Interaction of nanoparticles with edible plants and their possible implications in the food chain. J. Agric. Food Chem. 59, 3485–3498.

Rico, C.M., Morales, M.I., McCreary, R., Castillo-Michel, H., Barrios, A.C., Hong, J., Tafoya, A., Lee, W.Y., Varela-Ramirez, A., Peralta-Videa, J.R., Gardea-Torresdey, J.L., 2013. Cerium oxide nanoparticles modify the antioxidative stress enzyme activities and macromolecule composition in rice seedlings. Environ. Sci. Technol. 47, 14110–14118. http://dx.doi.org/10.1021/es4033887.

Rico, C.M., Peralta-Videa, J.R., Garadea-Torresdey, J.L., 2015. Chemistry, biochemistry of nanoparticles, and their role in antioxidant defense system in plants. In: Siddiqui, H.M., et al. (Ed.), Nanotechnology and Plant Sciences. Springer International Publishing, Switzerland, pp. 1–17.

Rizwan, M., Ali, S., Qayyum, M.F., Ok, Y.S., Adrees, M., Ibrahim, M., Zia-ur-Rehman, M., Farid, M., Abbas, F., 2017. Effect of metal and metal oxide nanoparticles on growth and physiology of globally important food crops: a critical review. J. Hazard. Mater. 322, 2–16. http://dx.doi.org/10.1016/j.jhazmat.2016.05.061.

Sharma, P., Bhatt, D., Zaidi, M.G.H., Saradhi, P.P., Khanna, P.K., Arora, S., 2012. Silver nanoparticle-mediated enhancement in growth and antioxidant status of *Brassica juncea*. Appl. Biochem. Biotechnol. 167, 2225–2233.

Shaw, A.K., Ghosh, S., Kalaji, H.M., Bosa, K., Brestic, M., Zivcak, M., Hossain, Z., 2014. Nano-CuO stress induced modulation of antioxidative defense and photosynthetic performance of syrian barley (*Hordeum vulgare* L.). Environ. Exp. Bot. 102, 37–47.

Shaw, A.K., Hossain, Z., 2013. Impact of nano-CuO stress on rice (*Oryza sativa* L.) seedlings. Chemosphere 93, 906–915.

Shen, C., Zhang, Q., Li, L., Bi, F., Yao, N., 2010. Induction of programmed cell death in Arabidopsis an rice by single-wall carbon nanotubes. Am. J. Bot. 97, 1602–1609. http://dx.doi.org/10.3732/ajb.1000073.

Shi, J., Abid, A.D., Kennedy, I.M., Hristova, K.R., Silk, W.K., 2011. To duckweeds (*Landoltia punctata*), nanoparticulate copper oxide is more inhibitory than the soluble copper in the bulk solution. Environ. Pollut. 159, 1277–1282.

Shweta, Tripathi, D.K., Singh, S., Singh, S., Dubey, N.K., Chauhan, D.K., 2016. Impact of nanoparticles on photosynthesis: challenges and opportunities. Mater. Focus 5 (5), 405–411.

Siddiqui, M.H., Al-Whaibi, M.H., Mohammad, F. (Eds.), 2015. Nanotechnology and Plant Sciences: Nanoparticles and Their Impact on Plants. Springer International Publishing, Switzerland.

Singh, S., Tripathi, D.K., Dubey, N.K., Chauhan, D.K., 2016. Effects of nano-materials on seed germination and seedling growth: striking the slight balance between the concepts and controversies. Mater. Focus 5 (3), 195–201.

Singh, S., Vishwakarma, K., Singh, S., Sharma, S., Dubey, N.K., Singh, V.K., Liu, S., Tripathi, D.K., Chauhan, D.K., 2017. Understanding the plant and nanoparticle interface at transcriptomic and proteomic level: a concentric overview. Plant Gene. http://dx.doi.org/10.1016/j.plgene.2017.03.006.

Song, G., Gao, Y., Wu, H., et al., 2012. Physiological effect of anatase TiO_2 nanoparticles on *Lemna minor*. Environ. Toxicol. Chem. 31, 2147–2152.

Tan, X.M., Lin, C., Fugetsu, B., 2009. Studies on toxicity of multi-walled carbon nanotubes on suspension rice cells. Carbon 47, 3479–3487.

Thwala, M., Musee, N., Sikhwivhilu, L., Wepner, V., 2013. The oxidative toxicity of Ag and ZnO naoparticles towards the aquatic plant *Spirodela punctuta* and the role of testing media parameters. Environ. Sci. 15, 1830–1843.

Tripathi, D.K., Singh, S., Singh, S., Pandey, R., Singh, V.P., Sharma, N.C., Prasad, S.M., Dubey, N.K., Chauhan, D.K., 2017a. An overview on manufactured nanoparticles in plants: uptake, translocation, accumulation and phytotoxicity. Plant Physiol. Biochem. 110, 2–12. http://dx.doi.org/10.1016/j.plaphy.2016.07.030.

Tripathi, D.K., Singh, V.P., Singh, S., Prasad, S.M., Chauhan, D.K., Dubey, N.K., 2016a. Effect of silicon and silicon nanoparticle (SiNp) on seedlings of maize cultivar and hybrid differing in arsenate tolerance. Front. Environ. Sci. 4, 46. http://dx.doi.org/10.3389/fenvs.2016.00046.

Tripathi, D.K., Singh, S., Singh, S., Srivastava, P.K., Singh, V.P., Singh, S., et al., 2017c. Nitric oxide alleviates silver nanoparticles (AgNps)-induced phytotoxicity in *Pisum sativum* seedlings. Plant Physiol. Biochem. 110, 167–177. http://dx.doi.org/10.1016/j.plaphy.2016.06.015.

Tripathi, D.K., Singh, S., Singh, V.P., Prasad, S.M., Chauhan, D.K., Dubey, N.K., 2016b. Silicon nanoparticles more efficiently alleviate arsenate toxicity than silicon in maize cultivar and hybrid differing in arsenate tolerance. Front. Environ. Sci. 4, 46.

Tripathi, D.K., Singh, S., Singh, V.P., Prasad, S.M., Dubey, N.K., Chauhan, D.K., 2017e. Silicon nanoparticles more effectively alleviated UV-B stress than silicon in wheat (*Triticum aestivum*) seedlings. Plant Physiol. Biochem. 110, 70–81.

Tripathi, D.K., Singh, V.P., Prasad, S.M., Chauhan, D.K., Dubey, N.K., 2015. Silicon nanoparticles (SiNp) alleviate chromium (VI) phytotoxicity in *Pisum sativum* (L.) seedlings. Plant Physiol. Biochem. 96, 189–198.

Tripathi, D.K., Tripathi, A., Guar, S., Singh, S., Singh, Y., Vishwakarma, K., et al., 2017b. Uptake, accumulation and toxicity of silver nanoparticle in autotrophic plants, and heterotrophic microbes: a concentric review. Front. Microbiol. 8, 7. http://dx.doi.org/10.3389/fmicb.2017.00007.

Tripathi, D.K., Mishra, R.K., Singh, S., Singh, S., Vishwakarma, K., Sharma, S., Singh, V.P., Singh, P.K., Prasad, S.M., Dubey, N.K., Pandey, A.C., 2017d. Nitric oxide ameliorates zinc oxide nanoparticles phytotoxicity in wheat seedlings: implication of the ascorbate–glutathione cycle. Front. Plant Sci. 8.

Ursache-Oprisan, M., Focanice, E., Creanga, D., Caltun, O., 2011. Sunflower chlorophyll levels after magnetic nanoparticle supply. Afr. J. Biotechnol. 10, 7092–7098.

Wang, H., Kou, X., Pei, Z., Xiao, J.Q., Shan, X., Xing, B., 2011. Physiological effects of magnetite (Fe_3O_4) nanoparticles on perennial ryegrass (*Lolium* perenne L.) and pumpkin (*Cucurbita mixta*) plants. Nanotoxicology 5, 30–42.

Wani, S.H., Sah, S.K., Hussain, M.A., Kumar, V., Balachandra, S.M., 2016. Transgenic approach for abiotic stress tolerance in crop plants. In: Al-Khayri, J.M., Jain, S.M., Johnson, D.V. (Eds.), Advances in Plant Breeding Strategies, Vol. 2: Agronomic, Abiotic and Biotic Stress Traits. Springer International Publishing, Switzerland, pp. 345–396. http://dx.doi.org/10.1007/978-3-319-22518-0.

Wei, H., Wang, E., 2013. Nanoparticles with enzyme-like characteristics (nanozymes): next-generation artificial enzymes. Chem. Soc. Rev. 42 (14), 6060–6093.

Winkel-Shirley, B., 2002. Biosynthesis of flavonoids and effects of stress. Curr. Opin. Plant Biol. 5, 218–222.

Yang, L., Watts, D.J., 2005. Particle surface characteristics may play an important role in phytotoxicity of alumina nanoparticles. Toxicol. Lett. 158, 122–132.

Yasur, J., Rani, P., 2013. Environmental effects of nanosilver: impact on castor seed germination, seedling growth, and plant physiology. Environ. Sci. Pollut. Res. 20, 8636–8648. http://dx.doi.org/10.1007/s11356-013-1798-3.

Zhao, L., Peng, B., Hernandez-Viezcas, J.A., Rico, C., Sun, Y., Peralta-Videa, J.R., Tang, X., Niu, G., Jin, L., Varela-Ramirez, A., Zhang, J.-Y., Gardea-Torresdey, J.L., 2012. Stress response and tolerance of *Zea mays* to CeO_2 nanoparticles: cross talk among H_2O_2, heat shock protein and lipid peroxidation. ACS Nano 6, 9615–9622.

Further Reading

He, D., Jones, A.M., Garg, S., Pham, A.N., Waite, T.D., 2011. Silver nanoparticle–reactive oxygen species interactions: application of a charging–discharging model. J. Phys. Chem. C 115, 5461–5468.

CHAPTER

18

Nanoparticles and Organic Matter: Process and Impact

Rajendran Velmurugan, Aran Incharoensakdi
Chulalongkorn University, Bangkok, Thailand

18.1 INTRODUCTION

Technological advancement and industrial needs are mostly dependent on fossil fuel as a primary energy source (Hall, 1980). As a result of the use of fossil fuels, environmentally harmful gases have been released into the atmosphere, which are contributing factors to global warming. The production of alternative energy has gained great importance because the supply of fossil fuels is limited, as evidenced by rapidly declining oil reserves (Hall, 1980; Blanch, 2012). As an alternative source, plant biomass has a high utilization potential and is highlighted as one of the most important future energy sources. The advantage of plant biomass is the utilization of CO_2, which can be returned to the atmosphere without additional CO_2 (Blanch, 2012). Significant progress in obtaining fuels and biochemicals from biomass has been achieved, such as fermentative ethanol from plants and microbial carbohydrates, biodiesel from oil crops, protein from plants and microbial extracts, and hydrocarbons from lignins. However, the methods developed for obtaining fuels and biochemicals from biomass are still limited by technical and economic barriers. Thus a novel approach is necessary to optimize biomass utilization for biochemical and biofuel production.

There are various technologies developed for the biomass conversion process utilizing biological and chemical agents (Blanch, 2012). Among the newly developed technologies, nanotechnology has gained much attention and achieved various biomass process applications. In the processing of biomass, nanotechnology has been applied to various processes such as modification of feed stocks, extraction of biomolecules, improvement of chemical and biological reactions, and obtaining desired products. Nanocatalysts were mostly obtained from metal oxides and silica nanomaterials. Metal oxides are one of the most acceptable and broadly employed nanomaterials, either as active catalyst or as supporting material. Certain groups of metals, particularly transition metals, have attracted much attention because of their outer electron configuration. Because of the structural

Nanomaterials in Plants, Algae, and Microorganisms
http://dx.doi.org/10.1016/B978-0-12-811487-2.00018-9

arrangement of metals in mixed metal oxide systems, mixed oxides have been applied in various successful conversion processes (Zhang et al., 2011). The configuration of the environment of the cation toward the biomolecule is represented as the advantage of a mixed metal oxide system. The surface-activated metal nanoparticles are produced by making a strong interaction between inorganic and organic molecules (Abu-Reziq et al., 2002; Lai et al., 2011). Linking of organic compounds onto metal units produces a material with generally higher surface area than the traditional porous materials. Based on the presence of functional groups, nanomaterials have been further grouped as acid, base, or bifunctional nanomaterials. On the other hand, base nanocatalysts were prepared by directly adding metal oxides or functionalization with silicon oxide or potassium fluoride (Lopez et al., 2005; Chiang et al., 2015; Ghotbabadi et al., 2016). Bifunctional nanocatalysts are catalysts holding both acidic and basic characteristics, developed for performing series of cascade reactions. Moreover, much interest has been focused on the synthesis, property, and possible benefit of nanomaterials in the biomass utilization processes. In this chapter, advances in nanomaterial-mediated extraction of plant and microalgae-derived biomolecules, theoretical background, and influence of important parameters are included, together with the advantages and drawbacks of individual nanomaterials.

18.2 PLANT COMPONENTS: NATURE AND USES

18.2.1 Carbohydrates

Carbohydrates are the most abundant structural component present in plants and microorganisms. Most carbohydrates are present in polymers such as cellulose, starch, and hemicelluloses. Cellulose is a structural component of plants that is associated with hemicelluloses and lignin (Fig. 18.1), whereas starch is present in potatoes, cereals, and microorganisms as a storage component. The denser and crystalline domains of cellulose with hindered access to the ether bonds cause complications in enzymatic and chemical access, while hemicelluloses have amorphous structures allowing easy chemical access. Cellulose is mainly used in the paper and textile industries, while starch is consumed mainly in the food and ethanol industries (Lynd et al., 2002). The monosaccharides extracted from plant materials are the actual raw material for current industrial applications. Glucose is used in various applications including food and fermentative production of various biochemicals. Xylose is another major pentose sugar, which is hydrogenated to xylitol for food applications and also for the production of furfural by acid hydrolysis.

18.2.2 Proteins and Amino Acids

Proteins and amino acids are biomolecules that are necessary for the structural and regulatory function of living things. Proteins are composed of amino acids that are attached to one another by peptide bonds. The nine essential amino acids are histidine, isoleucine, leucine, lysine, methionine, phenylalanine, threonine, tryptophan, and valine. Based on diversity, proteins and amino acids are varied with significant importance in commercial applications. To meet the protein and amino acid needs, three ways of production have been developed,

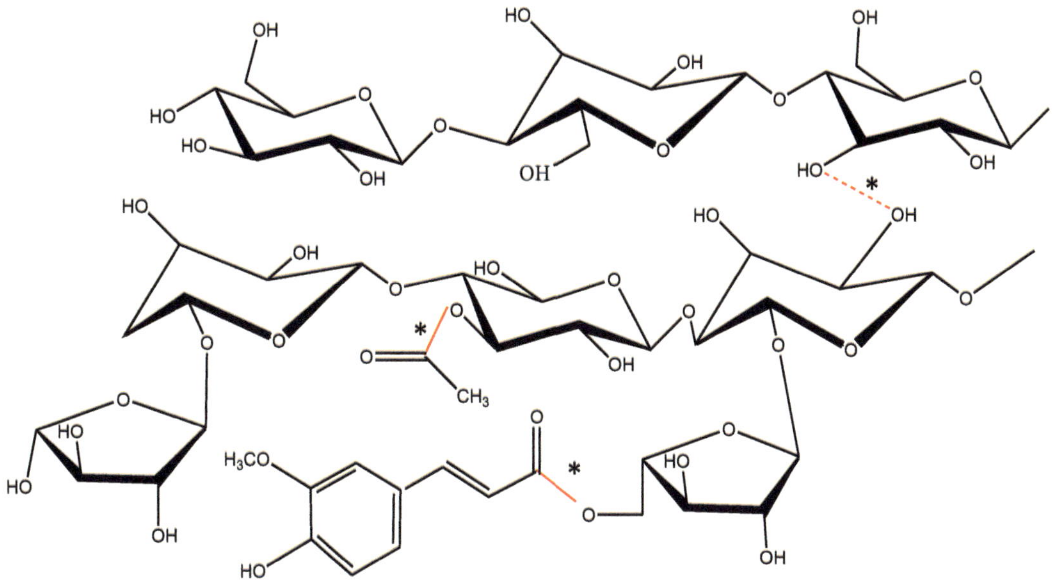

FIGURE 18.1 Chemical structure of lignocellulose present in plant biomass. *, indicates the susceptible bond for chemical treatment.

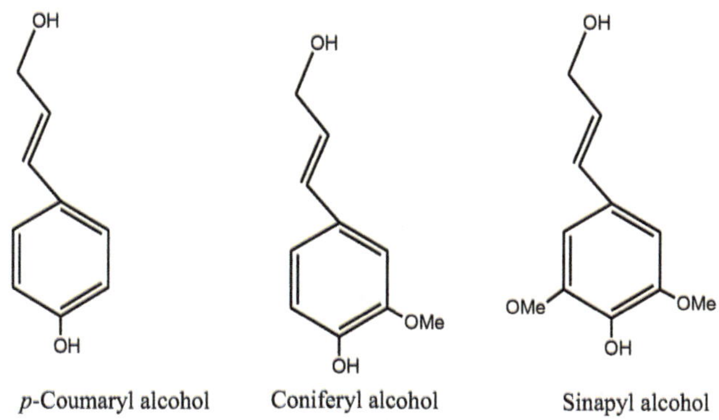

FIGURE 18.2 Structural units of lignin.

namely, extraction, chemical synthesis, and fermentation. In general, the presence of other compounds in protein extract, such as lignin and carbohydrates, is the main problem that may result in protein degradation or modification (Chaffin et al., 1998).

18.2.3 Lignins

Lignin is a polymer of propyl phenol units, namely, coniferyl alcohol and sinapyl alcohol, with a minor quantity of *p*-coumaryl alcohol (Fig. 18.2). The complex of these components is cross-linked together through carbon–carbon, ester, and ether linkages. Lignin is an abundant natural biopolymer making up to 10%–25% of plant biomass. As a renewable raw material, lignin has the potential to extend to any product currently sourced from petrochemical substances. Various applications of lignin include emulsifiers, dyes, binding agents, thermosets, paints, and fuels (Vanholme et al., 2010).

18.2.4 Lipids

Lipids exist in plants with many different physical and chemical properties. The simple lipids are mostly available in storage tissues, which are easy for extraction. In contrast, complex lipids are difficult to extract because of their membrane localization as well as their association with other biomolecules such as proteins and carbohydrates. These interactions are formed through covalent bonds, van der Waals interaction, hydrogen bonds, hydrophobic interaction, and ionic bond formation. To obtain a successful extraction method, it is necessary to develop an agent that not only solubilizes the lipids but also diminishes the interactions of other components with lipids.

18.3 COMPLICATIONS IN ORGANIC MATTER CONVERSION

18.3.1 Structural Complexity

This recalcitrancy of plant materials is considered as the major limitation for the efficient depolymerization of its structural components. Thus to harness the maximum benefit of plant biomass, an efficient deconstruction step is a prerequisite. Various chemical pretreatment strategies have been employed for the deconstruction of biomass for improved degradation; however, the hazardous nature of these chemicals has limited their wide adoption (Kumar et al., 2009). There is a general consensus for the need for greener and mild reaction protocols to avoid side reactions.

18.3.2 Crystallinity: Amorphous Modulation

Crystallinity has been considered as an important factor in determining the biological and chemical degradation of biomass. The amorphous nature is considered as more amenable in the hydrolysis process compared to crystalline biomass. Crystallinity of cellulose has been reduced by various physical and chemical treatments, which is necessary for improved enzymatic hydrolysis. However, the selective reduction of crystallinity with zero sugar losses has not yet been achieved.

18.3.3 Depolymerization

Plant cellulose consists of glucose monomers ranging from 800 to 10,000 units (Dieter et al., 2005). The increase in degree of polymerization increases the endoglucanase requirement and hydrolysis time. The reduction of degree of polymerization has been done with various physical, chemical, and biological pretreatments, but with not much success. Toward this end the potential depolymerization step to obtain desired products is required.

18.3.4 Economic Implications

To obtain biomolecules from plant and microbial biomass, there are numerous alternative methods available for a successful utilization process. Most of the techniques are not implemented because of their economic inviability. The heat, power, and chemical requirements for pretreatment should be low and it should be compatible with the thermally integrated process.

18.3.5 Water–Wastewater Prospects

The use of plant biomass treatment conditions, especially when employing chemical methods, results in a large amount of freshwater usage for washing organic and inorganic catalysts and degradatives. On the other hand, the release of degradative products such as sugars and phenolics along with wastewater leads to environmental pollution. The use of raw materials at high dry matter would reduce the quantity of wastewater generation and also reduce energy consumption during treatment.

18.3.6 By-Product Utilization

The conditions used for chemical extraction and hydrolysis result in the partial breakdown of plant materials to sugars, phenolics, acetic acid, and furan derivatives, which may cause harmful effects on downstream operation (Nakagame et al., 2010; Tengborg et al., 2001). Therefore to obtain a desired product it is necessary to improve the process that enables catalyst recovery.

18.4 NANOMATERIALS: A NEW CANDIDATE IN ORGANIC MATTER CONVERSION

Nanocatalysts are catalytically active materials comprising either particles smaller than 100nm or porous materials in which each pore has a diameter less than 100nm. In addition, nanocatalysts have unique features such as high reactivity, selectivity, lower process energy consumption, longer lifetime, and recyclability. The introduction of nanotechnology in plant biomass processing operation is a new option that potentially addresses challenges posed by chemical and biological treatments. Nanomaterials have been used to enhance the production of biochemicals such as carbohydrates, lignins, lipids, proteins, and pigments. Various nanomaterials such as metal oxides, nanophotocatalysts, electrospun nanomaterials, and mesoporous materials have been developed for plant biomass utilization (Lee and Juan, 2017). The developed catalysts are claimed to deliver a technique compatible with economic and environmental concerns. However, the appropriate processes have not been developed and require optimization to obtain a commercially viable technique.

18.5 CHARACTERISTICS OF NANOMATERIALS

The design of nanomaterials with optimal nanocatalytic properties including shape, size, as well as surface composition is vital to achieve efficient biomass utilization for biofuel and biochemical production (Gellman and Shukla, 2009; Lee and Juan, 2017).

18.5.1 Size

The reduction in catalyst size directly increases the reacting surface area. Thus exposure of more surface area facilitates the accessibility of substrates, which can increase catalytic activities.

18.5.2 Shape

The shape of particles (i.e., cubic, tetrahedral, and spherical) influences the catalytic activity and selectivity of nanocatalysts. For example, tetrahedral nanomaterial with more edges and corners can contain more surface-reacting sites than spherical or cubic nanomaterials.

18.5.3 Porosity

In general, porous material has additional active sites inside the porous system, which means that high porosity of nanomaterials directly increases the surface area. Therefore porosity with a highly accessible active site is crucial for substrate reactivity. Carbon-derived mesoporous nanomaterials are preferable for the biomolecule utilization process.

18.5.4 Surface Polarity

The adsorption or desorption process between nanoparticle surface-active sites and substrate determines the reactivity of nanomaterials with biomaterials. Normally, the surface polarity of nanomaterials consisting of biomaterials is characterized by their functional groups such as alkyl, carbonyl, carboxyl, ether, and hydroxyl with different ionic properties. Designing the surface of nanomaterials with high water tolerance and desirable polarity can enhance the reactivity.

18.5.5 Surface Composition

Reactivity of nanomaterials is determined by chemical composition of the surface. The chemical composition of the surface can be modified by a binary metal system, organic shell formation, or support interaction. The surface-activating molecule of the catalysts improves the activation energy of the overall reaction. Moreover, the catalysts with surface molecules have the potential to absorb the substrate. Therefore the functionalization of nanocatalysts by adding a capping agent (such as polymers or surfactants) is necessary.

18.6 FUNCTIONAL PROPERTIES OF NANOCATALYSTS FOR BIOMASS CONVERSION

Functional properties such as specificity, reactivity, durability, and recyclability of nanocatalysts determine their efficient use for commercialization. Developing a nanocatalyst with those requirements in biomass conversion enables the system to be more efficient (Lee and Juan, 2017).

18.6.1 Specificity

The nanomaterial should be specifically involved in the desired product conversion, whereas the other components of the nanocatalyst should be inactive to enable a successful conversion process. From this viewpoint, no unnecessary downstream process is needed. Moreover, the overall production costs can potentially be diminished.

18.6.2 Reactivity

The reactivity of nanomaterial is the parameter contributing to the conversion of available raw materials into products per unit time. Theoretically, the catalyst's reactivity can be determined by measuring the turnover frequency. The reactivity of a catalyst can also determine solid waste generation from raw materials.

18.6.3 Durability

In certain cases, the nature of the catalyst can be changed easily by changing the surface-active site. Therefore a highly stable catalyst enables recycling of the catalytic system without reducing its performance. The total number of recycles has been used as a measure of durability of catalysts.

18.6.4 Easy Recovery

The easy recovery of catalyst is necessary for developing an economically viable process. A large number of solid nanomaterials have been developed in biomass utilization. In general, nanoparticles can be separated by electrostatic repulsion forces by changing the pH (Seo et al., 2014). However, there are several difficulties in recovering and handling nanoparticles at large scale. Therefore the use of metal oxides and magnetic nanoparticles can facilitate the recovery of the whole catalytic system.

18.7 NANOPARTICLES: COMPONENTS DETERMINING THE FUNCTIONAL PROPERTIES

18.7.1 Acid Nanocatalyst

Conventional methods for biomass utilization use both homogeneous acidic and basic catalysts. Nevertheless, the application of acid catalysts can cause some undesirable reactions in biomolecule conversion leading to wastewater generation. Additionally, homogeneous catalysts have a corrosive property, which causes damage to the reactor. To solve these problems, heterogeneous catalysts have been developed and used for various applications. Heterogeneous catalysts have the potential to improve the process by enabling the separation of the catalyst and minimizing pollutant level. Various metal oxides in nanoscale have been developed for the organic matter; these include metal oxides, mixed metal oxides, acid-functionalized metal oxides, and metals supported on alumina or zeolite. Among them, acid-functionalized metal oxides showed

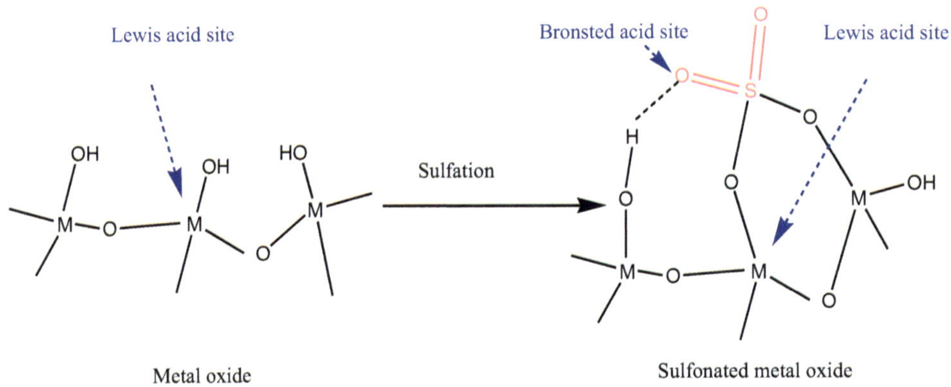

FIGURE 18.3 Sulfation of metal oxides and acid sites.

higher activity, which was ascribed to the formation of Bronsted acid and the availability of Lewis acid and Bronsted acid sites (Fig. 18.3). Heteropolyacids impregnated on silica, alumina, and activated carbon were used as solid acid catalysts for the production of biodiesel from canola oil (Laosiripojana et al., 2010). Boric acid-functionalized Fe_3O_4 was used for Mukaiyama aldol reaction, where the Fe_3O_4 was used for easy separation, while its boric acid portion was used for β-hydroxy carbonyl compounds, and the yields were in the range of 45%–98% at 80°C (Yazdani et al., 2015). Moreover, the reaction temperatures for biomass utilization were mild compared to the conventional acid-catalyzed reaction (Carrero et al., 2011). HZSM-5 (hydrogenated aluminosilicate zeolite) was used for the treatment of lipids extracted from *Nannochloropsis* sp. and the treated lipids had a drastic reduction in the content of heteroatoms such as nitrogen, oxygen, and sulfur (Li and Savage, 2013). Apart from the catalyst characteristics, the temperature used for the reaction has an influence on the yield of products. The treatment temperature at 400°C generates 75% of paraffinic oil suitable for liquid fuels, while the treatment temperature at 500°C generates 44% of hydrocarbon and 19% of hydrocarbon gas stream.

The sulfonated carbon nanocatalyst is considered a potential acid nanocatalyst containing the functional group $-SO_3H$ (Hara, 2010; Sani et al., 2013). The sulfonated carbon catalyst showed 1.15-fold higher conversion in esterification and threefold higher in transesterification than the Amberlyst-15 resin when oleic acid and triolein were used, respectively (Fu et al., 2013). The high catalytic activity of sulfonated carbon is caused by the synergistic action of $-SO_3H$ and active sites (Fu et al., 2013). Sulfonated TiO_2 showed higher catalytic activity with exposed facets for esterification of acetic acid and *n*-butanol (Zhao et al., 2014). This improved performance is caused by the acid strength of the TiO_2 particles, which was further enhanced with the loading of SO_4^{2-} groups on the surface of TiO_2. In addition, other catalysts such as zeolites and sulfonated SBA-15 (mesoporous silica) have been used in transesterification reactions (Velasquez-Orta et al., 2013).

18.7.2 Base Nanocatalyst

Metal base nanocatalysts including homogeneous base catalysts and heterogeneous base catalysts are developed for various chemical reactions such as transesterification and hydrogenation. The hydroxides of sodium and potassium and alkoxides are used as homogeneous base catalysts because of their higher catalytic efficiency and lower cost. However, drawbacks such as pollution, equipment corrosion, and product separation are highlighted in those methods. On the other hand, heterogeneous base catalysts have attracted considerable attention because of their favorable properties such as less corrosivity, easy recoverability, and recyclability (Lopez et al., 2005). The metal oxides CaMgO and CaMgO/Al_2O_3 showed highest methyl ester yield (85.3%) from *Nannochloropsis oculata* oil (Teo et al., 2014). Compared to CaMgO, the Al_2O_3-supported CaMgO catalyst has higher basic sites and is less sensitive to moisture content. On the other hand, the Mg–Zr catalyst was used to directly produce biodiesel from *Nannochloropsis* sp. (Li et al., 2011). Direct transesterification using Mg–Zr catalysts produced a methyl ester yield of 28%. However, most of the heterogeneous catalysts are expensive, which limits their wide use (Kafuku et al., 2010). A novel solid base nanocatalyst attached to the surfaces of Fe_3O_4/SiO_2 core–shell through triazabicyclodecene showed superior reactivity to those of commercial catalysts such as H_2SO_4, $AlCl_3$, Amberlyst-15, and $CuCl_2$ (Chiang et al., 2015) (Fig. 18.4). Potassium fluoride-impregnated clinoptilolite nanoparticles are a heterogeneous base catalyst, which catalyzed the one-pot three-component production of functionalized quinolines by condensation reactions between isatin, activated acetylenic compounds, and alcohols (Ghotbabadi et al., 2016).

18.7.3 Bifunctional Catalyst

In the conversion of biomass, bifunctional properties, such as acidic and basic characteristics, can be advantageous for a series of cascade reactions. Materials containing functionalized carbon, such as zirconia and titania, which are found to have high thermal stability, are indeed essential for applications such as biomass conversion. The use of acid- and base-functionalized mesoporous silica in an ionic liquid resulted in approximately 0.54 mol hydroxymethylfurfural per mol catalyst per hour at 393 K (Peng et al., 2012). The combination of amphoteric zirconia and surface sulfonation showed bifunctional catalytic activity for the production of hydroxymethylfurfural from glucose (Osatiashtiani et al., 2014). On the other hand, the heteropolyacid-based acid–base catalyst is involved in cascade isomerization/dehydration of glucose to hydroxymethylfurfural or lactic acid (Zhao et al., 2013). In addition, oxides and sulfides of Ni–Mo, Ni, Mo, and Pd have been investigated as bifunctional catalysts for the hydrocracking and hydrodeoxygenation of lignocellulosic biomass (Grilc et al., 2014). As a result of surface functionalization of sulfonated mesoporous silica (SBA-15) by trimethoxysilylpropyl thiopropane sulfonic acid, a hydroxymethylfurfural yield of 71% was observed because of the presence of a thiol group, which imparts significant hydrophobic properties (Shi et al., 2011). The bifunctional catalyst composed of protonated and vanadium-doped graphitic carbon nitride directly converted fructose to 2,5-diformylfuran via hydroxymethylfurfural in one step, which involves dehydration followed by subsequent oxidation of hydroxymethylfurfural (Chen et al., 2014). In biodiesel production, base catalysis is more suitable for transesterification of oils with low free fatty acids, whereas acid catalysis is suitable for esterification of free fatty acids. The bifunctional acid–base nanocatalysts are involved in the transesterification of lipids and the esterification of free fatty acids catalyzed by base and acid sites of the nanocatalyst, respectively (Fig. 18.5).

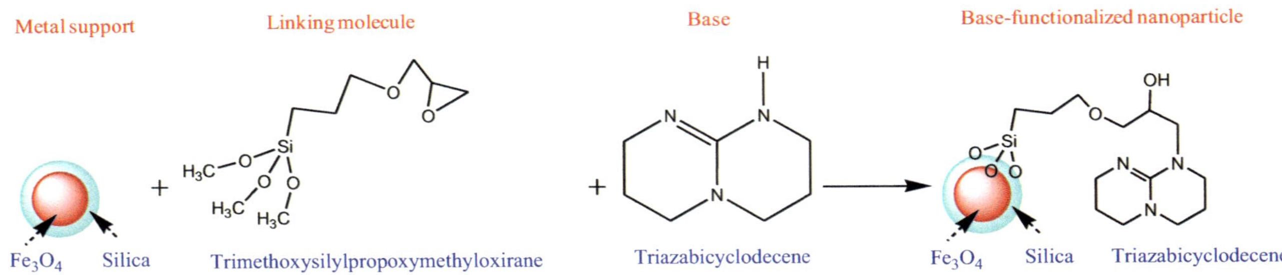

FIGURE 18.4 Base functionalization of a nanoparticle. *Modified from Chiang, Y.D., Dutta, S., Chen, C.T., Huang, Y.T., Lin, K.S., Wu, J., Suzuki, N., Yamauchi, Y., Wu, K.C.W., 2015. Functionalized* Fe_3O_4*@ silica core–shell nanoparticles as microalgae harvester and catalyst for biodiesel production. ChemSusChem 8, 789–794.*

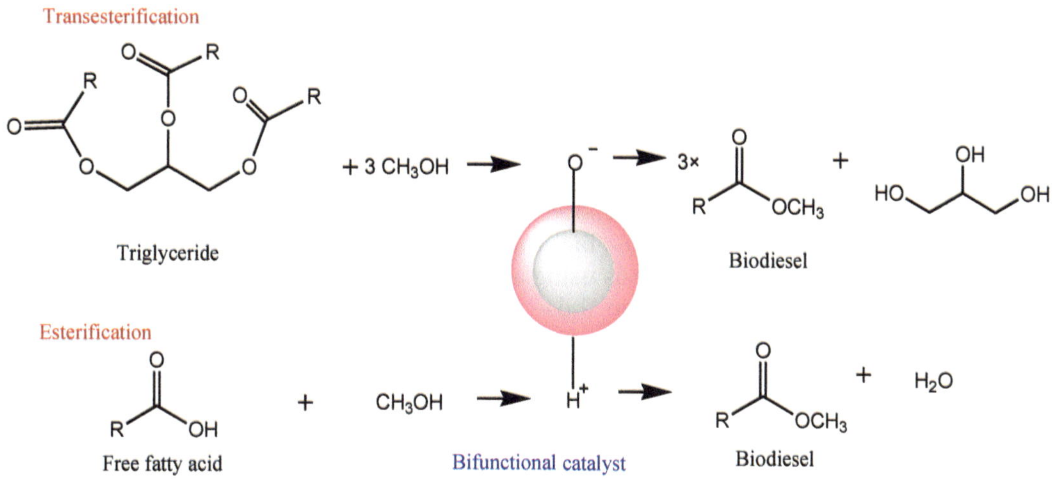

FIGURE 18.5 Metal nanoacid–base heterogeneous catalyst for the production of biodiesel. The internal basic sites (O^-) undergo transesterification and the external acid sites (H^+) undergo esterification. *Modified from Lima, A.L.D., Ronconib, C.M.C., Mota, J.A., 2016. Heterogeneous basic catalysts for biodiesel production. Catal. Sci. Technol. 6, 2877.*

18.7.4 Metal Oxides

Metal oxides are one of the most important and widely characterized solid catalysts. Metal oxides are considered as heterogeneous catalysts and are applied for acid–base and redox reactions. Certain groups of metals, particularly transition metals, have attracted much attention because of their outer electron configuration. They are applied widely in various reactions, which include oxidation, dehydration, dehydrogenation, and isomerization. The oxides of transition metals such as Nb_2O_5 (Tagusagawa et al., 2010), WO_3 (Tagusagawa et al., 2010), and TiO_2 (Kondo et al., 2005) are widely used as heterogeneous acid nanocatalysts. The mesopores of these transition metal oxides allow the substrate inside the metal pores for the catalyzed reaction. Mesoporous metal oxides are known for their special structural properties such as high surface area, variable pore size, and stability. The mesoporous CeO_2 was used for decomposition of methanol and showed an increase in transformation efficiency from 13% (nonmesoporous CeO_2) to 96% (Roggenbuck et al., 2007). The synergistic cooperation of Mn(III) and Mn(IV) along with a hematite phase has been shown to produce efficient oxidation of formylfuran carboxylic acid to furan dicarboxylic acid (Neatu et al., 2016).

18.7.5 Mixed Nanometal Oxides

Oxides consisting of two or more metal ions are generally called mixed metal oxides. Most of the mixed metal oxides are available in binary, ternary, or quaternary forms. The structural configuration of cations differs by the nature of neighboring cations and their coordination directs the type of interaction among the cations. The cation that comprises an active center can configure reactivity toward the biomolecule. There are various methods for the preparation of mixed metal oxides such as coprecipitation, the hydrothermal method, sol–gel, wet impregnation, mechanochemical synthesis, and microwave irradiation. Mixed metal oxides participate in chemical reactions because of their acid–base redox characteristics. Metal oxides are applied in various studies such as Ni/Mg–Al for reforming of ethanol, Mg–V–Al for hydroxylation of phenol, and various mixed metal oxides for deconstruction of biomass (Velmurugan and Incharoensakdi, 2016). Mesoporous Nb–W oxide was used for the Friedel–Crafts alkylation of anisole and the hydrolysis of carbohydrate. A specific concentration of Nb and W produces the specific mesoporous structure with acid properties. As expected, the turnover frequency of mesoporous Nb–W oxide was increased more than two times that of nonporous Nb–W oxide (Tagusagawa et al., 2010). However, the glucose and cellobiose yields of cellulose hydrolysis using a mesoporous Nb–W oxide-catalyzed reaction were about 8.5%, which is lower than those using conventional acids. The ternary mixed metal oxide (Zn–Ca–Fe) produced moderate catalytic activity in the hydrolysis of crystalline cellulose (Zhang et al., 2011). The highest cellulose hydrolysis of 42.6% and glucose selectivity of 69.2% were observed in a Zn–Ca–Fe oxide-catalyzed reaction. Compared to the fine particles, the nano-sized Zn–Ca–Fe oxide has more active sites and produces higher cellulose hydrolysis and glucose yield (Zhang et al., 2011). The application of $ZnO—ZrO_2$ for the production of hydroxymethylfurfural from cotton cellulose with the highest yield of 10% at 473 K has been reported (Yang et al., 2013). On the other hand, furfural conversion at approximately 100% condensation of furfural with acetone over Mg–Al was observed (Hora et al., 2014). The layered transition-mixed metal oxide $HNbMoO_6$ was proved for its hydrolytic potential for sucrose, cellobiose, starch, and cellulose. Comparatively, the layered $HNbMoO_6$ showed higher facile intercalation to saccharides and produced higher turnover frequency than Amberlyst-15. However, glucose yield from cellobiose was still only about 41%. The results proved that the hydrolysis is strongly influenced by the interaction of Lewis acid and Bronsted base sites (Takagaki, 2016).

18.7.6 Surface-Activated Metal Nanoparticles

Surface-activated nanoparticles can be prepared by linking organic molecules to inorganic materials. The nature of organic compounds, size, and functionality can be varied based on their applications. Linking of organic compound metal units produces a material with high surface area, which is generally higher compared to the commercial porous compounds such as zeolites and carbons. This attribute has made metal–organic frameworks ideal candidates for catalysis application. The increased surface area facilitates the selective inclusion of large molecules such as proteins and the utilization of the pores as reaction vessels. Moreover, the stability (thermal and chemical) of metal–organic complex frameworks facilitates further functionalization. These properties have led to their broad application as catalysts in several organic reactions. Metal oxide–organic compound frameworks have potential as heterogeneous catalysts, although their applications are still at the research level.

In most of the studies, the amino group-containing molecule was employed for providing the surface-modifying molecule. The high density of the surface functional group of NH_2, connecting to the other targeted biomolecules, can disperse well with high stability in aqueous solvents. A polyaminoamido dendron, an amino group-containing molecule, was used as a surface molecule for silica-coated magnetic nanoparticles, which improved the stability and solubility in solvent systems. Furthermore, when the dendronized particles were phosphonated before complexing with $[Rh(COD)Cl]_2$, the resulting product showed an improved hydroformylation reaction (Abu-Reziq et al., 2002). The *quasi*-homogeneous catalyst was prepared by coating the silane functionalized with different amine groups onto magnetite nanoparticles. This catalyst showed higher nitroaldol condensation reactivity compared to the heterogeneous base catalysts (Abu-Reziq and Alper, 2012). On the other hand, crown ether was used as a supporting material for magnetic nanoparticles and the whole catalyst was used for solid–liquid phase-transfer reactions. The magnetic nanoparticle-supported crown ethers showed similar activity to those of nonsupported crown ethers and can be recycled without loss of efficiency (Kawamura and Sato, 2007).

Ultrasmall palladium particles were used on magnetic nanoparticle supports to facilitate easy separation, in which immobilized auxiliaries are used as stabilizing ligands. The catalyst developed was active for Suzuki cross-coupling and Heck reactions and produced the highest yields of 83% and 56%, respectively (Yinghuai et al., 2007). In another study, nickel hydroxide was used as a Bronsted base for alcohol oxidation. The catalytic system efficiently oxidized primary and secondary alcohols (86%) to their carbonyls (Bhat and Bhat, 2015). The Morita–Baylis–Hillman reaction is a C—C bond formation strategy, which is achieved by a magnetic nanoparticle-supported quinuclidine. The supported catalyst demonstrated significant activity and recovery under an external magnetic field (Luo et al., 2007).

18.8 NANOPARTICLES ON ORGANIC MATTER

18.8.1 Protein Extraction

Proteins occupy a considerable part in plant materials and were extracted for a broad range of purposes, including enzyme isolation, to identify the genes encoding them, and to obtain proteins and amino acids for various foods and feed applications. Plant materials have a cellulose cell wall, which is further hardened by lignin meshwork. These components are degraded upon grinding and the release of such degraded products may affect the purity of protein. To extract the protein with high purity, special techniques are required to disrupt the cell walls and to protect the proteins from other contaminants. The extraction of protein usually involves solvent extraction. However, the chemicals used for protein extraction affect the protein structure and lead to a decrease of protein stability and unfolding of the protein. Nanomaterials have been used in various biological applications, which are distinguished by their unique properties. A gold nanoparticle functionalized with polymeric material was used as solid extraction sorbent for the isolation of proteins. Interestingly, the sorbent selectively captures proteins based on the difference in their pI values, and on the basis of the affinity of different proteins toward gold nanoparticles (Vergara-Barberan et al., 2016). On the other hand, cobalt oxide nanoparticles modified with cetyltrimethyl ammonium were used for protein extraction. The nanoparticles showed successful compatibility with a wide range of proteins such as insulin, chymotrypsinogen, and lysozyme (Shrivas and Wu, 2012). An iron oxide nanoparticle modified with oleate has been used for the separation of proteins from aqueous solution before matrix-assisted laser desorption ionization-mass spectrometry analysis. Interestingly, the nanoparticle extracts cytochrome *c* from the tryptic digestion products, which have a high content of salt and surfactant (Chang et al., 2007).

18.8.2 Fractionation of Carbohydrates

Nowadays, metal oxides have been used as a nanocatalyst for processing and extraction of carbohydrates (Crossley et al., 2010). As the metal oxides are nano-sized, they act more efficiently in the medium at the solid/liquid interface (Crossley et al., 2010). The successful use of metal oxides in obtaining carbohydrates can be a result of the destabilization of biomass by loosening the matrix, weakening intramolecular and intermolecular hydrogen bonds, causing a more open three-dimensional organization of structural molecules (Chen et al., 2011). Among the metal oxides, the oxides of aluminum, calcium, iron, magnesium, and zinc were used frequently in biomass destabilization because of their low production cost, easy recovery, and reactivity (Zi et al., 2015). Metal oxides in various forms have been used for a great variety of applications such as Al_2O_3 in catalytic hydrolysis of biomass (Zi et al., 2015), Zn–Ca–Fe for cellulose hydrolysis (Zhang et al., 2011), carbon-supported Ru for cellulose hydrolysis (Kobayashi et al., 2010), CaO in lignin isolation (Shi et al., 2012), Al_2O_3–MgO in cellulose extraction (Liu et al., 2014), and $SO_4^{2-}/TiO_2 - Fe_3O_4$ in the fast pyrolysis of biomass (Lu et al., 2014).

In this context, we quantified the effect of metal oxide on the recovery of glycogen from *Synechocystis* sp. PCC 6803 in our previous study (Velmurugan and Incharoensakdi, 2016). Metal oxides such as Al_2O_3, CaO, Fe_2O_3, MgO, and ZnO have the ability to liberate glycogen from *Synechocystis* sp. PCC 6803 biomass. The highest glycogen extraction was achieved with MgO followed by CaO, Fe_2O_3, Al_2O_3, and ZnO. It should be noted that the highest recovery values in mixed metal oxides were at least 1.3 times that of individual metal oxides. Improved degradation using mixed metal oxides is obtained because of the modification of the electronic properties between the metal oxides. In the case of binary metal oxides, CaO–MgO exhibited better activity on glycogen (79%) recovery. Compared to CaO–MgO and Al_2O_3–MgO, the ternary forms of those metal oxides showed less recovery in glycogen content. In general, metal oxide holds Bronsted and Lewis acids on surfaces and offers selectivity by varying the ratio of the molecules of metal oxides. Selectivity depends on the available acidic sites on the surfaces of metal oxides (Leliveld et al., 1999). In ternary forms, the accessibility of Bronsted and Lewis acids sites to the biomolecules might be reduced.

18.8.3 Production of Value-Added Products From Carbohydrates

Utilization of plant biomass, especially carbohydrate fraction, has gained much attention because of the demand in various applications. The carbohydrate fractions include cellulose (polymer of glucose), xylan (polymer of xylose), arabinan (polymer of arabinose), and oligosaccharides. To utilize the plant biomass, a specific method for dissolution of carbohydrate is necessary (Katahira et al., 2006). A number of processes for the conversion of polysaccharides into monosaccharides have been developed over the years. The vast majority of methods utilize either hydrolyzing enzymes or acids. In addition, there are other hydrolysis methods available, in which no enzymes or chemicals are applied. For instance, gamma-ray, electron-beam, ultrasound, and microwave irradiation may be used to hydrolyze lignocellulosic biomass. However, those processes are not well developed (Demirbas, 2004, 2005). Enzymatic hydrolysis is more expensive because of its cost associated with the isolation and purification of enzymes. Even though acid hydrolysis is less expensive, the disposal of wastewater generated is a major problem. Furthermore, the presence of noncarbohydrate fractions in lignocellulosic biomass complicates the hydrolysis of plant biomass. The hydrolysis of mechanically treated cellulose with zeolites was slow when compared with other catalysts such as sulfonated silica/carbon-like catalysts. Typically, less glucose yield was observed with zeolites, while up to 60% of glucose yield was achieved with the composite catalyst in aqueous medium at 423 K; however, the application of ionic liquids and microwave treatment improved the glucose yield with zeolites up to 37% (Van de Vyver et al., 2010). Hydrolysis of cellulose in water requires higher temperatures to break down the crystalline cellulose; however, glucose is not stable at higher temperatures. A promising strategy to avoid this problem, known as hydrolytic hydrogenation, is the combination of hydrolysis and hydrogenation, which simultaneously converts the released sugars to more stable sugar alcohols (Fukuoka and Dhepe, 2006). The weak magnetic solid catalyst $CaFe_2O_4$ was used to hydrolyze cellulose at 423 K and a maximum glucose yield of 49.8% was achieved (Zhang and Fang, 2012). Nanoparticles composed of metal oxides and organic molecules have been used as a new class of porous nanocatalysts for value-added products. Surface-activated magnetic solid acid (Fe_3O_4–porous silica–SO_3H) with a mesoporous structure showed efficient hydrolysis of about 96% from cellobiose compared with a 50% hydrolysis in amorphous cellulose (Lai et al., 2011). Hydrotalcite nanoparticles were activated with saturated aqueous $Ca(OH)_2$ for the hydrolysis of cellulose with a maximum hydrolysis yield of 47.4% at 423 K (Fang et al., 2011). The carbon-supported Ru (Ru/CMK-3) was used as a water-tolerant and reusable catalyst for the hydrolysis of cellulose. The synergistic action of CMK-3 and Ru converts the cellulose into glucose after 24 h (Kobayashi et al., 2010). The potential of heteropolyacid $H_3PW_{12}O_{40}$ was explored to catalyze the hydrolysis of cellulose under hydrothermal reaction conditions. The glucose yield of 50.5% was achieved at 453 K (Tian et al., 2010). In another study, hierarchical H-USY zeolite treated with oxalic acid showed an increased hydrolytic efficiency on hemicellulose (Zhou et al., 2013). Mesopores or macropores generated by oxalic acid increased total reducing sugar yield to 55.7%, which is caused by the improved accessibility of the acid sites and diffusion of particles.

18.8.4 Catalytic Pyrolysis of Lignin

In a nanomaterial system the reaction between lignin and nanoparticle generally proceeds in sequence starting from adsorption, activation of reactants, reaction, and finally desorption of products (Cullis, 1967). The adsorption of molecules is dependent on the characteristics of the adsorbing catalyst. According to Bielanski and Haber (1979)'s investigation, electron-rich surfaces of nanocatalysts donate electrons to the O_2 molecule to form radicals. The complete or partial oxidation of substrate always depends on the nature of oxygen species formed. In particular, the electrophilic species is mostly involved in complete oxidation, whereas the nucleophilic species is involved in partial oxidation. The electrophilic surface of the catalyst receives hydrogen from the lignin molecule and makes unstable active free radicals that initiate the complete oxidation of lignin (Corma et al., 2008). In some cases, electrophilic species may also attack the carbon–carbon bond and produce complete oxidation of lignin (Bielanski and Haber, 1991). Alternatively, partial oxidation involves abstraction of hydrogen and inclusion of nucleophilic oxygen into the lignin resulting in desirable products (Haber and Witko, 2003). Cu(II) 2,2-bipyridine has been used for the degradation of lignin from alkaline hydrogen peroxide-treated hybrid poplar resulting in the release of large amounts of alkali-soluble polymeric lignin (Li et al., 2015). The hydrogenolysis of a lignin over an Ru catalyst showed significant selectivity toward monomeric compounds in the presence of base. Notably, 100% production of monomeric aromatic compounds from lignin over Ni_7Au_3 catalyst was observed after adding NaOH as a promoter (Zhang et al., 2014). Metal oxides such as TiO_2 and ZnO showed higher lignin removal from corn stover in the presence of molecular oxygen and ammonia. The combined action of metal oxides and ammonia removed 70% of Klason lignin under optimum conditions (Yoo et al., 2013). In MoO_3-catalyzed hydrodeoxygenation of lignin compounds, it was observed that MoO_3 cleaved phenolic Ph—O—Me bonds over weaker aliphatic Ph—O—Me bonds (Prasomsri et al., 2014). The same phenomenon was observed in Mo_2C-catalyzed hydrodeoxygenation of anisole (Lee et al., 2014). The use of different supports and nitridation showed that the supports can modify the property of active sites and hence can increase selectivity and reactivity (Ghampson et al., 2012). Atensin et al. (2012) used a homogeneous transition metal triflate and supported Pd nanoparticles to catalyze the hydrogenolysis of ether and showed saturated alcohol production without any loss of aromaticity.

18.8.5 Extraction of Lipids

Lipids are generally defined as those components that are soluble in organic solvents but are insoluble in water. Traditional lipid extraction methods which are either modified Folch method or modified Bligh and Dyer method employ solvent extraction (Ranjith Kumar et al., 2015). Most of the conventional solvent extraction methods are known for their environmental issues and severe treatment conditions. Lipid extraction from plants is a difficult process because of the structural complexity of biomass. As one potential solution, nanoparticles have received particularly keen research interest in oil extraction.

In our previous study, metal oxides such as Al_2O_3, CaO, Fe_2O_3, MgO, and ZnO have been used to liberate the lipids from *Synechocystis* sp. PCC 6803 biomass. The metal oxides were used in various forms such as individual, binary, and ternary metal oxides for fractionation of lipids. The highest lipid extraction was achieved with MgO followed by CaO, Fe_2O_3, Al_2O_3, and ZnO. It should be noted that the highest recovery values in mixed metal oxides were at least 1.3 times that of individual metal oxides. In the case of binary metal oxides, CaO–MgO exhibited better activity on lipid recovery. Compared to CaO–MgO and Al_2O_3–MgO, the ternary forms of those metal oxides showed less recovery in lipid content (Velmurugan and Incharoensakdi, 2016). Nanoclays are considered as a potential catalyst for the extraction of lipids. The nanoclays were produced from Mg, Al, or Ca and were applied for lipid extraction from wet microalgae biomass; the conversion of the extracted lipid to biodiesel was nearly 100% (Lee et al., 2013a). The synthesis of novel aminoparticles with Fe^{3+}, Mn^{2+}, and Cu^{2+} was performed by sol–gel reaction with 3-aminopropyltriethoxysilane as a precursor. As a result, hydroxyl radicals were produced as in any Fenton-like reaction and caused cell damage leading to the effective release of lipids from wet microalgae (Lee et al., 2013b).

18.8.6 Transesterification of Lipids

Methyl esters are known as biodiesel, and are commonly produced by transesterification reaction. There are several developed nanomaterials used as heterogeneous catalysts for the transesterification reaction, which have been successfully used in the production of biodiesel (Table 18.1). CaO and MgO nanoparticles were applied for the transesterification of cooking oils and the result showed that the CaO nanoparticles had better reactivity than the MgO nanoparticles (Tahvildari et al., 2015). In another investigation, a mixture of FeO/CdO and FeO/SnO_2 nanoparticles was tested for its potential magnetic catalyst in transesterification. The nanocatalyst FeO/SnO_2 showed the highest yield of 84% in biodiesel production. Moreover, the catalyst showed significant activity toward esterification and transesterification of soybean oil (Alves et al., 2014). To improve the catalytic activity, an acid functional group was applied to magnetic particles for transesterification by heterogeneous nanocatalysts. The sulfonic- and sulfamic acid-functionalized crystalline Fe/Fe_3O_4 core–shell was synthesized and analyzed for catalytic activities in both acid-functionalized nanocatalysts. The sulfamic acid-coated nanocatalyst showed more activity compared to the sulfonic acid-coated nanocatalyst (Wang et al., 2015). Qiu et al. (2011) loaded ZrO_2 nanocatalyst with $C_4H_4O_6HK$ (potassium bitartrate) to improve transesterification and obtained a biodiesel yield of 98.03% from soybean oil. The nanocatalyst KF/CaO also showed 96.8% biodiesel yield from Chinese tallow seed oil (Wen et al., 2010). In another study, 95.2% biodiesel yield was obtained from pretreated Jatropha oil using Mg/Al oxides (Deng et al., 2011). On the other hand, the nanocatalyst Cs/Al/Fe_3O_4 was evaluated for catalytic efficacy of transesterification. The nanoparticle was prepared with various Cs/Al and Cs/Fe ratios to analyze the influence on catalytic performance and the result demonstrated molar ratios of 2.5:1 and 4:1, respectively, for the

TABLE 18.1 Nanomaterials Used for Effective Biodiesel Production

No.	Nanocatalyst	Feedstock	Yield (%)	References
1	CaO	Soybean oil	99	Reddy et al. (2006)
2	MgO	Soybean oil	99	Wang and Yang (2007)
3	MgO nanosheet	Sunflower and rapeseed oil	98	Verziu et al. (2008)
4	KF/Al_2O_3	Canola oil	97.7	Boz et al. (2009)
5	KF/CaO	Jatropha oil	96	Wen et al. (2010)
6	Hydrotalcite-derived particles with Mg/Al	Jatropha oil	95.2	Deng et al. (2011)
7	KF/CaO–Fe_3O_4	Stillingia oil	95	Hu et al. (2011)
8	Lithium-impregnated calcium oxide	Karanja oil and Jatropha oil	99	Kaur and Ali (2011)
9	ZrO_2 loaded with $C_4H_4O_6HK$	Soybean oil	98	Qiu et al. (2011)
10	MgO supported on titania	Soybean oil	95	Mguni et al. (2012)
11	ZnO nanorods	Olive oil	94.8	Molina (2013)
12	Hydrotalcite (Mg–Al)	Pungamia oil	90.8	Obadiah et al. (2012)
13	Cs/Al/Fe_3O_4	Sunflower oil	94.8	Feyzi et al. (2013)
14	Silicon-supported Ca/Fe_3O_4	Waste sunflower oil	97	Feyzi and Norouzi (2016)

highest biodiesel yield of 94.8% (Feyzi et al., 2013). In another approach, MgO nanocatalyst was prepared in nanosheet form and the highest biodiesel production of 98% from rapeseed and sunflower oil was obtained (Verziu et al., 2008). A silicon-supported Ca/Fe_3O_4 catalyst has also been prepared for biodiesel production. The developed nanocatalysts were effective in biodiesel production and resulted in 97% yield (Feyzi and Norouzi, 2016).

18.9 FURTHER PERSPECTIVES AND CONCLUSIONS

The utilization of biomass components such as cellulose, lignin, lipid, and protein is a preferred option for the synthesis of chemicals. However, the different chemical compositions of biomass versus reactivity of chemical catalyst pose challenges in the development of active and stable catalysts. The physical and thermal properties of nanoparticles are attractive features to utilize in various catalyzed reactions. As discussed in this chapter, the nanocatalyst has several productive applications including biochemicals and biofuels. In most of the nanocatalyst-mediated conversion processes, the reaction is carried out in hot aqueous conditions. To convert the biomass compounds efficiently, the functional group of catalysts should be designed to be compatible with the substrate. In addition, catalyst properties such as thermal stability, porosity, and reactivity should be analyzed before commercialization. Further research in developing the more stable functional group will speed up large-scale product conversion.

Acknowledgments

R.V. thanks the Graduate School and Faculty of Science, Chulalongkorn University (CU), for a postdoctoral fellowship from Rachadaphiseksomphot Endowment Fund. The research grants from CU on the Frontier Research Energy Cluster (CU-59-048-EN) and from Thailand Research Fund (IRG 5780008) to A.I. were also acknowledged.

References

Abu-Reziq, R., Alper, H., 2012. Magnetically separable base catalysts: heterogeneous catalysis vs. quasi-homogeneous catalysis. Appl. Sci. 2, 260–276.

Abu-Reziq, R., Avnir, D., Blum, J., 2002. Catalytic hydrogenolysis of aromatic ketones by a sol-gel entrapped combined Pd-[Rh(cod)Cl]$_2$ catalyst. J. Mol. Catal. A Chem. 187, 277–281.

Alves, M.B., Medeiros, F.C.M., Sousa, M.H., Rubim, J.C., Suarez, P.A.Z., 2014. Cadmium and tin magnetic nanocatalysts useful for biodiesel production. J. Braz. Chem. Soc. 25, 2304–2313.

Atensin, A.C., Ray, N.A., Stair, P.C., Marks, T.J., 2012. Etheric C-O bond hydrogenolysis using a tandem lanthanide triflate/supported palladium nanoparticle catalyst system. J. Am. Chem. Soc. 134, 14682–14685.

Bhat, P.B., Bhat, B.R., 2015. Nano Fe_3O_4@APTES@Ni$(OH)_2$ as a catalyst for alcohol oxidation. New J. Chem. 39, 273.

Bielanski, A., Haber, J., 1979. Oxygen in catalysis on transition metal oxides. Catal. Rev. 19, 1–41.

Bielanski, A., Haber, J., 1991. Oxygen in Catalysis. CRC Press, Boca Raton, FL, USA.

Blanch, H.W., 2012. Bioprocessing for biofuels. Curr. Opin. Biotechnol. 23, 390–395.

Boz, N., Degirmenbasi, N., Kalyon, D.M., 2009. Conversion of biomass to fuel: transesterification of vegetable oil to biodiesel using KF loaded nano-γ-Al_2O_3 as catalyst. Appl. Catal. B 89, 590–596.

Carrero, A., Vicente, G., Rodriguez, R., Linares, M., Del Peso, G., 2011. Hierarchical zeolites as catalysts for biodiesel production from *Nannochloropsis* microalga oil. Catal. Today 167, 148–153.

Chaffin, W.L., Lopez, J.L., Casanova, M., Gozalbo, M.C., Martinez, J.P., 1998. Cell wall and secreted proteins of *Candida albicans*: identification, function, and expression. Microbiol. Mol. Biol. Rev. 62, 130–180.

Chang, S.Y., Zhang, N.Y., Chen, C.S., Chen, C.D., Chen, Y.Y., Wang, C.R.C., 2007. Analysis of peptides and proteins affinity-bound to iron oxide nanoparticles by MALDI MS. 18, 910–918.

Chen, J., Guo, Y., Chen, J., Song, L., Chen, L., 2014. One-step approach to 2,5-diformylfuran from fructose by proton and vanadium containing graphitic carbon nitride. ChemCatChem 6, 3174–3181.

Chen, M.N., Zhang, D.Y., Thompson, L.T., Ma, Z.F., 2011. Catalytic properties of Ag promoted ZnO/Al_2O_3 catalysts for hydrogen production by steam reforming of ethanol. Int. J. Hydrog. Energy 36, 7516–7522.

Chiang, Y.D., Dutta, S., Chen, C.T., Huang, Y.T., Lin, K.S., Wu, J., Suzuki, N., Yamauchi, Y., Wu, K.C.W., 2015. Functionalized Fe_3O_4@ silica core–shell nanoparticles as microalgae harvester and catalyst for biodiesel production. ChemSusChem 8, 789–794.

Corma, A., Huber, G.W., Sauvanaud, L., OConnor, P., 2008. Biomass to chemicals: catalytic conversion of glycerol/water mixtures into acrolein, reaction network. J. Catal. 257, 163–171.

Crossley, S., Faria, J., Shen, M., Resasco, D.E., 2010. Solid nanoparticles that catalyze biofuel upgrade reactions at the water/oil interface. Science 327, 68–72.

Cullis, C.F., 1967. Heterogeneous catalytic oxidation of hydrocarbons. Ind. Eng. Chem. 59, 19–27.

Demirbas, A., 2004. Ethanol from cellulosic biomass resources. Int. J. Green Energy 1, 79–87.

Demirbas, A., 2005. Bioethanol from cellulosic material: a renewable motor fuel from ethanol. Energy Sources 27, 327–337.

Deng, X., Fang, Z., Liu, Y.H, and Yu, C.L., 2011. Production of biodiesel from Jatropha oil catalyzed by nanosized solid basic catalyst. Energy 36, 777–784.

Dieter, K., Brigitte, H., Hans-Peter, F., Andreas, B., 2005. Cellulose: fascinating biopolymer and sustainable raw material. Angew. Chem. Int. Ed. 44. http://dx.doi.org/10.1002/anie.200460587.

Fang, Z., Zhang, F., Zeng, H.Y., Guo, F., 2011. Production of glucose by hydrolysis of cellulose at 423 K in the presence of activated hydrotalcite nanoparticles. Bioresour. Technol. 102, 8017–8021.

Feyzi, M., Hassankhani, A., Rafiee, H.R., 2013. Preparation and characterization of $Cs/Al/Fe_3O_4$ nanocatalysts for biodiesel production. Energy Convers. Manag. 71, 62–68.

Feyzi, M., Norouzi, L., 2016. Preparation and kinetic study of magnetic $Ca/Fe_3O_4@SiO_2$ nanocatalysts for biodiesel production. Renew. Energy 94, 579–586.

Fu, X., Li, D., Chen, J., Zhang, Y., Huang, W., Zhu, Y., Yang, J., Zhang, C., 2013. A microalgae residue based carbon solid acid catalyst for biodiesel production. Bioresour. Technol. 146, 767–770.

Fukuoka, A., Dhepe, P.L., 2006. Catalytic conversion of cellulose into sugar alcohols. Angew. Chem. Int. Ed. 45, 5161–5163.

Gellman, A.J., Shukla, N., 2009. Nanocatalysis: more than speed. Nat. Mater. 8 (2), 87–88.

Ghampson, I.T., Sepulveda, C., Garcia, R., Radovic, L.R., Fierro, J.L.G., DeSisto, W.J., Escalona, N., 2012. Hydrodeoxygenation of guaiacol over carbon-supported molybdenum nitride catalysts: effects of nitriding methods and support properties. Appl. Catal. A 439–440, 111–124.

Ghotbabadi, H.S., Javanshir, S., Charati, F.R., 2016. Nano KF/clinoptilolite: an effective heterogeneous base nanocatalyst for synthesis of substituted quinolines in water. Catal. Lett. 146, 338–344.

Grilc, M., Likozar, B., Levec, J., 2014. Hydrodeoxygenation and hydrocracking of solvolysed lignocellulosic biomass by oxide, reduced and sulphide form of NiMo, Ni, Mo and Pd catalysts. Appl. Catal. B Environ. 150, 275–287.

Haber, J., Witko, M., 2003. Oxidation catalysis – electronic theory revisited. J. Catal. 216, 416–424.

Hall, D.O., 1980. Biomass for energy-a worldwide view. In: Campos-Lopez, E. (Ed.), Renewable Resources a Systematic Approach. Academic Press, USA, pp. 287–305.

Hara, M., 2010. Biomass conversion by a solid acid catalyst. Energy Environ. Sci. 3 (5), 601–607.

Hora, L., Kelbichova, V., Kikhtyanin, O., Bortnovskiy, O., Kubicka, D., 2014. Aldol condensation of furfural and acetone over Mg-Al layered double hydroxides and mixed oxides. 223, 138–147.

Hu, Sh., Guan, Y., Wang, Y., Han, H., 2011. Nano-magnetic catalyst $KF/CaO–Fe_3O_4$ for biodiesel production. Appl. Energy 88, 2685–2690.

Kafuku, G., Lee, K., Mbarawa, M., 2010. The use of sulfated tin oxide as solid superacid catalyst for heterogeneous transesterification of *Jatropha curcas* oil. Chem. Pap. http://dx.doi.org/10.2478/s11696-010-0063-1.

Katahira, S., Mizuike, A., Fukuda, H., Kondo, A., 2006. Ethanol fermentation from lignocellulosic hydrolysate by a recombinant xylose- and cellooligosaccharide-assimilating yeast strain. Appl. Microbiol. Biotechnol. 72, 1136–1143.

Kaur, M., Ali, A., 2011. Lithium ion impregnated calcium oxide as nano catalyst for the biodiesel production from karanja and jatropha oils. Renew. Energy 36, 2866–2871.

Kawamura, M., Sato, K., 2007. Magnetic nanoparticle supported crown ethers. Chem. Commun. 3404–3405.

Kobayashi, H., Komanoya, T., Hara, K., Fukuoka, A., 2010. Water-tolerant mesoporous-carbon-supported ruthenium catalysts for the hydrolysis of cellulose to glucose. ChemSusChem 3, 440–443.

Kondo, J.N., Yamashita, T., Nakajima, K., Lu, D., Hara, M., Domen, K., 2005. Preparation and crystallization characteristics of mesoporous TiO_2 and mixed oxides. J. Mater. Chem. 15, 2035–2040.

Kumar, P., Barrett, D.M., Delwiche, M.J., Stroeve, P., 2009. Methods for pretreatment of lignocellulosic biomass for efficient hydrolysis and biofuel production. Ind. Eng. Chem. Res. 48, 3713–3729.

Lai, D., Deng, L., Guo, Q., Fu, Y., 2011. Hydrolysis of biomass by magnetic solid acid. Energy Environ. Sci. 2011 (4), 3552.

Laosiripojana, N., Kiatkittipong, W., Sutthisripok, W., Assabumrungrat, S., 2010. Synthesis of methyl esters from relevant palm products in near-critical methanol with modified-zirconia catalysts. Bioresour. Technol. 101, 8416–8423.

Lee, H.V., Juan, J.C., 2017. Nanocatalysis for the conversion of non-edible biomass to biogasoline via deoxygenation reaction. In: Rai, M., da Silva, S.S. (Eds.), Nanotechnology for Bioenergy and Biofuel Production. http://dx.doi.org/10.1007/978-3-319-45459-7.

Lee, Y.C., Huh, Y.S., Farooq, W., Chung, J., Han, J.I., Shin, H.J., Jeong, S.H., Lee, J.S., Oh, Y.K., Park, J.Y., 2013a. Lipid extractions from docosahexaenoic acid (DHA)-richand oleaginous *Chlorella* sp. biomasses by organic-nanoclays. Bioresour. Technol. 137, 74–81.

Lee, Y.C., Huh, Y.S., Farooq, W., Han, J.I., Oh, Y.K., Park, J.Y., 2013b. Oil extraction by amino particle based H_2O_2 activation via wet microalgae harvesting. RSC Adv. 3, 12802–12809.

Lee, W.S., Wang, Z.S., Wu, R.J., Bhan, A., 2014. Selective vapor-phase hydrodeoxygenation of anisole to benzene on molybdenum carbide catalysts. J. Catal. 319, 44–53.

Leliveld, R.G., Ros, T.G., Dillon, A.J., Geus, J.W., 1999. Influence of Si/Al ratio on catalytic performance of (Co)Mo/sapponite catalysts. J. Catal. 185, 513–523.

Li, Y., Lian, S., Tong, D., Song, R., Yang, W., Fan, Y., Qing, R., Hu, C., 2011. One-step production of biodiesel from *Nannochloropsis* sp. on solid base Mg–Zr catalyst. Appl. Energy 88, 3313–3317.

Li, Z., Bansal, N., Azarpira, A., Bhalla, A., Chen, C.H., Ralph, J., Hegg, E.L., 2015. Chemical and structural changes associated with Cu-catalyzed alkaline-oxidative delignification of hybrid poplar. Biotechnol. Biofuels 8, 123.

Li, Z., Savage, P.E., 2013. Feedstocks for fuels and chemicals from algae: treatment of crude bio-oil over HZSM-5. Algal Res. 2, 154–163.

Lima, A.L.D., Ronconib, C.M.C., Mota, J.A., 2016. Heterogeneous basic catalysts for biodiesel production. Catal. Sci. Technol. 6, 2877.

Liu, L., Ju, M., Li, W., Jiang, Y., 2014. Cellulose extraction from *Zoysia japonica* pretreated by alumina-doped MgO in AMIMCl. Carbohydr. Polym. 113, 1–8.

Lopez, D.E., Goodwin, J.G., Bruce, D.A., Edgar Lotero, E., 2005. Transesterification of triacetin with methanol on solid acid and base catalysts. Appl. Catal. A 295, 97–105.

Lu, Q., Ye, X., Zhang, Z., Dong, C., Zhang, Y., 2014. Catalytic fast pyrolysis of cellulose and biomass to produce levoglucosenone using magnetic $SO_4{}^2/TiO_2 - Fe_3O_4$. Bioresour. Technol. 171, 10–15.

Luo, S., Zheng, X., Xu, H., Mi, X., Zhang, L., Cheng, J.P., 2007. Magnetic nanoparticle-supported Morita–Baylis–Hillman catalysts, Adv. Synth. Catal. 349, 2431–2434.

Lynd, L.R., Weimer, P.J., van Zyl, W.H., Pretorius, I.S., 2002. Microbial cellulose utilization: fundamental and biotechnology. Microbial. Mol. Biol. Rev. 66, 506–577.

Mguni, L.L., Meijboom, R., Jalama, K., 2012. Biodiesel production over nano-mgo supported on titania. World Acad. Sci. Eng. Technol. 64.

Molina, C.M.M., 2013. ZnO nanorods as catalyts for biodiesel product from olive oil (M.Sc. thesis). University of Louisville.

Nakagame, S., Chandra, R., Saddler, J., 2010. The effect of isolated lignins, obtained from a range of pretreated lignocellulosic substrates, on enzymatic hydrolysis. Biotechnol. Bioeng. 105, 871–879.

Neatu, F., Mari, R.S., Florea, M., Petrea, N., Pavel, O.D., Parvulescu, V.I., 2016. Selective oxidation of 5-hydroxymethyl furfural over non-precious metal heterogeneous catalysts. Appl. Catal. B Environ. 180, 751–757.

Obadiah, A., Kannan, R., Ravichandran, P., Ramasubbu, A., Vasanth Kumar, S., 2012. Nano hydrotalciteas a novel catalyst for biodiesel conversion. Dig. J. Nanomater. Biostruct. 7, 321–327.

Osatiashtiani, A., Lee, A.F., Brown, D.R., Melero, J.A., Moralese, G., Wilson, K., 2014. Bifunctional SO_4/ZrO_2 catalysts for 5-hydroxymethylfufural (5-HMF) production from glucose. Catal. Sci. Technol. 4, 333.

Peng, W.H., Lee, Y.Y., Wu, C., Wu, K.C.W., 2012. Acid–base bi-functionalized, large-pored mesoporous silica nanoparticles for cooperative catalysis of one-pot cellulose-to-HMF conversion. J. Mater. Chem. 22, 23181–23185.

Prasomsri, T., Shetty, M., Murugappan, K., Roman-Leshkov, Y., 2014. Insights into the catalytic activity and surface modification of MoO_3 during the hydrodeoxygenation of lignin-derived model compounds into aromatic hydrocarbons under low hydrogen pressures. Energy Environ. Sci. 7, 2660–2669.

Qiu, F., Li, Y., Yang, D., Li, X., Sun, P., 2011. Heterogeneous solid base nanocatalyst: preparation, characterization and application in biodiesel production. Bioresour. Technol. 102, 4150–4156.

Ranjith Kumar, R., Rao, P.H., Arumugam, M., 2015. Lipid extraction methods from microalgae: a comprehensive review. Front. Energy Res. 2, 61.

Reddy, C.R.V., Oshel, R., Verkade, J.G., 2006. Room-temperature conversion of soybean oil and poultry fat to biodiesel catalyzed by nanocrystalline calcium oxides. Energy Fuel 20, 1310–1314.

Roggenbuck, J., Schafer, H., Tsoncheva, T., Minchev, C., Hanss, J., Tiemann, M., 2007. Mesoporous CeO_2: synthesis by nanocasting, characterisation and catalytic properties. Microporous Mesoporous Mater. 101, 335–341.

Sani, Y.M., Daud, W.M.A.W., Aziz, A.A., 2013. Solid acid-catalyzed biodiesel production from microalgal oil—the dual advantage. J. Environ. Chem. Eng. 1, 113–121.

Seo, J.Y., Lee, K., Lee, S.Y., Jeon, S.G., Na, J.G., Oh, Y.K., Park, S.B., 2014. Effect of barium ferrite particle size on detachment efficiency in magnetophoretic harvesting of oleaginous *Chlorella* sp. Bioresour. Technol. 152, 562–566.

Shi, H., Fatehic, P., Xiao, H., Nia, Y., 2012. A process for isolating lignin of pre-hydrolysis liquor of kraft pulping process based on surfactant and calcium oxide treatments. Biochem. Eng. J. 68, 19–24.

Shi, X., Wu, Y., Yi, H., Rui, G., Li, P., Yang, M., Wang, G., 2011. Selective preparation of furfural from xylose over sulfonic acid functionalized mesoporous Sba-15 materials. Energies 4, 669–684.

Shrivas, K., Wu, H.F., 2012. Rapid and highly sensitive protein extraction via cobalt oxide nanoparticle based liquid–liquid micro-extraction coupled with MALDI mass spectrometry. Analyst 137, 890.

Tagusagawa, C., Takagaki, A., Iguchi, A., Takanabe, K., Kondo, J.N., Ebitani, K., 2010. Highly active mesoporous Nb-W oxide solid-acid catalyst. Angew. Chem. Int. Ed. 49, 1128–1132.

Tahvildari, K., Anaraki, Y.N., Fazaeli, R., Mirpanji, S., Delrish, E., 2015. The study of CaO and Mgo heterogenic nano catalyst coupling on transesterification reaction efficacy in the production of biodiesel from recycled cooking oil. J. Environ. Health Sci. Eng. 13, 73.

Takagaki, A., 2016. Kinetic analysis of aqueous phase cyclodehydration of 1,4-butanediol and erythritol over a layered niobium molybdate solid acid. Catal. Sci. Technol. 6, 791–799.

Tengborg, C., Galbe, M., Zacchi, G., 2001. Reduced inhibition of enzymatic hydrolysis of steam pretreated softwood. Enzyme Microb. Technol. 28 (9–10), 835–844.

Teo, S.H., Taufiq-Yap, Y., Ng, F., 2014. Alumina supported/unsupported mixed oxides of Ca and Mg as heterogeneous catalysts for transesterification of *Nannochloropsis* sp. microalga's oil. Energy Convers. Manag. 88, 1193–1199.

Tian, J., Wang, J., Zhao, S., Jiang, C., Zhang, X., Wang, X., 2010. Hydrolysis of cellulose by the heteropoly acid $H_3PW_{12}O_{40}$. Cellulose 17, 587–594.

Van de Vyver, S., Peng, L., Geboers, J., Schepers, H., de Clippel, F., Gommes, C.J., 2010. Sulfonated silica/carbon nanocomposites as novel catalysts for hydrolysis of cellulose to glucose. Green Chem. 12, 1560–1563.

Vanholme, R., Demedts, B., Morreel, K., Ralph, J., Boerjan, W., 2010. Lignin biosynthesis and structure. Plant Physiol. 153, 895–905.

Velasquez-Orta, S., Lee, J., Harvey, A., 2013. Evaluation of FAME production from wet marine and freshwater microalgae by in situ transesterification. Biochem. Eng. J. 76, 83–89.

Velmurugan, R., Incharoensakdi, A., 2016. Potential of metal oxides in fractionation of *Synechocystis* sp. PCC 6803 biomass for biofuel production. Algal Res. 19, 96–103.

Vergara-Barberan, M., Lerma-Garcia, M.J., Simo-Alfonso, E.F., Herrero-Martinez, J.M., 2016. Solid-phase extraction based on ground methacrylate monolith modified with gold nanoparticles for isolation of proteins. Anal. Chim. Acta 917, 37–43.

Verziu, M., Cojocaru, B., Hu, J., Richards, R., Ciuculesc, C., Filip, P., Parvulescu, V.I., 2008. Sunflower and rapeseed oil transesterification to biodiesel over different nanocrystalline MgO catalysts. Green Chem. 10, 373–381.

Wang, H., Covarrubias, J., Prock, H., Wu, X., Wang, D., Bossmann, S.H., 2015. Acid-functionalized magnetic nanoparticle as heterogeneous catalysts for biodiesel synthesis. J. Phys. Chem. C 119, 26020–26028.

Wang, L., Yang, J., 2007. Transesterification of soybean oil with nano-MgO or not in supercritical and subcritical methanol. Fuel 86, 328–333.

Wen, L., Wang, Y., Lu, D., Hu, S., Han, H., 2010. Preparation of KF/CaO nanocatalyst and its application in biodiesel production from Chinese tallow seed oil. Fuel 89, 2267–2271.

Yang, F., Li, G., Gao, P., Lv, X.-N., Sun, X., Liu, Z.-H., Fan, H., 2013. Mild hydrothermal degradation of cotton cellulose by using a mixed-metal-oxide ZnO–ZrO_2 catalyst. Energy Technol. 1, 581–586.

Yazdani, E., Azizi, K., Nakisa, A., Heydari, A., 2015. Boric acid-functionalized Fe_3O_4@SiO_2 as a novel superparamagnetically recoverable nano catalyst for mukaiyama-aldol reaction. Org. Chem. Res. 1, 27–36.

Yinghuai, Z., Peng, S.C., Emi, A., Zhenshun, S., Monalisa, Kemp, R.A., 2007. Supported ultra small palladium on magnetic nanoparticles used as catalysts for Suzuki cross-coupling and Heck reactions. Adv. Synth. Catal. 349, 1917–1922.

Yoo, C.G., Wang, C., Yu, C., Kim, T.H., 2013. Enhancement of enzymatic hydrolysis and Klason lignin removal of corn stover using photocatalyst-assisted ammonia pretreatment. Appl. Biochem. Biotechnol. 169, 1648–1658.

Zhang, F., Deng, X., Fang, Z., Zeng, H., Tian, X., Kozinski, J.A., 2011. Hydrolysis of microcrystalline cellulose over Zn-Ca-Fe oxide catalyst. Petrochem. Technol. 40, 43–48.

Zhang, F., Fang, Z., 2012. Hydrolysis of cellulose to glucose at the low temperature of 423 K with $CaFe_2O_4$-based solid catalyst. Bioresour. Technol. 124, 440–445.

Zhang, J., Asakura, H., van Rijn, J., Yang, J., Duchesne, P., Zhang, B., Chen, X., Zhang, P., Saeys, M., Yan, N., 2014. Highly efficient, NiAu-catalyzed hydrogenolysis of lignin into phenolic chemicals. Green Chem. 16, 2432–2437.

Zhao, H., Jiang, P., Dong, Y., Huang, M., Liu, B., 2014. A high-surface-area mesoporous sulfated nano-titania solid superacid catalyst with exposed (101) facets for esterification: facile preparation and catalytic performance. New J. Chem. 38, 4541–4548.

Zhao, Q., Wang, H., Zheng, H., Sun, Z., Shi, W., Wang, S., Wang, X., Jiang, Z., 2013. Acid–base bifunctional HPA nanocatalysts promoting heterogeneous transesterification and esterification reactions. Catal. Sci. Technol. 3, 2204–2209.

Zhou, L., Shi, M., Cai, Q., Wu, L., Hu, X., Yang, X., Chen, C., Xu, J., 2013. Hydrolysis of hemicelluloses catalyzed by hierarchical H-USY zeolites – the role of acidity and pore structure. Microporous Mesoporous Mater. 169, 54–59.

Zi, G., Yan, Z., Wang, Y., Chen, Y., Guo, Y., Yuan, F., Gao, W., Wang, Y., Wang, J., 2015. Catalytic hydrothermal conversion of carboxymethyl cellulose to value-added chemicals over metal–organic framework MIL-53(Al). Carbohydr. Polym. 115, 146–151.

CHAPTER

19

Ecological Risks of Nanoparticles: Effect on Soil Microorganisms

Praveen Kumar, Uday Burman, Ramesh K. Kaul

ICAR-Central Arid Zone Research Institute, Jodhpur, India

19.1 INTRODUCTION

Nanoparticles (NPs) are the building blocks of the new emerging field of nanotechnology that in broadest terms signifies the understanding and controlling of properties of matter at dimensions of roughly 1–100 nm. Two principal factors cause the properties of nanomaterials to differ significantly from other materials: increased relative surface area and quantum effects. As a particle decreases in size, the surface area increases and a greater proportion of atoms are found at the surface compared to those inside. As growth and catalytic chemical reactions occur at surfaces, nanoparticulate form is much more reactive than the same mass of material made up of larger particles (Nel et al., 2006; Ray et al., 2009; Ivask et al., 2014). NPs of different sizes and size distribution patterns have also been shown to follow different kinetics (Lankveld et al., 2010). In tandem with surface area effects, quantum effects can begin to dominate the properties of matter as size is reduced to the nanoscale. These can affect the optical, electrical, and magnetic behavior of materials. As a consequence, interaction of nanomaterials with microorganisms could be different from their reaction with bulk particles.

World nanomaterial production is estimated to be 11 million tons with a market value of €20 billion and the worth of products underpinned by nanotechnology to be about €2 trillion (European Commission, 2012). This industry is traditionally dominated by metal, metal oxides, carbon black, and synthetic amorphous silica. However, in the past years, many new nanomaterial-related applications have been developed such as UV filters in sun creams, anti-odor textiles, and lithium-ion batteries. Major technological breakthroughs can be achieved by using nanomaterials, therefore nanomaterials have been identified as a key enabling technology (European Commission, 2012).

Ecological concerns associated with NPs have been growing over time as ultimately soil is likely to be a major receptor of NPs (Gottschalk et al., 2009; Keller et al., 2013; Sun et al., 2014).

http://dx.doi.org/10.1016/B978-0-12-811487-2.00019-0

NPs can also enter the soil through industrial spills, landfill sites, or use of sewage sludge. Nanoscale-zerovalent iron is intentionally added to soil for decontamination (Simonin and Richaume, 2015). It is important that potential adverse effects are detected early, as soil microorganisms provide essential ecosystem services, including nutrient cycling, crop production, and climate regulation. Many NPs, specifically those of copper, silver, and zinc, have antimicrobial properties (Neal, 2008; Fabrega et al. 2009a,b; Dinesh et al., 2012). Thus these particles may be beneficial for medical applications, but their introduction into the natural environment could put beneficial soil microbial communities at risk (Science for Environment Policy, 2016).

Microorganisms constitute only <1% of the total mass of soil (de Neergaard, 2009), yet they play vital roles in supporting diverse ecosystem services. The metabolic diversity of microorganisms contributes to biochemical cycling of nutrients and maintains the structure and functions of the soil ecosystem. Soil bacteria, archaea, fungi, and actinomycetes constitute the basis of soil food webs, thereby determining the diversity of higher trophic levels (Aislabie and Deslippe, 2013). Soil microorganisms play an important role in organic matter decomposition, weathering of minerals, and ultimately soil formation. Microorganisms mineralize nutrients stored in organic matter and make them available for primary producers (Tian et al., 1992; Tian, 2010). Microbial decomposition also creates humus, which retains soil moisture and enhances the soil's cation exchange capacity. Moreover, the products of microbial metabolism are known to enhance the hydrophobicity and wettability of soils (Aislabie and Deslippe, 2013). Soil aggregation is primarily caused by polysaccharides and glycoproteins secreted by microorganisms (Haynes and Francis, 1993; Wright and Upadhyaya, 1998; Six et al., 2000; Cates et al., 2016). Microbial decomposition of biomass is therefore a critical soil process that supports plant growth, and in turn supports all the higher forms of life. By decomposing the organic matter and effecting its subsequent transformation, microorganisms determine the carbon storage capacity of soils (Conrad, 1996). The principles of microbial decomposition have found applications in bioremediation of polluted soil environments by harnessing the ability of microorganisms to transform or detoxify chemical substances into environmentally safe forms (Aislabie and Deslippe, 2013). Addition of nutrients and regulation of redox conditions can enhance natural detoxification-stimulating activity of indigenous microorganisms (Bollag et al., 1994). Cyanobacteria are the main contributors of oxygen production by bacteria on earth today (Madigan et al., 2010). Besides this vital function, microorganisms also support the different life forms by assisting and promoting the growth of primary producers by their activities in soil. Several soil bacteria are capable of fixing atmospheric nitrogen into ammoniacal form that is usable by plants. These bacteria, some of which live freely in the soil, while others live in association with plant roots, contribute to 60% of the nitrogen fixed on earth, thereby contributing significantly to the nitrogen economy (Jetten, 2008). Similarly, bacteria and fungi can also mobilize several immobile nutrients, which can be utilized for plant growth. Furthermore, soil microbes produce a variety of phytohormones, including auxins, gibberellins, and also antibiotics, which promote plant growth and protect plants from soilborne diseases (Asghar et al., 2002; Zahir et al., 2008). Several microbial metabolic products such as antibiotics, enzymes, organic acids, alcohols, and pigments are industrially important catalysts and find utility in biotechnology and green chemistry applications (Singh et al., 2017).

19.2 EFFECT OF NANOPARTICLES ON MICROORGANISMS

It has been reported that Ag NPs caused bactericidal effects and morphological alterations in cells in two soil bacteria, *Bacillus cereus* and *Pseudomonas stutzeri*, but no such effect was observed with Al_2O_3 (Fajardo et al., 2014). Ag NPs have also been reported to reduce enzyme activity, while bacterial growth and biomass were decreased by Cu- and Zn-based NPs (Frenk et al., 2013; Peyrot et al., 2014, Shen et al., 2015). Furthermore, Simonin and Richaume (2015) concluded that metal NPs are more toxic than organic (carbon-based) NPs. This was supported by Colman et al. (2013) who reported differences in the bacterial community structure of sewage sludge in the presence of Ag NPs. However, much higher concentrations of carbon NPs than generally predicted to be found in the environment are required to have negative effects (Rodrigues et al., 2013). Like the type of NP, soil properties can also play a determinant role in the toxic effects of NPs. Soil pH and soil organic material can influence the release of toxic ions from nanomaterials and thus their toxic effects on microorganisms (Klaine et al., 2008; Reidy et al., 2013). Toxicity of ZnO NPs on *Folsomia candida* was increased in acidic soil (Waalewijn-Kool et al., 2013). Some evidence suggests that the effects of nanomaterials, especially those that release ions, can be altered by organic matter (Klitzke et al., 2015; Shoults-Wilson et al., 2011; Benoit et al., 2013); however, in other studies, organic matter appeared to have little impact on the effects of Ag nanomaterial (Schlich and Hund-Rinke, 2015; Grillo et al., 2015). As a whole, in soils having low pH, organic matter content and cation exchange capacity obstruct sorption of Ag to soil colloids and as a result its mobility, toxicity, bioavailability, and consequently risk of its toxicity are enhanced, but in soil of higher pH organic matter, content and cation exchange capacity facilitate Ag sorption, thus restricting toxicity (Benoit et al., 2013; Klitzke et al., 2015; Schlich and Hund-Rinke, 2014). The toxic effects of different NPs on specific microorganisms are given in Table 19.1.

NPs gain entry into microbial cells mainly through endocytosis[1] and spread from cell to cell. In this way, they latch on to a preexisting transport mechanism of the cell (Elsaesser and Howard, 2012). Mechanisms at the nano–bio interface elicit the toxic effects of NPs, which may disrupt the growth and functioning of microorganisms. These can be classified into physical and chemical mechanisms, depending on the type of interaction manifested by the NP (Fig. 19.1).

[1] **Endocytosis:** The term endocytosis (*endo* + *cytosis*) was coined by the Nobel Prize-winning biochemist Christian deDuve in 1963 to denote a process in which cells transport molecules into the cells, wherein the molecules are engulfed by a portion of the plasma membrane and are budded off as vesicles. The process of endocytosis includes internalization of particulate matter (phagocytosis) and uptake of fluids (pinocytosis).

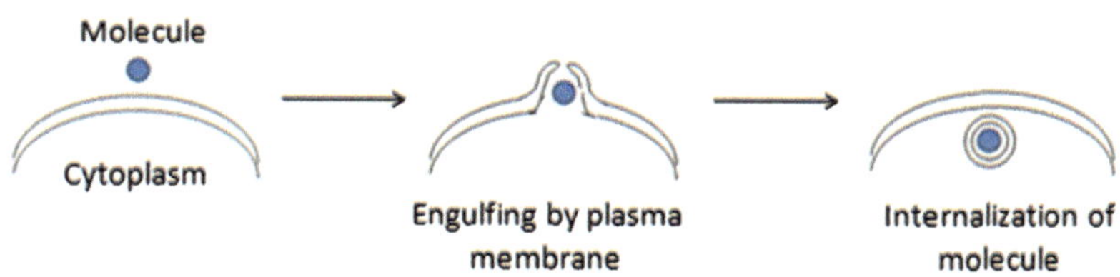

TABLE 19.1 Toxicity Mechanisms of Common Nanoparticles

Nanoparticle	Mechanism	Microorganisms Inhibited	References
METALLIC NANOPARTICLES			
Ag^+	• Adsorption to cells and intracellular invasion because of size and shape characteristics	*Escherichia coli* *Staphylococcus aureus*	Feng et al. (2000) Kim et al. (2011)
	• Reaction of free Ag^+ with thiol groups of proteins and inhibition of enzymes	*E. coli* *S. aureus*	Feng et al. (2000) Kim et al. (2011)
	• Binding to nucleic acids and impairment of DNA replication	*Pseudomonas putida* *Salmonella typhimurium* *Vibrio cholera*	Morones et al. (2005)
	• Generation of free radicals and associated cell membrane damage	*E. coli* *S. aureus*	Feng et al. (2000) Kim et al. (2011)
	• Dissipation of membrane potential	*Candida albicans* *Saccharomyces cerevisiae*	Nasrollahi et al. (2011)
	• Apoptosis or programmed cell death	*E. coli*	Bao et al. (2015)
CeO_2	• Cell membrane damage and disruption of cell wall	*E. coli* *S. typhimurium* *Enterococcus faecalis* *Bacillus subtilis*	Thill et al. (2006) Krishnamoorthy et al. (2014)
	• Growth inhibition in response to the size of particles	*E. coli* *B. subtilis*	Pelletier et al. (2010)
Cu^{2+}/CuO	• Action of metal ions at the cell surface leading to membrane damage	*E. coli* *Brevibacillus laterosporus* *Chryseobacterium indoltheticum* *Pantoea ananatis*	Karlsson et al. (2013) Palza (2015) Concha-Guerrero et al. (2014)
	• Reaction with amino acids and protein interference	*E. coli*	Karlsson et al. (2013) Palza (2015)
	• Dissipation of cell membrane potential leading to filamentous cell growth and loss of viability	*E. coli*	Chatterjee et al. (2014)
	• DNA degradation	*E. coli* *Lactobacillus brevis*	Chatterjee et al. (2014) Deryabin et al. (2013) Kaweeteerawat et al. (2015)
	• ROS generation and oxide reduction reactions	*B. laterosporus* *C. indoltheticum* *P. ananatis* *Vibrio fischeri*	Concha-Guerrero et al. (2014) Heinlaan et al. (2008)

TABLE 19.1 Toxicity Mechanisms of Common Nanoparticles—cont'd

Nanoparticle	Mechanism	Microorganisms Inhibited	References
TiO_2	• Photocatalytic production of reactive oxygen species (ROS) leading to oxidative stress, cell membrane disruption, and DNA strand breaks	*E. coli* *B. subtilis* *S. aureus* *Pseudomonas aeruginosa* *V. fischeri* *Bacteroides fragilis* *B. subtilis* *Enterococcus hirae* *S. typhimurium*	Barnes et al. (2013) Tsuang et al. (2008) Heinlaan et al. (2008) Adams et al. (2006) Kim et al. (2013) Gogniat and Dukan (2007)
	• Lipid modification and disruption of lipopolysaccharide wall layer	*E. coli*	Sunada et al. (1998) Maness et al. (1999)
	• Structural deformation of proteins	*Rhodopseudomonas acidophila*	Chen et al. (2005)
Zn^{2+}	• Shape-dependent cellular uptake and toxicity	Various bacteria	Sirelkhatim et al. (2015)
	• ROS generation and oxidative toxicity	*E. coli* *V. fischeri* *B. subtilis* *Campylobacter jejuni* *S. aureus*	Adams et al. (2006) Heinlaan et al. (2008) Arakha et al. (2015) Xie et al. (2011) Premanathan et al. (2011)
	• Cell membrane disorganization and increased membrane permeability	*Streptococcus agalactiae* *S. aureus* *E. coli* *C. jejuni*	Huang et al. (2008) Reddy et al. (2007) Brayner et al. (2006) Xie et al. (2011)
	• Membrane lipid modification	*E. coli* *Klebsiella pneumoniae*	Jiang et al. (2010) Reddy et al. (2014)
	• Protein interference and inhibition of protein function	*Bacillus licheniformis* *V. cholera*	Sinha and Khare (2014) Chatterjee et al. (2010)
	• Apoptosis	*S. aureus*	Premanathan et al. (2011)
NONMETALLIC NANOPARTICLES			
Carbon-based nanomaterials (C_{60}/fullerene/carbon nanotubes)	• Physical characteristics such as size and composition	*E. coli* *Pseudomonas fluorescens*	Kang et al. (2008) Riding et al. (2012)
	• Mechanical stress on the cell wall or membrane and cytotoxicity	*E. coli* *Shewanella oneidensis*	Fortner et al. (2005) Tang et al. (2007) De Windt et al. (2006)
	• Oxidative stress by ROS production, membrane lipid modification, and DNA damage	*Pseudomonas putida* *S. typhimurium* *B. subtilis* *P. aeruginosa*	Fang et al. (2007) Sera et al. (1996) Olivi et al. (2013)
Silica/SiO_2	• Oxidative stress and associated membrane disintegration	*E. coli* *B. subtilis*	Adams et al. (2006)
	• Interaction with bacterial lipopolysaccharides and destabilization of outer membrane	*E. coli*	Capeletti et al. (2014)

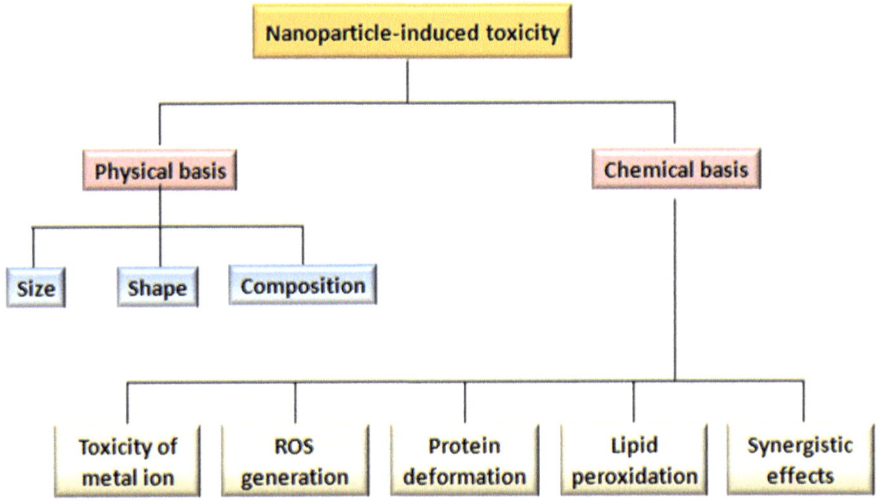

FIGURE 19.1 Mechanisms of nanoparticle-mediated toxicity to biological systems. *ROS*, reactive oxygen species.

19.3 PHYSICAL BASIS OF TOXICITY

Physical interactions at the NP and biological interface depend on characteristics such as size and surface properties of NPs and may lead to membrane disintegration, interference in transport processes, and protein dysfunction (Elsaesser and Howard, 2012; Tripathi et al., 2017). A common phenomenon observed in the NP–cell interaction is the adsorption of NPs onto the cell surface and the subsequent entrapment of cells by aggregates of NPs because NPs have a tendency to form larger agglomerates because of their rapid Brownian motion (Safe Work Australia Research Report, 2009; Wang et al., 2016b). The adhesion of NPs to cells is analogous to the spin supercurrents in physics, and is dependent on the size and form of NPs (Boldyreva, 2014). This leads to structural deformation and possible disruption of cell surface appendages, hardening of the cell, and physical injury (Zhang et al., 2012a). NP adsorption on the cell surface also leads indirectly to cell death by inhibiting nutrient uptake by cells and reducing the light availability for phototrophs (Wang et al., 2016b). Moreover, adsorption of cellular proteins onto NPs could induce changes in their conformation and also cause their aggregation (Billsten et al., 1997; Chen and von Mikecz, 2005). Thus the starting material and its properties greatly influence the fate and effects of NPs. Size and shape and their composition of the NP, which play a major role in NP toxicity, are discussed next.

19.3.1 Size of the Nanoparticle

The small size (1–100 nm) of NPs enables them to penetrate living tissues, thereby enhancing their bioavailability and toxicity. Besides internalization, subcellular distribution of particles is also size dependent (Shang et al., 2014). Several studies on the uptake of gold, silica, and iron oxide NPs have suggested an optimum size of 25–50nm for efficient cellular uptake (Chithrani et al., 2006; Lu et al., 2009; Huang et al., 2010; Zhang et al., 2012b). Ivask et al. (2014) studied the size-dependent cytotoxicity of Ag NPs to *Escherichia coli* cells and found that small Ag NPs of 10nm dimension were highly cytotoxic and were associated with intracellular Ag^+ release. Similar enhanced antimicrobial activities associated with decreased particle sizes have also been reported for TiO_2, ZnO, and CeO_2 NPs (Lin et al., 2014; Pelletier et al., 2010; Sirelkhatim et al., 2015). Once inside the cell, NPs can alter the catalytic activity of enzymes as well as the stability of protein structures (Fertsch-Gapp et al., 2011; Sajid et al., 2015; Vertegel et al., 2004). The destabilization of proteins caused by Si NPs of 6, 9, and 15nm has been attributed to the decreased curvature of such NPs, which leads to a greater unfolding of the protein (Lundqvist et al., 2004).

19.3.2 Shape of the Nanoparticle

Shape is another factor that influences uptake efficiency and toxicity of NPs. The design and geometries of NPs determine their ability to adhere to the microbial cells and effect their residence time within the cell (Albanese et al., 2012). The presence of sharp surface structures, as in the case of nanostar forms of NPs, would lead to disruption of cell membranes, irrespective of their surface compositions, charges, materials, and sizes (Chu et al., 2014). Furthermore, changes in shape and size of NPs also alter surface area and thus properties associated with surface area, which in turn affect cytotoxicity (Lee et al., 2014).

Shape also influences the cellular uptake of NPs. Uptake of rod- and wire-shaped particles is generally higher than sphere-, cylinder-, and cube-shaped particles as they penetrated bacterial cell walls more easily (Yang et al., 2009; Albanese et al., 2012). However, because of the length, rods and wires sometimes do not penetrate entry into the cells completely, leading to higher cell membrane damage (Stoehr et al., 2011). Shape-dependent interaction of Ag NPs with bacterial cells has been reported by Stoehr et al. (2011) and Sriram et al. (2012) with truncated triangular silver nanoplates displaying the strongest biocidal action against *E. coli* (Pal et al., 2007).

19.3.2.1 Composition of the Nanoparticle

The composition of nanomaterials alters their crystalline form and therefore, like size and shape, also affects their uptake. Furthermore, coupled with other associated changes, NPs of different composition generate different degrees of intracellular responses (Sohaebuddin et al., 2010). NP type affects the expression of cell surface receptors and thereby the internalization characteristics (Buzea et al., 2007). As stated by Albanese et al. (2012), carbon nanotubes possess 1000-fold greater endocytosis rate than Au NPs of similar diameter because of the inherent properties of the NPs. Surface charge is another criterion that dictates NP toxicity. For example, cationic NPs form complexes with the negatively charged DNA, plasma membrane, and molecules (Fröhlich, 2012). While polymeric NPs with positive surface charges are more cytotoxic, greater generation of reactive oxygen species (ROS) by negatively charged Si NPs also increases their cytotoxicity (Park et al., 2013; Shao et al., 2015). This might be caused by the greater stability of Si NPs with highly negative surface charge at physiological pH

(Kim et al., 2014). Exposure of *Bacillus* sp. to uncoated and coated Ag NPs also revealed that toxicity was highly dependent on surface charge (El Badawy et al., 2011).

Characterization of NPs is therefore necessary to attribute the toxic effects to a certain property of the NP or to the NP itself (Elsaesser and Howard, 2012). It further provides dynamic insights into the behavior of NPs within biological systems and facilitates better comprehension of their toxicity mechanisms.

19.4 BIOCHEMICAL MECHANISMS OF NANOPARTICLE-INDUCED TOXICITY

NP interaction with microorganisms may evoke perturbations in the normal metabolism and functioning of cells and cellular constituents, and prove to be toxic to organisms. The major means of NP-induced toxicity comprises a plethora of events initiated by the release of toxic metal ions, its interference in major cellular events, free radical generation, and subsequent cellular damage and cell death.

19.4.1 Release of Toxic Metal Ions and Subsequent Interaction

NP toxicity has mostly been associated with the formation of antimicrobial metal ions from the surface of NPs. Antibacterial effects of ZnO and Ag NPs is attributed to the release of Zn^{2+} and Ag^+ ions, while toxicity of CuO NPs is caused by the release of toxic Cu^{2+} ions and the toxic effect of CuO NPs themselves (Wang et al., 2016a; Feng et al., 2000; Fasim et al., 2002; Aziz et al., 2015, 2016). With respect to Ag NPs the formation of Ag^+ has been proposed as the main mechanism of toxicity (Feng et al., 2000). Similarly, polymeric nanocomposites, which use metal ions as active agents, also release free metal ions into the cellular environment leading to cytotoxicity (Palza, 2015).

The mechanisms of chemical toxicity by the released metal ions are varied. Ag^+ ions inhibit cellular respiratory enzyme(s) leading to the generation of ROS and cell damage (Hwang et al., 2008). Also, upon exposure to ionic silver, bacterial DNA conglomeration defense mechanisms are activated, severely compromising the bacterial replication ability (Seil and Webster, 2012). The biocidal effect of the ions released from CuO NPs has been attributed to the cytotoxicity induced by their interaction directly with the cellular membrane as well as to the formation of copper–peptide complexes (Palza, 2015). Although the toxicity mechanisms differ with respect to their metal characteristics, in general they have been grouped into five categories (Lemire et al., 2013).

19.4.2 Oxidative Stress and Antioxidant Depletion

Metal ions such as Cr, As, Hg, Cd, and even Fe and Cu at toxic levels are known to generate intracellular ROS, driven by the Fenton reaction (Birben et al., 2012). Aerobic respiration gives rise to partially reduced forms of molecular oxygen (O_2), such as hydrogen peroxide (H_2O_2) and superoxide ($O_2^{-\bullet}$). This H_2O_2 oxidizes the reduced metal ion (e.g., Fe^{2+}) to produce hydroxide (OH^-) and the highly reactive hydroxyl radical ($OH^\bullet$), and this is known as the Fenton reaction (Eq. 19.1). $O_2^{-\bullet}$ further reduces the oxidized metal released by the Fenton reaction (Eq. 19.2), leading to metal-catalyzed autoxidation, known as the metal-catalyzed Haber–Weiss reaction (Lemire et al., 2013):

$$H_2O_2 + Metal^{2+} \rightarrow Metal^{3+} + OH^- + OH^\bullet \quad (19.1)$$

$$O_2^{-\bullet} + Metal^{3+} \rightarrow Metal^{2+} + O_2 \quad (19.2)$$

Moreover, the $OH^\bullet$ generated either by the Fenton reaction or by the metal ions themselves can react with biomolecules to form a series of additional free radicals such as thiyl radical ($RS^\bullet$; Eq. 19.3) and carbon-centered radical [$(R)_3C^\bullet$; Eq. 19.4]:

$$\rightarrow OH^\bullet + RSH \rightarrow H_2O + RS^\bullet \quad (19.3)$$

$$\rightarrow OH^\bullet + (R)_3CH \rightarrow H_2O + (R)_3C^\bullet \quad (19.4)$$

These radicals may react with other thiol molecules to generate H_2O_2 and $O_2^{-\bullet}$, leading to additional ROS generation (Birben et al., 2012).

Changes in the normal redox state of cells lead to imbalances between the systemic manifestation of ROS and the ability of biological systems to readily detoxify the reactive intermediates (Bhattacharya, 2014). These perturbations are reflected as oxidative stress that leads to free radical-mediated damage to proteins, membranes, and DNA.

Metals also aggravate the toxic effects by depleting cellular antioxidants such as glutathione (GSH) by forming covalent bonds with sulfur, which causes oxidation of thiol (Fig. 19.2). Arsenic compounds are known to bind to sulfhydryl (—SH) groups of the GSH-dependent enzymes such as glutathione peroxidase, glutathione-*S*-transferases, and GSH reductase, and thereby inhibiting them (Schiller et al., 1977; Thomas, 2009). Depletion in cellular thiol pools has also been detected in cells of *E. coli* when exposed to doses of Ag, Cd, Zn, and Cr (Harrison et al., 2009). Thiol oxidation and antioxidant consumption increases the vulnerability of proteins to metal species or ROS and also prevents repair of the oxidized protein thiols by cellular thiol–disulfide exchange enzymes (Lemire et al., 2013).

The entire process is then followed by a cascade of events including oxidation of vitamins, membrane damage leading to altered Ca^{2+} homoeostasis, activation of Ca^{2+}-dependent proteases, end nucleases, lipases, depletion of ATP, NADP(H), and ultimately cell destruction (Stohs and Bagchi, 1995).

19.4.3 Enzyme Activity Interference

Metal ions catalyze the oxidation of amino acid side chains of proteins, forming carbonyl derivatives and thus inactivating enzymes (Amici et al., 1989; Lemire et al., 2013). In the

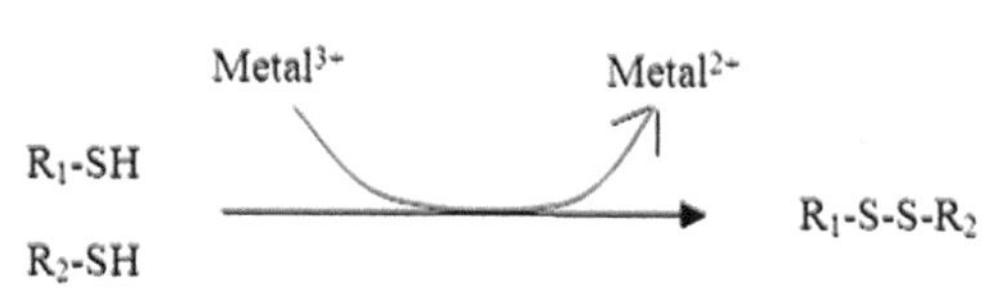

FIGURE 19.2 Metal-catalyzed oxidation of thiols, the cellular pool of antioxidants.

yeast *Saccharomyces cerevisiae*, chromium toxicity induces oxidative damage to the cytosolic enzymes involved in glycolysis and other catabolic reactions (Sumner et al., 2005).

Mononuclear metalloenzymes are usually targets of oxidative stress. These enzymes, which are dependent on metal ions for their structure and function, bind to Fe^{2+} in vivo and are damaged by H_2O_2 through Fe-catalyzed Fenton chemistry (Lemire et al., 2013). Micromolar concentrations of hydrogen peroxide in *E. coli* have been found to cause oxidation of cysteine residues of two mononuclear enzymes, peptide deformylase and threonine dehydrogenase, to sulfinic or sulfonic forms (Anjem and Imlay, 2012). This causes rapid damage to the enzymatic system (Fig. 19.3). However, structural and metabolic arrangements exist at the cellular level to avoid damage to other active-site residues.

Another mechanism involves the site-specific inactivation of the bacterial Fe–S dehydratases by toxic metals. However, the toxicity mechanism differs from ROS toxicity in being oxygen independent (Lemire et al., 2013). Metal ions convert the native [4Fe–4S] cluster of the dehydratases to a [3Fe–4S] form, which could be further degraded to the apoenzyme (Fig. 19.4). This phenomenon has been demonstrated with *E. coli* cells, wherein the extracts of cells treated with Ag^+ showed a decrease in the activity of the enzyme 6-phosphogluconate dehydratase, while the noncluster-containing enzymes, malate dehydrogenase and β-galactosidase, remained unaffected (Chahal et al., 2013). Calderón et al. (2009) has correlated the inactivation of [4Fe–4S] cluster-containing enzymes, mediated by metals as well as oxidative stress, with growth inhibition of *E. coli*.

Ionic/molecular mimicry is another mechanism of enzyme inhibition that can result in metal poisoning wherein the metal ion is structurally similar to the enzyme cofactor, or a complex formed between the metal cation and a cellular ligand structurally mimics an endogenous substrate, but is physiologically disruptive. An example of the former mechanism is the inhibition of ribonucleotide reductase, an Fe-containing enzyme responsible for the synthesis of

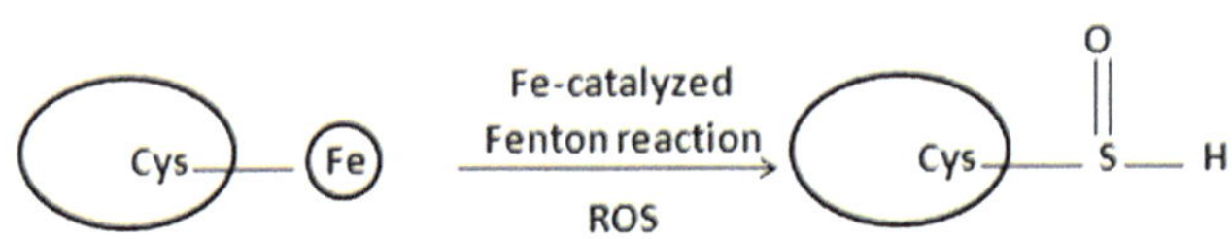

FIGURE 19.3 Metal-catalyzed oxidative inactivation of enzyme. *ROS*, reactive oxygen species. *Adapted from Lemire, J.A., Harrison, J.J., Turner, R.J., 2013. Antimicrobial activity of metals: mechanisms, molecular targets and applications. Nat. Rev. Microbiol. 11, 371–384.*

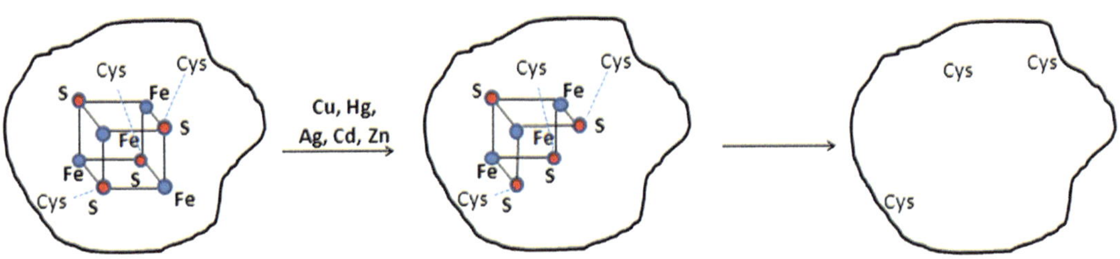

FIGURE 19.4 Metal-mediated degradation of the Fe–S cluster of dehydratases. *Adapted from Lemire, J.A., Harrison, J.J., Turner, R.J., 2013. Antimicrobial activity of metals: mechanisms, molecular targets and applications. Nat. Rev. Microbiol. 11, 371–384.*

deoxyribonucleotides, by the substitution of gallium (Ga) for iron. Ga has similar ionic radius to Fe, but has no oxidation state that is essential for enzyme activity (Nikolova et al., 2016). This Ga^{3+}–Fe^{3+} competition interferes with several biological processes requiring iron cofactors. Similarly, the methylmercury–cysteine complex mimics the amino acid methionine, leading to interference in the uptake and incorporation of methionine into specific proteins (Fig. 19.5).

Several instances of such competition between many divalent endogenous and toxic metal ions have been reported in biological systems. Arsenate and vanadate mimic phosphate and inhibit ATP synthesis (DeMaster and Mitchel, 1973). Some di- and trivalent metal cations such as Pb^{2+}, Cr^{3+}, and vanadate have been reported to mimic calcium and other metals such as iron. In addition, metal–anion complexes mimic and inhibit bicarbonates and chlorides (DeMaster and Mitchel, 1973). Metal complexes or cations can only mimic early steps in the metabolic processes of transport and metabolism, yet the process provides a possible explanation of the toxicity mechanism of at least a few metals.

19.4.4 Impairment of Membrane Function

The bactericidal effect of NPs could also be caused either by their adsorption or interaction through electrostatic interactions (Thill et al., 2006; Pal et al., 2007). Dissipation of the chemiosmotic force or siphoning off electrons from the electron transport chain is another mechanism of metal poison of microorganisms. Ag in *Vibrio cholerae* dissipates the chemiosmotic membrane potential through leakage of protons from the membrane (Dibrov et al., 2002). Moreover, Ag^+ inhibits the enzyme NADH:quinone oxidoreductase (Fig. 19.6), which spawns a redox-driven transmembrane Na^+ potential by translocating Na^+ in the respiratory chain of microorganisms (Fadeeva et al., 2011). Reduction of metal ions by quinones also steals electrons away from the respiratory chain (Oves et al., 2016). Loss of membrane integrity and the associated K^+ efflux have been reported to be induced extensively by copper and cadmium.

Brayner et al. (2006) observed an increase in membrane permeability of *E. coli* cells in the presence of ZnO NPs, leading to their cellular internalization. The NPs of CeO_2 are positively charged at neutral pH and thus display a strong electrostatic attraction toward bacterial outer membranes and cause cytotoxicity by altering the oxidoreduction mechanism (Thill et al., 2006).

This interaction differs depending on whether the bacterium is Gram positive or Gram negative, because of their distinct characteristics. Ag NPs, through the release of Ag^+ ions, inhibit microorganisms by causing cell membrane perturbation and release of intracellular contents (Jung et al., 2008). However, the Gram-negative *E. coli* was found to be more sensitive

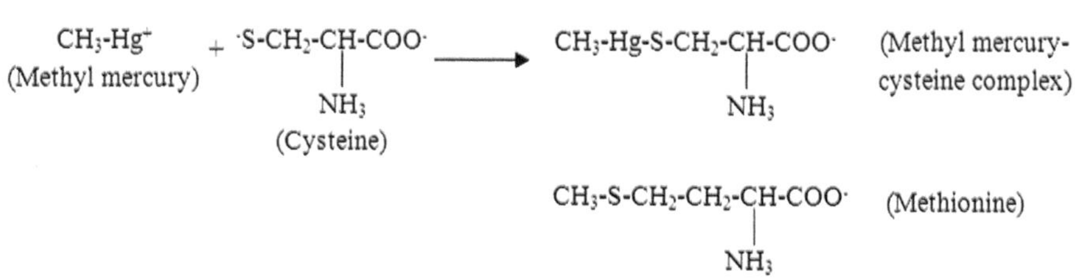

FIGURE 19.5 The hydrophobic chain of the methylmercury–cysteine complex, similar to that of methionine. *Adapted from Clarkson, T.W., 1993. Molecular and ionic mimicry of toxic metals. Annu. Rev. Pharmacol. Toxicol. 32, 545–571.*

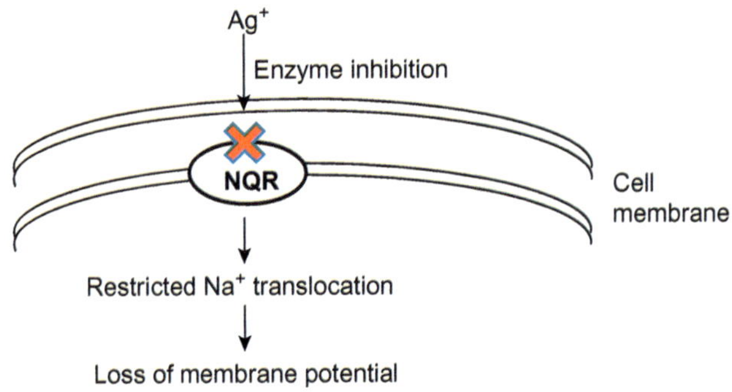

FIGURE 19.6 Membrane disintegration caused by Ag^+. *NQR*, NADH:quinone oxidoreductase. *Adapted from Lemire, J.A., Harrison, J.J., Turner, R.J., 2013. Antimicrobial activity of metals: mechanisms, molecular targets and applications. Nat. Rev. Microbiol. 11, 371–384.*

than the Gram-positive *Staphylococcus aureus*. This may be because of the trapping of Ag^+ ions by the negatively charged peptidoglycan in the walls of Gram-positive bacteria (Kawahara et al., 2000). Similar variability in susceptibility of bacteria to ZnO NPs has been reported by Arakha et al. (2015) who attributed the greater NP tolerance of Gram-positive bacteria to the thicker peptidoglycan wall layer. They also reported that the interaction of ZnO NPs, having positive surface potential, with a bacterial membrane of negative surface potential enhanced ROS production and exerted mechanical stress on the membrane depolarizing them. NPs of Pd, TiO_2, and C_{60} (fullerene) also cause membrane disruption in both Gram-positive and Gram-negative bacteria (De Windt et al., 2006; Fang et al., 2007; Heinlaan et al., 2008).

19.4.5 Interference With Nutrient Assimilation

As stated previously, metal ions enter the cell via unspecific or specific transporters and thus competitively inhibit binding sites of these transporters used for assimilation of essential ions. Sulfur starvation of *S. cerevisiae* cells caused by chromium by interfering with sulfate uptake and depleting the intracellular S metabolite pools is one such example. This is because of the affinity of Cr^{6+} to the sulfate transporters Sul1 and Sul2 (Pereira et al., 2008). Similarly, Ga binds to transferrin, blocks the cellular uptake and subsequent incorporation into ferritin, and inhibits cell growth (Chitambar and Narasimhan, 1991). This type of Ga induced Fe starvation in *Pseudomonas aeruginosa* also results from its binding with Fe-responsive transcriptional regulator *pvdS*, which controls genes involved in the synthesis of pyoverdine, the pyoverdine receptor, and other Fe-responsive genes (Kaneko et al., 2007). This causes deregulation of the expression of genes involved in the selective uptake of Fe and also interferes with Fe sensing (Fig. 19.7).

19.4.6 Genotoxicity or DNA Damage

Toxic doses of most metals, including Mn, Cr, Co, Cd, Mo, As, and organomercury complexes, are mutagenic in bacteria. The ROS created by the Fenton reaction disrupts DNA replication and leads to DNA strand breaks (Fig. 19.8), as has been described with *E. coli* (Linley et al., 2012). Cu-catalyzed Fenton chemistry is, however, an exception as it is restricted to the

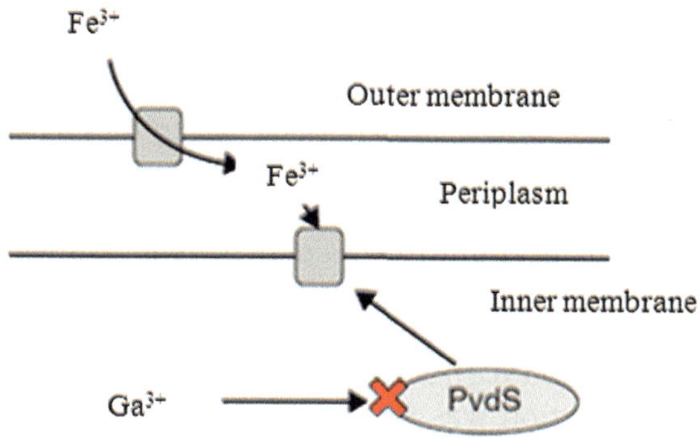

FIGURE 19.7 Metal interference in nutrient assimilation. *Adapted from Lemire, J.A., Turner, R.J., 2016. Mechanisms underlying the antimicrobial capacity of metals. In: de Bruijn, F.J. (Ed.) Stress and Environmental Regulation of Gene Expression and Adaptation in Bacteria, vol. 1. John Wiley & Sons, New Jersey, pp. 215–224.*

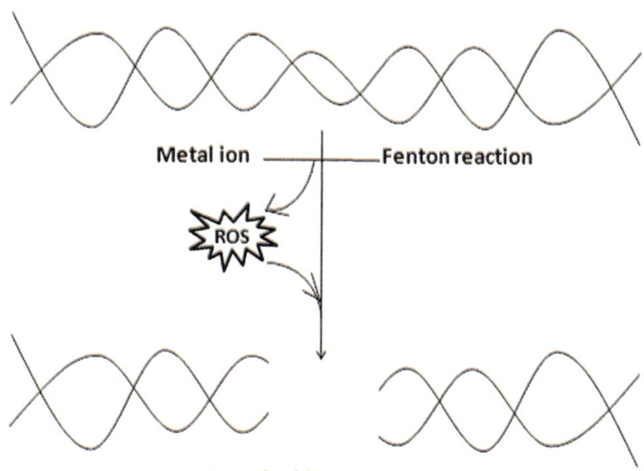

FIGURE 19.8 DNA damage induced by metal ions. *ROS*, reactive oxygen species. *Adapted from Lemire, J.A., Harrison, J.J., Turner, R.J., 2013. Antimicrobial activity of metals: mechanisms, molecular targets and applications. Nat. Rev. Microbiol. 11, 371–384.*

periplasmic space and the free radicals are quarantined in the cytosol, i.e., away from DNA (Macomber et al., 2007). However, Cu toxicity has been associated with destruction of extracellular DNA following cell lysis, which inhibits the horizontal gene transfer process (Warnes et al., 2012).

The genotoxicity of metals such as Cd in bacteria has been attributed to their direct interaction with DNA through the binding at T, A, and G bases as well as their ability to inhibit DNA repair (Hossain and Huq, 2002; Rossman et al., 1992).

Because of the strong toxicity, especially to bacteria and yeast, metal ions are being used widely as antimicrobial agents (Lemire et al., 2013). However, both physico-chemical properties of the medium and chemical properties of the metals can influence their toxicity.

19.4.7 Reactive Oxygen Species Generation

Aerobic cellular respiration, coupled with oxidation–reduction reactions with metal ions, leads to the formation of ROS with the potential to damage DNA, proteins, lipids, and carbohydrates, thereby disrupting the key cellular processes, as described in the previous section. Metal oxides such as MnO, WO_3, Fe_2O_3, and ZnO generate electrons in the presence of light, and react with molecular oxygen and surface hydroxyl groups to form hydroxyl and superoxide radicals causing oxidative stress (Niazi and Gu, 2009). Production of ROS, exceeding the capacity of ROS scavengers, on interaction of ZnO NPs with cell membranes of *E. coli* and *Bacillus subtilis* has been reported by Arakha et al. (2015). Moreover, higher ROS production was found in the presence of ZnO NPs with positive surface charge potential compared to that with a negative surface charge. Similar light-dependent ROS generation has also been reported as a cause of antibacterial activity of SiO_2 and TiO_2 (Adams et al., 2006).

ROS production has also been found to be associated with NPs as diverse as C_{60} or fullerenes, single-walled nanotubes, quantum dots, and ultrafine particles (Niazi and Gu, 2009). C_{60}-induced oxidative toxicity in *E. coli*, *Salmonella* sp. and *Shewanella* sp. (Fortner et al., 2005; Sera et al., 1996), cadmium telluride quantum dot-induced ROS generation in *E. coli*, *B. subtilis*, *P. aeruginosa*, and *S. aureus* (Dumas et al., 2009), and carbon nanotube-induced oxidative stress in *P. aeruginosa* (Olivi et al., 2013) are a few examples.

ROS-induced production of singlet oxygen, oxidative DNA damage-related mutagenicity, and cytotoxicity have been proposed to be the major mechanisms of oxidative toxicity (Sera et al., 1996; Niazi and Gu, 2009). Overproduction of ROS results in the failure of cells to maintain normal physiological redox-regulated functions and leads to oxidative modification of proteins (Fu et al., 2014). Induction of DNA-strand breaks and modification of nucleic acids lead to genotoxic effects and ultimately cell death.

19.4.8 Interactions With Proteins

As stated by Elsaesser and Howard (2012), NPs, which are of the same size as protein molecules, are able to interact with proteins, change their configuration, and interfere with cell signaling processes. The interaction of TiO_2 NPs with photosynthetic bacterial light-harvesting antenna complex LH2 was studied by Chen et al. (2005). The interaction was found to induce structural deformation of the LH2 complex. Sinha and Khare (2014) proposed that conformational changes in proteins subsequent to adsorption, electrostatic, hydrophobic, and hydrogen bonding interactions between proteins and NPs, lead to their inactivation. They found that exposure to Ag and ZnO NPs caused a 70% and 30% decrease, respectively, in the activity of nonhalophilic *Bacillus licheniformis* protease because of the loss of α-helical content of the proteases in the presence of NPs. Cytotoxicity of NPs of Al_2O_3, TiO_2, and ZnO on bacterial cells was found to be largely caused by harmful changes in bacterial proteins and phospholipids (Jiang et al., 2010). The NPs caused a decrease in the intensity ratio of β-sheets/α-helices, thereby changing the protein structure drastically. Similarly, the polystyrene/fullerene composite films are known to manifest bacteriostatic and fungistatic activities because of the interaction of fullerenes with functional groups of the amino acids composing bacterial protein (Alekseeva et al., 2014).

19.4.9 Lipid Peroxidation

Metals also affect the integrity of cytoplasmic membranes by causing peroxidation of membrane lipids especially in eukaryotes containing high levels of polyunsaturated fatty acids (PUFAs). Lipid peroxidation and the associated membrane disruption are a result of heavy metal-induced ROS. A single OH· can react with methylene groups of PUFAs and lead to a cyclic reaction (Fig. 19.9), with malondialdehyde as one of the final products, which is responsible for cell membrane damage (Sharma et al., 2012). The level of unsaturation in fatty acid of the plasma membrane of *S. cerevisiae* has been correlated with the organism's copper sensitivity by elevating levels of lipid peroxidation in the cells (Howlett and Avery, 1997).

Interestingly, lipid peroxidation is also found to be an important toxicity mechanism in bacteria, induced as a result of NPs. The generation of ROS triggers the peroxidation of lipids, particularly in the bacterial lipopolysaccharide layer. Studies in *E. coli* have suggested that TiO_2 exposure led to the destruction of endotoxin, an integral component of lipopolysaccharide of the outer membrane, under photocatalytic conditions (Sunada et al., 1998) and is the underlying mechanism causing mortality of *E. coli* K-12 cells irradiated in the presence of the TiO_2 photocatalyst (Maness et al., 1999). Fang et al. (2007) reported a ROS-mediated, altered membrane lipid composition induced by C_{60} in *Pseudomona putida*, characterized by an increase in cyclopropane fatty acids at the expense of unsaturated fatty acids, which ultimately affected membrane fluidity. The antibacterial activity of MgO NPs is also attributed to the ROS-induced lipid peroxidation in bacteria (Tang and Lv, 2014). Another mechanism of lipid modification involving the breakage of the phosphodiester bond of L-α-phosphatidylethanolamine, forming phosphate monoesters and a highly disordered alkyl chain on exposure to ZnO NPs, has also been reported (Jiang et al., 2010). Lipid peroxidation thus leads to phospholipid chain breakage and increases fluidity and permeability in membranes.

Although several toxicity mechanisms have been attributed to specific NPs, NPs mostly exert their antimicrobial activity through a synergism of these effects, as detailed in Table 19.1. This accounts for a broad spectrum of antimicrobial activities of NPs, which could be combined with the conventional processes of microbial control. This synergism has also led to the designing of novel organic and organic–inorganic hybrid NPs. Lopes et al. (2012) studied the antimicrobial properties of the organic NP SDS/DDAB (vesicles composed of sodium dodecyl sulfate/didodecyl dimethylammonium bromide) in *Vibrio fischeri* and *Salmonella typhimurium* and found

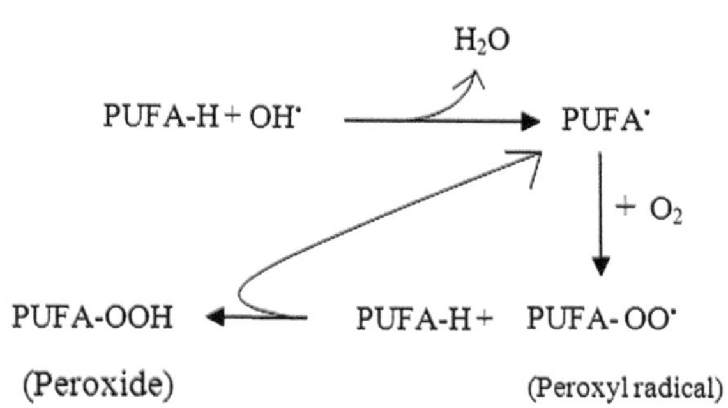

FIGURE 19.9 Reactive oxygen species-mediated peroxidation of membrane lipids. *PUFA*, polyunsaturated fatty acid.

that SDS/DDAB could induce genotoxicity in these bacteria. Beyth et al. (2010) reported the antimicrobial property of quaternary ammonium polyethylenimine (QPEI) NPs. The in vitro inhibition of carcinogenic *Mutans streptococci* biofilm was attributed to the cationic binding and electrostatic interaction between the QPEI and the microbial cell wall that disrupted the membrane function and caused leakage of K^+ ions, DNA, and RNA, which ultimately led to programmed cell death in the bacteria (Abachi et al., 2016). Organic–inorganic hybrid NPs, ONPs/Ag, and ONPs/Au (organic nanoparticles (ONPs) of lipoic acid and its doped derivatives) have also been demonstrated to be antibacterial as they could inhibit several bacteria, including *S. aureus*, *B. cereus*, *E. coli*, and *Salmonella typhi* (Aguilar et al., 2015; Prasad et al., 2016).

19.5 CONCLUSION AND FUTURE PERSPECTIVES

NPs, especially of metals, metal oxides, zeolites, and clays, are synthesized on a large scale every year and by passing through various pathways in the environment they ultimately end up in soil. These particles can adversely affect the population and/or activity of a range of soil microorganisms through their physical effect on membranes, generation of ROS, and production of metal ions. As microorganisms perform a range of functions related to decomposition and mineralization in soil, adverse effects on them can alter ecosystem services provided by the soil. Since production of NPs is set to increase in the future, there is the possibility of their adverse effects on soil microbial communities. However, in soil the NPs are exposed to a number of ligands and binding forces. These forces can alter the bioavailability and subsequent interaction of NPs with microorganisms. Therefore efforts have to be made in future to develop NPs with minimum toxicity to microorganisms and to identify soil management approaches that create conditions in the soil that can inactivate NPs and minimize their adverse effects on microorganisms.

References

Abachi, S., Lee, S., Vasantha Rupasinghe, H.P., 2016. Molecular mechanisms of inhibition of *Streptococcus* species by phytochemicals. Molecules 21, 215–246. http://dx.doi.org/10.3390/molecules21020215.

Adams, L.K., Lyon, D.Y., Alvarez, P.J., 2006. Comparative ecotoxicity of nanoscale TiO_2, SiO_2 and ZiO_2 water suspensions. Water Res. 40, 3527–3532.

Aguilar, C.A.H., Jiménez, A.B.P., Silva, A.R., Kaur, N., Thangarasu, P., Ramos, J.M.V., Singh, N., 2015. Organic-inorganic hybrid nanoparticles for bacterial inhibition: synthesis and characterization of doped and undoped ONPs with Ag/Au NPs. Molecules 20, 6002–6021.

Aislabie, J., Deslippe, J.R., 2013. Soil microbes and their contribution to soil services. In: Dymond, J.R. (Ed.), Ecosystem Services in New Zealand – Conditions and Trends. Manaaki Whenua Press, Lincoln, New Zealand, pp. 143–161.

Albanese, A., Tang, P.S., Chan, W.C.W., 2012. The effect of nanoparticle size, shape, and surface chemistry on biological systems. Annu. Rev. Biomed. Eng. 14, 1–16.

Alekseeva, O.V., Bagrovskaya, N.A., Noskov, A.V., 2014. Effect of C_{60} filling on structure and properties of composite films based on polystyrene. Arab. J. Chem. http://dx.doi.org/10.1016/j.arabjc.2014.09.008.

Amici, A., Levine, R.L., Tsai, L., Stadtman, E.R., 1989. Conversion of amino acid residues in proteins and amino acid homopolymers to carbonyl derivatives by metal-catalyzed oxidation reactions. J. Biol. Chem. 264 (6), 3341–3346.

Anjem, A., Imlay, J.A., 2012. Mononuclear iron enzymes are primary targets of hydrogen peroxide stress. J. Biol. Chem. 287 (19), 15544–15556.

Arakha, M., Saleem, M., Mallick, B.C., Jha, S., 2015. The effects of interfacial potential on antimicrobial propensity of ZnO nanoparticle. Sci. Rep. 5:9578.

Asghar, H.N., Zahir, Z.A., Arshad, M., Khalid, A., 2002. Relationship between in vitro production of auxins by rhizobacteria and their growth-promoting activities in *Brassica juncea* L. Biol. Fertil. Soils 35, 231–237.

Aziz, N., Faraz, M., Pandey, R., Shakir, M., Fatma, T., Varma, A., et al., 2015. Facile algae-derived route to biogenic silver nanoparticles: synthesis, antibacterial, and photocatalytic properties. Langmuir 31, 11605–11612.

Aziz, N., Pandey, R., Barman, I., Prasad, R., 2016. Leveraging the attributes of *Mucor hiemalis*-derived silver nanoparticles for a synergistic broad-spectrum antimicrobial platform. Front. Microbiol. 7, 1984. http://dx.doi.org/10.3389/fmicb.2016.01984.

Bao, H., Yu, X., Xu, C., Li, X., Li, Z., Wei, D., Liu, Y., 2015. New toxicity mechanism of silver nanoparticles: promoting apoptosis and inhibiting proliferation. PLoS One 10 (3), e0122535.

Barnes, R.J., Molina, R., Xu, J., Dobson, P.J., Thompson, I.P., 2013. Comparison of TiO_2 and ZnO nanoparticles for photocatalytic degradation of methylene blue and the correlated inactivation of gram-positive and gram-negative bacteria. J. Nanopart. Res. 15, 1432–1439.

Benoit, R., Wilkinson, K.J., Sauve, S., 2013. Partitioning of silver and chemical speciation of free Ag in soils amended with nanoparticles. Chem. Cent. J. 7, 75. http://dx.doi.org/10.1186/1752-153X-7-75.

Beyth, N., Yudovin-Farber, I., Perez-Davidi, M., Domb, A.J., Weiss, E.I., 2010. Polyethyleneimine nanoparticles incorporated into resin composite cause cell death and trigger biofilm stress in vivo. Proc. Natl. Acad. Sci. U.S.A. 107 (51), 22038–22043.

Bhattacharya, S., 2014. Reactive oxygen species and cellular defense system. In: Rani, V., Yadav, U.S. (Eds.), Free Radicals in Human Health and Disease. Springer, India, pp. 17–29.

Billsten, P., Freskgard, P.O., Carlsson, U., Jonsson, B.H., Elwing, H., 1997. Adsorption to silica nanoparticles of human carbonic anhydrase II and truncated forms induce a molten-globule-like structure. FEBS Lett. 402, 67–72.

Birben, E., Sahiner, U.M., Sackesen, C., Erzurum, S., Kalayci, O., 2012. Oxidative Stress and antioxidant defense. World Allergy Organ J. 5 (1), 9–19.

Boldyreva, L.B., 2014. The physical aspect of the effects of metal nanoparticles on biological systems. Spin supercurrents. Nanomater. Nanosci. 2, 1–7.

Bollag, J.-M., Mertz, T., Otjen, L., 1994. Role of microorganisms in soil bioremediation. In: Anderson, T.A., Coats, J.R. (Eds.), Bioremediation through Rhizosphere Technology. ACS Symposium Series. American Chemical Society, Washington, DC, pp. 2–10.

Brayner, R., Ferrari-Iliou, R., Brivois, N., Djediat, S., Benedetti, M.F., Fievet, F., 2006. Toxicological impact studies based on *Escherichia coli* bacteria in ultrafine ZnO nanoparticles colloidal medium. Nano Lett. 6, 866–870.

Buzea, C., Pacheco, I.I., Robbie, K., 2007. Nanomaterials and nanoparticles: sources and toxicity. Biointerphases 2 (4), MR17–71.

Calderón, I.L., Elías, A.O., Fuentes, E.L., Pradenas, G.A., Castro, M.E., Arenas, F.A., Pérez, J.M., Vásquez, C.C., 2009. Tellurite-mediated disabling of [4Fe-4S] clusters of *Escherichia coli* dehydratases. Microbiology 155 (6), 1840–1846.

Capeletti, L.B., de Oliveira, L.F., de Almeida Goncaçalves, K., de Oliveira, J.F.A., Saito, A., Kobarg, J., dos Santos, J.H.Z., Cardoso, M.B., 2014. Tailored silica–antibiotic nanoparticles: overcoming bacterial resistance with low cytotoxicity. Langmuir 30, 7456–7464.

Cates, A.M., Ruark, M.D., Hedtcke, J.L., Posner, J.L., 2016. Long-term tillage, rotation and perennialization effects on particulate and aggregate soil organic matter. Soil Tillage Res. 155, 371–380.

Chahal, H.K., Boyd, J.M., Outten, F.W., 2013. Fe-S cluster biogenesis in archaea and bacteria. In: Culotta, V., Scott, R.A. (Eds.), Metals in Cells. John Wiley & Sons, UK, pp. 445–472.

Chatterjee, A.K., Chakraborty, R., Basu, T., 2014. Mechanism of antibacterial activity of copper nanoparticles. Nanotechnology 25 (13), 135101 12p.

Chatterjee, T., Chakraborti, S., Joshi, P., Singh, S.P., Gupta, V., Chakrabarti, P., 2010. The effect of zinc oxide nanoparticles on the structure of the periplasmic domain of the *Vibriocholerae* ToxR protein. FEBS J. 277, 4184–4194.

Chen, M., von Mikecz, A., 2005. Formation of nucleoplasmic protein aggregates impairs nuclear function in response to SiO_2 nanoparticles. Exp. Cell Res. 305, 51–62.

Chen, X.-H., Zhang, L., Weng, Y.-X., Du, L.-C., Ye, M.-P., Yang, G.-Z., Fujii, R., Rondonuwu, F.S., Koyama, Y., Wu, Y.-S., Zhang, J.P., 2005. Protein structural deformation induced lifetime shortening of photosynthetic bacteria light-harvesting complex LH2 excited state. Biophys. J. 88 (6), 4262–4273.

Chitambar, C.R., Narasimhan, J., 1991. Targeting iron-dependent DNA synthesis with gallium and transferrin-gallium. Pathobiology 59 (1), 3–10.

Chithrani, B.D., Ghazani, A.A., Chan, W.C., 2006. Determining the size and shape dependence of gold nanoparticle uptake into mammalian cells. Nano Lett. 6 (4), 662–668.

Chu, Z., Zhang, S., Zhang, B., Zhang, C., Fang, C.-Y., Rehor, I., Cigler, P., Chang, H.-C., Lin, G., Liu, R., Li, Q., 2014. Unambiguous observation of shape effects on cellular fate of nanoparticles. Sci. Rep. 4, 4495.

Clarkson, T.W., 1993. Molecular and ionic mimicry of toxic metals. Annu. Rev. Pharmacol. Toxicol. 32, 545–571.

Colman, B.P., Arnaout, C.L., Anciaux, S., Gunsch, C.K., Hochella, M.F., Kim, B., Lowry, G.V., McGill, B.M., Reinsch, B.C., Richardson, C.J., Unrine, J.M., Wright, J.P., Yin, L., Bernhardt, E.S., 2013. Low concentrations of silver nanoparticles in biosolids cause adverse ecosystem responses under realistic field scenario. PLoS One 8 (2), e57189. http://dx.doi.org/10.1371/journal.pone.0057189.

Concha-Guerrero, S.I., Brito, E.M.S., Piñón-Castillo, H.A., Tarango-Rivero, S.H., Caretta, C.A., Luna-Velasco, A., Duran, R., Orrantia-Borunda, E., 2014. Effect of CuO nanoparticles over isolated bacterial strains from agricultural soil. J. Nanomater. 2014, 13:148743.

Conrad, R., 1996. Soil microorganisms as controllers of atmospheric trace gases (H_2, CO, CH_4, OCS, N_2O, and NO). Microbiol. Rev. 60 (4), 609–640.

De Windt, W., Boon, N., Van den Bulcke, J., Rubberecht, L., Prata, F., Mast, J., Hennebel, T., Verstraete, W., 2006. Biological control of the size and reactivity of catalytic Pd(0) produced by *Shewanella oneidensis*. Antonie Leeuwenhoek 90, 377–389.

DeMaster, E.G., Mitchel, R.A.A., 1973. A comparison of arsenate and vanadate as inhibitors or uncouplers of mitochondrial and glycolytic energy metabolism. Biochemistry 12, 3616–3621.

de Neergaard, A., 2009. Soil biology and microbiology. In: Verheye, W.H. (Ed.), Land Use, Land Cover and Soil Sciences - Volume VI: Soils and Soil Sciences – 1. Eolss Publishers, Oxford, UK, pp. 184–204.

Deryabin, D.G., Aleshina, E.S., Vasilchenko, A.S., Deryabina, T.D., Efremova, L.V., Karimov, I.F., Korolevskaya, L.B., 2013. Investigation of copper nanoparticles antibacterial mechanisms tested by luminescent *Escherichia coli* strains. Nanotechnol. Russ. 8 (5), 402–408.

Dibrov, P., Dzioba, J., Gosink, K.K., Hase, C.C., 2002. Chemiosmotic mechanism of antimicrobial activity of Ag^+ in *Vibrio cholerae*. Antimicrob. Agents Chemother. 46, 2668–2670.

Dinesh, R., Anandaraj, M., Srinivasan, V., Hamza, S., 2012. Engineered nanoparticles in the soil and their potential implications to microbial activity. Geoderma 173–174, 19–27.

Dumas, E.M., Ozenne, V., Mielke, R.E., Nadeau, J.L., 2009. Toxicity of CdTe quantum dots in bacterial strains. IEEE Trans. Nanobiosci. 8, 58–64.

El Badawy, A.M., Silva, R.G., Morris, B., Scheckel, K.G., Suidan, M.T., Tolaymat, T.M., 2011. Surface charge dependent toxicity of silver nanoparticles. Environ. Sci. Technol. 45, 283–287.

Elsaesser, A., Howard, C.V., 2012. Toxicology of nanoparticles. Adv. Drug Deliv. Rev. 64, 129–137.

European Commission, 2012. Communication from the Commission to the European Parliament, the Council and the European Economic and Social Committee. Second Regulatory Review on Nanomaterials. Brussels, 3.10.2012 COM(2012) 572 final http://eur-lex.europa.eu/legal-content/EN/TXT/PDF/?uri=CELEX:52012DC0572&from=EN.

Fabrega, J., Renshaw, J.C., Lead, J.B., 2009b. Interactions of silver nanoparticles with *Pseudomonas putida* biofilms. Environ. Sci. Technol. 43, 9004–9009.

Fabrega, J., Fawcett, S.R., Renshaw, J.C., Lead, J.R., 2009a. Silver nanoparticle impact on bacterial growth: effect of pH, concentration, and organic matter. Environ. Sci. Technol. 43, 7285–7290.

Fadeeva, M.S., Bertsova, Y.V., Euro, L., Bogachev, A.V., 2011. Cys377 residue in NqrF subunit confers Ag^+ sensitivity of Na^+-translocating NADH: quinone oxidoreductase from *Vibrio harveyi*. Biochemistry 76, 186–195.

Fajardo, C., Saccà, M.L., Costa, G., Nande, M., Martin, M., March 1, 2014. Impact of Ag and Al_2O_3 nanoparticles on soil organisms: in vitro and soil experiments. Sci. Total Environ. 473–474, 254–261. http://dx.doi.org/10.1016/j.scitotenv.2013.12.043. Epub December 27, 2013.

Fang, J., Lyon, D.Y., Wiesner, M.R., Dong, J., Alvarez, P.J., 2007. Effect of a fullerene water suspension on bacterial phospholipids and membrane phase behavior. Environ. Sci. Technol. 41, 2636–2642.

Fasim, F., Ahmed, N., Parsons, R., Gadd, G.M., 2002. Solubilization of zinc salts by a bacterium isolated from the air environment of a tannery. FEMS Microbiol. Lett. 213, 1–6.

Feng, Q.L., Wu, J., Chen, G.Q., Cui, F.Z., Kim, T.N., Kim, J.O., 2000. A mechanistic study of the antibacterial effect of silver ions on *Escherichia coli* and *Staphylococcus aureus*. J. Biomed. Mater. Res. 52, 662–668.

Fertsch-Gapp, S., Semmler-Behnke, M., Wenk, A., Kreyling, W.G., 2011. Binding of polystyrene and carbon black nanoparticles to blood serum proteins. Inhal. Toxicol. 23, 468–475.

Fortner, J.D., Lyon, D.Y., Sayes, C.M., Boyd, A.M., Falkner, J.C., Hotze, E.M., Alemany, L.B., Tao, Y.J., Guo, W., Ausman, K.D., Colvin, V.L., Hughes, J.B., 2005. C_{60} in water: nanocrystal formation and microbial response. Environ. Sci. Technol. 39, 4307–4316.

Frenk, S., Ben-Moshe, T., Dror, I., Berkowitz, B., Minz, D., 2013. Effect of metal oxide nanoparticles on microbial community structure and function in two different soil types. PLoS One. 8 (12), e84441. https://doi.org/10.1371/journal.pone.0084441.

Fröhlich, E., 2012. The role of surface charge in cellular uptake and cytotoxicity of medical nanoparticles. Int. J. Nanomed. 7, 5577–5591.

Fu, P.P., Xia, Q., Hwang, H.-M., Ray, P.C., Yu, H., 2014. Mechanisms of nanotoxicity, generation of reactive oxygen species. J. Food Drug Anal. 22 (1), 64–75.

Gogniat, G., Dukan, S., 2007. TiO_2 photocatalysis causes DNA damage via Fenton reaction-generated hydroxyl radicals during the recovery period. Appl. Environ. Microbiol. 73 (23), 7740–7743.

Gottschalk, F., Sonderer, T., Scholz, R.W., Nowack, B., 2009. Modeled environmental concentrations of engineered nanomaterials (TiO_2, ZnO, Ag, CNT, fullerenes) for different regions. Environ. Sci. Technol. 43, 9216–9222.

Grillo, R., Rosa, A.H., Fraceto, L.F., 2015. Engineered nanoparticles and organic matter: a review of the state-of-the-art. Chemosphere 119, 608–619. http://dx.doi.org/10.1016/j.chemosphere.2014.07.049.

Harrison, J.J., Tremaroli, V., Stan, M.A., Chan, C.S., Vacchi-Suzzi, C., Heyne, B.J., Parsek, M.R., Ceri, H., Turner, R.J., 2009. Chromosomal antioxidant genes have metal ion-specific roles as determinants of bacterial metal tolerance. Environ. Microbiol. 11 (10), 2491–2509.

Haynes, R.J., Francis, G.S., 1993. Changes in microbial biomass C, soil carbohydrate composition and aggregate stability induced by growth of selected crop and forage species under field conditions. Eur. J. Soil Sci. 44, 665–675.

Heinlaan, M., Ivask, A., Blinova, I., Dubourguier, H.C., Kahru, A., 2008. Toxicity of nanosized and bulk ZnO, CuO and TiO_2 to bacteria *Vibrio fischeri* and crustaceans *Daphnia magna* and *Thamnocephalus platyurus*. Chemosphere 71, 1308–1316.

Hossain, Z., Huq, F., 2002. Studies on the interaction between $Cd^{(2+)}$ ions and nucleobases and nucleotides. J. Inorg. Biochem. 90, 97–105.

Howlett, N.G., Avery, S.V., 1997. Induction of lipid peroxidation during heavy metal stress in *Saccharomyces cerevisiae* and influence of plasma membrane fatty acid unsaturation. Appl. Environ. Microbiol. 63 (8), 2971–2976.

Huang, J., Bu, L., Xie, J., Chen, K., Cheng, Z., Li, X., Chen, X., 2010. Effects of nanoparticle size on cellular uptake and liver MRI with polyvinylpyrrolidone-coated iron oxide nanoparticles. ACS Nano 4 (12), 7151–7160.

Huang, Z., Zheng, X., Yan, D., Yin, G., Liao, X., Kang, Y., Yao, Y., Huang, D., Hao, B., 2008. Toxicological effect of ZnO nanoparticles based on bacteria. Langmuir 24, 4140–4144.

Hwang, E.T., Lee, J.H., Chae, Y.J., Kim, Y.S., Kim, B.C., Sang, B.I., Gu, M.B., 2008. Analysis of the toxic mode of action by silver nano-particles using stress-specific bioluminescent bacteria. Small 4, 746–750.

Ivask, A., Kurvet, I., Kasemets, K., Blinova, I., Aruoja, V., Suppi, S., Vija, H., Käkinen, A., Titma, T., Heinlaan, M., Visnapuu, M., Koller, D., Kisand, V., Kahru, A., 2014. Size-dependent toxicity of silver nanoparticles to bacteria, yeast, algae, crustaceans and mammalian cells in vitro. PLoS One 9 (7), e102108.

Jetten, M.S., 2008. The microbial nitrogen cycle. Environ. Microbiol. 10 (11), 2903–2909.

Jiang, W., Yang, K., Vachet, R.W., Xing, B., 2010. Interaction between oxide nanoparticles and biomolecules of the bacterial cell envelope as examined by infrared spectroscopy. Langmuir 26 (23), 18071–18077.

Jung, W.K., Koo, H.C., Kim, K.W., Shin, S., Kim, S.H., Park, Y.H., 2008. Antibacterial activity and mechanism of action of the silver ion in *Staphylococcus aureus* and *Escherichia coli*. Appl. Environ. Microbiol. 74 (7), 2171–2178.

Kaneko, Y., Theondel, M., Olakanmi, O., Britigan, B.E., Singh, P.K., 2007. The transition metal gallium disrupts *Pseudomonas aeruginosa* iron metabolism and has antimicrobial and antibiofilm activity. J. Clin. Investig. 117, 877–888.

Kang, S., Herzberg, M., Rodrigues, D.F., Elimelech, M., 2008. Antibacterial effects of carbon nanotubes: size does matter. Langmuir 24, 6409–6413.

Karlsson, H.L., Cronholm, P., Hedberg, Y., Tornberg, M., De Battice, L., Svedhem, S., Wallinder, I.O., 2013. Cell membrane damage and protein interaction induced by copper containing nanoparticles–importance of the metal release process. Toxicology 313 (1), 59–69.

Kawahara, K., Tsuruda, K., Morishita, M., Uchida, M., 2000. Antibacterial effect of silver-zeolite on oral bacteria under anaerobic conditions. Dent. Mater. 16 (6), 452–455.

Kaweeteerawat, C., Chang, C.H., Roy, K.R., Liu, R., Li, R., Toso, D., Fischer, H., Ivask, A., Ji, Z., Zink, J.I., Zhou, Z.H., Chanfreau, G.F., Telesca, D., Cohen, Y., Holden, P.A., Nel, A.E., Godwin, H.A., 2015. Cu Nanoparticles have different impacts in *Escherichia coli* and *Lactobacillus brevis* than their microsized and ionic analogues. ACS Nano 9 (7), 7215–7225.

Keller, A.A., McFerran, S., Lazareva, A., Suh, S., 2013. Global life cycle releases of engineered nanomaterials. J. Nanopart. Res. 15, 1692–1709.

Kim, K.-M., Kim, H.M., Lee, W.-J., Lee, C.-W., Kim, T.-I., Lee, J.-K., Jeong, J., Paek, S.-M., Oh, J.-M., 2014. Surface treatment of silica nanoparticles for stable and charge-controlled colloidal silica. Int. J. Nanomed. 9 (Suppl. 2), 29–40.

Kim, S., Ghafoor, K., Lee, J., Feng, M., Hong, J., Lee, D.-U., Park, J., 2013. Bacterial inactivation in water, DNA strand breaking, and membrane damage induced by ultraviolet-assisted titanium dioxide photocatalysis. Water Res. 47, 4403–4411.

Kim, S.-H., Lee, H.-S., Ryu, D.-S., Choi, S.-J., Lee, D.-S., 2011. Antibacterial activity of silver-nanoparticles against *Staphylococcus aureus* and *Escherichia coli*. Korean J. Microbiol. Biotechnol. 39 (1), 77–85.

Klaine, S.J., Alvarez, P.J.J., Batley, G.E., Fernandes, T.F., Handy, R.D., Lyon, D.Y., Mahendra, S., McLaughlin, M.J., Lead, J.R., 2008. Nanomaterials in the environment: behavior, fate, bioavailability, and effects. Environ. Toxicol. Chem. 27, 1825–1851. http://dx.doi.org/10.1897/08-090.1.

Klitzke, S., Metreveli, G., Peters, A., Schaumann, G.E., Lang, F., 2015. The fate of silver nanoparticles in soil solution-sorption of solutes and aggregation. Sci. Total Environ. 535, 54–60. http://dx.doi.org/10.1016/j.scitotenv.2014.10.108.

Krishnamoorthy, K., Veerapandian, M., Zhang, L.-H., Yun, K., Kim, S.J., 2014. Surface chemistry of cerium oxide nanocubes: toxicity against pathogenic bacteria and their mechanistic study. J. Ind. Eng. Chem. 20 (5), 3513–3517.

Lankveld, D.P., Oomen, A.G., Krystek, P., Neigh, A., Troost-de Jong, A., Noorlander, C.W., Van Eijkeren, J.C., Geertsma, R.E., De Jong, W.H., 2010. The kinetics of the tissue distribution of silver nanoparticles of different sizes. Biomaterials 31 (32), 8350–8361.

Lee, J.H., Ju, J.E., Kim, B.I., Pak, P.J., Choi, E.K., Lee, H.S., Chung, N., 2014. Rod-shaped iron oxide nanoparticles are more toxic than sphere-shaped nanoparticles to murine macrophage cells. Environ. Toxicol. Chem. 33 (12), 2759–2766.

Lemire, J.A., Harrison, J.J., Turner, R.J., 2013. Antimicrobial activity of metals: mechanisms, molecular targets and applications. Nat. Rev. Microbiol. 11, 371–384.

Lemire, J.A., Turner, R.J., 2016. Mechanisms underlying the antimicrobial capacity of metals. In: de Bruijn, F.J. (Ed.), Stress and Environmental Regulation of Gene Expression and Adaptation in Bacteria, vol. 1. John Wiley & Sons, New Jersey, pp. 215–224.

Lin, X., Li, J., Ma, S., Liu, G., Yang, K., Tong, M., Lin, D., 2014. Toxicity of TiO_2 nanoparticles to *Escherichia* coli: effects of particle size, crystal phase and water chemistry. PLoS One 9 (10), e110247.

Linley, E., Denyer, S.P., McDonnell, G., Simons, C., Maillard, J.Y., 2012. Use of hydrogen peroxide as a biocide: new consideration of its mechanisms of biocidal action. J. Antimicrob. Chemother. 67, 1589–1596.

Lopes, I., Ribeiro, R., Antunes, F.E., Rocha-Santos, T.A.P., Rasteiro, M.G., Soares, M.V.M., Gonçalves, F., Pereira, R., 2012. Toxicity and genotoxicity of organic and inorganic nanoparticles to the bacteria *Vibrio fischeri* and *Salmonella typhimurium*. Ecotoxicology 21 (3), 637–648.

Lu, F., Wu, S.H., Hung, Y., Mou, C.Y., 2009. Size effect on cell uptake in well-suspended, uniform mesoporous silica nanoparticles. Small 5 (12), 1408–1413.

Lundqvist, M., Sethson, I., Jonsson, B.H., 2004. Protein adsorption onto silica nanoparticles: conformational changes depend on the particles' curvature and the protein stability. Langmuir 20 (24), 10639–10647.

Macomber, L., Rensing, C., Imlay, J.A., 2007. Intracellular copper does not catalyze the formation of oxidative DNA damage in *Escherichia coli*. J. Bacteriol. 189, 1616–1626.

Madigan, M.T., Martinko, J.M., Stahl, D.A., Clark, D.P., 2010. Brock Biology of Microorganisms, 13 Har/Psc edition. Pearson Benjamin-Cummings, San Francisco, 1043 pages.

Maness, P.-C., Smolinski, S., Blake, D.M., Huang, Z., Wolfrum, E.J., Jacoby, W.A., 1999. Bactericidal activity of photocatalytic TiO_2 reaction: toward an understanding of its killing mechanism. Appl. Environ. Microbiol. 65 (9), 4094–4098.

Morones, J.R., Elechiguerra, J.L., Camacho, A., Holt, K., Kouri, J.B., Ramirez, J.T., Yacaman, M.J., 2005. The bactericidal effect of silver nanoparticles. Nanotechnology 16, 2346–2353.

Nasrollahi, A., Pourshamsian, K., Mansourkiaee, P., 2011. Antifungal activity of silver nanoparticles on some of fungi. Int. J. Nano Dimens. 1 (3), 233–239.

Neal, A.L., 2008. What can be inferred from bacterium–nanoparticle interactions about the potential consequences of environmental exposure to nanoparticles? Ecotoxicology 17, 362–371.

Nel, A., Xia, T., Madler, L., Li, N., 2006. Toxic potential of materials at nanolevel. Science 311, 622–627.

Niazi, J.H., Gu, M.B., 2009. Toxicity of metallic nanoparticles in microorganisms- a review. In: Kim, Y., Platt, U., Gu, M.B., Iwahashi, H. (Eds.), Atmospheric and Biological Environmental Monitoring. Springer, pp. 193–206.

Nikolova, V., Angelova, S., Markova, N., Dudev, T., 2016. Gallium as a therapeutic agent: a thermodynamic evaluation of the competition between Ga^{3+} and Fe^{3+} ions in metalloproteins. J. Phys. Chem. B 120 (9), 2241–2248.
Olivi, M., Zanni, E., De Bellis, G., Talora, C., Sarto, M.S., Palleschi, C., Flahaut, E., Monthioux, M., Rapino, S., Uccelletti, D., Fiorito, S., 2013. Inhibition of microbial growth by carbon nanotube networks. Nanoscale 5 (19), 9023–9029.
Oves, M., Khan, M.S., Qari, A.H., Felemban, M.N., Almeelbi, T., 2016. Heavy metals: biological importance and detoxification strategies. J. Bioremediat. Biodegrad. 7, 334.
Pal, S., Tak, Y.K., Song, J.M., 2007. Does the antibacterial activity of silver nanoparticles depend on the shape of the nanoparticle? A study of the gram-negative bacterium *Escherichia coli*. Appl. Environ. Microbiol. 73 (6), 1712–1720.
Palza, H., 2015. Antimicrobial polymers with metal nanoparticles. Int. J. Mol. Sci. 16 (1), 2099–2116.
Park, Y.-H., Bae, H.C., Jang, Y., Jeong, S.H., Lee, H.N., Ryu, W.-I., Yoo, M.G., Kim, Y.-R., Kim, M.-K., Lee, J.K., Jeong, J., Son, S.W., 2013. Effect of the size and surface charge of silica nanoparticles on cutaneous toxicity. Mol. Cell. Toxicol. 9 (1), 67–74.
Pelletier, D.A., Suresh, A.K., Holton, G.A., McKeown, C.K., Wang, W., Gu, B., Mortensen, N.P., Allison, D.P., Joy, D.C., Allison, M.R., Brown, S.D., Phelps, T.J., Doktycz, M.J., 2010. Effects of engineered cerium oxide nanoparticles on bacterial growth and viability. Apppl. Environ. Microbiol. 76 (24), 7981–7989.
Pereira, Y., Lagniel, G., Godat, E., Baudouin-Cornu, P., Junot, C., Labarre, J., 2008. Chromate causes sulfur starvation in yeast. Toxicol. Sci. 106 (2), 400–412.
Peyrot, C., Wilkinson, K.J., Desrosiers, M., Sauvé, S., 2014. Effects of silver nanoparticles on soil enzyme activities with and without added organic matter. Environ. Toxicol. Chem. 33 (1), 115–125.
Prasad, R., Pandey, R., Barman, I., 2016. Engineering tailored nanoparticles with microbes: quo vadis? Wiley Interdiscip. Rev. 8, 316–330. http://dx.doi.org/10.1002/wnan.1363.
Premanathan, M., Karthikeyan, K., Jeyasubramanian, K., Manivannan, G., 2011. Selective toxicity of ZnO nanoparticles toward Gram-positive bacteria and cancer cells by apoptosis through lipid peroxidation. Nanomed. Nanotechnol. 7 (2), 184–192.
Ray, P.C., Yu, H., Fu, P.P., 2009. Toxicity and environmental risks of nanomaterials: challenges and future needs. J. Environ. Sci. Health C Environ. Carcinog. Ecotoxicol. Rev. 27 (1), 1–35.
Reddy, K.M., Feris, K., Bell, J., Wingett, D.G., Hanley, C., Punnoose, A., 2007. Selective toxicity of zinc oxide nanoparticles to prokaryotic and eukaryotic systems. Appl. Phys. Lett. 90, 213902.
Reddy, L.S., Nisha, M.M., Joice, M., Shilpa, P.N., 2014. Antimicrobial activity of zinc oxide (ZnO) nanoparticle against *Klebsiella pneumonia*. Pharm. Biol. 52 (11), 1388–1397.
Reidy, B., Haase, A., Luch, A., Dawson, K.A., Lynch, I., 2013. Mechanisms of silver nanoparticle release, transformation and toxicity: a critical review of current knowledge and recommendations for future studies and applications. Materials 6, 2295–2350. http://dx.doi.org/10.3390/ma6062295.
Riding, M.J., Martin, F.L., Trevisan, J., Llabjani, V., Patel, I.I., Jones, K.C., Semple, K.T., 2012. Concentration dependent effects of carbon nanoparticles in gram negative bacteria determined by infrared spectroscopy with multivariate analysis. Environ. Pollut. 163, 226–234.
Rodrigues, D.F., Jaisi, D.P., Elimelech, M., 2013. Toxicity of functionalized single-walled carbon nanotubes on soil microbial communities: implications for nutrient cycling in soil. Environ. Sci. Technol. 47, 625–633.
Rossman, T.G., Roy, N.K., Lin, W.C., 1992. Is Cadmium Genotoxic? IARC scientific publications, pp. 367–375.
Safe Work Australia Research Report, 2009–2010. Engineered Nanomaterials: Evidence on the Effectiveness of Workplace Controls to Prevent Exposure. Commonwealth of Australia, Barton, p. 132.
Sajid, M., Ilyas, M., Basheer, C., Tariq, M., Daud, M., Baig, N., Shehzad, F., 2015. Impact of nanoparticles on human and environment: review of toxicity factors, exposures, control strategies, and future prospects. Environ. Sci. Pollut. Res. 22, 4122–4143.
Schiller, C.M., Fowler, B.A., Woods, J.S., 1977. Effects of arsenic on pyruvate dehydrogenase activation. Environ. Health Perspect. 19, 205–207.
Schlich, K., Hund-Rinke, K., 2015. Influence of soil properties on the effect of silver nanomaterials on microbial activity in five soils. Environ. Pollut. 196, 321–330.
Science for Environment Policy, July 15, 2016. In: SCU (Ed.), Nanoparticles' Ecological Risks: Effects on Soil Microorganisms, European Commission DG Environment News Alert Service. The University of the West of England, Bristol. Issue 463.
Seil, J.T., Webster, T.J., 2012. Antimicrobial applications of nanotechnology: methods and literature. Int. J. Nanomed. 7, 2767–2781.

Sera, N., Tokiwa, H., Miyata, N., 1996. Mutagenicity of the fullerene C_{60}-generated singlet oxygen dependent formation of lipid peroxides. Carcinogenesis 17, 2163–2169.

Shang, L., Nienhaus, K., Nienhaus, G.U., 2014. Engineered nanoparticles interacting with cells: size matters. J. Nanobiotechnol. 12, 5. http://dx.doi.org/10.1186/1477-3155-12-5.

Shao, X.R., Wei, X.Q., Song, X., Hao, L.Y., Cai, X.X., Zhang, Z.R., Peng, Q., Lin, Y.F., 2015. Independent effect of polymeric nanoparticle zeta potential/surface charge, on their cytotoxicity and affinity to cells. Cell Prolif. 48 (4), 465–474.

Sharma, P., Jha, A.B., Dubey, R.S., Pessarakli, M., 2012. Reactive oxygen species, oxidative damage, and antioxidative defense mechanism in plants under stressful conditions. J. Bot. 2012:217037. http://dx.doi.org/10.1155/2012/217037.

Shen, Z., Chen, Z., Hou, Z., et al., 2015. Ecotoxicological effect of zinc oxide nanoparticles on soil microorganisms. Front. Environ. Sci. Eng. 9, 912–918. http://dx.doi.org/10.1007/s11783-015-0789-7.

Shoults-Wilson, W.A., Reinsch, B.C., Tsyusko, O.V., Bertsch, P.M., Lowry, G.V., Unrine, J.M., 2011. Effect of silver nanoparticle surface coating on bioaccumulation and reproductive toxicity in earthworms (*Eisenia fetida*). Nanotoxicology 5, 432–444. http://dx.doi.org/10.3109/17435390.2010.537382.

Simonin, M., Richaume, A., 2015. Impact of engineered nanoparticles on the activity, abundance, and diversity of soil microbial communities: a review. Environ. Sci. Pollut. Res. 22, 13710–13723. http://dx.doi.org/10.1007/s11356-015-4171-x.

Singh, R., Kumar, M., Mittal, A., Mehta, P.K., 2017. Microbial metabolites in nutrition, healthcare and agriculture. 3 Biotech 4. http://dx.doi.org/10.1007/s13205-016-0586-4.

Sinha, R., Khare, S.K., 2014. Differential interactions of halophilic and non-halophilic proteases with nanoparticles. Sustain. Chem. Process 2, 4. http://dx.doi.org/10.1186/2043-7129-2-4.

Sirelkhatim, A., Mahmud, S., Seeni, A., Kaus, N.H.M., Ann, L.C., Bakhori, S.K.M., Hasan, H., Mohamad, D., 2015. Review on zinc oxide nanoparticles: antibacterial activity and toxicity mechanism. Nano Micro Lett. 7 (3), 219–242.

Six, J., Paustian, K., Elliott, E.T., Combrink, C., 2000. Soil structure and organic matter i. Distribution of aggregate-size classes and aggregate-associated carbon. Soil Sci. Soc. Am. J. 64, 681–689.

Sohaebuddin, S.K., Thevenot, P.T., Baker, D., Eaton, J.W., Tang, L., 2010. Nanomaterial cytotoxicity is composition, size, and cell type dependent. Part. Fibre Toxicol. 7, 22. http://dx.doi.org/10.1186/1743-8977-7-22.

Sriram, M.I., Kalishwaralal, K., Barathmanikanth, S., Gurunathani, S., 2012. Size-based cytotoxicity of silver nanoparticles in bovine retinal endothelial cells. Nanosci. Methods 1, 56–77. http://dx.doi.org/10.1080/17458080.2010.547878.

Stoehr, L.C., Gonzalez, E., Stampfl, A., Casals, E., Duschl, A., Puntes, V., Oostingh, G.J., 2011. Shape matters: effects of silver nanospheres and wires on human alveolar epithelial cells. Part. Fibre Toxicol. 8, 36. http://dx.doi.org/10.1186/1743-8977-8-36.

Stohs, S.J., Bagchi, D., 1995. Oxidative mechanisms in the toxicity of metal ions. Free Radic. Biol. Med. 18 (2), 321–336.

Sumner, E.R., Shanmuganathan, A., Sideri, T.C., Willetts, S.A., Houghton, J.E., Avery, S.V., 2005. Oxidative protein damage causes chromium toxicity in yeast. Microbiology 151 (6), 1939–1948.

Sun, T.Y., Gottschalk, F., Hungerbühler, K., Nowack, B., 2014. Comprehensive probabilistic modelling of environmental emissions of engineered nanomaterials. Environ. Pollut. 185, 69–76.

Sunada, K., Kikuchi, Y., Hashimoto, K., Fujishima, A., 1998. Bactericidal and detoxification effects of TiO_2 thin film photocatalysts. Environ. Sci. Technol. 32, 726–728.

Tang, Y.J., Ashcroft, J.M., Chen, D., Min, G., Kim, C.H., Murkhejee, B., Larabell, C., Keasling, J.D., Chen, F.F., 2007. Charge associated effects of fullurene derivatives on microbial structural integrity and central metabolism. Nano Lett. 7, 754–760.

Tang, Z.X., Lv, B.F., 2014. MgO nanoparticles as antibacterial agent: preparation and activity. Braz. J. Chem. Eng. 31, 591–601.

Thill, A., Zeyons, O., Spalla, O., Chauvat, F., Rose, J., Auffan, M., Flank, A.M., 2006. Cytotoxicity of CeO_2 nanoparticles for *Escherichia coli*. Physico-chemical insight of the cytotoxicity mechanism. Environ. Sci. Technol. 40, 6151–6156.

Thomas, D.J., 2009. Unraveling arsenic—glutathione connections. Toxicol. Sci. 107, 309–311.

Tian, L., 2010. Understanding Soil Organic Matter Mineralization in Agroecosystems: Soil Enzyme Perspectives (Ph.D. dissertation). North Carolina State University, Raleigh, North Carolina.

Tian, G., Kang, B.T., Brussaard, L., 1992. Biological effects of plant residues with contrasting chemical compositions under humid tropical conditions - decomposition and nutrient release. Soil Biol. Biochem. 24, 1051–1060.

Tripathi, D.K., Tripathi, A., Shweta, S.S., Singh, Y., Vishwakarma, K., Yadav, G., Sharma, S., Singh, V.K., Mishra, R.K., Upadhyay, R.G., Dubey, N.K., 2017. Uptake, accumulation and toxicity of silver nanoparticle in autotrophic plants, and heterotrophic microbes: a concentric review. Front. Microbiol. 8.

Tsuang, Y.H., Sun, J.S., Huang, Y.C., Lu, C.H., Chang, W.H., Wang, C.C., 2008. Studies of photokilling of bacteria using titanium dioxide nanoparticles. Artif. Organs 32, 167–174.

Vertegel, A.A., Siegel, R.W., Dordick, J.S., 2004. Silica nanoparticle size influences the structure and enzymatic activity of adsorbed lysozyme. Langmuir 20, 6800–6807.

Waalewijn-Kool, P.L., Ortiz, M.D., Lofts, S., van Gestel, C.A.M., 2013. The effect of pH on the toxicity of zinc oxide nanoparticles to folsomia candida in amended field soil. Environ. Toxicol. Chem. 32, 2349–2355.

Wang, D., Lin, Z., Wang, T., Yao, Z., Qin, M., Zheng, S., 2016a. Where does the toxicity of metal oxide nanoparticles come from: the nanoparticles, the ions, or a combination of both? J. Hazard. Mater. 308, 328–334.

Wang, Y., Zhu, X., Lao, Y., Lv, X., Tao, Y., Huang, B., Wang, J., Zhou, J., Cai, Z., 2016b. TiO_2 nanoparticles in the marine environment: physical effects responsible for the toxicity on algae *Phaeodactylum tricornutum*. Sci. Total Environ. 565, 818–826.

Warnes, S.L., Highmore, C.J., Keevil, C.W., 2012. Horizontal transfer of antibiotic resistance genes on abiotic touch surfaces: implications for public health. mBio 3. http://dx.doi.org/10.1128/mBio.00489-12.

Wright, S., Upadhyaya, A.A., 1998. Survey of soils for aggregate stability and glomalin, a glycoprotein produced by hyphae of arbuscular mycorrhizal fungi. Plant Soil 198, 97–107.

Xie, Y., He, Y., Irwin, P.L., Jin, T., Shi, X., 2011. Antibacterial activity and mechanism of action of zinc oxide nanoparticles against Campylobacter jejuni. Appl. Environ. Microbiol. 77 (7), 2325–2331.

Yang, H., Liu, C., Yang, D., Zhang, H., Xi, Z., 2009. Comparative study of cytotoxicity, oxidative stress and genotoxicity induced by four typical nanomaterials: the role of particle size, shape and composition. J. Appl. Toxicol. 29, 69–78.

Zahir, Z.A., Munir, A., Asghar, H.N., Shaharoona, B., Arshad, M., 2008. Effectiveness of rhizobacteria containing ACC deaminase for growth promotion of peas (*Pisum sativum*) under drought conditions. J. Microbiol. Biotechnol. 18, 958–963.

Zhang, W., Hughes, J., Chen, Y., 2012a. Impacts of hematite nanoparticle exposure on biomechanical, adhesive, and surface electrical properties of *Escherichia coli* cells. Appl. Environ. Microbiol. 78, 905–3915.

Zhang, Y., Xu, D., Li, W., Yu, J., Chen, Y., 2012b. Effect of size, shape, and surface modification on cytotoxicity of gold nanoparticles to human HEp-2 and canine MDCK cells. J. Nanomater.:375496. http://dx.doi.org/10.1155/2012/375496.

Further Reading

Auld, D.S., 1995. Removal and replacement of metal ions in metallopeptidases. Methods Enzymol. 248, 228–242.

Cabelli, D.E., Riley, D., Rodriguez, J.A., Valentine, J.S., Zhu, H., 2000. Models of superoxide dismutases. In: Meunier, B. (Ed.), Biomimetic Oxidations Catalyzed by Transition Metal Complexes. Imperial College Press, London, pp. 461–508.

da Silva, F.J.J.R., Williams, R.J.P., 2001. The Biological Chemistry of the Elements. Oxford Univ. Press.

Dai, Y., Wensink, P.C., Abeles, R.H., 1999. One protein, two enzymes. J. Biol. Chem. 274, 1193–1195.

Eady, R.R., Robson, R.L., Richardson, T.H., Miller, R.W., Hawkins, M., 1987. The vanadium nitrogenase of *Azotobacter chroococcum*. Purification and properties of the VFe protein. Biochem. J. 244 (1), 197–207.

Gadd, G.M., 1992. Metals and microorganisms: a problem of definition. FEMS Microbiol. Lett. 100, 197–204.

Gallinari, P., di Marco, S., Jones, P., Pallaoro, M., Steinkuhler, C., 2007. HDACs, histone deacetylation, and gene transcription: from molecular biology to cancer therapeutics. Cell Res. 17, 195–211.

Gantt, S.L., Gattis, S.G., Fierke, C.A., 2006. Catalytic activity and inhibition of human histone deacetylase 8 is dependent on the identity of the active site metal ion. Biochemistry 45, 6170–6178.

Gomis-Rüth, F.X., Stöcker, W., Huber, R., Zwilling, R., Bode, W., 1993. Refined 1.8 A X-ray crystal structure of astacin, a zinc-endopeptidase from the crayfish *Astacusastacus* L. Structure determination, refinement, molecular structure and comparison with thermolysin. J. Mol. Biol. 229 (4), 945–968.

Haferburg, G., Kothe, E., 2007. Microbes and metals: interactions in the environment. J. Basic Microbiol. 47, 453–467.

Hernick, M., Fierke, C., 2010. Mechanisms of metal-dependent hydrolases in metabolism. In: Comprehensive Natural Products II: Chemistry and Biology, vol. 8. Elsevier Ltd., pp. 547–581.

Holland, D.R., Hausrath, A.C., Juers, D., Matthews, B.W., 1995. Structural analysis of zinc substitutions in the active site of thermolysin. Protein Sci. 4, 1955–1965.

Holm, R.H., Kennepohl, P., Solomon, E.I., 1996. Structural and functional aspects of metal sites in biology. Chem. Rev. 96, 2239–2314.

Hu, Y., Lee, C.C., Ribbe, M.W., 2012. Vanadium nitrogenase: a two-hit wonder? Dalton Trans. 41 (4), 10.

Key, H.M., Dydio, P., Clark, D.S., Hartwig, J.F., 2016. Abiological catalysis by artificial haem proteins containing noble metals in place of iron. Nature 534, 534–537.

Nichol, B.M., 1971. Metals as constituents of enzymes. Nutr. Rev. 29 (4), 97–100.

Perezgasga, L., Sánchez-Sánchez, L., Aguila, S., Vazquez-Duhalt, R., 2012. Substitution of the catalytic metal and protein PEGylation enhances activity and stability of bacterial phosphotriesterase. Appl. Biochem. Biotechnol. 166 (5), 1236–1247.

Roane, T.M., Pepper, I.L., 2000. Microorganisms and metal pollution. In: Maier, R.M., Pepper, I.L., Gerba, C.B. (Eds.), Environmental Microbiology. Academic Press, London, pp. 403–423.

Sabbioni, E., Girardi, F., Marafante, E., 1976. Replacement of metal in metalloenzymes. A lead-alkaline phosphatase. Biochemistry 15 (2), 271–276.

Singh, R., Gautam, N., Mishra, A., Gupta, R., 2011. Heavy metals and living systems: an overview. Indian J. Pharmacol. 43 (3), 246–253.

Solnick, J., Schauer, D.B., 2001. Emergence of diverse *Helicobacter* species in the pathogenesis of gastric and enterohepatic diseases. Clin. Microbiol. Rev. 14, 59–97.

Stadtman, E.R., 1993. Oxidation of free amino acids and amino acid residues in proteins by radiolysis and by metal-catalyzed reactions. Annu. Rev. Biochem. 62, 797–821.

Stern, S.T., McNeil, S.E., 2008. Nanotechnology safety concerns revisited. Toxicol. Sci. 101 (1), 4–21.

Valdez, C.E., Alexandrova, A.N., 2012. Why urease is a di-nickel enzyme whereas the CcrA β-lactamase is a di-zinc enzyme. J. Phys. Chem. B 116, 10649–10656.

Valdez, C.E., Smith, Q.A., Nechay, M.R., Alexandrova, A.N., 2014. Mysteries of metals in metalloenzymes. Acc. Chem. Res. 47 (10), 3110–3117.

Waldron, K.J., Robinson, N.J., 2009. How do bacterial cells ensure that metalloproteins get the correct metal? Nat. Rev. Microbiol. 7 (1), 25–35.

Xu, Y., Feng, L., Jeffrey, P.D., Shi, Y., Morel, F.M., 2008. Structure and metal exchange in the cadmium carbonic anhydrase of marine diatoms. Nature 452, 56–61.

Yang, W., Peters, J.I., Williams, R.O., 2008. Inhaled nanoparticles—a current review. Int. J. Pharm. 356, 239–247.

Yang, L., Liao, R.-Z., Ding, W.-J., Liu, K., Yu, J.-G., Liu, R.-Z., 2012. Why calcium inhibits magnesium-dependent enzyme phosphoserine phosphatase? A theoretical study. Theor. Chem. Acc. 131, 1275.

CHAPTER

20

Application of Nanotechnology to Enhance the Nutrient Quality of Food Crops and Agricultural Production

Pragya Mishra[1], Anita Singh[1], Raghvendra R. Mishra[2], SuYean Ong[3], Sheo M. Prasad[1]

[1]University of Allahabad, Allahabad, India; [2]Banaras Hindu University, Varanasi, India; [3]University Sains Malaysia, Penang, Malaysia

20.1 INTRODUCTION

Nanoscience and nanotechnology, being the novel frontiers of the 21st century, have shown a significant role in various fields such as computer electronics, communication, energy production, medicine, and the food industry. At the commencement of the second technical revolution, nanotechnology is considered as a novel technology (Ozimek et al., 2010) and is one of the most important tools in modern agriculture. Furthermore, agrifood nanotechnologies have worthwhile potency for the future. Despite this, nanotechnology could also benefit the agricultural food supply chains by offering potential applications to farmers especially during postharvest management, for the food industry (processing, preservation, and packaging), as well as for consumers via novel agrifood production (FAO/WHO, 2010). Nanotechnology comprises the prefix "nano," a Greek word meaning "dwarf." According to Pehanich (2006), nanotechnology is the study of matter at the dimensions of approximately 1–100 nm with more charge and larger surface area, higher stability and solubility, and a behavior that is different to their bulk-sized equivalent counterparts (Chowdappa and Gowda, 2013). The Royal Society explains nanotechnologies in terms of the design, characterization, production, and application of nanostructures, nanodevices, and nanosystems by understanding and controlling the shape and size of particles of any matter at the nanometer scale (RSRAE, 2004). Nanobiotechnology is considered as an application of nanotechnology in the biological fields. At present, nanotechnology has proved to be a multidisciplinary field that ensures

Nanomaterials in Plants, Algae, and Microorganisms
http://dx.doi.org/10.1016/B978-0-12-811487-2.00020-7

development of current methodology, technology, and facility as well as advanced opportunities in different dynamics of science such as engineering, physics, chemistry, and biology.

With inadequate availability of natural resources such as land, water, and soil fertility, demand for global food has been continually increasing because of alterations in climate, urbanization, and other anthropogenic activities, leading to environmental issues such as soil runoff, decreased fertility of soil, and accumulation of agrochemicals. There is a tremendous need to feed an estimated population growth from 6 to 9 billion by 2050 (Chen and Yada, 2011). Because of the limited reserves of natural gas and petroleum, the cost of chemical fertilizers, pesticides, and other production inputs has been drastically increased (Ditta, 2012). All the aforementioned facts have led researchers to overcome these constraints by using precision farming practices and effective nanotechnological applications to agriculture. Nowadays, agriculture as a source of food has become progressively important in a world with depleting resources and an ever-increasing global population. It is assumed that with this cumulative world population, there is a need to use the application of modern technologies such as nanotechnology and nanobiotechnology in the agricultural sector as well as in food sciences. With the commencement of nanotechnology, there is an excellent opportunity to tackle multifaceted technical issues related to the agricultural supply chains and have potential for revolutionary changes.

The United States Department of Agriculture roadmap published in Sep. 2003 (USDA, 2003) has addressed the application of nanotechnology in the agricultural and food industries for the first time. Previously, nanotechnologies have revolutionized health (Agrahari et al., 2016) and beauty care, the textile and automobile industries, information and communication technology, and energy sectors. Some of the examples in the market include antibacterial dressings, transparent sunscreen lotions, stain-resistant fabrics, scratch-free paints for cars, and self-cleaning windows. It is expected that nanotechnology will also remodel the entire food industry by altering the approach toward food production, processing, packaging, transportation, and consumption. However, the least influenced sectors are agronomy and environmental science, which are still in their infancy among the diverse applications of nanotechnology (Gonzalez-Melendi et al., 2008).

In the agricultural sector the key challenges to benefit from various applications of nanotechnology are arable areas with low productivity, large nonarable areas resulting in reduction of arable lands, wastage of inputs (water, fertilizers, pesticides), as well as final products, which obviously have negative impacts on food security. The main emphasis of nanotechnology agricultural research includes nanogenetic manipulation of agricultural crops, agricultural diagnostics, drug delivery and nanotechnology, precisely controlled release of nanofertilizers and nanocomplexes, nanobiosensors, nanopesticides, nanoherbicides, and nanobiofarming (Agrawal and Rathore, 2014; Singh and Prasad, 2016) as well as nanonutrients to alleviate the stress response in agricultural crops and maintain the quality and growth of plants (Wang et al., 2014; Tripathi et al., 2015, 2016). In spite of this, some reports also demonstrated the adverse impacts of nanoparticles on plants that not only deplete the growth and development of plants but also severely damage their metabolic system (Peralta-Videa et al., 2014, 2016; Tripathi et al., 2015, 2016; Cox et al., 2016; Shweta et al., 2016; Singh et al., 2016, 2017; Tiwari et al., 2017; Tripathi et al., 2017a,b,c,d,e). Nanotechnology will affect our lives tremendously over the next decade by its dynamic influence in various fields, such as food, medicine, and pharmaceutical sciences. In parallel, it can also elevate agricultural production, for which its applications include:

1. Agrochemicals (nanoformulations for applying pesticides and fertilizers for crop improvement);
2. Nanosensors/nanobiosensors (identification of diseases and residues of agrochemicals);
3. Nanodevices (genetic manipulation of plants);
4. Disease diagnostics applications among plants;
5. Nanotechnological application in husbandry and poultry (production, health care, animal breeding); and
6. Postharvest management.

In addition, applications of nanotechnology in the agricultural field also include nanotechnology-aided delivery systems for agriculture chemicals, field-sensing systems to monitor the environmental stresses and crop conditions, and improvement of plant traits against environmental stress and diseases.

20.2 NANOBIOTECHNOLOGICAL MATERIALS AND THEIR SYNTHESIS

Nanomaterials are generally classified on the basis of characterization, fabrication, and manipulation of structures, materials, or objects with a particle size less than 100nm (approximately 1–100nm in length) (Azeredo, 2009; Ozimek et al., 2010; Duncan, 2011; Cushen et al., 2012; Coles and Frewer, 2013) in at least one dimension. The European Union (European Commission, 2011) defines nanomaterial as a natural, incidental, or manufactured material containing particles, in an unbound state or as an agglomerate, whereby for 50% or more of the particles in the number size distribution, one or more external dimensions is in the size range of 1–100nm. Characteristically, nanoscale means size dimensions between approximately 1 and 100nm because at this scale the physical, chemical, and biological properties of materials differ with respect to the same properties at a larger scale (Fig. 20.1). A single nanometer (nm) is 1 billionth of a meter. Any form of a material with one or more dimensions in the nanoscale is known as nanomaterial. Nanoparticles are the most notorious nanomaterials; they have high reactivity because of their surface effects (Fiorani, 2005) because there is a high proportion of atoms located on their surface and consequently these particles can be subjected to phase transformation (Gilbert et al., 2003). At the nanoscale, materials that have one dimension but are extended in the two other dimensions are layers, for example, graphene, thin films, and surface coatings. On the other hand, materials that are nanoscale in two dimensions but extended in one dimension include nanowires and nanotubes. Meanwhile, materials with three dimensions are particles, for instance, precipitates, dendrimers, fullerenes, colloids, and tiny particles of semiconductor materials (quantum dots).

Because of increased relative surface area and quantum effects, nanomaterial-specific properties are derived. Likewise, any material that was intentionally produced to achieve specific properties with specific composition at the nanoscale is termed manufactured/engineered nanomaterial. Such engineered nanomaterials have different properties when compared

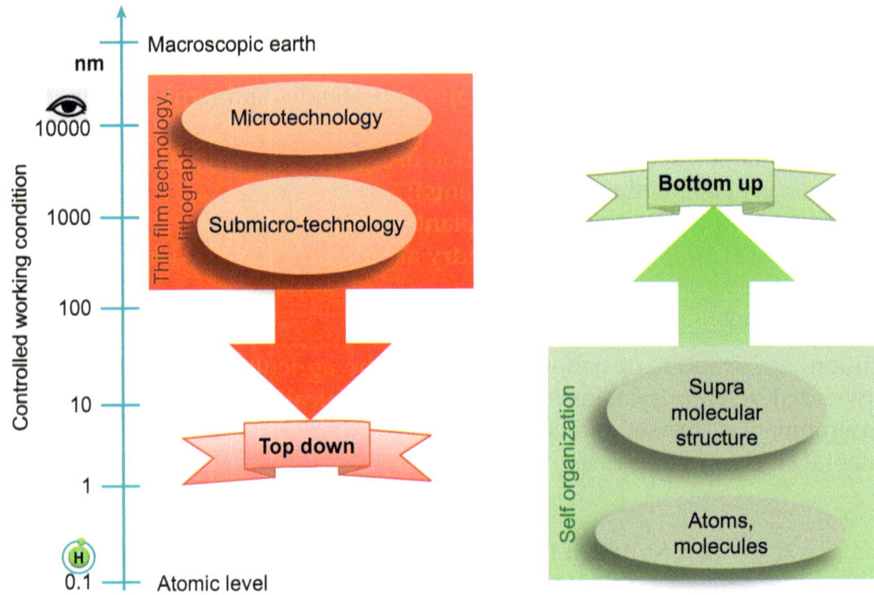

FIGURE 20.1 Nanoscale and manufacturing of nanomaterials. *Modified from Mura, S., Seddaiu, G., Bacchini, F., Roggero, P.P., Greppi, G.F., 2013. Advances of nanotechnology in agro-environmental studied. Ital. J. Agro, 8–18.*

with their conventional counterparts. Any discrete entity that has all three dimensions in the nanoscale is considered a nanoparticle.

There are different forms of nanomaterial: nanoparticles (discrete entities with three dimensions), nanotubes (cylindrical lattice arrangement of material), fullerenes (spherical molecular arrangement), nanofibers (length-to-diameter ratio of at least 3:1, and are in the nano range), nanowhiskers (fine fibers in the nano range, and dimensions of 5–20 nm in cross-section with lengths of several μm), and nanosheets (material where only one dimension is in the nano range) (Ozimek et al., 2010; Cushen et al., 2012).

There are several processes through which nanoparticles are synthesized physically, chemically, and biologically. The techniques involved in the physical synthesis of nanoparticles are vapor evaporation, condensation, laser fluence, and laser ablation. Some new techniques also came into existence that synthesize size-controlled metallic nanoparticles in the presence of both solution and gas matrices; these are known as pulsed laser ablation in aqueous media (Hodak et al., 2000; Singh and Gopal, 2007). The characterization of the synthesized nanoparticles is usually done by UV–visible absorption, transmission electron microscopy, small-angle X-ray diffraction analysis, and wide-angle X-ray diffraction analysis, which gives an idea of the structural property of nanoparticles (Jung et al., 2006). The chemical syntheses of nanoparticles are based upon reduction of metal ions in solution followed by formation of small metal clusters or aggregates. Some of the common techniques utilized for chemical synthesis of nanoparticles are chemical reduction, thermal decomposition, microemulsion technique, and photochemical/electrochemical method.

A new environmentally friendly approach toward the development of nanoparticles is based on the use of microorganisms and plants that have great potential for nanoparticle

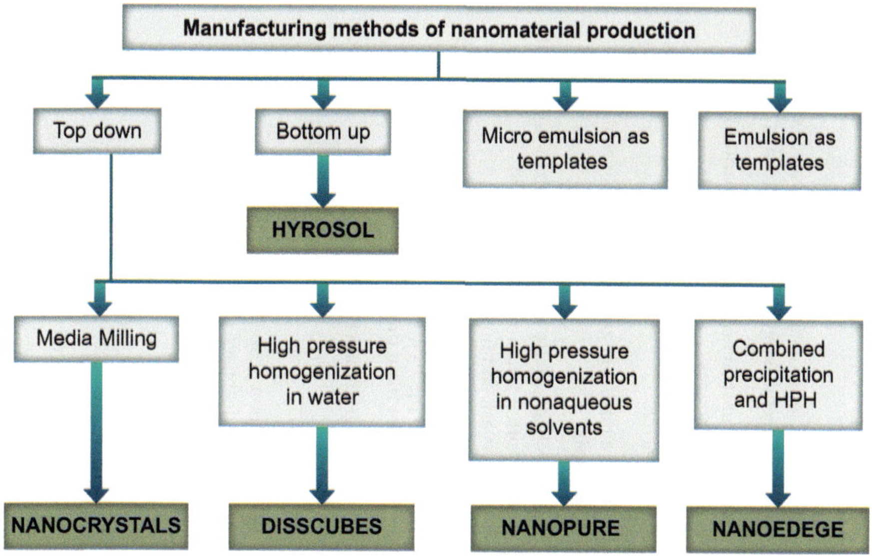

FIGURE 20.2 Nanomaterial manufacturing methods and different nanoparticles. *HPH*, high pressure homogenization. *Source: http://www.pharmainfo.net.*

synthesis. These biomolecules were applied as a replacement for chemicals used for reduction during nanoparticle synthesis. The applications for biological synthesis of nanoparticles are biocontrol (suppression of disease), biofertilization (enhancement of nutrient availability), and phytostimulation (production of plant hormones) (Martinez-Viveros et al., 2010; Bhattacharyya and Jha, 2012). In addition, plants in the form of inactivated tissue, extracts, and living plants are also used as a suitable alternative to physical and chemical methods for the biological synthesis of metal nanoparticles (especially gold and silver nanoparticles).

Some of the compounds such as hydrogen, sodium borohydride, hydrazine, hydrogen peroxide, hydroxylamine, ethylenediaminetetraacetic acid, and dimethylene sulfoxide were also used for the reduction of metal ions for the synthesis of nanoparticles (Singh and Prasad, 2016).

Whether the mode of nanoparticle synthesis involves the physical, chemical, or biological method, there are two main approaches to manufacturing nanomaterials, as shown in Fig. 20.2.

1. Top down (milling/homogenization): In this category, the manufacturing of nanomaterial involves breaking down larger particles of matter into particles of only nanometers in dimension, by using milling, nanolithography, or precision engineering.
2. Bottom up (controlled precipitation): This manufacturing category is the alternative production method of nanomaterials, which includes crystallization, layer-by-layer deposition, solvent extraction/evaporation, self-assembly, microbial synthesis, and biomass reactions (Azeredo, 2009; Cushen et al., 2012).

Regardless of the mode of synthesis, the synthesized nanoparticles will then be incorporated into the sectors of food production as shown in Table 20.1.

TABLE 20.1 List of Different Nanoparticles Used at the Production Site (Agricultural Sector) and Marketing Site (Food Sector)

	Nanoparticle	Size	Function	Advantage	References
At the production site (agricultural sector)	Silver nanoparticles	Diameter of 1–5nm	Antifungal and antibacterial agents	Controlling phytopathogens	Park (2006)
			Alternative to pesticides	Control of sclerotium forming phytopathogenic fungi	Jo et al. (2009) and Kim et al. (2012)
	Gold nanoparticles	–	Antibacterial activity	Effective against clinically isolated pathogens *Staphylococcus aureus* and *Escherichia coli*	MubarakAli et al. (2011)
	Surface-modified hydrophobic nanosilica	–		Controls a range of agricultural insect pests	Brecht et al. (2004)
	Copper nanoparticles	–	Antimicrobial activity	Antimicrobial activity against Gram-positive, Gram-negative bacteria, yeast, and fungi	Esteban-Tejeda et al. (2009)
			Copper-based biopesticides	Provides efficient delivery system for copper-based fungicides for plant protection	Brunel et al. (2013)
	Porous hollow silica nanoparticles	Thickness 15nm and a pore diameter of 4–5nm	Shielding that protects against pesticides and improves photostability	Prevents degradation by UV light	
	Single-walled carbon nanotubes	–	Functionalizes with poly-3-aminobenzenesulfonic acid for high dispersibility	Root elongation of six different crop species [cabbage (*Brassica oleracea*), carrot (*Daucus carota*), cucumber (*Cucumis sativus*), lettuce (*Lactuca sativa*), onion (*Allium cepa*), and tomato (*Solanum lycopersicum*)]	Canas et al. (2008)
	Multiwalled carbon nanotubes	–	Increases water uptake by seeds	Seed germination and growth of tomato plants	Khodakovskaya et al. (2009)

	Magnetic nanoparticles coated with tetramethylammonium hydroxide	–	As a stabilizing agent	Chlorophyll *a* level was increased in maize plants	Esquivel (2007)
	ZnO	25 nm		Growth and development of peanuts	Prasad et al. (2012)
	Nano-TiO_2	–	Improves light absorbance, promotes the activity of Rubisco activase, and enhances nitrogen metabolism	Enhances germination and growth of spinach seeds	Mahmoodzadeh et al. (2013)
At the marketing site (food sector)	Nanotubes from hydrolyzed milk protein α-lactalbumin	–	Carrier for nanoencapsulation of nutrients	Supplements pharmaceuticals	Bugusu et al. (2009)
	Casein micelles	–	Nanovehicles	Entrapment, protection, and delivery of sensitive hydrophobic nutraceuticals	Semo et al. (2007)
	Recrystallized amylose	10–20 nm	Encapsulates bioactive substances in microregions	Encapsulation	Sekhon (2014)
	Crystallites of triglycerides	10–100 nm	Formation of large clusters, flocs, and fat crystal networks	Texture improvement	IOM (2009)
	Nanosilver	–	Antimicrobial, antiodorant	Food packaging	WWICS (2009)
	Nanosilica	–	Used in food contact surfaces	Food packaging	Sekhon (2014)
	Nanoselenium	–	Additive to a green tea product	Health benefits (enhances uptake of selenium)	Sustech GMBH and Co. (2003)
	Nanocalcium and nanomagnesium	–	Health supplements		Sekhon (2014)
	Nanoiron	–	Decontaminant	Treatment of contaminated water	Sekhon (2014)

Nanotechnology has the potential to transform the agricultural sector as well as food industry by changing the way the food is produced, processed, packaged, transported, and consumed through the use of new tools and nanoscale devices with novel properties. As a result, a "smart" agrifood system is coined.

20.3 APPLICATION OF NANOBIOTECHNOLOGY AT THE PRODUCTION SITE (AGRICULTURAL SECTOR)

20.3.1 Maintenance of Plant and Other Food Crops

Progress in manufacturing has resulted in the fabrication of nanomaterials with a diverse range of sizes and shapes, which further established the basis for engineering to create unique properties targeted toward specified applications at the production site with special reference to the agricultural sector to protect the food crops (Fig. 20.3). Previously, it was reported that various fields such as medicine, environmental science, and food processing have employed the successful and safe use of nanomaterials. However, in the agriculture sector, especially for plant protection and production, the usages are limited and are becoming the thrust area in the research community.

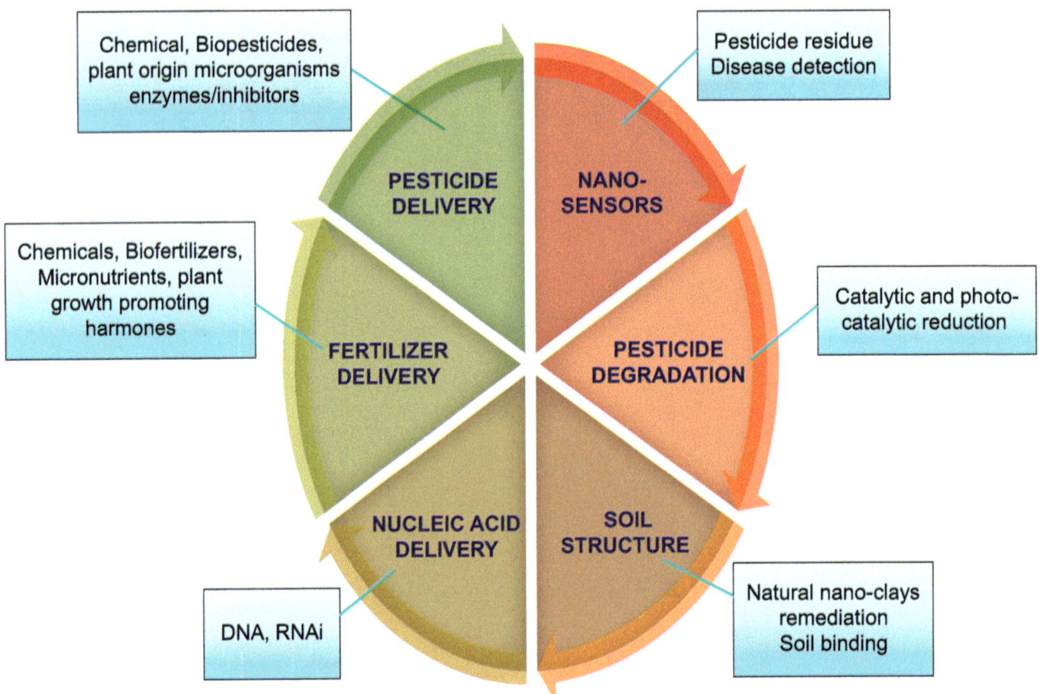

FIGURE 20.3 Application of nanotechnology in plant protection and nutrition. *Modified from Ghormade, V., Deshpande, M.V., Paknikar, K.M., 2011. Perspectives for nano-biotechnology enabled protection and nutrition of plants. Biotechnol. Adv. 29, 792–803.*

Preliminary studies have shown the potential of nanomaterials for improving seed germination and growth, plant protection, pathogen detection, and pesticide/herbicide residue detection. Nanoparticles enter plants through the cuticle, stomata, stigma, root tips, cortex, lateral root junction, and by wounding. As a result, improved root length, root area, chlorophyll content, enzyme activities, and protein content, which enhance the prevention of membrane damage and stress tolerance, are achieved. It has also been reported that nanomaterials have a significant role in plant protection by:

1. Increasing dispersion and wettability, and the affinity to target pesticides (Bergeson, 2010; Jianhui et al., 2005);
2. Providing better penetration of herbicides through cuticles and tissues of weeds (Gonzalez-Melendi et al., 2008; Perez-de-Luque and Rubiales, 2009);
3. Altering viral capsids to achieve different configurations and deliver specific nucleic acids, enzymes, or antimicrobial peptides acting against parasites (Torney et al., 2007).

20.3.2 Protection From Pathogens

The major limiting factor in cost-effective crop production is pests, including insects, mites, nematodes, and pathogens. To counteract them, pesticides are frequently used, which has resulted in the development of pest and disease resistance, accumulation of residues in produce, and environmental pollution. Therefore, there is an urgent need for an alternative approach to control pests and pathogens. Application of nanotechnology in crop protection holds a significant promise in the management of insects and pathogens by providing diagnostic tools for early detection and by controlling and targeting delivery of agrochemicals to counteract them. Nanoparticles remain bound to the cell wall of pathogens and cause deformity because of high-energy transfer leading to their death. Among a wide variety of bacterial and fungal pathogens the most common bacterial agents are *Erwinia carotovora*, *Pseudomonas* spp., *Corynebacterium*, and *Xanthomonas campestris*, which attack most vegetables. Fungal pathogens causing spoilage of vegetables are species belonging to genera *Alternaria, Aspergillus, Cladosporium, Colletotrichum, Phomopsis, Fusarium, Penicillium, Phoma, Phytophthora, Pythium, Rhizopus* spp., *Botrytis cinerea, Ceratocystis fimbriata, Rhizoctonia solani, Sclerotinia sclerotiorum*, and some mildews. These organisms are host specific and nonhost specific, causing huge economic losses, and some produce toxic metabolites and adversely affect human health.

They enter the plant tissue through mechanical or chilling injuries, and cause overwhelming losses (Tournas, 2005). So, the application of nanotechnology to solve this problem overcomes the environmental losses from the use of pesticides.

20.3.3 As a Nanonutrient to Maintain Nutrient Quality of Food Crops

Nutrients are required by all plants for growth and productivity despite the fact that the conditions of soil and water are vastly different from place to place. Soil composition, soil pH, humus content, and other factors determine the biological availability of nutrients in the land. Plants growing on the land and in the water pick up nutrients from the soil and water, respectively. The soil/water that is poor in nutrients or in a biologically inactive state result in nutrient deficiencies that eventually lead to loss of crop yield. Therefore, it is desirable to remedy imbalances in micronutrient availability to crops by using a nanonutrient that is economical, efficient, and easy to use. For example, a new patented nanonutrient called "Nualgi" has been developed. This product has all the plant growth micronutrients in nanoform and is biologically available for plant absorption. This can be used as a foliar spray/drip irrigation for all crops to remedy micronutrient deficiencies and to increase crop yields. Nualgi can also be used to grow diatom algae in any water including sewage-polluted water, which is also an environmental challenge. It is recommended that 100 g of Nualgi, which is dissolved in 100 L of water, can be sprayed onto 1 acre of crops once in 15 days to solve most problems such as purification of water. The dosage may be increased based on the severity of the problems. Previous studies have demonstrated the role of nanomaterials in enhancing the seed germination rates in tomato (Khodakovskaya et al., 2009), spinach (Zheng et al., 2005), and rice (Lin and Xing, 2007), at least in the early stages of growth, and consequently increase the crop dry matter yield. For slow and effective release of nutrients in the environment, zeolites are used. These are a group of naturally occurring minerals having a honeycomb-like layered crystal structure loaded with nitrogen and potassium in their network of interconnected tunnels and cages with other slowly dissolving ingredients containing phosphorus, calcium, and a complete set of minor and trace nutrients.

20.3.4 As a Nanofertilizer

India is the second largest consumer of fertilizers in the world and it imports almost its entire supply of nonurea fertilizers. The fertilizer was formulated using microbial enzymes to break down the respective salts into nano-form. Since it is a complete biosource, nanofertilizer is also eco-friendly and improves soil aggregation, moisture retention, and carbon build-up. There is no health hazard and it is suitable for all varieties of crops including food grains and vegetables as well as for horticulture. Fertilizers have an important role in enhancing food production in developing countries, especially after the introduction of high-yielding and fertilizer-responsive crop varieties. In spite of this, it is known that yields of many crops have begun to reduce as a result of imbalanced fertilization and decrease in soil organic matter. Moreover, excessive application of nitrogen and phosphorus fertilizers has an adverse effect on the groundwater and also leads to eutrophication in aquatic ecosystems. Furthermore, the fact that fertilizer use efficiency is only about 20%–50% for nitrogen and 10%–25% for phosphorus fertilizers implies that food production will need to be much more efficient than ever before (Shaviv, 2000; Chinnamuthu and Boopathi, 2009). The application of excess volume of fertilizers has an enormous impact on the ecology as runoff (Tilman et al., 2002). Thus there is an urgent need to solve the excessive usage of pesticides and fertilizers by:

1. Finding alternatives to current pesticide and fertilizer deployment;
2. Rapidly and locally detecting the presence of pathogens and pests, as well as pesticides and nutrient levels; and
3. Developing methods for either agrochemical removal or degradation to promote soil health.

Because of the limited availability of land and water resources, development of agriculture can only be achieved by increasing the productivity via effective use of modern technologies. Among these, nanotechnology has the potential to revolutionize agricultural systems, biomedicine, environmental engineering, safety and security, water resources, energy conversion, and numerous other areas (Baruah and Dutta, 2009). With the increase in the world population and the demand for food, new agricultural practices have been developed to improve food production through the use of more effective pesticides and fertilizers. However, such modern technologies have caused uncontrolled release of undesired substances into the environment, with the potential to contaminate the soil and groundwater. Today, nanotechnology represents a promising approach to improve agricultural production and remediate the polluted sites.

20.3.5 As a Genetic Tool

Nanotechnology in combination with other technologies such as biotechnology can make genetic manipulation of plants easier. In this technique, nanomaterials such as nanofibers or nanocapsules are used as vectors for new genetic materials instead of conventional viral vectors. These new vehicles could carry a larger number of genes, as well as substances that are able to trigger gene expression (Miller and Senjen, 2008; Nair et al., 2010) or to control the release of genetic material over time. In such a manner, nanoparticles can be used as an alternative for transporting exogenous DNA into the cells, and, moreover, they can also be an effective way to produce transgenic plants. The possibility of linking nanoparticles with proteins, lipids, and ligands has enabled better transfection by making the crossing of membranes and the breaking of the endosomal barriers more efficient. Furthermore, with the addition of magnetic particles, it is possible to amend the intracellular kinetics of nanoparticle–DNA complexes.

Some examples of nanomaterial application as genetic tools include the application of nanofiber arrays to quickly and efficiently deliver genetic material into cells, as well as potential applications in drug delivery, crop engineering, and environmental monitoring. In spite of this, other examples include carbon nanofibers that are surface modified with plasmid DNA to achieve controlled biochemical manipulations in cells through their integration (Miller and Kinnear, 2007) and PEGylated chitosan nanoparticles that are quite versatile to control the release of genetic material; their transfection efficiency can also be modified because of PEGylation. In addition, the fluorescent-labeled starch nanoparticle was also reported as a plant transgenic vehicle in which the nanoparticle biomaterial was designed in such a way that it binds and transports genes across the cell wall of plant cells by inducing instantaneous pore channels in the cell wall, cell membrane, and nuclear membrane with the help of ultrasound.

DNA-coated silver nanoparticle (coated with plasmid) treatments have been demonstrated to penetrate the isolated protoplast of petunia and carry plasmatic DNA into the nucleus by incubation with ethylene glycol (Rad et al., 2013). Currently, a new method for direct delivery of DNA into intact plant cells, which is known as gene gun or particle bombardment, is being used. In this technique, gold particles are used for bombardment because they adsorb DNA readily and are nontoxic to cells. It was reported that in intact tobacco and maize tissues, plasmid DNA transferred by the gene gun method using gold-capped nanoparticles was successfully expressed. This method has great advantage because the delivery of both DNA and effector molecules into the specific sites is simultaneous and thus results in site-targeted delivery and expression of chemicals and genes, respectively (Nair et al., 2010).

20.4 APPLICATIONS OF NANOBIOTECHNOLOGY AT THE MARKETING SITE (FOOD SECTOR)

A wide range of industries and markets have shown increasing interest in the use of nanomaterials because of their well-known benefits (Chen and Hu, 2005) in maintaining safety and quality of food, while satiating increased consumer demand and also fulfilling the food safety guidelines and regulations given by the government. Nanotechnology offers a wide range of applications, ranging from production, processing, safety, packaging, and transportation to the delivery of food products to the ultimate consumers. Food becomes nanofood when nanoparticles, nanotechnology techniques, or tools are used during cultivation, production, processing, or packaging of the food. The application of nanobiotechnology in the food and dairy industries involves the encapsulation and delivery of substances in the targeted sites, increasing the flavor, introducing antibacterial nanoparticles into food, enhancing shelf-life, sensing contamination, improving food storage, tracking, tracing, and brand protection. The most profitable sectors that achieved a competitive advantage and positioning through the use of nanotechnology are food and dairy manufacturers, agricultural producers, and consumers (Sekhon, 2014).

Nowadays, the food industry is constantly seeking newer technologies that can help to improve hygiene, extend the shelf-life of products, prevent food-borne illnesses, and detect and control chemical as well as physical contamination. To achieve these goals, nanotechnology is one of the best technologies to provide the solution to improve the processes of production, packaging, distribution and storage of the food, and their nutritional components. The potential of food nanotechnology seems unlimited and offers numerous promising applications for improved food production, processing, packaging, and storage (Fig. 20.4).

The major issue during the processing of any food product is the loss of nutrients. The loss of nutrients in food can be prevented by the use of nanoencapsulated food additives and supplements that signify greater advantage over the conventional use of similar substances in food. They function as nanocarrier systems and the technique used for their development is known as microencapsulation. Some of the existing nanoencapsulated additives are vitamins, antioxidants, colors, flavors, and preservatives. Examples of nanonutrients and supplements are omega-3 fatty acids, vitamins A, E, and D_2, isoflavones, and β-carotene. Apart from this, nano-size carriers based on nanoencapsulated substances in liposomes, micelles, or protein have also been developed to retain the nutrient with controlled release when required, thereby enhancing the bioavailability of certain compounds such as flavors and odors (Chaudhry and Castle, 2011; Cushen et al., 2012).

In food-packaging production, one of the major challenges is the migration and permeability of the packaging material. To solve such a problem, polymer nanocomposites are proven to be very useful and have desired characteristics, such as gas and water vapor barrier properties, mechanical strength, thermal and chemical stability, recyclability, biodegradability, dimensional stability, heat resistance, and optical clarity (Sorrentino et al., 2007; Lee,

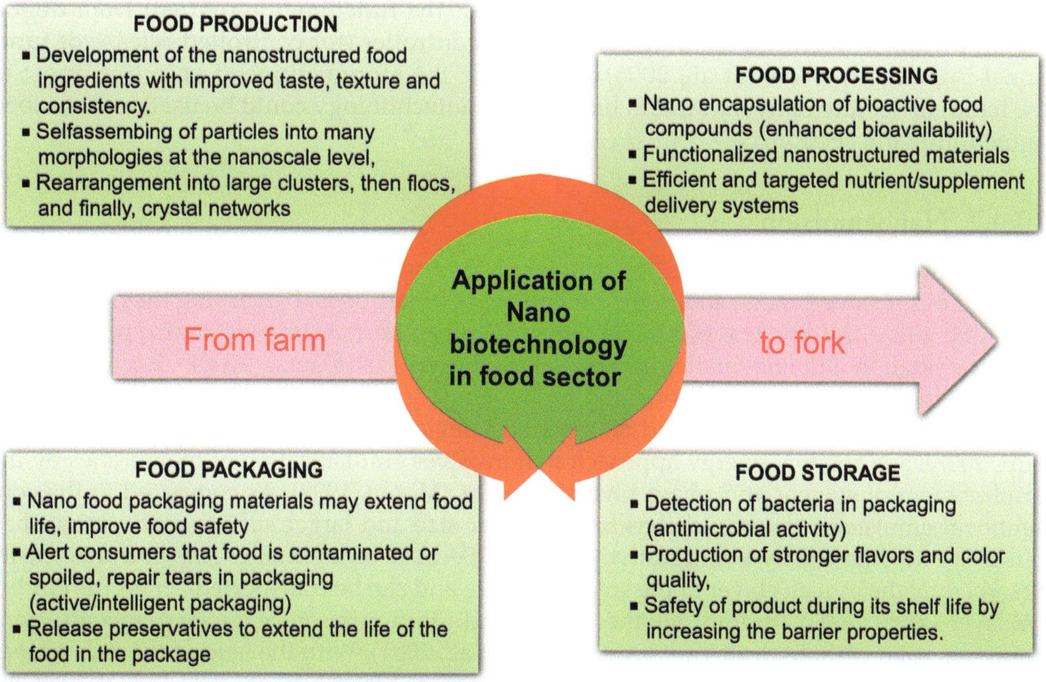

FIGURE 20.4 Application of nanobiotechnology in different food sectors. *Modified from Ranjan, S., Dasgupta, N., Chakraborty, A.R., Melvin Samuel, S., Ramalingam, C., Shanker, R., Kumar, A., 2014. Nanoscience and nanotechnologies in food industries: opportunities and research trends. J. Nano. Res. 16, 2464.*

2010; Duncan, 2011). This packaging supports the preservation of fresh foods, extends their shelf-life, and reduces packaging waste associated with processed foods. On the other hand, the nanoparticles (nanoplatelets, silica, carbon nanotubes, graphene, starch nanocrystals, cellulose-based nanofibers, chitin or chitosan nanoparticles, etc.) were used as reactive agents, which can be incorporated into packaging in the form of nanosensors for quality and safety control of the packed products. These nanosensors have the potential to detect the presence of gases, aromas, physical, chemical, and biological contaminants, or to respond accordingly to the changes in the surrounding environmental conditions (Azeredo, 2009; Duncan, 2011). Hence because it is proven to have a dynamic role in food preservation, "intelligent" or "smart" nanopackaging is now used to monitor the condition and surrounding environment of the packaged food (Silvestre et al., 2011; Rhim et al., 2013).

In the marketing sector with respect to the food industry, the application of nanotechnology involves several methods. In the food sector, many useful functional ingredients were utilized to give the particular product functionality and these ingredients include drugs, vitamins, antimicrobials, antioxidants, flavorings, colorants, and preservative agents. All of them have different molecular, chemical, and physical properties with specific molecular weights and also physical states. Hence these functional ingredients require some form of delivery system because they are rarely utilized in their pure form (Qureshi et al., 2012). The application of nanotechnology includes a "nanoparticulate delivery system" (Chen et al., 2006) as a vehicle (carrying the functional ingredient to the targeted site of action), as a shield (to protect the functional ingredient from degradation during different processing stages), as a controller (for controlled release of functional ingredient) (Weiss et al., 2006). To achieve the aforementioned objective, most of the nanoparticulate delivery system based on nanotechnology could be used as described in the following section.

20.4.1 Nanoemulsions

There are several applications for nanoemulsions such as in salad dressing, in making flavored oils, in sweeteners, and in the processing of other foods. By varying the condition of heat, pH, and ultrasonic waves, the flavors of food can be changed and it also prevents the further oxidation and enzymatic reactions of food products (Kumar, 2000). Two approaches are used to prepare nanoemulsions: the high-energy approach that involves high-pressure valve homogenizers, an ultrasound method, and devices having high speed (Letchford and Burt, 2007), and the low-energy approach that involves emulsification of membranes by the displacement of solvent and phase inversion point (Moore, 1999). As compared to the conventional emulsions, nanoemulsions have smaller size and larger surface area so that they can be easily digested by the digestive enzymes. Because of its smaller size, it can be easily absorbed by the mucous layer of the small intestine and can be transported across the epithelial layer (Zarif, 2003). To improve the texture and uniformity of ice cream, nanoemulsions of proteins and carbohydrates are used (Hogan et al., 2001). With the help of the nanoencapsulation technique, a potential carrier of nutrients, e.g., biomolecules such as milk proteins and micelles and carbohydrates such as dextrin, can be synthesized (Fernandez et al., 2009). In addition, the nanoemulsions also play an important role as antimicrobial agent toward Gram-positive organisms, whereby they are used for decontamination of food-packaging articles (Wang et al., 2012). Nanoemulsions having an antimicrobial role can be synthesized from nonionic surfactants, soybean oil, and tributyl phosphate (Sanguansri and Augustin, 2006). They also help in maintaining the flavor of the functional compounds of food products by reducing the degradation action on enzymes, temperature, and other oxidation processes (Bhattacharya et al., 2007).

20.4.2 Nanosensors

Nanosensors can detect any change in color and gas production caused by food spoilage. Several gases such as hydrogen, hydrogen sulfide, nitrogen oxides, sulfur dioxide, and ammonia can be assessed by the sensors (Kang et al., 2007). As compared to the conventional sensors, nanosensors have high sensitivity and selectivity. The sensors are made up of several metals such as palladium, platinum, and gold (Mannino and Scampicchio, 2007). These nanosensors embedded with carbon black and polyaniline help in the detection of foodborne pathogens (Sozer and Kokini, 2009). Nanosensors are installed at the packaging plant to detect the presence of microorganisms responsible for causing harm to food, so there is no need to send food samples to the laboratory for further quality tests. There are different types of nanosensors such as array biosensors, nanoparticle-based sensors, electronic noses, nano-test strips, and nanocantilevers (Tang et al., 2009). To sense the chemical sensor, electronic noses are attached to a data-processing system (Vidhyalakshmi et al., 2009). Nano-gas sensors and nano-smart dust can also be used to detect different kinds of environmental pollution (Biswal et al., 2012). Applications of nanobiosensors are efficiently used to detect the presence of viruses and bacteria.

20.4.3 Nanocomposites

Nanocomposites basically provide a highly versatile chemical functionality that helps in the maintenance of freshness and reduces the risk of causing any microbial infestation of the food products. Therefore, they increase the shelf-life of the food products. Manufacturing industries can use nanocomposites to layer their bottles to prevent leakage. For packaging applications, bionanocomposites, which are made up of starch and cellulose derivatives, polylactic acid, polycaprolactone, polyhydroxybutyrate, and polybutylene succinate, are efficiently used as layering materials (Baeumner, 2004). Guard IN is another kind of nanocomposite that scavenges ethylene gas and helps in the ripening of vegetables and fruits (Gupta and Moulik, 2008). Being eco-friendly and biodegradable, nanocomposites are easily used in the food-packaging industries. Top Screen DS13 is a kind of nanocomposite that can be easily recyclable (Mason et al., 2006). NanoCeram PAC is also an eco-friendly nanocomposite that helps in the rapid absorption of unpleasant components and reduces the chance of foul odors (Bordes et al., 2009). Applications of lactase or cholesterol reductase in packaging materials provide these enzymes to consumers who have low levels of these enzymes (Brody, 2003). An enzyme immobilization technique for coating packaging materials provides a larger surface area and faster transfer rates and is considered as the most effective type of nanocomposite for use in the food-packaging industry (Burdo, 2005).

20.4.4 For Food Packaging

Nanopackaging is done to fulfill the objective of improving barrier properties and antimicrobial release, maintaining flavors, protecting foods, and extending shelf-life, which subsequently increase the facility of transportation, safety, and functionality. Some of the examples of nanopackaging potential applications are:

Nanocoatings: A waxy coating is used widely for some food such as fruits and vegetables. Nanoscale edible coatings, which are as thin as 5 nm wide, having moisture, lipid, and gas barrier properties, are also available (Rhim, 2004).

Nanolaminates: Two or more layers of materials that are physically or chemically bonded to each other and in the range of nanometer dimensions. Some of the materials used for the preparation of nanolaminates are polyelectrolytes (proteins, polysaccharides), charged lipids (phospholipids, surfactants), and colloidal particles (micelles, vesicles, droplets).

Food-packaging methods are used to ensure that the quality of the food is kept intact and that food is packaged in a way that is safe for consumption. Packaging mainly aims at providing physical protection to prevent the food from external shocks and vibration, microbial infestation, and temperature by providing barrier protection from scavenging oxygen and other spoilage-causing gases. The packaging materials are preferably made of biodegradable materials to reduce environmental pollution. This idea has been turned into reality because of the introduction of nanotechnology in the food-packaging industry. High barrier plastic with antimicrobial properties and contaminant detection are some of the methods taken to ensure better food packaging.

20.5 CONCLUSION

The potential uses and benefits of nanobiotechnology for the production site (agriculture) and marketing site (food sector) are enormous. Nanotechnology enhances the use of water, pesticides, and fertilizers by plants more efficiently and therefore, offers potential benefits to the farmers through agricultural produce and food production. In the food industry, innovative products were developed in terms of food processing, preservation, and packaging. These include creating biofuels, nanocomposites for plastic film coatings used in food packaging, antimicrobial nanoemulsions for decontamination of food, and nanobiosensors for identification of pathogen contamination. It is predicted that multidisciplinary approaches could potentially improve food production by incorporating new emerging technologies and disciplines such as chemical biology integrated with nanotechnologies to tackle the existing biological bottlenecks that are currently limiting further developments. Therefore, numerous nanoparticulate agroformulations with higher bioavailability and efficacy and better selectivity could be made available for farmers as well as for food processors in the near future. Despite the rapid developments in food nanotechnology, little is known about the occurrence, fate, and toxicity of nanoparticles. The potential benefits of nanotechnology for agriculture, food, fisheries, and aquaculture need to be balanced in a way that does not harm the environmental components such as soil, water, and air, and the occupational health of workers.

Acknowledgments

The authors are also thankful to the Head, Department of Botany, University of Allahabad, Allahabad, and also the Director, Center of Food Technology, University of Allahabad for providing necessary laboratory facilities during research work. The author A. Singh is thankful to the CSIR [No. 13(8913-A)/2017-pool] for providing fund as Senior Research Associate.

References

Agrahari, R., Bhatia, A.K., Sexna, A., Pant, G., Singh, R.P., Mishra, R.R., Mishra, V.K., et al., 2016. Advance applications of nanotechnology in medicine. Int. J. Sci. Eng. Res. 7 (4), 1284–1315.

Agrawal, S., Rathore, P., 2014. Nanotechnology pros and cons to agriculture: a review. Int. J. Curr. Microbiol. Appl. Sci. 3 (3), 43–55.

Azeredo, M.C.H., 2009. Nanocomposites for food packaging applications. Food Res. Int. 42, 1240–1253.

Baeumner, A., 2004. Nanosensors identify pathogens in food. Food Technol. 58 (8), 51–55.

Baruah, S., Dutta, J., 2009. Nanotechnology applications in pollution sensing and degradation in agriculture: a review. Environ. Chem. Lett. 7, 161–204.

Bergeson, L.L., 2010. Nanosilver pesticide products: what does the future hold? Environ. Qual. Manag. 19, 73–82.

Bhattacharya, S., Jang, J., Yang, L., Akin, D., Bashir, R., 2007. BioMEMS and nanotechnology-based approaches for rapid detection of biological entities. J. Rapid Methods Auto. Microbiol. 15 (1), 1–32.

Bhattacharyya, P., Jha, D., 2012. Plant growth-promoting rhizobacteria (PGPR): emergence in agriculture. World J. Microbiol. Biotechnol. 28, 1327–1350.

Biswal, S.K., Nayak, A.K., Parida, U.K., Nayak, P.L., 2012. Applications of nanotechnology in agriculture and food sciences. Int. J. Sci. Innov. Discov. 2 (1), 21–36.

Bordes, P., Pollet, E., Avérous, L., 2009. Nano-biocomposites: biodegradable polyester/nanoclay systems. Prog. Polymer Sci. 34 (2), 125–155.
Brecht, M.O., Datnoff, L.E., Kucharek, T.A., Nagata, R.T., 2004. Influence of silicon and chlorothalonil on the suppression of gray leaf spot and increase plant growth in St. Augustine grass. Plant Dis. 88, 338–344.
Brody, A.L., 2003. Nano food packaging technology. Food Technol. 57 (12), 52–54.
Brunel, F., El Gueddari, N.E., Moerschbacher, B.M., 2013. Complexation of copper (II) with chitosan nanogels: toward control of microbial growth. Carbohydr. Polym. 92, 1348–1356.
Bugusu, B., Mejia, C., Magnuson, B., Tafazoli, S., 2009. Global regulatory food policies on nanotechnology. Food Technol. 63 (5), 24–29.
Burdo, O.G., 2005. Nanoscale effects in food-production technologies. J. Eng. Phys. Thermophys 78 (1), 90–96.
Canas, J.E., Long, M., Nations, S., Vadan, R., Dai, L., Luo, M., Ambikapathi, R., Lee, E.H., Ols zyk, D., 2008. Effects of functionalized and nonfunctionalized single – walled carbon nanotubes on root elongation of select crop species. Environ. Toxicol. Chem. 27, 1922–1931.
Chaudhry, Q., Castle, L., 2011. Food applications of nanotechnologies: an overview of opportunities and challenges for developing countries. Trends Food Sci. Technol. 22, 595–603.
Chen, F., Hu, X., 2005. Study on red fermented rice with high concentration of monacolin k and low concentration of citrinin. Int. J. Food Microbiol. 103, 331–337.
Chen, H., Yada, R., 2011. Nanotechnologies in agriculture: new tools for sustainable development. Trends Food Sci. Technol. 22, 585–594.
Chen, H., Weiss, J., Shahidi, F., 2006. Nanotechnology in nutraceuticals and functional foods. Food Technol. 60 (3), 30–36.
Chinnamuthu, C.R., Boopathi, P.M., 2009. Nanotechnology and agroecosystem. Madras Agric. J. 96 (1–6), 17–31.
Chowdappa, P., Gowda, S., 2013. Nanotechnology in crop protection: status and scope. Pest Manage. Hortic. Ecosyst. 19 (2), 131–151.
Coles, D., Frewer, L.J., 2013. Nanotechnology applied to European food production – a review of ethical and regulatory issues. Trends Food Sci. Technol. 34 (1), 32–43.
Cox, A., Venkatachalam, P., Sahi, S., Sharma, N., 2016. Silver and titanium dioxide nanoparticle toxicity in plants: a review of current research. Plant Physiol. Biochem. 107, 147–163.
Cushen, M., Kerry, J., Morris, M., Cruz-Romero, M., Cummins, E., 2012. Nanotechnologies in the food industry – recent developments, risks and regulation. Trends Food Sci. Technol. 24, 30–46.
Ditta, A., 2012. How helpful is nanotechnology in agriculture. Adv. Nat. Sci. Nanosci. Nanotechnol. 3, 033002.
Duncan, V.T., 2011. Applications of nanotechnology in food packaging and food safety: barrier materials, antimicrobials and sensors. J. Colloid Interface Sci. 363, 1–24.
Esquivel, D.M.S., 2007. Magnetic nanoparticles in social insects: are they the geomagnetic sensors? In: Entomological Society of America. Annual Meeting; San Diego, CA. December 9–12.
Esteban-Tejeda, L., Malpart ida, F., Es teban-Cubillo, A., Pecharromán, C., Moya, J.S., 2009. Antibacterial and antifungal activity of a soda-lime glass containing copper nanoparticles. Nanotechnology 20, 505–701.
European Commission, 2011. Commission recommendation of 18October 2011 on the definition of nanomaterial, 2011/696/EU. In: Official Journal, L 275/38, 20/10/2011 available from: http://eurex.europa.eu/LexUriServ/LexUriServ.do?uri=OJ: L:2011:275:0038:0040:EN: PDF.
FAO/WHO, 2010. Food and Agriculture Organization of the United Nations and World Health Organization. Expert Meeting on the Application of Nanotechnologies in the Food and Agriculture Sectors: Potential Food Safety Implications Meeting Report. Rome, Italy. Available from: www.fao.org/docrep/012/i1434e/i1434e00.pdf.
Fernandez, A., Torres-Giner, S., Lagaron, J.M., 2009. Novel route to stabilization of bioactive antioxidants by encapsulation in electrospun fibers of zein prolamine. Food Hydrocoll. 23 (5), 1427–1432.
Fiorani, D., 2005. Nanostructure Science and Technology. Surface Effects in Magnetic Nanoparticles. XIV. 300 pp.
Ghormade, V., Deshpande, M.V., Paknikar, K.M., 2011. Perspectives for nano-biotechnology enabled protection and nutrition of plants. Biotechnol. Adv. 29, 792–803.
Gilbert, B., Zhang, H., Huang, F., Finnegan, M.P., Waychunas, G.A., Banfield, J.F., 2003. Special phase transformation and crystal growth pathways observed in nanoparticles. Geochem.Trans. 4, 20–28.
Gonzalez-Melendi, P., Fernandez-Pacheco, R., Coronado, M.J., 2008. Nanoparticles as smart treatment-delivery systems in plants: assessment of different techniques of microscopy for their visualization in plant tissues. Ann. Bot. (Lond.) 101, 187–195.
Gupta, S., Moulik, S.P., 2008. Biocompatible microemulsions and their prospective uses in drug delivery. J. Pharm. Sci. 97 (1), 22–45.

Hodak, J.H., Henglein, A., Giersig, M., Hartland, G.V., 2000. Laser induced inter-diffusion in AuAg core-shell nanoparticles. J. Phys. Chem. B 104, 11708–11718.
Hogan, S.A., McNamee, B.F., Dolores O'Riordan, E., O'Sullivan, M., 2001. Microencapsulating properties of sodium caseinate. J. Agri. Food Chem. 49 (4), 1934–1938.
IOM (Institute of Medicine), 2009. Nanotechnology in Food Products: Workshop Summary. The National Academies Press, Washington, DC. Available from: http://www.nap.edu/openbook.php?record_id=12633.
Jianhui, Y., Kelong, H., Yuelong, W., Suqin, L., 2005. Study on anti-pollution nanopreparation of dimethomorph and its performance. Chin. Sci. Bull. 50, 108–112.
Jo, Y.K., Kim, B.H., Jung, G., 2009. Antifungal activity of silver ions and nanoparticles on phytopathogenic fungi. Plant Dis. 93, 1037–1043.
Jung, J., Oh, H., Noh, H., Ji, J., Kim, S., 2006. Metal nanoparticles generation using a small ceramic heater with a local heating area. J. Aerosol Sci. 37, 1662–1670.
Kang, S., Pinault, M., Pfefferle, L.D., Elimelech, M., 2007. Single-walled carbon nanotubes exhibit strong antimicrobial activity. Langmuir 23 (17), 8670–8673.
Khodakovskaya, M., Dervishi, E., Mahmood, M., Xu, Y., Li, Z., Watanabe, F., Biris, A.S., 2009. Carbon nanotubes are able to penetrate plant seed coat and dramatically affect seed germination and plant growth. ACS Nano 3, 3221–3227.
Kim, S.W., Jung, J.H., Lamsal, K., Kim, Y.S., Min, J.S., Lee, Y.S., 2012. Antifungal effects of silver nanoparticles (AgNPs) against various plant pathogenic fungi. Mycobiology 40, 53–58.
Kumar, M.N.V.R., 2000. A review of chitin and chitosan applications. React. Funct. Polym. 46 (1), 1–27.
Lee, W., 2010. The anodization of aluminum for nanotechnology applications. J. Miner. Met. Mater. Soc. 62 (6), 57–63.
Letchford, K., Burt, H., 2007. A review of the formation and classification of amphiphilic block copolymer nanoparticulate structures: micelles, nanospheres, nanocapsules and polymersomes. Eur. J. Pharm. Biopharm. 65 (3), 259–269.
Lin, D., Xing, B., 2007. Phytotoxicity of nanoparticles: inhibition of seed germination and root growth. Environ. Poll. 150, 243–250.
Mahmoodzadeh, H., Nabavi, M., Kashefi, H., 2013. Effect of nanoscale titanium dioxide particles on the germination and growth of canola (*Brassica napus*). J. Ornam. Hortic. Plants 3, 25–32.
Mannino, S., Scampicchio, M., 2007. Nanotechnology and food quality control. Vet. Res. Commun. 31 (1), 149–151.
Martinez-Viveros, O., Jorquera, M., Crowley, D.E., Gajardo, G., Mora, M.L., 2010. Mechanisms and practical considerations involved in plant growth promotion by rhizobacteria. J. Soil Sci. Plant Nutr. 10 (3), 293–319.
Mason, T.G., Wilking, J.N., Meleson, K., Chang, C.B., Graves, S.M., 2006. Nanoemulsions: formation, structure, and physical properties. J Phys. Condens. Matter 18 (41), 635–666.
Miller, G., Kinnear, S., 2007. Nanotechnology: a new wave of assaults on our food. Clean Food Org. 4, 52–57.
Miller, G., Senjen, R., 2008. Out of the laboratory and on to our plates nanotechnology in food and agriculture. In: Friends of the Earth, Australia, Europe and U.S.A. available from: http://www.foeeurope.
Moore, S., 1999. Nanocomposite achieves exceptional barrier in films. Mod. Plast. 76 (2), 31–32.
MubarakAli, D., Thajuddin, N., Jeganathan, K., Gunasekaran, M., 2011. Plant extract mediated synthesis of silver and gold nanoparticles and its antibacterial activity against clinically isolated pathogens. Colloids Surf. B Biointerfaces 85, 360–365.
Mura, D., Seddaiu, N., Bacchini, K., Roggero, K., Greppi, M., 2013. Advances of nanotechnology in agro-environmental studied. Ital. J. Agro, 8–18.
Nair, R., Varguese, S.H., Nair, B.G., Maekawa, T., Yoshida, Y., Kumar, D.S., 2010. Nanoparticulate material delivery to plants. Plant Sci. 179, 154–163.
Ozimek, L., Pospiech, E., Narine, S., 2010. Nanotechnologies in food and meat processing. Acta Sci. Pol. Technol. Aliment. 9 (4), 401–412.
Park, K.H., 2006. Preparation Method Antibacterial Wheat Flour by Using Silver Nanoparticles. Korea Patent Application: WPI ACC NO: 2006–489267/200650.
Pehanich, M., 2006. Small gains in processing and packaging. Food Process. 11, 46–48.
Peralta-Videa, J.R., Hernandez-Viezcas, J.A., Zhao, L., Diaz, B.C., Ge, Y., Priester, J.H., Holden, P.A., Gardea-Torresdey, J.L., 2014. Cerium dioxide and zinc oxide nanoparticles alter the nutritional value of soil cultivated soybean plants. Plant Physiol. Biochem. 80, 128–135.

Peralta-Videa, J.R., Huang, Y., Parsons, J.G., Zhao, L., Lopez-Moreno, L., Hernandez-Viezcas, J.A., Gardea-Torresdey, J.L., 2016. Plant-based green synthesis of metallic nanoparticles: scientific curiosity or a realistic alternative to chemical synthesis? Nanotechnol. Environ. Eng. 1 (1), 4.

Perez-de-Luque, A., Rubiales, D., 2009. Nanotechnology for parasitic plant control. Pest Manag. Sci. 65, 540–545.

Prasad, T.N.V.K.V., Sudhakar, P., Sreenivasulu, Y., Latha, P., Munaswamy, V., Reddy, K.R., Sreeprasad, T.S.P., Sajanlal, R., Pradeep, T., 2012. Effect of nanoscale zinc oxide particles on the germination, growth and yield of peanut. J. Plant Nutr. 35 (6), 905–927.

Qureshi, M.A., Karthikeyan, S., Punita, K., Khan, P.A., Uprit, S., Mishra, U.K., 2012. Application of nanotechnology in food and dairy processing: an overview Pak. J. Food Sci. 22 (1), 23–31.

Rad, S.J., Naderi, R., Alizadeh, H., Yaraghi, A.S., 2013. Silver-nanoparticle as a vector in gene delivery by incubation. Int. Res. J. Appl. Life Sci. 2, 21–33.

Ranjan, S., Dasgupta, N., Chakraborty, A.R., Melvin Samuel, S., Ramalingam, C., Shanker, R., Kumar, A., 2014. Nanoscience and nanotechnologies in food industries: opportunities and research trends. J. Nano. Res. 16, 2464.

Rhim, J.W., Park, H.M., Ha, C.S., 2013. Bio-nanocomposites for food packaging applications. Prog. Polymer Sci. 38, 1629–1652.

Rhim, J.W., 2004. Increase in water vapor barrier property of biopolymer-based edible films and coatings by compositing with lipid materials. J. Food Sci. Biotechnol. 13, 528–535.

RSRAE (The Royal Society & Royal Academy of Engineering), 2004. Nanoscience and Nanotechnologies: Opportunities and Uncertain Ties. RS Policy Document 19/04. Royal Society, London. Available from: http://www.royalsoc.ac.uk.

Sanguansri, P., Augustin, M.A., 2006. Nanoscale materials development-a food industry perspective. Trends Food Sci. Technol. 17 (10), 547–556.

Sekhon, B.S., 2014. Nanotechnology in agri-food production: an overview. Nanotechnol. Sci. Appl. 7, 31–53.

Semo, E., Kesselman, E., Danino, D., Livney, Y.D., 2007. Casein micelle as a natural nano-capsular vehicle for nutraceuticals. Food Hydrocoll. 21, 936–942.

Shaviv, A., 2000. Advances in controlled release of fertilizers. Adv. Agron. 71, 1–49.

Shweta, Tripathi, D.K., Singh, S., Singh, S., Dubey, N.K., Chauhan, D.K., 2016. Impact of nanoparticles on photosynthesis: challenges and opportunities. Mater. Focus 5 (5), 405–411.

Silvestre, C., Duraccio, D., Cimmino, S., 2011. Food packaging based on polymer nanomaterials. Prog. Polymer Sci. 36, 1766–1782.

Singh, S.C., Gopal, G., 2007. Zinc nanoparticles in solution by laser ablation technique. Bull. Mater. Sci. 30 (3), 291–293.

Singh, A., Prasad, S.M., 2016. Nanotechnology and its role in agro-ecosystem: a strategic perspective. Int. J. Environ. Sci. Technol. http://dx.doi.org/10.1007/s13762-016-1062-8.

Singh, S., Tripathi, D.K., Dubey, N.K., Chauhan, D.K., 2016. Effects of nano-materials on seed germination and seedling growth: striking the slight balance between the concepts and controversies. Mater. Focus 5 (3), 195–201.

Singh, S., Vishwakarma, K., Singh, S., Sharma, S., Dubey, N.K., Singh, V.K., Liu, S., Tripathi, D.K., Chauhan, D.K., 2017. Understanding the plant and nanoparticle interface at transcriptomic and proteomic level: a concentric overview. Plant Gene. doi.org/10.1016/j.plgene.2017.03.006.

Sorrentino, A., Gorrasi, G., Vittoria, V., 2007. Potential perspectives of bio- nanocomposites for food packaging applications. Trends Food Sci. Technol. 18 (2), 84–95.

Sozer, N., Kokini, J.L., 2009. Nanotechnology and its applications in the food sector. Trends Biotechnol. 27 (2), 82–89.

Sustech, GMBH and Co. 2003. Patent Application EP20030748025, Sweet Containing Calcium, Germany; Sustech GMBH Co. 2004. International Patent Application PCT/EP2003/010213 Coated Chewing Gum, Germany.

Tang, D., Sauceda, J.C., Lin, Z., 2009. Magnetic nanogold microspheres-based lateral-flow immunodipstick for rapid detection of aflatoxin B2 in food. Biosens. Bioelectron. 25 (2), 514–518.

Tilman, D., Cassman, K.G., Matson, P.A., Naylor, R., Polasky, S., 2002. Agricultural sustainability and intensive production practices. Nature 418, 671–677.

Tiwari, M., Sharma, N.C., Fleischmann, P., Burbage, J., Venkatachalam, P., Sahi, S.V., 2017. Nanotitania exposure causes alterations in physiological, nutritional and stress responses in tomato (*Solanum lycopersicum*). Front. Plant Sci. 8.

Torney, F., Trewyn, B.G., Lin, V.S.-Y., Wang, K., 2007. Mesoporous silica nanoparticles deliver DNA and chemicals into plants. Nature Nanotechnol. 2, 295–300.

Tournas, V.H., 2005. Spoilage of vegetable crops by bacteria and fungi and related health hazards. Crit. Rev. Microbiol. 31, 33–44.

Tripathi, D.K., Singh, V.P., Prasad, S.M., Chauhan, D.K., Dubey, N.K., 2015. Silicon nanoparticles (SiNp) alleviate chromium (VI) phytotoxicity in *Pisum sativum* (L.) seedlings. Plant Physiol. Biochem. 96, 189–198.

Tripathi, D.K., Singh, S., Singh, V.P., Prasad, S.M., Chauhan, D.K., Dubey, N.K., 2016. Silicon nanoparticles more efficiently alleviate arsenate toxicity than silicon in maize cultiver and hybrid differing in arsenate tolerance. Front. Environ. Sci. 4. article 46.

Tripathi, D.K., Singh, S., Singh, S., Pandey, R., Singh, V.P., Sharma, N.C., Prasad, S.M., Dubey, N.K., Chauhan, D.K., 2017a. An overview on manufactured nanoparticles in plants: uptake, translocation, accumulation and phytotoxicity. Plant Physiol. Biochem. 110, 2–12.

Tripathi, D.K., Mishra, R.K., Singh, S., Singh, S., Vishwakarma, K., Sharma, S., Singh, V.P., Singh, P.K., Prasad, S.M., Dubey, N.K., Pandey, A.C., 2017b. Nitric oxide ameliorates zinc oxide nanoparticles phytotoxicity in wheat seedlings: implication of the ascorbate–glutathione cycle. Front. Plant Sci. 8.

Tripathi, D.K., Singh, S., Singh, S., Srivastava, P.K., Singh, V.P., Singh, S., Prasad, S.M., Singh, P.K., Dubey, N.K., Pandey, A.C., Chauhan, D.K., 2017c. Nitric oxide alleviates silver nanoparticles (AgNps)-induced phytotoxicity in Pisum sativum seedlings. Plant Physiol. Biochem. 110, 167–177.

Tripathi, D.K., Tripathi, A., Shweta, S.S., Singh, Y., Vishwakarma, K., Yadav, G., Sharma, S., Singh, V.K., Mishra, R.K., Upadhyay, R.G., Dubey, N.K., 2017d. Uptake, accumulation and toxicity of silver nanoparticle in autotrophic plants, and heterotrophic microbes: a concentric review. Front. Microbiol. 8.

Tripathi, D.K., Singh, S., Singh, V.P., Prasad, S.M., Dubey, N.K., Chauhan, D.K., 2017e. Silicon nanoparticles more effectively alleviated UV-B stress than silicon in wheat (*Triticum aestivum*) seedlings. Plant Physiol. Biochem. 110, 70–81.

USDA (United States Department of Agriculture), 2003. Nanoscale Science and Engineering for Agriculture and Food Systems. United States Department of Agriculture, Washington, DC, USA.

Vidhyalakshmi, R., Bhakyaraj, R., Subhasree, R.S., 2009. Encapsulation 'the future of probiotics'—a review. Adv. Biol. Res. 3 (3–4), 96–103.

Wang, Y., Zhang, Q., Zhang, C.-L., Li, P., 2012. Characterisation and cooperative antimicrobial properties of chitosan/nano-ZnO composite nanofibrous membranes. Food Chem. 132 (1), 419–427.

Wang, S., Wang, F., Gao, S., 2014. Foliar application with nano-silicon alleviates Cd toxicity in rice seedlings. Environ. Sci. Pollut. Res. Int. 22 (4), 2837–2845.

Weiss, J., Takhistov, P., Mc'clements, J.D., 2006. Functional materials in food nanotechnology. J. Food Sci. 71 (9), 107–116.

Woodrow Wilson International Centre for Scholars, 2009. The Nanotechnology Consumer Inventory. Available at: www.nanotechproject.org/inventories/consumer/.

Zarif, L., 2003. Nanocochleate cylinders for oral and parenteral delivery of drugs. J. Liposome Res. 13 (1), 109–110.

Zheng, L., Hong, F., Lu, S., Liu, C., 2005. Effect of nanoTiO2 on strength of naturally aged seeds and growth of spinach. Biol. Trace Elem. Res. 104, 83–91.

CHAPTER

21

Potential Applications and Avenues of Nanotechnology in Sustainable Agriculture

Kanchan Vishwakarma[1], *Neha Upadhyay*[1], *Nitin Kumar*[1], *Durgesh K. Tripathi*[1,2,4], *Devendra Kumar Chauhan*[2], *Shivesh Sharma*[1], *Shivendra Sahi*[3]

[1]Centre for Medical Diagnostic and Research, Motilal Nehru National Institute of Technology, Allahabad, India; [2]University of Allahabad, Allahabad, India; [3]Western Kentucky University, Bowling Green, KY, United States; [4]Banaras Hindu University, Varanasi, India

21.1 INTRODUCTION

Materials having dimensions smaller than 100 nm can be classified as nanomaterials (NMs). The advancements in nanotechnology alongside biotechnology have resulted in enhancement of the applications of NMs in different areas. Dendrimers (carbon based, metallic, and steel oxide based) and biocomposites (Nair et al., 2010; Rizwan et al., 2017) are currently under development. Utilization of these materials includes water purification, food processing, packaging, and environmental remediation (Byrappa et al., 2008; Zhang and Webster, 2009; Zuverza-Mena et al., 2017).

NMs comprise different varieties of constituents. Considering market volume, the most available categories contain inorganic nonmetallic NMs, carbon-based NMs, metallic nanoparticles (NPs), and organic polymeric materials. NMs can occur in different arrangements and are engineered according to the applications. In relation to public exposure and industrial impact, the aforementioned NMs are of greater relevance (Mazzaglia et al., 2017). However, new varieties of NMs are either under development or at early stages of market advancement. As per the records, the worldwide extent of NMs being marketed annually is approximately 11.5 million tons, having a market value of nearly $20 billion

Nanomaterials in Plants, Algae, and Microorganisms
http://dx.doi.org/10.1016/B978-0-12-811487-2.00021-9

(Mazzaglia et al., 2017). Carbon nanofibers and carbon nanotubes (CNTs) are presently marketed annually in amounts of thousands of tons. The marketed amount of nanosilver annually is estimated to be about 20 tons. Applications of NMs differ significantly, from commodity purposes in everyday goods to specialized applications. Apart from marketing, the estimation of release of NPs into different environmental ecosystems is required to quantify mass loadings as well as to identify their utilization in essential environmental implications. In Fig. 21.1, the low estimate signifies the lower estimate of production in 2010 including fewer release assumptions during manufacturing, use, and disposal. In such a situation, most of the engineered NPs end up in landfills in much higher amounts than those used in production and emission (Keller et al., 2013). In 2010, the release of engineered NMs from various sectors was collectively estimated to be the highest, at approximately 240,000 metric tons per year in landfills, whereas it was the lowest in air (Fig. 21.1).

In cosmetics, the major NMs include titanium dioxide, zinc oxide, and synthetic amorphous silica. Among biomedical applications, gold nanoparticles (Au NPs) and silver nanoparticles (Ag NPs) as medical diagnostics and antimicrobials are being used, respectively. Additionally, NMs can be used in paints and coatings, sunlight and gas cells, catalysts, etc. The financial divisions with major utilization of NMs are aerospace, transportation, agrifood, medicine, and nanobiotechnology. With the distribution of NPs, plant communities experience notable exposure to manufactured/engineered NMs and show different positive/negative responses (Shweta et al., 2016; Singh et al., 2016; Tripathi et al., 2015, 2016, 2017a,b,c,d,e; Cox et al., 2017; Du et al., 2017; Venkatachalam et al., 2017; Medina-Velo et al., 2017).

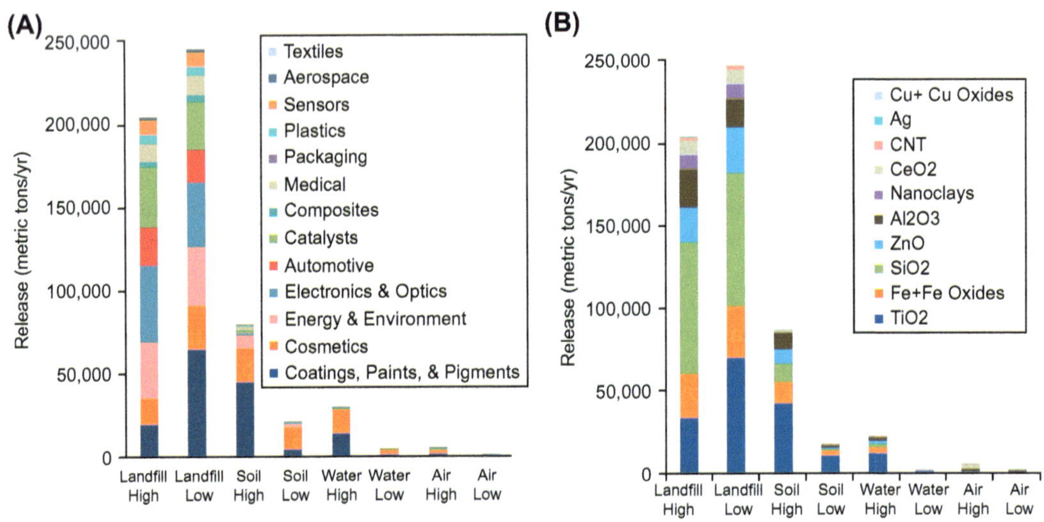

FIGURE 21.1 Estimated global releases of engineered nanomaterials into different compartments, by application, based on 2010 production estimates by (A) application and (B) engineered nanomaterial type. Low indicates low production and release assumption estimates, and high refers to high production and release assumptions. *Adapted with permission from Keller, A.A., McFerran, S., Lazareva, A., Suh, S., 2013. Global life cycle releases of engineered nanomaterials. J. Nanopart. Res. 15, 1692.*

21.1.1 Global Applications of Nanomaterials

Applications of nanotechnology in industrial products continue to be explored in spite of the fact that maximum applications are restricted to the bulk utilization of NMs. Examples of industrial applications of NMs comprise Ag NPs in food packaging, zinc oxide and titanium dioxide nanoparticles in sunscreen, disinfectants, and clothing, carbon nanotubes (CNTs) for stain-resistant textiles, etc. (Kah, 2015; Rizwan et al., 2017). By 2011, the project on emerging nanotechnologies assessed the availability of more than 1300 NM-based products and new products entering the market at a rate of three to four products per week (Gogos et al., 2012).

In developing countries, applications of nanotechnology include treatment of disease and avoidance of health issues. Other industrial applications of nanotechnology include purification, water filtration, environmental cleaning purposes such as desalination of water, wastewater treatment, and other nanoremediations. Industrial applications of NMs include military items, construction materials, nanofabrication of nanowires, nanorods, etc.

Nanobiotechnology is a biology-oriented field of research, a hybrid of nanoscience and biotechnology with equivalent contributions from both fields (Derrick et al., 2007). Applications of nanotechnology in agriculture are shown in Fig. 21.2.

The discipline corresponding to applications of nanotechnology in medical science is known as nanomedicine (Kah, 2015). It comprise the applications of nanomaterials, nanoelectronic biosensors, biological devices, etc. in medical science (Rao, 2016). The major problems for nanomedicine include the toxicity of these materials and their environmental impact.

The properties of NMs can be modified by conjugating them to biological molecules for different applications. The dimensions of NMs are comparable to those of biomolecular structures; hence, NMs can be beneficial for both in vitro and in vivo applications. The unification

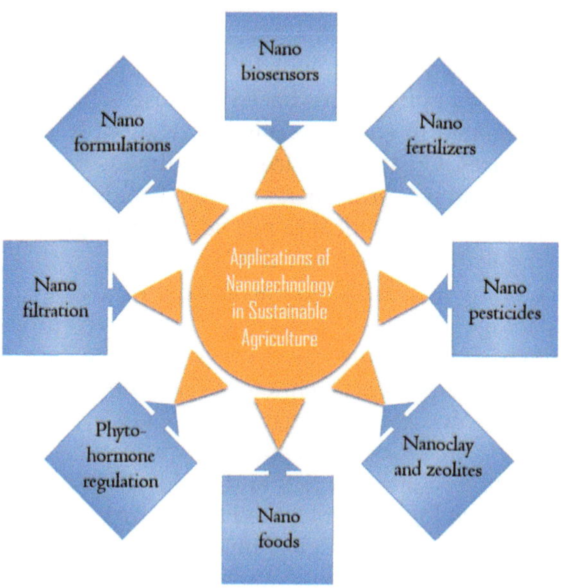

FIGURE 21.2 Nanotechnology in plant science and sustainable agriculture.

of NMs and biology has led to the development of analytical tools, diagnostic devices, physical therapy applications, and drug delivery vehicles.

As it is mentioned earlier, nanobiotechnology refers to the amalgamation of nanotechnology and biology (Kah, 2015). It highlights the merging of biological research with different fields of nanotechnology resulting in products such as nanodevices, NPs, nanovesicles, etc. (Zhang et al., 2016a,b). However, similar to nanotechnology and biotechnology, the field of nanobiotechnology represents certain potential ethical issues related to it. The most important objectives that recurrently originate in nanobiotechnology are focused on the application of nanotools to related biological problems and further refining these applications. Novel nanotools are developed by refining the already available nanotools. The imaging of tissues and biomolecules is also a crucial issue for nanobiotechnology researchers. Other topics of concern related to nanobiotechnology include the applications of nanophotonics for controlling molecular phenomena in cells and application of cantilever array sensors (Sastry et al., 2002). The use of microorganisms for synthesis of efficient NPs has created interest among researchers. These microbial phenomena have revealed new possibilities for novel applications of nanobiotechnology, such as biosynthesis of metallic nanomaterials. Microorganisms are capable of changing the oxidation state of metals ions. As compared to chemical and physical methods, these microbial processes can be accomplished in the aqueous phase under mild and environmentally suitable conditions (Liu and Lal, 2015).

Green nanotechnology corresponds to applications of nanotechnology in enhancement of environmental sustainability of processes causing adverse effects on the environment. The process of green nanotechnology includes production of green nanoproducts and applications of novel nanoproducts in the establishment of sustainability (Mansur et al., 2017). Green nanotechnology is defined as the development of clean technologies for minimizing potential human health and environmental risks associated with manufacture and applications of nanoproducts and replacement of present products with new environmentally friendly nanoproducts (Guo, 2012; Morganti et al., 2016).

In the last few decades, researchers have been on a quest for the development of new and enhanced energy technologies with the ability to improve life. To switch from the current generation of technology, there is development of energy applications based on nanotechnology. An important subfield of nanotechnology interrelated to energy is nanofabrication. It is the process of designing and creating nanoscale devices. Creating devices with dimensions less than 100 nm opens up new possibilities to capture, store, and transfer energy. The distinctive level of control that can be provided by nanofabrication is critical in providing the capability of deciphering many problems related to energy technologies (Guo, 2012).

Furthermore, researchers have started developing techniques to utilize nanotechnology to improve consumer-related products. The advantages achieved from the design of such products include improved efficiency of heating and lighting, increased energy storage capacity, and reduction in pollution from the use of energy. Nanotechnology has effects in the area of consumer goods. Several products that involve nanomaterials are already available. Applications of nanotechnology in textiles include embedded "wearable electronics." These new products have the potential particularly in the area of cosmetics, and find applications in heavy metal industry (Liu and Lal, 2015). Nanotechnology is anticipated to be a driving force of technology and business and has the potential to provide higher performance materials and intelligent systems alongside new production methods.

CNTs consist of cylinders having single or multiple layers of graphene. The diameter of single- and multiwalled CNTs ranges between 0.8–2.0 nm and 5.0–20.0 nm, respectively. The length of CNTs range from smaller than 100 nm to 0.5 m (Baughman et al., 2002). Their walls can be made up of either metal or semiconducting material depending on the coordination of the lattice in reference to the tube axis, known as chirality. The cross-sectional area of multiwalled CNTs has an elastic modulus of magnitude 1 TPa and a tensile strength of 100 GPa, which is nearly 10 times greater than any other industrial fiber. Single-walled CNTs have thermal conductivity of the order of 3500 W/m K, which is higher than that of diamond (Gogos et al., 2012).

By 2013, production of CNTs increased to 1000 tons per 12 months. CNTs find applications in energy storage, boat hulls, automotive materials, water filters, carrying items, skinny-movie electronics, actuators, coatings, and electromagnetic shields (Gogos et al., 2012). Organized CNT structures analogous to forests, yarns, and normal sheets were synthesized in smaller volumes (Gogos et al., 2012). A couple of reviews have emphasized the possibility of utilizing CNTs as building blocks to fabricate three-dimensional macroscopic (>1 mm) all-carbon devices. Lalwani et al. (2013) discovered a novel crosslinking method for the fabrication of macroscopic, porous, freestanding, carbon scaffold materials using both single- and multiwalled CNTs as building blocks. These scaffold materials exhibit macro-, micro-, and nano-structured pores, and their porosity can be regulated for explicit applications. These three-dimensional carbon scaffolds can be utilized for the assembly of supercapacitors, excessive-efficiency catalysis, discipline emission transistors, photovoltaics, implants, and biomedical devices as well as in agricultural applications. For instance, single-walled CNTs influence the stress-responsive gene expression in tomato plant to regulate plant development (Lahiani et al., 2015).

21.2 NANOTECHNOLOGY FOR SUSTAINABLE DEVELOPMENT OF CROPS

Nanotechnology opens up a wide scope of novel functions within the field of agriculture to meet the demands of the growing population. NPs have distinctive physicochemical properties, i.e., high surface area, tunable pore size, high reactivity, and particle morphology. Management of most suitable nutrients for sustainable crop construction is a priority-based subject of study in agriculture. In this regard, nanonutrition has proved to be the most interesting discipline of study and concerns the availability of nanosized nutrients for sustainable crop production (Fig. 21.3). Making use of this technology, the effectivity of micro- as well as macronutrients of crops can be expanded (Mazzaglia et al., 2017). Within the literature, various NPs and NMs have been efficaciously used for higher nutrient uptake of crops as compared to traditional fertilizers. NMs also offer potential as delivery vehicles and as sensors for key processes in crop production (Khot et al., 2012; Grillo et al., 2016; Wang et al., 2016). The assessment summarizes these NPs and NMs into macro-, micro-, and nanocarrier-founded fertilizers and plant-growth-enhancing NPs with uncertain mechanisms; describing their position in growth and yield of crops, concentration/rate of application, mechanism of action if identified, toxic effects, if any, and gaps in the present research.

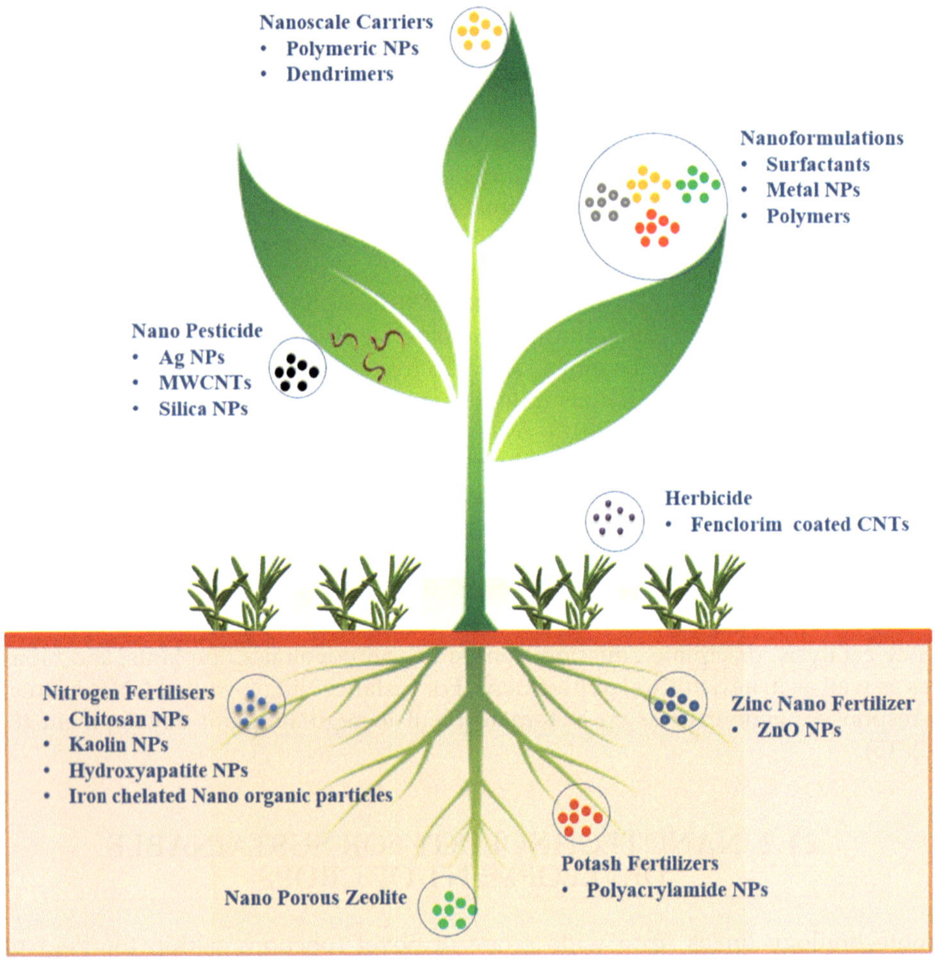

FIGURE 21.3 Application of nanotechnology in crop protection by utilizing various formulations of nanoscale products.

The application of nanotechnology has widely been utilized in various agriculture-based fields and numerous reports are available on the application of NMs in the area of plant protection (Kah, 2015; Servin et al., 2015). NMs have also been efficiently utilized in the area of crop biotechnology for the protection of agricultural crops. They also function as magic bullets for an efficient release of active chemicals and genes for improvement of crop heath (Pérez-de-Luque and Rubiales, 2009). An example of this type of delivery system is reported by Torney et al. (2007) with the development of mesoporous silica NPs that have been formulated with chemicals and function as a gene carrier for the transfer of genes to tobacco and corn plants. The cell wall then absorbs the NPs and the target genes are smoothly transferred to the host plant in an ecofriendly way.

NMs also function in the development of novel and effective bioformulations such as insecticides or pesticides to develop a sustainable crop protection system (Barik et al., 2008; Gajbhiye et al., 2009). The use of silica NPs has been reported previously for gene transfer in bacterial cells and this technique may be utilized for the development of several novel biocontrol agents (Torney, 2009). Additionally, the application of nanoemulsions was also reported for the development of pesticides against phytopathogens (Wang et al., 2007). Barik et al. (2008) reported the use of silica NPs for the development of an insecticide that was proved to be absorbed into the cuticular layer of insects to restrict their growth. However, because of the small size of NPs, modification of NMs is simple and effectively applied to control the parasites of plant and crop pathogens (Ulrichs et al., 2005).

NPs of zinc oxide, copper oxide, and cerium oxide have also been exploited in sweet potato for their bioavailability (Anderson et al., 2017; Bradfield et al., 2017). The widely studied NP includes nanosilver that has strong bacteriostatic and bactericidal properties (Yang et al., 2009). It is also characterized as having a broad spectrum of antimicrobial properties because of their large surface area (Suman et al., 2010). It was also reported that AgNPs alter cell membrane properties (Pal et al., 2007) and inhibit protein expression related to ATP generation (Yamanaka et al., 2005). The accurate mechanism of the foregoing process still needs to be discovered. The use of Ag NPs as an antifungal agent was reported under in vitro studies where the concentration of Ag NPs at 100 ppm proved to be effective toward restricting the growth of fungus on potato dextrose agar medium (Kim et al., 2012). The use of aluminum NPs was also explored for their insecticidal activity and proved to be effective against *Sitophylus oryzae* L. and *Rhyzopertha dominica* (F.), insect pests in stored food (Teodoro et al., 2010). In addition, cerium oxide NPs were accumulated in *Raphanus sativus* and the growth of plant was improved via enhancement of nutrients in soil (Zhang et al., 2016a,b, 2017).

An additional outstanding property of nanotechnology is the addition of NPs with various fertilizers and pesticides to create an effective formulation that promotes plant growth. This involves the release of bioformulations including chemicals and NPs into plant systems to increase their efficacy (Scrinis and Lyons, 2007). The controlled delivery of these formulations is completed with the aid of different courses of action such as diffusion and biodegradation (Vidhyalakshmi et al., 2009). The process for the development of effective formulation of NPs and fertilizers or pesticides is termed as nanoencapsulation, which also modifies the application of pesticides or herbicides and the formulation that is developed is known as nanocapsules.

Furthermore, these nanocapsules also assist the effective delivery of herbicides via tissues and cuticles, which permits smooth and continuous release of the active chemicals (Pérez-de-Luque and Rubiales, 2009). Therefore, these formulations have been utilized as the successful carrier materials for instant and prolonged transfer of herbicides or pesticides to agricultural crops. Furthermore, in comparison with commercially existing pesticides, NP-based fertilizers and pesticides are more beneficial, economic, and ecofriendly for the control of plant pathogens (Park et al., 2006; Kumar and Yadav, 2009; Prasad et al., 2011; Swamy and Prasad, 2012; Prasad and Swamy, 2013). Therefore, studies related to the development of NP-based bioformulation for plant disease suppression may expand the area for NP-based research in plant crop improvement and pest management.

21.2.1 Nanobiosensors

The application of nanotechnology is extremely broad and additionally incorporates the utilization of agricultural substances such as nutrients, chemicals, and water during agricultural processes, and nanosensors that have been proved to be more advantageous (Fig. 21.3). Nanobiosensors are nanosensors immobilized with various bioreceptors, which are specific for the target compounds. A nanobiosensor is developed on a nanoscale to detect and analyze results at the atomic level and has been applied for the detection of chemicals such as glucose, urea, herbicides, pesticides, or other compounds, analysis of enzymes and metabolic products, as well as detection of numerous pathogens (Rai and Ingle, 2012). Nanobiosensors have also been utilized to control soil nutrients and applied for the reduction of environmental pollution (Ingale and Chaudhari, 2013).

Nanobiosensors possess several advancements by enhancing sensitivity, subsequently decreasing the reaction time to detect microbial infections in crops (Small et al., 2001), and therefore accelerating agricultural production and reducing food safety issues. A fiber-optic biosensor developed by coating the antibodies onto the fiber-optic probes has been utilized for detecting plant pathogens and food toxins as well as for the detection of *Escherichia coli* with the help of a flow injection system. A novel biosensor was developed by Hashimoto et al. (2008) for the fast and sensitive detection of soilborne illness by incorporating two biosensors by immobilizing two individual microorganisms of the same quantity on an electrode.

The use of NPs as nanobiosensors for the detection of crop pathogens is currently at an initial stage and can be utilized for the detection of plant pathogens to enhance crop productivity. However, some reports are available on the utilization of NPs as diagnostic tools for the detection of microorganisms causing crop diseases (Servin et al., 2015). Yao et al. (2009) utilized fluorescent silica NPs in conjugation with an antibody for the detection of *Xanthomonas axonopodis*, a Gram-negative proteobacteria that causes bacterial spot disease in Solanaceae plants. AuNPs are proven to be a potential marker and could be used as a biosensor because of their optical properties, which can be utilized for pathogen detection. An immunosensor was developed by using AuNPs for the detection of Karnal bunt disease in wheat (Singh et al., 2010).

A number of nanobiosensors that utilize NMs including nanowires (Wang et al., 2014a), silicon NPs (Wang et al., 2014b), CNTs (Philip, 2009), and Au NPs (Rico et al., 2011) have been reported in recent studies for the detection of foodborne and plant pathogens. The application of nanobiosensors in the field of plant pathogen diagnosis still needs to be explored, most importantly under field applications. Nanobiosensors, when applied to the agricultural field, can be used for the detection of plant fungi, viruses, and other phytopathogens as well as for the analysis of the concentration of nutrients in soil (Brock et al., 2011; Jones, 2014). The application of a nanobiosensor, i.e., nano smart dust, was reported by Mousavi and Rezaei (2011) for the rapid analysis of environmental pollution. The application of nanotechnology-based devices will enhance the use of nanobiosensors for diagnosis of infections in crops and assist in minimizing the use of pesticides (Sharon et al., 2010). Because of the presence of extremely low concentrations of pollutants and detectable pathogens, the diagnosis is complicated (Misra et al., 2013).

21.2.2 Nanopesticides

Pesticide consists of four main classes but their function is similar. They either kill the unwanted substances or make the plants resistant. These categories comprise herbicides performing on undesirable weeds and grasses, insecticides on quite a lot of insects, fungicides for molds and algae, and disinfectants to manage bacteria by stopping them unfolding. Therefore, preparing this pesticide within a type of shell or in a closed envelope to have gradual and effective release will decrease the soil run-off rate and enhance solubility (Chinnamuthu and Murugesa Boopathi, 2009). An emulsion coating of pesticide was once developed with the appropriate mixing of oil and water leading to the formation of a water-insoluble product named β-cypermethrin and compared with an uncoated product for balance and effectiveness in spraying (Wang et al., 2007). *Spodoptera littoralis*, a primary pest that affects mostly the plant kingdom establishing its resistivity to virtually all pesticides, was controlled against hydrophobic nanosilica-covered tomato without any hindrance and showed optimistic response to the destruction of pests at 300–350 ppm (El-bendary and El-Helaly, 2013). When citric acid molecules were mixed with multiwalled CNT-designated pesticides such as Zineb and Mancozeb and trapped in aqueous solution, it led to the encapsulation of the pesticides, which resulted in major damage to the fungi *Alternaria alternata* (Sarlak et al., 2014). Yang et al. (2009) analyzed the effect of polyethylene glycol-coated NPs filled with essential oil against *Tribolium castaneum* (Yang et al., 2009). A pesticide formulation was further developed by loading pesticide validamycin to porous hollow silica NPs (Liu et al., 2006a,b,c). However, several other NMs, for example, Ag NPs, zinc oxide NPs, and titanium oxide NPs, have been explored for their insecticidal properties and successfully utilized in controlling silkworm disease and rice pests (Goswami et al., 2010).

Pesticides and herbicides, which are used in agriculture, enhance growth and yield. Presently, the poor features of traditional pesticides and herbicides in agriculture are controversial. The major dangers of pesticides include reduction in nitrogen fixation, development of pathogen and pest resistance, contribution to bioaccumulation of pesticides, reduction in soil biodiversity, pollinator decline, and destruction of habitats for birds (Ghormade et al., 2011). Accordingly, the use of nanopesticides unravel these concerns to some extent; their usage with herbicides reduces the quantity of herbicide required for weed eradication (Fig. 21.3). As an important ingredient and sensible delivery method, herbicides are released within the soil according to the soil's condition (Gruère et al., 2011). Ag NPs have a pesticidal effect against pathogenic fungi, stated to have inhibitory results on conidial germination of the genus *Raffaelea*, which causes mortality in oak trees (Nair et al., 2010). Furthermore, avermectin (a pesticide that blocks neurotransmission in insects) has a very short life span with a half-life of 6 h, considering the fact that it cannot be exposed UV rays; however, its life span has been reported to increase by using porous Ag NPs with a shell thickness of 15 nm and pore diameter of 4–5 nm (Ghormade et al., 2011). Ag NPs, which encapsulate avermectin and prevent its degradation and potential under UV rays, have increased life span for roughly 30 days (Ghormade et al., 2011). Gruère et al. (2011) reported a nanosurfactant based on soybean micelles. In the case where a crop is susceptible to glyphosate, it is found that it can be glyphosate resistant as well when applied with nanotechnology-derived surfactant. Application of nanoformulations open novel methods to increase the stability of biological substances. The function of NPs in pest inhibition and their potential for insect pest management were mentioned earlier as well (Bhattacharyya et al., 2010). Green synthesis of Ag NPs by *Tinospora cordifolia* showed highest pesticidal influence toward the top louse *Pediculus humanus* and fourth instar larvae of *Anopheles subpictus* and *Culex quinquefasciatus* (Jayaseelan et al., 2011). Most likely, the nanoformulations are expected to enhance the solubility of poorly soluble lively parts, unlock the active elements in a designated fashion, and resist untimely degradation (Kah et al., 2013). Nanopesticides greatly managed the supply of pesticides, low chemical dose, and high optimistic outcome. Ag NPs at a 100 mg/kg reduced mycelial growth and conidial germination on cucurbits and pumpkins against powdery mildew (Lamsal et al., 2011a,b). Phenolic suspension of hydrophobic alumina-silicate NPs is tremendously robust toward grasserie disease in *Bombyx mori* leaves (Goswami et al., 2010). Nanoencapsulated pesticides adsorbed on the plant floor help in sustained release for longer as compared to conventional pesticides that draws off in the rain (Scrinis and Lyons, 2007).

Such old approaches to inhibit pathogens and pests have negative effects on the atmosphere. Consequently, the incorporation of NPs is gaining ever-escalating demand for sustainable agriculture without disturbing the environment. Studies carried under nonsterile conditions make it clear that the increase in crop development/yield is the influence of diminished disease occurrence. This is feasible from both the antipathogenic activity of the NP itself and because of key defensive pathways and metabolites within the plant.

21.2.3 Nano-Based Smart Delivery Systems: Nanoscale Carriers and Nanofertilizers

Fertilizers play an important role where ancient chemical fertilizers are replaced with nano- and biofertilizers with their efficiency and environment-friendly nature. The primary use of adding fertilizers is to facilitate the faster uptake of nutrients from the soil leading to better crop yield. The symbiotic exchange between soil and the plant system is very efficient. When fertilizers are applied in slow and efficient way in an encapsulated form, the uptake of nutrients by plants increase drastically, and the required energy is restored (Mazzaglia et al., 2017). This can be achieved when fertilizers are in encapsulated form. The main element nitrogen is needed in abundance and uptake of this causes many problems for plants. A different type of fertilizer for different crop problems is also reviewed. Nanofertilizers produced from NMs to be assessed on crops are either available commercially by registered companies for applications other than to be used for plant fertility or produced in very small quantities in-house. Large-scale production of nanofertilizers is still dubious. However, a number of countries are moving ahead with the concept of nanofertilizers (Dimkpa and Bindraban, 2017).

Nanoscale carriers would be used for the efficient supply of pesticides, fertilizers, plant growth regulators, herbicides, etc (Prasad et al., 2012). The processes applied via these carriers for better storage, efficient supply, and managed release comprise entrapment and encapsulation, polymers and dendrimers, floor ionic and vulnerable bond attachments, and others (Sawant et al., 2006). These mechanisms help to fortify their steadiness against degradation within the environment and ultimately minimize the amount to be applied, which reduces chemical compound run-off and alleviates environmental issues. In a review by Zeeb and McClements (2017), it was found that there has been a large increase in the research carried out on the development and characterization of NP-based colloidal delivery systems for bioactive food components such as microemulsions, nanoliposomes, nanoemulsions, nanogels, and solid lipid and biopolymeric NPs.

Nanofertilizers have proved to be yet another landmark within the historical past of crop creation by way of nanotechnology. As it is already discussed, nanofertilizers are the NMs that can provide nutrients to plants or they assist to augment the activities of conventional fertilizers. Replacement of traditional fertilizers with nanofertilizers is beneficial to release nutrients into the soil steadily and in a controlled way, thus preventing water pollution (Naderi and Danesh-Shahraki, 2013). There are many problems with the use of chemical fertilizers, however; low use effectivity is the prominent one, which is costly and causes environmental pollution (Wilson et al., 2008). Nanomaterials have significant properties including large surface area and small size, in the range of 1–100 nm. These might be utilized as nanocoatings, such as sulfur coatings (≤100 nm layer) for surface protection and controlled release, eventually boosting their efficiency (Santoso et al., 1995). Nanofertilizers have proven to be more efficient in comparison with other fertilizers because they decrease nitrogen loss because of leaching, emissions, and long-term incorporation by soil microorganisms (Liu et al., 2006a,b,c). In addition, controlled release of fertilizers may also enhance soil by minimizing the toxic effects associated with the overapplication of normal chemical fertilizers (Suman et al., 2010). There are also reviews about the use of nanoencapsulated slow release of fertilizers (DeRosa et al., 2010). Not too long ago, biodegradable, polymeric chitosan NPs (~78 nm) had been used for controlled discharge of NPK fertilizer sources equivalent to urea, calcium phosphate, and potassium chloride (Corradini et al., 2010). Different NMs such as kaolin and polymeric biocompatible NPs might also be utilized for this purpose (Wilson et al., 2008). Hydroxyapatite NPs of 16 nm in size, synthesized by Liu and Lal (2014), exhibited a fertilizing effect on soybean. There are so many reports where application of nanofertilizers reflects a positive effect in terms of good crop yield as well as environmental pollution.

The application of NPs increased growth rate and seed germination by 33% and 20%, respectively, compared to regular P fertilizer. This indicated that the roots of soybean could absorb hydroxyapatite NPs as an effective P nutrient source. Soil amended with metallic copper NPs significantly increased 15-day lettuce seedling growth by 40% and 91%, respectively (Shah and Belozerova, 2009). Some studies focused on the characteristics of NPs also revealed that NPs could enter plant cells and transport DNA and chemicals inside the cell (Torney et al., 2007; Ghafariyan et al., 2013; Ambrogio et al., 2013). These studies provide a platform on which we can assume that NPs can also deliver nutrients to the plants as fertilizers. The nanoorganic iron-chelated fertilizers demonstrated high absorption, increase in photosynthesis, and expansion in leaf surface area (INIC, 2009). Moreover, nanofertilizers have a great impact on the soil because they can reduce the toxicity of the soil and decrease the frequency of fertilizer application (Naderi and Danesh-Shahraki, 2013). DeRosa et al. (2010) reported that in nanofertilizers, nutrients can be encapsulated by NMs, coated with a thin protective film, or delivered as emulsions or NPs. Nano- and subnanocomposites control the release of nutrients from the fertilizer capsule (Liu et al., 2006a,b,c). Urea-modified hydroxyapatite NP-encapsulated *Gliricidia sepium* nanocomposite exhibited a slow and sustained release of nitrogen over time at three different pH values (Kottegoda et al., 2011). Manikandan et al. reported that nanoporous zeolite used on N fertilizer might be used as an alternate strategy to enhance the effectiveness of N used in crop production systems (Manikandan and Subramanian, 2014). Nanofertilizers because of their characteristic features have a great role in sustainable agriculture (El-Ramady, 2014). Thus from the previously mentioned findings we can say that the use of nanofertilizers leads to an increased efficiency of the micro- and macroelements, reduces the toxicity of the soil, and reduces the frequency of application of conventional fertilizers.

21.2.3.1 Nitrogen Fertilizers

There are number of reasons of nitrogen loss from the surroundings which may include volatilizaion, denitrification, leaching and run-off. Caution should be taken when making use of this N fertilizer on crops. Steady application of N fertilizer to the precrops will not benefit from the superior utilization of the whole amount added. Hence, it is vital to interchange with a hand fertilizer to constantly make the vitamins and minerals easily available to plants. The element nitrogen is mainly for plant development and animal nutrition, and is the nutrient taken up in the greatest amount by all plants. Sources of nitrogen include ammonia, diammonium phosphate, ammonium nitrate, ammonium sulfate, calcium cyanamide, calcium nitrate, sodium nitrate, and urea (Booze-Daniels and Schmidt, 1997). It is mainly used because it is easy to handle and can be applied quickly. However, this may cause environmental issues (Derrick et al., 2007). When it comes to nitrogen fertilizers, its gradual liberation is beneficial to plants. This is because a farmer can use fertilizers less often because the nutrients are released slowly and steadily. Uptake of nitrogen can be increased by applying 25 kg Mg Oha-1, which elevates the constructive uptake (El-bendary and El-Helaly, 2013).

21.2.3.2 Potash Fertilizers

Potash fertilizers are utilized by plants as K^+ ions without forming any organic compound in the plant. They also help in photosynthesis and control water storage and stomatal opening in leaves. For the steady release of these fertilizers, a polyacrylamide-based coating of pellets was used. In the process, potash and clay were mixed together and dried for 1 h, and then coated with toothpaste for proper attachment to the polymer. This polymer was then dipped in polyacrylamide polymer. The study revealed the difference in dissolution with and without the coating; when the potash used is less, the release is also decreased and can be maintained with less water (Zhang and Wang, 2005). The main cash crop of India and Sri Lanka is rice, and studies were done on rice using nanopotash fertilizer as the source of potassium in rice, which showed an increase in the number of grains per panicle and the amount, compared to muriate of potash, which was less. Studies were also carried out on wheat and corn with nanocoated fertilizer; the slow release in crops increased the yield rate and effective recovery of N fertilizer (Subbarao et al., 2013).

21.2.3.3 Nanoporous Zeolite

Nanoporous zeolite helps in the steady release of fertilizer to plants, and allows proper utilization of nutrients from the fertilizers instead of minimal uptake. Because of its large surface area, many molecules can be arranged on it and are released and taken by plants whenever required (Naderi and Danesh-Shahraki, 2013). Zeolite (clinoptilolite) was made by nano-sized particles using a ball mill. One gram was taken in a centrifuge and stirred along with 1.5 M of 50 mL of calcium sulfate solution for 8 h, filtered and washed with deionized water and air dried. Furthermore, the solid:liquid ratio was maintained up to 1:10 for the purpose of synthesis (Selva Preetha et al., 2014). Nitrogen fertilizers are important but because of their high solubility, they cause severe damage to plants and the surroundings. Hence, nanoporous zeolite was used along with urea. This showed better uptake and steady release of nitrogen-efficient urea (Manikandan and Subramanian, 2014). Aluminum zeolites are also used because they are highly porous and allow retention of the soil (Prasad et al., 2014).

21.2.3.4 *Zinc Nanofertilizer*

The micronutrient zinc is deficient on earth. The quantity of zinc consumption by way of everyday meals is considerably less; consequently, through utilizing zinc-founded fertilizer, there are fewer chances of indirect delivery to humans. Identical NPs can be used to coat zinc to achieve a diffused and soluble zinc (Milani et al., 2010). When the pH rises, the solubility of zinc decreases (Bickel and Killorn, 2001). Equal ratios of surface area and size of NPs should be cautiously designed. If not, complete solubility of zinc can be affected. This is proven by taking the ratio of nano-ZnO and bulk ZnO (Malik and Kumar, 2014).

21.2.4 Nanoherbicide

Another area to consider is when there are standing crops grown together with weeds. To kill these weeds, herbicides are used but traditional herbicides when sprayed can affect the crops too causing a significant loss in crop yield (Deva and Kadiri, 2016). With the use of nanoherbicides, which are in the 1–100nm range, an attempt can be made to mingle with the soil particles to destroy the weeds from their roots without affecting the main crops. Because the NPs are target distinctive, they may be able to kill the weeds and achieve higher yields (Bickel and Killorn, 2001). Additionally, there are regulations implemented for the controlled release of herbicides that limit injury to humans. More generally, herbicides comprising triazine, ametryn, and atrazin have been nanoencapsulated to achieve 84% efficient release to plants (Grillo et al., 2012). The adsorptive stripping voltammetry process was developed to produce the herbicide fenclorim with CNTs at pH 4 and with adsorption tactics on the electrodes (Raveendranath Babu et al., 2014). Atrazine is a broadly used herbicide that can kill the weeds and undesirable grass growing close the crops. Continuous use of herbicides makes soil free from nutrients and causes crops to become immune; however, use of modified silver with NPs and carboxymethyl cellulose makes degradation of herbicides easier. In a review by Varma and Manika (2017), interaction of NPs (Ag, CNTs, TiO_2, and ZnO) with the fungus *Piriformospora indica* (*Hymenomycetes, Basidiomycota*) was focused on its herbicidal correlation with plant growth.

21.2.5 Nanofiltration in Agriculture

In many regions around the world including North Africa and the Middle East, the scarcity of water has adversely affected agricultural sectors. The increasing demand of water for domestic and industrial purpose aggravates the competition for limited resources; in addition, increasing volumes of water are being supplied to high priority sectors rather than agriculture. A sustainable and cost-effective method to deal with water scarcity is to control wastage of water and implement competent irrigation techniques; however, the implementation of such improvements is slow and also may not be appropriate for areas affected by persistent scarcity of water. An alternative way for dealing with this problem is the utilization of non-conventional water sources such as recycling, desalination, harvesting of rainwater, and reuse of brackish water (Ghermandi and Messalem, 2009).

Nanotechnology-based methods including application of nanofilters have been successfully used to facilitate water (salt water and brackish) treatment processes for effective irrigation. The characteristics of nanofilters lie between reverse osmosis and ultrafiltration, which can be used widely for several applications, generally in wastewater treatment and water-softening processes. Nanofilters have pore lengths of 0.5–1 nm. These filters facilitate the fractionation and purification of mixtures of monovalent or multivalent salts and organic solute particles of molecular weight 100–1000 Da under operating pressures of 1 and 4 MPa (Savage and Diallo, 2005).

Nanofiltration (NF) is a quite advanced membrane filtration process used mainly for removing solids, including bacteria, viruses, fungi, nematodes, and other parasites, from surface as well as fresh groundwater. In agriculture, relevant practices require that water used for irrigation purposes should be devoid of particles (>50–100 μm), have a low salt concentration to avoid toxicity generated by ions, and increased soil salinity. Hence, it is important to treat irrigation water properly to get rid of all undesirable contaminants that are capable of decreasing crop productivity (Rasouli et al., 2013), damaging the quality of crops (Bernstein et al., 2011), reducing the choice of crops (Levy and Tai, 2013), and harming the suitability of crops (Liu et al., 2013).

NF can generally be implemented for elimination of organic matter, fine particles, turbidity, harmful microorganisms, and selective ions without the use of disinfectants (Mrayed et al., 2011; Riera et al., 2013). This membrane-based methodology has shown potential to facilitate salty water desalination because it has: (1) a high rejection degree for divalent ions that may constitute a potentially commercial substitute for irrigation with brackish water; (2) an effective tolerance for fouling situations (Guttman, 1968; Sotto et al., 2013); (3) the ability to remove certain specific pollutants, which further facilitate usage of concentrate steam without enhancing fouling of the membrane; (4) the ability to process at reduced applied pressure thus utilizing less energy and operating costs; and (5) the ability to retain ions crucial for plant growth (Nederlof et al., 2005; Al-Amoudi and Lovitt, 2007; Zhao et al., 2012).

A solar system equipped with nanofilters is also used in hot and arid countries for the production of desalinated water for irrigation purposes. The outcomes of the Josefowitz Oasis Project indicated that 25% less irrigation and fertilizer are required for crops grown with desalinated water. In addition, crop yield is also increased.

21.3 NANOTECHNOLOGY IN PLANT NUTRITION AND HEALTH

In the era of expanding global population, the chief objective for agricultural scientists today is to enhance the yield and productivity of crops. Rapid diminution of agricultural land and restricted energy resources required for agricultural purposes has led to further aggravation of the problem. Hence, there is a need to develop new and information-based techniques and technologies to enhance crop yield. Fig. 21.4 describes the utilization of engineered nanomaterials for enhanced plant nutrition by targeted supplying of cargos.

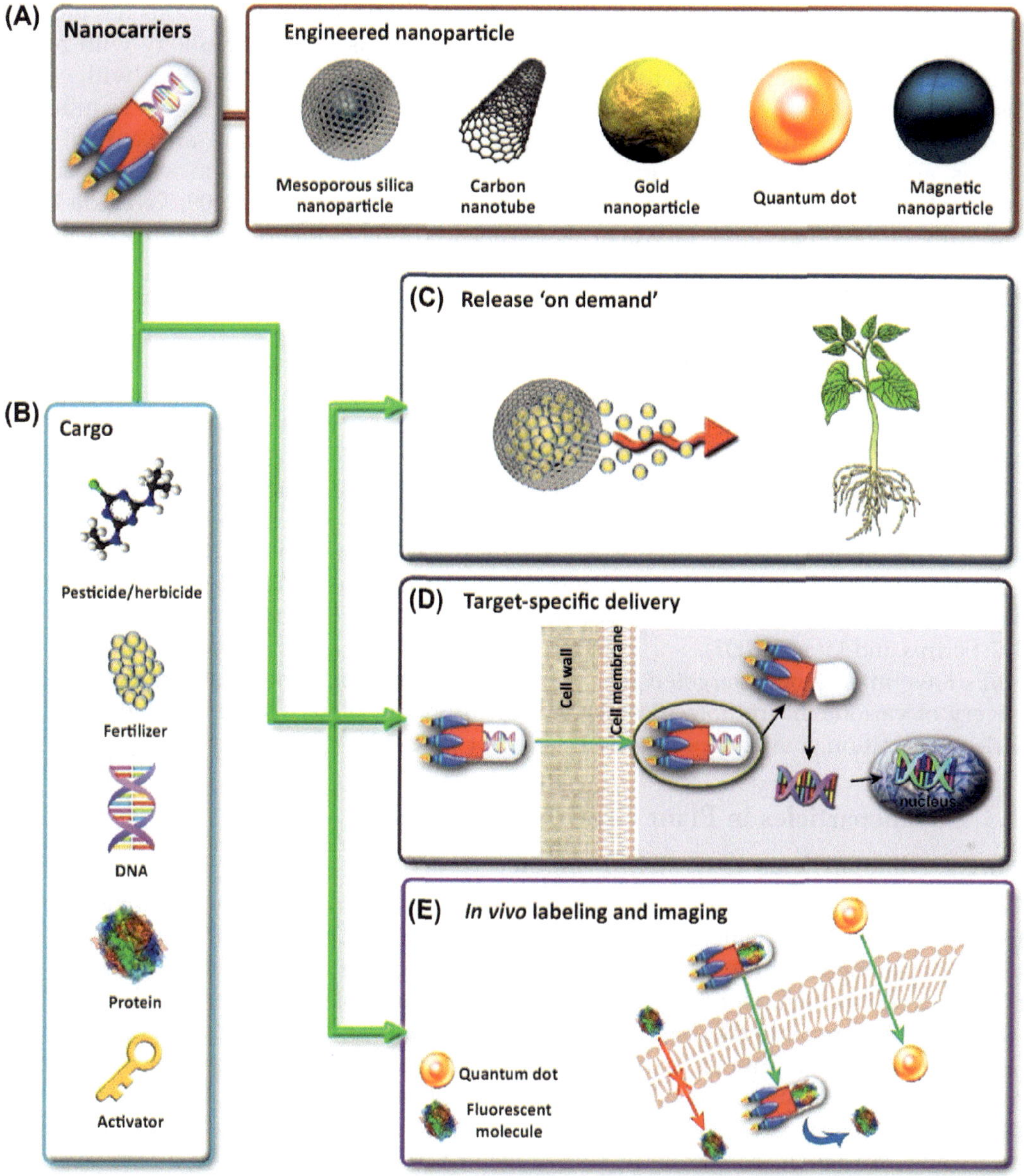

FIGURE 21.4 Applications of nanotechnology in plant sciences. (A) Engineering nanoparticles used as nanocarriers for delivery of various exogenous cargos (B), including agrochemicals and bioactive molecules. (C) Nanocarrier-mediated release of agrochemicals in a controlled manner. (D) Nanocarrier-mediated delivery of bioactive molecules into plant cells. (E) Nanocarrier-mediated intracellular fluorescent agents (e.g., quantum dots or fluorescent protein) for delivery of intracellular labeling and imaging. *Adapted with permission from Wang, P., Lombi, E., Zhao, F.J., Kopittke, P.M., 2016. Nanotechnology: a new opportunity in plant sciences. Trends Plant Sci. 21, 699–712. http://dx.doi.org/10.1016/j.tplants.2016.04.005.*

Nanobiotechnology and nanotechnology are the evolving scientific domains having huge potential to refurbish agricultural sectors and applied areas. Nanotechnology with respect to agronomy is focusing on targeted farming involving the utilization of NMs with characteristic properties to enhance the growth and yield of crops. There is an urgent requirement to raise food quality, worldwide production of food, analysis of plant- and animal-based diseases, and examination of the growth of plants as well as reduction in waste to achieve sustainable agricultural conditions (Gruère et al., 2011; Frewer et al., 2009; Pérez-de-Luque et al., 2013; Prasad et al., 2014; Biswal et al., 2012; Ditta, 2012; Sonkaria et al., 2012). There is a great necessity to utilize nanostructures in the agricultural sector for improving the efficiency and sustainability of practices carried out in agriculture by giving less input and obtaining reduced wastage as compared to conventional methods.

Nanotechnology contributes tremendously in terms of agriculture growth, enhanced reactivity, improved bioavailability, and bioactivity along with the greater surface impact of NMs. Engineered nanoparticles (ENPs) are able to enter into plant cells and leaves, and can transport DNA and chemicals into plant cells (Galbraith, 2007; Torney et al., 2007). This area of research provides a platform for biotechnology to target specific gene manipulation and expression in the specific cells of plants (Singh et al., 2017). Research is going on to target drug delivery for disease diagnosis applicable for both flora and fauna. Research in the field of nanotechnology is required to discover novel applications to target-specific delivery of chemicals, proteins, and nucleotides for the genetic transformation of crops (Torney et al., 2007; Scrinis and Lyons, 2007).

NPs have mainly been targeted for controlled release of agrochemicals and site-targeted delivery of various macromolecules needed for improved plant disease resistance, efficient nutrient utilization, and enhanced growth (Nair et al., 2010).

21.3.1 Nanoparticles in Plant Growth Enhancement

21.3.1.1 Nanoparticles as Growth Promoter

The effect of NMs on germination and growth have been reported by researchers to promote their usage in agricultural applications (Table 21.1). When NPs interact with plants, various physiological and physical changes occur, depending on NP properties. The efficacy of NPs is concentration dependent and varies from plant to plant.

Because of inimitable mechanical, thermal, chemical, and electrical properties, carbon NMs have attained an important place among NPs. The significant increment in seed germination rate and vegetative biomass of tomato seeds after exposure of multiwalled CNTs was observed (Khodakovskaya et al., 2009) (Table 21.1). According to this study, cell wall penetration by NPs caused an increase in uptake of water by seeds, which led to the enhancement in germination percentage. Srinivasan and Saraswathi (2010) concluded that single-walled CNTs act as nanotransporters for efficient DNA and dye molecule delivery into plant cells. In another study, it has been reported that multiwalled CNTs mediated efficient water and nutrient (Ca, Fe) uptake to facilitate the germination of seeds and growth of plants (Villagarcia et al., 2012; Tiwari et al., 2014) (Table 21.1). The enhancement in root elongation in rape, ryegrass, and corn by multiwalled CNTs has been reported by Lin and Xing (2007). However, according to Canas et al. (2008), single-walled CNT (0.16, 0.9, and 5 g/L) exposure to cucumber seedlings and onion enhanced their root growth (Table 21.1).

TABLE 21.1 Effect of Different Nanomaterials on Plants

Plant	Plant Parts	Nanomaterial	Effect	References
Tomato	Seeds	MWCNTs	Enhanced seed germination and vegetative biomass	Khodakovskaya et al. (2009)
Tomato	Tissue	SWCNTs	DNA-dye molecule delivery	Srinivasan and Saraswathi (2010)
Tomato	Roots	MWCNTs	Efficient water and nutrient uptake	Villagarcia et al. (2012)
Rape, rye grass, corn	Root	MWCNTs	Root elongation	Lin and Xing (2007)
Cucumber, onion	Root	SWCNTs	Root elongation	Canas et al. (2008)
Cabbage, maize	Root	Ag NPs	Root elongation	Pokhrel and Dubey (2013)
Arabidopsis thaliana	Seeds	Au NPs	Enhanced seed germination	Kumar et al. (2013)
Boswellia ovalifoliata	Seeds	Au NPs	Enhanced seed germination	Savithramma et al. (2012)
Bacopa monnieri	Whole plant	Ag NPs	Increased carbohydrate, protein, and decreased total phenol content	Krishnaraj et al. (2012)
Pearl millet	Whole plant	Zn NPs	Enhanced shoot length, chlorophyll, and yield	Tarafdar et al. (2014)
Petroselinum crispum	Seeds	Nanoanatase	Enhanced shoot length, chlorophyll, and yield	Dehkourdi and Mosavi (2013)
Foeniculum vulgare Mill.	Seeds	TiO_2	Enhanced seed germination	Feizi et al. (2013)
Zea mays L.	Seeds	SiO_2	Enhanced seed germination	Suriyaprabha et al. (2012)
Changbai larch	Seeds	SiO_2	Enhanced seed germination and chlorophyll biosynthesis	Bao-shan et al. (2004)
Brassica juncea	Seeds	Au NPs	Increase in net productivity and yield	Arora et al. (2012)
Trigonella foenum	Seeds	Ag NPs	Enhanced growth, diosgenin content, and secondary metabolite production	Jasim et al. (2016)

Different studies inferred that metal NPs facilitate plant growth and development. Ag NPs were found to increase root length in cabbage and maize plants in comparison with $AgNO_3$ (Pokhrel and Dubey, 2013; Kumar et al., 2013) (Table 21.1). A significant role of Au NPs in seed germination and the antioxidant system of *Arabidopsis thaliana* has been reported (Kumar et al., 2013). In addition, the improved seed germination by Au NPs in *Boswellia ovalifoliata* has been observed (Savithramma et al., 2012) (Table 21.1). In *Bacopa monnieri*, biologically synthesized

Ag NPs were found to induce synthesis of carbohydrate and protein, and decrease the total phenol content (Krishnaraj et al., 2012). Biologically synthesized Zn NPs effectively enhanced shoot (10.8%), chlorophyll content (18.4%), and grain yield (29.5%) in pearl millet (Tarafdar et al., 2014) (Table 21.1).

The effect of Fe NPs on spinach in hydroponic solution has been reported by Almeelbi and Bezbaruah (2014). The role of NPs in plant growth and biomass has been reported by many authors. The increase in Fe content by up to 11–21 times in spinach leaves, stems, and roots has been investigated. When seeds were soaked in NPs or sprayed with nanosized TiO_2, plant growth was found to be promoted (Zheng et al., 2005). Nanoanatase-treated parsley seeds were observed to have enhanced germination, root and shoot length, and chlorophyll content (Dehkourdi and Mosavi, 2013). TiO_2 NPs improved the germination rates of salvia seeds (Feizi et al., 2013) (Table 21.1). Suriyaprabha et al. (2012) reported that germination of seed is increased by nano-SiO_2 because it provided better nutrient availability to maize seeds. Changbai larch (*Larix olgensis*) seedlings, when treated with nano-SiO_2, had improved seedling growth and chlorophyll biosynthesis (Bao-shan et al., 2004). Nano-SiO_2 increases the photosynthetic rate, transpiration rate, electron transport rate, and other physiological parameters, which improved plant growth and development (Al-Whaibi, 2014; Xie et al., 2011). The growth-stimulating function of NPs on plants seems to be significant. Dedicated field research is required to study the promoting effects of these NPs on the yields of some important crops.

In a study carried out by Arora et al. (2012), it has been observed that treatment of *Brassica juncea* seedlings with Au NPs of 10 ppm can successfully improve their growth, leading to an increase in net productivity in terms of seed yield. However, increasing Au NP concentration did not significantly alter the response relevant to agronomical traits. Furthermore, the observed enhancement in growth and yield parameters of the treated seedlings is related to their redox status. The results obtained in the present findings are novel from an application viewpoint.

The effect of biologically synthesized NPs on plant growth and diosgenin content of fenugreek (*Trigonella foenum*) was investigated by Jasim et al. (2016) (Table 21.1). The results observed were considerable because the seedlings treated with the NPs were found to have not only enhanced growth but also enhanced diosgenin content. This study suggests that Ag NPs can be used as elicitors for enhancement of secondary metabolite production in medicinal plants.

21.3.1.2 Nanoparticles in Disease Suppression

The primary reasons for the progression of disease in plants are bacteria, viruses, fungi, and nematodes. They cause a decline in yield as well as deprive plant foodstuffs. The literature has mentioned the positive effect of NPs against pathogens responsible for low crop growth. Jo et al. (2009) concluded that Ag NPs at a concentration of 200 mg/L inhibited the colony formation of pathogenic fungi by 50% in ryegrass. Similarly, Lamsal et al. (2011a,b) found that use of Ag NP treatment improved disease inhibition. Synergistic use of Ag NPs and the fungicide fluconazole successfully suppressed *Candida albicans*, *Phoma glomerata*, and *Trichoderma* spp. (Gajbhiye et al., 2009). Jo et al. (2009) showed that efficient quantity for the suppression of colony formation using silver on *Bipolaris sorokiniana* was better when compared to *Magnaporthe grisea* with 50% at optimum concentration. A considerable decrease in mycelia growth was seen in spores treated with Ag NPs (Kasprowicz et al., 2010). Cui et al. (2012), in a similar fashion, demonstrated a method of antibacterial activity in Au NPs in opposition to *E. coli* and found that there is a decline in membrane potential, reduction in ATPase activity, and suppression of ATP at the cellular stage. Ag NPs synthesized from tea extracts also show activity against *Vibrio harveyi* (Vaseeharan et al., 2010).

NMs that act either as supplements to plants or enhance the performance of usual fertilizers are known as nanofertilizers. Substitution of conventional fertilizers with nanofertilizers is advantageous because they function to discharge nutrients in the soil in a gradual and regulated fashion, thus minimizing water pollution (Naderi and Danesh-Shahraki, 2013). Innumerable reports have discussed the purpose of nanofertilizers related to superior crop yield and reduced environmental contamination. The application of NPs augmented seed germination by 33% and 20% correspondingly, with reference to standard P fertilizer. Soil enriched with metallic Cu NPs appreciably improved 15-day lettuce seedling growth by 40% and 91% correspondingly (Shah and Belozerova, 2009). Additionally, nanofertilizers impose immense impact on soil since they are capable of decreasing the toxicity in the soil as well as the regularity of fertilizer application (Naderi and Danesh-Shahraki, 2013). DeRosa et al. (2010) stated that nutrients can also be encapsulated in nanofertilizers after covering with a thin defensive film or distributed as emulsions or NPs. Nano- and subnanocomplexes manage the discharge of nutrients from the fertilizer shell (Liu et al., 2006a,b,c). Nanofertilizers, because of their attributes, have an immense role in sustainable agriculture.

21.3.2 Nanofoods

The agrifood industries have been investing huge amount of money into nanotechnological research. Food is nanofood when NPs or nanotechnology techniques or tools are used during cultivation, production, processing, or packaging of the food. This does not mean atomically modified food or food produced by nanomachines. Agrifood is not the only user of nanotechnology, and there are many other innovative ways to improve the quality and sustainability of products and processing by this technology. Its potential is extensively discussed in scientific articles and the practical implementation, as well as the economic significance, has been predicted in market forecasts (Walz et al., 2017).

Nanotechnology has been widely used for improving crop productivity and quality. In this respect, various experiments have been conducted to evaluate NPs and their effects on growth and development in a variety of crops. In crop plants including corn, wheat, ryegrass, tomato, radish, lettuce, spinach, pumpkin, corn, and cucumber, various positive morphological effects of NPs (mainly metal based and carbon based), such as improved germination rate, root and shoot length, biomass of seedlings, have been observed. In one study, iron oxide NPs have been observed to improve agronomical traits including the grain yield of soybean. In a study carried out by Kole et al. (2013), the effect of the carbon-based NP fullerol on yield of a medicinally rich vegetable crop, bitter melon, has been evaluated. In this study, it has been observed that fullerol treatment resulted in an increased biomass yield of 54%, a fruit number of 59%, and a fruit weight of 70%, which further led to an improvement of up to 128% in fruit yield. The content of two anticancerous phytomedicines cucurbitacin-B and lycopene were also reported to be enhanced.

India is a country with a wide array of climatic conditions that further support the production of a wide variety of fruits and vegetables. After China, India is the largest producer of fruits and vegetables. In spite of this, there is a large gap between demand and supply, which is attributed to losses after harvesting. For example, tropical fruits such as banana, papaya, and mango are in high demand in markets for exportation purposes but maintenance of the freshness of these fruits during export is an important task because these fruits are spoiled and damaged quickly. Lack of refrigeration facilities further intimidates the challenge of keeping fruits fresh. About 40% of produce is lost in tropical countries during postharvest handling.

These losses are caused by the absence of proper processing, preservation, and packaging facilities for the perishables. Currently, the packaging industries have started to utilize innovative methodologies such as nanotechnology to prevent losses because of postharvesting of perishable fruits and vegetables. In this respect, Tamil Nadu Agricultural University is currently developing nanomatrices that can be incorporated with packaging material for efficient preservation of mango to increase the shelf-life of mango fruits. This was an integrative effort of Indian, Sri Lankan, and Canadian partners to implement the nanoapplication of a natural product hexanal to delay ripening of mango and banana. Hexanal is reported to inhibit the ethylene-mediated ripening of fruits. Likewise, the Industrial Technology Institute, Sri Lanka, has developed a biowax that has the potential to reduce postharvest damage. By using these types of nanotechnology-based strategies, it is expected that an efficient and effective packaging system will be implemented to increase the shelf-life of highly perishable vegetables and fruits. Nanotechnology-based GuardIN Fresh (Fayetteville, AR, USA) benefits from perishable produce and floral products by scavenging the ethylene gas that hastens ripening. Nanotechnology is the basis of many novel and functional foods, and food colors, favors, and textures can all be manipulated and altered at the nanoscale. In 2004, researchers had been able to alter rice color from purple to green. Cellular "injection" with carbon nanofibers containing foreign DNA has also been used to genetically modify golden rice.

21.4 CONCLUSION AND FUTURE PROSPECTS

There has been continuous use of agrochemicals to enhance agricultural productivity but it has led to the contamination of top soil, groundwater, and food. It is required to increase agricultural development and reduce the pollution of necessary sources, but considering the damage related to ecosystems, new technologies need to be devised. Nanotechnology, in this respect, is becoming a widely accepted field for sustainable agricultural development. Promising outcomes and applications are already being established in the areas of delivery of pesticides, biopesticides, fertilizers, and genetic material for plant transformation. Hence with the utilization of nanotechnology, reduction in dose and controlled delivery of fertilizers and pesticides can be ensured. The main prediction is that the application of NPs to stabilize biocontrol preparations will prove to go a long way in diminishing environmental hazards. The tools of nanotechnology can be employed to address the urgent issues of plant growth protection and maintenance of sustainability. Furthermore, nanotechnology can attempt to provide and fundamentally modernize technologies to be used in environmental detection, sensing, and remediation.

Acknowledgment

The authors are thankful to the Director of MNNIT, Allahabad, for providing necessary facilities and the "Design and Innovation Centre," a project sponsored by the Ministry of Human Resource Development, government of India, for support to execute this study.

References

Al-Amoudi, A., Lovitt, R.W., 2007. Fouling strategies and the cleaning system of NF membranes and factors affecting cleaning efficiency. J. Membr. Sci. 303, 4–28.

Almeelbi, T., Bezbaruah, A., 2014. Nanoparticle-sorbed phosphate: iron and phosphate bioavailability studies with *Spinacia oleracea* and *Selenastrum capricornutum*. ACS Sustain. Chem. Eng. 2, 1625–1632.

Al-Whaibi, M.H., 2014. Role of nano-SiO_2 in germination of tomato (*Lycopersicum esculentum* seeds Mill.). Saudi J. Biol. Sci. 2, 13–17.

Ambrogio, M.W., Frasconi, M., Yilmaz, M.D., Chen, X., 2013. New methods for improved characterization of silica nanoparticle-based drug delivery systems. Langmuir 29 (49), 15386–15393.

Anderson, A., McLean, J.E., Jacobson, A.R., Britt, D., 2017, 1–48. CuO and ZnO nanoparticles modify interkingdom cell signaling processes relevant to crop production: a review. J. Agric. Food Chem.

Arora, S., Sharma, P., Kumar, S., Nayan, R., Khanna, P.K., Zaidi, M.G.H., 2012. Gold-nanoparticle induced enhancement in growth and seed yield of *Brassica juncea*. Plant Growth Regul. 66, 303–310.

Bao-shan, L., shao-qi, D., Chun-hui, L., Li-jun, F., Shu-chun, Q., Min, Y., 2004. Effect of TMS (nanostructured silicon dioxide) on growth of Changbai larch seedlings. J. For. Res. 15, 138–140.

Barik, T.K., Sahu, B., Swain, V., 2008. Nanosilica-from medicine to pest control. Parasitol. Res. 103, 253–258.

Baughman, R.H., Zakhidov, A.A., De Heer, W.A., 2002. Carbon nanotubes–the route toward applications. Science 297, 787–792.

Bernstein, N., Ioffe, M., Luria, G., Bruner, M., Nishri, Y., Philosoph-Hadad, S., et al., 2011. Effects of K and N nutrition on function and production of *Ranunculus asiaticus*. Pedosphere 21, 288–301. http://dx.doi.org/10.1016/S1002-0160(11)60129-X.

Bhattacharyya, A., Bhaumik, A., Rani, P.U., Mandal, S., Epidi, T.T., 2010. Nano-particles-A recent approach to insect pest control. Afr. J. Biotechnol. 9 (24), 3489–3493.

Bickel, A., Killorn, R., November 2001. "Spatial Response of Corn to Banded Zinc Sulfate Fertilizer in Iowa". North central Extension Industry Soil Fertility Conference, Potash and Phosphate Institute, 605/692-6280 Vol. 17, 14–15.

Biswal, S.K., Nayak, A.K., Parida, U.K., Nayak, P.L., 2012. Applications of nanotechnology in agriculture and food sciences. Int. J. Sci. Innov. Discov. 2, 21–36.

Booze-Daniels, J.N., Schmidt, R.E., 1997. The Use of Slow Release Nitrogen.

Bradfield, S.J., Kumar, P., White, J.C., Ebbs, S.D., 2017. Zinc, copper, or cerium accumulation from metal oxide nanoparticles or ions in sweet potato: yield effects and projected dietary intake from consumption. Plant Physiol. Biochem. 110, 128–137.

Brock, D.A., Douglas, T.E., Queller, D.C., Strassmann, J.E., 2011. Primitive agriculture in a social amoeba. Nature 469, 393–396.

Byrappa, K., Dayananda, A.S., Sajan, C.P., Basavalingu, B., Shayan, M.B., Soga, K., Yoshimura, M., 2008. Hydrothermal preparation of ZnO: CNT and TiO_2: CNT composites and their photocatalytic applications. J. Mater. Sci. 43, 2348–2355.

Canas, J.E., Long, M., Nations, S., Vadan, R., Dai, L., Luo, M., Ambikapathi, R., Lee, E.H., Olszyk, D., 2008. Effects of functionalized and nonfunctionalized single-walled carbon nanotubes on root elongation of select crop species. Environ. Toxicol. Chem. 27 (9), 1922–1931.

Chinnamuthu, C.R., Murugesa Boopathi, P., 2009. Nanotechnology and agroecosystem. Madras Agric. J. 96, 1–6.

Corradini, E., De Moura, M.R., Mattoso, L.H.C., 2010. A preliminary study of the incorporation of NPK fertilizer into chitosan nanoparticles. Express Polym. Lett. 4 (8), 509–515.

Cox, A., Venkatachalam, P., Sahi, S., Sharma, N., 2017. Reprint of: silver and titanium dioxide nanoparticle toxicity in plants: a review of current research. Plant Physiol. Biochem. 110, 33–49.

Cui, Y., Zhao, Y., Tian, Y., Zhang, W., Lu, X., Jiang, X., 2012. The molecular mechanism of action of bactericidal gold nanoparticles on *Escherichia coli*. Biomaterials 33, 2327–2333.

Dehkourdi, E.H., Mosavi, M., 2013. Effect of anatase nanoparticles (TiO_2) on parsley seed germination (*Petroselinm crispum*) invitro. Biol. Trace Elem. Res. 155, 283–286.

DeRosa, M.C., Monreal, C., Schnitzer, M., Walsh, R., Sultan, Y., 2010. Nanotechnology in fertilizers. Nat. Nanotechnol. 5, 91.

Derrick, M., Oosterhuis, Donald, D.H., 2007. Evaluation of slow release of nitrogen and potassium fertilizer for cotton production". Afr. J. Agric. Res. 3, 068–073.

Deva, S., Kadiri, L., 2016. A review on impact of nanoparticles on plant growth and development. Adv. Life Sci. 5 (1), 16–21.

Dimkpa, C.O., Bindraban, P.S., 2017. Nanofertilizers: new products for the industry? J. Agric. Food Chem. http://dx.doi.org/10.1021/acs.jafc.7b02150.

Ditta, A., 2012. How helpful is nanotechnology in agriculture? Adv. Nat. Sci. Nanosci. Nanotechnol. 3, 033002.

Du, W., Tan, W., Peralta-Videa, J.R., Gardea-Torresdey, J.L., Ji, R., Yin, Y., Guo, H., 2017. Interaction of metal oxide nanoparticles with higher terrestrial plants: physiological and biochemical aspects. Plant Physiol. Biochem. 110, 210–225.

El-bendary, H.M., El-Helaly, A.A., 2013. First record nanotechnology in agricultural: silica nano-particles a potential new insecticide for pest control. Appl. Sci. Rep. ISSN: 2310-9440.

El-Ramady, H.R., 2014. Integrated nutrient management and postharvest of crops. In: Sustainable Agriculture Reviews. Springer International Publishing, pp. 163–274.

Feizi, H., Kamali, M., Jafari, L., Moghaddam, P.R., 2013. Phytotoxicity and stimulatory impacts of nanosized and bulk titanium dioxide on fennel (*Foeniculum vulgare* Mill). Chemosphere 91 (4), 506–511.

Frewer, L.J., Norde, W., Fischer, A.R.H., Kampers, F.W.H., 2009. Nanotechnology in the Agri-Food Sector: Implications for the Future. Wiley-VCH, Weinheim, Germany.

Gajbhiye, M., Kesharwani, J., Ingle, A., Gade, A., Rai, M., 2009. Fungus-mediated synthesis of silver nanoparticles and their activity against pathogenic fungi in combination with fluconazole. Nanomedicine 5, 382–386.

Galbraith, D.W., 2007. Nanobiotechnology: silica breaks through in plants. Nat. Nanotechnol. 2, 272–273.

Ghafariyan, M.H., Malakouti, M.J., Dadpour, M.R., Stroeve, P., Mahmoudi, M., 2013. Effects of magnetite nanoparticles on soybean chlorophyll. Environ. Sci. Technol. 47, 10645–10652.

Ghermandi, A., Messalem, R., 2009. Solar-driven desalination with reverse osmosis: the state of the art. Desalin. Water Treat. 7 (1–3), 285–296.

Ghormade, V., Deshpande, M.V., Paknikar, K.M., 2011. Perspectives for nano-biotechnology enabled protection and nutrition of plants. Biotechnol. Adv. 29, 792–803.

Gogos, A., Knauer, K., Bucheli, T.D., 2012. Nanomaterials in plant protection and fertilization: current state, foreseen applications, and research priorities. J. Agric. Food Chem. 60, 9781–9792. http://dx.doi.org/10.1021/jf302154y.

Goswami, A., Roy, I., Sengupta, S., Debnath, N., 2010. Novel applications of solid and liquid formulations of nanoparticles against insect pests and pathogens. Thin Solid Films 519, 1252–1257.

Grillo, R., Abhilash, P.C., Fraceto, L.F., 2016. Nanotechnology applied to bio-encapsulation of pesticides. J. Nanosci. Nanotechnol. 16, 1231–1244.

Grillo, R., dos Santos, N.Z.P., Maruyama, C.R., Rosa, A.H., de Lima, R., Fraceto, L.F., 2012. Poly (ε-caprolactone) nanocapsules as carrier systems for herbicides: physico-chemical characterization and genotoxicity evaluation. J. Hazard. Mater. 231, 1–9.

Gruère, G., Narrod, C., Abbott, L., 2011. Agricultural, Food, and Water Nanotechnologies for the Poor. International Food Policy Research Institute, Washington, DC.

Guo, K.W., 2012. Green nanotechnology of trends in future energy: a review. Int. J. Energy Res. 36 (1), 17.

Guttman, L., 1968. A general non-metric technique for finding the smallest space for a configuration of points. Psychometrica 33, 479–506.

Hashimoto, Y., Nakamura, H., Asaga, K., Karube, I., 2008. A new diagnostic method for soil-borne disease using a microbial biosensor. Microb. Environ. 23, 35–39.

Ingale, A.G., Chaudhari, A.N., 2013. Biogenic synthesis of nanoparticles and potential applications: an eco-friendly approach. J. Nanomed. Nanotechol. 4 (165), 1–7.

Iran Nanotechnology Initiative Council (INIC), 2009. First nano-organic iron chelated fertilizer invented in Iran [webpage on theInternet]. Iran Nanotechnology Initiative Council, Tehran, Iran. Available from: http://www.iranreview.org/content/Documents/Iranians_Researchers_Produce_Nano_Organic_Fertilizer.htm.

Jasim, B., Thomas, R., Mathew, J., Radhakrishnan, E.K., 2016. Plant growth and diosgenin enhancement effect of silver nanoparticles in Fenugreek (*Trigonella foenum-graecum* L.). Saudi Pharm. J. 25 (3), 443–447.

Jayaseelan, C., Rahuman, A.A., Rajakumar, G., Kirthi, A.V., Santhoshkumar, T., Marimuthu, S., et al., 2011. Synthesis of pediculocidal and larvicidal silver nanoparticles by leaf extract from heartleaf moonseed plant, *Tinospora cordifolia* Miers. Parasitol. Res. 109 (1), 185–194.

Jo, Y.K., Kim, B.H., Jung, G., 2009. Antifungal activity of silver ions and nanoparticles on phytopathogenic fungi. Plant Dis. 93 (10), 1037–1043.

Jones, P.B.C., 2014. A Nanotech Revolution in Agriculture and the Food Industry. Information Systems for Biotechnology, Blacksburg, VA. 2006 http://www.isb.vt.edu/articles/jun0605.htm.

Kah, M., 2015. Nanopesticides and nanofertilizers: emerging contaminants or opportunities for risk mitigation? Front. Chem. 3.

Kah, M., Beulke, S., Tiede, K., Hofmann, T., 2013. Nanopesticides: state of knowledge, environmental fate, and exposure modeling. Crit. Rev. Environ. Sci. Technol. 43, 1823–1867.

Kasprowicz, M.J., Kozioł, M., Gorczyca, A., 2010. The effect of silver nanoparticles on phytopathogenic spores of *Fusarium culmorum*. Can. J. Microbiol. 56, 247–253.

Keller, A.A., McFerran, S., Lazareva, A., Suh, S., 2013. Global life cycle releases of engineered nanomaterials. J. Nanopart. Res. 15, 1692.

Khodakovskaya, M., Dervishi, E., Mahmood, M., Xu, Y., Li, Z., Watanabe, F., Biris, A.S., 2009. Carbon nanotubes are able to penetrate plant seed coat and dramatically affect seed germination and plant growth. ACS Nano 3, 3221–3227.

Khot, L.R., Sankaran, S., Maja, J.M., Ehsani, R., Schuster, E.W., 2012. Applications of nanomaterials in agricultural production and crop protection: a review. Crop Prot. 35, 64–70.

Kim, S.W., Jung, J.H., Lamsal, K., Kim, Y.S., Min, J.S., Lee, Y.S., 2012. Antifungal effects of silver nanoparticles (AgNPs) against various plant pathogenic fungi. Mycobiology 40, 53–58.

Kole, C., Kole, P., Randunu, K.M., Choudhary, P., Podila, R., Ke, P.C., Rao, A.M., Marcus, R.K., 2013. Nanobiotechnology can boost crop production and quality: first evidence from increased plant biomass, fruit yield and phytomedicine content in bitter melon (*Momordica charantia*). BMC Biotechnol. 13, 37.

Kottegoda, N., Munaweera, I., Madusanka, N., Karunaratne, V., 2011. A green slow-release fertilizer composition based on urea-modified hydroxyapatite nanoparticles encapsulated wood. Curr. Sci. (Bangalore) 101, 73–78.

Krishnaraj, C., Jagan, E.G., Ramachandran, R., Abirami, S.M., Mohan, N., Kalaichelvan, P.T., 2012. Effect of biologically synthesized silver nanoparticles on *Bacopa monnieri* (Linn.) Wettst. plant growth metabolism. Process Biochem. 47, 651–658.

Kumar, V., Yadav, S.K., 2009. Plant-mediated synthesis of silver and gold nanoparticles and their applications. J. Chem. Technol. Biotechnol. 84 (2), 151–157.

Kumar, V., Guleria, P., Kumar, V., Yadav, S.K., 2013. Gold nanoparticle exposure induces growth and yield enhancement in *Arabidopsis thaliana*. Sci. Total Environ. 461, 462–468.

Lahiani, M.H., Chen, J., Irin, F., Puretzky, A.A., Green, M.J., Khodakovskaya, M.V., 2015. Interaction of carbon nanohorns with plants: uptake and biological effects. Carbon 81, 607–619.

Lalwani, G., Kwaczala, A.T., Kanakia, S., Patel, S.C., Judex, S., Sitharaman, B., 2013. Fabrication and characterization of three-dimensional macroscopic all-carbon scaffolds. Carbon 53, 90–100.

Lamsal, K., Kim, S.W., Jung, J.H., Kim, Y.S., Kim, K.S., Lee, Y.S., 2011a. Application of silver nanoparticles for the control of Colletotrichum species in vitro and pepper anthracnose disease in field. Mycobiology 39, 194–199.

Lamsal, K., Kim, S.W., Jung, J.H., Kim, Y.S., Kim, K.S., Lee, Y.S., 2011b. Inhibition effects of silver nanoparticles against powdery mildews on cucumber and pumpkin. Mycobiology 39 (1), 26–32.

Levy, D., Tai, G.C.C., 2013. Differential response of potatoes (*Solanum tuberosum* L.) to salinity in an arid environment and field performance of the seed tubers grown with fresh water in the following season. Agric. Water Manag. 116, 122–127.

Lin, D., Xing, B., 2007. Phytotoxicity of nanoparticles: inhibition of seed germination and root growth. Environ. Pollut. 150 (2), 243–250.

Liu, F., Wen, L.X., Li, Z.Z., Yu, W., Sun, H.Y., Chen, J.F., 2006a. Porous hollow silica nanoparticles as controlled delivery system for water-soluble pesticide. Mater. Res. Bull. 41, 2268–2275.

Liu, R., Lal, R., 2014. Synthetic apatite nanoparticles as a phosphorus fertilizer for soybean (*Glycine max*). Sci. Rep. 4, 5686.

Liu, R., Lal, R., 2015. Potentials of engineered nanoparticles as fertilizers for increasing agronomic productions. Sci. Total Environ. 514, 131–139.

Liu, X.M., Feng, Z.B., Zhang, F.D., Zhang, S.Q., He, X.S., 2006b. Preparation and testing of cementing and coating nano-subnanocomposites of slow/controlled-release fertilizer. Agric. Sci. China 5 (9), 700–706.

Liu, X., Feng, Z., Zhang, S., Zhang, J., Xiao, Q., Wang, Y., 2006c. Preparation and testing of cementing nano-subnano composites of slow or controlled release of fertilizers. Sci. Agric. Sin. 39, 1598–1604.

Liu, Y., Li, X., Xing, Z., Zhao, X., Pan, Y., 2013. Responses of soil microbial biomass and community composition to biological oil crusts in the revegetated areas of the Tengger Desert. Appl. Soil Ecol. 65, 52–59.

Malik, S., Kumar, A., 27 June 2014. "Approach For Nano-Particle Synthesis: Using As Nano-Fertilizer". International Journal of Pharmaceutical Research and Bioscience, ISSN: 2277-8713 Volume 3 (3), 519–527.

Manikandan, A., Subramanian, K.S., 2014. Fabrication and characterisation of nanoporous zeolite based N fertilizer. Afr. J. Agric. Res. 9 (2), 276–284.

Mansur, A.A., Ramanery, F.P., Borsagli, F.M., de Oliveira, L.C., Mansur, H.S., 2017. Eco-friendly nano-photocatalysts based on Bi-doped ZnS quantum dots/polysaccharide hybrids for environmental applications: towards green nanotechnology. Adv. Sci. Eng. Med. 9 (3), 179–187.

Mazzaglia, A., Fortunati, E., KennyJ, M., Torre, L., Mariano, G., 2017. Nanomaterials in plant protection. In: Axelos, M.A.V., Van de Voorde, M. (Eds.), Nanotechnology in Agriculture and Food Science, pp. 115–134.

Medina-Velo, I.A., Barrios, A.C., Zuverza-Mena, N., Hernandez-Viezcas, J.A., Chang, C.H., Ji, Z., Zink, J.I., Peralta-Videa, J.R., Gardea-Torresdey, J.L., 2017. Comparison of the effects of commercial coated and uncoated ZnO nanomaterials and Zn compounds in kidney bean (*Phaseolus vulgaris*) plants. Journal of Hazardous Materials 332, 214–222.

Milani, N., McLaughlin, M.J., Hettiaratchchi, G.M., Beak, D.G., Kirby, J.K., Stacey, S., August 2010. "Fate of nanoparticulate zinc oxide fertilisers in soil: solubility, diffusion and solid phase speciation". Soil solutions for Changing World 1–6.

Misra, A.N., Misra, M., Singh, R., Nunziata, M.L., Morganti, G., Cardillo, M., Chianese, A., 2013. Nanotechnology in agriculture and food industry. Int. J. Pure Appl. Sci. Technol. 16 (2), 1.

Morganti, P., Palombo, M., Carezzi, F., Nunziata, M.L., Morganti, G., Cardillo, M., Chianese, A., 2016. Green nanotechnology serving the bioeconomy: natural beauty masks to save the environment. Cosmetics 3 (4), 41–419.

Mousavi, S.R., Rezaei, M., Zou, I., Leslie, G., 2011. Nanotechnology in agriculture and food production. J. Appl. Environ. Biol. Sci. 1 (10), 414–419.

Mrayed, S.M., Sanciolo, P., Zou, I., Leslie, G., 2011. An alternative membrane treatment process to produce low-salt and high-nutrient recycled water suitable for irrigation purposes. Desalination 274 (19), 144–149.

Naderi, M.R., Danesh-Shahraki, A., Nair, B.G., Maekawa, T., Yoshida, Y., Kumar, D.S., 2013. Nanofertilizers and their roles in sustainable agriculture. Int. J. Agric. Crop Sci. 5 (19), 2229–2232.

Nair, R., Varghese, S.H., Nair, B.G., Maekawa, T., Yoshida, Y., Kumar, D.S., 2010. Nanoparticulate material delivery to plants. Plant Sci. 179 (3), 154–163.

Nederlof, M.M., Van Paassen, J.A.M., Jong, R., 2005. Nanofiltration concentrate disposal: experiences in The NetherlandsEscherichia coli. Desalination 178, 303–312.

Pal, S., Tak, Y.K., Song, J.M., Choi, S.H., 2007. Does the antibacterial activity of silver nanoparticles depend on the shape of the nanoparticle? A study of the gram-negative bacterium *Escherichia coli*. Appl. Environ. Microbiol. 73 (3), 1712–1720.

Park, H.J., Kim, S.H., Kim, H.J., Choi, S.H., 2006. A new composition of nanosized silica-silver for control of various plant diseases. Plant Pathol. J. 22 (3), 295–302.

Pérez-de-Luque, A., Rubiales, D., 2009. Nanotechnology for parasitic plant control. Pest Manag. Sci. 65 (2), 540–545.

Pérez-de-Luque, A., Hermosín, M.C., 2013. Nanotechnology and its use in agriculture. In: Bagchi, D., Bagchi, M., Moriyama, H., Shahidi, F. (Eds.), Bio-nanotechnology: A Revolution in Food, Bomedical and Health Sciences. Wiley-Blackwell, West Sussex, UK, pp. 299–405.

Philip, D., Dubey, B., 2009. Biosynthesis of Au, Ag and Au–Ag nanoparticles using edible mushroom extract. Spectrochim. Acta Part A Mol. Biomol. Spectrosc. 73 (2), 374–381.

Pokhrel, L.R., Dubey, B., Patel, A., Dalwadi, P., Prasad, R., Patel, P., Selvaraj, K., 2013. Evaluation of developmental responses of two crop plants exposed to silver and zinc oxide nanoparticlesNicotiana tobaccum. Sci. Total Environ. 10 (41) 321–332.

Prasad, K.S., Pathak, D., Patel, A., Dalwadi, P., Prasad, R., Patel, P., Selvaraj, K., 2011. Biogenic synthesis of silver nanoparticles using *Nicotiana tobaccum* leaf extract and study of their antibacterial effect. Afr. J. Biotechnol. 10 (41), 8122.

Prasad, R., Swamy, V.S., Varma, A., 2013. Antibacterial activity of silver nanoparticles synthesized by bark extract of *Syzygium cumini*. J. Nanopart. 2013, 13746–13752.

Prasad, R., Bagde, U.S., Varma, A., 2012. Intellectual property rights and agricultural biotechnology: an overview. Afr. J. Biotechnol. 11 (6), 13746–13752.

Prasad, R., Kumar, V., Prasad, K.S., 2014. Nanotechnology in sustainable agriculture: present concerns and future aspects. Afr. J. Biotechnol. 13 (6), 705–713.

Rai, M., Ingle, A., 2012. Role of nanotechnology in agriculture with special reference to management of insect pests. Appl. Microbiol. Biotechnol. 94 (2), 287–293.

Rao, V.D., Pouya, A.K., Karimian, N., 2016. Personalized nanoMedicine: not just a tool but towards an excellence. Transl. Biomed. 7, 2–255.

Rasouli, F., Pouya, A.K., Karimian, N., 2013. Wheat yield and physico-chemical properties of a sodic soil from semi-arid area of Iran as affected by applied gypsum. Geoderma 193, 246–255.

Raveendranath Babu, T., Raja Sekhar Reddy, S., Sujana, P., Peralta-Videa, J.R., Gardea-Torresdey, J.L., 2014. Determination of the herbicide fenclorim by adsorptive stripping voltammetry at carbon nano tubes Paste electrodes (CNTPE). J. Adv. Sci. 4, 109–3498.

Rico, C.M., Majumdar, S., Duarte-Gardea, M., Peralta-Videa, J.R., Gardea-Torresdey, J.L., 2011. Interaction of nanoparticles with edible plants and their possible implications in the food chain. J. Agric. Food Chem. 59, 3485–3498.

Riera, F.A., Suarez, A., Muro, C., Ok, Y.S., Adrees, M., Ibrahim, M., Zia-ur-Rehman, M., Farid, M., Abbas, F., 2013. Nanofiltration of UHT flash cooler condensates from a dairy factory: characterization and water reuse potential. Desalination 309, 52–63.

Rizwan, M., Ali, S., Qayyum, M.F., Ok, Y.S., Adrees, M., Ibrahim, M., Zia-ur-Rehman, M., Farid, M., Abbas, F., 2017. Effect of metal and metal oxide nanoparticles on growth and physiology of globally important food crops: a critical review. J. Hazard. Mater. 322 (1), 2–16.

Santoso, D., Lefroy, R.D., Blair, G.J., 1995. Sulfur and phosphorus dynamics in an acid soil/crop system. Soil Res. 33 (1), 113–124.

Sarlak, N., Taherifar, A., Salehi, F., 2014. Synthesis of nanopesticides by encapsulating pesticide nanoparticles using functionalized carbon nanotubes and application of new nanocomposite for plant disease treatment. J. Agric. Food Chem. 62 (21), 4833–4838.

Sastry, M., Rao, M., Ganesh, K.N., 2002. Electrostatic Assembly of Nanoparticles and Biomacromolecules Acc. Chem. Res. 35 (10), 847–855.

Savage, N., Diallo, M.S., Bhumi, G., 2005. Nanomaterials and water purification: opportunities and challengesBoswellia ovalifoliolata. J. Nanopart. Res. 7 (4), 331–342.

Savithramma, N., Ankanna, S., Bhumi, G., Kale, A., Tolcheva, E., Levchenko, T.S., Torchilin, V.P., 2012. Effect of nanoparticles on seed germination and seedling growth of *Boswellia ovalifoliolata* an endemic and endangered medicinal tree taxon. Nano Vis. 2 (4), 61–68.

Sawant, R.M., Hurley, J.P., Salmaso, S., Kale, A., Tolcheva, E., Levchenko, T.S., Torchilin, V.P., 2006. "SMART" drug delivery systems: double-targeted pH-responsive pharmaceutical nanocarriers. Bioconjug. Chem. 17 (4), 943–949.

Scrinis, G., Lyons, K., Sharmila Rahale, C., 2007. The emerging nano-corporate paradigm: nanotechnology and the transformation of nature, food and agri-food systems. Int. J. Sociol. Agric. Food 15, 22–44.

Selva Preetha, P., Subramanian, K.S., Sharmila Rahale, C., De la Torre-Roche, R., Hamdi, H., White, J.C., Bindraban, P., Dimkpa, C., 2014. Sorption characteristics of nanozeolite based slow release sulphur fertilizer. Int. J. Dev. Res. 4, 225–228.

Servin, A., Elmer, W., Mukherjee, A., De la Torre-Roche, R., Hamdi, H., White, J.C., Bindraban, P., Dimkpa, C., 2015. A review of the use of engineered nanomaterials to suppress plant disease and enhance crop yield. J. Nanopart. Res. 17, 1–21.

Shah, V., Belozerova, I., Kumar, R., 2009. Influence of metal nanoparticles on the soil microbial community and germination of lettuce seeds. Water Air Soil Pollut. 197 (4), 143–148.

Sharon, M., Choudhary, A.K., Kumar, R., Singh, S., Dubey, N.K., Chauhan, D.K., 2010. Nanotechnology in agricultural diseases and food safety. J. Phytol. 2 (4), 405–411.

Shweta, Tripathi, D.K., Singh, S., Singh, S., Dubey, N.K., Chauhan, D.K., 2016. Impact of nanoparticles on photosynthesis: challenges and opportunities. Mater. Focus 5 (5), 405–411.

Singh, D., Singh, S.C., Kumar, S., Lal, B., Singh, N.B., 2010. Effect of titanium dioxide nanoparticles on the growth and biochemical parameters of *Brassica oleracea*. In: Symposium of International Conference on Food and Agricultural Applications of Nanotechnologies. São Pedro, SP, Brazil.

Singh, S., Tripathi, D.K., Dubey, N.K., Chauhan, D.K., Dubey, N.K., Singh, V.K., Liu, S., Tripathi, D.K., Chauhan, D.K., 2016. Effects of nano-materials on seed germination and seedling growth: striking the slight balance between the concepts and controversies. Mater. Focus 5 (3), 195–201. http://dx.doi.org/10.1016/j.plgene.2017.03.006.

Singh, S., Vishwakarma, K., Singh, S., Sharma, S., Dubey, N.K., Singh, V.K., Liu, S., Tripathi, D.K., Chauhan, D.K., 2017. Understanding the plant and nanoparticle interface at transcriptomic and proteomic level: a concentric overview. Plant Gene 67 (10), 4708–4716. http://dx.doi.org/10.1016/j.plgene.2017.03.006 .

Small, J., Call, D.R., Brockman, F.J., Straub, T.M., Chandler, D.P., 2001. Direct detection of 16S rRNA in soil extracts by using oligonucleotide microarrays. Appl. Environ. Microbiol. 67 (10), 4708–4716.

Sonkaria, S., Ahn, S.H., Khare, V., 2012. Nanotechnology and its impact on food and nutrition: a review. Recent Pat. Food Nutr. Agric. 4, 8–18.

Sotto, A., Arsuaga, J.M., VanderBruggen, B., 2013. Sorption of phenolic compounds on NF/RO membrane surfaces: influence on membrane performance. Desalination 309, 64–73.

Srinivasan, C., Saraswathi, R., Sirisha, D., 2010. Nano-agriculture-carbon nanotubes enhance tomato seed germination and plant growth. Curr. Sci. 99, 273–275.

Subbarao, C.V., Kartheek, G., Sirisha, D., 2013. Slow release of potash fertilizer through polymer coating. Int. J. Appl. Sci. Eng. 11, 25–1191.

Suman, P.R., Jain, V.K., Varma, A., Rajendran, V., Kannan, N., 2010. Role of nanomaterials in symbiotic fungus growth enhancementZea mays. Curr. Sci. 99, 1189–1191.

Suriyaprabha, R., Karunakaran, G., Yuvakkumar, R., Rajendran, V., Kannan, N., 2012. Silica nanoparticles for increased silica availability in maize (*Zea mays*. L) seeds under hydroponic conditions. Curr. Nanosci. 8 (3), 902–908.

Swamy, V.S., Prasad, R., Mahawar, H., Rathore, I., 2012. Green synthesis of silver nanoparticles from the leaf extract of *Santalum album* and its antimicrobial activity. J. Optoelectron. Biomed. Mater. 4 (3), 53–59.

Tarafdar, J.C., Raliya, R., Mahawar, H., Rathore, I., 2014. Development of zinc nanofertilizer to enhance crop production in pearl millet (*Pennisetum americanum*). Agric. Res. 3, 257–262.

Teodoro, S., Micaela, B., David, K.W., Villegas, J., Montoya, L.C., Garcia, S.B., 2010. Novel use of nano-structured alumina as an insecticideZea mays. Pest Manag. Sci. 66, 577–579.

Tiwari, D.K., Dasgupta-Schubert, N., Cendejas, L.V., Villegas, J., Montoya, L.C., Garcia, S.B., 2014. Interfacing carbon nanotubes (CNT) with plants: enhancement of growth, water and ionic nutrient uptake in maize (*Zea mays*) and implications for nanoagriculture. Appl. Nanosci. 4, 577–591.

Torney, F., 2009. Nanoparticle mediated plant transformation: emerging technologies in plant science research. In: Proceedings of the Interdepartmental Plant Physiology Major Fall Seminar Series, (IPPM'09), Iowa State University, vol. 696.

Torney, F., Trewyn, B.G., Lin, V.S., Wang, K., Dubey, N.K., Chauhan, D.K., 2007e. Mesoporous silica nanoparticles deliver DNA and chemicals into plantsTriticum aestivum. Nat. Nanotechnol. 2, 295–300.

Tripathi, D.K., Singh, S., Singh, V.P., Prasad, S.M., Dubey, N.K., Chauhan, D.K., 2017e. Silicon nanoparticles more effectively alleviated UV-B stress than silicon in wheat (*Triticum aestivum*) seedlings. Plant Physiol. Biochem. 110, 70–81.

Tripathi, D.K., Singh, V.P., Prasad, S.M., Chauhan, D.K., Dubey, N.K., Sharma, S., Singh, V.P., Singh, P.K., Prasad, S.M., Dubey, N.K., Pandey, A.C., 2015b. Silicon nanoparticles (SiNp) alleviate chromium (VI) phytotoxicity in *Pisum sativum* (L.) seedlings. Plant Physiol. Biochem. 96, 189–198.

Tripathi, D.K., Mishra, R.K., Singh, S., Singh, S., Vishwakarma, K., Sharma, S., Singh, V.P., Singh, P.K., Prasad, S.M., Dubey, N.K., Pandey, A.C., 2017b. Nitric oxide ameliorates zinc oxide nanoparticles phytotoxicity in wheat seedlings: implication of the ascorbate–glutathione cycle. Front. Plant Sci. 8, 2–12.

Tripathi, D.K., Singh, S., Singh, S., Pandey, R., Singh, V.P., Sharma, N.C., Prasad, S.M., Dubey, N.K., Chauhan, D.K., Pandey, A.C., Chauhan, D.K., 2017a. An overview on manufactured nanoparticles in plants: uptake, translocation, accumulation and phytotoxicityPisum sativum. Plant Physiol. Biochem. 110, 2–12.

Tripathi, D.K., Singh, S., Singh, S., Srivastava, P.K., Singh, V.P., Singh, S., Prasad, S.M., Singh, P.K., Dubey, N.K., Pandey, A.C., Chauhan, D.K., 2017c. Nitric oxide alleviates silver nanoparticles (AgNps)-induced phytotoxicity in *Pisum sativum* seedlings. Plant Physiol. Biochem. 110, 167–177.

Tripathi, D.K., Singh, S., Singh, V.P., Prasad, S.M., Chauhan, D.K., Dubey, N.K., 2016. Silicon nanoparticles more efficiently alleviate arsenate toxicity than silicon in maize cultivar and hybrid differing in arsenate tolerance. Front. Environ. Sci. 4, 46.

Tripathi, D.K., Tripathi, A., Shweta, S.,S., Singh, Y., Vishwakarma, K., Yadav, G., Sharma, S., Singh, V.K., Mishra, R.K., Upadhyay, R.G., Dubey, N.K., Lee, Y., Chauhan, D.K., 2017d. Uptake, accumulation and toxicity of silver nanoparticle in autotrophic plants, and heterotrophic microbes: a concentric review. Front. Microbiol. 8, 1–16.

Ulrichs, C., Mewis, I., Goswami, A., 2005. Crop Diversification Aiming Nutritional Security in West Bengal: Biotechnology of Stinging Capsules in Nature's Water-Blooms. Ann Tech Issue of State Agri Technologists Service Assoc, pp. 1–18.

Varma, A., Manika, K., 2017. Role of nanoparticles on plant growth with special emphasis on *Piriformospora indica*: a review. In: Nanoscience and Plant – soil Systems. Springer International Publishing, pp. 387–403.

Vaseeharan, B., Ramasamy, P., Chen, J.C., 2010. Antibacterial activity of silver nanoparticles (AgNps) synthesized by tea leaf extracts against pathogenic *Vibrio harveyi* and its protective efficacy on juvenile *Feneropenaeus indicus*. Lett. Appl. Microbiol. 50, 352–356.

Venkatachalam, P., Jayaraj, M., Manikandan, R., Geetha, N., Rene, E.R., Sharma, N.C., et al., 2017. Zinc oxide nanoparticles (ZnONPs) alleviate heavy metal-induced toxicity in *Leucaena leucocephala* seedlings: a physiochemical analysis. Plant Physiol. Biochem. 110, 59–69. http://dx.doi.org/10.1016/j.plaphy.2016.08.022.

Vidhyalakshmi, R., Bhakyaraj, R., Subhasree, R.S., Biris, A.S., Khodakovskaya, M.V., 2009. Encapsulation "the future of probiotics"-a review. Adv. Biol. Res. 3 (15), 96–103.

Villagarcia, H., Dervishi, E., de Silva, K., Biris, A.S., Khodakovskaya, M.V., 2012. Surface chemistry of carbon nanotubes impacts the growth and expression of water channel protein in tomato plants. Small 8 (15), 2328–2334.

Walz, E., Gräf, V., Greiner, R., 2017. Potential economic impact of engineered nanomaterials in agriculture and the food sector. In: Axelos, M.A.V., Van de Voorde, M. (Eds.), Nanotechnology in Agriculture and Food Science, pp. 363–376.

Wang, L., Li, X., Zhang, G., Dong, J., Eastoe, J., 2007. Oil-in-water nanoemulsions for pesticide formulations. J. Colloid Interface Sci. 314, 230–235.

Wang, P., Lombi, E., Zhao, F.J., Kopittke, P.M., Naidu, R., 2016b. Nanotechnology: a new opportunity in plant sciences. Trends Plant Sci. 21, 699–712. http://dx.doi.org/10.1016/j.tplants.2016.04.005.

Wang, T., Jin, X., Chen, Z., Megharaj, M., Naidu, R., Lu, G.M., 2014a. Green synthesis of Fe nanoparticles using eucalyptus leaf extracts for treatment of eutrophic wastewater. Sci. Total Environ. 466 (1), 210–213.

Wang, T., Lin, J., Chen, Z., Megharaj, M., Naidu, R., Tao, G., 2014b. Green synthesized iron nanoparticles by green tea and eucalyptus leaves extracts used for removal of nitrate in aqueous solution2Indocalamus barbatus. J. Clean. Prod. 83, 413–419.

Wilson, M.A., Tran, N.H., Milev, A.S., Kannangara, G.K., Volk, H., Lu, G.M., 2008. Nanomaterials in soilsEscherichia coli. Geoderma 146 (1), 291–302.

Xie, Y., Li, B., Zhang, Q., Zhang, C., Lu, K., Tao, G., 2011. Effects of nano-TiO_2 on photosynthetic characteristics of *Indocalamus barbatus*. J. Northeast Agric. Univ. 39, 22–25.

Yamanaka, M., Hara, K., Kudo, J., Cheng, T.C., Chang, C.Y., Chiu, C.Y., Liao, C.Y., Hsu, J.J., Lin, Z.P., 2005. Bactericidal actions of a silver ion solution on *Escherichia coli*, studied by energy-filtering transmission electron microscopy and proteomic analysis. Appl. Environ. Microbiol. 71, 7589–7593 (Trans Tech Publications).

Yang, F.L., Li, X.G., Zhu, F., Lei, C.L., 2009. Structural characterization of nanoparticles loaded with garlic essential oil and their insecticidal activity against *Tribolium castaneum* (Herbst) (Coleoptera: Tenebrionidae). J. Agric. Food Chem. 57, 10156–10162.

Yao, K.S., Li, S.J., Tzeng, K.C., Cheng, T.C., Chang, C.Y., Chiu, C.Y., Liao, C.Y., Hsu, J.J., Lin, Z.P., 2009a. Fluorescence silica nanoprobe as a biomarker for rapid detection of plant pathogens. Adv. Mater. Res. 79, 513–516 (Trans Tech Publications).

Zeeb, B., McClements, D.J., 2017. Nanoparticle-based delivery systems for nutraceuticals: Trojan Horse hydrogel beads. In: Axelos, M.A.V., Van de Voorde, M. (Eds.), Nanotechnology in Agriculture and Food Science, pp. 134–150.

Zhang, L., Webster, T.J., Shi, H., Ma, X., 2009b. Nanotechnology and nanomaterials: promises for improved tissue regenerationRaphanus sativus. Nano Today 4 (10), 66–80.

Zhang, P., Liu, G., Chen, X., Schwab, P., Wang, Q., Ebbs, S.D., Ma, X., 2016a. Nanobiotechnology: cell membrane-based delivery systemsRaphanus sativus. Nano Today 13, 7–9.

Zhang, Q.C., Wang, G.H., 2005. Studies on nutrient uptake of rice and characteristics of soil microorganisms in a long-term fertilization experiments for irrigated rice. J. Zhejiang Univ. Sci. B 6, 147.

Zhang, W., Dan, Y., Shi, H., Ma, X., Bonilla-Bird, N., López-Moreno, M.L., Komárek, M., Peralta-Videa, J.R., Gardea-Torresdey, J.L., 2016b. Effects of aging on the fate and bioavailability of cerium oxide nanoparticles to radish (*Raphanus sativus* L.) in soil. ACS Sustain. Chem. Eng. 4 (10), 5424–5431.

Zhang, W., Musante, C., White, J.C., Schwab, P., Wang, Q., Ebbs, S.D., Ma, X., 2017. Bioavailability of cerium oxide nanoparticles to *Raphanus sativus* L. in two soils. Plant Physiol. Biochem. 110, 185–193.

Zhao, S., Zou, L., Mulcahy, D., 2012. Brackish water desalination by a hybrid forward osmosis – nanofiltration system using divalent draw solute. Desalination 284, 175–181. http://dx.doi.org/10.1016/j.desal.2011.08.053.

Zheng, L., Hong, F., Lu, S., Liu, C., 2005. Effect of nano-TiO_2 on strength of naturally aged seeds and growth of spinach. Biol. Trace Elem. Res. 105, 83–91.

Zuverza-Mena, N., Martínez-Fernández, D., Du, W., Hernandez-Viezcas, J.A., Bonilla-Bird, N., López-Moreno, M.L., Komárek, M., Peralta-Videa, J.R., Gardea-Torresdey, J.L., 2017. Exposure of engineered nanomaterials to plants: insights into the physiological and biochemical responses-A review. Plant Physiol. Biochem. 110, 236–264.

Further Reading

Conley, D.J., Paerl, H.W., Howarth, R.W., Boesch, D.F., Seitzinger, S.P., Havens, K.E., Lancelot, C., Likens, G.E., 2009. Controlling eutrophication: nitrogen and phosphorus. Science 323, 1014–1015.

Correll, D.L., 1998. The role of phosphorus in the eutrophication of receiving waters: a review. J. Environ. Qual. 27, 261–266.

http://www.iranreview.org/content/Documents/Iranians_Researchers_Produce_Nano_Organic_Fertilizer.htm.

Khiyami, M.A., Almoammar, H., Awad, Y.M., Alghuthaymi, M.A., Abd-Elsalam, K.A., 2014. Plant pathogen nanodiagnostic techniques: forthcoming changes? Biotechnol. Biotechnol. Equip. 28 (5), 775–785.

Manik, A., Subramanian, K.S., 2014. Fabrication and characterization of nanoporous zeolite based N fertilizer. Afr. J. Agric. Res. 9 (2), 276–284.

Martínez-Fernández, D., Vítková, M., Michálková, Z., Komárek, M., 2016. Engineered nanomaterials for phytoremediation of metal/metalloids contaminated soils: implications for plant physiology. In: Ansari, A.A., Gill, S.S., Gill, R., Lanza, G.R., Lee, N. (Eds.). Ansari, A.A., Gill, S.S., Gill, R., Lanza, G.R., Lee, N. (Eds.), Phytoremediation: Management of Environmental Contaminants, vol. V. Springer, New York, USA.

Prasad, R., Jain, V.K., Varma, A., 2010. Role of nanomaterials in symbiotic fungus growth enhancement. Curr. Sci. 99 (9), 1189–1191.

Rai, V., Acharya, S., Dey, N., 2012. Implications of Nanobiosensors in Agriculture, pp. 315–324.

Rameshaiah, G.N., Jpallavi, S., 2015. Nano fertilizers and nano sensors – an attempt for developing smart agriculture. Int. J. Eng. Res. Gen. Sci. 3, 2091–2730.

CHAPTER

22

Nanoencapsulation of Essential Oils: A Possible Way for an Eco-Friendly Strategy to Control Postharvest Spoilage of Food Commodities From Pests

Akash Kedia[1], Nawal Kishore Dubey[2]

[1]Government General Degree College at Mangalkote, Burdwan, India; [2]Banaras Hindu University, Varanasi, India

22.1 INTRODUCTION

Postharvest damage of agricultural food commodities by storage pests can cause around 30%–35% of the yearly loss. The major culprits of such losses are insects and fungi. Furthermore, the secretion of mycotoxins by different fungi on stored food commodities poses a serious health concern to consumers (Prakash et al., 2012a). Various outbreaks of mycotoxin poisoning through contaminated foods have been reported occasionally from different parts of the world (Reddy and Raghavender, 2007; Wagacha and Muthomi, 2008), exploring the severity of the problem. The application of synthetic pesticides to minimize such losses has been reported to cause serious risk to both health and the environment depending on their toxicity, level of contamination, and the duration of exposure (Kohler and Triebskorn, 2013), leaving behind the urgent need to search for safer alternatives for the management of storage losses. In this context, a careful systematic search for phytochemicals is needed by the agricultural industries and government organizations to develop plant-based pesticides because they are biodegradable and could be better alternatives of synthetic pesticides for food protection. Additionally, the bioactivity of plant products is caused by synergistic effects of many active components leading to multiple modes of action and less resistance development during their pesticidal effect (Kedia et al., 2015a).

Nanomaterials in Plants, Algae, and Microorganisms
http://dx.doi.org/10.1016/B978-0-12-811487-2.00022-0

Essential oils (EOs) from higher plants are emerging as the better alternative for conventional pesticides and many articles and patents for their use as effective pesticides have been published (Lai et al., 2006; Dayan et al., 2009). EOs are aromatic, volatile oily liquids consisting of a mixture of various terpenes, phenols, and many other minor compounds having broad antimicrobial and insecticidal activities against a number of fungi, bacteria, and insects including postharvest pests (Burt, 2004; Kedia et al., 2015a; Prakash et al., 2015). Furthermore, many EOs are included in the Generally Recognized As Safe list by the US Food and Drug Administration and Environment Protection Agency and they have long been used in traditional medicine and pharmaceutical preparations from times immemorial (Prakash et al., 2012b). Antiviral, antimutagenic, antioxidant, antidiabetic, antiinflammatory, and antiprotozoal properties of different EOs have been well investigated (Raut and Karuppayil, 2014). Based on a wide spectrum of biological activity, EOs and their different components could be exploited as safe eco-friendly products for use in several fields including postharvest management of food commodities. From the last few years, several studies have centered on their potential as food additives for the postharvest protection of food items (Prakash et al., 2015).

However, the major inconvenience of using EOs for postharvest applications is their volatility and unstable and fragile nature. EOs are normally sensitive to degradation by factors such as oxidation, volatilization, heating, light, etc. during application (Asbahani et al., 2015). Different reports are available stating the decreased effect of EOs when applied during in vivo trials for the protection of food commodities (Varma and Dubey, 2001; Kedia et al., 2015b). This decreased efficacy of EOs could be caused by the degradation of the active component (Ilboudo et al., 2010) because the hydrogenated components, mono- and sesquiterpenes, present in EOs are vulnerable to degradation causing a decline in bioactivity (Kim et al., 2003). Temperature and light also play a major role in enhancing the oxidation process (Isman, 2000).

Hence there should be a suitable strategy to enhance the action, durability, and controlled release of EOs. Microencapsulation is such a technique in which a shell material encloses small particles of core material offering protection to these core materials from inauspicious environmental conditions while avoiding the drawbacks of volatility (Bertolini et al., 2001). Microencapsulation of EOs provides various advantages, namely, it minimizes reactivity with the environmental factors, decreases the evaporation rate, promotes handling ability, causes uniform distribution with very small amounts, and is delivered safely at the right time (Gibbs et al., 1999). The purpose behind microencapsulation is to entrap volatile components as microcarriers for their protection within shell material, reduce the rate of evaporation, ease handling, and control the release of volatiles during application (Baranauskiene et al., 2007). These criteria would be effective in improving efficacy, mode of application, and minimizing environmental damage (Moretti et al., 2002). The diameter of microencapsulates can vary from the nanorange to several microns, and presently there is increased concern about devising nanoencapsulates. In the past, these formulations for controlled release through nanoencapsulation have been found to enhance the insecticidal effects of pesticides on field crops (Latheeef, 1995) and work against stored-product pests in warehouses (Arthur, 1999) and antibiotics with enhanced antibacterial activity (Drulis-Kawa and Dorotkiewicz-Jach, 2010), providing an opportunity for future pesticide formulations. Currently, the nanoencapsulation technique has received considerable attention in the field of medicine for the formulation of therapeutic drugs and has established great

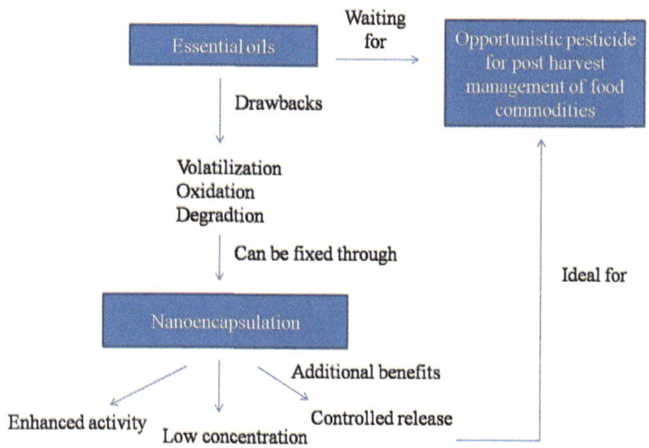

FIGURE 22.1 Figure shows limitations and the solution for formulation of essential oils as postharvest protectant of food commodities during storage.

potential in drug delivery systems through controlled release properties (Reis et al., 2006). The application of nanoparticles in agriculture can maximize the effects of pesticides at low concentrations such as in the field of medicine for enhanced drug delivery (Abreu et al., 2012). However, the nanoencapsulation process has a huge opportunity in the food and agriculture field (Fig. 22.1), but little information exists on their use in postharvest protection of food commodities (Yang et al., 2009). Table 22.1 comprises a list of EO/EO components encapsulated through different techniques in different wall materials and their potential for various applications including postharvest management of food items against storage pests.

In this chapter, different techniques for EO micro-/nanocapsule preparation, range of wall materials, and the properties that should be tested after encapsulation are discussed so that they can be applied on a large scale for postharvest management of food items against storage pests.

22.2 TECHNIQUES FOR ESSENTIAL OIL ENCAPSULATION

The selection of an appropriate encapsulation technique for preparing EO encapsulates is a crucial step because many properties such as size of the particle, surface area of the particle, shape of the particle, solubility of shell material, encapsulation efficiency of shell material, and releasing mechanisms of volatiles from microcapsules alter on changing the encapsulation technique. The choice of technique must be considered keeping in mind the required size and physicochemical properties of particles and the nature of the core and wall material of the capsule. Many techniques for microencapsulation have been developed so far and are used currently for different purposes. However, the techniques for emulsification, coacervation, nanoprecipitation, inclusion complexation, and supercritical fluid have greater potency to produce microcapsules in the range of 10–1000 nm (Ezhilarasi et al., 2013). This section contains a short description of encapsulation processes that are commonly used and can be

TABLE 22.1 Some Essential Oil/Essential Oil Components Encapsulated Through Various Techniques in Different Wall Materials and Their Properties

Essential Oil/Essential Oil Components	Method of Encapsulation	Wall Material	Properties Checked	Result Obtained	References
Artemisia arborescens	Liposome	Phosphatidyl-choline	Stability for over 1 year, antiviral activity	Up to 6 months neither oil leakage nor vesicle size alteration was found. However, after 1 year of storage, vesicle fusion was observed. The encapsulated product showed enhanced in vitro antiherpetic activity	Sinico et al. (2005)
Carum copticum	Phase separation	Chitosan	Fumigation toxicity	The oil-loaded nanogels showed higher fumigation toxicity against *Sitophilus granarius* and *Tribolium confusum* adults during storage as compared to EO alone	Ziaee et al. (2014)
Cinnamomum camphora	Compound coacervation	Gelatin blended with gum arabic with added polystyrene	Oil/wall volume ratio, emulsification stirring speed, concentration of cross-linking agent, treated time, and oil release properties	The encapsulation efficiency of 99.6% was observed at 0.75 oil/wall volume ratio; homogenization stirring speed decreases size of capsule; sustained oil release increases on increased concentration of glutaraldehyde and treatment time; addition of polystyrene improved the constant release of EO	Chang et al. (2006)
Cinnamomum verum	Molecular inclusion/coprecipitation	β-Cyclodextrin	Effect of the ratio on the inclusion efficiency; components variation after encapsulation	The retention of EO was found maximum (94.18%) at the oil-to-cyclodextrin ratio of 10:90; the qualitative and quantitative composition of the EO in the total oil extracts was found similar to the starting oil	Petrović et al. (2013)
Cymbopogom citratus	Simple coacervation	Polyvinyl alcohol cross-linked with glutaraldehyde	EO component analysis and antimicrobial assay	The encapsulation process did not affect the antimicrobial activity or any deterioration in the EO composition	Leimann et al. (2009)
Elettaria cardamomum	Spray drying	Mesquite gum	Total oil retention, surface oil, moisture content, and bulk and particle density in the microcapsules	The greatest encapsulation efficiency (83.6%) was observed for an oil:gum ratio of 1:4, with a surface oil concentration of 2590 mg/kg powder; moisture content was found similar for all ratios tested	Beristain et al. (2001)

Lippia graveolens	Spray drying	β-Cyclodextrin	Antimicrobial and antioxidant property	The encapsulated EOs showed similar chemical profile, similar antimicrobial activity, but four- to eightfold increased antioxidant activity as compared with nonencapsulated EOs	Arana-Sánchez et al. (2010)
Lippia sidoides	Spray drying	Maltodextrin DE10 and gum arabic	Shape and size of microparticles, moisture content, powder recovery, and antifungal activity	The best result obtained at maltodextrin:gum arabic ratio of 0:1 (m/m) and carrier:EO ratio of 4:1 (m/m) with enhanced antifungal activity	Fernandes et al. (2008)
Mentha piperita	Spray drying	Modified starch	Composition analysis, and release of volatiles at different water activity	The compositions of pure, emulsified, and encapsulated EO were almost the same; the aroma binding capacity of different modified starch matrices was found to depend on the water activity because the release of aromas during storage increased with increasing water activity	Baranauskiene et al. (2007)
Origanum dictamnus EO components	Liposomes	Phosphatidyl choline	Antioxidant and antimicrobial assay	All test components showed enhanced antimicrobial and antioxidant activities after their encapsulation	Liolios et al. (2009)
Pomegranate seed EO	Spray drying	Skimmed milk	Moisture content, particle size, bulk density, and hygroscopicity	The maximum encapsulation efficiency (95.6%) was found at a ratio of core to wall material, 1/9; feed solids concentration, 30% (w/w); inlet air temperature, 187°C; and drying air flow rate, 22.80 m^3/h	Goula and Adamopoulos (2012)
Rosemary EO	Phase separation	Ethylcellulose	Confocal laser scanning microscopy was used to observe surface properties, oil content in the microcapsules, and the presence of microcapsules attached to textile materials	Reducing the stirring speed increases the size of microcapsules; microcapsules had on average 40% of empty space fully occupied by EO; EO was found present even at the elevated temperature after grafting on textile materials	Voncina et al. (2009)

(Continued)

TABLE 22.1 Some Essential Oil/Essential Oil Components Encapsulated Through Various Techniques in Different Wall Materials and Their Properties—*cont'd*

Essential Oil/Essential Oil Components	Method of Encapsulation	Wall Material	Properties Checked	Result Obtained	References
Thymol and cinnamaldehyde	Molecular inclusion and subsequent freeze drying	β-Cyclodextrin	Effect of interactions with water on stability of nanocapsules	No component released from the nanoparticle at relative humidity < 84%, but it increased abruptly from 84% relative humidity coincidentally with the abrupt increase of sorbed water	Cevallos et al. (2010)
Vetiver EO	Complex coacervation	Gelatin/gum arabic cross-linked with glutaraldehyde or transglutaminase	The effect of type of cross-linking, physical state, and the type of release media	Increased resistivity of nanoparticles on increasing cross-linking; the release of core material decreased in the dissolution media as: anhydrous ethanol > Tween 80 > sodium dodecyl sulfate, and the wet microparticles showed faster release than freeze-dried ones	Prata et al. (2008)
Zanthoxylum limonella	Coacervation	Gelatin cross-linked with glutaraldehyde	Oil loading percentage, gelatin concentration, and degree of cross-linking, surface characteristics of microcapsules	Degree of cross-linking with increased concentration of glutaraldehyde reduces release rate; absence of any significant interaction between oil and polymer	Maji et al. (2007)

EO, essential oil.

applied for EOs. It is, however, not a complete list because it is a growing field where new inventions are regularly added.

22.2.1 Spray-Drying Method

Spray drying is a simple, common, easy-to-perform, and economically effective method for microencapsulation of chemically reactive volatile components (Gharsallaoui et al., 2007). The solution of coating material is prepared in deionized water, cooled and mixed with EO in different ratios, followed by homogenization with the help of a high-speed homogenizer. The homogenized emulsion is then spray dried in a spray dryer and atomized to create contact between drop and hot air. The droplet is maintained at a low temperature through rapid evaporation of water allowing simultaneous entrapment of EO. During this drying process, a film surrounding the droplet surface is formed increasing the concentration of core material and subsequently the formation of a porous, dry particle. The microcapsules then fall to bottom and are collected (Baranauskiene et al., 2007). Upon addition of water, depending on the porosity of the particle, the core material is released immediately. However, a slower release can be maintained by mixing more hydrophobic or readily available cross-linked carrier materials (Zuidam and Shimoni, 2010).

A literature survey revealed the suitability of this method for encapsulation of EOs. Bylaitė et al. (2001) prepared a solution of whey protein, skimmed milk powder, and their mixtures as a coating matrix (30 wt%) at 50°C in deionized water. After dissolving, maltodextrins were added (1:9 ratio) at 80°C. Caraway EO (15 wt%) was then emulsified into the solution and homogenized at 20,000 rpm for 5 min. The emulsion was then spray dried in a spray dryer after setting the parameters as follows: 180±5°C inlet temperature, 90±5°C outlet temperature, and 750–800 mm/H_2O pressure. In a similar way, Baranauskiene et al. (2007) developed microencapsulants of *Mentha piperita* EO by preparing the solutions of modified starches as coating matrices (30% w/w) in 40°C deionized water. The solution was then cooled, mixed overnight, and then emulsified with EO (15.25% w/w). Homogenization was achieved in an ultrahomogenizer (13,500 rpm for 7 min) and then dried in a spray dryer after setting the parameters as follows: 200°C±10°C inlet temperature, 120°C±10°C outlet temperature, and 400 mm/H_2O pressure. The microencapsulants from these studies showed effective protection of EOs against volatilization and degradation. Furthermore, it was found that on increasing water activity, the leakage of volatiles from the encapsulate also increased during storage. These properties of encapsulated EOs are ideal for effective pesticides and can be further utilized for postharvest protection of food items.

However, in this technique the process is not fully controlled and needs further development. The drying step needs a trial-and-error procedure; however, improvement can be made after studying the molecular interactions of water/wall, water/core, and wall/core.

22.2.2 Spray Cooling/Chilling Method

The method of spray cooling/chilling is another frequently used and economical encapsulation process for converting EOs into free-flowing powders. The process is almost the same as spray drying except no water is evaporated. The EO is emulsified into dissolved wall materials, atomized to dissipate droplets from the feedstock, and then the droplets are directly mixed with a cooling medium to solidify it into powder form (Zuidam and Shimoni, 2010). Usually, the melting point of the shell material lies at 34–42°C for spray chilling and is higher for spray cooling.

Moretti et al. (2002) prepared an aqueous dispersion of gelatin (10% wt/wt) in purified water and then added a suitable amount of *Rosmarinus officinalis* and *Thymus herba-barona* EOs. The mixture was emulsified with the help of a high shear mixer (turbine) at 1200 rpm and Na_2SO_4 was added to obtain the coacervate phase. To this, glutaraldehyde solution was added while stirring at 750 rpm for 3h (final pH 8.5°C). The hardened microparticles were then filtered, rinsed with cold water, and then dehydrated by freeze drying. The results from this study showed high encapsulation yields (over 98%) with both EOs, and the product showed interesting larvicidal effects favoring their application in integrated pest control strategies.

This technique is mainly used to encapsulate water-soluble, heat-sensitive, active ingredients and is less researched for encapsulation of EOs. The resultant microcapsules are insoluble in water and free of their core materials above the melting temperature of the wall material. Thus a kind of temperature-controlled release system is achieved by this process. However, the crucial step in this process is to control the crystallization process, otherwise unwanted release properties or particle softening will occur (Schrooyen et al., 2001).

22.2.3 Liposomal Preparation

Liposomal preparation is another simple and cost-effective way to prepare microcapsules of EOs. Liposomes are closed lipid bilayer membranes composed mainly of phospholipids (lecithin) and cholesterol. The principle of liposome formation is simply the hydrophilic–hydrophobic interactions between water and phospholipids. This method provides the stability of water-insoluble substances such as EOs in an aqueous environment in the inner hydrophobic core of a micelle. The method is promising for delivering water-insoluble active components in an aquatic medium and widely researched as formulations for applications of EOs as antibiotics in pharmaceuticals (Sherry et al., 2013).

In one of the methods used by Liolios et al. (2009), common lipid compositions containing egg L-α-phosphatidylcholine and cholesterol were used for the shell material. The liposomes were developed by mechanical shaking technique (thin film method). The lipids were melted in chloroform/methanol at a ratio of 3:1 and the organic solvent was moved out with the help of a rotary evaporator. The EO components, namely, carvacrol or carvacrol/thymol (6/1) or carvacrol/γ-terpinene (3/1), were then dissolved in methanol, mixed with a thin film of lipids, and organic solvents were vaporized under nitrogen stream. A solution of phosphate buffer saline solution (pH 7.4) was mixed with the resultant film, which was then vortexed and sonicated. The suspension was then left to hydrate (2h) and was ultracentrifuged (30,000 rpm for 60 min) to remove unincorporated compounds. The pellet was collected and dispersed in distilled water to reconstitute multilamellar vesicles. All tested compounds in the study performed enhanced antimicrobial activities after their encapsulation. Similarly, in the study of Moghimipour et al. (2012), liposomal encapsulates of *Eucalyptus camaldulensis* leaf EO performed enhanced antifungal activity favoring its application for food preservation.

The use of liposomal-microencapsulated products is quite limited in food preservation because of their unstable nature during storage, low yield, leakage of active ingredients during storage, and high cost of preparation (Zuidam et al., 2003). However, these are suitable materials for active drug delivery and are widely researched by pharmaceutical organizations.

22.2.4 Coacervation/Phase Separation

The process of coacervation/phase separation involves phase separation, which first sorts out polyelectrolyte(s) from a solution and then agglomerates into a liquid, coacervate phase around the active ingredient (Korus, 2001). It is important that the core material should be well matched with the introduced polymer and be less soluble or insoluble in the coacervation medium. The process of coacervation can be simple (single polymer) or complex (two or more polymers). In both cases the mixture must be agitated continuously and a suitable stabilizer must be introduced to prevent coagulation of the microcapsule products (Arshady, 1999). Furthermore, an enzymatic cross-linker such as glutaraldehyde or transglutaminase can be added to cross-link the hydrocolloid shell (Zuidam and Shimoni, 2010).

King (1995) prepared the coacervate of EOs by adding ethanol dropwise to the 1%–10% gelatin solution, which resulted in the separation of two phases, with one phase having a higher gelatin concentration. After adding the gellan gum (anionic polysaccharide), the gelatin (amphoteric hydrocolloid) formed complex coacervates. On lowering the pH gradually, microcapsules formed from coacervate material depositing around the oil droplets. Dong et al. (2011) prepared microcapsules of peppermint EO through the coacervation technique using gum arabic/gelatin as wall material and transglutaminase as a hardening agent. The particles with thin walls showed high encapsulation efficiency with fast release, while larger particles showed slower release at high temperature. However, the products showed higher stability upon storage and the rate of release was faster in hot medium and slower in cold medium. A similar finding was observed by Maji et al. (2007) with nanocapsules of *Zanthoxylum limonella* EO prepared through coacervation. On increasing the polymer content, the oil-loading capacity and oil content in microcapsules were found to be decreased. Furthermore, the oil retention capacity improved after the reaction of gelatin with glutaraldehyde. An increased concentration of glutaraldehyde resulted in an increased degree of cross-linking and subsequently decreased the oil release rate.

However, the method looks simple, but it is actually complicated and costly. The optimization of wall material concentration is baffling because it affects the yield of microcapsules (Nakagawa et al., 2004). The process also may cause vaporization of active ingredients, dissipation of active components into the solvent, and needs toxic chemicals (glutaraldehyde) to steady the unstable coacervates (Flores et al., 1992).

22.2.5 Molecular Inclusion

The process of molecular inclusion involves entrapment of EO/EO components within a circular molecule of β-cyclodextrin. The circular molecule of cyclodextrin is prepared by the cleavage of starch with the enzymatic action of cyclodextrin glucosyltransferase and subsequently the ends are linked with α(1-4) bondage (Hedges and McBride, 1999). The product protects unstable active components of EO and has a huge opportunity in the food industry because the product is obtained in the nanorange. Being lipophobic in nature, a wide range of aromatic components of EO is entrapped easily within the central hydrophobic cavity of the molecule.

The methods used for entrapping EO/EO components within a cyclodextrin molecule include stirring of aqueous solution of cyclodextrin with desired EO with the help of a powerful mixer and filtering off the precipitated complex. Cevallos et al. (2010) successfully prepared inclusion complexes of thymol and cinnamaldehyde, the major components of thyme and cinnamon EOs, respectively, with β-cyclodextrin. EO components were introduced to a saturated solution of β-cyclodextrin at 55°C and stirred for 4h. The mixture was then cooled at ambient temperature and then put overnight (2°C). The precipitated complexes were then filtered and freeze dried after submergence in liquid nitrogen. In this study, they found that up to 84% relative humidity, no components were released but above that level the release increased suddenly with the sudden increase of sorbed water. From the study they concluded that water sorption significantly affects the complex stability, which in turn is regulated by the water sorption isotherm. Bhandari et al. (1998a) in a similar way successfully encapsulated lemon oil in β-cyclodextrin using the foregoing method followed by spray drying, and found a very fine product with excellent flowability.

This technique has received global attention and is now researched widely because the product can be obtained within the desired range and the process is easily handled. The only limitation is the high cost of cyclodextrins, which limits its application for further large-scale trials for the postharvest management of food items.

22.2.6 Cocrystallization

The process of cocrystallization for EO nanoencapsulation is relatively simple, economical, and flexible. The cocrystallization of EO components can be done with supersaturated sucrose syrup above 120°C temperature and 95–97°Bx moisture. The crystal structure of sucrose then can be modified to form nanocrystals incorporating the EO components within them (Bhandari et al., 1998b). In the study of Beristain et al. (1996), they successfully encapsulated orange peel oil through the cocrystallization process and found good entrapment potential similar to spray-dried and extruded products. However, to protect the product from oxidization during storage, the introduction of a potent antioxidant was necessary.

Relatively few studies have reported the practice of cocrystallization to prepare microcapsules of EOs. Because the process requires high temperature, this method is not suitable for heat-sensitive volatile components of EOs.

22.2.7 Emulsification

The technique of emulsification produces nanoemulsions, colloidal dispersions comprising two immiscible liquids, of which one entraps the other forming a droplet in the size range of 50–1000nm. These nanoemulsions are then dried to a powder state using a spray dryer or freeze dryer. The technique offers great potential to encapsulate a broad range of active ingredients including EOs in a simple and economical way. The process needs the input of high-energy emulsification processes such as microfluidization, high-speed homogenization, or ultrasonication because the particle size decreases on increasing the energy input (Walstra, 1996).

Natrajan et al. (2015) prepared nanocapsules of *Curcuma longa* and *Cymbopogon citratus* EOs using a mixture of alginate and chitosan as wall material through the process of emulsification. At first, 20mL (0.3 and 0.6mg/mL) alginate solution was added to 1% (w/v) Tween-80 and the pH was adjusted between 5 and 5.5. Following this, 0.6mL ethanolic EO solution (20mg/mL) was then mixed dropwise and the solution was subjected to sonication for 15min. Thereafter, 4mL calcium chloride solution (0.67mg/mL) was mixed into it and the emulsion was vortex mixed for 30min. To this emulsion, a further 4mL of chitosan solution [0.3 or 0.6mg/mL in 1% (v/v) acetic acid] was added and vortex mixed for the next 30min. The final emulsion was left overnight and subjected to rotary evaporation at 40°C for 20min to achieve a fine dried powder of nanocapsules. From this study it was found that at 0.3mg/mL concentration for alginate and 0.6mg/mL concentration for chitosan, nanocapsules within the range of 300nm with good stability were produced. The oil-loaded nanocapsules were also found more hemocompatible and antiproliferative in nature as compared with the nonencapsulated EOs, favoring their use in biomedical and pharmaceutical fields.

The emulsification method through microfluidization yields less surface oil on products, but on increasing energy input it results in overprocessing caused by recoalescence. However, ultrasonication and high-pressure homogenization resulted in reduced droplet size with minimum recoalescence on increasing energy input. The size of the nanoparticles, their distribution, stability on medium, and other parameters were altered by various emulsion preparation techniques and handling conditions, namely, speed, pressure, temperature, emulsifiers, and concentration.

22.2.8 Nanoprecipitation

The process of nanoprecipitation or solvent displacement is like emulsification and is caused by spontaneous emulsification of the inner organic phase comprising the dissolved polymer, active ingredient (EO/EO component), and organic solvent in the aqueous external phase. The most common polymers used for this process are polycaprolactone, polylactide, polylactide-*co*-glycolide, and polyalkylcyanoacrylate (Reis et al., 2006).

Reports on encapsulation of EO/EO components through nanoprecipitation show the formation of nanocapsules within a range of 300nm and enhanced bioactivity as compared with the bare EO/EO components (Anand et al., 2010; Suwannateep et al., 2011). The process is efficient for forming nanocapsules in much lower range with good stability and higher encapsulation efficiency. However, the process relies on a good drying technique and only polymer-based shell materials can be implemented. The choice of proper solvent and nonsolvent phase is also a crucial step because it depends on the choice of bioactive component. The solvents must be water miscible to produce enough diffusion rate for spontaneous emulsification.

22.2.9 Supercritical Fluid Technique

A supercritical fluid is a substance showing properties in between liquids and gases such as modest viscosity, modest density, high solvation, high diffusion, and high mass transfer above its thermodynamic critical point. Out of a number of compounds that exhibited the properties of supercritical fluid such as carbon dioxide, propane, nitrogen, water, etc., carbon dioxide is most commonly used for nanoencapsulation because it is eco-friendly, utilizes less organic solvent, and is applied below 30°C (Gouin, 2004). The technique is relatively new, somewhat similar to that of spray drying, and is used for encapsulation of thermally sensitive compounds, giving the opportunity for rapid nanoencapsulation of heat-sensitive volatile components of EOs.

This process utilizes the mixing of the bioactive compounds, the polymer, and supercritical fluid, and subsequent evaporation of supercritical fluid in the spraying process (Reis et al., 2006). The main advantage of this process is the minimal use of organic solvent; however, it needs a high initial investment to buy the equipment.

22.3 CARRIERS/WALL MATERIALS FOR ENCAPSULATION

The selection of a suitable wall material is a crucial step in the formation of micro-/nano-capsules because it determines the stability and characteristics of the resulting microcapsules. The wall material must be able to create a network for entrapping the active ingredient. A number of coating materials such as gums, carbohydrates, gelatins, lipids, inorganic materials, polymers, and proteins can be used for this purpose; however, carbohydrate-based matrixes were used more frequently for EO microcapsules by different researchers (Rodea-González et al., 2012). Each type of material has its own merits and demerits. The choice of a suitable wall material may depend upon the desired property of the product, type of core material, encapsulation process, economic condition, and suitability according to the safety guidelines of the US Food and Drug Administration or the European Food Safety Authority (Madene et al., 2006). To encapsulate the EO/EO compounds, the wall materials must not have any reactivity with the EO/EO compounds, they must be easy to handle, they must protect the core material against adverse environment conditions (light, pH, oxygen, heat, etc.), they must possess desired release properties, and they should have an effective redispersion property to release the active ingredients at right time and place (Carvalho et al., 2015).

This section aims to provide existing knowledge about the materials used as wall materials for encapsulating EO/EO components in different studies and their potential for making EO-based microcapsules for use in large scale.

22.3.1 Carbohydrate Based Materials

Carbohydrates are used extensively as wall material for encapsulation of EOs most commonly in spray-dried encapsulations. To bind EOs in a desired range the choice of carbohydrates must be governed by the following properties: low cost, high diversity, modest viscosities, good solubility, good availability, and widespread use in foods as a safe component, all of which makes them the perfect choice for encapsulation of EOs.

22.3.1.1 Starch and Starch-Based Materials

Starch (amylose and amylopectin) and starch-based ingredients (maltodextrin, modified starches, and β-cyclodextrin) are widely used materials for encapsulation of EOs. Amylose comprises a linear chain of α(1-4)-linked D-glucose units usually within the range of 500–6000, while amylopectin comprises a linear D-glucose chain with side molecules of α(1-6)-linked D-glucose units at approximately every 20–30 linear chain units, usually a total of up to 2 million D-glucose units. Both amylose and amylopectin have the ability to interact intramolecularly and intermolecularly to form uniform structures (Parker and Ring, 2001). Dextrins (maltodextrin and cyclodextrin) are any product obtained by degradation of starch by different processes such as heat, acid, enzyme, etc. Dextrin, also referred as starch gum, has lower tensile strength than starch and thus dries faster and is thicker than starch. As a result, dextrins are suitable materials to encapsulate water-insoluble flavors (Wurzburg, 2006). Maltodextrins are the starch hydrolysates generated by acid or enzyme catalysis with molecular weight below 4000 g/mol. Maltodextrin is cost effective, shows low viscosity at high solid ratio, and can be found in different molecular weights. Cyclodextrins, a type of cyclic oligosaccharide, are obtained by the acidic or enzymatic degradation of amylose, amylopectin, or glycogen. The typical ringed structure of cyclodextrins contains 6–8 glucose monomer units with a cavity diameter of 0.7–0.8 nm. Usually for EO encapsulation, β-cyclodextrin (a ring molecule with seven sugar units and inner diameter of 0.6 nm) is preferred over α- or γ-cyclodextrin because the solubility of β-cyclodextrin in water at 25°C is much lower than α- or γ-cyclodextrin (Blanchard and Katz, 2006).

Tian et al. (2013) prepared an inclusion complex of amylose corn starch–cinnamaldehyde and observed the reduced release rate of cinnamaldehyde from the complex in comparison to free cinnamaldehyde (57.5%–28.4%) as analyzed from Fourier-transform infrared spectroscopy and thermogravimetric analysis. They further suggested that the desired rate of release of cinnamaldehyde in food systems can be achieved using high-amylose corn starch as wall material with an ultrasound treatment process. Baranauskiene et al. (2007) observed the effect of different commercially available modified food starches as wall materials for flavor retention of the *M. piperita* EO and found increased emulsification and encapsulation efficacy for all *n*-octenyl succinic anhydride-modified starches compared to hydrolyzed starches (dextrins). Furthermore, water activity was found to control the binding efficiency of different modified starch matrices because by increasing the water activity of the product, the leakage of aromas was found to be higher. Petrović et al. (2013) observed that the retention of cinnamon EO volatiles reached 94.18% when the EO was encapsulated within β-cyclodextrin at a ratio of 10:90. The qualitative and quantitative composition of the volatiles before and after nanoencapsulation was found to be similar and thus can be used for food preservation.

These modified starches have been used for coating and encapsulation of various active ingredients including EOs using different technologies such as spray drying, fluidized bed spray drying, and most commonly molecular inclusion with β-cyclodextrin. The major disadvantage of using maltodextrin is poor emulsifying capacity, low retentivity, and dependence on dextrose equivalent (DE) value for good encapsulation. Anandaraman and Reineccius (1986) investigated that maltodextrin with high DE value saved entrapped orange peel oil from oxidation, suggesting the effect of DE in the practicality of the wall system. Presently, molecular inclusion of EOs within β-cyclodextrin is a growing field in research because it is emerging as a suitable carrier, which transforms liquid compounds into crystalline form, covers unpleasant smells or tastes of some compounds, improves physical and/or chemical stability, and reduces evaporation (Marques, 2010).

22.3.1.2 Cellulose and Its Derivatives

Cellulose is a β-D-glucose linear chain linked with β-(1-4)-glycosidic bonds. The cellulose molecules form rigid structures by arranging themselves close together because of lack of side chains. Compared to starch, cellulose is more crystalline and is insoluble in water. Methylcellulose is a well-known derivative of cellulose used for the encapsulation process. It is promptly soluble in cold water and converts into gel on heating at 50°C or above. It shows good film-forming property and is used for the formation of nanofilms (Chang and

Zhang, 2011). Ethyl methylcellulose also behaves like methylcellulose because, like methylcellulose, it is soluble in cold water and produces gels on heating, but the gels are comparably weaker than methylcellulose. In contrast, another derivative of cellulose, ethylcellulose, is water insoluble and hence has huge application for coating materials for controlled release applications.

Voncina et al. (2009) encapsulated rosemary EO in ethylcellulose using a phase separation method. The microcapsules were then grafted onto cotton textile substrate and the prepared textile materials were analyzed by confocal laser fluorescence microscopy. The result of the study showed the microcapsules had on average 40% of empty space fully occupied by rosemary oil. Furthermore, rosemary oil in grafted textile was present even at the elevated temperature. This property of ethylcellulose-encapsulated microcapsules of EOs can be utilized for postharvest applications against temperature-resistant pests.

22.3.1.3 Plant Gums

Plant gums are complex macromolecular substances of various chemical structures and/or chain structure. Many of the plant-based gums can be used for encapsulation purposes but the most commonly used plant gums for EO encapsulation are gum arabic and mesquite gum. Both the gums are highly branched complexes having L-arabinose and D-galactose as the main components and 4-*O*-methyl-D-glucuronate and L-rhamnose as minor components. The major difference is the ratio of these components: 2:4:1:1 for mesquite gum and 4:2:1:1 for gum arabic, respectively (Román-Guerrero et al., 2009). However, they also possess a small amount of protein, which also helps in emulsification and film formation (Trejo-Espino et al., 2010). Gum arabic, mainly extracted from *Acacia seyal* or *Acacia senegal*, is highly soluble in water and maximum viscosity is achieved between pH 6 and 7. It is a splendid emulsifier and creates a strong protective covering around EO droplets (Krishnan et al., 2005). Mesquite gum is extracted from *Prosopis* spp., forms dense films during encapsulation, and sometimes shows superiority over gum arabic.

These natural gums either alone or in combinations are excellent wall materials for EO encapsulation. Bertolini et al. (2001) prepared microcapsules of β-pinene and citral (EO components) using gum arabic as wall material and observed a greater stability of capsules with similar external morphologies and with no apparent cracks or porosity. Beristain et al. (2001) reported that cardamom-based microcapsules can be successfully formed through spray drying using mesquite gum with high encapsulation efficiency (83.6%) at a ratio of 4:1 gum:oil. In a similar way, a mixture of gum arabic and mesquite gum showed high encapsulating efficiency (93.5%) of orange peel oil as observed by Beristain and Vernon-Carter (1995).

The reports on desired emulsifying ability and good encapsulation capability of plant gums can be utilized as an alternative encapsulating medium to synthetic ones. Being plant-based materials these microcapsules can readily be utilized for application in food safety.

22.3.1.4 Animal Polysaccharides

The major animal polysaccharide used for encapsulation purposes is chitosan, a polymer of D-glucosamine and *N*-acetyl-D-glucosamine units linked by a β-(1-4)-glycosidic bond. Chitosan is a deacetylated form of chitin extracted mainly from crustacean shells (Tan et al., 2005).

Chitosan is a well-studied wall material for EO encapsulation and is supported by a plethora of literature (Pedro et al., 2009). Zivanovic et al. (2005) encapsulated a number of EOs in chitosan films and observed an enhanced antimicrobial effect of chitosan–oregano EO film against *Listeria monocytogenes* and *Escherichia coli*. Similarly, oregano oil loaded in chitosan films was found suitable for meat preservation because it showed sustained release of carvacrol from EO products (Chami et al., 2005). Hsieh et al. (2006) prepared chitosan-based citronella EO microparticles by a modified emulsification technique and showed that the smallest particle showed the highest release rate. Anchisi et al. (2006) developed chitosan-entrapped *M. piperita* EO beads using the techniques of coacervation and ionotropic gelation and found the beads stable and resistant. Hussain and Maji (2008) encapsulated *Z. limonella* EO in a mixture of chitosan and gelatin. The rate of release of EO from the product was found to vary with different percentages of chitosan after evaluating chitosan–gelatin ratio. Similarly, Hosseini et al. (2013) obtained regular, spherical, and 40–80nm nanoparticles of chitosan-encapsulated oregano EO as confirmed by scanning electron and atomic force microscopy. The in vitro release behavior from the same study also showed slow release of EO during storage, making it suitable for food preservation.

22.3.2 Protein-Based Materials

Compared to carbohydrate, protein-based wall materials have not been studied extensively because of their dissimilar chemical groups, amphiphilic nature, interaction with core material, and high molecular weight (Kim et al., 1996). However, food or milk proteins such as whey protein, casein, soy protein, etc. showed desirable solubility, viscosity, and film-forming capacity, making them suitable for encapsulation. During emulsification these molecules were quickly adsorbed at the fresh-formed EO–water interface entrapping the EO droplets and protecting them against recoalescence (Dickinson, 2011). Casein is present in milk as spherical micelles and can be sorted out from the molecularly dispersed whey proteins by ultracentrifugation. Caseins are extremely heat stable and are insoluble at pH 4–5. However, their solubility can be altered by increasing or decreasing the pH. Whey protein becomes insoluble at its isoelectric point (pH 5) but remains soluble at other pHs.

Whey protein-based microcapsules protected orange EO against oxidation when prepared through spray drying (Kim and Morr, 1996). Sheu and Rosenberg (1995) used a mixture of whey proteins and carbohydrate as wall material where whey protein acted as film-forming material and carbohydrate (maltodextrin) served as matrix-forming material. In the food industry, β-lactoglobulin is the most preferred whey protein because it showed desirable emulsifying and foaming properties (Jouenne and Crouzet, 2000).

Other proteins such as soy protein and gelatin derivatives can also be used to form stable emulsions with EOs; however, their solvability in cold water and their high costs limit their potential application.

22.3.3 Lipid-Based Materials

Lipid-based wall materials are used as liposomal preparation of microcapsules utilizing phospholipids, usually phosphatidylcholine. On addition of water, they self-assemble into organized structures and bilayers. While mixing, some external energy is provided to form the liposomes from the bilayer (Taylor et al., 2005).

The encapsulation capability of liposomes can be affected by many factors, namely, type of phospholipids used for shell material, cholesterol content of lipid, the molar ratio of EOs to lipids, the methods of preparation, and the type of EO used. Liposomes are able to protect the fluidity of EOs at 4–5°C for at least 6 months (Sherry et al., 2013). The lipid-encapsulated microcapsules of EOs are promising agents to enhance the antimicrobial property of EOs, to study the interaction of EOs on cell membranes, and as a substitute to cure several diseases.

22.3.4 Other Materials

Other materials such as polyvinyl alcohol, a hydrophilic polymer, can also be used for preparing EO-based microcapsules because it forms hydrogel on addition of glutaraldehyde along with methanol, sulfuric acid, and acetic acid. Polyvinyl alcohol bears a simple chemical structure, is easy to handle, and shows potential in pharmaceutical and biomedical fields (Leimann et al., 2009).

22.4 CHARACTERIZATION OF MICRO-/NANOCAPSULES

The main purpose of EO encapsulation for protection of food items is the controlled release and protection of flavors. The controlled release of the active ingredients from EOs at the right place and time will improve the storage effectivity of food items (Baranauskiene et al., 2007). Microcapsules have various properties including shape, size, charge, thickness of wall, mechanical strength, crystallinity, flowability, persistence, permeability, and release properties. These properties are based on wall material and method of preparation and can be modified according to the desired product. To understand the behavior of encapsulates in different environments, information about their properties is very important. In this section the properties of EO encapsulates that should be analyzed before large-scale application are discussed. Furthermore, it is recommended to conduct an in vivo bioassay of encapsulants against a variety of storage pests including fungi and insects before their practical applications for postharvest protection of food items.

22.4.1 Total and Surface Oil Determination (Encapsulation Efficiency)

The measurement of total EO that has been encapsulated and the EO present at the surface of prepared microcapsules is important as these data can alter the degree of bioactivity of microcapsules. Furthermore, encapsulation efficiency depends on the nature of wall material and the process of encapsulation. The general way of calculating total oil is by the distillation of encapsulated dried powders (10 g) in hydrodistillation apparatus (Clevenger apparatus) for 3 h. The weight of oil retrieved from the sample can be calculated by multiplying by the specific gravity of that particular oil (Bylaitë et al., 2001). To calculate surface oil, 10 g of encapsulated dried powder is taken in a Soxhlet extraction apparatus and surface oil is washed by using pentane. The extract is collected and the total EO in the sample can be determined by gas chromatography (GC) (Bylaitë et al., 2001).

22.4.2 Surface Characterization

The surface characterization of microcapsules is done to observe shape, size, agglomeration, and damage, if any. These properties can alter the result during storage and are generally observed with the help of an electron microscope. Yang et al. (2009) prepared garlic EO-loaded nanocapsules of polyethylene observed under a transmission electron microscope at 80 kV supported by a Gatan 832 CCD camera. The transmission electron microscope image showed almost round particles (within 300 nm average size) with good dispersion, without any apparent cracks or porosity, which indicated good protection of EO.

22.4.3 Analysis of EO Composition Before and After Encapsulation

The bioactivity of EO depends on the concentration of its active components. Any change in component concentration may decrease its activity. Hence measurement of EO composition is necessary before and after encapsulation to observe the role of encapsulation on enhanced bioactivity. Various encapsulation processes and the interaction between wall material and EO can alter the composition of encapsulated EO. The most common method for EO component analysis is GC and gas chromatography-mass spectrometry (GC-MS) analysis. Yang et al. (2009) observed the EO composition before and up to 5 months after encapsulation to compare the concentration of various components entrapped within capsules through GC-MS. The result showed no significant chemical variations between pre- and postencapsulated EO stored up to 5 months.

22.4.4 Release of Volatiles

Application of EO-based microcapsules for postharvest management of food items needs release of volatiles at the proper time and in the right environmental conditions. The rate of release depends on many factors, namely, nature of wall materials, nature of volatiles, and storage conditions. Keeping this point in mind, microcapsules can be designed for release of volatiles, which depends on time, temperature, or humidity. For time-, temperature-, and humidity-controlled microcapsule preparation, generally, porous, heat-sensitive, and water-soluble wall materials, respectively, are taken.

Baranauskiene et al. (2007) observed the time- and humidity-dependent release properties of peppermint EO encapsulated within modified starches. For time-dependent studies, they placed 0.5 g encapsulated powder into a sample flask (125 mL) kept inside a water bath (25°C). The volatiles accumulated in the headspace were removed under nitrogen flow at 120 mL/min. Then, the sample was collected at timed intervals to trap released volatiles. The trapped volatiles were then desorbed from Tenax with the help of a thermal desorption device, after which the desorbed volatiles were directed toward GC. For humidity (water activity)-dependent studies, they placed spray-dried microcapsules after washing with pentane into an HPLC bottle fitted into a headspace containing 2 mL of saturated salt solution to obtain the water activity yield. After definite time periods, the vials were shifted to an automated headspace autosampler for analysis through GC. From this study, the release of volatiles was found to be increased on increasing the time period and water activity. The authors concluded that the increased rate of water activity was found to be responsible for causing structural changes of coating matrices, leading to increased diffusion rate of volatiles. These properties of encapsulated EOs are useful in postharvest management because the increase in humidity leads to increased deterioration of food items by storage pests.

22.4.5 Evaluation of Storage Stability

For proper functioning of microcapsules, they should be stable over time. Various factors such as agglomeration, lack of binding capacity, light, temperature, etc. can destroy the microcapsule structure or degrade the active ingredient, ultimately leading to decreased bioactivity. In a simple way, the stability of a nanoemulsion can be checked by placing it in an oven at 50°C. After different time intervals, it can be visually examined for surface oil layers, which indicate poor emulsion stabilization by the carrier. Bylaitė et al. (2001) observed the formation of oxidation compounds over time during storage of capsules at various temperatures (50°C and room temperature) and light conditions (presence and absence of light). They removed surface oil by washing with pentane, kept the microcapsules in separate, tightly capped 40-mL bottles, and placed them in the previously mentioned environmental conditions. The samples along with the nonencapsulated EO stored in the same conditions were then analyzed through GC. Their study revealed that the microcapsules of *Carum carvi* EO using milk protein-based matrices as wall material can be protected against oxidation and are found stable at least up to 26 weeks after their preparation.

22.5 CONCLUSION AND FUTURE PROSPECTS

EO-based micro-/nanocapsules have potential for application in the postharvest management of food items, and the strategy will help to mitigate adverse impacts of synthetic pesticides/preservatives on the environment and to human health. This is a growing field where new information is regularly added through ongoing research. The main difficulties that need to be addressed before commercialization are the issues of stability, safety, and production of cost-effective products in the quantities required to control postharvest loss. However, detailed research is required to develop nanoencapsulated EOs as pesticides for large-scale application.

References

Abreu, F.O., Oliveira, E.F., Paula, H.C., de Paula, R.C., 2012. Chitosan/cashew gum nanogels for essential oil encapsulation. Carbohydr. Polym. 89 (4), 1277–1282.

Anand, P., Nair, H.B., Sung, B., Kunnumakkara, A.B., Yadav, V.R., Tekmal, R.R., Aggarwal, B.B., 2010. Design of curcumin loaded PLGA nanoparticles formulation with enhanced cellular uptake, and increased bioactivity in vitro and superior bioavailability in vivo. Biochem. Pharmacol. 79 (3), 330–338.

Anandaraman, S., Reineccius, G.A., 1986. Stability of encapsulated orange peel oil. Food Technol. 40, 88–93.

Anchisi, C., Meloni, M.C., Maccioni, A.M., 2006. Chitosan beads loaded with essential oils in cosmetic formulations. J. Cosmet. Sci. 57, 205–214.

Arana-Sánchez, A., Estarrón-Espinosa, M., Obledo-Vázquez, E.N., Padilla-Camberos, E., Silva-Vázquez, R., Lugo-Cervantes, E., 2010. Antimicrobial and antioxidant activities of Mexican oregano essential oils (*Lippia graveolens* HBK) with different composition when microencapsulated in β-cyclodextrin. Lett. Appl. Microbiol. 50 (6), 585–590.

Arshady, R., 1999. Microspheres, microcapsules and liposomes. In: Arshady, R. (Ed.). Arshady, R. (Ed.), Preparation and Chemical Applications, vol. I. Citus Books, London, UK, pp. 279–322.

Arthur, F.H., 1999. Evaluation of an encapsulated formulation of cyfluthrin to control *Sitophilus oryzae* (L.) on stored wheat. J. Stored Prod. Res. 35, 159–166.

Asbahani, A.E., Miladi, K., Addi, E.H., Bitar, A., Casabianca, H., Mousadik, A.E., Hartmann, D.J., Jilale, A., Renaud, F.N.R., Elaissari, A., 2015. Antimicrobial activity of nano-encapsulated essential oils: comparison to non-encapsulated essential oils. J. Colloid Sci. Biotechnol. 4 (1), 39–48.

Baranauskiene, R., Bylaitë, E., Zukauskaite, J., Venskutonis, R.P., 2007. Flavor retention of peppermint (*Mentha piperita* L.) essential oil spray-dried in modified starches during encapsulation and storage. J. Agric. Food Chem. 55 (8), 3027–3036.

Beristain, C.I., Garcîa, H.S., Vernon-Carter, E.J., 2001. Spray-dried encapsulation of Cardamom (*Elettaria cardamomum*) essential oil with mesquite (*Prosopis juliflora*) gum. LWT Food Sci. Technol. 34, 398–401.

Beristain, C.I., Vazquez, A., Garcia, H.S., Vernon-Carter, E.J., 1996. Encapsulation of orange peel oil by co-crystallization. LWT Food Sci. Technol. 29 (7), 645–647.

Beristain, C.I., Vernon-Carter, E.J., 1995. Studies on the interaction of arabic (*Acacia senegal*) and mesquite (*Prosopis juliflora*) gum as emulsion stabilizing agents for spray dried encapsulated orange peel oil. Dry. Technol. 29, 645–667.

Bertolini, A.C., Siani, A.C., Grosso, C.R.F., 2001. Stability of monoterpenes encapsulated in gum arabic by spray-drying. J. Agric. Food Chem. 49 (2), 780–785.

Bhandari, B.R., D'Arcy, B.R., Thi Bich, L.L., 1998a. Lemon oil to β-cyclodextrin ratio effect on the inclusion efficiency of β-cyclodextrin and the retention of oil volatiles in the complex. J. Agric. Food Chem. 46 (4), 1494–1499.

Bhandari, B.R., Datta, N., D'Arcy, B.R., Rintoul, G.B., 1998b. Co-crystallization of honey with sucrose. LWT Food Sci. Technol. 31 (2), 138–142.

Blanchard, P.H., Katz, F.R., 2006. Starch hydrolysates. In: Stephen, A.M., Phillips, G.O., Williams, P.A. (Eds.), Food Polysaccharides and Their Applications, second ed. Taylor & Francis, Boca Raton, FL, pp. 119–145.

Burt, S., 2004. Essential oils: their antibacterial properties and potential applications in foods – a review. Int. J. Food Microbiol. 94, 223–253.

Bylaitë, E., Venskutonis, P.R., Maþdþierienë, R., 2001. Properties of caraway (*Carum carvi* L.) essential oil encapsulated into milk protein-based matrices. Eur. Food Res. Technol. 212, 661–670.

Carvalho, I.T., Estevinho, B.N., Santos, L., 2015. Application of microencapsulated essential oils in cosmetic and personal healthcare products – a review. Int. J. Cosmet. Sci. 2015, 1–11.

Cevallos, P.A.P., Buera, M.P., Elizalde, B.E., 2010. Encapsulation of cinnamon and thyme essential oils components (cinnamaldehyde and thymol) in β-cyclodextrin: effect of interactions with water on complex stability. J. Food Eng. 99 (1), 70–75.

Chami, N., Bennis, S., Chami, F., Aboussekhra, A., Remmal, A., 2005. Study of anticandidal activity of carvacrol and eugenol in vitro and in vivo. Oral Microbiol. Immunol. 20, 106–111.

Chang, C., Zhang, L., 2011. Cellulose-based hydrogels: present status and application prospects. Carbohydr. Polym. 84 (1), 40–53.

Chang, C.P., Leung, T.K., Lin, S.M., Hsu, C.C., 2006. Release properties on gelatin-gum arabic microcapsules containing camphor oil with added polystyrene. Colloids Surf. B 50 (2), 136–140.

Dayan, F.E., Cantrell, C.L., Duke, S.O., 2009. Natural products in crop protection. Bioorg. Med. Chem. 17, 4022–4034.

Dickinson, E., 2011. Double emulsions stabilized by food biopolymers. Food Biophys. 6 (1), 1–11.

Dong, Z., Ma, Y., Hayat, K., Jia, C., Xia, S., Zhang, X., 2011. Morphology and release profile of microcapsules encapsulating peppermint oil by complex coacervation. J. Food Eng. 104 (3), 455–460.

Drulis-Kawa, Z., Dorotkiewicz-Jach, A., 2010. Liposomes as delivery systems for antibiotics. Int. J. Pharm. 387 (1), 187–198.

Ezhilarasi, P.N., Karthik, P., Chhanwal, N., Anandharamakrishnan, C., 2013. Nanoencapsulation techniques for food bioactive components: a review. Food Bioprocess Technol. 6, 628–647.

Fernandes, L.P., Turatti, I.C., Lopes, N.P., Ferreira, J.C., Candido, R.C., Oliveira, W.P., 2008. Volatile retention and antifungal properties of spray-dried microparticles of *Lippia sidoides* essential oil. Dry. Technol. 26 (12), 1534–1542.

Flores, R.J., Wall, M.D., Carnahan, D.W., Orofino, T.A., 1992. An investigation of internal phase losses during the microencapsulation of fragrances. J. Microencapsul. 3, 287–307.

Gharsallaoui, A., Roudaut, G., Chambin, O., Voilley, A., Saurel, R., 2007. Applications of spray-drying in microencapsulation of food ingredients: an overview. Food Res. Int. 40, 1107–1121.

Gibbs, F., Kermasha, S., Alli, I., Mulligan, C.N., Bernard, 1999. Encapsulation in the food industry: a review. Int. J. Food Sci. Nutr. 50 (3), 213–224.

Gouin, S., 2004. Microencapsulation: industrial appraisal of existing technologies and trends. Trends Food Sci. Technol. 15 (7–8), 330–347.

Goula, A.M., Adamopoulos, K.G., 2012. A method for pomegranate seed application in food industries: seed oil encapsulation. Food Bioprod. Process. 90 (4), 639–652.

Hedges, A., McBride, C., 1999. Utilization of β-cyclodextrin in food. Cereal Foods World 44 (10), 700–704.

Hosseini, S.F., Zandi, M., Rezaei, M., Farahmandghavi, F., 2013. Two-step method for encapsulation of oregano essential oil in chitosan nanoparticles: preparation, characterization and in vitro release study. Carbohydr. Polym. 95, 50–56.

Hsieh, W.C., Chang, C.P., Gao, Y.L., 2006. Controlled release properties of chitosan encapsulated volatile citronella oil microcapsules by thermal treatments. Colloids Surf. B 53, 209–214.

Hussain, M.R., Maji, T.K., 2008. Preparation of genipin cross-linked chitosan-gelatin microcapsules for encapsulation of *Zanthoxylum limonella* oil (ZLO) using salting-out method. J. Microencapsul. 1–7.

Ilboudo, Z., Dabiré, L.C.B., Nébié, R.C.H., Dicko, I.O., Dugravot, S., Cortesero, A.M., Sanon, A., 2010. Biological activity and persistence of four essential oils towards the main pest of stored cowpeas, *Callosobruchus maculatus* (F.) (Coleoptera: Bruchidae). J. Stored Prod. Res. 46, 124–128.

Isman, M.B., 2000. Plant essential oils for pest and disease management. Crop Prot. 19, 603–608.

Jouenne, E., Crouzet, J., 2000. Effect of pH on retention of aroma compounds by β-lactoglobulin. J. Agric. Food Chem. 48 (4), 1273–1277.

Kedia, A., Prakash, B., Mishra, P.K., Dwivedy, A.K., Dubey, N.K., 2015b. *Trachyspermum ammi* L. essential oil as plant based preservative in food system. Ind. Crops Prod. 69, 104–109.

Kedia, A., Prakash, B., Mishra, P.K., Singh, P., Dubey, N.K., 2015a. Botanicals as eco friendly biorational alternatives of synthetic pesticides against *Callosobruchus* spp. (Coleoptera: Bruchidae) -a review. J. Food Sci. Technol. 52, 1239–1257.

Kim, S.I., Roh, J.Y., Kim, D.H., Lee, H.S., Ahn, Y.J., 2003. Insecticidal activities of aromatic plant extracts and essential oils against *Sitophilus oryzae* and *Callosobruchus chinensis*. J. Stored Prod. Res. 39 (3), 293–303.

Kim, Y.D., Morr, C.V., 1996. Microencapsulation properties of gum arabic and several food proteins: spray-dried orange oil emulsion particles. J. Agric. Food Chem. 44 (5), 1314–1320.

Kim, Y.D., Morr, C.V., Schenz, T.W., 1996. Microencapsulation properties of gum arabic and several food proteins: liquid orange oil emulsion particles. J. Agric. Food Chem. 44 (5), 1308–1313.

King, A.K., 1995. Encapsulation of food ingredients: a review of available technology, focusing on hydrocolloids. In: Risch, S.J., Reineccius, G.A. (Eds.), Encapsulation and Controlled Release of Food Ingredients. American Chemical Society, Washington, DC, pp. 26–41.

Kohler, H.R., Triebskorn, R., 2013. Wildlife ecotoxicology of pesticides: can we track effects to the population level and beyond? Science 341, 759–765.

Korus, J., 2001. Microencapsulation of flavours in starch matrix by coacervation method. Pol. J. Food Nutr. Sci. 10 (1), 17–23.

Krishnan, S., Kshirsagar, A.C., Singhal, R.S., 2005. The use of gum arabic and modified starch in the microencapsulation of a food flavoring agent. Carbohydr. Polym. 62 (4), 309–315.

Lai, F., Wissing, S.A., Müller, R.H., Fadda, A.M., 2006. *Artemisia arborescens* L. essential oil-loaded solid lipid nanoparticles for potential agricultural application: preparation and characterization. AAPS PharmSciTech. 7 (1), E1–E9 Article 2.

Latheeef, M.A., 1995. Influence of pyrethroid formulations and sprayer nozzle size on tobacco budworm mortality on cotton. Southwest. Entomol. 20, 49–54.

Leimann, F.V., Gonçalves, O.H., Machado, R.A., Bolzan, A., 2009. Antimicrobial activity of microencapsulated lemongrass essential oil and the effect of experimental parameters on microcapsules size and morphology. Mater. Sci. Eng. 29 (2), 430–436.

Liolios, C.C., Gortzi, O., Lalas, S., Tsaknis, J., Chinou, I., 2009. Liposomal incorporation of carvacrol and thymol isolated from the essential oil of *Origanum dictamnus* L. and *in vitro* antimicrobial activity. Food Chem. 112, 77–83.

Madene, A., Jacquot, M., Scher, J., Desobry, S., 2006. Flavour encapsulation and controlled release – a review. Int. J. Food Sci. Technol. 41, 1–21.

Maji, T.K., Baruah, I., Dube, S., Hussain, M.R., 2007. Microencapsulation of *Zanthoxylum limonella* oil (ZLO) in glutaraldehyde crosslinked gelatin for mosquito repellent application. Bioresour. Technol. 98, 840–844.

Marques, H.M.C., 2010. A review on cyclodextrin encapsulation of essential oils and volatiles. Flavour Fragr. J. 25, 313–326.
Moghimipour, E., Aghel, N., Mahmoudabadi, A.Z., Ramezani, Z., Handali, S., 2012. Preparation and characterization of liposomes containing essential oil of *Eucalyptus camaldulensis* leaf. Jundishapur J. Nat. Pharm. Prod. 7, 117–122.
Moretti, M.D., Sanna-Passino, G., Demontis, S., Bazzoni, E., 2002. Essential oil formulations useful as a new tool for insect pest control. AAPS PharmSciTech 3 (2), 64–74.
Nakagawa, K., Iwamoto, S., Nakajima, M., Shono, A., Satoh, K., 2004. Microchannel emulsification using gelatine and surfactant-free coacervate microencapsulation. J. Colloid Interface Sci. 278, 198–205.
Natrajan, D., Srinivasan, S., Sundar, K., Ravindran, A., 2015. Formulation of essential oil-loaded chitosan–alginate nanocapsules. J. Food Drug Anal. 23 (3), 560–568.
Parker, R., Ring, S.G., 2001. Aspects of the physical chemistry of starch. J. Cereal Sci. 34, 1–17.
Pedro, A.S., Cabral-Albuquerque, E., Ferreira, D., Sarmento, B., 2009. Chitosan: an option for development of essential oil delivery systems for oral cavity care? Carbohydr. Polym. 76, 501–508.
Petrović, G.M., Stojanović, G.S., Radulović, N.S., 2013. Encapsulation of cinnamon oil in β-cyclodextrin. J. Med. Plants Res. 4 (14), 1382–1390.
Prakash, B., Kedia, A., Mishra, P.K., Dubey, N.K., 2015. Plant essential oils as food preservatives to control moulds, mycotoxin contamination and oxidative deterioration of agri-food commodities – potentials and challenges. Food Control 47, 381–391.
Prakash, B., Singh, P., Kedia, A., Dubey, N.K., 2012b. Assessment of some essential oils as food preservatives based on antifungal, antiaflatoxin, antioxidant activities and in vivo efficacy in food system. Food Res. Int. 49, 201–208.
Prakash, B., Singh, P., Kedia, A., Singh, A., Dubey, N.K., 2012a. Efficacy of essential oil combination of *Curcuma longa* L. and *Zingiber officinale* Rosc. as a postharvest fungitoxicant, aflatoxin inhibitor and antioxidant agent. J. Food Saf. 32, 279–288.
Prata, A.S., Zanin, M.H., Ré, M.I., Grosso, C.R., 2008. Release properties of chemical and enzymatic crosslinked gelatin-gum arabic microparticles containing a fluorescent probe plus vetiver essential oil. Colloids Surf. B 67 (2), 171–178.
Raut, J.S., Karuppayil, S.M., 2014. A status review on the medicinal properties of essential oils. Ind. Crops Prod. 62, 250–264.
Reddy, B.N., Raghavender, C.R., 2007. Outbreaks of aflatoxicoses in India. Afr. J. Food Agric. Nutr. Dev. 7 (5), 1–15.
Reis, C.P., Neufeld, R.J., Ribeiro, A.J., Veiga, F., 2006. Nanoencapsulation I. Methods for preparation of drug-loaded polymeric nanoparticles. Nanomedicine 2, 8–21.
Rodea-González, D.A., Cruz-Olivares, J., Román-Guerrero, A., Rodríguez-Huezo, M.E., Vernon-Carter, E.J., Pérez-Alonso, C., 2012. Spray-dried encapsulation of chia essential oil (*Salvia hispanica* L.) in whey protein concentrate-polysaccharide matrices. J. Food Eng. 111, 102–109.
Román-Guerrero, A., Orozco-Villafuerte, J., Pérez-Orozco, J.P., Cruz-Sosa, F., Jiménez-Alvarado, R., Vernon-Carter, E.J., 2009. Application and evaluation of mesquite gum and its fractions as interfacial film formers and emulsifiers of orange peel-oil. Food Hydrocoll. 23 (3), 708–713.
Schrooyen, P.M.M., van der Meer, R., De Kruif, C.G., 2001. Microencapsulation: its application in nutrition. Proc. Nutr. Soc. 60, 475–479.
Sherry, M., Charcosset, C., Fessi, H., Greige-Gerges, H., 2013. Essential oils encapsulated in liposomes: a review. J. Liposome Res. 23 (4), 268–275.
Sheu, T.Y., Rosenberg, M., 1995. Microencapsulation by spray drying ethyl caprylate in whey protein and carbohydrate wall systems. J. Food Sci. 60 (1), 98–103.
Sinico, C., De Logu, A., Lai, F., Valenti, D., Manconi, M., Loy, G., Bonsignore, L., Fadda, A.M., 2005. Liposomal incorporation of *Artemisia arborescens* L. essential oil and in vitro antiviral activity. Eur. J. Pharm. Biopharm. 59 (1), 161–168.
Suwannateep, N., Banlunara, W., Wanichwecharungruang, S.P., Chiablaem, K., Lirdprapamongkol, K., Svasti, J., 2011. Mucoadhesive curcumin nanospheres: biological activity, adhesion to stomach mucosa and release of curcumin into the circulation. J. Control. Release 151 (2), 176–182.
Tan, L.H., Chan, L.W., Heng, P.W., 2005. Effect of oil loading on microspheres produced by spray drying. J. Microencapsul. 22, 253–259.
Taylor, T.M., Weiss, J., Davidson, P.M., Bruce, B.D., 2005. Liposomal nanocapsules in food science and agriculture. Crit. Rev. Food Sci. Nutr. 45 (7–8), 587–605.
Tian, Y., Zhu, Y., Bashari, M., Hu, X., Xu, X., Jin, Z., 2013. Identification and releasing characteristics of high-amylose corn starch–cinnamaldehyde inclusion complex prepared using ultrasound treatment. Carbohydr. Polym. 91 (2), 586–589.

Trejo-Espino, J.L., Rodríguez-Monroy, M., Vernon-Carter, E.J., Cruz-Sosa, F., 2010. Emulsifying properties of the gum produced by *Prosopis laevigata* (Humb. & Bonpl. ex Willd) M.C. Johnst. (Mesquite) cells suspension culture in bioreactor. Rev. Mex. Ing. Quím. 9 (3), 251–260.

Varma, J., Dubey, N.K., 2001. Efficacy of essential oils of *Caesulia axillaris* and *Mentha arvensis* against some storage pests causing biodeterioration of food commodities. Int. J. Food Microbiol. 68, 207–210.

Voncina, B., Kreft, O., Kokol, V., Chen, W.T., 2009. Encapsulation of rosemary oil in ethylcellulose microcapsules. Text. Polym. J. 1 (1), 13–19.

Wagacha, J.M., Muthomi, J.W., 2008. Mycotoxin problem in Africa: current status, implications to food safety and health and possible management strategies. Int. J. Food Microbiol. 124, 1–12.

Walstra, P., 1996. Emulsion stability. In: Becher, P. (Ed.), Encyclopedia of Emulsion Technology. Marcel Dekker, New York, pp. 1–62.

Wurzburg, O.B., 2006. Modified starches. In: Stephen, A.M., Phillips, G.O., Williams, P.A. (Eds.), Food Polysaccharides and Their Applications, second ed. Taylor & Francis, Boca Raton, FL, pp. 87–118.

Yang, F.L., Li, X.G., Zhu, F., Lei, C.L., 2009. Structural characterization of nanoparticles loaded with garlic essential oil and their insecticidal activity against *Tribolium castaneum* (Herbst) (Coleoptera: Tenebrionidae). J. Agric. Food Chem. 57 (21), 10156–10162.

Ziaee, M., Moharramipour, S., Mohsenifar, A., 2014. Toxicity of *Carum copticum* essential oil-loaded nanogel against *Sitophilus granarius* and *Tribolium confusum*. J. Appl. Entomol. 138 (10), 763–771.

Zivanovic, S., Chi, S., Draughon, A.F., 2005. Antimicrobial activity of chitosan films enriched with essential oils. J. Food Sci. 70, M45–M51.

Zuidam, N.J., Shimoni, E., 2010. Overview of microencapsulation use in food products or processes and methods to make them. In: Zuidam, N.J., Nedovic, V.A. (Eds.), Encapsulation Technique for Active Food Ingredients and Food Processing. Springer, New York, pp. 3–29.

Zuidam, N.J., Van Winden, E., De Vrueh, R., Crommelin, D.J.A., 2003. Stability, storage and sterilization of liposomes. In: Torchilin, V.P., Weissig, V. (Eds.), Liposomes. Oxford University Press, Oxford, pp. 149–165.

Index

Note: 'Page numbers followed by "f" indicate figures, "t" indicate tables and "b" indicate boxes.'

D

E

F

Q

R

S

Edwards Brothers Inc.
Ann Arbor MI. USA
November 27, 2017